FORTSCHRITTE DER CHEMIE ORGANISCHER NATURSTOFFE

PROGRESS IN THE CHEMISTRY OF ORGANIC NATURAL PRODUCTS

BEGRÜNDET VON · FOUNDED BY

L. ZECHMEISTER

HERAUSGEGEBEN VON · EDITED BY

W. HERZ   **H. GRISEBACH**   **G. W. KIRBY**
TALLAHASSEE, FLA.   FREIBURG i. BR.   LOUGHBOROUGH, LEICS.

NEUNUNDZWANZIGSTER BAND
TWENTY-NINTH VOLUME

VERFASSER · AUTHORS

E. GLOTTER · D. GOLDSMITH · D. GROSS · J. R. HANSON · S. HUNECK
F. JOHNSON · D. LAVIE · E. PREMUZIC · W. RÜDIGER

MIT 18 ABBILDUNGEN · WITH 18 FIGURES

1971

WIEN · SPRINGER-VERLAG · NEW YORK

DAS WERK IST URHEBERRECHTLICH GESCHÜTZT

DIE DADURCH BEGRÜNDETEN RECHTE, INSBESONDERE DIE DER ÜBERSETZUNG,
DES NACHDRUCKES, DER ENTNAHME VON ABBILDUNGEN, DER FUNKSENDUNG,
DER WIEDERGABE AUF PHOTOMECHANISCHEM ODER ÄHNLICHEM WEGE
UND DER SPEICHERUNG IN DATENVERARBEITUNGSANLAGEN, BLEIBEN,
AUCH BEI NUR AUSZUGSWEISER VERWERTUNG, VORBEHALTEN

THIS WORK IS SUBJECT TO COPYRIGHT

ALL RIGHTS ARE RESERVED, WHETHER THE WHOLE OR PART OF THE MATERIAL
IS CONCERNED, SPECIFICALLY THOSE OF TRANSLATION, REPRINTING, RE-USE
OF ILLUSTRATIONS, BROADCASTING, REPRODUCTION BY PHOTOCOPYING
MACHINE OR SIMILAR MEANS, AND STORAGE IN DATA BANKS

© 1971 BY SPRINGER-VERLAG / WIEN

LIBRARY OF CONGRESS CATALOG CARD NUMBER AC 39-1015

PRINTED IN AUSTRIA

ISBN 3-211-81024-2 Springer-Verlag Wien–New York
ISBN 0-387-81024-2 Springer-Verlag New York–Wien

Inhaltsverzeichnis

Contents

Vorkommen, Struktur und Biosynthese natürlicher Piperidinverbindungen. Von D. GROSS, Institut für Biochemie der Pflanzen, Deutsche Akademie der Wissenschaften, Halle (Saale), Weinberg, DDR ... 1

Einführung ... 2

I. Einfache Piperidinderivate ... 3
  1. Piperidin und Methylpiperidine ... 3
  2. Piperidin- und Piperideincarbonsäuren ... 4
  3. Biosynthese der Pipecolinsäure ... 6

II. Aliphatisch substituierte Piperidinbasen ... 8
  1. N-Substituierte Piperidine (*Piper*-Alkaloide) ... 8
  2. α-Substituierte Piperidine ... 9
    a) *Conium*-Alkaloide ... 9
    b) *Punica*-Alkaloide ... 12
    c) Strukturähnliche *Sedum*-, *Lobelia*- und *Haloxylon*-Basen ... 14
    d) *Withania*-Alkaloide ... 16
    e) Febrifugin und Isofebrifugin ... 17
    f) Nigrifactin ... 18

III. Aliphatisch α,α′-disubstituierte Piperidine ... 18
  1. Pinidin ... 18
  2. *Lobelia*- und *Sedum*-Alkaloide ... 19
  3. *Cassia*- und *Prosopis*-Alkaloide ... 22
  4. *Caria*- und *Azima*-Alkaloide ... 23
  5. Lythraceen-Alkaloide ... 24

IV. Heterocyclisch substituierte Piperidine ... 24
  1. Anabasin- und Tetrahydroanabasinalkaloide ... 24
  2. Lobinalin ... 29
  3. *Ormosia*-Alkaloide ... 31
  4. Lamprolobin, Aphyllinsäuremethylester und Leontiformin ... 31
  5. *Nuphar*-Alkaloide ... 32
  6. Piperidinhaltige Indolalkaloide (Secamine, Secodine und Nitrarin) ... 33

V. Monoterpenoide Piperidinalkaloide ... 34

VI. Verschiedenartige Piperidinstrukturen ... 36
  1. Alkaloide ... 36
  2. Betalaine ... 38
  3. Antibiotica ... 40

VII. Schlußbetrachtung ... 41

Literaturverzeichnis ... 42

Gallenfarbstoffe und Biliproteide. Von W. RÜDIGER, Botanisches Institut der Universität München, BRD 60

    I. Einleitung... 61

   II. Nomenklatur 62

  III. Chemische Untersuchungsmethoden 64
       1. Farbreaktionen 64
           Die Gmelin-Reaktion und ihre Erweiterung 64
           Die Jaffe-Schlesinger-Reaktion und ihre Erweiterung 66
           Die Diazoreaktion 66
       2. Abbaureaktionen 68
           Abbau mit Permanganat 69
           Abbau mit Chromsäure und Chromat 70

   IV. Physikalische Untersuchungsmethoden 73
       1. Elektronenspektren 73
       2. Optische Aktivität 76
       3. Massenspektren 79
       4. NMR-Spektren 84
       5. Chromatographie 86

    V. Bilirubin 89
       1. Bilirubin-Konjugate 90
       2. Bilirubin-Proteide 91

   VI. Umwandlungsprodukte des Bilirubins 94
       1. Bilane und Bilene-(b) *(Urobilinoide)* 94
       2. Biladiene-(a,b) 99

  VII. Bilatriene 104
       1. Biliverdin und Mesobiliverdin 104
       2. Biliverdin und Biliverdin-Proteide bei Vertebraten 104
       3. Bilatriene bei Invertebraten 108

VIII. Gallenfarbstoffe mit Äthylidengruppe 111
       1. Aplysia-Farbstoffe 111
       2. Phycobiliproteide 114
           Phycobiline 118
       3. Phytochrom 124

Literaturverzeichnis... 128

The Chemistry of Glutarimide Antibiotics. By F. JOHNSON, The Dow Chemical Company, Eastern Research Laboratory, Wayland, Massachusetts, USA... 140

    I. Introduction 140

   II. The Chemistry of the Glutarimide Antibiotics 141
       1. Nomenclature 141
       2. Isolation and Determination 145

3. The Structure of Cycloheximide and Its Isomers 146
   a) The Gross Structures of Cycloheximide, Isocycloheximide and Naramycin-B 146
   b) Absolute Configuration 151
   c) Fine Structure... 152
   d) Miscellaneous Chemistry 168
4. The Streptovitacins and E-73..................................... 174
5. Inactone 179
6. Actiphenol (C-73) 180
7. Streptimidone and Protomycin 181
8. Fermicidin, Niromycin-A and Niromycin-B 186

III. Synthesis... 186
   1. Cycloheximide, Naramycin-B and Isocycloheximide 187
   2. α-Epiisocycloheximide 191
   3. Actiphenol... 193
   4. Homologs and Analogs of Cycloheximide and Other Related Substances 193

IV. Biosynthesis 197

References.. 202

Chemie und Biosynthese der Flechtenstoffe. Von S. HUNECK, Institut für Biochemie der Pflanzen, Deutsche Akademie der Wissenschaften, Halle (Saale), Weinberg, DDR.................................... 209

I. Einleitung... 209

II. Methoden zum Nachweis und zur Strukturaufklärung der Flechtenstoffe 210
   A. Dünnschichtchromatographie................................... 210
   B. Papierchromatographie 211
   C. Gaschromatographie 211
   D. Infrarotspektroskopie... 211
   E. Ultraviolettspektroskopie 211
   F. NMR-Spektroskopie 212
   G. Massenspektrometrie 213
   H. Röntgenstrukturanalyse 215
   I. Chemische Methoden.. 215

III. Einteilung der Flechtenstoffe 216

IV. Strukturaufklärung und Synthese der Flechtenstoffe 216
   1. Produkte des Primärstoffwechsels 216
   2. Acetogenine... 220
   3. Phenylalanin-Derivate 269
   4. Vitamine 273
   5. Enzyme 273

V. Biosynthese der Flechtenstoffe 273

VI. Aus Mycobionten isolierte Verbindungen 285

VII. Chemotaxonomie der Flechten 285

VIII. Antibiotische und weitere biologische Wirkungen der Flechtenstoffe.. 287

Literaturverzeichnis.. 288

The Cucurbitanes, a Group of Tetracyclic Triterpenes. By D. LAVIE and E. GLOTTER, Department of Chemistry, The Weizmann Institute of Science, Rehovot, Israel.................................... 307

    I. Introduction 308

   II. The Carbon Skeleton up to 1960 309

  III. Nomenclature 310

  IV. Structure Determination and Chemistry of Cucurbitacins........... 311

      1. Cucurbitacins B, D, E and I 311
         1.1. Interrelationship Between Cucurbitacins B (**5**), D (Elatericin A) (**6**), E (Elaterin) (**2**) and I (Elatericin B) (**7**).................. 311
         1.2. The Skeleton 312
         1.3. The Side Chain 313
         1.4. The Ring A Substituents; the α-Hydroxy-ketone (**5**), (**6**) and the Diosphenols (Enolized α-Diketone) (**2**), (**7**)................ 314
         1.5. The 19-Methyl Group and Ring C Carbonyl................ 316
         1.6. The Ring B Double Bond 317
         1.7. The 16-Hydroxy Group 322
         1.8. The Alkaline Treatment of Elaterin 323

      2. Cucurbitacins A (**61**) and C (**69**) 325
         2.1. Structure Determination................................ 325
         2.2. Interrelationship Between Cucurbitacins A, B and C 329

      3. Stereochemistry of Cucurbitacins 329

      4. Stereochemistry of Ring A Ketols.............................. 332

      5. Interrelationship Between the Cucurbitane and Lanostane Series..... 334

      6. Synthesis of a 32-nor-Cucurbitane Skeleton...................... 336

   V. Cucurbitacins G, H, L, J, K, Dihydrocucurbitacin B and 22-Deoxocucurbitacin D 337

      1. Cucurbitacins G and H (**108**) 337
      2. Cucurbitacins J, K (**109**) and L (**110**) 337
      3. Dihydrocucurbitacin B (**111**) 338
      4. 22-Deoxocucurbitacin D (**112**) 339
      5. "β-Elaterin" 339

  VI. Isocucurbitacin B (**119**), 22-Deoxoisocucurbitacin D (**120**) and Tetrahydrocucurbitacin I (**121**) 341

      1. Isocucurbitacin B (**119**) 341
      2. 22-Deoxoisocucurbitacin D (**120**) 341
      3. Tetrahydrocucurbitacin I (**121**) 342

 VII. Bryodulcosigenin (**123**), Bryosigenin (**124**), Bryogenin (**125**), Gratiogenin (**126**) and 16-Hydroxygratiogenin (**127**) 343

VIII. Cucurbitacin F (**134a**), O (**135a**), P (**136a**), and Q (**137a**)........... 346

  IX. Cucurbitacins of Unknown Structure 348

X. Biogenetic Aspects 348

XI. Physical Methods in the Structure Elucidation of the Cucurbitacins.. 350

XII. Biological Properties of the Cucurbitacins 351

XIII. Tables... 352
    1. Occurrence of the Cucurbitacins in Nature 352
    2. Physical Constants of the Cucurbitacins 356

References 357

Biogenetic-type Synthesis of Terpenoid Systems. By D. GOLDSMITH, Department of Chemistry, Emory University, Atlanta, Georgia, USA 363

Introduction 363
  I. Theory of Polyene Cyclization 364
 II. Acid Catalyzed Cyclization 366
III. Oxidative Cyclization 369
 IV. Arene and Alkyl Sulfonates, Acetals and Allylic Alcohols 378
  V. Cyclopropyl Ketones, Enols, and Tertiary Alcohols 384
 VI. Carbonium Ion Catalyzed Cyclization.............................. 390
VII. Radical Cyclization 390
References 391

The Biosynthesis of the Diterpenes. By J. R. HANSON, Chemical Laboratory, The University of Sussex, Falmer, Brighton, England 395

  I. Introduction 395

 II. The Biogenesis of the Diterpenes.................................. 396
    1. The Bicyclic Diterpenes 397
    2. The Tricyclic Diterpenes 399
    3. The Tetracyclic Diterpenes 401
    4. The Macrocyclic Diterpenes....................................... 403

III. Biosynthetic Evidence 403
    1. The Bi- and Tricyclic Diterpenes.................................. 403
    2. The Tetracyclic Diterpenes 406

 IV. Conclusion.. 412

References 413

Chemistry of Natural Products Derived from Marine Sources. By E. PREMUZIC, World Life Research Institute, Colton, California, USA 417

  I. Introduction.. 417
 II. Steroids... 418
III. Sapogenins of Marine Origin...................................... 425

IV. Bile Alcohols and Bile Acids 431
   V. Terpenes and Related Hydrocarbons............................ 435
  VI. Halogen-Containing Compounds 446
 VII. Non-Proteinoid Nitrogen-Containing Substances................. 451
VIII. Quinonoid and Related Pigments 460
  IX. Carbohydrates 467
   X. Related Topics 469
Addendum... 469
References 472

Namenverzeichnis. Author Index..................................... 489

Sachverzeichnis. Subject Index..................................... 510

Vorkommen, Struktur und Biosynthese natürlicher Piperidinverbindungen

Von D. GROSS, Halle (Saale)

Inhaltsübersicht

	Seite
Einführung	2
I. Einfache Piperidinderivate	3
1. Piperidin und Methylpiperidine	3
2. Piperidin- und Piperideincarbonsäuren	4
3. Biosynthese der Pipecolinsäure	6
II. Aliphatisch substituierte Piperidinbasen	8
1. N-Substituierte Piperidine (*Piper*-Alkaloide)	8
2. α-Substituierte Piperidine	9
a) *Conium*-Alkaloide	9
b) *Punica*-Alkaloide	12
c) Strukturähnliche *Sedum*-, *Lobelia*- und *Haloxylon*-Basen	14
d) *Withania*-Alkaloide	16
e) Febrifugin und Isofebrifugin	17
f) Nigrifactin	18
III. Aliphatisch α,α'-disubstituierte Piperidine	18
1. Pinidin	18
2. *Lobelia*- und *Sedum*-Alkaloide	19
3. *Cassia*- und *Prosopis*-Alkaloide	22
4. *Caria*- und *Azima*-Alkaloide	23
5. Lythraceen-Alkaloide	24
IV. Heterocyclisch substituierte Piperidine	24
1. Anabasin- und Tetrahydroanabasinalkaloide	24
2. Lobinalin	29
3. *Ormosia*-Alkaloide	31
4. Lamprolobin, Aphyllinsäuremethylester und Leontiformin	31
5. *Nuphar*-Alkaloide	32
6. Piperidin-haltige Indolalkaloide (Secamine, Secodine und Nitrarin)	33
V. Monoterpenoide Piperidinalkaloide	34
VI. Verschiedenartige Piperidinstrukturen	36
1. Alkaloide	36
2. Betalaine	38
3. Antibiotika	40
VII. Schlußbetrachtung	41
Literaturverzeichnis	42

Einführung

Eine große Anzahl heterocyclischer Naturstoffe leitet sich vom Ringsystem des Piperidins (1) ab. Als Substituenten einfach oder mehrfach substituierter Piperidinbasen finden sich Methyl-, Carboxyl-, Hydroxyl- und Aminogruppen sowie aliphatische Seitenketten unterschiedlicher Länge. Die Substitution erfolgt bevorzugt an den C-Atomen 2, 3 und 6 sowie am Heteroatom. In zahlreichen Fällen ist der Piperidinring in α- oder β-Stellung direkt oder über eine C-Brücke mit einem weiteren Heterocyclus verbunden, z. B. einem Piperidin-, Piperidein-, Pyridin-, Indol-, Chinolizidin- oder Furanrest. Darüber hinaus kann der Piperidinring zum 2,6-Dioxopiperidin (Glutarimid) oxydiert oder zum Piperidein dehydriert sein.

Während Naturstoffe mit Pyridinstruktur im Tier- und Pflanzenreich weit verbreitet sind und einzelnen von ihnen wie NAD oder Pyridoxalphosphat als Coenzymen des Primärstoffwechsels besondere Bedeutung zukommt, handelt es sich bei den natürlichen Piperidinverbindungen im allgemeinen um sekundäre Pflanzenstoffe (vgl. *18*, *191*). Dabei sind einige wie z. B. die Pipecolinsäure (7) sporadisch auf verschiedene Pflanzenfamilien verteilt. Andere Piperidinbasen wie die *Conium*- oder *Piper*-Alkaloide weisen dagegen eine ausgesprochene Artspezifität auf. Im Gegensatz zu den meisten Pyridinalkaloiden finden sich die Piperidinbasen oft mit strukturell andersartig gebauten Alkaloiden vom Chinolizidin- oder Tropantyp vergesellschaftet, was in den meisten Fällen durch eine enge biogenetische Verwandtschaft bedingt sein dürfte.

Trotz enger struktureller Ähnlichkeit liegt den natürlich vorkommenden Pyridin- und Piperidinverbindungen kein gemeinsames Entstehungsprinzip zugrunde. Die Biosynthese des Pyridinringes ist sehr intensiv bearbeitet und in ihren Grundzügen weitgehend aufgeklärt (vgl. *119*, *120*). Dagegen sind umfassende Untersuchungen für den Piperidinring bisher nur für die Pipecolinsäure und für einige Alkaloide bekannt geworden (vgl. *66*, *123*, *191*). Der Grund dafür dürfte in dem oft schwierig zu beschaffenden oder zu kultivierenden bzw. für Biosyntheseversuche wenig geeigneten Pflanzenmaterial zu suchen sein.

In einer früheren Übersicht sind bereits Struktur und Biosynthese natürlich vorkommender Pyridinverbindungen besprochen worden (*120*). Die in dieser Arbeit unberücksichtigt gebliebenen Naturstoffe mit Piperidinstruktur weisen vielfach eine enge strukturelle Verwandtschaft auf und werden in vorliegender Zusammenfassung behandelt. Dabei wird insbesondere auf Vorkommen, Konstitution und Biosynthese eingegangen. Methodische Details der Isolierung sowie der Struktur- und Konfigurationsaufklärung sind der zitierten Originalliteratur zu entnehmen.

Literaturverzeichnis: SS. 42—59

I. Einfache Piperidinderivate

1. Piperidin und Methylpiperidine

Piperidin (1), der Grundkörper der nachfolgend zu besprechenden Naturstoffklasse, soll nach älteren Angaben in *Nicotiana tabacum* L., *Petrosimonia monandra* Bunge, *Piper nigrum* L. und *Psilocaulon absimile* N. E. Brown enthalten sein (vgl. *39*). Ein neuerer Nachweis stammt aus der Arbeitsgruppe um SANDBERG, wonach Piperidin neben einigen Piperidinalkaloiden in der Chenopodiacee *Haloxylon salicornicum* (Moq.-Tand.) Boiss. vorkommt (*212*).

Δ^1-Piperidein (2) stellt als Produkt des Lysinstoffwechsels die Vorstufe verschiedener vom Lysin (19) abgeleiteter Piperidinalkaloide dar (vgl. Schema 1, 3 und 5).

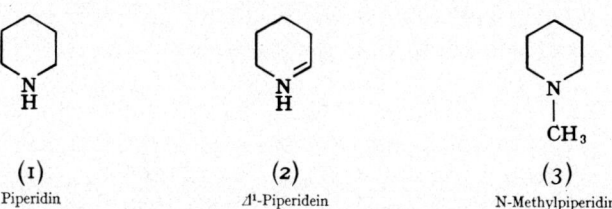

(1) Piperidin   (2) Δ^1-Piperidein   (3) N-Methylpiperidin

Darüber hinaus finden sich auch einige methylsubstituierte Piperidine in höheren Pflanzen. YURASHEWSKII und STEPANOVA haben N-Methylpiperidin (3) in den Chenopodiaceen *Girgensohnia oppositiflora* Fenzl und *G. diptera* Bunge aufgefunden (*351*). Das entsprechende N-Oxid (4) isolierten 1970 BRANDÄNGE und LÜNING (*43*) aus *Vandopsis longicaulis* Schltr., einer Orchidacee, in der auch die aus dem Tierreich bereits bekannte N-Methylpyridinium-Verbindung enthalten ist (*43*). Als weitere pflanzliche Piperidine sind (+)-α-Pipecolin (5) aus *Pinus sabiniana* Dougl. und anderen *Pinus*-Arten (*319*) sowie das in *Nanophyton erinaceum*

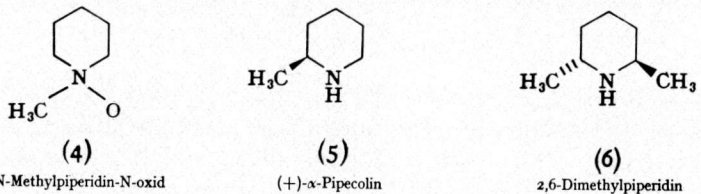

(4) N-Methylpiperidin-N-oxid   (5) (+)-α-Pipecolin   (6) 2,6-Dimethylpiperidin

(Pallas) Bunge (*39*) und *Anabasis salsa* Paulsen (*350*) entdeckte 2,6-Dimethylpiperidin (6) zu nennen. Während der letzteren Verbindung 2 R : 6 R-Konfiguration zukommt (*148*), konnte für (+)-α-Pipecolin (5) S-Konfiguration ermittelt werden (*258*).

Zur Biosynthese des Piperidins (1) und der erwähnten Methylpiperidine liegen noch keine experimentellen Befunde vor. Möglicherweise stellt die Aminosäure Lysin (19) den Precursor der Basen (1)—(4) dar. Für α-Pipecolin (5) diskutiert LEETE eine Bildung aus drei Molekülen Acetat (*184*) über 5-Oxohexansäure (*187 a*).

2. Piperidin- und Piperideincarbonsäuren

Aus der Pyridinreihe sind mehrere natürlich vorkommende Mono- oder Dicarbonsäuren wie z. B. Nicotinsäure (78) oder Chinolinsäure bekannt. Dagegen hat man in der Natur als Piperidincarbonsäure bisher nur die L(—)-Pipecolinsäure (7) und einige von ihr abgeleitete Verbindungen gefunden.

Die als cyclische Iminosäure aufzufassende Pipecolinsäure ist im Tierreich (z. B. *214, 236*, vgl. *39*) und in der Pflanzenwelt weit verbreitet: *Trifolium repens* L. (*218*), *Phaseolus vulgaris* L. (*118, 352, 353*), *P. angularis* W. F. Wight (*140*) und *P. aureus* Roxb. (*100*), *Leucaena glauca* Benth. (*141, 153*), *Pyrus malus* L. (*152*), *Strophanthus scandens* Griff. (*276*), *Acacia*-Arten (*60, 112, 304*), *Albizzia*-Arten (*174*), *Morus alba* L. (*171*), *Nicotiana tabacum* L. (*324*), *Thea sinensis* L. (*239*), *Humulus lupulus* L. (*130, 131*), *Sedum acre* L. (*127*), *Mimosa pudica* L. (*323*), verschiedenen Gramineen (*97, 130, 131, 150, 151, 165, 244*) und in einer Reihe weiterer höherer Pflanzen (*130, 131, 154, 224, 242, 243, 244*, vgl. *39*).

Entsprechend dem natürlich vorkommenden Trigonellin, das durch N-Methylierung der Nicotinsäure entsteht, findet man in einigen höheren Pflanzen das N-Methylbetain der Pipecolinsäure. Diese als Homostachydrin (8) bezeichnete linksdrehende Verbindung wurde 1958 von

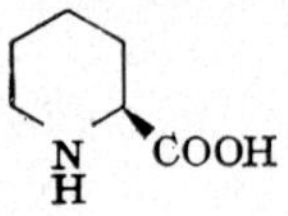

(7) Pipecolinsäure   (8) Homostachydrin

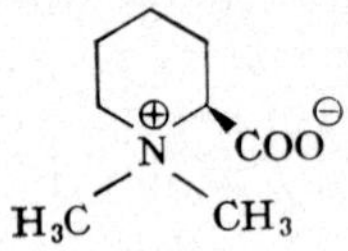

WIEHLER und MARION aus den Samen von *Medicago sativa* L. Grimm *(Leguminosae)* isoliert (*341*) und wenig später von PAILER und KUMP in *Achillea moschata* Wulf. und *A. atrata* L. *(Compositae)* nachgewiesen (*241*). Die Aufklärung der absoluten Konfiguration erfolgte 1959 unabhängig durch zwei Arbeitsgruppen (*27, 261*). In Biosyntheseuntersuchungen, die im gleichen Jahr von ROBERTSON und MARION an Luzerne vorgenommen worden sind, konnte überraschenderweise kein Einbau von Lysin-(2-^{14}C) in Homostachydrin beobachtet werden (*262*). Es ist denk-

Literaturverzeichnis: SS. 42—59

Vorkommen, Struktur und Biosynthese natürlicher Piperidinverbindungen 5

bar, daß diese Fütterungsversuche zu einem Zeitpunkt durchgeführt worden sind, in dem sich das Pflanzenmaterial in einem für eine optimale Homostachydrinsynthese ungeeigneten physiologischen Zustand befand.

Von den pflanzlichen Pipecolinsäure-Abkömmlingen sind zwei hydroxylierte Verbindungen zu nennen. (—)-4-Hydroxypipecolinsäure (9) wurde 1955 von VIRTANEN et al. in verschiedenen *Acacia*-Arten sowie in *Lysiloma bahamense* Benth., *Albizzia lophantha* Benth. und *Strelitzia reginae* Banks aufgefunden (*335, 337*). FOWDEN hat diese hydroxylierte Iminosäure 1958 in *Armeria maritima* Willd. (*96*) und später auch in verschiedenen *Acacia*-Arten nachgewiesen (*98, 304*). CLARK-LEWIS und Mitarb. berichten über das Vorkommen in anderen *Acacia*-Arten und haben trans-Konfiguration für die von ihnen isolierte Substanz ermittelt (*58, 59, 60*). Weitere Angaben zum Vorkommen von (9) stammen von GMELIN (*112*) sowie von KRAUS und REINBOTHE (*174*). Nach SCHENK und SCHÜTTE ist 4-Hydroxypipecolinsäure (9) neben der erstmals von diesen Autoren aufgefundenen 4-Aminopipecolinsäure (10) in *Strophanthus scandens* Griff. enthalten (*275, 276*).

OH	NH₂	HO,,,
(9)	(10)	(11)
4-Hydroxypipecolinsäure	4-Aminopipecolinsäure	5-Hydroxypipecolinsäure

Weiter verbreitet scheint die 5-Hydroxypipecolinsäure (11) zu sein. Diese von VIRTANEN und KARI 1954 in *Rhapis excelsa* Henry ex Rehder entdeckte Verbindung (*336*) findet sich nach Angaben anderer Autoren in zahlreichen höheren Pflanzen: *Baikiaea plurijuga* Harms (*117*), *Salix fragilis* L. (*37a*), *Leucaena glauca* Benth. (*141, 153*), verschiedenen *Acacia*-Arten (*112, 304, 337*), *Albizzia*-Arten (*174*), *Morus alba* L. (*171*), *Strelitzia reginae* Banks (*337*) und anderen höheren Pflanzen. WITKOP und FOLTZ haben für die natürlich vorkommende 5-Hydroxypipecolinsäure (11) trans-Konfiguration nachgewiesen (*345*).

Neben der Pipecolinsäure (7) und den vorstehend aufgeführten Hydroxy- und Aminoderivaten (9), (10), (11) finden sich in der Natur einige Piperideinmonocarbonsäuren.

Δ^1-Piperidein-2-carbonsäure (12) und Δ^1-Piperidein-6-carbonsäure (13) sind Stoffwechselprodukte des Lysins (19) und sollen hier nicht näher besprochen werden.

Die als Baikiain (14) bezeichnete Δ^4-Piperidein-2-carbonsäure wurde 1950 von KING und Mitarb. aus *Baikiaea plurijuga* Harms (*164*) isoliert und ist später auch in einigen anderen Pflanzen nachgewiesen worden.

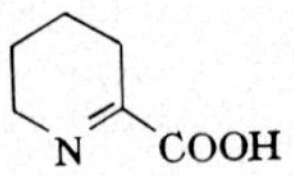

(12) Δ¹-Piperidein-2-carbonsäure
(13) Δ¹-Piperidein-6-carbonsäure 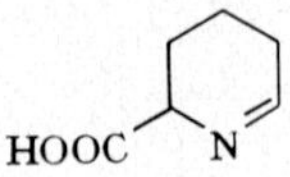
(14) Baikiain 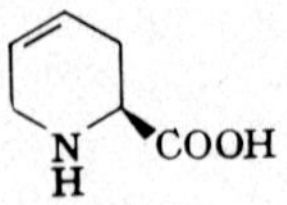

Bekannter und von biogenetischem Interesse sind die *Areca*-Alkaloide, die sich von der Δ³-Piperidein-3-carbonsäure ableiten (vgl. *39*). Diese schon Ende des vorigen Jahrhunderts von JAHN in *Areca catechu* L. aufgefundenen Pflanzenbasen Arecaidin (15) und Arecolin (16) sowie Guvacin (17) und Guvacolin (18) zeigen wegen ihrer β-ständigen Carboxylgruppe möglicherweise biogenetische Verwandtschaft zur Nicotinsäure (78). Leider liegen noch keine Biosyntheseergebnisse vor, so daß der Bildungsweg dieser Piperideinalkaloide noch ungeklärt ist.

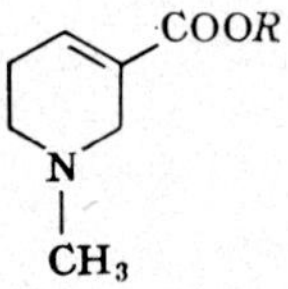

(15) R = H Arecaidin
(16) R = CH₃ Arecolin
(17) R = H Guvacin 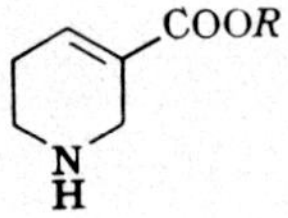
(18) R = CH₃ Guvacolin

3. Biosynthese der Pipecolinsäure

Die Biosynthese der Pipecolinsäure (7) ist an verschiedenen biologischen Objekten eingehend untersucht worden. Als Resultat dieser Arbeiten hat sich ergeben, daß Pipecolinsäure offenbar in allen Organismen aus der Aminosäure Lysin (19) gebildet wird. Bei der Entstehung der Pipecolinsäure aus Lysin geht ein Stickstoffatom verloren, so daß sich die Frage ergibt, ob der Heterostickstoff der α- oder der ε-Aminogruppe des Lysins entstammt. Beide Biosynthesemöglichkeiten sind in Schema 1 dargestellt.

Für den tierischen Organismus wurde an Ratten (*42, 127, 269, 270*) und an Leberhomogenaten (*42, 194*) ein Einbau von ^{14}C-markiertem Lysin in Pipecolinsäure nachgewiesen. ROTHSTEIN und MILLER konnten bereits 1954 durch Versuche mit Lysin-(^{15}N) zeigen, daß der Heterostickstoff der Pipecolinsäure dem ε-Aminostickstoff des Lysins entspricht (*268*). Somit ist eine oxydative α-Desaminierung des Lysins zu ε-Amino-α-ketocapronsäure (20) anzunehmen. Nachfolgende Cyclisierung zu Δ¹-Piperidein-2-carbonsäure (12) würde nach Hydrierung zu Pipecolinsäure (7) führen.

Literaturverzeichnis: SS. 42—59

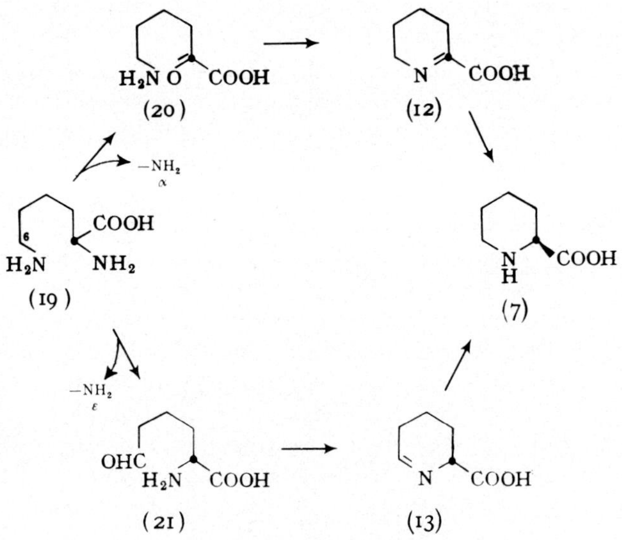

Vorkommen, Struktur und Biosynthese natürlicher Piperidinverbindungen 7

Schema 1. Biosynthese der Pipecolinsäure (7)

Neuere Versuche von GUPTA und SPENSER mit Lysin-(6-^{14}CT$_2$) an Ratten stehen im Einklang mit diesem Biosyntheseweg (*127*). Die isolierte Pipecolinsäure wies dasselbe ^{14}C/T-Verhältnis wie das der applizierten Vorstufe auf. Der Gegenversuch mit Lysin-(2-T,6-^{14}C) ergab einen T-Verlust (*127a*).

SCHWEET et al. haben 1954 für eine Lysin-Mangelmutante von *Neurospora crassa* gezeigt, daß Lysin-(1-^{14}C) in Pipecolinsäure (7) und ε-Amino-α-ketocapronsäure (20) eingebaut wird (*303*). Dieser Lysineinbau ist kürzlich von GUPTA und SPENSER für *N. crassa* bestätigt worden (*127*). Somit wird in diesem Pilz derselbe Biosyntheseweg der Pipecolinsäure wie im tierischen Organismus beschritten.

Für die höhere Pflanze ist die Inkorporation von ^{14}C-markiertem Lysin in Pipecolinsäure mehrfach beschrieben (*41, 98, 99, 118, 127, 199, 229, 323*). Nach Angaben von SCHÜTTE und SEELIG entstammt der Heterostickstoff der Pipecolinsäure bei *Phaseolus vulgaris* L. der α-Aminogruppe des Lysins (*301*), so daß α-Aminoadipinsäure-δ-semialdehyd (21) als Intermediärprodukt anzunehmen ist. Dieser Befund steht in Übereinstimmung mit Arbeiten an *Aspergillus nidulans*, wonach die Aminogruppe von ^{15}N-markierter α-Aminoadipinsäure in Pipecolinsäure eingebaut wird (*15*). Untersuchungen von GUPTA und SPENSER an *Phaseolus vulgaris* L. und an *Sedum acre* L. haben jedoch gezeigt, daß in diesen Pflanzen Lysin-(6-^{14}CT$_2$) mit weitgehend intaktem ^{14}C/T-Verhältnis in

Pipecolinsäure eingeht (*127*), während Lysin-(2-T,6-^{14}C) unter T-Verlust inkorporiert wird (*127a*).

Auf Grund dieses Befundes diskutieren die Autoren einen Biosyntheseweg über ε-Amino-α-ketocapronsäure (20). Die widersprüchlich erscheinenden Angaben der genannten Arbeitsgruppen lassen noch keine endgültige Entscheidung zu, welches Bildungsprinzip für die Pipecolinsäure in der höheren Pflanze realisiert ist. Vielleicht sind mehrere Entstehungsmechanismen vorhanden.

Die hydroxylierten Pipecolinsäuren (9) und (11) können entweder durch stellungs- und stereospezifische Hydroxylierung der Pipecolinsäure oder durch Cyclisierung einer bereits hydroxylierten offenkettigen Vorstufe wie z. B. Hydroxylysin entstehen. Für die erste Möglichkeit sprechen Versuche von FOWDEN an *Acacia homalophylla* A. Cunn. *ex* Benth. (*98*) und von SCHENK et al. an *Strophanthus scandens* Griff. (*276*). In *Strophanthus* wurde ein direkter Übergang von Pipecolinsäure (7) in 4-Hydroxypipecolinsäure (9) und 4-Aminopipecolinsäure (10) nachgewiesen, wobei 4-Oxopipecolinsäure als Vorstufe der Aminoverbindung postuliert wird (*276*). FOWDEN diskutiert, daß die Entstehung der 4- bzw. 5-Hydroxypipecolinsäure möglicherweise durch Wasseranlagerung an Baikiain (14) erfolgt (*101*). Dafür könnte das gemeinsame Vorkommen von Baikiain und hydroxylierten Pipecolinsäuren in *Acacia*-Arten sprechen.

Andererseits ist an Rattenleberhomogenaten (*194*) und Heuschrecken (*321*) nachgewiesen, daß δ-Hydroxylysin-(6-^{14}C) in 5-Hydroxypipecolinsäure (11) eingebaut wird. Somit könnte die Hydroxylierung auch auf der Stufe des Lysins erfolgen. Demgegenüber stehen aber andere, an Ratten durchgeführte Experimente, bei denen die ^{14}C-Aktivität von appliziertem δ-Hydroxylysin-(6-^{14}C) nur auf indirektem Weg in 5-Hydroxypipecolinsäure gelangt (*245a*).

II. Aliphatisch substituierte Piperidinbasen

1. N-Substituierte Piperidine (*Piper*-Alkaloide)

Die Wurzeln verschiedener *Piper*-Arten enthalten neben Piperidin (1) einige artspezifische Alkaloide. Es handelt sich um Piperidide, deren Säurekomponente (z. B. Piperinsäure) amidartig an einen Piperidinring gebunden ist. Man kennt darüber hinaus auch einige *Piper*-Alkaloide wie Piperlonguminin und Peepuloidin, bei denen an Stelle des Piperidins ein Isobutylamin- bzw. Pyrrolidylrest vorhanden ist.

Zu den Piperidin-haltigen Pflanzenbasen gehören das für den scharfen Pfeffergeschmack verantwortliche Piperin (22) und das Piperettin (23) (vgl. *39*). GREWE und Mitarb. haben 1970 gezeigt, daß früher als Chavicin bezeichnetes *cis,cis*-Piperin (vgl. *39*) nicht mit diesem identisch ist und keinen Piperidinabkömmling darstellt und daß neben *trans,trans*-Piperin

Literaturverzeichnis: SS. 42—59

(22) kein weiteres Isomere in *Piper nigrum* L. vorkommt (*115a*). Die gleiche Arbeitsgruppe hat aus *Piper nigrum* einige Nebenalkaloide isoliert und charakterisiert, von denen Piperolein A (24) und Piperolein B (24a) Piperidin-haltig sind. Als weiteres Alkaloid dieser Gruppe ist das 1968 von Joshi und Mitarb. aus *Piper longum* L. isolierte Piplartin (25) zu nennen (*156*). Diese Base enthält einen Trimethoxycinnamoylrest und ist mit dem von indischen Autoren bereits früher aufgefundenen Piperlongumin (*16, 53, 54*) identisch. Die Lage der Δ^3-Doppelbindung im 2-Oxopiperideinrest ist durch neuere Untersuchungen gesichert (*156*).

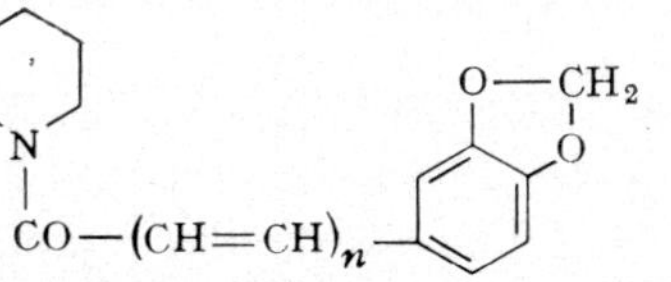

(22) n = 2 Piperin *(trans,trans)*
(23) n = 3 Piperettin

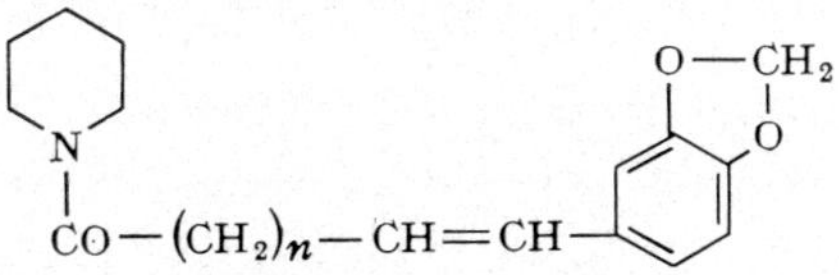

(24) n = 4 Piperolein A
(24a) n = 6 Piperolein B

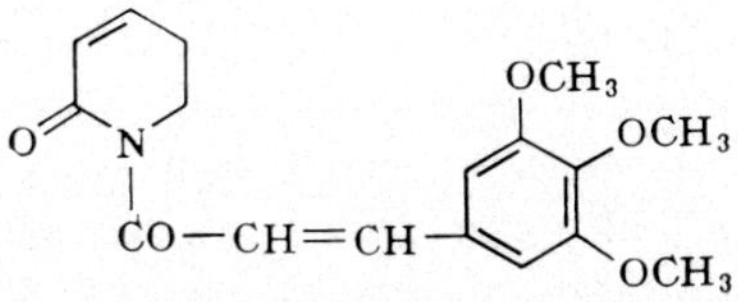

(25) Piplartin
(Piperlongumin)

Zur Biosynthese der *Piper*-Alkaloide liegen noch keine experimentellen Befunde vor. Es ist anzunehmen, daß die Säurekomponente des Piplartins (25) aus dem Zimtsäurestoffwechsel hervorgeht. Bei den Alkaloiden (22)—(24a) sollten zusätzliche C_2-Einheiten zur Kettenverlängerung an den C_6—C_3-Körper angeknüpft werden.

2. α-Substituierte Piperidine

a) Conium-Alkaloide

Die schon seit dem Altertum bekannte Giftwirkung von *Conium maculatum* L. *(Umbelliferae)* ist vor allem dem DL-Coniin (26) zuzuschreiben. Weitere *Conium*-Alkaloide sind (+)- und (—)-N-Methylconiin (27), γ-Conicein (28), (+)-Conhydrin (29) und das isomere Pseudo-

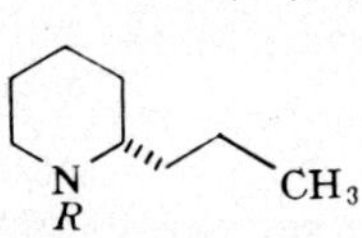

(26) R = H Coniin.
(27) R = CH_3 N-Methylconiin

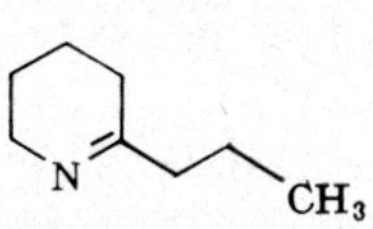

(28) γ-Conicein

conhydrin (30). Diesen Pflanzenbasen liegt ein Piperidin- bzw. $\varDelta^1$-Piperideinring mit einer α-ständigen unverzweigten C_3-Seitenkette zugrunde. Conhydrin (29) und Pseudoconhydrin (30) besitzen eine Hydroxylgruppe am C-1' des Propylrestes bzw. am C-5 des Piperidins.

(29) Conhydrin   (30) Pseudoconhydrin

Die Struktur dieser Pflanzenbasen ist schon seit längerem bekannt (vgl. *39*). Lediglich γ-Conicein (28) hat 1961/1962 durch BEYERMAN et al. (*31*) sowie durch BÜCHEL und KORTE (*47*) eine Korrektur erfahren, wonach sich dieses Alkaloid nicht vom $\varDelta^2$-, sondern vom $\varDelta^1$-Piperidein (2) ableitet. Die stereochemische Aufklärung des (2S : 1'R)-Conhydrins (29) und des (2S : 5S)-Pseudoconhydrins (30) verdanken wir im wesentlichen den Untersuchungen von HILL (*144, 145*), SICHER und TICHY (*305, 306*), BALENOVIC und STIMAC (*22*) sowie YANAI und LIPSCOMB (*349*). Pseudoconhydrin (30) gehört neben Cassin (56), Carnavallin (57), den *Prosopis*-Alkaloiden (58)—(62) und Febrifugin (45) zu den Hydroxypiperidinalkaloiden, unterscheidet sich jedoch von diesen durch die *trans*-Stellung des Propylrestes und der Hydroxylgruppe. Seit der klassischen Coniinsynthese durch LADENBURG vor etwa 100 Jahren sind verschiedene Darstellungsmethoden für die *Conium*-Alkaloide beschrieben worden (vgl. z. B. *74, 93, 246* und dort zitierte Literatur). Das Vorkommen der vorstehend genannten Alkaloide ist fast ausschließlich auf die Art *Conium maculatum* L. beschränkt.

<small>Das Auftreten von Coniin in *Parietaria officinalis* L. *(Urticaceae)* (*251*) bedarf möglicherweise einer Bestätigung.</small>

Nach pflanzenphysiologischen Untersuchungen von FAIRBAIRN und Mitarb. liegen Coniin (26) und γ-Conicein (28) in der Pflanze teilweise in gebundener Form vor (*87, 88*). Die durch klimatische oder tageszeitliche Schwankungen bedingten qualitativen und quantitativen Unterschiede im Alkaloidspektrum von *Conium maculatum* L. weisen darauf hin, daß die *Conium*-Alkaloide einem starken Stoffwechsel unterliegen und teilweise ineinander übergehen können (*65, 89, 90*). Der wechselseitige Übergang von γ-Conicein und Coniin ist durch ^{14}C-Markierung bestätigt worden (*88, 188*).

Zur Biosynthese der *Conium*-Alkaloide existieren zahlreiche Untersuchungen. Nach der Hypothese von ROBINSON entsteht Coniin aus Lysin, das nach oxydativer Desaminierung und Decarboxylierung in $\varDelta^1$-Piperidein übergeht und mit Acetoacetat unter nachfolgender De-

Literaturverzeichnis: SS. 42—59

carboxylierung und Reduktion Coniin ergibt (266). Hinweise für dieses Bildungsprinzip stammen von SCHIEDT und HÖSS (277, 278) sowie von CROMWELL und ROBERTS (67), wonach Lysin und seine Folgeprodukte Cadaverin und Δ^1-Piperidein in γ-Conicein und Coniin eingebaut werden sollen. Demgegenüber stehen Arbeiten von LEETE, der keine oder nur eine verschwindend niedrige Inkorporation dieser Verbindungen finden konnte (181, 187). Nach LEETE wird Coniin entsprechend Schema 2 aus 4 Acetat-Einheiten über eine Poly-β-ketosäure gebildet (181, 182). Nach Applikation von ^{14}C-markiertem Acetat konnte er den Radiokohlenstoff in den entsprechenden C-Atomen des Coniins nachweisen. Kürzlich hat LEETE gezeigt, daß Octansäure-(1-^{14}C) mit hohen Einbauraten in Coniin eingebaut wird, wobei die ^{14}C-Aktivität zu über 85% im C-Atom 6 des Coniins lokalisiert war (187, 187a). Das würde für einen direkten Einbau dieser C_8-Carbonsäure sprechen. Allerdings ergab der Gegenversuch mit Octansäure-(8-^{14}C) nicht eindeutig die erwartete ^{14}C-Verteilung, so daß weitere Experimente abgewartet werden müssen. Inkorporationsversuche mit 5-Oxooctansäure-(6-^{14}C) und dem entsprechenden Aldehyd haben gezeigt, daß diese Verbindungen mit hohen Einbauraten in Coniin inkorporiert werden, wobei die Radioaktivität ausschließlich im C-Atom 1' lokalisiert ist (187a, 190a). Es wird diskutiert, daß aus vier Acetat-Einheiten gebildete 5-Oxooctansäure zum Aldehyd reduziert wird, der einer Transaminierung und nachfolgenden spontanen Cyclisierung zum γ-Conicein unterliegt.

Als direkte Vorstufe des Coniins wird γ-Conicein diskutiert, das als Primäralkaloid entstehen und von dem sich Coniin, N-Methylconiin und Pseudoconiin ableiten sollen. Der direkte Übergang γ-Conicein → Coniin ist mit ^{14}C-markierten Verbindungen nachgewiesen (88, 188). Für eine

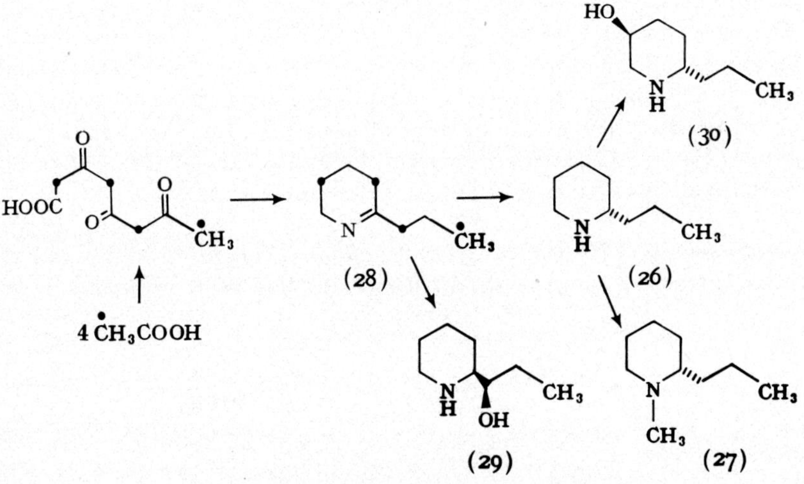

Schema 2. Biosynthese der *Conium*-Alkaloide

zentrale Rolle des γ-Coniceins sprechen auch $^{14}CO_2$-Kurzzeitversuche von DIETRICH und MARTIN (*72, 73*). Diese Arbeitsgruppe konnte bei ihren Versuchen 3-Formyl-4-hydroxy-2 H-pyran aus *Conium maculatum* L. isolieren (*170*). Möglicherweise stellt diese Verbindung ein Intermediärprodukt der Alkaloidbiosynthese dar.

b) *Punica-Alkaloide*

Das in der Rinde von *Punica granatum* L. *(Punicaceae)* enthaltene Isopelletierin (**31**) besitzt dasselbe C-Gerüst wie Coniin (**26**), unterscheidet sich von diesem aber durch die Ketogruppe in der C_3-Seitenkette. Isopelletierin wurde schon 1878/1880 von TANRET neben drei weiteren Alkaloiden aus *P. granatum* L. isoliert und ist damals als „Pelletierin" bezeichnet worden (vgl. *39*). Spätere Untersuchungen haben ergeben, daß es sich dabei und auch bei dem von HESS bearbeiteten Alkaloid um Isopelletierin gehandelt hat (*32, 79, 111*). Der Vorschlag von GILMAN und MARION (*111*), das heutige Isopelletierin in Pelletierin zurückzubenennen, ist in der Literatur nicht einheitlich durchgeführt worden. In der vorliegenden Arbeit wird für Verbindung (**31**) der Name Isopelletierin beibehalten.

Die Aufklärung der absoluten Konfiguration durch BEYERMAN und Mitarb. hat R-Konfiguration für das Asymmetriezentrum ergeben (*32, 34*). Synthesen des (—)-Isopelletierins stammen ebenfalls aus dieser Arbeitsgruppe (*32, 33*). Darüber hinaus kann Isopelletierin unter „zellmöglichen Bedingungen", d. h. *in vitro* im neutralen pH-Bereich, aus Δ^1-Piperidein (**2**) und Acetondicarbonsäure (*82, 83, 84*) bzw. Acetoacetat (*282, 344*) erhalten werden.

In *Punica granatum* findet sich Isopelletierin in der L- und DL-Form. Es wird von N-Methylisopelletierin (**32**) und einigen noch strukturunbekannten Alkaloiden begleitet (*177*, vgl. *39*). Als weiterer Inhaltsstoff wurde 1967 von ROBERTS et al. 2-(2'-Propenyl)-Δ^1-piperidein (**32 a**) isoliert und charakterisiert (*260*). Dieses Piperideinderivat wird als Vorstufe der *Punica*-Basen diskutiert. Weiterhin sind Isopelletierin (**31**) bzw. sein N-Methylderivat (**32**) von FRANCK (*103*) sowie von MARION und CHAPUT (*208*) in *Sedum*-Arten *(Crassulaceae)*, von MORTIMER et al. in *Duboisia myoporoides* R. Br. *(Solanaceae)* (*219, 220*) und von der Arbeitsgruppe um SCHWARTING in einer anderen Solanacee *Withania somnifera* Dunal

(**31**) R = H   Isopelletierin   (**32 a**)
(**32**) R = CH$_3$ N-Methylisopelletierin

Literaturverzeichnis: SS. 42—59

(*162*, *302*) nachgewiesen worden. In diesen Pflanzen finden sich außerdem typische *Sedum*-Alkaloide wie Sedamin (38) und Sedridin (33), Pyridin- oder Piperidinbasen wie Nicotin oder Anabasin (69) bzw. die charakteristischen *Withania*-Alkaloide wie Anaferin (43), Cuskhydrin, Tropin oder Pseudotropin. Die Zusammensetzung im Alkaloidspektrum deutet bereits eine mögliche biogenetische Verwandtschaft an.

Die Biosynthese des Isopelletierins (31) und seines N-Methylderivates ist 1968 etwa gleichzeitig von drei Arbeitsgruppen untersucht worden. Bei diesen von LIEBISCH *et al.* (*192*) sowie von O'DONOVAN und KEOGH (*161a*, *235*) an *Punica granatum* und von GUPTA und SPENSER an *Sedum sarmentosum* Bunge (*126*, *126a*) durchgeführten Arbeiten hat sich übereinstimmend ergeben, daß der Piperidinring aus dem Lysinstoffwechsel hervorgeht. Nach Verfütterung von Lysin-(2-^{14}C) und Lysin-(6-^{14}C) war die Radioaktivität ausschließlich im C-Atom 2 bzw. 6 des Piperidinringes lokalisiert (*126*, *161a*). Das bedeutet, daß der Lysineinbau nicht über ein symmetrisches Zwischenprodukt verläuft, wenn auch andererseits von außen appliziertes Cadaverin-(1,5-^{14}C) zur Isopelletierinbildung verwertet wird (*192*). Im Fütterungsexperiment mit Lysin-(6-^{14}C, 4,5-T) blieb das ^{14}C/T-Verhältnis von applizierter Vorstufe und isoliertem Alkaloid konstant (*126*). Die Autoren postulieren daher ε-Amino-α-ketocapronsäure (20) als mögliche Zwischenstufe und schließen α-Aminoadipinsäure-δ-semialdehyd (21) als Intermediärprodukt aus. Die N-Methylgruppe des N-Methylisopelletierins (32) entstammt Methionin (*192*), die C$_3$-Seitenkette von (31) und (32) dem Acetatstoffwechsel (*161a*, *235*). Tritierte Pipecolinsäure zeigte keine Inkorporation (*192*). Entsprechend Schema 3 dürften Lysin (19) und Acetat, möglicherweise Acetoacetyl-CoA, als Precursoren feststehen.

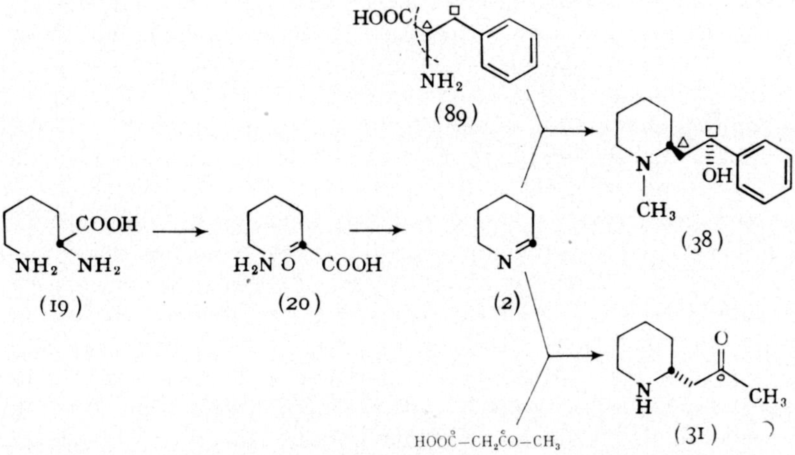

Schema 3. Biosynthese des Isopelletierins (31) und des Sedamins (38)

c) Strukturähnliche Sedum-, Lobelia- und Haloxylon-Basen

In den Gattungen *Sedum, Lobelia* und *Haloxylon* finden sich einige Piperidinalkaloide, die eine enge strukturelle Verwandtschaft zu den vorstehend genannten *Conium-* und *Punica*-Alkaloiden zeigen.

Zu diesen Verbindungen gehört das 1955 von BEYERMAN und MULLER (*35*) und etwas später von SCHÖPF und UNGER (*294*) aus *Sedum acre* L. isolierte (+)-Sedridin (*33*). Diese synthetisch zugängliche Substanz (*29, 63, 95, 287*) besitzt 2 S : 8 S-Konfiguration (*28, 34, 94*) und ist nach einer von SCHÖPF und Mitarb. 1957 vorgeschlagenen Nomenklatur (*288*) als (+)-8-Methylnorlobelol zu bezeichnen. FRANCK hat diese Base ebenfalls in *Sedum acre* L. nachweisen können (*102, 103*).

(**33**) Sedridin   (**34**) 8-Äthylnorlobelol   (**35**) Halosalin

SCHÖPF *et al.* ist 1957 die Isolierung und Strukturaufklärung eines Nebenalkaloids aus *Lobelia inflata* L. gelungen, das als (2 R : 8 S)-(+)-8-Äthylnorlobelol (I) (*34*) charakterisiert werden konnte (*288*). Dieses *Lobelia*-Alkaloid unterscheidet sich vom Sedridin (*33*) und den später zu besprechenden Norallosedamin (*36*) und Allosedamin (*37*) durch eine verlängerte Seitenkette und durch die entgegengesetzte Konfiguration am C-Atom 2.

Letzteres trifft auch für das (—)-Halosalin (*35*) zu, ein von SANDBERG und Mitarb. 1967 in *Haloxylon salicornicum* (Moq.-Tand.) Boiss. und in einigen anderen Chenopodiaceen aufgefundenes Alkaloid mit einer C_5-Seitenkette (*212, 274*). Als Nebenalkaloide wurden Anabasin (*69*), Piperidin (*1*), Haloxin, Aldotripiperidein u. a. nachgewiesen. Halosalin ist kürzlich konfigurativ aufgeklärt worden (*213*) und als (2 R : 8 R)-(—)-8-Propylnorlobelol zu bezeichnen.

Darüber hinaus kennt man einige Alkaloide, bei denen ein Piperidinring in 2-Stellung über eine C_2-Brücke mit einem Phenylrest verbunden ist. Dazu gehört das (+)-Norallosedamin [(2 S : 8 R)-(+)-8-Phenylnorlobelol I] (*36*). Diese von SCHÖPF *et al.* aus *Lobelia inflata* L. isolierte und strukturell aufgeklärte Verbindung (*285, 286, 288*) wird von (—)-Allosedamin [(2 S : 8 R)-(—)-8-Phenyllobelol I] (*37*) begleitet. Die Autoren diskutieren, daß beide Alkaloide im pflanzlichen Stoffwechsel durch Methylierung bzw. Entmethylierung ineinander übergehen können und somit in einem engen biogenetischen Zusammenhang stehen. Diese Annahme wird durch die Tatsache unterstützt, daß beide Verbindungen dieselbe absolute Konfiguration (Allosedaminreihe) aufweisen.

Literaturverzeichnis: SS. 42—59

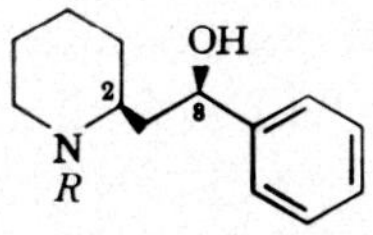

(36) R = H Norallosedamin
(37) R = CH₃ Allosedamin

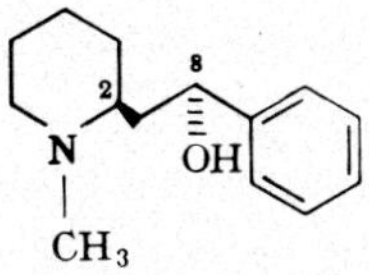

(38) Sedamin

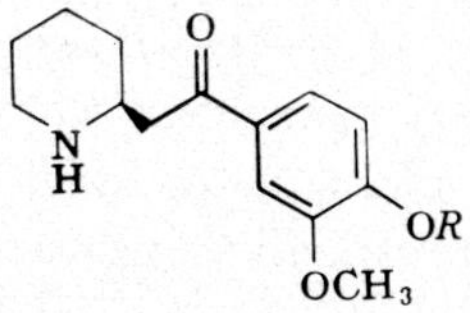

(39) R = H Pleurospermin
(40) R = CH₃ O-Methylpleurospermin

Als diastereomere Verbindung wurde (—)-Sedamin [(—)-8-Phenyllobelol II] (38) 1951 von MARION et al. (*209*) und später auch von FRANCK (*102, 103*) in *Sedum acre* L. aufgefunden. Diese Pflanzenbase liegt wahrscheinlich in der L-Form und als Racemat vor und besitzt (2S:8S)-Konfiguration (*28, 30, 286*).

(—)-Sedamin (38) und (—)-Allosedamin (37) sind am C-2 des Piperidinringes gleich, am C-Atom 8 der Seitenkette jedoch entgegengesetzt konfiguriert.

Ein ähnlich gebautes Alkaloid findet sich in der Lauracee *Cryptocarya pleurosperma* White and Francis (*110, 198*). Diese als Pleurospermin (39) bezeichnete Verbindung liegt in der Pflanze als Racemat vor.

Das O-Methylderivat des Pleurospermins (40) wurde 1968 von HART et al. aus *Boehmeria platyphylla* Don. als eines der ersten Urticaceen-Alkaloide isoliert (*132, 135*) und ist 1969 von FARNSWORTH et al. in *B. cylindrica* (L.) Sw. nachgewiesen worden (*90a*). Vom biogenetischen Standpunkt sind zwei Nebenalkaloide (*90a, 135*) interessant. Es handelt sich um Cryptopleurin (41) und um ein neuartiges Secophenanthrochinolizidinalkaloid (42) aus *B. platyphylla* und *B. cylindrica*, die sehr wahrscheinlich aus Pleurospermin bzw. seinem O-Methylderivat und einer C_6C_2-Einheit hervorgehen.

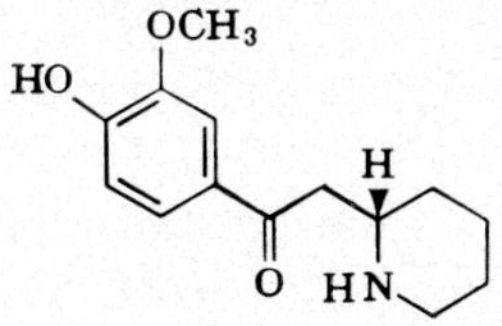

(39) Pleurospermin 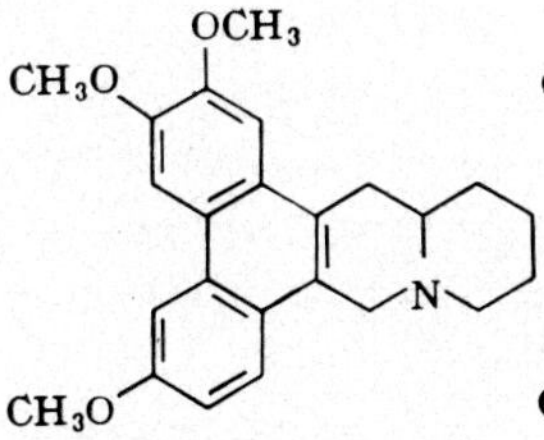   (41) Cryptopleurin 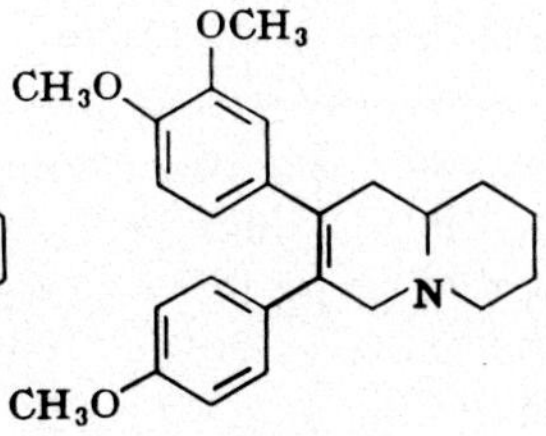   (42)

Von den vorstehend genannten Alkaloiden dieser Gruppe ist bisher das Sedamin (38) aus *Sedum acre* L. biogenetisch untersucht. Nach Verabreichung von Lysin-(2-^{14}C) wurde der Radiokohlenstoff im C-Atom 2 des isolierten Sedamins nachgewiesen (*124, 125*). Das bedeutet, daß Lysin über ein unsymmetrisches Zwischenprodukt in Sedamin eingebaut wird (Schema 3). Bei *Haloxylon salicornicum* wird Lysin-(6-^{14}C) in Halosalin (35) inkorporiert (*234*).

In neueren Untersuchungen ist von GUPTA und SPENSER an *Sedum acre* gezeigt worden, daß im Gegensatz zur Pipecolinsäure-Biosynthese (vgl. S. 7) Lysin-(2-T,6-^{14}C) wie bereits für Lysin-(6-^{14}CT$_2$) nachgewiesen mit intaktem T/^{14}C-Verhältnis in Sedamin eingebaut wird (*127a*). Auf Grund dieser Befunde dürften ε-Amino-α-ketocapronsäure (20) und α-Aminoadipinsäure-δ-semialdehyd (21) als Zwischenprodukte auszuschließen sein. Die Autoren diskutieren als mögliche Vorstufen des Sedamins N$_δ$-Methyl- oder N$_δ$-Acetyl-lysin, die zum entsprechenden N-monosubstituierten Cadaverinderivat decarboxyliert und nach oxydativer α-Desaminierung zur N-substituierten Piperideinium-Verbindung cyclisieren sollen.

Die C$_6$C$_2$-Seitenkette des Sedamins (38) entstammt Phenylalanin, das unter Decarboxylierung und Desaminierung in Sedamin eingebaut wird. Es ist denkbar, daß sich zuerst eine Carbonylgruppe am C-Atom 8 ausbildet, die durch stereospezifische Reduktion zu den an dieser Position unterschiedlich konfigurierten Alkaloiden der Sedamin- und Allosedaminreihe führt. Die N-Methylgruppe des Sedamins kommt aus dem C$_1$-Stoffwechsel (Methionin) (*125*).

d) *Withania*-Alkaloide

Der Arbeitskreis um SCHWARTING hat 1962/1964 aus der Solanacee *Withania somnifera* Dunal neben den schon strukturbekannten Alkaloiden Cuskhygrin, Isopelletierin (31), Tropin und Pseudotropin zwei optisch inaktive Basen aufgefunden, die als Anaferin (43) und Anahygrin (44) bezeichnet werden (*178, 267, 302*). Anahygrin soll neben Lupinenalkaloiden in *Genista transcaucasica* Schischk. gefunden worden sein (*40*). Bei diesen Verbindungen ist ein Piperidinring über eine Propanonkette mit einem N-Methylpyrrolidin bzw. einem zweiten Piperidinring verbunden. Sie zeigen somit Ähnlichkeit zu den vorstehend aufgeführten *Conium*-, *Punica*-, *Sedum*- und *Lobelia*-Alkaloiden.

Die Annahme, daß Anaferin in der *meso*-Form vorliegt (*302*), ist von SCHÖPF *et al.* (*279a, 286a*) kürzlich bestätigt worden. EL-OLEMY und SCHWARTING haben für die L(+)-Form des Anaferinracemats S,S-Konfiguration und für die D(−)-Form R,R-Konfiguration nachweisen können (*83a*).

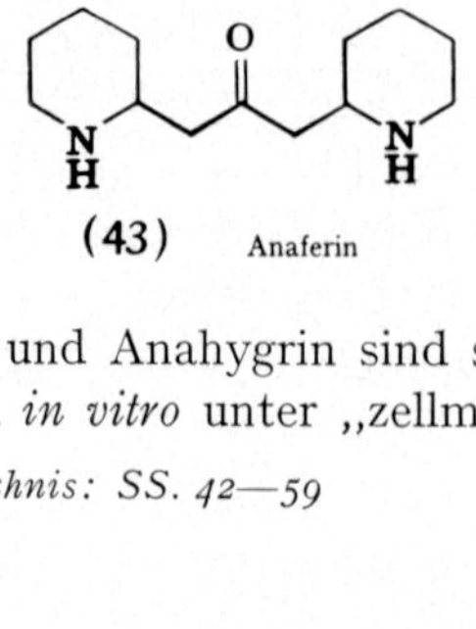

(43) Anaferin 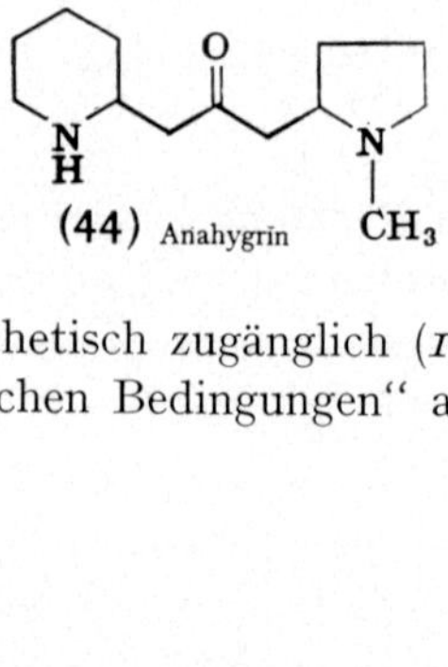   (44) Anahygrin

Anaferin und Anahygrin sind synthetisch zugänglich (*178, 267*) und können auch *in vitro* unter „zellmöglichen Bedingungen" aus Δ^1-Piperi-

Literaturverzeichnis: SS. 42—59

dein (2) und Acetondicarbonsäure bzw. Δ^1-Piperidein, Acetondicarbonsäure und N-Methyl-2-hydroxypyrrolidin dargestellt werden (*82, 83, 84, 280*).

Biogenetisch ist interessant, daß diese Alkaloide mit Tropanbasen vergesellschaftet vorkommen. Es ist auffällig, daß der Piperidinring in keinem Fall N-methyliert ist. Man kann annehmen, daß sich die Tropanbasen aus dem Ornithin- und die Piperidinalkaloide aus dem Lysinstoffwechsel ableiten. Biosyntheseexperimente von KEOGH und O'DONOVAN (*161a, 235*) ergaben, daß Lysin-(2-^{14}C) von *Withania somnifera* in Anaferin (43) eingebaut wird, wobei die Radioaktivität im C-Atom 2 der beiden Piperidinringe lokalisiert ist. Somit wird Lysin unsymmetrisch inkorporiert. Die C$_3$-Brücke geht aus Acetat hervor. Möglicherweise stellt Isopelletierin (31) eine Vorstufe des Anaferins dar, da es mit hohen Einbauraten in Anaferin inkorporiert wird.

e) Febrifugin und Isofebrifugin

Beim (+)-Febrifugin (45) und Isofebrifugin (46) ist ein hydroxylierter Piperidinring über eine C$_3$-Brücke mit einem 4-Chinazolon verknüpft. Diese Alkaloide haben wegen ihrer Antimalariawirkung besonderes Interesse gefunden und sind Gegenstand intensiver Untersuchung. Wegen ihrer Toxizität sind sie aber bisher nicht zur klinischen Anwendung gelangt.

Sie wurden 1946/1948 aus den Wurzeln der Saxifragacee *Dichroa febrifuga* Lour., einer als Ch'an San bezeichneten chinesischen Droge, und aus *Hydrangea umbellata* Rheder isoliert (*1, 57, 168, 169, 175*, vgl. *39*). Die Konstitutionsaufklärung erwies sich als recht schwierig. Die 1950 von KOEPFLI und Mitarb. vorgeschlagene Struktur (*167*) konnte jedoch

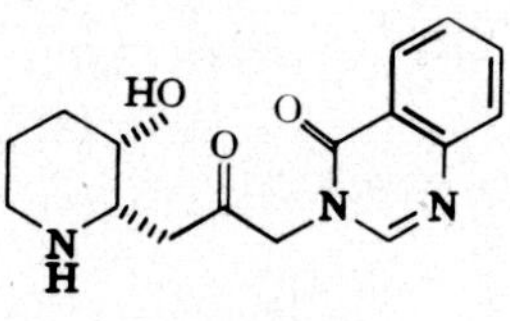

(45) Febrifugin 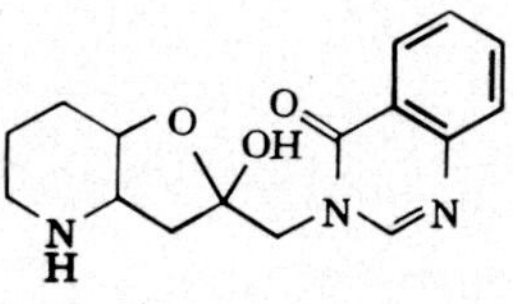   (46) Isofebrifugin

zwei Jahre später durch BAKER *et al.* (*21*) bestätigt werden. Die gleiche Arbeitsgruppe hat mehrere Verfahren zur Synthese von DL- und D-Febrifugin ausgearbeitet und eine Reihe von Derivaten dargestellt (*19, 20, 21*). Von HILL und EDWARDS ist 1962 die absolute Konfiguration des D-Febrifugins (45) ermittelt worden, wonach den beiden *cis*-ständig angeordneten Substituenten des Piperidinringes S-Konfiguration zukommt (*147*).

Zur Biosynthese des Febrifugins liegen bisher nur Hypothesen vor. ROBINSON postuliert Anthranilsäure, Ameisensäure, Ammoniak, einen C_3-Baustein und Lysin als Vorstufen (266). Demgegenüber diskutiert LEETE, daß nach Ausbildung eines 4-Chinazolonringes eine aus 4 Acetateinheiten bestehende Seitenkette angefügt wird, die in geeigneter Weise mit Ammoniak zum Piperidinring cyclisiert (184). Nach MACHOLAN soll sich aus Aminoacetessigsäure und einem oxydierten Aminovaleraldehyd ein Aminoketon ausbilden, das mit einem entsprechenden Chinazolon zum Febrifugin umgesetzt wird (205).

f) Nigrifactin

Als eine weitere Verbindung dieser Substanzklasse sei das 1969 von TERASHIMA et al. aus *Streptomyces* Stamm FFD-101 isolierte Nigrifactin (46a) genannt (320), ein Δ^1-Piperidein mit einer α-ständigen C_7-Seitenkette mit drei konjugierten Doppelbindungen, für das zwei Synthesen beschrieben worden sind (122, 240). Die Biosynthese dürfte auf dem Polyacetatweg erfolgen.

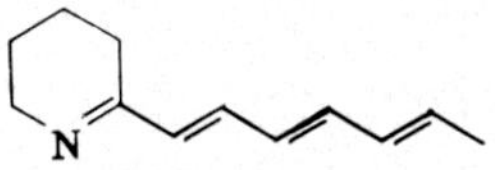

(46a) Nigrifactin

III. Aliphatisch α, α'-disubstituierte Piperidine

1. Pinidin

Das zu den 2,6-disubstituierten Piperidinalkaloiden gehörende Pinidin (47) ist 1955 von TALLENT et al. aus *Pinus sabiniana* Dougl., *P. jeffreyi* A. Murr. und *P. torreyana* Parry ex Torr. isoliert (319) und etwas später als

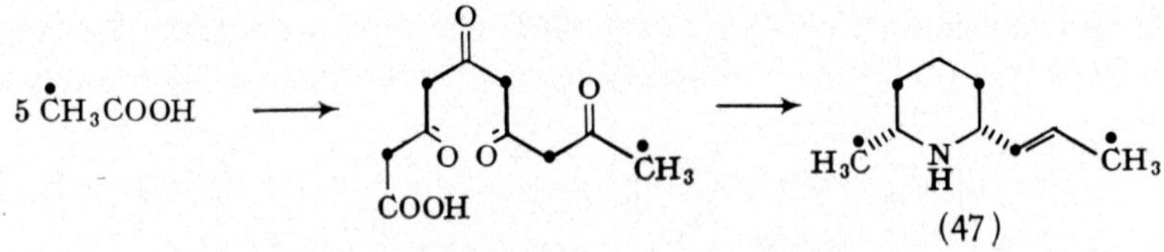

5 $\overset{\bullet}{C}H_3COOH$ ⟶ ⟶ (47)

Schema 4. Biosynthese von Pinidin

(—)-*cis*-2-Methyl-6-(2-propenyl)-piperidin identifiziert worden (318). Die Aufklärung der absoluten Konfiguration stammt von HILL und Mitarb. (146). Den beiden Substituenten am C-2 und C-6 des Pinidins kommt R-Konfiguration zu; die Doppelbindung der Seitenkette ist *trans*-ständig. LEETE und JUNEAU haben 1969 zeigen können, daß Lysin keine Vorstufe des Pinidins darstellt und daß 5 Acetateinheiten zum Aufbau dieses C_9-Alkaloids verwendet werden (190). Nach Applikation von Acetat-(1-^{14}C) an *Pinus jeffreyi* fand sich der Radiokohlenstoff in den

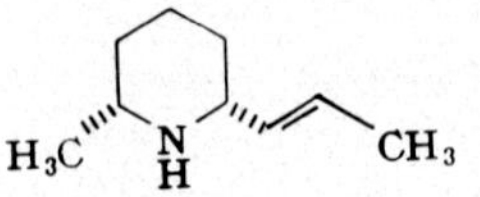

(47) Pinidin

Literaturverzeichnis: SS. 42—59

Vorkommen, Struktur und Biosynthese natürlicher Piperidinverbindungen

C-Atomen 2,4,6 und 9 des isolierten Pinidins. Diese ^{14}C-Verteilung ist in voller Übereinstimmung mit dem in Schema 4 dargestellten Polyacetatweg.

2. *Lobelia*- und *Sedum*-Alkaloide

Eine ungewöhnlich hohe Zahl disubstituierter Piperidine findet sich in den Gattungen *Lobelia*, insbesondere in *Lobelia inflata* L., und *Sedum*. Diese Verbindungen tragen vielfach Trivialnamen, leiten sich aber im wesentlichen von drei Grundstrukturen ab, die nach SCHÖPF *et al.* als Lobelidiole (48), Lobelionole (49) und Lobelidione (50) bezeichnet werden (*288*). Die große Variationsbreite dieser Alkaloide kommt durch unter-

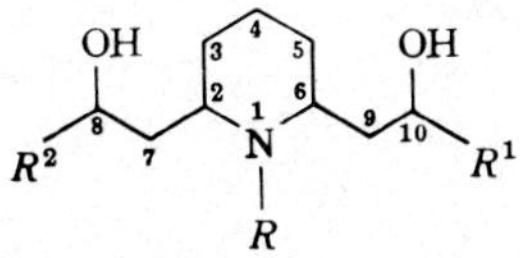

(48) Lobelidiol 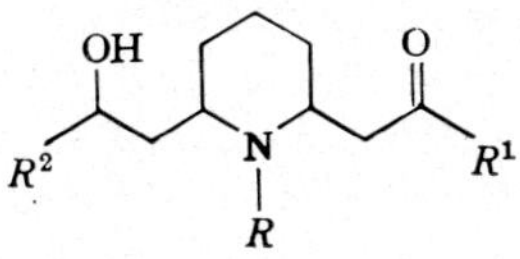   (49) Lobelionol

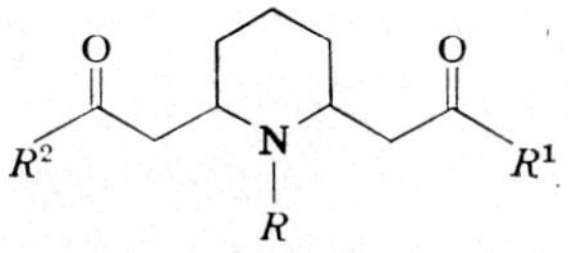

(50) Lobelidion

schiedliche Substitution (R = H oder CH$_3$; R^1 und R^2 im allgemeinen Methyl-, Äthyl- oder Phenylreste, wobei symmetrische oder unsymmetrische Substitution vorliegen kann) und durch das Auftreten von Stereoisomeren (Diastereomere werden durch römische Ziffern I oder II unterschieden) zustande. In Einzelfällen kann der Piperidinring auch als Piperidein auftreten.

Von den zahlreichen Alkaloiden dieser Gruppe, die vor allem in den Laboratorien von FRANCK, SCHÖPF und TSCHESCHE bearbeitet worden sind, sollen hier nur einige typische Vertreter genannt werden. Die restlichen Alkaloidstrukturen müssen den in dieser Übersicht aufgenommenen Originalarbeiten (*13, 102, 103, 104, 105, 288, 292, 293, 328*) und darin zitierten Referenzen sowie anderen Zusammenfassungen (*18, 39*) entnommen werden.

Als Vertreter der Lobelionole sei das als Hauptalkaloid aus *Lobelia inflata* L. isolierte (—)-Lobelin [(2 S : 6 R : 8 S)-(—)-8,10-Diphenyllobelionol)] (51) genannt (*292, 293*), das sterisch in die Reihe des Sedamins (38) gehört. Da in dieser Pflanze als Nebenalkaloide Allosedamin (37) und seine Norverbindung (36) vorkommen, die gegenüber dem Lobelin (51) eine entgegengesetzte sterische Anordnung der Hydroxylgruppe besitzen

(Allosedaminreihe), sollten diese monosubstituierten Alkaloide und Lobelin auf getrennten Biosynthesewegen entstehen.

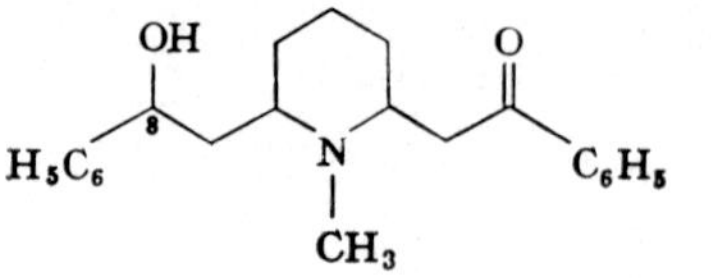

(51) Lobelin

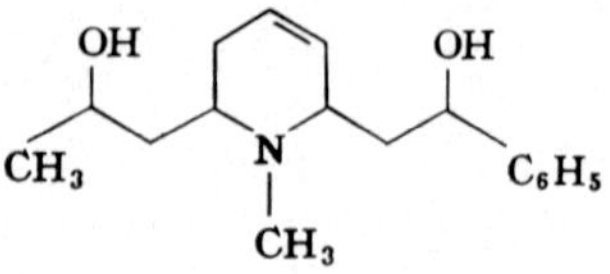

(52) Sedinin

Das von FRANCK aus *Sedum acre* L. isolierte Sedinin [*trans*-8-Methyl-10-phenyl-4,5-dehydrolobelidiol] (52) (*102, 103, 104, 105*) gehört zu den unsymmetrisch disubstituierten Piperidinalkaloiden.

Aus der Reihe der Lobelidione soll das von TSCHESCHE und Mitarb. (*328*) in *Lobelia syphilitica* L. und anderen *Lobelia*-Arten aufgefundene *cis*-8,10-Diäthylnorlobelidion (53) vorgestellt werden.

Ein Alkaloid vom Typ des Lobelidiols ist das (—)-*cis*-8,10-Diphenyllobelidiol (54) aus *Isotoma longiflora* Presl., eine von ARTHUR und CHAN entdeckte Verbindung (*13*). Die Aufklärung der absoluten Konfiguration ist SCHÖPF und Mitarb. gelungen (*292*).

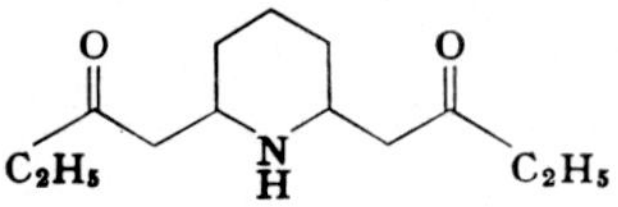

(53) *cis*-8,10-Diäthylnorlobelidion

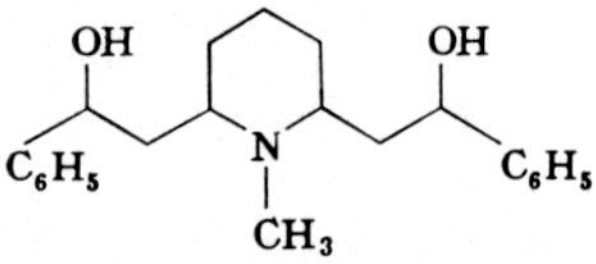

(54) *cis*-8,10-Diphenyllobelidiol

Zur Biogenese der *Lobelia*-Alkaloide liegen neuere Untersuchungen für das Lobelin (51) vor. KEOGH und O'DONOVAN haben 1970 die früheren Angaben anderer Autoren über den Einbau von Lysin-(U-^{14}C) (*309a*) und von Phenylalanin-(2-^{14}C) (89) (*24a*) bestätigen können und darüber hinaus nach Applikation von Lysin-(2-^{14}C) und Phenylalanin-(2,3-^{14}C) an *Lobelia inflata* die ^{14}C-Verteilung im isolierten Lobelin ermittelt (*161b, 235*). Es ließ sich zeigen, daß der Piperidinring nicht aus Acetat, sondern aus Lysin über ein symmetrisches Intermediärprodukt gebildet wird. Die beiden C_6C_2-Seitenketten gehen gleichwertig aus Phenylalanin hervor. Die Autoren diskutieren, daß zwei aus Phenylalanin entstehende C_6C_2-Einheiten — möglicherweise Benzoylessigsäure (89a) — mit Δ^1-Piperidein zu einem Lobelidion (50) kondensieren. Reduktion der einen Carbonylgruppe und N-Methylierung führen zu Lobelin (51) (Schema 4a). Dieser Biosyntheseweg ähnelt der Sedaminbildung (vgl. Schema 3), zeigt aber durch das symmetrische Lysinfolgeprodukt und durch die Verwertung von zwei C_6C_2-Einheiten zur α,α'-Substitution charakteristische Unterschiede. Das gemeinsame Vorkommen von 2- und

Literaturverzeichnis: SS. 42—59

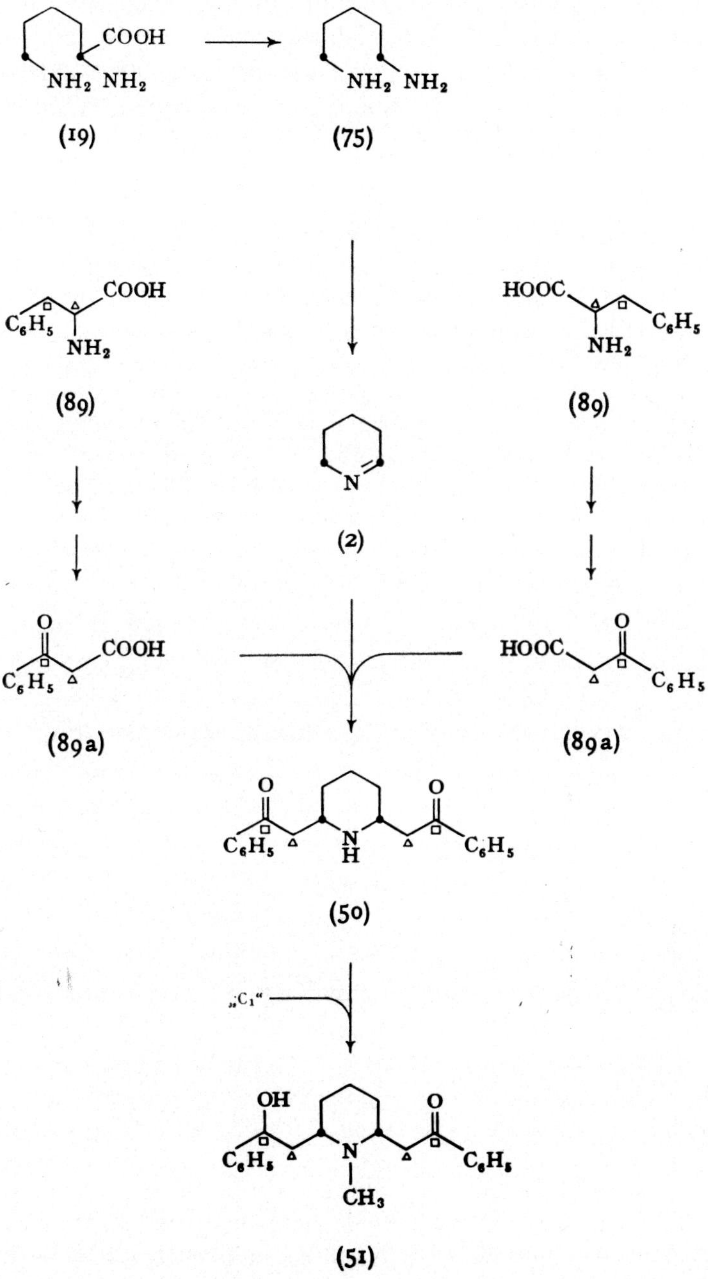

Vorkommen, Struktur und Biosynthese natürlicher Piperidinverbindungen

Schema 4a. Biosynthese des Lobelins (51)

2,6-substituierten Piperidinen in einer Pflanze könnte andeuten, daß bei gleicher stereochemischer Anordnung die monosubstituierten Alkaloide als Vorstufen der disubstituierten Verbindungen in Frage kommen. Bei unterschiedlicher sterischer Anordnung sollten die mono- und disubstituierten Alkaloide auf getrennten Bildungswegen entstehen.

3. Cassia- und Prosopis-Alkaloide

Von den aliphatisch disubstituierten Piperidinbasen mit einer längeren Seitenkette seien die kürzlich von MacConnell et al. im Gift der Feuerameise *Solenopsis saevissima* aufgefundenen Solenopsine A, B und C (55) sowie die Dehydrosolenopsine B und C (55a) genannt, deren Struktur durch Synthese bestätigt werden konnte (*204*). Es sind die ersten alkylierten Piperidinderivate, die in einem tierischen Gift nachgewiesen worden sind. Die absolute Konfiguration ist noch unbekannt, möglicherweise besitzen (55) und (55a) auch die spiegelbildliche Anordnung der Substituenten. Die Doppelbindung in der Seitenkette der Dehydrosolenopsine B und C ist *cis*-ständig. Die Substituenten am Piperidinring der Solenopsine und Dehydrosolenopsine sind im Gegensatz zu den nachfolgend zu besprechenden pflanzlichen Piperidinen dieses Strukturtyps *trans*-ständig angeordnet.

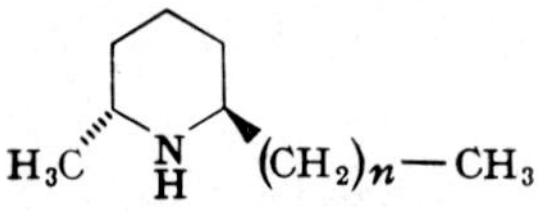

(55) n = 10 Solenopsin A
n = 12 Solenopsin B
n = 14 Solenopsin C

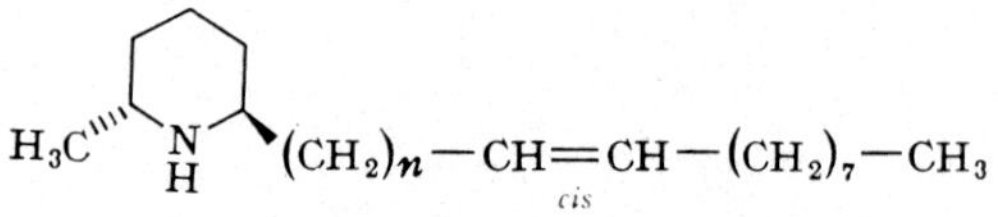

(55a) n = 3 Dehydrosolenopsin B
n = 5 Dehydrosolenopsin C

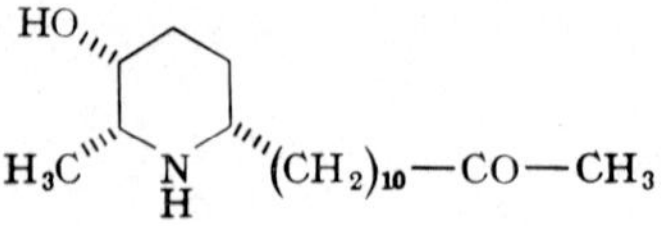

(56) Cassin

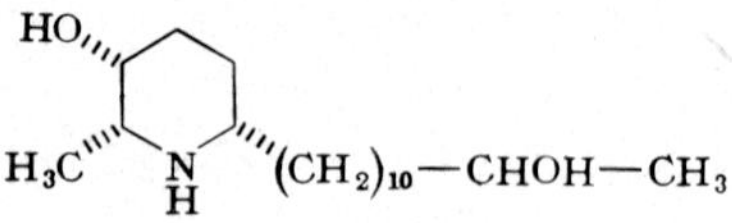

(57) Carnavallin

Nach Untersuchungen von Highet aus dem Jahr 1964 enthält die Leguminose *Cassia excelsa* Schrad. ein als Cassin (56) bezeichnetes Piperidinalkaloid, das eine C_{12}-Seitenkette aufweist (*142*). Auf Grund einer Korrektur der ursprünglich angegebenen Formel (*143*) und durch Arbeiten anderer Autoren zur Konstitution und Konfiguration (*202, 257*) liegt (—)-Cassin die Struktur eines 2(R)-Methyl-3(R)-hydroxy-6(S)-[11-oxododecyl]-piperidins zugrunde.

Literaturverzeichnis: SS. 42—59

LYTHGOE und VERNENGO haben 1967 aus *Cassia carnaval* Spreg. drei weitere Piperidinbasen isoliert, von denen das Carnavallin (57) an Stelle der Ketofunktion des Cassins (56) eine Hydroxylgruppe trägt, deren sterische Anordnung allerdings noch nicht gesichert ist *(202)*. Ähnlich gebaut sind die von der Arbeitsgruppe um GOUTAREL *(250)* in der afrikanischen Mimosacee *Prosopis africana* (Guill. et Per.) Taub. nachgewiesenen Pflanzenbasen Prosopin (58) und Prosopinin (59) sowie (60), (61) und (62). Diese *Prosopis*-Alkaloide unterscheiden sich untereinander im wesentlichen durch verschiedenartige Oxydationsgrade am C-10′ und C-11′ in der Seitenkette. Es wird angenommen, daß diesen Piperidinbasen all-*cis*-Konfiguration zukommt.

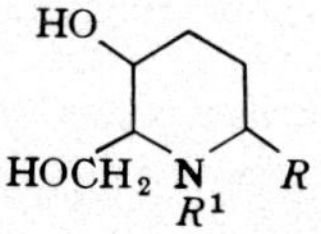

(58) $R = (CH_2)_{10}-CHOH-CH_3$ $R^1 = H$ Prosopin
(59) $R = (CH_2)_9-CO-CH_2CH_3$ $R^1 = H$ Prosopinin
(60) $R = (CH_2)_{11}CH_3$ $R^1 = H$
(61) $R = (CH_2)_{11}CH_3$ $R^1 = CH_3$
(62) $R = (CH_2)_{10}-CO-CH_3$ $R^1 = H$

4. *Caria-* und *Azima*-Alkaloide

Strukturähnlichkeit zu den vorstehend aufgeführten *Cassia-* und *Prosopis*-Alkaloïden weist das aus *Caria papaya* L. *(Cariacaceae)* isolierte Carpain (63) auf, für das in älteren Arbeiten eine monomere Piperidinstruktur angenommen wurde (vgl. *39*).

Spätere massenspektrometrische Untersuchungen von SPITELLER-FRIEDMANN und SPITELLER haben jedoch ergeben, daß es sich beim Carpain um ein dimeres Alkaloid mit einer bis-Lactonstruktur handelt *(311)*, dem nach TICHY und SICHER *(322)*, GOVINACHARI et al. *(115)* sowie COKE und RICE *(62)* all-*cis*-Konfiguration zukommt.

Carpain (63) und Cassin (56) besitzen eine spiegelbildliche Anordnung der Substituenten am Piperidinring *(257)*. Das als Nebenalkaloid

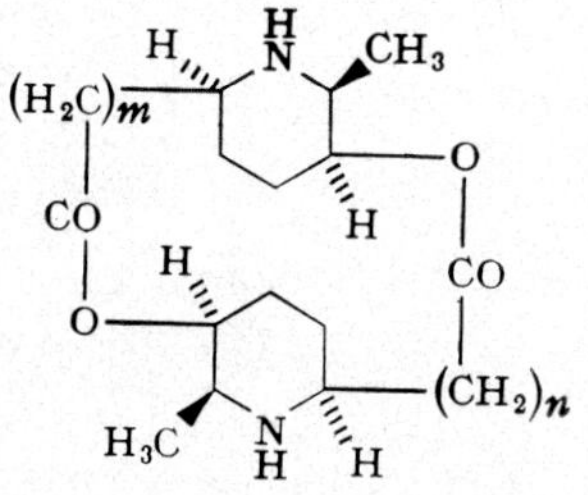

(63) $m = n = 7$ Carpain
(64) $m = n = 5$ Azimin
(65) $n = 7, m = 5$ Azcarpin

in *Caria papaya* L. vorkommende Pseudocarpain unterscheidet sich vom Carpain durch die entgegengesetzte Anordnung der einen der beiden Methylgruppen (C-2′) *(115)*. RALL und Mitarb. haben 1967/1968 aus der Salvadoracee *Azima tetracantha* Lam. neben Carpain (63) die beiden dimeren Piperidinalkaloide Azimin (64) und Azcarpin (65) isoliert und in ihrer absoluten Konfiguration aufgeklärt (*249, 308*). Azimin ist ein bis-Lacton, das aus zwei Molekülen Aziminsäure [2(S)-Methyl-3(S)-hydroxy-6(R)-(5'-carboxypentyl)piperidin] aufgebaut ist. Beim Azcarpin (65) handelt es sich um ein unsymmetrisches bis-Lacton aus Aziminsäure und Carpaminsäure [2(S)-Methyl-3(S)-hydroxy-6(R)-(7'-carboxyheptyl)-piperidin].

Die bisher nur für Carpain (63) vorliegenden Biosyntheseuntersuchungen haben gezeigt, daß Acetat besser eingebaut wird als Lysin oder Mevalonat (*26*). Von LEETE wird daher ein Polyacetatweg diskutiert, wobei 14 Acetateinheiten zum Aufbau des Carpains benötigt werden (*184*). Für die analog gebauten Verbindungen dieses Strukturtyps (64) und (65) ist ein ähnlicher Biosyntheseweg zu postulieren.

5. Lythraceen-Alkaloide

Aus der Gruppe der Lythraceen-Alkaloide, die im allgemeinen zur Gruppe der Chinolizidine gehören, kennt man zwei Piperidin-haltige Vertreter. Es handelt sich um die von FUJITA und Mitarb. 1967 aus *Lythrum anceps* Makino isolierten Basen Lythranin (66) und Lythranidin (67), die wie das ähnlich gebaute Lythramin (68) eine Biphenylstruktur aufweisen (*108*) und kürzlich konfigurativ auch aufgeklärt werden konnten (3,11 S : 5,9 R) (*107* und dort zitierte Literatur).

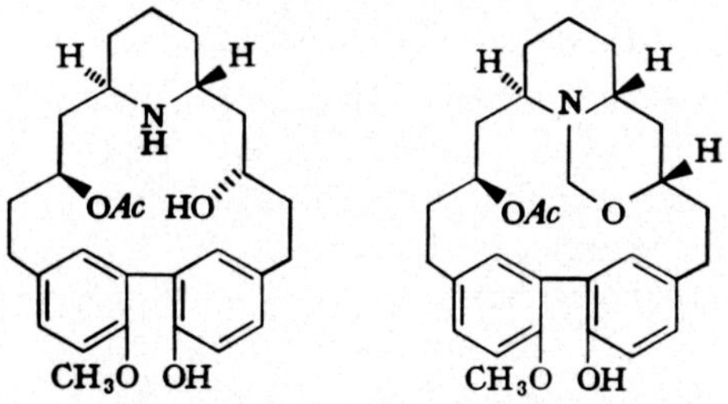

(66) Lythranin
(67) Lythranidin Ac = H (Desacetyllythranin)

(68) Lythramin

Das gemeinsame Vorkommen mit Chinolizidinalkaloiden, für die ein spezifischer Einbau von Lysin und Phenylalanin kürzlich nachgewiesen worden ist (*172, 173*), könnte eine Bildung des Piperidinringes aus Lysin andeuten. Entsprechende Biosyntheseversuche stehen jedoch noch aus.

IV. Heterocyclisch substituierte Piperidine

1. Anabasin- und Tetrahydroanabasinalkaloide

Zu den Pflanzenbasen, bei denen ein Piperidin- oder Piperideinring direkt oder über eine kürzere C-Kette mit einem weiteren Heterocyclus

Vorkommen, Struktur und Biosynthese natürlicher Piperidinverbindungen 25

verbunden ist, gehören (—)-Anabasin (69) und sein N-Methylderivat (70) sowie (—)-Anatabin (71) und N-Methylanatabin (72) (vgl. *39*). Für das C-Atom 2' des natürlich vorkommenden Anabasins ist durch oxydativen Abbau zur sterisch zugeordneten L-(—)-N-Methylpipecolinsäure S-Konfiguration ermittelt worden (*201*). Eine entsprechende sterische Anordnung ließ sich auch für (—)-N-Methylanabasin (70), (—)-Anatabin (71) und (—)-N-Methylanatabin (72) ableiten.

(69) R = H Anabasin
(70) R = CH₃ N-Methylanabasin

(71) R = H Anatabin
(72) R = CH₃ N-Methylanatabin

Anabasin findet sich u. a. in *Mackinlaya sublata* Philipson und *M. macrosciadea* F. Muell. (*92*), in *Anabasis aphylla* L. (*45, 46, 273*, vgl. *39*), in *Leontice alberti* Regel (*159*), in *Duboisia myoporoides* R. Br. und in verschiedenen *Nicotiana*-Arten (*114, 220, 309,* vgl. *39*). Es wurde außerdem von der Arbeitsgruppe um SANDBERG in der Chenopodiacee *Haloxylon salicornicum* (Moq.-Tand.) Boiss. (*212, 274*) sowie von SCHLUNEGGER und STEINEGGER erstmals in einer Leguminose *Priestleya elliptica* DC. (*279*) nachgewiesen. Vielfach wird Anabasin von Pyridin- oder Piperidinbasen oder von typischen Chinolizidinalkaloiden wie Lupanin oder Spartein begleitet. Das ist z. B. der Fall in *Anabasis aphylla* L., *Leontice alberti* Regel und *Priestleya elliptica* DC.

Das 1967/1968 von Taschkenter Autoren (*223, 272*) aus Samen von *Anabasis aphylla* L. isolierte und strukturell aufgeklärte Anabasamin (73) sowie das von KISAKI und Mitarb. (*166*) in *Nicotiana tabacum* aufgefundene Anatallin (74) stellen neuartige, bezüglich ihrer Ringverknüpfung besonders interessante Strukturtypen dar.

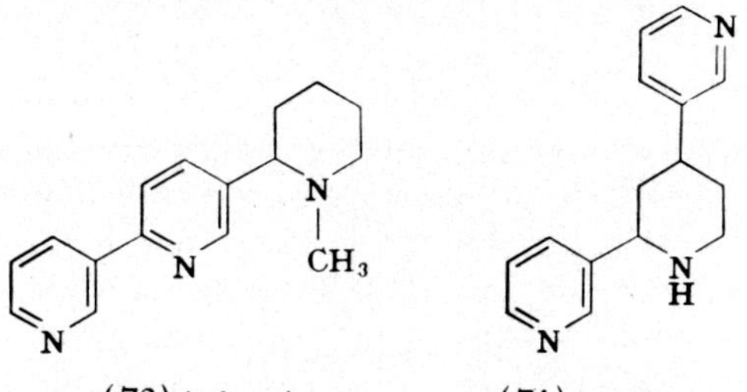

(73) Anabasamin

(74) Anatallin

Anabasamin enthält einen N-Methylanabasinteil, der in 6-Stellung einen Pyridinring als Substituenten trägt, während beim Anatallin ein β-ständiger Pyridinkern in Position 4' des Anabasins angefügt ist. Anatallin kann leicht zu dem als Nicotellin bekannten *Nicotiana*-Alkaloid dehydriert werden.

Von den Alkaloiden dieser Gruppe ist bisher nur das Anabasin biogenetisch eingehend untersucht worden (Übersichtsreferate vgl. *119, 120, 183, 184, 221, 248*). Die Ergebnisse dieser zahlreichen Arbeiten können hier nur zusammenfassend abgehandelt werden.

Der Piperidinring des Anabasins leitet sich aus dem Lysinstoffwechsel ab (Weg A in Schema 5), wie durch Einbauversuche an *Nicotiana glauca* und *N. rustica* sowie an *Anabasis aphylla* gezeigt werden konnte (*116, 179, 180, 310*). Nach Verfütterung von Lysin-(2-^{14}C) war die gesamte Radioaktivität im C-Atom 2' des isolierten Anabasins lokalisiert, so daß

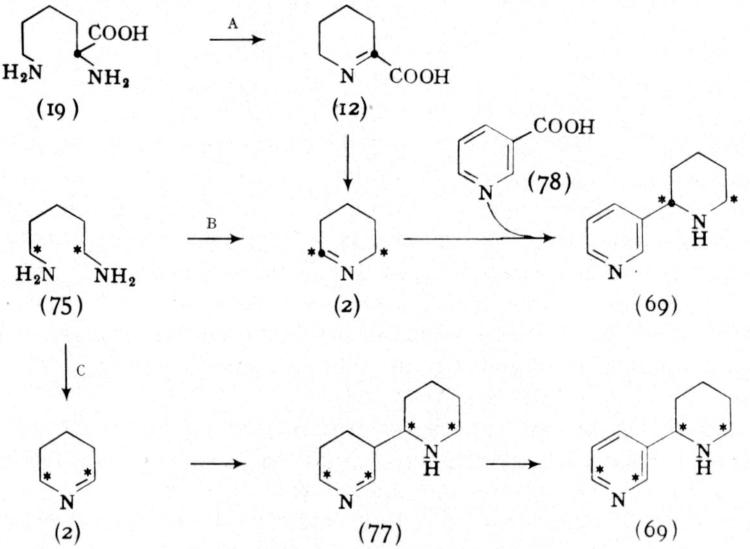

Schema 5. Biosynthese des Anabasins

ein Biosyntheseweg über unsymmetrische Zwischenstufen wie Δ^1-Piperidein-2-carbonsäure (**12**) oder Δ^1-Piperidein (**2**) angenommen werden muß. Ein hoher und spezifischer Einbau von Δ^1-Piperidein-(6-^{14}C) wurde kürzlich von Leete nachgewiesen (*186*). Auf Grund von Versuchen mit Lysin-(2-^{14}C, ε-^{15}N) entstammt der Heterostickstoff des Piperidinringes der ε-Aminogruppe des Lysins (*189*). Auch Cadaverin (**75**), das Decarboxylierungsprodukt des Lysins, zeigt entsprechend Weg B in Schema 5 Inkorporation in Anabasin (*180*).

Der Pyridinring des Anabasins geht bei *Nicotiana* und *Anabasis* nicht aus einem vorgebildeten Piperidinring hervor, sondern entsteht wie der des Nicotins aus Nicotinsäure (78) unter Verlust ihrer Carboxylgruppe und des Wasserstoffatoms an C-6 (*310*). Obwohl verschiedentlich eine Wechselbeziehung Nicotin-Anabasin diskutiert wird, dürfte nach LEETE eine Ringerweiterung des Nicotins zum Anabasin, wobei die N-Methylgruppe des Nicotins das 6. Ringkohlenstoffatom des Anabasins ergibt, nicht möglich sein (*185*). Durch *in vitro*-Versuche an Erbsen- und Lupinenextrakten ist ein weiterer Syntheseweg zum Aufbau des Anabasins bekannt geworden, bei dem beide Ringsysteme des Anabasins aus Cadaverin (75) entstehen (Schema 5, Weg C). Erbsen enthalten kein Anabasin, sind aber imstande, appliziertes Cadaverin (75) zu ω-Aminovaleraldehyd (76) zu oxydieren (*138, 139, 206*), der spontan zum Δ^1-Piperidein (2) cyclisiert. Nach Dimerisierung zu Tetrahydroanabasin (77) (*283*) erfolgt Dehydrierung zu Anabasin (*138*). Dieser ungewöhnliche Bildungsweg ist von MOTHES und Mitarb. durch Einsatz von Cadaverin-(1,5-^{14}C) im *in vitro*-System bestätigt worden (*222*), konnte aber nicht in intakten Pflanzen nachgewiesen werden.

Die bisher zur Biosynthese des Anatallins (74) und Anatabins (71) vorliegenden Untersuchungen haben ergeben, daß diese beiden Alkaloide nicht aus Lysin oder 6-Hydroxylysin entstehen, so daß der Lysinweg zur Bildung des Piperidin- bzw. Δ^3-Piperideinringes dieser Basen wahrscheinlich auszuschließen ist (*166*). Andererseits deutet das gemeinsame Vorkommen von Nicotellin und Anatallin in *Nicotiana* möglicherweise auf ein Entstehungsprinzip, bei dem Pyridin- und Piperidinring aus einheitlichen Vorstufen hervorgehen und eine ein- oder wechselseitige Umwandlung denkbar ist.

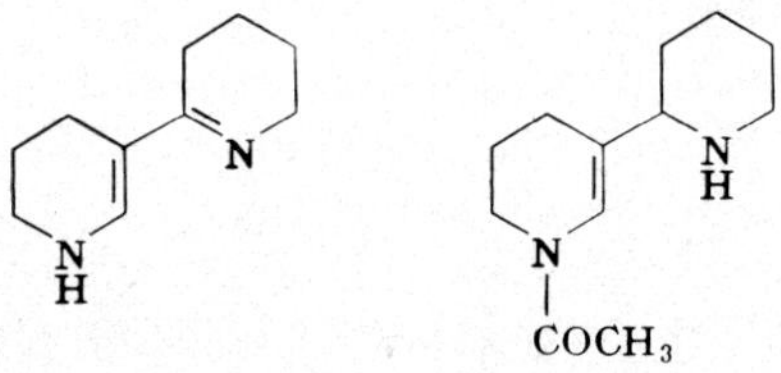

(79) Hystrin   (80) Ammodendrin

In der Gattung *Genista*, die durch das Vorkommen von Chinolizidinalkaloiden ausgezeichnet ist, stellt *Genista hystrix* Lge. offenbar eine Ausnahme dar. In dieser Art sind an Stelle von Chinolizidinbasen zwei Dipiperidylalkaloide enthalten. Neben dem schon länger bekannten Ammodendrin (80) (vgl. *39*) wurde 1967/1968 von STEINEGGER und Mitarb. das inzwischen auch synthetisch zugängliche Hystrin (79) entdeckt (*312, 313, 315a, 316*). Es soll nativ vorliegen und biogenetisch nicht aus N-Acetylhystrin hervorgehen (*315*).

Zu den Pflanzenbasen, die sich von der Enaminform des Anabasins ableiten und bei denen der Δ^2-Piperideinring des Tetrahydroanabasins durch Essigsäure, *cis*- bzw. *trans*-Zimtsäure oder α-Truxillsäure acyliert ist, gehört das racemisch vorkommende Ammodendrin (80), das 1935 von ORECHOFF und PROSKURNINA aus der Leguminose *Ammodendron conollyi* Bunge ex Boiss. isoliert und etwas später als N-Acetyl-Δ^2-tetrahydroanabasin charakterisiert wurde *(237, 238)*. Ammodendrinsynthesen sind von SCHÖPF und Mitarb. beschrieben *(281, 284)*. Das von PROSKURNINA und MERLIS 1949 ebenfalls in *A. conollyi* aufgefundene Isoammodendrin *(211, 247)* und das 1963 von RIBAS und VEGA in *Retama sphaerocarpa* Boiss. nachgewiesene Sphaerocarpin *(254)* sind nach Untersuchungen von DOMINGUEZ *et al.* *(78)* mit dem (+)-Antipoden des Ammodendrins identisch.

Verschiedene Arbeitsgruppen haben (+)- bzw. (±)-Ammodendrin auch in anderen Leguminosen aufgefunden, z. B. *Retama raetam* Webb. et Bert. *(91)*, *Genista lusitanica* L. *(317)* und *G. hystrix* Lge. *(312, 313, 315, 315a)*. In der südafrikanischen *Liparia parva* Vog. ex Walp. und *L. sphaerica* L. wird (+)-Ammodendrin von typischen Chinolizidinalkaloiden wie (−)-Lupanin und (+)-Spartein begleitet *(314)*, während DL-Ammodendrin in *Coelidium fourcadei* Compton neben Isotripiperidein und α-Aldotripiperidein auftritt *(12)*.

Ähnlich gebaut ist das Adenocarpin (81), ein N-*trans*-Cinnamoyl-Δ^2-tetrahydroanabasin, das in der L- und DL-Form vorkommt. Das als Orensin bezeichnete Racemat kann nach SCHÖPF *et al.* *(284)* synthetisch dargestellt werden. Adenocarpin bzw. Orensin wurden vor etwa 20 Jahren von der Arbeitsgruppe um RIBAS *(252, 255,* vgl. *39)* und kürzlich von BERNASCONI und STEINEGGER *(25)* in verschiedenen *Adenocarpus*-Arten aufgefunden. SCHÜTTE und Mitarb. haben durch Verfütterung von Cadaverin-(1,5-^{14}C) *(75)* an *Adenocarpus viscosus* L. zeigen können, daß die Radioaktivität des isolierten Adenocarpins ausschließlich im Basenteil lokalisiert war *(299)*. Gemeinsam mit der Tatsache, daß Tetrahydroanabasin *(77)* wesentlich besser als Cadaverin inkorporiert wird, schließen die Autoren auf die Biosynthesefolge Cadaverin *(75)* → Tetrahydroanabasin *(77)* → Adenocarpin (81).

(81) R = *trans*-Cinnamoyl   Adenocarpin/Orensin
(82) R = *cis*-Cinnamoyl   Isoorensin

Das zu den racemischen Alkaloiden gehörende Isoorensin (82) enthält denselben basischen Rest wie Adenocarpin und Orensin, im Gegensatz dazu jedoch einen *cis*-Cinnamoylrest, wie Untersuchungen zur Struktur

Literaturverzeichnis: SS. 42—59

und zur Synthese durch SCHÖPF et al. (*289, 290, 291*) sowie von RIBAS-MARQUES und VIDAL (*256*) gezeigt haben. Die bei der Hydrolyse des Isoorensins (82) gefundene *trans*-Zimtsäure entsteht sekundär durch *cis* → *trans*-Umlagerung. Isoorensin ist bisher nur in verschiedenen *Adenocarpus*-Arten nachgewiesen worden (*25, 252, 255,* vgl. *39*).

Bei dem in den gleichen Pflanzenarten vorkommenden (+)- und (−)-Santiaguin (83) (*25, 252, 253, 255,* vgl. *39*) handelt es sich um ein von der α-Truxillsäure abgeleitetes Dimeres des (+)-Adenocarpins. Das Racemat kann aus α-Truxillsäurechlorid und Isotripiperidein synthetisch dargestellt werden. Phenylalanin (89), Zimtsäure und Adenocarpin (81) sind spezifische biogenetische Vorstufen (*233*).

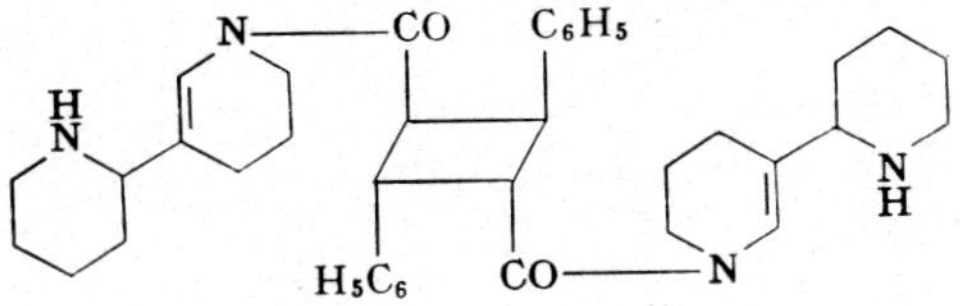

(83) Santiaguin

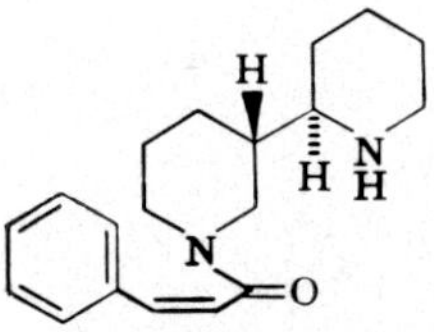

(84) Astrophyllin

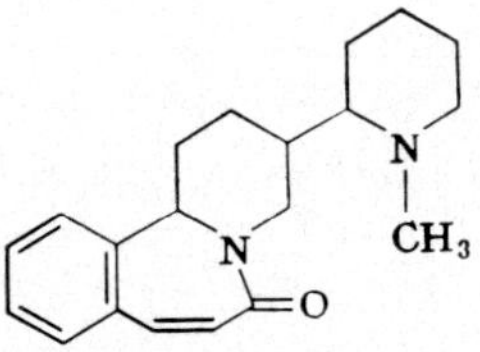

(85) Astrocasin

Als Hexahydroanabasinabkömmling sei das 1965 von LLOYD in der Euphorbiacee *Astrocasia phyllanthoides* Robinson et Millsp. aufgefundene Astrophyllin (84) genannt (*195, 196*), das als Dihydroisoorensin eine enge Verwandtschaft zum Isoorensin (82) aufweist. Astrophyllin ist als N-*cis*-Cinnamoyl-3(S)-[2'(R)-piperidyl]-piperidin charakterisiert. Die absolute Konfiguration wird von der des (+)-α,β-Dipiperidyls abgeleitet. Biogenetisch nahestehend dürfte das in der gleichen Pflanze vorkommende Astrocasin (85) sein (*195, 196*), bei dem durch Ringschluß zwischen der o-Stellung des Zimtsäurerestes und dem C-6 des Hexahydroanabasins ein siebengliedriger Lactamring entstanden ist.

2. Lobinalin

Als weiteres heterocyclisch substituiertes Piperideinalkaloid sei das Lobinalin (86) genannt. Dieser Verbindung liegt ein *trans*-N-Methyldekahydrochinolin zugrunde, das in den Positionen 5 und 7 je einen Phenylrest und in Stellung 6 einen Δ^1-Piperideinring trägt.

Lobinalin ist vor über 30 Jahren von MANSKE in *Lobelia cardinalis* L. und später von anderen Autoren (*55, 157, 158*) in weiteren *Lobelia*-Arten nachgewiesen worden. ROBISON und Mitarb. haben 1964/1966 Lobinalin in seiner Struktur und relativen Konfiguration aufgeklärt (*263, 264, 265*). Die Konstitution wurde 1967 durch die Arbeitsgruppe um MANSKE massenspektrometrisch bestätigt (*61*). Danach soll die freie Base als ein Gemisch mehrerer tautomerer Formen vorliegen.

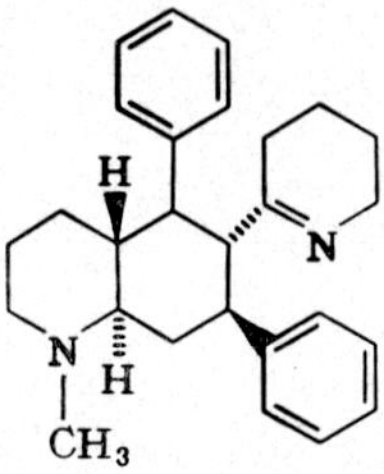

(86) Lobinalin

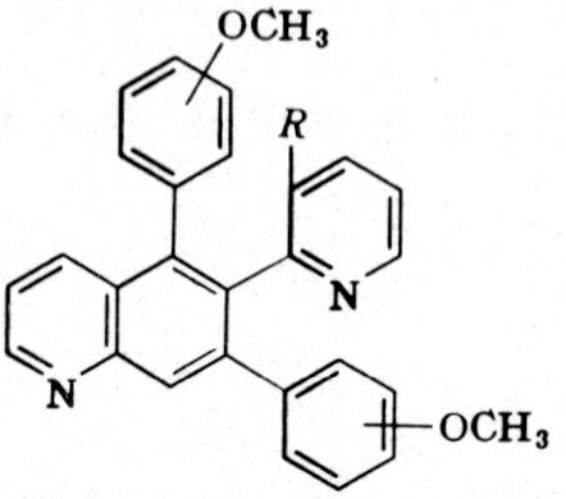

(87) R = H Syphilobin A
(88) R = OH Syphilobin F

TSCHESCHE und Mitarb. haben aus *Lobelia syphilitica* L. die beiden strukturähnlichen Verbindungen Syphilobin A (87) und Syphilobin F (88) isoliert (*327*), die einen Chinolinring, zwei methoxylierte Phenylreste und einen hydroxylierten Pyridinring aufweisen.

Die Biosynthese des Lobinalins (86) ist von GUPTA und SPENSER untersucht worden (*124*). Mit ^{14}C-markiertem Phenylalanin (89) sowie mit Lysin-(2-^{14}C) und Lysin-(6-^{14}C) (19) durchgeführte Arbeiten haben ergeben, daß Lobinalin bei *Lobelia cardinalis* L. aus Lysinäquivalenten und zwei aus Phenylalanin hervorgehenden C_6C_2-Einheiten aufgebaut wird (Schema 6).

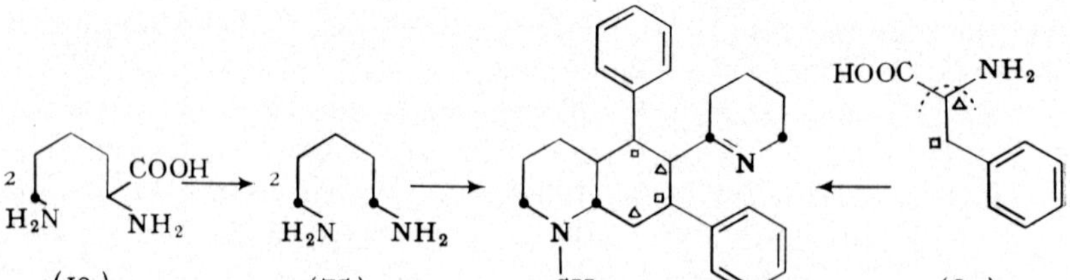

(19)   (75)   (86)   (89)

Schema 6. Biosynthese des Lobinalins

Aus der Radioaktivitätsverteilung des isolierten Lobinalins ist zu schließen, daß Lysin über ein symmetrisches Intermediärprodukt wie Cadaverin (75) in die beiden heterocyclischen Ringe inkorporiert wird.

Somit ordnet sich Lobinalin zwanglos in das Biogeneseschema der α-substituierten *Sedum*- und *Lobelia*-Alkaloide ein und ist aus biogenetischen Überlegungen heraus nicht als Abkömmling des Hydrochinolins, sondern als Dimerisierungsprodukt zweier monosubstituierter Piperidinbasen aufzufassen.

Tschesche und Mitarb. diskutieren Lobinalin als Vorstufe von Syphilobin A (87) und Syphilobin F (88) *(327)*. Das würde bedeuten, daß die beiden Pyridinringe dieser Alkaloide aus Lysin → Cadaverin → Δ^1-Piperidein hervorgehen. Dieser Biosyntheseweg des Pyridinringsystems ist bisher in der intakten höheren Pflanze nicht nachgewiesen.

3. *Ormosia*-Alkaloide

Die in *Ormosia*-Arten *(Papilionaceae)* vorkommenden Alkaloide gehören in die Gruppe der Chinolizidinbasen (vgl. *38*). Man kennt aber aus *Ormosia jamaicensis* Urb. und *O. panamensis* Benth. sowie aus *Piptanthus nanus* M. Pop. isolierte Verbindungen, die ein tetracyclisches sparteinähnliches Chinolizidingerüst mit einem angefügten Piperidinring enthalten. Es handelt sich um Ormosanin (90) und das am C-6 epimere Piptanthin (91) sowie das durch Dimerisierung von Ormosajin entstandene Ormojin (92), deren Isolierung und Strukturaufklärung von mehreren Arbeitsgruppen durchgeführt worden sind *(69, 70, 81, 136, 137, 197, 225, 226, 330, 342, 343)*. Das gemeinsame Vorkommen dieser Pflanzenbasen mit anderen charakteristischen Chinolizidinalkaloiden könnte auf eine Biosynthese aus Lysin hinweisen (vgl. *297*).

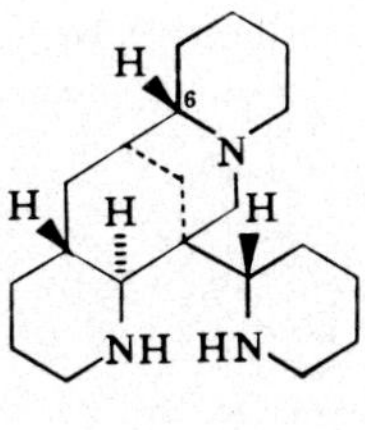

(90) Ormosanin 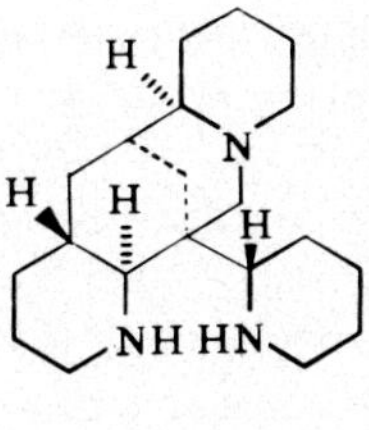   (91) Piptanthin 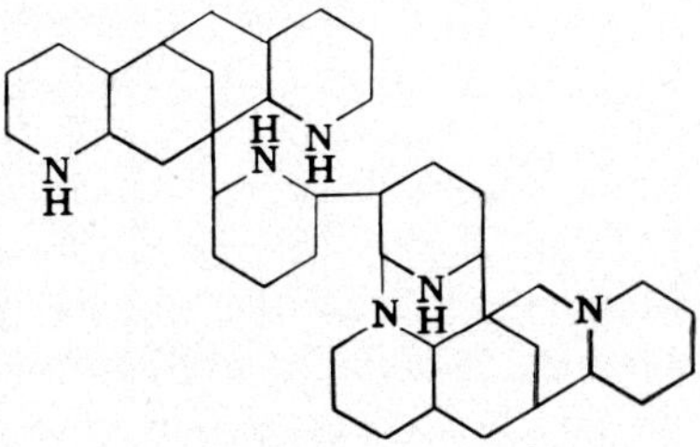   (92) Ormojin

4. Lamprolobin, Aphyllinsäuremethylester und Leontiformin

Von den Piperidinring-haltigen Chinolizidinalkaloiden sei das 1968 von Hart et al. aus der Leguminose *Lamprolobium fructicosum* Benth. isolierte (—)-Lamprolobin (93) genannt *(133, 134)*. Diese Pflanzenbase wurde als (1R:5R:10R)-1-(Glutarimidomethyl)-chinolizidin charakterisiert und ist in seiner DL-Form synthetisch zugänglich *(113, 340)*.

Andere Strukturtypen stellen Aphyllinsäuremethylester (94), eine 1966 von ASLANOV und Mitarb. in Samen von *Anabasis aphylla* L. aufgefundene Chinolizidinbase (*14*), und das von MOLLOV und IVANOV 1970 in *Leontice leontopetalum* L. entdeckte Leontiformin (95) (*217*) dar.

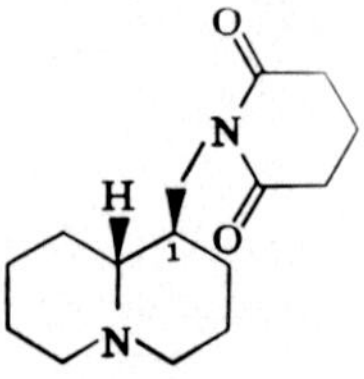

(93) Lamprolobin 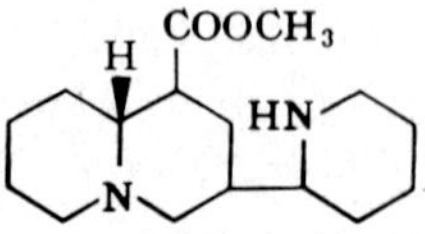   (94) Aphyllinsäuremethylester 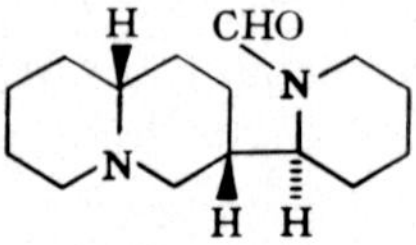   (95) Leontiformin

Diese Verbindungen finden sich vergesellschaftet mit typischen Chinolizidinalkaloiden, für die Lysin (19) als Vorstufe gilt. Daher ist zu diskutieren, daß auch (94) aus dieser Aminosäure entsteht, möglicherweise durch Aufspaltung eines Sparteingerüstes. Der 2,6-Dioxopiperidinring des Lamprolobins (93) könnte aus Glutar- oder Glutaminsäure hervorgehen. Entsprechende Biosyntheseuntersuchungen stehen noch aus.

5. *Nuphar*-Alkaloide

Den typischen, teilweise schwefelhaltigen Alkaloiden wie Nupharidin oder Thionupharidin aus der Gattung *Nuphar (Nymphaceae)* liegt das Chinolizidinskelett zugrunde (vgl. *347*). Daneben sind auch sesquiterpenoide Pflanzenbasen mit einem dreifach substituierten Piperidinring aufgefunden worden, der eine isoprenoide Seitenkette, eine Methylgruppe und einen Furanring als Substituenten trägt. Zu diesen Alkaloiden gehört das 1957/1959 von ARATA *et al.* aus *Nuphar japonicum* DC. isolierte (—)-Nupharamin (96) (*6, 7*), dessen absolute Konfiguration einige Jahre später ermittelt werden konnte (*9, 160, 161*). Auf Grund dieser Untersuchungen kommt den Substituenten an C-2 und C-6 R-Konfiguration zu. Die Methylgruppe am C-Atom 3 besitzt S-Konfiguration.

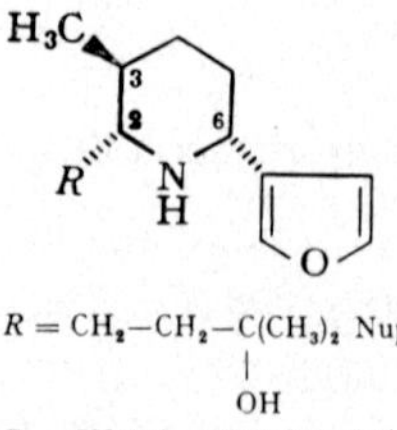

(96) $R = CH_2-CH_2-C(CH_3)_2$ Nupharamin
          |
          OH
(97) $R = CH_2CH=C(CH_3)CH_2OH$ Nuphamin
(98) $R = CH_2CH=C(CH_3)_2$ Anhydronupharamin

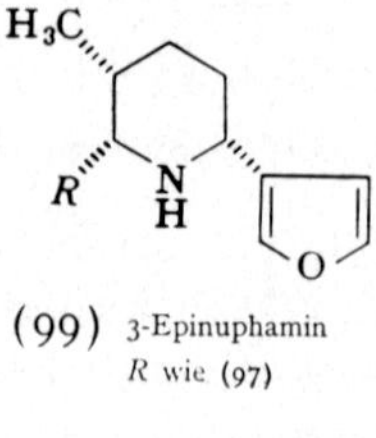

(99) 3-Epinuphamin
     R wie (97)

Literaturverzeichnis: SS. 42—59

Das von der Arbeitsgruppe um ARATA 1965 in *N. japonicum* gefundene und strukturell aufgeklärte (—)-Nuphamin (97) weist eine ungesättigte Seitenkette auf (*8, 9, 10*) und ist auch in *Nuphar luteum subsp. variegatum* enthalten (*346 a*). Das später entdeckte Anhydronupharamin (98) trägt keine Hydroxylgruppe in der ungesättigten Seitenkette (*11*). Kürzlich von WONG und LALONDE in *Nuphar luteum* nachgewiesenes 3-Epinuphamin (99) besitzt am C-3 eine entgegengesetzt konfigurierte Methylgruppe (*346*).

Bei dem von BARCHET und FORREST aus *Nuphar variegatum* Engelm. *ex* Clinton isolierten und als Nuphenin bezeichneten Alkaloid handelt es sich um ein Stereoisomeres des Anhydronupharamins (98), das sich von diesem durch die axial-äquatoriale oder äquatorial-axiale Anordnung der Wasserstoffatome an C-2 und C-3 unterscheidet (*23*).

An *Nuphar luteum* (L.) Smith durchgeführte Biosyntheseexperimente haben ergeben, daß das Hauptalkaloid Thiobinupharidin mit großer Wahrscheinlichkeit aus Isopreneinheiten aufgebaut wird (*300*). Es ist daher anzunehmen, daß auch die Piperidin-haltigen *Nuphar*-Alkaloide auf einem isoprenoiden Bildungsweg entstehen. Dafür sprechen die typischen Methylverzweigungen. Möglicherweise stellen in *Nuphar* die Piperidin-haltigen Alkaloide Vorstufen der Chinolizidinbasen dar.

6. Piperidin-haltige Indolalkaloide (Secamine, Secodine und Nitrarin)

Aus der umfangreichen und strukturell so mannigfaltigen Gruppe der Indolalkaloide kennt man seit kurzem einige biogenetisch interessante Verbindungen, die neben dem charakteristischen Indolteil einen Piperidin- bzw. Δ^2-Piperideinring besitzen. Zu diesen Verbindungen gehören die 1968 von der Arbeitsgruppe um G. F. SMITH isolierten dimeren Indolalkaloide Dihydrosecamin, Tetrahydrosecamin und Secamin aus der Apocynacee *Rhazia stricta* Decaisne und aus *R. orientalis* A. DC. (*85, 86*). Diese als Secamine bezeichneten Pflanzenbasen stellen einen neuartigen Strukturtyp dar.

Wenig später gelang der gleichen Forschergruppe die Entdeckung entsprechender monomerer Basen, die Secodine genannt wurden. Bei diesen Substanzen befindet sich am C-2 des Indolkerns eine Äthylgruppe oder ein verzweigter C_3-Rest. Das C-Atom 3 des Indols ist über eine Äthylenbrücke mit dem Heteroatom eines äthyl-substituierten Piperidin- oder Δ^2-Piperideinringes verbunden. Es handelt sich um 2-Äthyl-3-[2-(3-äthylpiperidin)äthyl]indol (100) aus *Tabernamontana cumminsii* (*68*) sowie um vier in den oben genannten *Rhazia*-Arten vorkommende Secodinalkaloide: 16,17,15,20-Tetrahydrosecodin (101), 16,17,15,20-Tetrahydrosecodin-17-ol (102), 16,17-Dihydrosecodin (103) (*44, 64*) und 16,17-Dihydrosecodin-17-ol (104) (*24*). Diese Alkaloidtypen stellen sicher

Zwischen- oder Nebenprodukte der zu den Indolalkaloiden führenden Biosynthesekette dar. Auf Grund sehr eingehender Untersuchungen (vgl. *118a, 121*) ist bekannt, daß für die meisten Indolalkaloide Tryptamin und eine Monoterpeneinheit als Ausgangsstufen verwertet werden. Das dürfte auch für die Secamine und Secodine zutreffen. Die 5 C-Atome

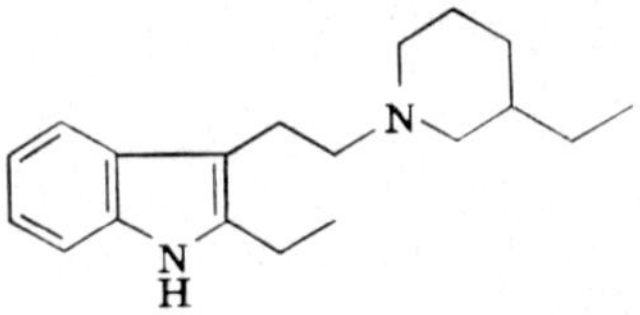

(100) 2-Äthyl-3-[2'(3''-äthylpiperidin)äthyl]indol

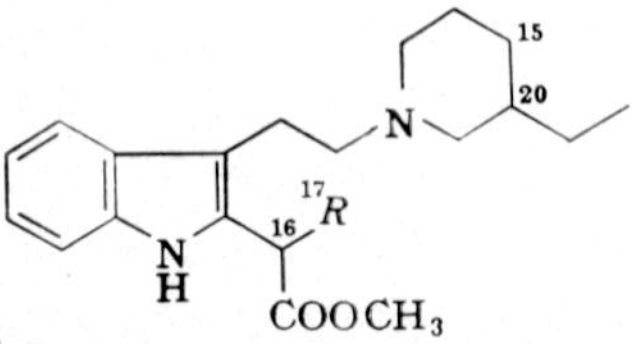

(101) R = CH$_3$ 16,17,15,20-Tetrahydrosecodin
(102) R = CH$_2$OH 16,17,15,20-Tetrahydrosecodin-17-ol

des Piperidinringes und die Äthylseitenkette der Verbindungen (100) bis (104) sollten dabei aus der Monoterpeneinheit hervorgehen. Der Heterostickstoff des Piperidinringes würde dem Aminostickstoff des

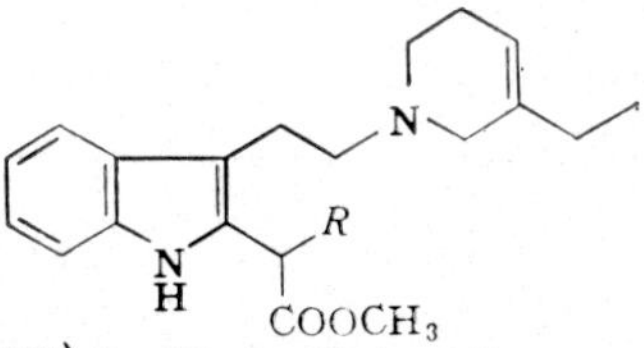

(103) R = CH$_3$ 16,17-Dihydrosecodin
(104) R = CH$_2$OH 16,17-Dihydrosecodin-17-

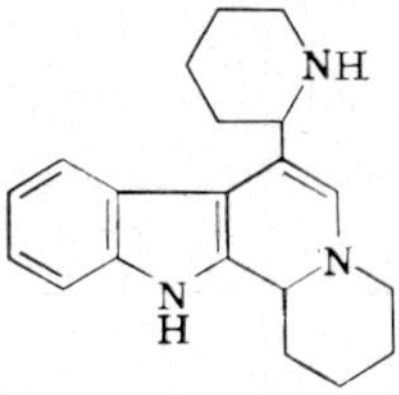

(105) Nitrarin

Tryptamins entsprechen. Somit unterscheidet sich das Bildungsprinzip dieses Piperidinringes stark von den anderen, zum Piperidinheterocyclus führenden Biosynthesewegen.

Schließlich sei noch ein von NORMATOV und YUNUSOV aus *Nitraria schoberi* L. *(Zygophyllaceae)* isoliertes Indolalkaloid Nitrarin (105) genannt (*230*), das sich gegenüber den erwähnten Secaminen und Secodinen durch ein andersartiges C-Gerüst auszeichnet.

V. Monoterpenoide Piperidinalkaloide

Vor etwa 10 Jahren haben drei Arbeitsgruppen etwa gleichzeitig aus der Apocynacee *Skytanthus acutus* Meyen das Alkaloid Skytanthin (106) isoliert und strukturell aufgeklärt (*5, 50, 75, 76*). Durch stereospezifische Umwandlung zu konfigurativ bekannten Verbindungen und durch Synthesen konnte die absolute Konfiguration der natürlich vorkommenden Stereoisomeren als α-, β- und δ-Skytanthin ermittelt werden (*49, 80*). Als Begleitalkaloide finden sich in der gleichen Pflanze β-Skytanthin-

Literaturverzeichnis: SS. 42—59

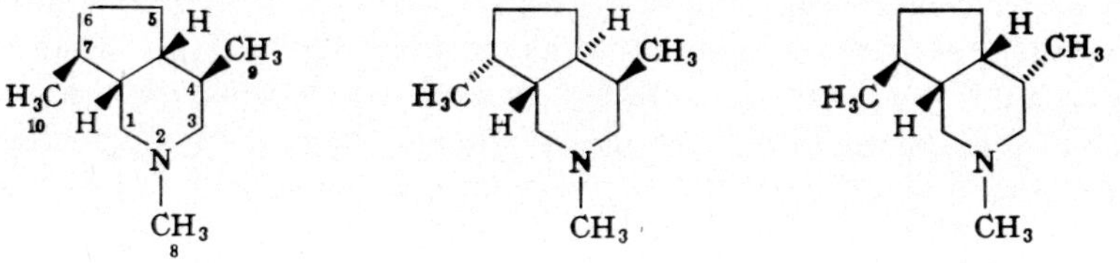

(106) Skytanthin α-, β-, δ-

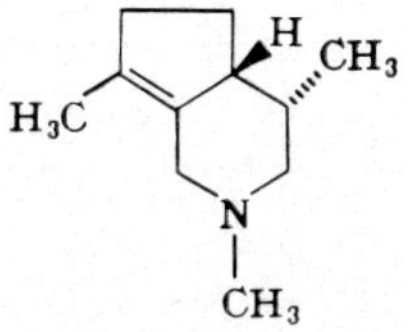

(107) Δ⁷-Dehydroskytanthin

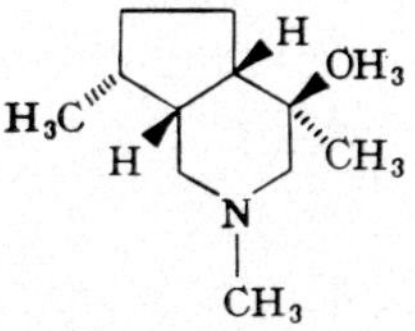

(108) 4-Hydroxyskytanthin

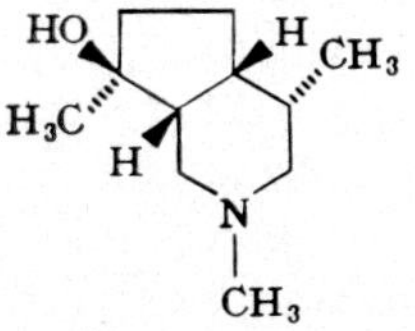

(109) 7-Hydroxyskytanthin

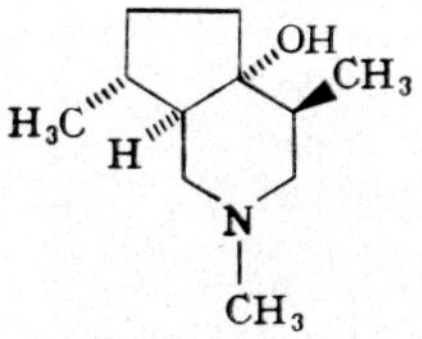

(110) 4a-Hydroxyskytanthin

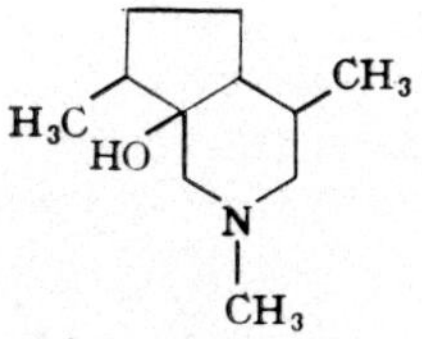

(111) 7a-Hydroxyskytanthin

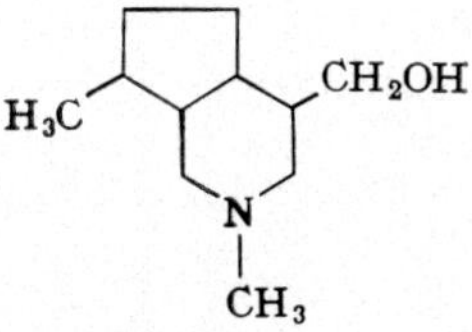

(112) Tecostanin

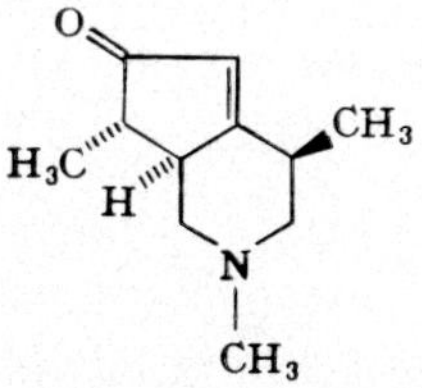

(112 a) Tecomanin

N-oxid (vgl. (106)) (317a) sowie Δ⁷-Dehydroskytanthin (107), 4- und 7-Hydroxyskytanthin (108) (109) (3, 5, 48, 207). Strukturverwandte Verbindungen sind die aus *Tecoma stans* Juss. *(Bignoniaceae)* isolierten Basen N-Normethylskytanthin, 4a- und 7a-Hydroxyskytanthin (110), (111) (71) sowie Tecostanin (112) (128, 129) und Tecomanin (112a) (155). Es handelt sich bei diesen Alkaloiden um Verbindungen, denen ein N-methylierter Piperidinring mit einem ankondensierten Methylcyclopentanring zugrunde liegt.

Die Biosynthese des Skytanthins ist erstmals 1964 von der italienischen Arbeitsgruppe um MARINI-BETTOLO untersucht worden (51, 52, 200, 207).

Es wurde nachgewiesen, daß Mevalonsäure-(2-^{14}C) wesentlich besser in Skytanthin inkorporiert wird als Phenylalanin oder Acetat. Spätere Experimente von AUDA und Mitarb. haben den spezifischen Einbau von Mevalonsäure bestätigt (*17*). Diese Autoren konnten durch gezielten chemischen Abbau zeigen, daß bei 15 Monate alten *Skytanthus*-Pflanzen 44% der ^{14}C-Aktivität im C-Atom 6 und etwa je 25% in Position 3 und 9 des isolierten Skytanthins lokalisiert waren. Dagegen ergaben dreijährige Pflanzen einen gleichwertigen Einbau des C-Atoms 2 der Mevalonsäure in die C-Atome 3 und 6 des Skytanthins. Es ist noch offen, ob dieser unterschiedliche Mevalonat-Einbau durch verschiedenartige Randomerisierung der Mevalonsäurefolgeprodukte zustande kommt, die vom Pflanzenalter abhängig ist, oder ob möglicherweise ein Gemisch radioisomerer Skytanthine zum chemischen Abbau eingesetzt worden ist, wie es von APPEL diskutiert wird (*4*). Es dürfte aber erwiesen sein, daß das C-Gerüst der Alkaloide vom Typ des Skytanthins aus zwei Mevalonsäurebausteinen, wahrscheinlich über iridoide Zwischenstufen, gebildet wird. Das trifft auch für das Actinidin zu, ein Alkaloid aus *Actinidia polygama* (Sieb. et Zucc.) Maxim. mit einer dem N-Normethylskytanthin entsprechenden Pyridinstruktur (vgl. *121*). Es ist noch ungeklärt, ob die Pyridin- und Piperidinringe dieser Alkaloidgruppe ineinander übergehen oder ob sie auf parallel angelegten Biosynthesewegen entstehen.

VI. Verschiedenartige Piperidinstrukturen

1. Alkaloide

Unter den Alkaloiden aus der Familie *Himantandraceae* (vgl. *259*) findet man einige Vertreter mit einem Piperidinring, der über eine *trans*-ständige Äthylidenbrücke mit einem Dekalin, das zusätzlich einen γ-Lactonring aufweist, verbunden ist. Zu diesen Verbindungen gehören das von PINHEY *et al.* strukturell (*245*) und von FRIDRICHSON und MATHIESON (*106*) konfigurativ aufgeklärte Himbacin (**114**), sein N-Norderivat (+)-Himbelin (**113**) und das stereoisomere (+)-Himandravin. Letzteres unterscheidet sich vom Himbelin durch entgegengesetzte Konfiguration an einem oder beiden asymmetrischen C-Atomen des Piperidinringes.

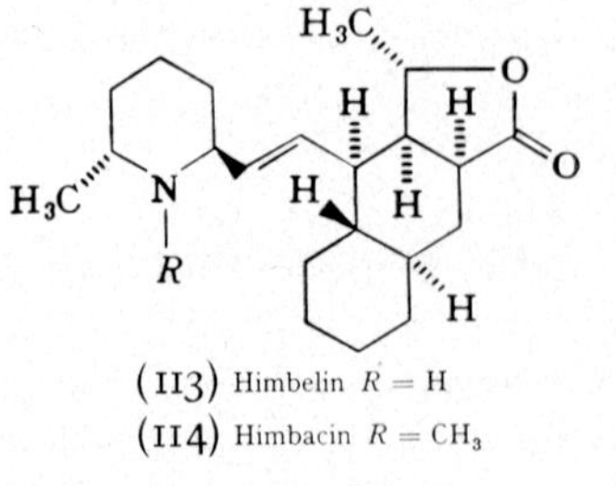

(**113**) Himbelin $R = H$
(**114**) Himbacin $R = CH_3$

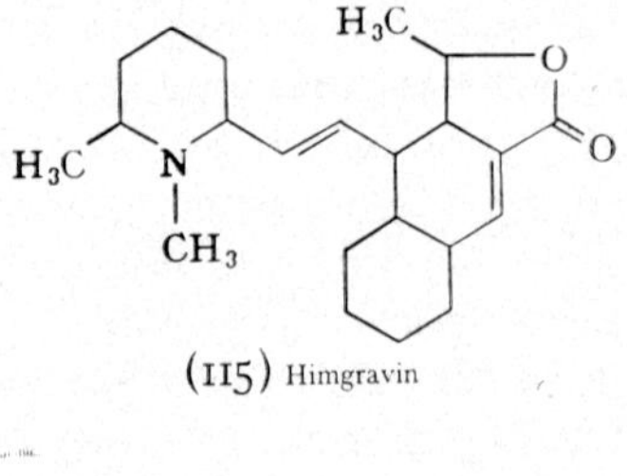

(**115**) Himgravin

Literaturverzeichnis: SS. 42—59

Das Dehydroderivat des Himbacins wird als Himgravin (115) bezeichnet, dessen Struktur zusätzlich durch NMR-Messung gesichert wurde (*2*, *245*). Für diese Pflanzenbasen diskutiert LEETE als Vorstufe eine Poly-β-ketosäure, die nach zweifacher C-Methylierung Verbindung (113) ergibt (*184*).

Alkaloide mit einer Piperidonstruktur sind das 1959/1961 von NAKANO et al. bearbeitete Julocrotin (116) aus der Euphorbiacee *Julocroton montevidensis* Klotzsch (*227, 228*) und das kürzlich im Arbeitskreis von DOEPKE aufgefundene Campedin (117) aus *Campanula medium* L. (*77*).

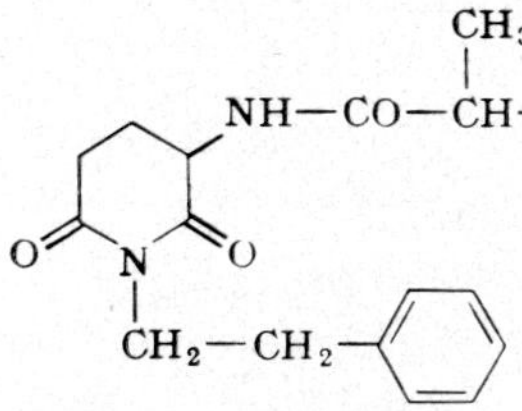

(116) Julocrotin

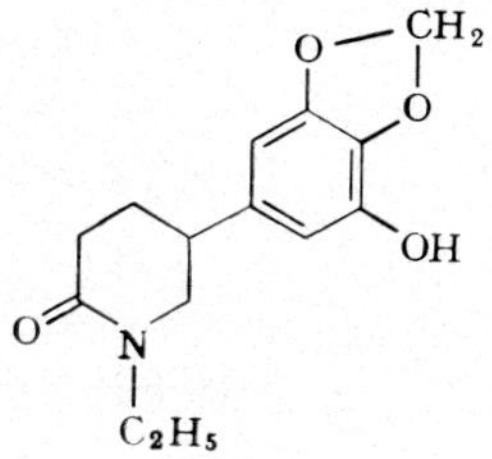

(117) Campetin

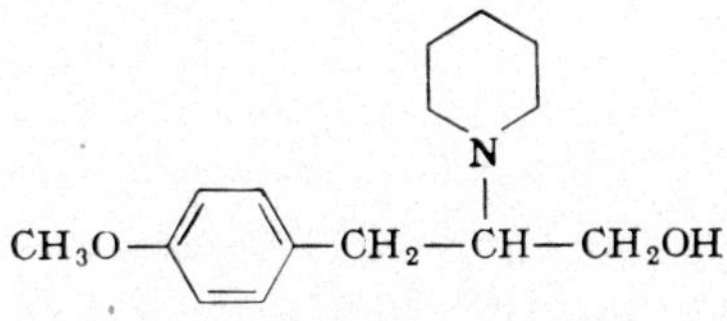

(118) Alangin A

Alangin A (118), ein Alkaloid aus *Alangium lamarckii* Thwaits *(Alangiaceae)*, ist als 3'-(p-Methoxyphenyl)-2-(1'-piperidyl)-n-propanol identifiziert worden (*36*).

Schließlich sei auf die stickstoffhaltigen Steroide verwiesen, von denen einige einen methylierten Piperidin- bzw. Piperideinring enthalten. Stellvertretend seien genannt als Spirosolan das Solasodin (119) aus verschiedenen *Solanum*-Arten und als 22,26-Epiminocholestan das Tomatillidin (120) aus *Solanum tomatillo* Phil. f. (*37*) sowie aus der *Veratrum*-Gruppe das Veralkamin (121) (*325, 326*) (ausführliche Übersichten vgl. *176, 295, 296, 298*). Sie stellen keine eigentlichen Piperidinalkaloide dar und sollen daher nicht näher behandelt werden. Corsevin (122) aus

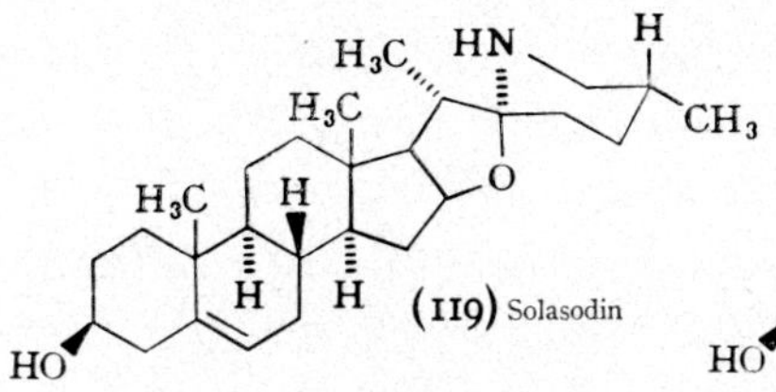

(119) Solasodin

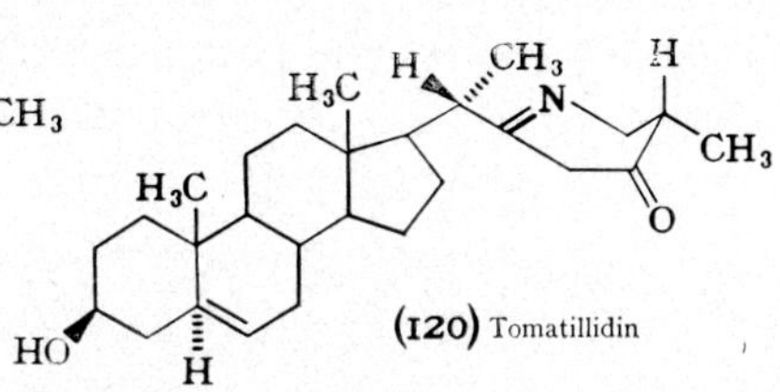

(120) Tomatillidin

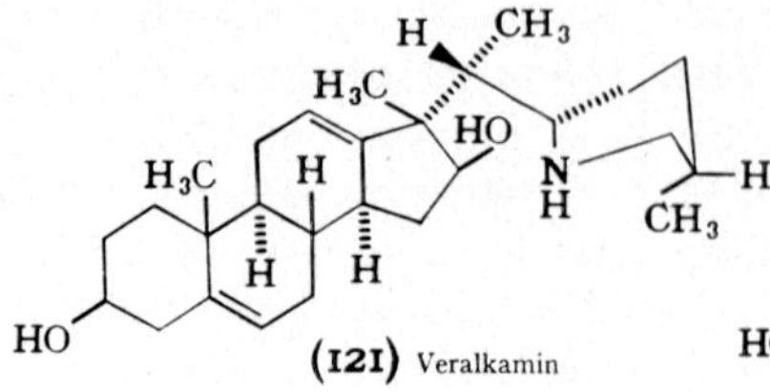

(121) Veralkamin

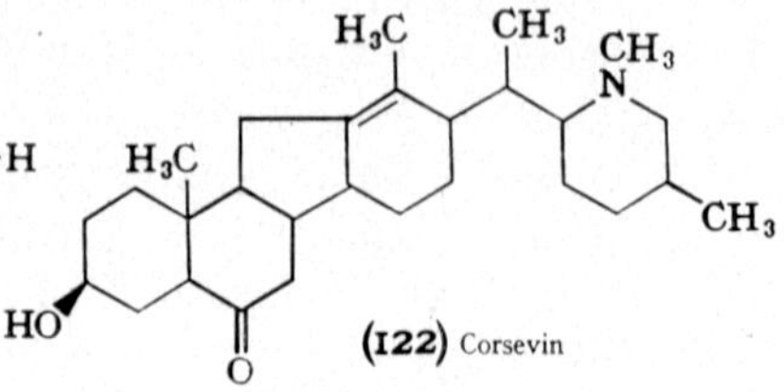

(122) Corsevin

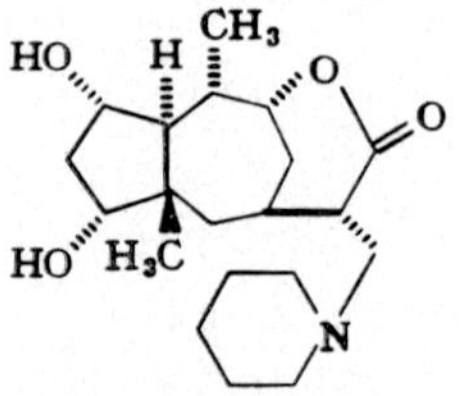

(122a) Pulchellidin

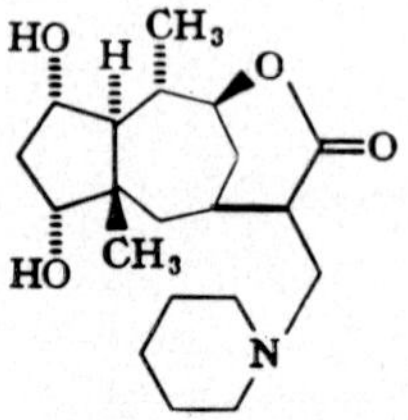

(122b) Neopulchellidin

der Liliacee *Korolkowia severzowii* Rgl. (*232*) und die von YANAGITA *et al.* bearbeiteten Sesquiterpenalkaloide Pulchellidin (122a) und Neopulchellidin (122b) (*303a, 348b, c, d*) sind ebenfalls Piperidinring-haltig. Letztere sind aus *Gaillardia pulchella* Foug. isoliert und strukturell sowie konfigurativ aufgeklärt worden. Die Autoren diskutieren Lysin (19) als mögliche Vorstufe des Piperidinringes.

2. Betalaine

Weiterhin sind die Betalaine aufzuführen, die als rotviolette (Betacyane) und gelbe (Betaxanthine) stickstoffhaltige Farbstoffe bei anthocyanfreien Centrospermen verbreitet sind. Diese Naturstoffgruppe wird durch Betanin (123), das Pigment der Roten Rübe, und durch Indicaxanthin (124) der Kaktusfeige *Opuntia ficus-indica* Mill. repräsentiert (Übersicht über Betalaine (vgl. *203*)).

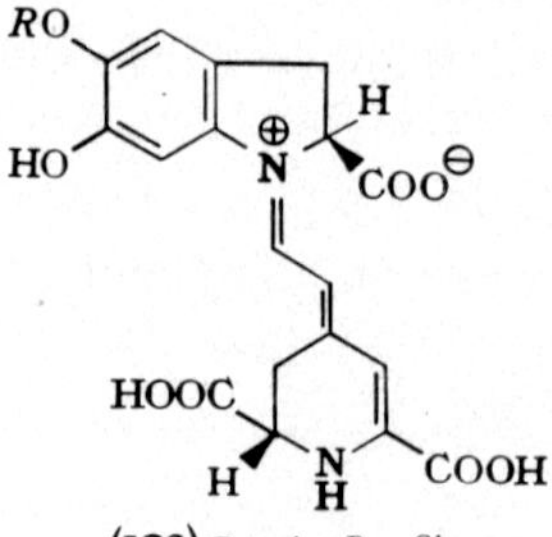

(123) Betanin, R = Glucose
Betanidin, R = H

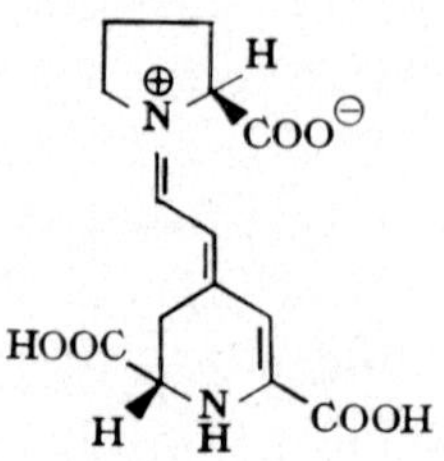

(124) Indicaxanthin

Literaturverzeichnis: SS. 42—59

Die erst vor kurzem abgeschlossene Konstitutionsaufklärung dieser alkaloidartigen Strukturen hat sich als schwierig erwiesen. Beim Betanidin, dem Aglukon des Betanins (123), ist ein 5,6-Dihydroxydihydroindol-2-carbonsäurerest mit einer Pyridin-2,6-dicarbonsäureeinheit [Betalaminsäure (129)] verbunden. Im Indicaxanthin (124) ist der Dihydroindolteil durch Prolin ersetzt.

Nach einer Hypothese von WYLER, MABRY und DREIDING sollen der Cyclodopateil (128) und die Betalaminsäure (129) des Betanidins (123) biogenetisch aus je einem Molekül 3,4-Dihydroxyphenylalanin (Dopa) (126) entstehen (*348*). Als Zwischenstufen werden zwei Moleküle Dopachinon (127) diskutiert, von denen eines nach WOODWARD-Spaltung und erneuter Cyclisierung Betalaminsäure ergibt (Schema 7). Für diese Annahme sprechen Biosyntheseversuche, in denen Dopa und verwandte Verbindungen Inkorporation in Betanidin (123) (*149, 193, 215*), Amaranthin (*109*) und Indicaxanthin (124) (*216*) zeigten.

MILLER et al. (*215*) sowie LIEBISCH et al. (*193*) konnten Tyrosin (125) und Dopa (126) als spezifische Precursoren bestätigen und nachweisen,

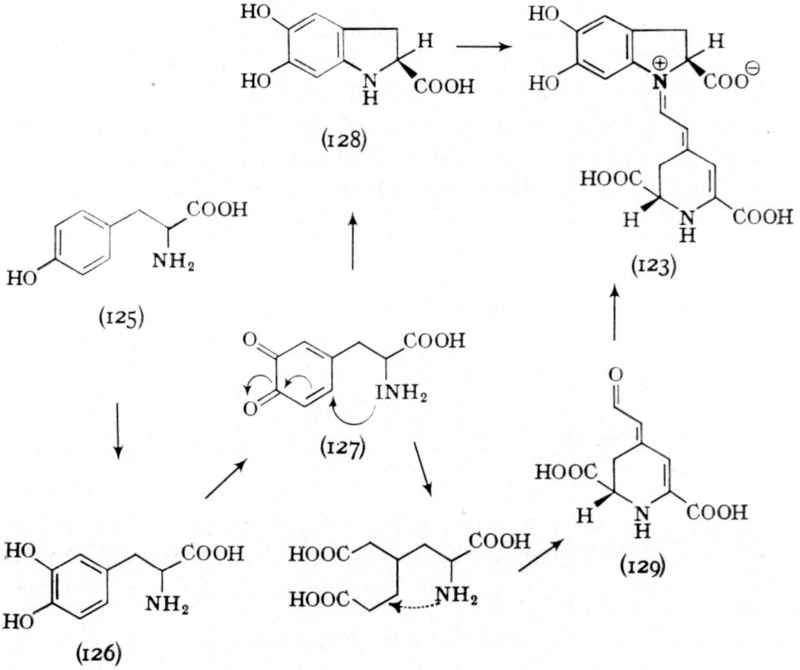

Schema 7. Biosynthese des Betanidins

daß diese Phenylpropankörper sowohl in den Cyclodopa- als auch in den Betalaminsäureteil des Betanins (123) eingebaut werden, wenn auch mit unterschiedlichen spezifischen Inkorporationsraten. Nach Ansicht dieser Autoren verläuft die Bildung des Betanins entsprechend dem in Schema 7 dargestellten Biosyntheseweg.

3. Antibiotika

Aus der Gruppe der Antibiotika sind ebenfalls einige Vertreter bekannt, die einen Piperidin- bzw. Piperidonring enthalten.

Von den B-Streptograminen seien Staphylomycin S, Vernamycin B_δ, Mykamycin sowie Ostreogrycin B, B_1 und B_2 genannt, bei denen eine 4-Oxopipecolinsäure (130) amidartig in einen Polypeptidverband eingefügt ist (*331, 334, 338, 339*). Interessanterweise enthalten die genannten Verbindungen gleichzeitig einen 3-Hydroxypicolinsäurerest, so daß ein Pyridin- neben einem Piperidinring vorliegt. Aspartocin, ein von *Streptomyces griseus var. spiralis* und von *St. violaceus* gebildetes Polypeptid-Antibiotikum, ist aus sechs Aminosäuren und aus Pipecolinsäure (7) aufgebaut (*210*). In dem zur Gruppe der Sideromycine gehörenden Ferrimycin A_1 findet sich ein vollständig substituierter 2-Piperidonring (vgl. *231*).

Eine weitere Gruppe der Antibiotika leitet sich vom Cycloheximid (131) ab (*307, 332*). Diese Verbindungen besitzen einen Glutarimidring (2,6-Dioxopiperidin), der am C-Atom 4 über eine C_2-Brücke mit einem 2,4-Dimethylcyclohexanonrest verknüpft ist. Durch strukturelle Veränderungen ergibt sich eine Reihe weiterer Cycloheximid-Antibiotika. Diese wenigen, ausgewählten Beispiele mögen zeigen, daß der Piperidinring in einer Vielzahl von Naturstoffen enthalten ist.

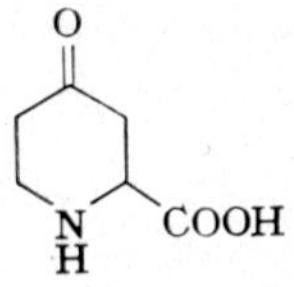

(130) 4-Oxopipecolinsäure

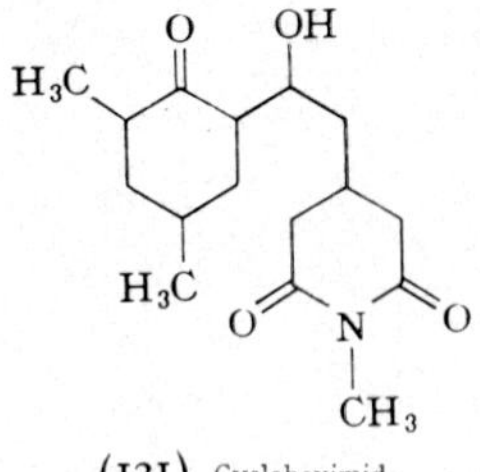

(131) Cycloheximid

Für Cycloheximid und Streptimidon liegen bereits erste Biosynthesebefunde vor (*163, 332, 333*). Durch Versuche an *St. noursei* ist gezeigt worden, daß die Methylgruppen aus Methionin und die restlichen C-Atome aus dem Acetat-Malonat-Stoffwechsel hervorgehen (Schema 8). Auf Grund der ermittelten Radioaktivitätsverteilung nach Verfütterung von Acetat-(1-^{14}C), Malonat-(1,3-^{14}C) und Kohlendioxid-(^{14}C) ist anzunehmen, daß 6 Moleküle Malonat zum Aufbau des Cycloheximids verwendet werden, wobei eine Malonateinheit ohne Decarboxylierung eingebaut

Literaturverzeichnis: SS. 42—59

wird. Mevalonat, Lysin, Glutaminsäure und Citrat ergaben keine Inkorporation. Ähnliche Ergebnisse wurden für Streptimidon an *St. rimosus f. paromomycinus* erhalten, so daß auch hier eine Kondensation von 6 Malonatresten vorliegen dürfte.

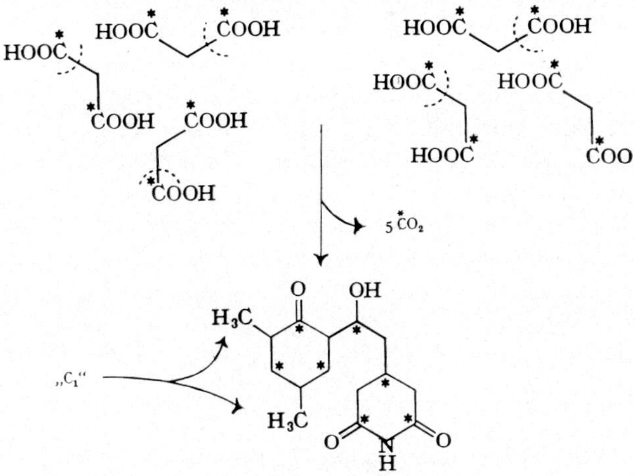

Schema 8. Biosynthese der Cycloheximid-Antibiotika

Schlußbetrachtung

Aus der Piperidinreihe sind bis heute über 130 verschiedene Naturstoffe bekannt, die sich bevorzugt in höheren Pflanzen, vereinzelt auch im Tierreich und bei verschiedenen Mikroorganismen finden. Diese Piperidinverbindungen unterscheiden sich durch die Art der Substitution am Piperidinkern und durch die konfigurative Anordnung der Substituenten. Darüber hinaus gibt es auch einige komplizierter gebaute Naturstoffe, die in ihrem Molekül einen Piperidinring enthalten.

Bei den pflanzlichen Naturstoffen mit Piperidinstruktur erscheint auffällig, daß sie vielfach mit weiteren Alkaloidtypen vergesellschaftet vorkommen. Das gemeinsame Auftreten dieser andersartigen Strukturen ist in den meisten Fällen in einer biogenetischen Verwandtschaft begründet.

Die bisher vorliegenden Untersuchungen zur Biosynthese haben gezeigt, daß sich die Natur zur Bildung des Piperidinringes mehrerer Prinzipien bedient, wobei unterschiedliche Precursoren verwertet werden.

Eine Reihe natürlicher Piperidinverbindungen leitet sich von der Aminosäure Lysin (19) ab, die einer oxydativen Desaminierung unter nachfolgender intramolekularer Cyclisierung unterliegt. Das entstehende reaktionsfähige Azomethin fungiert als Vorstufe von Piperidinabkömmlingen wie Pipecolinsäure (7), Anabasin (69) oder Isopelletierin (31). Bei diesen Substanzen erfolgt der Lysineinbau über unsymmetrische Zwischenstufen, wobei der Aminogruppenstickstoff des Lysins den Heterostickstoff des Piperidins ergibt (vgl. Schema 1, 3 und 5). Demgegenüber wird bei der Biosynthese von Lobelin (51) und Lobinalin (86) ein symmetrisches Intermediärprodukt durchlaufen (vgl. Schema 4a und 6).

Eine Anzahl weiterer Piperidinverbindungen wird aus Acetat über eine Polyketidkette aufgebaut, wie Untersuchungen an den *Conium*-Alkaloiden oder am Pinidin (47) ergeben haben (vgl. Schema 2 und 4). Schließlich ist für die monoterpenoiden Alkaloide vom Typ des Skytanthins (106) Mevalonsäure als Vorstufe nachgewiesen.

Einzelheiten über den Reaktionsmechanismus und den stereochemischen Ablauf der genannten Biosynthesewege sind nicht bekannt. Man weiß auch noch sehr wenig über Enzymologie und Regulation der einzelnen zum Piperidinheterocyclus führenden Reaktionsschritte.

Literaturverzeichnis

1. ABLONDI, F., S. GORDON, J. MORTON II and J. H. WILLIAMS: An Antimalarial Alkaloid from *Hydrangea*. II. Isolation. J. Organ. Chem. (USA) **17**, 14 (1952).
2. ABRAHAM, R. J. and H. J. BERNSTEIN: Proton Magnetic Resonance Spectrum and Structure of Himgravine. Austral. J. Chem. **14**, 64 (1961).
3. ADOLPHEN, G., H. H. APPEL, K. H. OVERTON and W. D. C. WARNOCK: Hydroxyskytanthines I and II. Two Minor Alkaloids of *Skytanthus acutus* Meyen. Tetrahedron **23**, 3147 (1967).
4. APPEL, H. H.: Sobre la Existencia de Esquitantinas "Radioisomeras". Scientia (Valparaiso) **35**, 128 (1968) [Chem. Abstr. **71**, 124733b (1969)].
5. APPEL, H. H. and B. MÜLLER: Alkaloids from Skytanthus acutus. Scientia (Valparaiso) **28** (115), 5 (1961) [Chem. Abstr. **57**, 2332g (1962)].
6. ARATA, Y. and T. OHASHI: Constituents of *Rhizoma nupharis*. XII. Nupharamine. Yakugaku Zasshi **77**, 792 (1957) [Chem. Abstr. **51**, 17756b (1957)].
7. — — Constitution of Nupharamine. Yakugaku Zasshi **79**, 127 (1959) [Chem. Abstr. **53**, 10215 (1959)].
8. — — Nuphamine: A New Alkaloid from *Nuphar japonicum* DC. Chem. Pharm. Bull. (Tokyo) **13**, 392 (1965).
9. — — Constituents of *Nuphar japonicum*. XXII. Structure of Nuphamine. Chem. Pharm. Bull. (Tokyo) **13**, 1247 (1965) [Chem. Abstr. **64**, 6703 (1966)].
10. — — Constituents of *Nuphar japonicum*. XXIII. The Absolute Configuration of Nuphamine. Chem. Pharm. Bull. (Tokyo) **13**, 1365 (1965) [Chem. Abstr. **64**, 6703 (1966)].
11. ARATA, Y., T. OHASHI, M. YONEMITSU and S. YASUDA: Constituents of *Nuphar japonicum*. XXIV. Structure of a New Alkaloid, Anhydronupharamine. Yakugaku Zasshi **87**, 1094 (1967) [Chem. Abstr. **68**, 2912 (1968)].
12. ARNDT, R. R. and L. M. DU PLESSIS: The Alkaloids of *Coelidium fourcadei*. J. S. Afr. Chem. Inst. **21**, 54 (1968) [Chem. Abstr. **70**, 4342f (1969)].
13. ARTHUR, H. R. and R. P. K. CHAN: A New Alkaloid from *Isotoma longiflora*. J. Chem. Soc. (London) **1963**, 750.
14. ASLANOV, X. A., S. Z. MUKHAMEDZHANOV and A. S. SADYKOV: Chemical Studies of the *Anabasis aphylla* Seeds. Nauch. Tr. Tashkent. Gos. Univ. Nr. **286**, 71 (1966) [Chem. Abstr. **67**, 73730f (1967)].

15. ASPEN, A. J. and A. MEISTER: Conversion of α-Aminoadipic Acid to L-Pipecolic Acid by *Aspergillus nidulans*. Biochemistry **1**, 606 (1962).
16. ATAL, C. K. and S. S. BANGA: Structure of Piplartine, a New Alkaloid from *Piper longum*. Curr. Sci. (India) **32**, 354 (1963) [Chem. Abstr. **59**, 15329h (1963)].
17. AUDA, H., H. R. JUNEJA, E. J. EISENBRAUN, G. R. WALLER, W. R. KAYS and H. H. APPEL: Biosynthesis of Methylcyclopentane Monoterpenoids. I. *Skytanthus*-Alkaloids. J. Amer. Chem. Soc. **89**, 2476 (1967).
18. AYER, W. A. and T. E. HABGOOD: The Pyridine Alkaloids. In: R. H. F. MANSKE: The Alkaloids, Chemistry and Physiology. Vol. XI, p. 459. New York-London: Academic Press. 1968.
19. BAKER, B. R. and F. J. McEVOY: An Antimalarial Alkaloid from *Hydrangea* XXIII. Synthesis by the Pyridine Approach. II. J. Organ. Chem. (USA) **20**, 136 (1955).
20. BAKER, B. R., F. J. McEVOY, R. E. SCHAUB, J. B. JOSEPH and J. H. WILLIAMS: An Antimalarial Alkaloid from *Hydrangea*. XXI. Synthesis and Structure of Febrifugine and Isofebrifugine. J. Organ. Chem. (USA) **18**, 178 (1953).
21. BAKER, B. R., R. E. SCHAUB, F. J. McEVOY and J. H. WILLIAMS: An Antimalarial Alkaloid from *Hydrangea*. XII. Synthesis of 3-[β-Keto-γ-(3-hydroxy-2-piperidyl)propyl]-4-quinazolone, the Alkaloid. J. Organ. Chem. (USA) **17**, 132 (1952).
22. BALENOVIC, K. and N. STIMAC: Pseudoconhydrine of the Configuration at C-2 with that of α-Amino Acids. Croat. Chem. Acta **29**, 153 (1957) [Chem. Abstr. **53**, 15967e (1959)].
23. BARCHET, R. and T. P. FORREST: Alkaloids of *Nuphar variegatum*. Tetrahedron Letters **1965**, 4229.
24. BATTERSBY, A. R. and A. K. BHATNAGAR: Evidence from Synthesis and Isolation Concerning the Rearrangement Process in Indole Alkaloid Biosynthesis. Chem. Comm. **1970**, 193.
24a. BEECHAM, A. F., S. R. JOHNS and J. A. LAMBERTON: The absolute Configuration of (+)-9-Aza-1-methylbicyclo(3,3,1)nonan-3-one, an alkaloid from Euphorbia atoto. Austral. J. Chem. **20**, 2291 (1967).
25. BERNASCONI, R. und E. STEINEGGER: Die Alkaloide von *Adenocarpus mannii* Hooker, einer Leguminose des Kilimanjaro. Pharm. Acta Helv. **45**, 42 (1970).
26. BEVAN, C. W. L. and A. U. OGAN: Studies on West African Medicinal Plants. I. Biogenesis of Carpaine in *Carica papaya* Linn. Phytochemistry **3**, 591 (1964).
27. BEYERMAN, H. C.: The Absolute Configuration of (−)-Homostachydrine from Alfalfa. Rec. trav. chim. Pays-Bas **78**, 134 (1959).
28. BEYERMAN, H. C., J. EENSHUISTRA and W. EVELEENS: On the Absolute Configuration of Sedamine and Sedridine. Rec. trav. chim. Pays-Bas **76**, 415 (1957).
29. BEYERMAN, H. C., J. EENSHUISTRA, W. EVELEENS and A. ZWEISTRA: Studies on *Sedum* Alkaloids. V. Synthesis and Optical Resolution of Sedamine and Sedridine. Rec. trav. chim. Pays-Bas **78**, 43 (1959).
30. BEYERMAN, H. C., W. EVELEENS and Y. M. F. MULLER: On the Synthesis and Stereochemistry of Sedamine. Rec. trav. chim. Pays-Bas **75**, 63 (1956).
31. BEYERMAN, H. C., M. VAN LEEUWEN, J. SMIDT and A. VAN VEEN: γ-Coniceine of Conium maculatum L., a Revision of the Generally Accepted Structure. Rec. trav. chim. Pays-Bas **80**, 513 (1961).
32. BEYERMAN, H. C. and L. MAAT: Synthesis and Absolute Configuration of Tanret's (−)-Pelletierine. Rec. trav. chim. Pays-Bas **82**, 1033 (1963).
33. − − Resolution of Isopelletierine: A Second Synthesis of Pelletierine. Rec. trav. chim. Pays-Bas **84**, 385 (1965).

34. BEYERMAN, H. C., L. MAAT, A. VAN VEEN, A. ZWEISTRA and W. VON PHILIPSBORN: The Complete Configuration of Sedridine and of Pelletierine. Rec. trav. chim. Pays-Bas **84**, 1367 (1965).
35. BEYERMAN, H. C. und Y. M. F. MULLER: Über die Isolierung und Strukturaufklärung eines neuen Alkaloids aus *Sedum acre* L. Rec. trav. chim. Pays-Bas **74**, 1568 (1955).
36. BHAKUNDI, D. S., M. M. DHAR and M. L. DHAR: Structure of Alangine A, an Alkaloid from *Alangium lamarckii*. J. sci. ind. Res. (New Delhi) Sect. B **19**, 8 (1960) [Chem. Abstr. **54**, 21168h (1960)].
37. BIANCHI, E., C. DJERASSI, H. BUDZIKIEWICZ and Y. SATO: Structure of Tomatillidine. J. Organ. Chem. (USA) **30**, 754 (1965).
37a. BLUNDEN, G., S. B. CHALLEN and B. JAQUES: Accumulation of Anthoxanthines and Iminoacids in Leaf Galls of *Salix fragilis* L. Nature **212**, 514 (1966).
38. BOHLMANN, F. and D. SCHUMANN: Lupine Alkaloids. In: R. H. F. MANSKE, The Alkaloids, Chemistry and Physiology. Vol. IX, p. 175. New York-London: Academic Press. 1967.
39. BOIT, H.-G.: Ergebnisse der Alkaloid-Chemie bis 1960, S. 125ff. Berlin: Akademie-Verlag. 1961.
40. BOSTOGANASHVILL, V. S.: *Genista transcaucasica*. I. *Genista abchasica* Alkaloids. Tr. Farmakokhim. Akad. Nauk. Gruz. SSR, Ser. 1, **1967**, 196 [Chem. Abstr. **69**, 8900p (1968)].
41. BOULANGER, P. and R. OSTEUX: The destination of Lysine in the Animal and Vegetable Kingdoms. C. R. hebd. Seances Acad. Sci. **239**, 458 (1954) [Chem. Abstr. **49**, 2536h (1955)].
42. — — Stoffwechsel der DL-Pipecolinsäure bei den Vertebraten. Hoppe-Seyler's Z. physiol. Chem. **321**, 79 (1960).
43. BRANDÄNGE, S. and B. LÜNING: Studies on *Orchidaceae* Alkaloids. XVII. Alkaloids from Vandopsis longicaulis Schltr. Acta Chem. Scand. **24**, 353 (1970).
44. BROWN, R. T., G. F. SMITH, K. S. J. STAPLEFORD and D. A. TAYLOR: The Occurrence of 16,17,15,20-Tetrahydrosecodine in *Rhazya orientalis*. Chem. Comm. **1970**, 190.
45. BRUTKO, L. I. and P. S. MASSAGETOV: Lupinine Content in *Anabasis aphylla*. Med. Prom. SSSR **18**, 34 (1964) [Chem. Abstr. **62**, 10823d (1965)].
46. BRUTKO, L. I., P. S. MASSAGETOV and L. M. UTKIN: Content of Anabasine Isomers in Different Samples of *Anabasis aphylla*. Rast. Resur. **4** (3), 334 (1968) [Chem. Abstr. **70**, 17564e (1969)].
47. BÜCHEL, K. H. und F. KORTE: Acyl-lacton Umlagerung, XXIV. Synthese von Schierlingsalkaloiden nach dem Reaktionsprinzip der Acyl-lacton-Umlagerung. Chem. Ber. **95**, 2460 (1962).
48. CASINOVI, C. G., F. DELLE MONACHE, G. GRANDOLINI, G. B. MARINI-BETTOLO and H. H. APPEL: Two New Alkaloids from *Skytanthus acutus* Meyen. Chem. and Ind. **1963**, 984.
49. CASINOVI, C. G., F. DELLE MONACHE, G. B. MARINI-BETTOLO, E. BIANCHI e J. GARBARINO: Ricerche nella serie della Skytanthina. Nota II. Sintesi di tre stereoisomeri a configurazione nota della Skytanthina. Gazz. chim. ital. **92**, 479 (1962).
50. CASINOVI, C. G., J. A. GARBARINO and G. B. MARINI-BETTOLO: Structure of the Alkaloid of *Skytanthus acutus* Meyen. Chem. and Ind. **1961**, 253.
51. CASINOVI, C. G., G. GIOVANNOZZI-SERMANNI e G. B. MARINI-BETTOLO: Studi preliminari sulla biosintesi delle skitanthine. Gazz. chim. ital. **94**, 1356 (1964).
52. — — — — The Biosynthesis of Skytanthines. Rend. Accad. Naz. XL. Ser. IV, **16—17**, 89 (1965) [Chem. Abstr. **68**, 10289 (1968)].

53. CHATTERJEE, A. and C. P. DUTTA: The Structure of Piperlongumine, a New Alkaloid Isolated from the Roots of *Piper longum* Linn. Sci. Cult. (Calcutta) **29**, 568 (1963) [Chem. Zbl. **137**, (10), 1149 (1966)].
54. — — Alkaloids of *Piper longum* Linn. I. Structure and Synthesis of Piperlongumine and Piperlonguminine. Tetrahedron **23**, 1769 (1967).
55. CHAUBAL, M. G., R. M. BAXTER and G. C. WALKER: Paper Chromatography of Alkaloidal Extracts of *Lobelia* Species. J. Pharm. Sci. **51**, 885 (1962).
57. CHOU, T. Q., F. Y. FU and Y. S. KAO: Antimalarial Constituents of Chinese Drug Ch'an Shan, *Dichroa febrifuga* Lour. J. Amer. Chem. Soc. **70**, 1765 (1948).
58. CLARK-LEWIS, J. W. and J. DAINIS: Flavan Derivatives XIX. Austral. J. Chem. **20**, 2191 (1967).
59. CLARK-LEWIS, J. W. and P. I. MORTIMER: Occurence of 4-Hydroxypipecolic Acid in *Acacia* Species. Nature **184**, 1234 (1959).
60. — — The 4-Hydroxypipecolic Acid from *Acacia* Species, and its Stereoisomers. J. Chem. Soc. (London) **1961**, 189.
61. CLUGSTON, D. M., D. B. MACLEAN and R. H. F. MANSKE: The Examination of Lobinaline and Some Degradation Products by Mass Spectrometry. Canad. J. Chem. **45**, 39 (1967).
62. COKE, J. L. and W. Y. RICE: The Absolute Configuration of Carpaine. J. Organ. Chem. (USA) **30**, 3420 (1965).
63. COOKE, G. A. and G. FODOR: Diastereoisomeric 2-Aryl-6-methyl-3,4(1,2-piperido)-oxazines. Canad. J. Chem. **46**, 1105 (1968).
64. CORDELL, G. A., G. F. SMITH and G. N. SMITH: The Isolation of Monomeric Secodine-Type Alkaloids from *Rhazya* Species. Chem. Comm. **1970**, 189.
65. CROMWELL, B. T.: Separation, Micro-Estimation and Distribution of the Alkaloids of Hemlock (*Conium maculatum* L.) Biochem. J. **64**, 259 (1956).
66. — The Biogenesis of the Piperidine Alkaloids. In: J. B. PRIDHAM and T. SWAIN: Biosynthetic Pathways in Higher Plants, p. 147. London-New York: Academic Press. 1965.
67. CROMWELL, B. T. and M. F. ROBERTS: The Biogenesis of γ-Coniceine in Hemlock (*Conium maculatum* L.). Phytochemistry **3**, 369 (1964).
68. CROOKS, P. A., B. ROBINSON and G. F. SMITH: Isolation and Structure of an Indole Alkaloid of Biogenetic Interest from *Tabernamontana cumminsii*. Chem. Comm. **1968**, 1210.
69. DAVIES, A. P. and C. H. HASSALL: The Molecular Structure of Ormojine, an Alkaloid of *Ormosia jamaicensis*. Tetrahedron Letters **1966**, 6291.
70. DESLONGCHAMPS, P., J. S. WILSON and Z. VALENTA: The Structure of Piptanthine, a Catalytic Epimerization of *Ormosia* Alkaloids. Tetrahedron Letters **1964**, 3893.
71. DICKINSON, E. M. and G. JONES: Pyrindane Alkaloids from *Tecoma stans*. Tetrahedron **25**, 1523 (1969).
72. DIETRICH, S. M. C. and R. O. MARTIN: Biosynthesis of *Conium* Alkaloids Using Carbon-14-Dioxide. Interrelation of γ-Coniceine, Coniine and N-Methylconiine. J. Amer. Chem. Soc. **90**, 1921 (1968).
73. — — The Biosynthesis of *Conium* Alkaloids Using Carbon-14-Dioxide. The Kinetics of ^{14}C Incorporation into the Known Alkaloids and Some New Alkaloids. Biochemistry **8**, 4163 (1969).
74. DIETZSCH, K.: Über 2-Alkyl-Δ^1-piperideine und eine neue Synthese aus 1-Alkylcyclopentanolen mit Natriumazid. Z. Chem. **6**, 98 (1966).
75. DJERASSI, C., J. P. KUTNEY and M. SHAMMA: Alkaloid Studies. XXXII. Studies on *Skytanthus acutus* Meyen. The Structure of the Monoterpenoid Alkaloid Skytanthine. Tetrahedron **18**, 183 (1962).

76. DJERASSI, C., J. P. KUTNEY, M. SHAMMA, J. N. SHOOLERY and L. F. JOHNSON: Alkaloid Studies. XXVII. The Structure of Skytanthine. Chem. and Ind. **1961**, 210.
77. DÖPKE, W. und G. FRITSCH: Der Alkaloidgehalt von *Campanula medium*. Pharmazie **25**, 128 (1970).
78. DOMINGUEZ, J., I. RIBAS and J. VEGA: *Papilionaceae* Alkaloids. XXVII. Structure of Sphaerocarpine and its Identity with Isoammodendrine. An. Real. Soc. espan. Fisica Quim., Ser. B **52**, 43 (1956) [Chem. Abstr. **51**, 1212 (1957)].
79. DRILLIEN, G. et C. VIEL: Sur la structure de la pelletiérine, alcaloïde du grenadier. Bull. soc. chim. France **1963**, 2393.
80. EISENBRAUN, E. J., A. BRIGHT and H. H. APPEL: The Absolute Configuration and Partial Synthesis of α-, β-, γ- and δ-Skytanthine. Chem. and Ind. **1962**, 1242.
81. EISNER, U. and F. SORM: The Structure of Piptanthine. Collect. Czech. Chem. Comm. **24**, 2348 (1959).
82. EL-OLEMY, M. M. and A. E. SCHWARTING: Simulated Biosynthesis of Anahygrine. Experientia **21**, 249 (1965).
83. — — Simulated Biosynthesis of Some Pyrrolidine and Piperidine Alkaloids of *Whitania somnifera*. Abh. dtsch. Akad. Wiss. Berlin, Kl. Chem., Geol. Biol. **1966**, (3), 137.
83a. EL-OLEMY, M. M. and A. E. SCHWARTING: The Resolution and Absolute Configuration of the Racemic Isomer of Anaferine. J. Organ. Chem. (USA) **34**, 1352 (1969).
84. EL-OLEMY, M. M. A. E. SCHWARTING and W. J. KELLEHER: Simulated Biosynthesis of Hygrine, Cuscohygrine, Anahygrine, Anaferine and Isopelletierine. Lloydia **29**, 58 (1966).
85. EVANS, D. A., J. A. JOULE and G. F. SMITH: The Alkaloids of *Rhazia orientalis*. Phytochemistry **7**, 1429 (1968).
86. EVANS, D. A., G. F. SMITH, G. N. SMITH and K. S. J. STAPLEFORD: *Rhazia* Alkaloids: The Secamines, a New Group of Indole Alkaloids. Chem. Comm. **1968**, 859.
87. FAIRBAIRN, J. W. and A. A. E. R. ALI: The Alkaloids of Hemlock. (*Conium maculatum* L.). III. The Presence of Bound Forms on the Plant. Phytochemistry **7**, 1593 (1968).
88. — — The Alkaloids of Hemlock (*Conium maculatum* L.). IV. Isotopic Studies of the Bound Forms of Alkaloids in the Plant. Phytochemistry **7**, 1599 (1968).
89. FAIRBAIRN, J. W. and S. B. CHALLEN: The Alkaloids of Hemlock (*Conium maculatum* L.). Biochem. J. **72**, 556 (1959).
90. FAIRBAIRN, J. W. and P. K. SUWAL: The Alkaloids of Hemlock (*Conium maculatum* L.). II. Evidence for a Rapid Turnover of the Major Alkaloids. Phytochemistry **1**, 38 (1961).
90a. FARNSWORTH, N. R., N. K. HART, S. R. JOHNS and J. A. LAMBERTON: Alkaloids of *Boehmeria cyclindrica* (Family *Urticaceae*): Identification of a Cytotoxic Agent, Highly Active Against eagle's 9 KB Carcinoma of the Nasopharynx in Cell Culture, as Cyclopleurine. Austral. J. Chem. **22**, 1805 (1969).
91. FAUGERAS, G., R. PARIS and M. H. MEYRUEY: Alkaloids of the *Retama raetam*. Isolation of Cytisine from the Fruits. Ann. pharm. france **20**, 768 (1962) [Chem. Abstr. **58**, 10506h (1963)].
92. FITZGERALD, J. S., S. R. JOHNS, J. A. LAMBERTON and A. H. REDCLIFFE: 6,7,8,9-Tetrahydropyridoquinlazolines, a New Class of Alkaloids from *Mackinlaya* Species. Austral. J. Chem. **19**, 151 (1966).

93. FODOR, G. and E. BAUERSCHMIDT: A Direct Stereoselective Synthesis of Optically Active Conhydrines. J. Heterocycl. Chem. 5, 205 (1968).
94. FODOR, G. and D. BUTRUILLE: Novel Method of Configurational Correlation: (+) Sedridine and (S)-(+)-Octan-2-ol. Chem. and Ind. 1968, 1437. Tetrahedron 27, 2055 (1971).
95. FODOR, G. and G. A. COOKE: Stereospecific Synthesis of ε-Coniceine via Optically Active Sedridine. Tetrahedron, Suppl. 8 (I), 113 (1966).
96. FOWDEN, L.: Some Observations on a Hydroxypipecolic Acid from Thrift *(Armeria maritima)*. Biochem. J. 70, 629 (1958).
97. — δ-Acetylornithine: A Constituent of Some Common Grasses. Nature 182, 406 (1958).
98. — The Metabolism of Labelled Lysine and Pipecolic Acid by *Acacia phyllodes*. J. exp. Bot. 11, 302 (1960).
99. — The Metabolism of Simple Heterocyclic Nitrogen Compounds with Especial Reference to the Imimo Acids. Abh. dtsch. Akad. Wiss. Berlin, Kl. Chem., Geol. Biol. 1963, 17.
100. — Amino-acid Analogues and the Growth of Seedlings. J. exp. Bot. 14, 387 (1963).
101. — Amino Acid Biosynthesis. In: J. B. PRIDHAM and T. SWAIN: Biosynthetic Pathways in Higher Plants, p. 73. London-New York: Academic Press. 1965.
102. FRANCK, B.: Über Mauerpfeffer-Alkaloide. Angew. Chem. 70, 269 (1958).
103. — *Sedum*-Alkaloide. II. Alkaloide in *Sedum acre* und verwandten *Sedum*-arten. Chem. Ber. 91, 2803 (1958).
104. — *Sedum*-Alkaloide. III. Zur Konstitution des Sedinins. Chem. Ber. 92, 1001 (1959).
105. — *Sedum*-Alkaloide. IV. Struktur und Biosynthese des Sedinins. Chem. Ber. 93, 2360 (1960).
106. FRIDRICHSONS, J. and A. M. MATHIESON: The Direct Determination of Molekular Structure: The Crystal Structure of Himbacine Hydrobromide at $-150°$. Acta Cryst. 15, 119 (1962) [Chem. Abstr. 56, 12401i (1962)].
107. FUJITA, E. and K. FUJI: Lythraceous Alkaloids. Part IV. Structure and Absolute Configuration of Lythranine, Lythranidine, and Lythramine. J. Chem. Soc. (London) C 1971, 1651.
108. FUJITA, E., K. FUJI, K. BESSHO, A. SUMI and S. NAKAMURA: The Structures of Lythranine, Lythranidine and Lythramine, Novel Alkaloids from *Lythrum anceps* Makino. Tetrahedron Letters 1967, 4595.
109. GARAY, A. S. and G. H. N. TOWERS: Studies on the Biosynthesis of Amaranthin. Canad. J. Bot. 44, 231 (1966).
110. GELLERT, E.: The Constituents of *Cryptocarya pleurosperma* White and Francis. I. Pleurospermine: A New Alkaloid of the Leaves. Austral. J. Chem. 12, 90 (1959).
111. GILMAN, R. E. e L. MARION: La Pelletierine de Tanret. Bull. soc. chim. France 1961, 1993.
112. GMELIN, R.: Die freien Aminosäuren der Samen von *Acacia willardiana (Mimosaceae)*. Isolierung von Willardiin, einer neuen pflanzlichen Aminosäure, vermutlich L-Uracil-[β-(α-amino-propionsäure)]-(3). Hoppe Seyler's Z. physiol. Chem. 316, 164 (1959).
113. GOLDBERG, S. I. and A. H. LIPKIN: *Leguminosae* Alkaloids. VII. The Synthesis of (±)-Lamprolobine. J. Organ. Chem. (USA) 35, 242 (1970).
114. GONZALES, A. G. and D. D. RODRIGUEZ: Alkaloids in Plants of the Canary Island. VII. *Nicotiana glauca* and *N. paniculata*. An. Real. Soc. espan. Fisical. Quim., Ser. B 58, 431 (1962) [Chem. Abstr. 58, 3686 (1963)].
115. GOVINDACHARI, T. R., K. NAKARAGAN and N. VISWANATHAN: Carpaine and Pseudocarpaine. Tetrahedron Letters 1965, 1907.

115a. GREWE, R., W. FEIST, H. NEUMANN und S. KERSTEN: Über die Inhaltsstoffe des schwarzen Pfeffers. Chem. Ber. **103**, 3752 (1970).
116. GRIFFITH, T. and G. D. GRIFFITH: Biosynthesis of Piperidine Alkaloids. I. Phytochemistry **5**, 1175 (1966).
117. GROBBELAAR, N., J. K. POLLARD and F. C. STEWARD: New Soluble Nitrogen Compounds (Amino- and Imino-Acids and Amides) in Plants. Nature **175**, 703 (1955).
118. GROBBELAAR, J. K. and F. C. STEWARD: Pipecolic Acid in *Phaseolus vulgaris*: Evidence on its Derivation from Lysine. J. Amer. Chem. Soc. **75**, 4341 (1953).
118a. GRÖGER, D.: Indolalkaloide. In: K. MOTHES und H. R. SCHÜTTE: Biosynthese der Alkaloide, S. 459, 699. Berlin: VEB Deutscher Verlag der Wissenschaften. 1969.
119. GROSS, D.: Pyridinalkaloide. In: K. MOTHES und H. R. SCHÜTTE: Biosynthese der Alkaloide. S. 215. Berlin: VEB Deutscher Verlag der Wissenschaften. 1969.
120. — Naturstoffe mit Pyridinstruktur und ihre Biosynthese. Fortschr. Chem. org. Naturst. **28**, 109 (1970).
121. — Die Biosynthese iridoider Naturstoffe. Fortschr. Bot. **32**, 93 (1970).
122. GSCHWEND, H. W.: The Synthesis of Nigrifactin. Tetrahedron Letters **1970**, 2711.
123. GUPTA, R. N.: Biosynthesis of the Piperidine Alkaloids. Lloydia **31**, 318 (1968).
124. GUPTA, R. N. and I. D. SPENSER: Biosynthesis of the Piperidine Alkaloids. The C_6-C_2-Units of Sedamine and Lobinaline. Chem. Comm. **1966**, 893.
125. — — The Biosynthesis of Sedamine. Canad. J. Chem. **45**, 1275 (1967).
126. — — Biosynthesis of the Piperidine Alkaloids: Origin of the Piperidine Nucleus of N-Methylisopelletierine. Chem. Comm. **1968**, 85.
126a. — — Biosynthesis of N-Methylpelletierine. Phytochemistry **8**, 1937 (1969).
127. — — Biosynthesis of the Piperidine Nucleus. The Mode of Incorporation of Lysine into Pipecolinic Acid and into Piperidine Alkaloids. J. Biol. Chem. **244**, 88 (1969).
127a. — — Biosynthesis of the Piperidine Nucleus: The Occurrence of Two Pathways from Lysine. Phytochemistry **9**, 2329 (1970).
128. HAMMOUDA, Y., M. PLAT et J. LEMEN: Structure de la tecostanine: alcaloïde du *Tecoma stans* Juss. (3ᵉ mémoire). Monoterpénoïdes (2ᵉ mémoire). Bull. soc. chim. France **1963**, 2802.
129. — — — Isolement de la tecostanine. Alcaloides du *Tecoma stans* (Juss.). *Bignoniacees* (2ᵉ memoire). Ann. pharm. franc. **21**, 699 (1963).
130. HARRIS, G. and J. R. A. POLLOCK: Pipecolinic Acid, a Widely Occurring Amino Acid. Chem. and Ind. **1952**, 931.
131. — — Amino Acids and Peptides of Hops and Wort. II. Pipecolinic Acid, a New Amino Acid in Barley and Hops. J. Inst. Brewing **59**, 28 (1953) [Chem. Abstr. **49**, 4934 (1955)].
132. HART, N. K., JOHNS, S. R. and J. A. LAMBERTON: 3,4-Dimethoxy-ω-(2'-piperidyl)acetophenon, a New Alkaloid from *Boehmeria platyphylla* Don. (Family *Urticaceae*). Austral. J. Chem. **21**, 1397 (1968).
133. — — — Alkaloids of *Lamprolobium fructicosum* Benth. *(Leguminosae)*. Chem. Comm. **1968**, 302.
134. — — — The Alkaloids of *Lamprolobium fructicosum* Benth. (Family *Leguminosae*). Austral. J. Chem. **21**, 1619 (1968).
135. — — — Minor Alkaloids of *Boehmeria platyphylla (Urticaceae)*. II. Isolation of Cyclopleurine and a New Secophenanthroquinolizidine Alkaloid. Austral. J. Chem. **21**, 2579 (1968).

136. HASSALL, C. H. and E. M. WILSON: Alkaloids of *Ormosia jamaicensis* (Urb.). Chem. and Ind. **1961**, 1358.
137. — — The Isolation and Characterisation of the Alkaloids of *Ormosia jamaicensis* Urb. J. Chem. Soc. (London) **1964**, 2657.
138. HASSE, K. und P. BERG: Oxydation von Cadaverin zu Anabasin. Naturwiss. **44**, 584 (1957).
139. HASSE, K. und H. MAISACK: Δ^1-Pyrrolin und Δ^1-Piperidein aus Putrescin und Cadaverin durch enzymatische Oxydation. Naturwiss. **42**, 627 (1955).
140. HATANAKA, S.: Isolation of Pipecolinic Acid from *Phaseolus angularis*. Sci. Pap. Coll. Gen. Educ., Univ. Tokyo **17**, 219 (1967) [Chem. Abstr. **70**, 882i (1969)].
141. HEGARTY, M. P.: The Isolation and Identification of 5-Hydroxypiperidine-2-carboxylic Acid from *Leucaena glauca* Benth. Austral. J. Chem. **10**, 484 (1957).
142. HIGHET, R. J.: Alkaloids of *Cassia* Species. I. Cassine. J. Organ. Chem. (USA) **29**, 471 (1964).
143. HIGHET, R. J. and P. F. HIGHET: Alkaloids of *Cassia* Species. II. The Side Chain of Cassine. J. Organ. Chem. (USA) **31**, 1275 (1966).
144. HILL, R. K.: Stereochemistry of the Hemlock Alkaloids. I. Conhydrine. J. Amer. Chem. Soc. **80**, 1609 (1958).
145. HILL, R. K.: Stereochemistry of the Hemlock Alkaloids. II. Pseudoconhydrine. J. Amer. Chem. Soc. **80**, 1611 (1958).
146. HILL, R. K., T. H. CHAN and J. A. JOULE: The Stereochemistry of Pinidine. Tetrahedron **21**, 147 (1965).
147. HILL, R. K. and A. G. EDWARDS: The Absolute Configuration of Febrifugine. Chem. and Ind. **1962**, 858.
148. HILL, R. K. and J. W. MORGAN: The Absolute Configuration of trans-2,6-Dimethylpiperidine. J. Organ. Chem. (USA) **31**, 3451 (1966).
149. HÖRHAMMER, L., H. WAGNER und W. FRITZSCHE: Zur Biosynthese der Betacyane. I. Biochem. Z. **339**, 398 (1964).
150. HOSENEY, R. C. and K. F. FINNEY: Free Amino Acid Composition of Flours Milled from Wheats Harvested at Various Stages of Maturity. Crop. Sci. **7**, 3 (1967) [Chem. Abstr. **66**, 84837b (1967)].
151. HUANG, W. N. and N. P. YANG: Nitrogen Metabolism in Higher Plants. V. Secretion of Amino Acids from Plant Root Systems. [Chem. Abstr. **62**, 13525f (1965)].
152. HULME, A. C. and W. ARTHINGTON: New Amino-Acids in Young Apple Fruits. Nature **170**, 659 (1952).
153. HYLIN, J. W.: Biosynthesis of Mimosine. Phytochemistry **3**, 161 (1964).
154. INUKAI, F., Y. SUYAMA and N. MORI: Preparation of L-Pipecolic Acid. II. Isolation of L-Pipecolic Acid from Legumes of Peas. [Chem. Abstr. **66**, 83078t (1967)].
155. JONES, G., H. M. FALES and W. C. WILDMAN: The Structure of Tecomanine. Tetrahedron Letters **1963**, 397; Chem. Comm. **1971**, 994.
156. JOSHI, B. S., V. N. KAMAT and A. K. SAKSENA: On the Structure of Piplartine and a Synthesis of Dihydropiplartine. Tetrahedron Letters **1968**, 2395.
157. KACZMAREK, F. und E. STEINEGGER: Chromatographische Untersuchung der Basenfraktion von *Lobelia cardinalis*, insbesondere von Lobinalin und dem neuen *Cardinalis*-Alkaloid 2. Pharm. Acta Helv. **33**, 852 (1958).
158. — — Botanische Klassifizierung und Alkaloidvorkommen in der Gattung *Lobelia*. Pharm. Acta Helv. **34**, 413 (1959).
159. KAMALITDINOV, D., S. ISKANDAROV and S. Y. YUNUSOV: Alkaloids of *Leontice alberti*. Khim. Prir. Soedin. **5** (5), 409 (1969) [Chem. Abstr. **72**, 75653k (1970)].

160. KAWASAKI, I., S. MATSUTANI and T. KANEKO: The Absolute Configuration of Nupharamine. Bull. Chem. Soc. (Japan) **36**, 623 (1963) [Chem. Abstr. **60**, 3021g (1964)].
161. — — — Absolute Configuration of Nupharamine. Bull. Chem. Soc. (Japan) **36**, 1474 (1963) [Chem. Abstr. **60**, 5571 (1964)].
161a. KEOGH, M. F. and D. G. O'DONOVAN: Biosynthesis of Some Alkaloide of *Punica granatum* and *Witnania somnifera*. J. Chem. Soc. (London) C **1970**, 1792.
161b. KEOGH, M. F. and D. G. O'DONOVAN: Biosynthesis of Lobeline. J. Chem. Soc. (London) C **1970**, 2470.
162. KHANNA, K. L., A. E. SCHWARTING and J. M. BOBBITT: The Occurrence of Isopelletierine in *Withania somnifera*. J. Pharm. Sci. **51**, 1194 (1962).
163. KHARATYAN, S., M. PUZA, J. SPIZEK, L. DOLEZILOVA and Z. VANEK: Biogenesis of Cycloheximide and Related Compounds. Chem. and Ind. **1963**, 1038.
164. KING, F. E., T. J. KING and A. J. WARWICK: The Chemistry of Extractives from Hardwoods. III. Baikiain, an Amino-Acid Present in *Baikiaea plurijuga*. J. Chem. Soc. (London) **1950**, 3590.
165. KIRCHMEIER, O.: Zur Chemotaxonomie von Futterpflanzen: Pipecolinsäure, ein charakteristischer Bestandteil von Gras-Silagen. Naturwiss. **52**, 454 (1965).
166. KISAKI, T., S. MIZUSAKI and E. TAMAKI: Phytochemical Studies on Tobacco Alkaloids. XI. A New Alkaloid in *Nicotiana tabacum* Roots. Phytochemistry **7**, 323 (1968).
167. KOEPFLI, J. B., J. A. BROCKMAN and J. MOFFAT: The Structure of Febrifugine and Isofebrifugine. J. Amer. Chem. Soc. **72**, 3323 (1950).
168. KOEPFLI, J. B., J. F. MEAD and J. A. BROCKMAN: An Alkaloid with hight Antimalarial Activity from *Dichroa febrifuga*. J. Amer. Chem. Soc. **69**, 1837 (1947).
169. KOEPFLI, J. B., J. F. MEAD and J. A. BROCKMAN: Alkaloids of *Dichroa febrifuga*. I. Isolation and Degradatives Studies. J. Amer. Chem. Soc. **71**, 1048 (1949).
170. KOLEOSO, O. A., S. M. C. DIETRICH and R. O. MARTIN: Biosynthesis of *Conium* Alkaloids. Identification of a Novel Nonnitrogenous Base from *Conium maculatum* as 3-Formyl-4-hydroxy-2H-pyran. Biochemistry **8**, 4172 (1959).
171. KONDO, Y.: Pipecolic and 5-Hydroxypipecolic Acids in Mulberry Leaves. Sericult. Sci. (Japan) **26**, 345, 349 (1957) [Chem. Abstr. **52**, 17407h (1958)].
172. KOO, S. H., F. COMER and I. D. SPENSER: Biosynthesis of the *Lythraceae* Alkaloids: Mode of Incorporation of Phenylalanine. Chem. Comm. **1970**, 897.
173. KOO, S. H., R. N. GUPTA, I. D. SPENSER and J. T. WROBEL: Biosynthesis of the *Lythraceae* Alkaloids: Incorporation of Lysine. Chem. Comm. **1970**, 396.
174. KRAUS, G.-J. und H. REINBOTHE: Die Aminosäuren der Gattung *Albizzia Durazz*. Biochem. Physiol. Pflanzen **161**, 243 (1970).
175. KUEHL, F. A. Jr., C. F. SPENCER and K. FOLKERS: Alkaloids of *Dichroa febrifuga* Lour. J. Amer. Chem. Soc. **70**, 2091 (1948).
176. KUPCHAN, S. M. and A. W. BY: Steroid Alkaloids: The *Veratrum* Group. In: R. H. F. MANSKE: The Alkaloids, Chemistry and Physiology Vol. X, p. 193. New York-London: Academic Press. 1970.
177. KUWATA, S.: Pelletierine. I. Isolation of Pelletierine from Pomegranate Root Bark. Bull. Chem. Soc. (Japan) **33**, 1668 (1960) [Chem. Abstr. **55**, 12414 (1961)].
178. LEARY, J. D., J. M. BOBBITT, A. ROTHER and A. E. SCHWARTING: Structure and Synthesis of the Alkaloid Anahygrine. Chem. and Ind. **1964**, 283.
179. LEETE, E.: The Biogenesis of Nicotine and Anabasine. J. Amer. Chem. Soc. **78**, 3520 (1956).
180. — The Biogenesis of *Nicotiana* Alkaloids. VI. The Piperidine Ring of Anabasine. J. Amer. Chem. Soc. **80**, 4393 (1958).

181. LEETE, E.: The Biosynthesis of Coniine from Four Acetate Units. J. Amer. Chem. Soc. **85**, 3523 (1963).
182. — Biosynthesis of the Hemlock Alkaloids. The Incorporation of Acetate-1-^{14}C into Coniine and Conhydrine. J. Amer. Chem. Soc. **86**, 2509 (1964).
183. — Biosynthesis of Alkaloids. Science **147**, 1000 (1965).
184. — Alkaloid Biogenesis. In: P. BERNFELD: Biogenesis of Natural Compounds. First Edit., p. 739. Oxford-London-New York-Paris: Pergamon Press. 1963. Sec. Edit., p. 953. Oxford-London-Edinburgh-New York-Toronto-Sydney-Paris-Braunschweig: Pergamon Press. 1967.
185. — The Metabolism of Nicotine-2'-^{14}C in *Nicotiana glauca*. Tetrahedron Letters **1968**, 4433.
186. — Biosynthesis of the Nicotiana Alkaloids. XVI. The Incorporation of Δ^1-Piperideine-6-^{14}C into the Piperidine Ring of Anabasine. J. Amer. Chem. Soc. **91**, 1697 (1969).
187. — Biosynthesis of Coniine from Octanoic Acid in Hemlock Plants *(Conium maculatum)*. J. Amer. Chem. Soc. **92**, 3835 (1970).
187a. — Biosynthesis of the Hemlock and Related Piperidine Alkaloids. Accounts Chem. Res. **4**, 100 (1971).
188. LEETE, E. and N. ADITYACHAUDHURY: Biosynthesis of the Hemlock Alkaloids. II. The Conversion of γ-Coniceine to Coniine and ψ-Conhydrine. Phytochemistry **6**, 219 (1967).
189. LEETE, E., E. G. GROS and T. J. GILBERTSON: The Biosynthesis of Anabasine. Origin of the Nitrogen of the Piperidine Ring. J. Amer. Chem. Soc. **86**, 3907 (1964).
190. LEETE, E. and K. N. JUNEAU: Biosynthesis of Pinidine. J. Amer. Chem. Soc. **91**, 5614 (1969).
190a. LEETE, E. and J. O. OLSON: 5-Oxo-octanoic Acid and 5-Oxo-octanal, Precursors of Coniine. Chem. Comm. **1970**, 1651.
191. LIEBISCH, H. W.: Piperidinalkaloide. In: K. MOTHES und H. R. SCHÜTTE: Biosynthese der Alkaloide, S. 275. Berlin: VEB Deutscher Verlag der Wissenschaften. 1969.
192. LIEBISCH, H. W., N. MAREKOV und H. R. SCHÜTTE: Die Biosynthese der Alkaloide aus *Punica granatum*. Z. Naturforsch. **23b**, 1116 (1968).
193. LIEBISCH, H. W., B. MATSCHINER und H. R. SCHÜTTE: Beiträge zur Physiologie und Biosynthese des Betanins. Z. Pflanzenphysiol. **61**, 269 (1969).
194. LINDSTEDT, S. and G. LINDSTEDT: On the Formation of 5-Hydroxypipecolic Acid from δ-Hydroxy-DL-lysine. Arch. Biochem. Biophys. **85**, 565 (1959).
195. LLOYD, H. A.: Astrocasine: A New *Euphorbiaceae* Alkaloid. Tetrahedron Letters **1965**, 1761.
196. — Alkaloids of *Astrocasia phyllanthoides* II. Astrophylline. Tetrahedron Letters **1965**, 4537.
197. LLOYD, H. A. and E. C. HORNING: Alkaloids of *Ormosia panamensis* Benth. and Related Species. J. Amer. Chem. Soc. **80**, 1506 (1958).
198. LODER, J. W.: Synthesis of Pleurospermine, the Leaf Alkaloid of *Cryptocarya pleurosperma* White and Francis. Austral. J. Chem. **15**, 296 (1962).
199. LOWY, P. H.: The Conversion of Lysine to Pipecolic Acid by *Phaseolus vulgaris*. Arch. Biochem. Biophys. **47**, 228 (1953).
200. LUCHETTI, M. A.: Biosynthesis of Skytanthine in vitro. Ann. Ist. Super Sanita **1**, 563 (1965) [Chem. Abstr. **65**, 9349 (1966)].
201. LUKES, R., A. A. AROJAN, J. KOVAR und K. BLAHA: Zur Konfiguration stickstoffhaltiger Verbindungen. XV. Bestimmung der absoluten Konfiguration von Anabasin und Anatabin. Collect. Czech. Chem. Comm. **27**, 751 (1962).

202. LYTHGOE, D. and M. J. VERNENGO: Alkaloids from *Cassia carnaval* Speg.: Cassine and Carnavaline. Tetrahedron Letters **1967**, 1133.
203. MABRY, T. J. and A. S. DREIDING: The Betalains. In: T. J. MABRY, R. R. ALSTON and V. C. RUNNECKLES: Recent Advances in Phytochemistry. New York: Appleton-Century-Crofts. 1968.
204. MACCONNELL, J. G., M. S. BLUM and H. M. FALES: The Chemistry of Fire Ant Venom: Tetrahedron **26**, 1129 (1971); Science **168**, 840 (1970).
205. MACHOLAN, L.: Aminoketocarbonsäuren. V. Zur Frage der Biogenese von Chinazolinalkaloiden. Collect. Czech. Chem. Comm. **24**, 550 (1959).
206. MANN, P. J. G. and W. R. SMITHIES: Plant Enzyme Reactions Leading to the Formation of Heterocyclic Compounds. I. The Formation of Unsaturated Pyrrolidine and Piperidine Compounds. Biochem. J. **61**, 89 (1955).
207. MARINI-BETTOLO, G. B.: Skytanthines: A New Group of Natural Alkaloids. Ann. Ist. Super Sanita **4**, 489 (1968) [Chem. Abstr. **71**, 91703m (1969)[.
208. MARION, L. and M. CHAPUT: A New Occurrence of DL-Methylisopelletierine. Canad. J. Res. B **27**, 215 (1949).
209. MARION, L., R. LAVIGNE and L. LEMAY: The Structure of Sedamine. Canad. J. Res. **29**, 347 (1951).
210. MARTIN, J. H. and W. K. HAUSMANN: The Isolation and Identifikation of D-α-Pipecolinic Acid, α(L),β-Methylaspartic Acid and α,β-Diaminobutyric Acid from the Polypeptide Antibiotic Aspartocin. J. Amer. Chem. Soc. **82**, 2079 (1960).
211. MERLIS, V. M. and N. F. PROSKURNINA: Alkaloids of *Ammodendron conollyi*. III. Structure of Isoammodendrine. Zhurn. Obshchei Khimii (USSR) **20**, 1722 (1950) [Chem. Abstr. **45**, 1302a (1951)].
212. MICHEL, K.-H., SANDBERG, F., F. HAGLID and T. NORIN: Alkaloids of *Haloxylon salicornicum* (Moq.-Tand.) Boiss. Acta pharm. Suecica **4**, 97 (1967).
213. MICHEL, K.-H., F. SANDBERG, F. HAGLID, T. NORIN, R. P. K. CHAN and J. C. CRAIG: The Absolute Configuration of Halosaline. Acta Chem. Scand. **23**, 3479 (1970).
214. MICHL, M.: Über das Vorkommen von Pipecolinsäure in tierischen Giften. Mh. Chem. **88**, 701 (1957).
215. MILLER, H. E., H. ROESLER, A. WOHLPART, H. WYLER, M. E. WILCOX, H. FROHOFER, T. J. MABRY und A. S. DREIDING: Biogenese der Betalaine. Biotransformation von Dopa und Tyrosin in den Betalaminteilsäureteil des Betanins. Helv. chim. Acta **51**, 1470 (1968).
216. MINALE, L., M. PIATELLI and R. A. NICOLAUS: Pigments of Centrospermae. IV. On the Biogenesis of Indicaxanthin and Betanin in *Opuntia ficus-indica* Mill. Phytochemistry **4**, 593 (1965).
217. MOLLOV, N. M. and I. C. IVANOV: Leontiformine, A New 3-Piperidyl-(2)-quinolizidine, from *Leontice leontopetalum* L. Tetrahedron **26**, 3805 (1970).
218. MORRISON, R. I.: The Isolation of L-Pipecolinic Acid from *Trifolium repens*. Biochem. J. **53**, 474 (1953).
219. MORTIMER, P. I.: The Structure of Isopeletierine from *Duboisia myoporoides* R. Br. Austral. J. Chem. **11**, 82 (1958).
220. MORTIMER, P. I. and S. WILKINSON: The Occurrence of Nicotine, Anabasine and iso-Pelletierine in *Duboisia myoporoides*. J. Chem. Soc. (London) **1957**, 3967.
221. MOTHES, K. und H. R. SCHÜTTE: Die Biosynthese von Alkaloiden. I. Angew. Chem. **75**, 265 (1963).
222. MOTHES, K., H. R. SCHÜTTE, H. SIMON und F. WEYGAND: Die Bildung von Anabasin aus Cadaverin-[1,5-^{14}C] mit Hilfe von Extrakten aus Erbsenkeimlingen. Z. Naturforsch. **14b**, 49 (1959).

223. MUKHAMEDZHANOV, S. Z., X. A. ASLANOV, A. S. SADYKOV, V. B. LEONTEV and V. K. KIRYUKHIN: Structure of Anabasamine. Khim. Prir. Soedin. **4** (3), 158 (1968) [Chem. Abstr. **69**, 87277e (1968)].
224. MURAKAMI, T., F. INUGAI, M. NAGASAWA, H. INATOMI and N. MORI: Water-Soluble Constituents of Crude Drugs. III. Free Amino Acids Isolated from Ginges Rhizome. Yakugaku Zasshi **85**, 845 (1965) [Chem. Abstr. **64**, 10090a (1966)].
225. NAEGELI, P., R. NAEGELI, W. C. WILDMAN and R. J. HIGHET: Dehydrogenation of Three *Ormosia* Alkaloids. Tetrahedron Letters **1963**, 2075.
226. NAEGELI, P., W. C. WILDMAN and H. A. LLOYD: Structure Relationships in the *Ormosia* Alkaloids. Tetrahedron Letters **1963**, 2069.
227. NAKANO, T., C. DJERASSI, R. A. CORRAL and O. O. ORAZI: The Structure of Julocrotine. Tetrahedron Letters **1959**, 8.
228. — — — — Structure of Julocrotine. J. Organ. Chem. (USA) **26**, 1184 (1961).
229. NIGAM, S. N. and W. B. MCCONNEL: Studies on Wheat Plants Using Carbon-14 Compounds. Canad. J. Biochem. **41**, 1367 (1963).
230. NORMATOV, M. and S. Y. YUNUSOV: Alkaloids of *Nitraria schoberi*. The Structure of Nitrarine. Khim. Prir. Soedin. **4** (2), 139 (1968) [Chem. Abstr. **69**, 77570t (1968)].
231. NÜESCH, J. and F. KNÜSEL: Sideromycins. In: D. GOTTLIEB and P. D. SHAW: Antibiotics, Vol. I., p. 499. Berlin-Heidelberg-New York: Springer. 1967.
232. NURIDDINOV, R. N. and S. Y. YUNUSOV: The Constitution of Corsevine. Khim. Prir. Soedin. **3**, 398 (1967) [Chem. Abstr. **68**, 105400 (1968)].
233. O'DONOVAN, D. G. and P. B. CREEDON: The Biosynthesis of Santiaguin. J. Chem. Soc. (London) C **1971**, 1604.
234. — — Biosynthesis of the Alkaloids of *Haloxylon salicornicum*. Tetrahedron Letters **1971**, 1341.
235. O'DONOVAN, D. G. and M. F. KEOGH: Biosynthesis of Piperidine Alkaloids. Tetrahedron Letters **1968**, 265.
236. ONODERA, R. and M. KANDATSU: Occurrence of L-(−)-Pipecolic Acid in the Culture Medium of *Rumen ciliate protozoa*. Agric. Biol. Chem. (Tokyo) **33**, 113 (1969).
237. ORECHOFF, A. und N. PROSKURNINA: Über die Alkaloide von *Ammodendron conollyi* Bge. I. Ber. dtsch. chem. Ges. **68**, 1807 (1935).
238. — — Sur les alcaloïdes de *Ammodendron conollyi*. Constitution de l'ammodendrine. Bull. soc. chim. France **5**, 29 (1938).
239. OZAWA, M., N. SATO, H. INATOMI, Y. SUYAMA and F. INUKAI: Free Amino Acids in Plants. X. L-Pipecolic Acid in Tea Plant *(Thea sinensis)* Seeds. [Chem. Abstr. **72**, 19074 (1970)].
240. PAILER, M. und E. HASLINGER: Synthese von Nigrifactin. Mh. Chem. **101**, 508 (1970).
241. PAILER, M. und W. G. KUMP: Über die Untersuchung basischer Inhaltsstoffe von *Achillea*-Arten. Arch. Pharm. **293**, 646 (1960).
242. PARISH, D. H.: Amino Acids of Sugar Cane. I. Amino Acids of Cane Juice and the Effect of Nitrogenous Fertilization on the Levels of these Substances. J. Sci. Food. Agr. **16**, 240 (1965) [Chem. Abstr. **63**, 10617c (1965)].
243. — Composition of Cane Juice. VI. Amino Acids [Chem. Abstr. **63**, 18980a (1965)].
244. PHILLIPS, D. M.: Pipecolinic Acid (Pipecolic Acid) Chem. and Ind. **1953**, 127.
245. PINHEY, J. T., E. RITCHIE and W. C. TAYLOR: The Chemical Constituents of *Himantandra* (Galbulimima) Species. IV. The Structures of Himbacine, Himbeline, Himandravine, and Himgravine. Austral. J. Chem. **14**, 106 (1961).

245a. POLAN, C. E., W. G. SMITH, C. Y. NG, R. H. HAMMERSTEDT and L. M. HENDERSON: Metabolism of Hydroxylysine by Rats. J. Nutr. **91**, 143 (1967) [Chem. Abstr. **66**, 93031m (1967)].

246. PRASAD, K. B. und S. C. SHAW: Synthese von Pyridin-Alkaloiden. I. Chem. Ber. **98**, 2822 (1965).

247. PROSKURINIA, N. F. and V. M. MERLIS: Alkaloide of *Ammodendron conollyi* Bge. II. Isolation of the New Alkaloids, Isoammodendrine and Connoline. J. gen. Chem. (UdSSR) **19**, 1396 (1949) [Chem. Abstr. **44**, 1119 (1950)].

248. RAPOPORT, H.: The Biosynthesis of the Pyridine and Piperidine Alkaloids. The Tobacco Alkaloids. Abh. dtsch. Akad. Wiss. Berlin, Kl. Chem., Geol. Biol. **1966** (3), 111.

249. RALL, G. J. H., T. M. SMALBERGER, H. L. de WAAL and R. R. ARNDT: Dimeric Piperidine Alkaloids from *Azima tetracantha* Lam.: Azimine, Azcarpine and Carpaine. Tetrahedron Letters **1967**, 3465.

250. RATLE, G., X. MONSEUR, B. C. DAS, J. YASSI, Q. KHUONG-HUU et R. GOUTAREL: Le prosopine et la prosopinine, alcaloïdes du *Prosopis africana* (Guill. et Perr.) Taub. Bull. soc. chim. France **1966**, 2945.

251. REMZIYE, Salih Hisar: *Parietaria officinalis*. Türk. Ij. tecr. Biol. Derg. **11**, 172 (1951) [Chem. Abstr. **47**, 1893 (1953)].

252. RIBAS-MARQUES, I. and A. N. BLANCO: *Papilionaceae* Alkaloids. XXXIX. Absolute Configuration of Adenocarpine, Santiaguine, Ammodendrine, Anabasine, Anatabine, and its Methyl Derivatives. Partial Absolute Configuration of the α,β-Dipiperidines. An. Real. Soc. espan. Fisica. Quim., Ser. B. **57** (12), 781 (1962) [Chem. Abstr. **58**, 5738 (1963)].

253. RIBAS, I. and M. RIBAS: Papilionaceous Alkaloids. XLIV. Constribution to the Stereochemistry of Santiaguine. An. Real. Soc. espan. Fisica. Quim., Ser. B. **62**, 845 (1966) [Chem. Abstr. **66**, 76217j (1967)].

254. RIBAS, I. and J. VEGA: Alkaloids of the Leguminous *Papilionaceae*. XX. Alkaloids of the Fruits of *Retama sphaerocarpa*. Ion (Madrid) **13**, 148 (1953) [Chem. Abstr. **48**, 5195 (1954)].

255. RIBAS-MARQUES, I.: Die Alkaloide von *Adenocarpus*-Arten. Abh. dtsch. Akad. Wiss. Berlin, Kl. Chem., Geol. Biol. **1963**, (4), 149.

256. RIBAR-MARQUES, I. und A. VIDAL: Die alkalische Hydrolyse des Isoorensins. Naturwiss. **53**, 252 (1966).

257. RICE, W. Y. and J. L. COKE: Structure and Configuration of Alkaloids. II. Cassine. J. Organ. Chem. (USA) **31**, 1010 (1966).

258. RIPPERGER, H. und K. SCHREIBER: Die absolute Konfiguration von $(+)$-α-Pipecolin. Tetrahedron **21**, 1485 (1965).

259. RITCHIE, E. and W. C. TAYLOR: The Galbulimima Alkaloids. In: R. H. F. MANSKE: The Alkaloids, Chemistry and Physiology. Vol. IX, p. 529. New York-London: Academic Press. 1967.

260. ROBERTS, M. F., B. T. CROMWELL and D. E. WEBSTER: The Occurrence of 2-(2-Propenyl)-Δ^1-piperideine in the Leaves of Pomegranate (*Punica granatum* L.). Phytochemistry **6**, 711 (1967).

261. ROBERTSON, A. L. and L. MARION: Absolute Configuration of $(-)$-Homostachydrine. Canad. J. Chem. **37**, 829 (1959).

262. — — The Biogenesis of Alkaloids. XXI. The Biogenesis of $(-)$-Homostachydrine and the Occurrence of Trigonelline in Alfalfa. Canad. J. Chem. **37**, 1043 (1959).

263. ROBISON, M. M., B. F. LAMBERT, L. DORFMAN and W. G. PIERSON: The Stereochemistry and Synthesis of the Lobinaline Ring System. J. Org. Chem. (USA) **31**, 3220 (1966).

264. ROBISON, M. M., W. G. PIERSON, L. DORFMAN, B.F. LAMBERT and R. A. LUCAS: The Structure of Lobinaline. Tetrahedron Letters 1964, 1513.
265. — — — — The Skeletal Structure of Lobinaline. J. Organ. Chem. (USA) 31, 3206 (1966).
266. ROBINSON, R.: A Theory of the Mechanism of Phytochemical Synthesis of Certain Alkaloids. J. Chem. Soc. (London) 111, 876 (1917).
267. ROTHER, A., J. M. BOBBITT and A. E. SCHWARTING: Structure and Synthesis of the Alkaloid Anaferine. Chem. and Ind. 1962, 654.
268. ROTHSTEIN, M. and L. L. MILLER: Loss of the α-Amino Group in Lysine Metabolism to Form Pipecolic Acid. J. Amer. Chem. Soc. 76, 1459 (1954).
269. — — The Metabolism of L-Lysine-6-^{14}C. J. Biol. Chem. 206, 243 (1954).
270. — — The Conversion of Lysine to Pipecolic Acid in the Rat. J. Biol. Chem. 211, 851 (1954).
271. SAAYMAN, H. M. and D. G. ROUX: The Origins of Tannins and Flavonoids in Black-Wattle Barks and Heartwoods, and their Associated "Non-Tannin" Components. Biochem. J. 97, 794 (1965).
272. SADYKOV, A. S., S. Z. MUKHAMEDZHANOV and X. A. ASLANOV: Structure of Anabasamine, a New Base from *Anabasis aphylla* Seeds. Dokl. Akad. Nauk (Uzb. SSR) 24, 34 (1967) [Chem. Abstr. 68, 78473e (1968)].
273. SADYKOV, A. S. and B. TUMUR: The Investigation of the Alkaloids of *Anabasis aphylla*. Dokl. Akad. Nauk (Uzb. SSR) 1960 (1), 27 [Chem. Abstr. 56, 3563 (1962)].
274. SANDBERG, F., K.-H. MICHEL, B. STAF and M. TJERNBERG-NELSON: Screening of Plants of the Family *Chenopodiaceae* for Alkaloids. Acta pharm. Suecica 4, 51 (1967).
275. SCHENK, W. und H. R. SCHÜTTE: 4-Aminopipecolinsäure, eine neue basische Iminosäure in *Strophanthus scandens*. Naturwiss. 48, 223 (1961).
276. SCHENK, W., H. R. SCHÜTTE und K. MOTHES: Über den Stoffwechsel von Pipecolinsäure, 4-Hydroxypipecolinsäure, 4-Aminopipecolinsäure und 4-Oxopipecolinsäure in *Strophanthus scandens*. Flora (Jena) 152, 590 (1963).
277. SCHIEDT, U. und H. G. HÖSS: Zur Biosynthese des Coniins. Z. Naturforsch. 13b, 691 (1958).
278. — — Zur Biogenese der Alkaloide. III. Lysin als Vorstufe des Coniins. Hoppe Seyler's Z. physiol. Chem. 330, 74 (1962).
279. SCHLUNEGGER, E. und E. STEINEGGER: Erstmalige Isolierung von Anabasin aus einer Leguminose, *Priestleya elliptica* DC. Pharm. Acta Helv. 45, 147 (1970).
279a. SCHÖPF, C., G. BENZ, F. BRAUN, H. HINKEL, G. KRÜGER, R. ROKOHL und A. HUTZLER: Darstellung und Umwandlung des *meso*-1,3-Di[2-piperidyl]-propan-2-ons (Anaferin) und der entsprechenden racem. Verbindung. Liebig's Ann. Chem. 737, 1 (1970).
280. SCHÖPF, C., G. BENZ, F. BRAUN, H. HINKEL und R. ROKOHL: Die Kondensation von Δ^1-Piperidein mit Acetondicarbonsäure und Formaldehyd. Angew. Chem. 65, 161 (1953).
281. SCHÖPF, C. und F. BRAUN: Konstitution und Synthese des Ammodendrins. Naturwiss. 36, 377 (1949).
282. SCHÖPF, C., F. BRAUN, K. BURKHARDT, G. DUMMER und H. MÜLLER: Die Kondensation von Δ^1-Piperidein mit Acetessigsäure und Benzoylessigsäure zu Isopelletierin bzw. α-Phenylacyl-Piperidin. Liebig's Ann. Chem. 626, 123 (1959).
283. SCHÖPF, C., F. BRAUN und A. KOMZAK: Der Übergang von Δ^1-Piperidein in Tetrahydroanabasin unter zellmöglichen Bedingungen. Chem. Ber. 89, 1821 (1956).

284. Schöpf, C., F. Braun und K. Kreibich: Ammodendrin, Orensin und N-Benzoyl-Δ^2-tetrahydroanabasin aus Isotripiperidein. Liebigs's Ann. Chem. **674**, 87 (1964).
285. Schöpf, C., W. Bundschuh, G. Dummer, T. Kauffmann und R. Kress: Synthese und absolute Konfiguration von (+)-8-Phenylnorlobelol-I [(+)-Norallosedamin] aus *Lobelia inflata* L. Liebig's Ann. Chem. **628**, 101 (1959).
286. Schöpf, C., G. Dummer, W. Wüst und R. Rausch: Synthese und absolute Konfiguration von (−)-Sedamin und von (−)-8-Phenyllobelol-I [(−)-Allosedamin]. Liebig's Ann. Chem. **626**, 134 (1959).
286a. Schöpf, C., E. Gams, H. Hinkel, G. Krüger und M. Höhn: Die Konfiguration der stereomeren meso-1,3-Di[2-piperidyl]-propan-2-ole und das Verhalten von Tetrahydro-(1,3)-oxazin-Derivaten bei der Hydrolyse mit Säuren. Liebig's Ann. Chem. **737**, 24 (1970).
287. Schöpf, C., E. Gams, F. Koppernock, R. Rausch und R. Walbe: Δ^1-Piperidein und verwandte Verbindungen. XV. Eine einfache Synthese der optischen Antipoden des Sedridins und Allosedridins sowie die absolute Konfiguration des (+)- und (−)-Allosedridins. Liebig's Ann. Chem. **732**, 181 (1970).
288. Schöpf, C., T. Kauffmann, P. Berth, W. Bundschuh, G. Dummer, H. Fett, G. Habermehl, E. Wieters und W. Wüst: Über die stärker hydrophilen Nebenalkaloide aus *Lobelia inflata* L.: Ein Beitrag zur Biogenese der *Lobelia*-Alkaloide. Liebig's Ann. Chem. **608**, 88 (1957).
289. Schöpf, C. und W. Merkel: Konstitution und Synthese des Isoorensins. Abh. dtsch. Akad. Wiss. Berlin, Kl. Chem., Geol. Biol. **1966** (3), 133.
290. — — Konstitution und Synthese des Isoorensins. Naturwiss. **53**, 274 (1966).
291. — — Konstitution und Synthese des Isoorensins. Liebig's Ann. Chem. **701**, 180 (1967).
292. Schöpf, C. und E. Müller: Die absolute Konfiguration des (−)-Lobelins und seiner Reaktionsprodukte. Liebigs' Ann. Chem. **687**, 241 (1965).
293. Schöpf, C. und E. Schenkenberger: Über stereochemische Zusammenhänge zwischen Alkaloiden aus *Lobelia inflata* L.: ein Beitrag zur Biogenese des Lobelins. Liebig's Ann. Chem. **682**, 206 (1965).
294. Schöpf, C. und R. Unger: Physiological Strains of *Sedum acre* characterized by Different Alkaloids. Experientia **12**, 19 (1956) [Chem. Abstr. **50**, 15549 (1956)].
295. Schreiber, K.: Über die Biochemie der Steroidalkaloide. Abh. dtsch. Akad. Wiss. Berlin, Kl. Chem., Geol. Biol. **1966** (3), 65.
296. — Steroid Alkaloids: The Solanum Group. In: R. H. F. Manske: The Alkaloids, Chemistry and Physiology. Vol. X, p. 1. New York-London: Academic Press. 1968.
297. Schütte, H. R.: Chinolizidinalkaloide. In: K. Mothes und H. R. Schütte: Biosynthese der Alkaloide, S. 324. Berlin: VEB Deutscher Verlag der Wissenschaften. 1969.
298. Schütte, H. R.: Steroidalkaloide. In: K. Mothes und H. R. Schütte: Biosynthese der Alkaloide, S. 616. Berlin: VEB Deutscher Verlag der Wissenschaften. 1969.
299. Schütte, H. R., K. L. Kelling, D. Knöfel und K. Mothes: Zur Biosynthese von Adenocarpin in *Adenocarpus viscosus*. Phytochemistry **3**, 249 (1964).
300. Schütte, H. R. und J. Lehfeld: Über die Biosynthese der Alkaloide in *Nuphar luteum*. Arch. Pharm. **298**, 460 (1965).
301. Schütte, H. R. und G. Seelig: Zur Biosynthese der Pipecolinsäure in *Phaseolus vulgaris*. Z. Naturforsch. **22b**, 824 (1967).

302. SCHWARTING, A. E., J. M. BOBBITT, A. ROTHER, C. K. ATAL, K. L. KHANNA, J. D. LEARY and W. G. WALTER: The Alkaloids of *Withania somnifera*. Lloydia **26**, 258 (1963).
303. SCHWEET, R. S., J. T. HOLDEN and P. H. LOWY: The Metabolism of Lysine in *Neurospora*. J. Biol. Chem. **211**, 517 (1954).
303a. SEKITA, T. and S. INAYAMA: The Complete Structures of Pulchellidine and Pulchellin. A Crystallographic Study of 11,13-Dibromopulchellin. Tetrahedron Letters **1970**, 135.
304. SENEVIRATNE, A. S. and L. FOWDEN: The Amino Acids of the Genus *Acacia*. Phytochemistry **7**, 1039 (1968).
305. SICHER, J. and M. TICHY: Absolute Configuration of Conhydrine. Chem. and Ind. **1958**, 16.
306. — — Stereochemical Studies. XII. The Absolute Configuration of Conhydrine and ψ-Conhydrine. Coll. Czech. Chem. Comm. **23**, 2081 (1958).
307. SISLER, H. D. and M. R. SIEGEL: Cycloheximide and Other Glutarimide Antibiotics. In: D. GOTTLIEB and P. D. SHAW: Antibiotics. Vol. I, p. 283. Berlin-Heidelberg-New York: Springer. 1967.
308. SMALBERGER, T. M., G. J. H. RALL, H. L. DE WALL and R. R. ARNDT: The Structures and Configuration of Azimine and Azcarpine. Tetrahedron **24**, 6417 (1968).
309. SMITH, H. H. and D. V. ABASHIAN: Chromatographic Investigations on the Alkaloids Content of *Nicotiana* Species and Interspecific Combinations. Amer. J. Bot. **50**, 435 (1963).
309a. SMOGROVICOVA, H., P. NEMEC, I. KOMPIS, A. JINDRA und P. KOVACS: Celostatni biochemicky sjezd. **1966**, 4 (zitiert nach *161b*).
310. SOLT, M. L., R. F. DAWSON and D. R. CHRISTMAN: Biosynthesis of Anabasine and of Nicotine by Excised Root Cultures of *Nicotiana glauca*. Plant Physiol. **35**, 887 (1960).
311. SPITELLER-FRIEDMANN, M. und G. SPITELLER: Anwendung der Massenspektrometrie zur Strukturaufklärung von Alkaloiden. V. Die Struktur des Carpains. Mh. Chem. **95**, 1234 (1964).
312. STEINEGGER, E. und C. MOSER: Die Alkaloide von *Genista hystrix* Lge. 13. Mitteilung über Leguminosen-Alkaloide. Pharm. Acta Helv. **42**, 177 (1967).
313. STEINEGGER, E., C. MOSER und P. WEBER: Konstitution des neuen Alkaloides Hystrin aus *Genista hystrix* Lge. 15. Mitteilung über Leguminosen-Alkaloide. Phytochemistry **7**, 849 (1968).
314. STEINEGGER, E. und E. SCHLUNEGGER: Alkaloide der südafrikanischen *Liparia parva* Vog. ex Walp und *Liparia sphaerica* L. Pharm. Acta Helv. **45**, 369 (1970).
315. STEINEGGER, E. und D. SCHNYDER: Zur Alkaloidbildung bei *Genista hystrix* Lge. Pharm. Acta Helv. **45**, 157 (1970).
315a. STEINEGGER, E. und F. SCHNYDER: Alkaloidführung und Taxonomie von *Genista legionensis* (Pau) Lainz = *Genista hystrix* Lge. subsp. *legionensis* (Pau) P. Gibbs. Pharm. Acta Helv. **45**, 648 (1970).
316. STEINEGGER, E. und P. WEBER: Totalsynthese des neuen Leguminosen-Alkaloids Hystrin. Helv. chim. Acta **51**, 206 (1968).
317. STEINEGGER, E. und K. WICKY: Die Alkaloide von *Genista lusitanica* L. = *Echinospartum lusitanicum* (L.) Rothm. 11. Mitteilung über Leguminosen-Alkaloide. Pharm. Acta Helv. **40**, 610 (1965).
317a. STREETER, M. P., G. ADOLPHEN and H. H. APPEL: β-Skytanthine N-Oxide in *Skytanthus acutus* Meyen. Chem. and Ind. **1969**, 1631.

318. TALLENT, W. H. and E. C. HORNING: The Structure of Pinidine. J. Amer. Chem. Soc. 78, 4467 (1956).
319. TALLENT, W. H., V. L. STROMBERG and E. C. HORNING: Pinus Alkaloids. The Alkaloids of *P. sabiniana* Dougl. and Related Species. J. Amer. Chem. Soc. 77, 6361 (1955).
320. TERASHIMA, T., Y. KURODA and Y. KANEKO: Studies on a New Alkaloid of *Streptomyces*. Structure of Nigrifactin. Tetrahedron Letters 1969, 2535.
321. THOMPSON, J. F. and C. J. MORRIS: Conversion of 5-Hydroxylysine to 5-Hydroxypipecolic Acid in Honey Locust Leaves. Arch. Biochem. Biophys. 125, 362 (1968).
322. TICHY, M. and J. SICHER: The Configuration of Carpaine. Tetrahedron Letters 1962, 511.
323. TIWARI, H. P., W. R. PENROSE and I. D. SPENSER: Biosynthesis of Mimosine: Incorporation of Serine and α-Aminoadipic Acid. Phytochemistry 6, 1245 (1967).
324. TOMITA, H., S. MITUSAKI and E. TAMAKI: Chemical Studies on Ninhydrinpositive Compounds in Cured Tobacco Leaves. I. Identification of 2-Pyrrolidine Acetic Acid, a New Amino Acid, and 1-Pipecolic Acid. Agric. Biol. Chem. (Tokyo) 28, 451 (1964).
325. TOMKO, J., A. VASSOVA, G. ADAM, K. SCHREIBER and E. HÖHNE: Veralkamine, a Novel Type of Steroidal Alkaloid with a 17β-Methyl-18-nor-17-isocholestane Carbon Skeleton. Tetrahedron Letters 1967, 3907.
326. TOMKO, J., A. VASSOVA, G. ADAM und K. SCHREIBER: Über Veralkamin, ein neuer Steroidalkaloidtyp mit 17β-Methyl-18-nor-17-isocholestan-Kohlenstoffgerüst. Tetrahedron 24, 4865 (1968).
327. TSCHESCHE, R., D. KLÖDEN und H. W. FEHLHABER: Über die Alkaloide aus *Lobelia syphilitica* L. II. Syphilobin A und Syphilobin F. Tetrahedron 20, 2885 (1964).
328. TSCHESCHE, R., K. KOMETANI, F. KOWITZ und G. SNATZKE: Über die Alkaloide aus *Lobelia syphilitica* L., I. Chem. Ber. 94, 3327 (1961).
329. VÄHÄTALO, M. L. and A. J. VIRTANEN: Bound Homoserine in Fruits of Cowberry and Cranberry. Acta Chem. Scand. 11, 747 (1957).
330. VALENTA, Z., P. DESLONGCHAMPS, M. H. RASHID, R. H. WIGHTMAN and J. S. WILSON: *Ormosia* Alkaloids, I. Structure of Ormojanine and Ormosanine. Tetrahedron Letters 1963, 1559.
331. VANDERHAEGHE, H. and G. PARMENTIER: The Structure of Factor S of Staphylomycin. J. Amer. Chem. Soc. 82, 4414 (1960).
332. VANEK, Z., J. CUDLIN and M. VONDRACEK: Cycloheximide and Other Glutarimide Antibiotics. In: D. GOTTLIEB and P. D. SHAW: Antibiotics, Vol. II, p. 222. Berlin-Heidelberg-New York: Springer. 1967.
333. VANEK, Z., M. PUZA, J. CUDLIN, L. DOLEZILOVA and M. VONDRACEK: Metabolites of *Streptomyces noursei*. III. Incorporation of ^{14}C-Carbon Dioxide into Cycloheximide. Biochem. Biophys. Res. Comm. 17, 532 (1964).
334. VAZQUEZ, D.: The Streptogramin Family of Antibiotics. In: D. GOTTLIEB and P. D. SHAW: Antibiotics, Vol. I, p. 387. Berlin-Heidelberg-New York: Springer. 1967.
335. VIRTANEN, A. I. and R. GMELIN: On the Structure of 4-Hydroxypipecolic Acid Isolated from Green Plants. Acta Chem. Scand. 13, 1244 (1959).
336. VIRTANEN, A. I. and S. KARI: 5-Hydroxy-piperidine-2-carboxylic Acid in Green Plants. Acta Chem. Scand. 8, 1290 (1954).
337. — — 4-Hydroxy-piperidine-2-carboxylic Acid in Plants. Acta Chem. Scand. 9, 170 (1955).

338. WATANABE, K.: Untersuchungen über Mikamycin. 6. Mitt. Konstitutionelle Aminosäuren von Mikamycin B. J. Antibiotics (Tokyo) **14**, 1 (1961) [Chem. Zbl. **136** (22), 6923 (1965)].
339. — Untersuchungen über Mikamycin. 7. Mitt. Die Konstitution von Mikamycin B. J. Antibiotics (Tokyo) **14**, 14 (1961) [Chem. Zbl. **136** (22), 6923 (1965)].
340. WENKERT, E. and A. R. JEFFCOAT: Synthesis of Lamprolobine. J. Organ. Chem. (USA) **35**, 515 (1970).
341. WIEHLER, G. and L. MARION: (−)-Homostachydrine, a New Alkaloid Isolated from the Seeds of *Medicago sativa* L. Grimm. Canad. J. Chem. **36**, 339 (1958).
342. WILSON, E. M.: Structures of Ormosinine and Panamine. Chem. and Ind. **1965**, 472.
343. — The Identity of Piptamine and Ormosanine, and the Structures of Ormojanine, Ormosinine and Panamine. Tetrahedron **21**, 2561 (1965).
344. WISSE, J. H., H. DE KLONIA and B. J. VISSER: pH Dependence of the Synthesis of Isopelletierine from Δ^1-Piperidine. Rec. trav. Chim. Pays-Bas **83**, 1265 (1964).
345. WITKOP, B. and C. M. FOLTZ: The Configuration of 5-Hydroxypipecolic Acid from Dates. J. Amer. Chem. Soc. **79**, 192 (1957).
346. WONG, C. F. and R. T. LA LONDE: The Structure of 3-Epinuphamine, a New Alkaloid from *Nuphar luteum subsp. variegatum*. Phytochemistry **9**, 1851 (1970).
346a. — — Sesquiterpene Alkaloids of *Nuphar luteum subsp. variegatum*. Phytochemistry **9**, 2417 (1970).
347. WROBEL, J. T.: *Nuphar* Alkaloids. In: R. H. F. MANSKE: The Alkaloids, Chemistry and Physiology. Vol. IX, p. 441. New York-London: Academic Press. 1967.
348. WYLER, H., T. J. MABRY and A. S. DREIDING: Über die Konstitution des Randenfarbstoffes Betanin. VI. Helv. chim. Acta **46**, 1745 (1963).
348a. YAMADA, Y., K. HATANO and M. MATSUI: Synthesis of (±)-Lamprolobine and (±)-Epilamprolobine. Agric. Biol. Chem. (Tokyo) **34**, 1536 (1970).
348b. YANAGITA, M., S. INAYAMA and T. KAWAMATA: The Stereostructures of Pulchellidine and Pulchellin. Tetrahedron Letters **1970**, 131.
348c. — — — Neopulchellidine and Neopulchellin. Tetrahedron Letters **1970**, 3007.
348d. YANAGITA, M., S. INAYAMA, T. KAWAMATA, T. OKURA and W. HERZ: Pulchellidine, a Novel Sesquiterpene Alkaloid isolated from *Gaillardia pulchella* Foug. Tetrahedron Letters **1969**, 2073, 4170.
349. YANAI, H. S. and W. N. LIPSCOMB: The Structure of ψ-Conhydrine. Tetrahedron **6**, 103 (1959).
350. YUNUSOV, T. K., A. S. SADYKOV and O. S. OTROSHCHENKO: Alkaloids of *Anabasis salsa*. Nauch. Tr. Tashkent, Gos. Univ. Nr. **263**, 16 (1964) [Chem. Abstr. **63**, 2123 (1965)].
351. YURASHEVSKII, N. K. and N. L. STEPANOVA: Alkaloids of *Girgensohnia oppositiflora*. J. allg. Chem. **16**, 141 (1946) [Chem. Abstr. **40**, 6754 (1946)].
352. ZACHARIUS, R. M., J. F. THOMPSON and F. C. STEWARD: The Detection, Isolation and Identification of (−)-Pipecolic Acid as a Constituent of Plants. J. Amer. Chem. Soc. **74**, 2949 (1952).
353. — — — The Detection, Isolation and Identification of L(−)-Pipecolic Acid in the Non-Protein Fraction of Beans *(Phaseolus vulgaris)*. J. Amer. Chem. Soc. **76**, 2908 (1954).

(Eingelaufen am 18. Januar 1971)

Gallenfarbstoffe und Biliproteide

Von **W. Rüdiger**, Saarbrücken und München

Mit 12 Abbildungen

Inhaltsübersicht

	Seite
I. Einleitung	61
II. Nomenklatur	62
III. Chemische Untersuchungsmethoden	64
1. Farbreaktionen	64
Die Gmelin-Reaktion und ihre Erweiterung	64
Die Jaffe-Schlesinger-Reaktion und ihre Erweiterung	66
Die Diazoreaktion	66
2. Abbaureaktionen	68
Abbau mit Permanganat	69
Abbau mit Chromsäure und Chromat	70
IV. Physikalische Untersuchungsmethoden	73
1. Elektronenspektren	73
2. Optische Aktivität	76
3. Massenspektren	79
4. NMR-Spektren	84
5. Chromatographie	86
V. Bilirubin	89
1. Bilirubin-Konjugate	90
2. Bilirubin-Proteide	91
VI. Umwandlungsprodukte des Bilirubins	94
1. Bilane und Bilene-(b) (*Urobilinoide*)	94
2. Biladiene-(a,b)	99
VII. Bilatriene	104
1. Biliverdin und Mesobiliverdin	104
2. Biliverdin und Biliverdin-Proteide bei Vertebraten	104
3. Bilatriene bei Invertebraten	108
VIII. Gallenfarbstoffe mit Äthylidengruppe	111
1. Aplysia-Farbstoffe	111
2. Phycobiliproteide	114
Phycobiline	118
3. Phytochrom	124
Literaturverzeichnis	128

Literaturverzeichnis: SS. 128—139

I. Einleitung

Als „*Gallenfarbstoffe*" oder „*Biline*" (engl. „bile pigments") werden hier Verbindungen bezeichnet, die dasselbe Grundgerüst besitzen wie Bilirubin (1), der Hauptfarbstoff in der Galle des Menschen und der Wirbeltiere: Sie bestehen aus vier „Pyrrol"ringen, die über drei Ein-Kohlenstoff-Brücken miteinander verknüpft sind. Die äußeren Ringe enthalten Sauerstoff-Funktionen; sie besitzen Pyrrolon- (2) oder davon abgeleitete Strukturen. Die inneren Ringe sind echte Pyrrol- (3) oder Pyrrolenin (4)-Ringe.

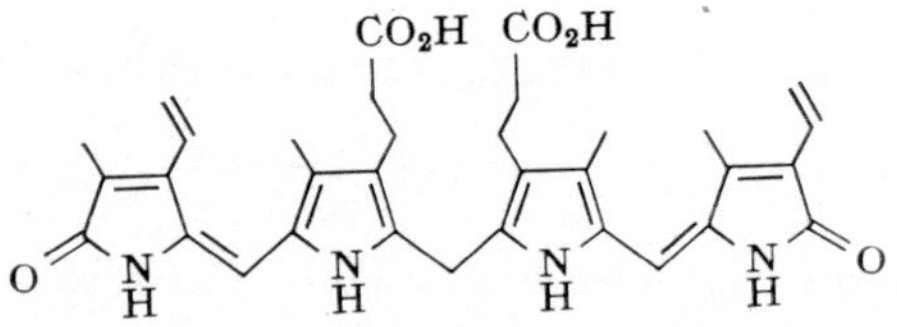

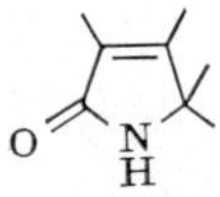

Diese Definition bedeutet einerseits eine Einengung des ursprünglichen Begriffs „Farbstoffe in der Galle", der ohnehin nicht eindeutig war, da auch dem Organismus von außen zugeführte Stoffe (verschiedenster Strukturen) zum Teil mit der Gallenflüssigkeit ausgeschieden werden. Zu den auf diese Weise in die Galle gelangenden Farbstoffen gehört z. B. das Phylloerythrin, das aus dem Chlorophyll der Nahrung stammt. Auf der anderen Seite sind Gallenfarbstoffe nach dieser Definition nicht auf das Vorkommen in der Galle beschränkt. Der derart erweiterte Begriff schließt eine Reihe von physiologisch wirksamen Pigmenten ein (z. B. bei niederen Tieren und Pflanzen), die durchweg in Protein-gebundener Form vorkommen.

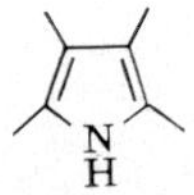

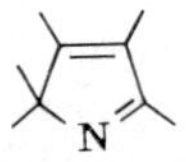

Für photosynthetisch aktive Chromoproteide von Algen, deren Chromophore Gallenfarbstoffe sind, führte O'H EOCHA (*146*) den Begriff *Biliproteide* (engl. *biliproteins*) ein. Dieser Name wird hier für alle Protein-Komplexe von Gallenfarbstoffen unabhängig von ihrem Vorkommen verwendet. Voraussetzung ist eine spezifische Bindung des Pigmentes an das Protein; diese kann kovalent oder nicht-kovalent sein.

In den letzten Jahren wurden wesentliche Fortschritte bei den Strukturuntersuchungen von Gallenfarbstoffen und Biliproteiden erzielt. Das ist einerseits auf verfeinerte chromatographische Trennverfahren, anderseits auf neue physikalische und chemisch analytische Untersuchungsmethoden zurückzuführen. Schließlich führten neue synthetische Methoden zu Gallenfarbstoffen und deren Bausteinen definierter Strukturen, die als Vergleichssubstanzen wesentlich für einige Strukturfragen bei Gallenfarbstoffen geworden sind. Im folgenden werden die heutigen Kenntnisse über die Strukturen der Gallenfarbstoffe und Biliproteide zusammengefaßt. Die Darstellung der historischen Entwicklung und älterer Ergebnisse findet man in früheren Werken und Übersichtsartikeln (58, 72, 116, 117, 201, 202, 203, 239).

II. Nomenklatur (der Biline)

Die natürlich vorkommenden Gallenfarbstoffe besitzen Trivialnamen, welche aber nicht von allen Autoren im gleichen Sinn, d. h. für die gleichen Verbindungen benutzt werden. Für rationelle Bezeichnungen sei hier eine Nomenklatur kurz erläutert, die im Prinzip auf FISCHER (58) und SIEDEL (203) zurückgeht und Vorschläge von LEMBERG (116) berücksichtigt.

Grundkörper ist das Tetrapyrran (5), das den Namen ,,Bilan" erhält. In der vorliegenden Arbeit wird die ursprünglich vorgeschlagene Zählweise (5a) beibehalten, da die im folgenden beschriebene Klassifizierung der Biline sich daraus zwanglos ergibt. Das ist bei der Zählweise entsprechend Formel (5b) nicht der Fall; diese Numerierung, die sich von der der Porphyrine und Corrine ableitet, ist bei manchen synthetischen Bilinoiden vorzuziehen*.

Die Klassifizierung der Gallenfarbstoffe sei an einem Schema (Formelübersicht 1) erläutert, das gleichzeitig die Umwandlung der einzelnen Typen ineinander beschreibt (Einzelheiten s. unter Nachweisreaktionen S. 64) sowie Trivialnamen einzelner Beispiele anführt. Als ,,Bilane" werden alle Derivate von (5) bezeichnet, die wie der Grundkörper drei Methylenbrücken enthalten. Besitzt ein Gallenfarbstoff eine Doppelbindung mehr (also eine Methin- statt einer Methylenbrücke), so gehört er zu den ,,Bilenen"**, wobei die Stellung der Doppelbindung nachstehend angegeben wird. So ist das Urobilin (12), das eine Methinbrücke in der Mitte besitzt, ein Bilen-(b) (6)***. Bei den Pigmenten mit zwei Methin-Brücken muß man Biladiene-(a,b) (7) von Biladienen-(a,c) (8) unterscheiden, wohin-

---

* Häufig wird eine gemischte Nomenklatur benutzt, bei der die Brückenatome mit den Ziffern 5, 10 und 15 als auch mit den Buchstaben a, b und c bezeichnet werden (vgl. 80).

** Die von SIEDEL gewählten Namen Bilien, Bilidien und Bilitrien entsprechen nicht den Regeln der chemischen Nomenklatur, wenn man vom Bilan ausgeht.

*** Die Methylen- bzw. Methinbrücken werden von FISCHER und von SIEDEL als α, β, γ bezeichnet. Da diese Buchstaben bei Porphyrinen in anderer Bedeutung vorkommen, ist die obige, von LEMBERG vorgeschlagene Bezeichnung vorzuziehen.

Literaturverzeichnis: SS. 128—139

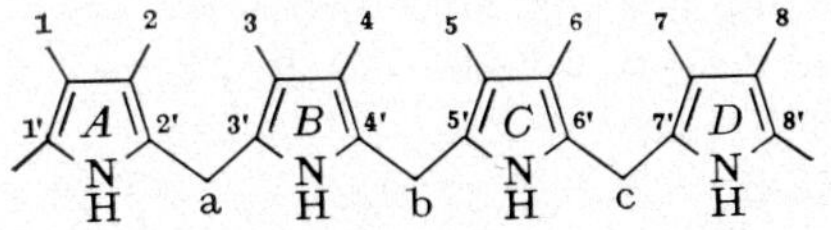

Gallenfarbstoffe und Biliproteide

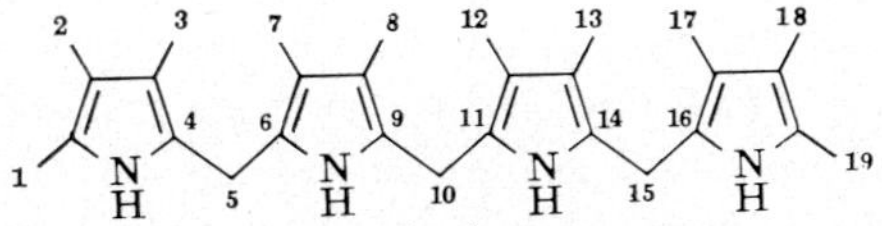

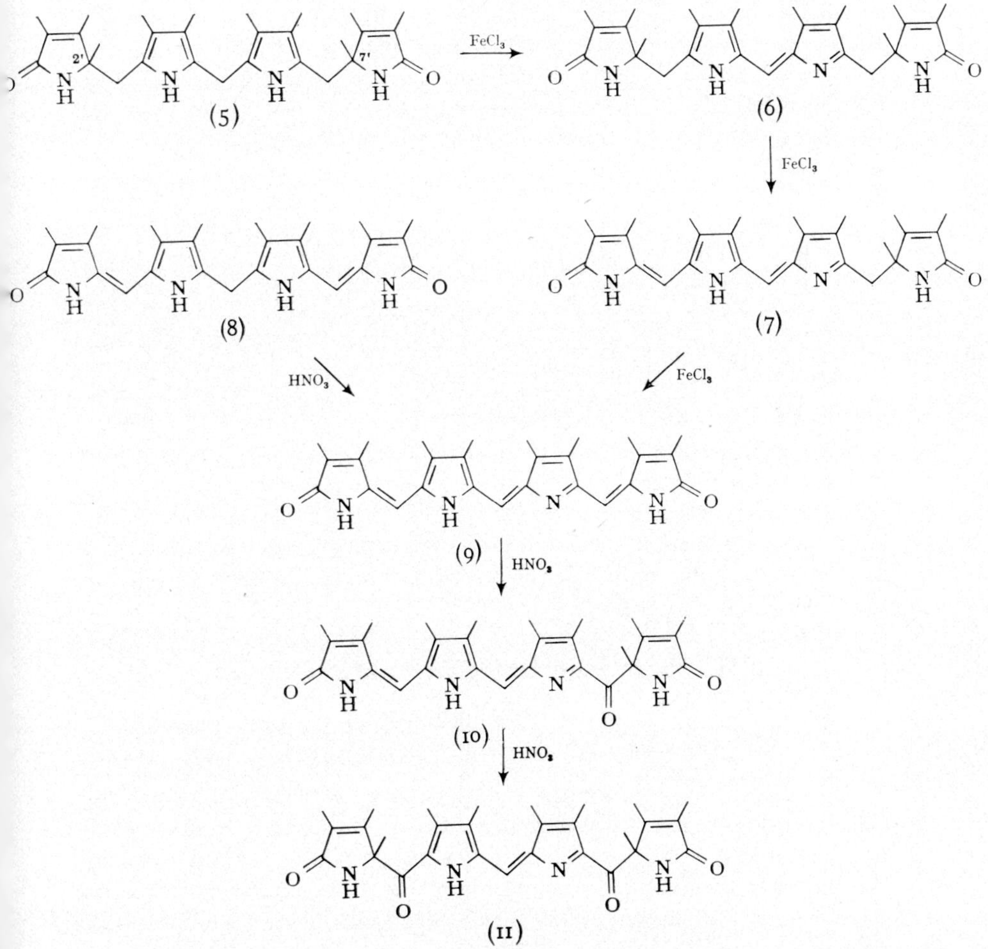

Formelübersicht 1

gegen die Bezeichnung Bilatrien (9) eindeutig ist. Als Beispiele für Brückensubstituierte Gallenfarbstoffe seien das Biladien-(a,b)-on(c) (10) und das Bilen-(b)-dion(a,c) (11) angeführt.

Der ausführliche Name für Bilirubin (1) ist 1′,8′-Dioxo-1,3,6,7-tetramethyl-2,8-divinyl-biladien-(a,c)-dipropionsäure-(4,5). Zur Vereinfachung schlug LEMBERG die Bezeichnung „Biladien-(a,c)" für diese Substanz vor, was zur Verwirrung Anlaß gibt, da jetzt eine Stoff*gruppe* und ein *einzelner* Stoff mit demselben Namen belegt werden. Diese Schwierigkeit läßt sich vermeiden, wenn man bei den *einzelnen* Bilinen zusätzlich zur Gruppenbezeichnung angibt, aus welchen Porphyrinen sie (formal) entstanden sind. Die Bezeichnungsweise ist bei einigen Gallenfarbstoffen bereits üblich; so heißt das Bilirubin (1), das aus Protoporphyrin-IX durch Öffnung an der α-Methinbrücke entstanden ist, genauer Bilirubin-IXα; nach der hier vorgeschlagenen Nomenklatur ist es als Protobiladien-(a,c)-IXα zu benennen; Urobilin (12), welches sich (formal) aus Mesoporphyrin IX ableitet, heißt Mesobilen-(b)-IXα.

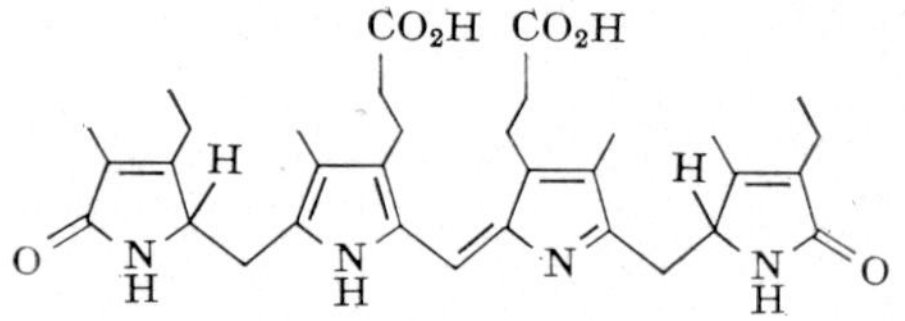

(12)

An der bezeichneten Methin-Brücke (hier α) soll der Porphyrinring *oxydativ* geöffnet worden sein, d. h. der Zusatz α soll den Austausch des Brücken-C-Atoms gegen zwei O-Atome einschließen. Die von GRAY (*79*) vorgeschlagene Bezeichnung „Bilenon" für den unsubstituierten Grundkörper *mit* den beiden Pyrrolon-*Sauerstoff-A*tomen ergibt keine wesentliche Vereinfachung, kann aber zu Verwechslungen mit Biladienonen (10) und Bilendionen (11) führen.

III. Chemische Untersuchungsmethoden

1. Farbreaktionen

Die Gmelin-Reaktion und ihre Erweiterung

Die 1826 von TIEDEMANN und GMELIN entdeckte Farbreaktion Bilirubin-reicher Körperflüssigkeiten mit Nitrit-haltiger Salpetersäure hat sich unter dem Namen „Gmelin-Reaktion" als üblicher Nachweis von Bilirubin durchgesetzt (zur Ausführung vgl. *203*). Den dabei nacheinander auftretenden Farbstufen konnte SIEDEL (*203*) die entsprechenden Strukturen (8)—(11) (Formelübersicht 1) zuordnen. Charakteristische Farbänderungen geben unter den Bedingungen der Gmelin-Reaktion außer

Literaturverzeichnis: SS. 128—139

den Biladienen-(a,c) noch die Bilatriene, während Gallenfarbstoffe niederer (Bilane, Bilene-(b), Biladiene-(a,b)) oder höherer Oxydationsstufen (Biladienone, Bilendione) nach vorhergehender Farbvertiefung (Salzbildung an Stickstoff) ausbleichen. Das trifft aber auch bei anderen N-haltigen Farbstoffen (z. B. Porphyrinen) zu. Wenn die Gmelin-Reaktion trotzdem zum Nachweis dieser Gallenfarbstoffe herangezogen wurde, war die Interpretation nicht eindeutig (vgl. *174*).

Der Nachweis von Gallenfarbstoffen aller Oxydationsstufen gelingt mit Hilfe der *erweiterten Gmelin-Reaktion* (vgl. Formelübersicht 1) (*174*): Mit Natriumamalgam werden alle Gallenfarbstoffe zu Bilanen reduziert (*58, 203*), Vinylgruppen werden dabei zu Äthylgruppen reduziert, so daß man sowohl von der Proto- als auch von der Meso-Reihe zu Mesobilan gelangt, was zur Vereinheitlichung der Reaktion beiträgt. Die Oxydation der Bilane mit Eisen(III)-chlorid ergibt je nach den Bedingungen verschiedene Resultate: In verdünnter methanol. Salzsäure entsteht als Hauptprodukt in der Kälte Bilen-(b) (*104, 192*), in der Wärme Biladien-(a,b) (*206*) und etwas Bilatrien und Biladienon. Biladien-(a,b) läßt sich in stärkerer Säure (1 ml konz. HCl auf 2 ml Methanol) (*234*) oder in Eisessig (*174*) glatt zu Bilatrien weiteroxydieren.

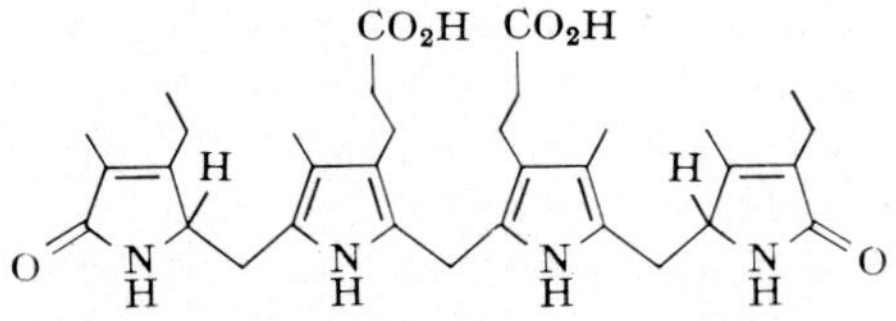

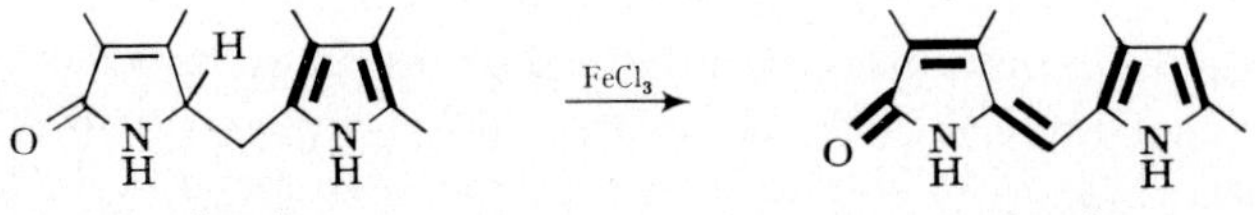

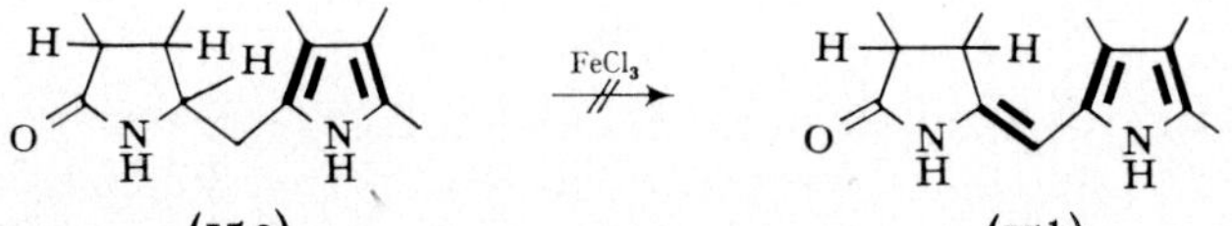

Bei einer standardisierten Oxydation, bei der das rohe Bilan (0,2 bis 2 mg) mit 10% $FeCl_3$ in konz. HCl (5 Tropfen) und Methanol (3 ml) 30 Min. auf 60° C erhitzt wird (*174*), erhält man Bilene-(b) neben Biladienen-(a,b) sowie wenig Bilatrien und Biladienon, die sich chromatographisch voneinander trennen. Die genaue Untersuchung des aus Mesobilan-IX α (13) erhaltenen Pigmentgemisches ergab das Vorhandensein von drei Bilenen, acht Biladienen-(a,b) und drei Bilatrienen (*214*). Die Vielzahl an farbigen, charakteristischen Derivaten eines ursprünglich einheitlichen Farbstoffes ermöglicht so die sichere Identifizierung als Gallenfarbstoff.

Die treibende Kraft bei der Oxydation mit $FeCl_3$ dürfte der Gewinn an Mesomerie-Energie infolge der Ausdehnung der Konjugation sein (in Formel (14a, b für einen mittleren und einen äußeren Ring formuliert). Wenn z. B. der äußere Ringe hydriert ist, fällt diese treibende Kraft weg (15a, b); tatsächlich erweist sich die Dehydrierung in diesem Fall als unmöglich oder jedenfalls erheblich erschwert (*235*).

Die Jaffe-Schlesinger-Reaktion und ihre Erweiterung

Die von JAFFE 1869 entdeckte grüne Fluoreszenz der Zink-Komplexe von Bilenen-(b) wurde 1903 von SCHLESINGER in die klinische Chemie eingeführt. Beim üblichen Nachweis wird eine schwach alkalische oder ammoniakalische Untersuchungslösung mit alkoholischer Zinkacetat-Lösung versetzt; bei hoher Bilan-(b)-Konzentration sieht man die grüne Fluoreszenz bereits im Tageslicht; empfindlicher ist der Nachweis im UV-Licht.

Die Zink-Komplexe von Biladienen-(a, b) und Biladien(a, b)-onen(c) besitzen eine hell- bis dunkelrote Fluoreszenz; die Reaktionsbedingungen entsprechen denen bei den Bilenen-(b). Biladienone entstehen durch Oxydation aus Bilatrienen oder Biladienen-(a,c) (vgl. Formelübersicht 1); als Oxydationsmittel zur Ausführung dieser Reaktion wird meist eine ammoniakalische Jod-Lösung verwendet.

Bilatriene ergeben unter den Bedingungen der JAFFE-SCHLESINGER-Reaktion (ohne Oxydationsmittel) keine sichtbare Fluoreszenz; nach COLE, CHAPMAN und SIEGELMAN (*33*) liegen die Emissionsmaxima der Zink-Komplexe im nahen Infrarotgebiet (710—730 nm). Bei entsprechender Meßmethodik läßt sich somit auch die Fluoreszenz der Bilatrien-Zink-Komplexe zu deren Nachweis heranziehen.

Die JAFFE-SCHLESINGER-Reaktion kann auch als Mikromethode z. B. bei chromatographisch getrennten Gallenfarbstoffen angewendet werden. Die Chromatogramme werden mit 0,2 proz. äthanol. Zinkacetat-Lösung besprüht und im UV-Licht betrachtet; dabei werden die Fluoreszenzfarben der Bilene-(b), Biladiene-(a,b) und Biladien(a,b)-one(c) (nach deren Trennung) sichtbar (*174*, *212*).

Die Diazoreaktion

Seitdem HIJMANS VAN DEN BERGH 1916 die von EHRLICH 1883 entdeckte Farbreaktion von Bilirubin mit diazotierter Sulfanilsäure (16a) standardisierte, ist diese Reaktion die klassische Methode zur Bilirubin-

Literaturverzeichnis: SS. 128—139

Bestimmung geworden. Die langandauernde Kontroverse um die Frage, ob die „direkte" (d. h. rasche) Diazoreaktion — im Gegensatz zur „indirekten" (d. h. langsameren) Reaktion des freien Bilirubins — auf spezifische Bilirubin-Verbindungen oder nur auf feiner verteiltes Bilirubin zurückzuführen ist, darf heute als im Sinne der ersten Alternative entschieden gelten; jedoch spielt bei der Reaktionsgeschwindigkeit wohl die bessere Wasserlöslichkeit der Bilirubin-Konjugate eine Rolle (vgl. 95) (zur Struktur dieser Konjugate vgl. S. 90). Auch andere aromatische Diazonium-Salze reagieren in ähnlicher Weise; für die chromatographische Trennung der entstehenden Azofarbstoffe wird häufig die Umsetzung mit den unpolaren Äthylestern der diazotierten Anthranilsäure (Formel (16b)) vorgezogen.

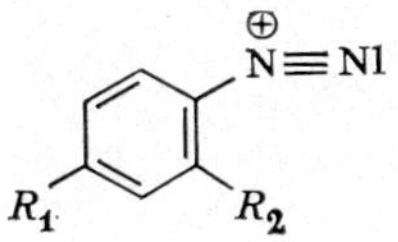

(16a) $R_1 = SO_3^\ominus;\ R_2 = H$
(16b) $R_1 = H;\ R_2 = CO_2C_2H_5$

Die Kupplung mit Diazoniumsalzen ist typisch für einfache Pyrrolderivate mit freier α- oder β-Position bzw. mit einem leicht eliminierbaren Substituenten (z. B. Carboxyl). In der Reihe der Gallenfarbstoffe ist die Reaktion auf Biladiene-(a,c) beschränkt. Da diese keine freie α- oder β-Position besitzen, muß das Molekül im Verlauf der Reaktion gespalten werden. FISCHER und HABERLAND (54) zeigten, daß Biladiene-(a,c) (8) zu denselben Azofarbstoffen (17) führen wie die entsprechenden Dipyrromethene der Struktur (14b); demnach erfolgt die Spaltung an der (mittleren) Methylenbrücke.

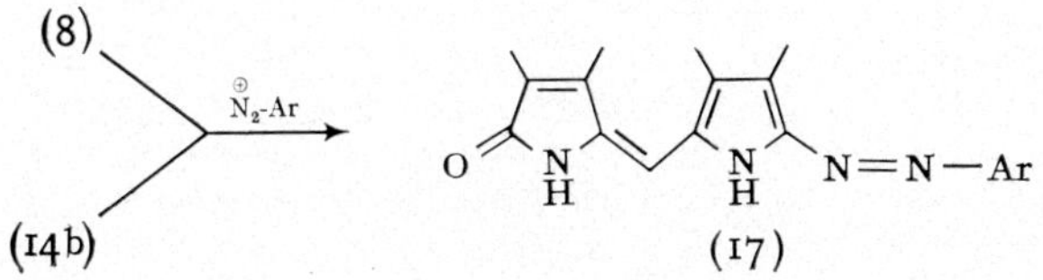

Die Spaltung erfolgt sowohl links als auch rechts des Methylen-C-Atoms; bei unsymmetrisch substituierten Biladienen (z. B. Bilirubin IXα (1)) bilden sich zwei Azofarbstoffe, die sich von der linken bzw. rechten Molekülhälfte ableiten. Den Mechanismus untersuchten TREIBS und FRITZ (223) am Beispiel einer Reihe von Dipyrromethanen (Formelübersicht 2); dabei wurde der Einfluß von Substituenten sowohl am Pyrrolkörper als auch am Phenylkern des Diazoniumsalzes studiert. Der elektrophile Angriff erfolgt an einem mit der Methylen-Brücke verknüpften α-C-Atom, da dort die Elektronendichte besonders hoch ist.

Das entstandene Zwischenprodukt (18a) zerfällt in den Azofarbstoff (17a) und ein Carbenium-Ion (18b), welches auch als Immonium-Ion formuliert werden kann. (18b) reagiert (sogar schneller als das Ausgangsmaterial) mit weiterem Diazoniumsalz zum Azofarbstoff (17b) unter Freisetzung von Formaldehyd, der zum Teil mit Dimedon abgefangen wurde. Ein analoger Mechanismus wurde bei der säurekatalysierten Spaltung von Bilanen formuliert (*91*). In diesem Fall kann eine Rekombination der Bruchstücke erfolgen (vgl. S. 104).

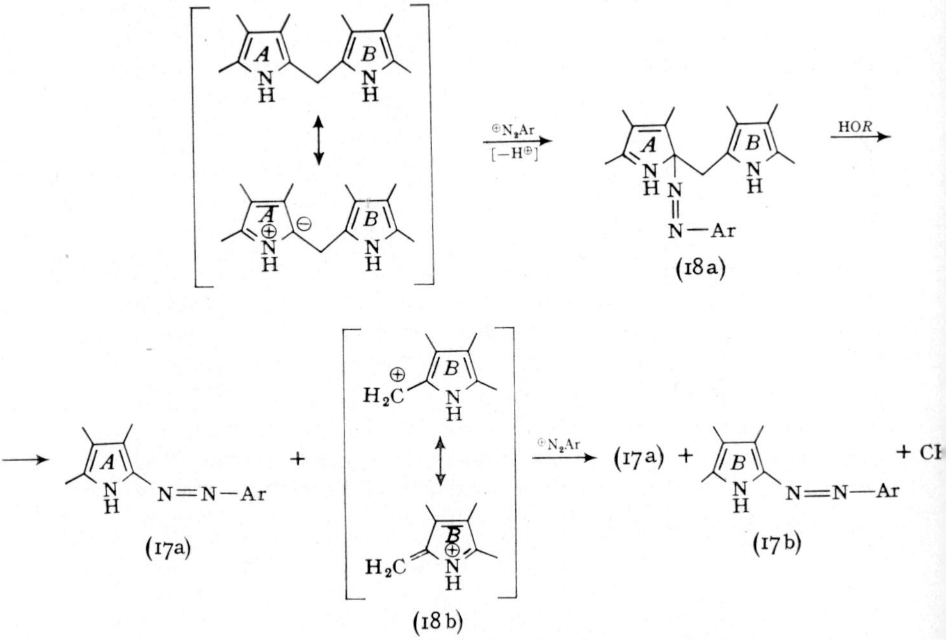

Formelübersicht 2

2. Abbaureaktionen

Methoden des *reduktiven* Abbaus (mit Jodwasserstoffsäure und mit Resorcin), die bei der ursprünglichen Strukturaufklärung der Gallenfarbstoffe eine entscheidende Rolle spielten (*58*, dort S. 640), sind später nicht weiterentwickelt worden und haben in neuerer Zeit keinen wesentlichen Beitrag zu Strukturuntersuchungen bei Gallenfarbstoffen geliefert. Dagegen haben verfeinerte Methoden des *oxydativen* Abbaus (zusammen mit verbesserten chromatographischen Verfahren) Strukturuntersuchungen bei neuen Gallenfarbstoffen sehr erleichtert.

Literaturverzeichnis: SS. 128—139

Abbau mit Permanganat

Der Abbau mit Permanganat wurde zunächst für Strukturuntersuchungen von Porphyrinen entwickelt (*138*) und dann für bestimmte Fragestellungen im Gebiet der Gallenfarbstoffe angewendet (*79, 132, 135, 161, 162, 218*). Das Prinzip ist die Bildung von Pyrroldicarbonsäure-(2,5) (**19**) aus Pyrrolkernen, die in beiden α-Stellungen ungesättigte Gruppen (z. B. Methin-Brücken) tragen oder bei denen solche ungesättigten Gruppen im Verlauf der Oxydation entstehen (**20**). Gesättigte Substituenten in den Positionen bleiben erhalten, während ungesättigte Gruppen oxydiert werden (z. B. Vinyl zu Carboxyl).

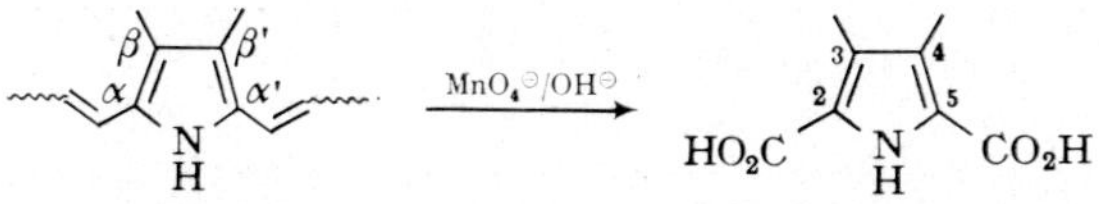

(20) → MnO$_4^\ominus$/OH$^\ominus$ → (19)

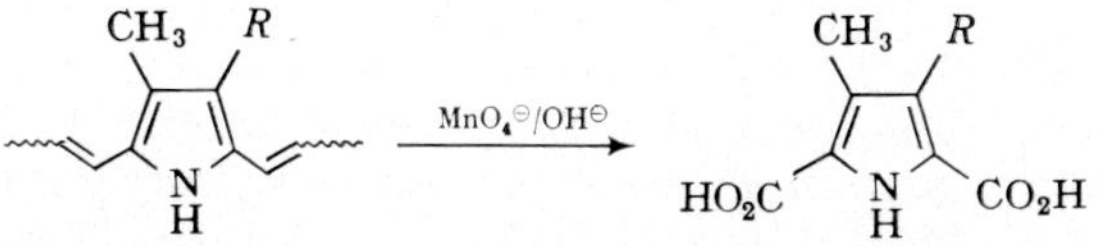

(23) R = CH=CH$_2$
(24) R = CH$_2$CH$_2$CO$_2$H

(21) R = CO$_2$H
(22) R = CH$_2$CH$_2$CO$_2$H

Die Oxydation wurde ursprünglich in 2 n Na$_2$CO$_3$ durchgeführt (*79*) später dann in 0,1 n HCl (*132*); jedoch besteht in saurer Lösung die Gefahr der Überoxydation unter Zerstörung der Säure (**22**). Die nach Zerstören des Permanganatüberschusses aus der stark sauren Oxydationslösung mit Äther extrahierten Säuren werden dann durch Papierchromatographie getrennt und mit alkalischer, diazotierter Sulfanilsäure nachgewiesen (R_F-Werte in Tabelle 1).

Tabelle 1. R_F-*Werte von Pyrroldicarbonsäuren-(2,5)*

Fließmittel	A		B	
Substanz	(21)	(22)	(21)	(22)
Nach (*132*)	0,45—0,47	0,82—0,84	0,22—0,24	0,05—0,07
Nach (*138*)	0,33	0,86	0,43	0,28
Nach (*161*)	0,34—0,40	0,80—0,85	0,26—0,31	0,03—0,06
Nach (*135*)	0,51	0,82	0,37	0,04

Absteigende Papierchromatographie auf Whatman No. 1.
Fließmittel A: n-Butanol/Eisessig/Wasser (4 : 1 : 5).
Fließmittel B: Äthanol/Ammoniak (d = 0,88)/Wasser (20 : 1 : 4).

Da bei den Gallenfarbstoffen nur die mittleren Pyrrolkerne C-Atome in beiden α-Positionen besitzen, können sich Pyrroldicarbonsäuren-(2,5) nur von diesen mittleren Kernen ableiten. Die äußeren Kerne, die Sauerstoff-Funktionen an einer α-Position enthalten, werden durch Permanganat oxydativ zerstört (*166*). Die β-Substituenten der entstandenen Pyrroldicarbonsäuren entsprechen also den β-Substituenten der mittleren Pyrrolkerne. Die Frage, welche Substituenten ein Gallenfarbstoff an den beiden Kernen trägt, hat eine Rolle bei Untersuchungen zur Biogenese dieser Pigmente gespielt (vgl. *162*, *176*, *218*).

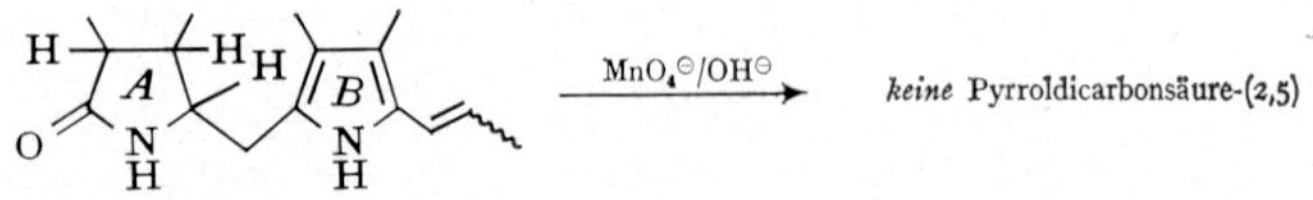

$$\text{(25)} \quad \xrightarrow{MnO_4^{\ominus}/OH^{\ominus}} \quad keine \text{ Pyrroldicarbonsäure-(2,5)}$$

Hydrierte Endringe von Gallenfarbstoffen (**25**) (vgl. Stercobilin S. 96) verhindern die Bildung von Pyrroldicarbonsäuren-(2,5) (*79*); offensichtlich ist hier die Ausbildung einer Struktur mit zwei ungesättigten α-Substituenten (vgl. (**20**)) durch das Oxydationsmittel nicht möglich.

Der Abbau und der Nachweis der Abbauprodukte wird durch die Gegenwart von Protein nicht gestört; deshalb können auch Chromoproteide direkt mit der Methode untersucht werden (*137*).

Abbau mit Chromsäure und Chromat

Der Chromsäureabbau ist zur Charakterisierung der β-Substituenten einer Reihe von Pyrrolfarbstoffen (außer Gallenfarbstoffen von Chlorinen, Corrinen und Porphyrinen) herangezogen worden. (Literaturübersicht in (*178*)). Aus Pyrrol-Ringen entstehen dabei Maleinimide, aus β,β'-Dihydropyrrol-Ringen die entsprechenden Succinimide. Einen wesentlichen Fortschritt bei der Untersuchung der Gallenfarbstoffe brachte die

Tabelle 2. *Halbwertszeiten von Pyrrolpigmenten beim Chromatabbau bei pH 1,7 (nach 178)*

Pigment	$t^1/_2$
Aplysioviolin (**99**)	0,5 Min.
Mesobiladien-(a,b)-IXα (**67**)	3 Min.
Protobilatrien-IXα (**81**)	4,5 Min.
Mesobilatrien-IXα (**81a**)	16 Min.
Mesobilen-(b)-IXα (**12**)	20 Min.
Phaeophorbid a	5 Std.
Protoporphyrin IX	7 Std.
Cyanocobalamin	7 Tage

Literaturverzeichnis: SS. 128—139

Beobachtung, daß diese Pigmente von Chromsäure bzw. Chromat unter milderen Bedingungen (d. h. bei höheren pH-Werten) abgebaut werden als andere Pyrrolfarbstoffe. Die Bestimmung der Abbaugeschwindigkeit unter Standardbedingungen kann somit zur Unterscheidung der Gallenfarbstoffe von anderen Pyrrolfarbstoffen dienen (Tabelle 2).

Zum einen bleiben unter diesen milden Bedingungen des Abbaus ungesättigte Seitenketten (z. B. Vinylgruppen) erhalten; bei früheren Abbauversuchen mit Chromsäure wurden Pyrrolkerne mit ungesättigten Seitenketten oxydativ zerstört. Zum anderen lassen sich bei pH 1,7 („Chromatabbau") Pyrroldialdehyde-(2,5) (26) fassen, die erst in stärker saurem Medium (in 2 n H_2SO_4 „Chromsäureabbau") in Maleinimide (27) übergehen.

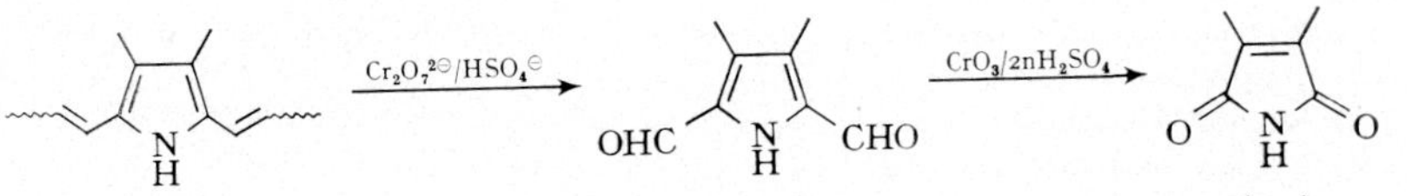

Die Pigmente werden entweder in Lösung (Aceton bzw. Dimethylsulfoxid) mit der wäßrigen Oxydationslösung vermischt (CrO_3 in 2 n H_2SO_4 bzw. Dichromat in 1 proz. $KHSO_4$); die Abbauprodukte müssen dann mit Essigester extrahiert werden. Bei kleineren Pigment-Mengen empfiehlt sich der Abbau auf der Dünnschichtplatte (durch Aufsprühen der Oxydationslösung), wobei sich die Extraktion der Abbauprodukte erübrigt (*178*). Die Abbauprodukte werden durch Dünnschichtchromatographie getrennt und mit Dinitrophenylhydrazin (Pyrrolaldehyd) bzw. Chlor/Benzidin (Imide) (*178*) oder t-Butylhypochlorit/Kaliumjodid/ Stärke (Imide) (*101*) sichtbar gemacht. Die R_F-Werte einiger typischer Abbauprodukte sind in Tabelle 3 zusammengestellt.

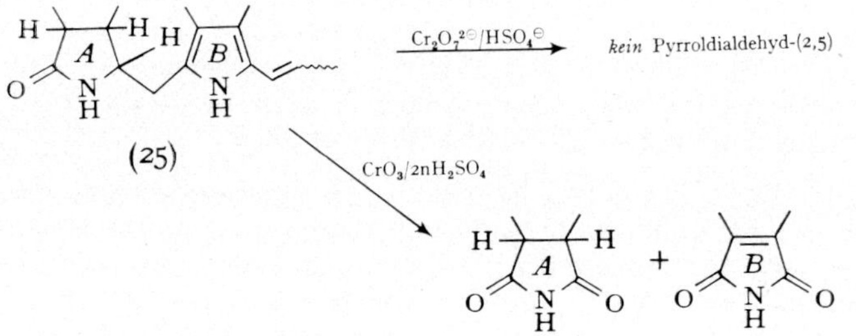

Die Entstehung der Dialdehyde entspricht in ihrer Bedeutung der Entstehung der Dicarbonsäuren beim Permanganat-Abbau; jedoch

Tabelle 3. *Dünnschichtchromatographie von Imiden und Pyrroldialdehyden auf Kieselgel G mit CCl_4/Äthylacetat/Cyclohexan (5:3:1) (nach 178)*

Substanz	R_F
Succinimid	0,06
Methylsuccinimid	0,10
Methyl-äthyliden-succinimid	0,21
Methyl-äthyl-succinimid	0,22
2-Methyl-3-(2-methoxycarbonyl-äthyl)-succinimid = Dihydrohämatinsäure-imid-methylester	0,08
Maleinimid	0,24
Methylmaleinimid	0,28
Methyl-äthyl-maleinimid	0,41
Methyl-vinyl-maleinimid	0,43
Methyl-(2-methoxycarbonyl-äthyl)-maleinimid = Hämatinsäure-imid-methylester	0,26
Pyrroldialdehyd-(2,5)	0,23
3-Methyl-pyrroldialdehyd-(2,5)	0,26
3-Methyl-4-äthyl-pyrroldialdehyd-(2,5)	0,38
3-Methyl-4-vinyl-pyrroldialdehyd-(2,5)	0,40
3-Methyl-4-(2-methoxycarbonyl-äthyl)-pyrroldialdehyd-(2,5)	0,25

bietet der Chromat-Abbau zwei Vorteile: 1. Die Ausbeute an Abbauprodukten ist erheblich höher; man kommt deshalb mit weniger des zu untersuchenden Farbstoffs aus (0,5 μg Gallenfarbstoff gegenüber mindestens 6 mg beim Permanganat-Abbau). 2. Aus den endständigen Ringen entstehen gleichzeitig Maleinimide, so daß man eine doppelte Kontrolle für den Nachweis der β-Substituenten an inneren und äußeren Ringen besitzt. Auch hier verläuft die Reaktion bei Anwesenheit eines hydrierten Endringes anormal; man erhält keinen Pyrroldialdehyd, wohl aber (in stärker saurem Medium) die entsprechenden Maleinimide (27) bzw. — aus dem hydrierten Ring — Succinimide (28). Über welche Zwischenstufen die Oxydation in diesem Falle verläuft, ist noch unbekannt.

Auch Chromsäure- und Chromatabbau werden durch die Gegenwart von Protein nicht gestört, man kann deshalb Chromoproteide (Biliproteide) direkt untersuchen. Die Variation beim Chromsäureabbau erlaubt hier sogar Aussagen über die Bindung an das Protein: Ester- und Amidbindungen bleiben bei Raumtemperatur erhalten, d. h. die derart mit dem Protein verknüpften Pyrrolringe bleiben in Form der Imide an das Protein gebunden. Diese Bindungen werden jedoch in 2 n H_2SO_4 bei 100° verseift. Da der Chromsäureabbau (abgesehen von dieser Verseifung) bei beiden Temperaturen die gleichen Abbauprodukte liefert, kann man die Abbauprodukte der vorher gebundenen Pyrrolkerne selektiv durch Abbau bei 100° freisetzen (Anwendung vgl. S. 106, 121).

Literaturverzeichnis: SS. 128—139

IV. Physikalische Untersuchungsmethoden

1. Elektronenspektren

Gallenfarbstoffe besitzen durchweg „spektralreine" Farben, d. h. sie haben trotz komplizierter Molekülstruktur nur *eine* Absorptionsbande im sichtbaren Spektralbereich, während z. B. Porphyrine, deren Molekülstruktur eine höhere Symmetrie aufweist, vier Banden im entsprechenden Spektralbereich besitzen. Da eine allgemeine Theorie für die Elektronenspektren der Gallenfarbstoffe noch fehlt, seien hier einige empirisch abgeleitete Gesichtspunkte herausgestellt.

Tabelle 4. *Langwellige Absorptionsmaxima von Gallenfarbstoffen der Meso-Reihe IX α in Methanol (nm)*

Pigment	Lit.	neutral (freie Base)	H^+ (Hydrochlorid)	Zn^{2+} (Komplex)
Bilan*	(*172*)	(216)	—	—
Bilen-(a)*	(*48*)	400	450	—
Bilen-(b)	(*156*)	448	492	507
Biladien-(a,b)**		555	565	625
Bilatrien	(*156*)	640	685	682

* Bei zweikernigen Modellsubstanzen gemessen.
** W. Rüdiger und W. Klose, unveröffentlicht.

Die Lage der Absorptionsbanden im sichtbaren Bereich ist für die Gallenfarbstoffe der Meso-Reihe, die gesättigte Alkyl-Substituenten in allen β-Positionen besitzen, in Tabelle 4 zusammengestellt. Methylen-Brücken können in erster Näherung als Alkyl-Substituenten aufgefaßt werden; die durch Methylen-Brücken voneinander „getrennten" Pyrrol- bzw. Pyrrolon-Ringe beeinflussen sich in ihren spektralen Eigenschaften nicht*. So wird für das Mesobilan eine Überlagerung der Elektronenspektren von Alkyl-substituiertem Pyrrol und Pyrrolon erwartet. Diese besitzen *kein* Maximum im sichtbaren Bereich, Pyrrole ein solches auch nicht im UV-Bereich (bis 220 nm); 3,4-Dimethylpyrrolon (29) besitzt eine Bande bei 220 nm (*167*). Mesobilan hat ebenfalls kein Maximum im sichtbaren Spektralbereich; die Banden im UV-Bereich sind noch nicht exakt bestimmt worden, da dieses Bilin leicht der Autoxydation unterliegt. Die zweikernigen Modell-Substanzen (30a) und (30b) besitzen in Übereinstimmung mit der Theorie ein Maximum bei 216 nm (*172*).

* Das läßt sich nicht ohne weitere auf andere Aromate übertragen; 3,4-Dimethyl-5-benzyl-Δ^3-pyrrolon-(2), ein Analogon von (30) mit einem Phenyl- statt eines Pyrrolkernes, besitzt ein Maximum bei zirka 330 nm (*166*).

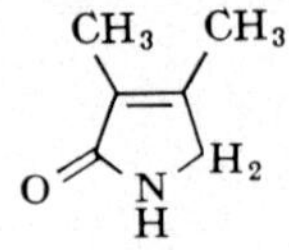

(29)

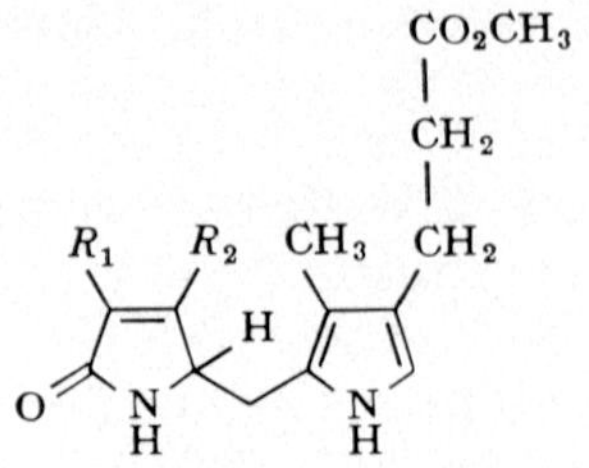

(30a) $R_1 = CH_3;\ R_2 = C_2H_5$
(30b) $R_1 = C_2H_5;\ R_2 = CH_3$

Durch Einführung einer Methin-Brücke entsteht ein neuer Chromophor, der die beiden durch diese Brücke verbundenen Ringe einschließt. Zwei durch eine Methin-Brücke verbundene Pyrrol-Ringe bilden ein Dipyrromethen vom „klassischen Typ" (vgl. Formel (31a)) (*48*); dazu gehört das Mesobilen-(b) oder Urobilin (*12*). Werden durch eine Methin-Brücke ein Pyrrol- und ein Pyrrolon-Ring miteinander verknüpft, so entsteht ein Dipyrromethen vom „Neo-Typ" (vgl. (32)) (*48*). Dieser Typ sollte in dem (bisher nicht isolierten) Bilen-(a) verwirklicht sein. Dem gleichen Farbstoff-Typ gehören auch Biladiene-(a,c) an (*48*), die denselben Chromophor, getrennt durch die mittlere Methylen-Brücke, in beiden Molekül-Hälften enthalten. Im Biladien-(a,b) verbinden zwei benachbarte Methin-Brücken einen Pyrrolon- und zwei Pyrrol-Ringe zu einer Tripyrren-Struktur, während im Bilatrien alle vier Ringe einen Chromophor vom Tetrapyrren-Typ bilden. Diese beiden Strukturtypen besitzen außer der langwelligen Bande im UV-Bereich, die sich beim Übergang von der freien Base zum Hydrochlorid bzw. zum Zink-Komplex nur wenig verschiebt. Diese liegt (in Methanol) für das Mesobiladien-(a,b) bei 320—325 nm, für das Mesobilatrien bei 365—370 nm. Die Lage aller Banden ist Lösungsmittel-abhängig.

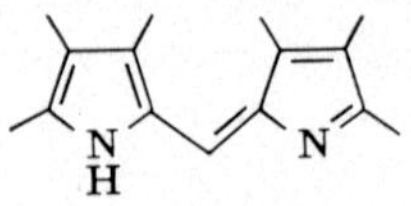

(31a) 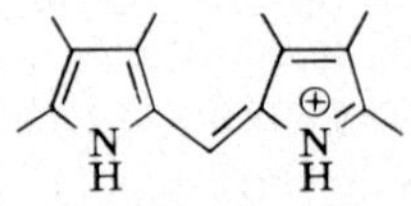   (31b)

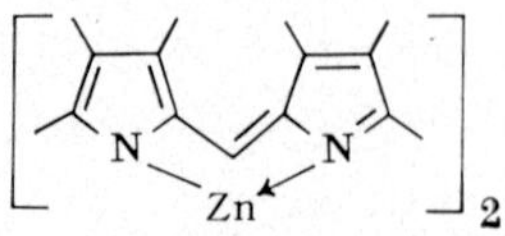

(31c) 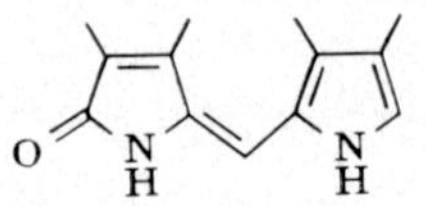   (32)

Literaturverzeichnis: SS. *128—139*

Wie aus den Daten der Tabelle 4 hervorgeht, wird die langwellige Absorptionsbande nicht nur durch Ausdehnung der Konjugation (beim Übergang von einem Farbstoff-Typ zum anderen), sondern auch (bei ein und demselben Farbstoff-Typ) beim Übergang von der freien Base (z. B. (31a)) zum Hydrochlorid (z. B. (31b)) bzw. Zink-Komplex (z. B. (31c)) bathochrom verschoben. Charakteristisch für die Zugehörigkeit zu einem Farbstoff-Typ ist die Lage des Maximums beim Hydrochlorid, während individuelle Unterschiede zwischen verschiedenen Gallenfarbstoffen desselben Typs besser in Spektren der freien Base oder des Zink-Komplexes zum Ausdruck kommen. So tritt bei Einführung von zusätzlichen, mit dem Chromophor konjugierten Doppelbindungen in den Seitenketten (Vinylgruppen der Proto-Reihe an Stelle von Äthylgruppen der Meso-Reihe) bei den freien Basen eine größere Rotverschiebung (10 bis 20 nm) als bei den Hydrochloriden (maximal 5 nm) ein.

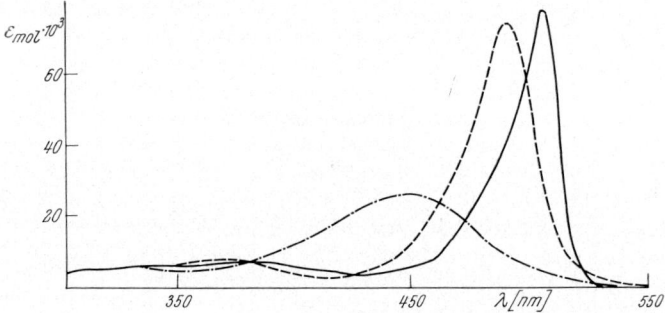

Abb. 1. Elektronenspektren von Mesobilen-(b)-IX α (12) als Dimethylester in Methanol: freie Base (— — — —), Hydrochlorid (— · — · —), Zink-Komplex (———) (nach (156))

Beim Übergang von der Base zum Hydrochlorid (bzw. zum Metall-Komplex) findet man außer der bathochromen Verschiebung eine Erhöhung der molaren Extinktion (Abb. 1). Diese macht man sich zur Bestimmung der zugehörigen pK_a-Werte mit Hilfe der spektrophotometrischen pH-Titration zunutze (76). Dazu mißt man die Absorptionskurven bei verschiedenen pH-Werten und trägt die Extinktionswerte bei einer bestimmten Wellenlänge (z. B. bei einem Absorptionsmaximum) oder besser das Verhältnis der Extinktionen bei zwei Wellenlängen (177) in Abhängigkeit vom pH-Wert auf; der pK_a-Wert des Hydrochlorids läßt sich dann graphisch ermitteln (Abb. 2). Die Hydrochloride werden mit zunehmender Ausdehnung des Chromophors stärkere Säuren (pK_a-Werte für Mesobilen-(b)-IXα: 7,4; Mesobiladien-(a,b)-IXα: 4,0; Mesobilatrien-IXα: zirka 3,0), d. h. sie geben das zusätzliche Proton zunehmend leichter ab, was durch eine stärker werdende Delokalisierung des bindenden Elektronenpaars erklärt wurde (76).

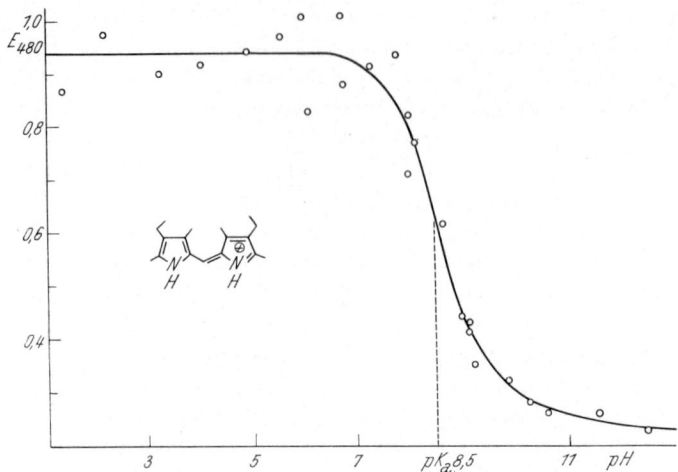

Abb. 2. Spektrophotometrische pH-Titration von 3, 5, 3', 5'-Tetramethyl-4, 4'-diäthylpyrromethen (W. RÜDIGER und W. KLOSE, unveröffentlicht)

2. Optische Aktivität

Asymmetriezentren sind bei Gallenfarbstoffen bisher nur in den Endringen, nicht jedoch in den mittleren (Pyrrol-)Ringen gefunden worden. Optisch aktive Seitenketten, wie man sie in Porphyrinen kennt (z. B. Porphyrine a und c), wurden bei Bilenen bisher nicht beschrieben.

Bilane ((5) in Formelübersicht 1) besitzen zwei *α-ständige* asymmetrische C-Atome in den Positionen 2' und 7', die bei der Oxydation zu Bilenen-(b) erhalten bleiben. Die entsprechenden Biladiene-(a,b) besitzen noch eines, Biladiene-(a,c) keines dieser Asymmetriezentren mehr. *1,2,7,8-Tetrahydrobilane* (Struktur (5c)) enthalten zusätzliche *β-ständige* asymmetrische C-Atome in den Positionen 1,2,7 und 8, die man nicht nur in den entsprechenden Bilenen-(b), sondern auch noch in den davon abgeleiteten Biladienen und Bilatrienen wiederfinden sollte. Derartige Biladiene und Bilatriene sind als Naturprodukte noch nicht aufgefunden worden; durch chemische Umwandlung aus Naturstoffen erhaltene Biladiene-(a,b) vom 1,2- bzw. 7,8-Dihydro-Typ (Mesobilirhodin, vgl. S. 102); stabiles Mesobiliviolin, vgl. S. 101) sind entweder racemisch oder in bezug auf die optische Aktivität noch nicht untersucht worden. Jedoch kennt

Literaturverzeichnis: SS. 128—139

man die optische Aktivität von einem Biladien-(a,b) (Aplysioviolin, vgl. S. 112) und einem Bilatrien (Phycobiliverdin, vgl. S. 118) mit einem Asymmetriezentrum in Position 1.

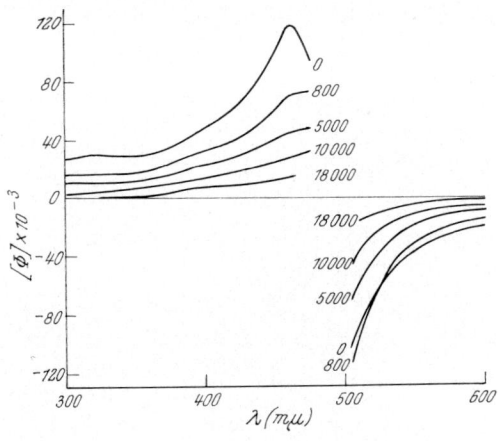

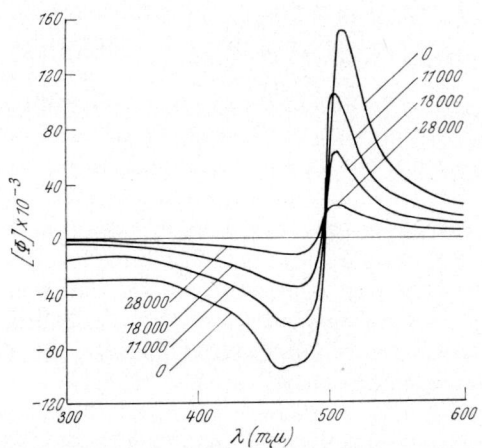

Abb. 3. ORD-Spektren von Bilenen-(b) in $CHCl_3$; die Zahlen geben Mol Trichloressigsäure pro Mol Bilen an. Oben: Stercobilinhydrochlorid; unten: d-Urobilinhydrochlorid (nach (128))

Am längsten bekannt (56) ist die optische Aktivität des *Stercobilins*, eines Bilens-(b) vom 1,2,7,8-Tetrahydro-Typ. Die Drehung des Hydrochlorids beträgt $(\alpha)_D^{CHCl_3} = -4000°$ (232). Die optische Aktivität wurde zunächst auf die durch die vier zusätzlichen H-Atome hervorgerufenen (also β-ständigen (77, 170)) Asymmetriezentrum zurückgeführt, da *Uro*-

bilin, das entsprechende, an den Endringen nicht hydrierte Bilen-(b), nur in optisch inaktiver Form bekannt war (*48, 55, 56, 60*). Erst nach dem Auffinden von optisch aktiven d-Urobilin (*199*) (Hydrochlorid: $(\alpha)_D^{CHCl_3} = +5000°$ (*77, 231*) wurde der Einfluß der α-ständigen asymmetrischen C-Atome auf die optische Aktivität der Gallenfarbstoffe als wesentlich, der der α-ständigen als unwesentlich erkannt (vgl. *34*). Die entgegengesetzten Vorzeichen der (gleich großen) Drehwerte machten ferner wahrscheinlich, daß das Stercobilin die entgegengesetzte Chiralität von d-Urobilin besitzt; das wurde durch Messung der ORD-Kurven (*73, 128*) bestätigt, da diese sich (fast) wie ein Bild und Spiegelbild zueinander verhalten (Abb. 3). Die hohen Cotton-Effekte weisen auf einen inhärent dissymmetrischen (*127, 129*) („verdrillten") Dipyrromethen-Chromophor in beiden Farbstoffen hin; da Substanzen, welche Wasserstoff-Brücken lösen (Methanol, Trifluoressigsäure, Trichloressigsäure) die Drehwerte erniedrigen, wurde auf eine durch intramolekulare Wasserstoffbrücken stabilisierte Helix-Struktur geschlossen (*128*). Als Alternative wurde eine Verdrillung auf Grund sterischer Hinderung diskutiert (*34*). Auch durch die Bildung des Zink-Komplexes werden die Drehwerte erniedrigt (*35, 128*). Aus Betrachtungen an einem Atommodell mit Stuart-Briegleb-Kalotten folgt, daß eine Helix sich nur bei gleicher Konfiguration der Asymmetriezentren 2' und 7' in einem Farbstoff ausbilden kann; der Drehsinn der Helix des 2'R—7'R'-Pigments ist entgegengesetzt demjenigen des 2'S—7'S-Pigments (Bezeichnung der Asymmetriezentren nach (*24*)). Bei Bilanen gibt es keinen „verdrillten" Chromophor, da an der mittleren Methylen-Brücke freie Drehbarkeit besteht. Tatsächlich sind die Drehwerte der durch Reduktion der optisch aktiven Bilene erhältlichen Bilane erheblich niedriger als die der Ausgangspigmente (*55, 56, 121*); durch Reoxydation wird die volle optische Aktivität wiederhergestellt (vgl. *58*, S. 696).

Von besonderem theoretischen Interesse ist der hohe Drehwert des Bilirubin-Serumalbumin-Komplexes, da das freie Bilirubin kein Asymmetriezentrum besitzt und daher optisch inaktiv ist. Der Einfluß des Proteins wird auf S. 91 diskutiert.

Durch Vergleich mit synthetisch erhaltenen Bilenen-(b) und Tetrahydrobilenen-(b) bekannter Konfiguration (vgl. S. 98) sind das natürliche Stercobilin als 1 S, 2 S, 2' S, 7 S, 7' S, 8 S-1,2,7,8-Tetrahydrobilen-(b), das d-Urobilin als 2'R,7'R-Bilen-(b) erkannt worden (*165*). Das natürliche, optisch inaktive Urobilin (untersucht im Fall Ori (*230*)) ist ein Racemat aus den 2'R,7'R- und 2'S,7'S-Formen (und keine „meso"-Form 2'R,7'S bzw. 2'S,7'R), da es sich durch fraktionierte Kristallisation partiell in die beiden optisch aktiven Komponenten trennen läßt (*230*). In den meisten der daraufhin untersuchten Fällen liegt jedoch kein echtes Racemat vor, da die linksdrehende Komponente ein 1,2-Dihydrobilen-(b) ist (230). Warum für das Hydrochlorid dieser Komponente, die die 2' S,7' S-Konfiguration besitzen sollte, nur Drehwerte bis $(\alpha)_D^{CHCl_3} = -3000°$ gefunden werden, ist bisher nicht bekannt; wahrscheinlich waren die Präparate mit rechtsdrehendem d-Urobilin verunreinigt. In einem 1,2-Dihydrobilen-(b), das durch katalytische Reduktion und Reoxydation eines natürlichen Biladiens-(a,b) erhalten worden war, wurde der niedrige Drehwert (Hydrochlorid: $(\alpha)_D^{CHCl_3} = +2470°$) durch Racemisierung bei der chemischen Reaktion an C-2' erklärt (36); C-7' muß R-Konfiguration besitzen. Das aus demselben Biliproteid stammende Aplysioviolin (vgl. S. 111) ist nach seinem ORD-Spektrum (Abb. 9 S. 114, positive Cotton-Effekt-Kurve beim Hauptabsorptionsmaximum) ein 7' R-Biladien-(a,b). Der Einfluß des β-ständigen Asymmetriezentrums (an C-1) auf die optische Aktivität ist hier noch unbekannt; für das aus einem verwandten Biliproteid erhaltene Bilatrien Phycobiliverdin, das nur ein Asymmetriezentrum an C-1 besitzt, wurde ein negativer Cotton-Effekt beim langwelligen Maximum (615 nm) und eine geringe optische Aktivität ($(\alpha)_D^{CHCl_3} = 660°$) gefunden (33). Ein aus Aplysioviolin erhaltenes Abbauprodukt, welches noch das Asymmetriezentrum an C-1 besitzt, zeigt nur eine geringe optische Drehung (Abb. 9, S. 114).

3. Massenspektren

Massenspektren wurden bisher durchweg nur zur Untersuchung oder Identifizierung einzelner Gallenfarbstoffe herangezogen (33, 90, 118, 188, 214). Aus den bisherigen Daten kann man aber die charakteristische Fragmentierung dieser Pigmente bereits ableiten, wenn man (genauer untersuchte) zweikernige Modellverbindungen mit heranzieht.

Bei *Bilatrienen* ist der Molekül-Peak vorherrschend (meist der „base peak"); man beobachtet in geeigneten Fällen die ausgeprägte Fragmentierung von Seitenketten (z. B. Propionsäure-Gruppen (vgl. Abb. 4)). Das konjugierte System der durch drei Methin-Brücken miteinander verknüpften Pyrrol- und Pyrrolon-Ringe ist also ein stabiles System. Eine gewisse Fragmentierung an der mittleren Brücke (14, 15, 33, 89, 184, 214) dürfte auf deren Reduktion zu einer Methylen-Brücke im Verlauf einer Disproportionierung* unter den Meßbedingungen beruhen, da die Massenzahlen der entstehenden Fragmente mit denen der entsprechenden Biladiene-(a,c) identisch sind (135, 214, vgl. aber 89). Bilatriene lassen sich chemisch leicht zu Biladienen-(a.c) reduzieren (59); Modellversuche mit einem Bilatrien weisen auf eine solche Reaktion auch unter den Meßbedingungen der Massenspektrometrie hin (46).

---

* Bei Bilatrienen treten daneben um zwei Masseneinheiten ärmere Fragmente auf (14).

Gallenfarbstoffe mit Methylen-Brücken unterliegen einer Fragmentierung an diesen Brücken. Mechanismen für diese Reaktion sind für Dipyrromethane angegeben worden (90); sie seien an einem Beispiel erläutert (Formelübersicht 3): Aus dem Dipyrromethan (33) können einerseits unter Wanderung eines Protons die Ionen (34) und (35a) entstehen. Die Richtung dieser Umlagerung wird durch mehrere Faktoren kontrolliert. Einmal erleichtert die den Methylen-Brücken benachbarte Äthylgruppe die Protonenübertragung; bei der entsprechenden Verbindung mit

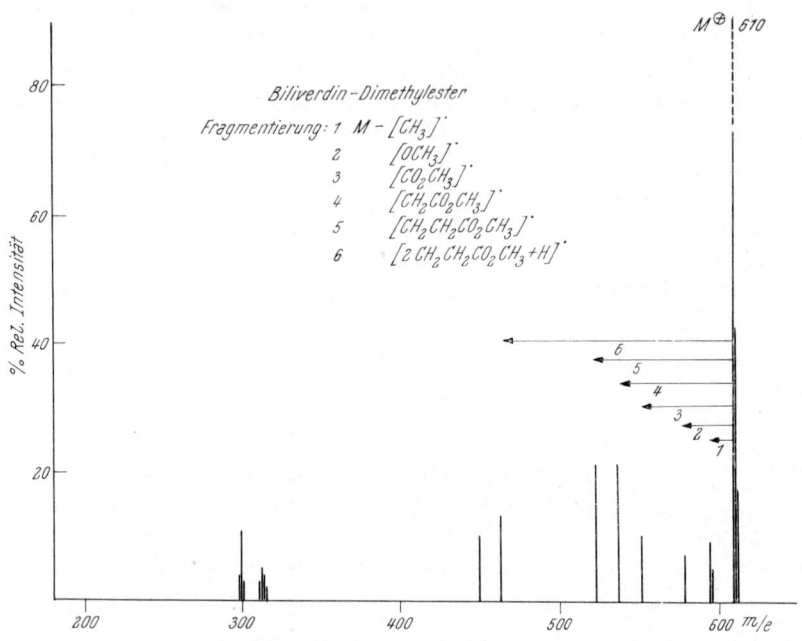

Abb. 4. Massenspektrum von Protobilatrien-IX α-dimethylester (184)

Methylgruppe ist die Reaktion weniger ausgeprägt. Weiterhin stabilisiert die kernständige Carbonylgruppe das entstehende Ion. Die entsprechende Spaltung, bei der die Methylengruppe an Ring B verbleibt (Fragmente (34a) und (35)), ist daher viel weniger ausgeprägt. Anderseits beobachtet man eine Fragmentierung ohne Umlagerung (Reaktion 3 und 4), bei der die Fragmente (36)—(37a) entstehen können. Im Fall von (33) ist nur (37a) ausgeprägt; die anderen Ionen sind aber bei analogen Dipyrromethanen beobachtet worden.

Bei *Biladienen-(a,c)* werden zwei Hauptfragmente gefunden, die von der Spaltung der mittleren (Methylen-)Brücke herrühren (Abb. 5). Im Falle der Dicarbonsäure wurden die Fragmentierungsreaktionen 2 und 4, nicht jedoch 1 und 3 beobachtet (90); bei gleichen β-Substituenten

Literaturverzeichnis: SS. 128—139

Formelübersicht 3

in beiden Molekül-Hälften beträgt die Massendifferenz beider Hauptfragmente 13 (entspr. (35) und (37)) (*90*). Im Falle der Dimethylester wurde eine Massendifferenz von 14 gefunden (*135*), was bei gleicher Entstehungsart beider Fragmente auch zu erwarten ist (Reaktionen 1 und 2, bzw. 3 und 4). Wenn die beiden Molekülhälften unsymmetrisch substituiert sind, ist der Unterschied in den Massenzahlen der beiden Hauptfragmente entsprechend größer; diese Tatsache hat eine Rolle bei Modellversuchen zur Biogenese der Gallenfarbstoffe gespielt (vgl. *135*, *218*).

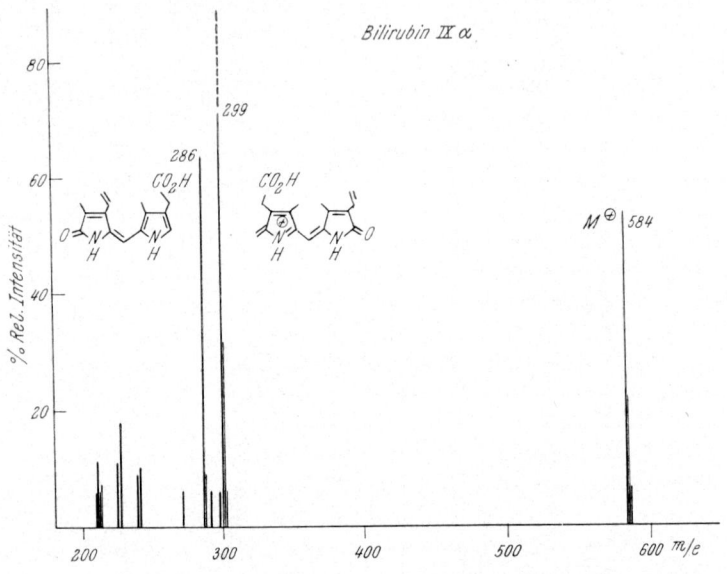

Abb. 5. Massenspektrum von Protobiladien-(a, c)-IX α (*90*, *184*)

Bei *Biladienen-(a,b)* ergibt die Fragmentierung an der Methylen-Brücke ein einkerniges und ein dreikerniges Spaltstück. Die Methylen-Gruppe bleibt bei dem Tripyrrol-Fragment (Entstehung nach Reaktion 4), welches meist der „base-peak" im Massenspektrum von Biladienen-(a,b) ist (Abb. 6). Bei 7,8-Dihydrobiladienen (stabiles Mesobiliviolin vgl. S. 101) ist die Fragmentierungsreaktion 1 vorherrschend (*214*).

Die Massenpeaks sind bei Biladienen-(a,b) nur wenig ausgeprägt, außerdem sind stets Satelliten mit den Massenzahlen $M + 2$ und $M - 2$ zugegen, die wahrscheinlich durch Redox-Disproportionierung des Farbstoffs im Massenspektrometer entstehen (vgl. *118*). Das ist bei den *Bilenen-(b)* in noch höherem Maße der Fall (Abb. 6); daher ist es manchmal schwierig, bei diesen Verbindungen das wahre Molekulargewicht zu bestimmen (*78*, *118*, *172*). Erfolgversprechende Ansätze finden sich

Literaturverzeichnis: SS. 128—139

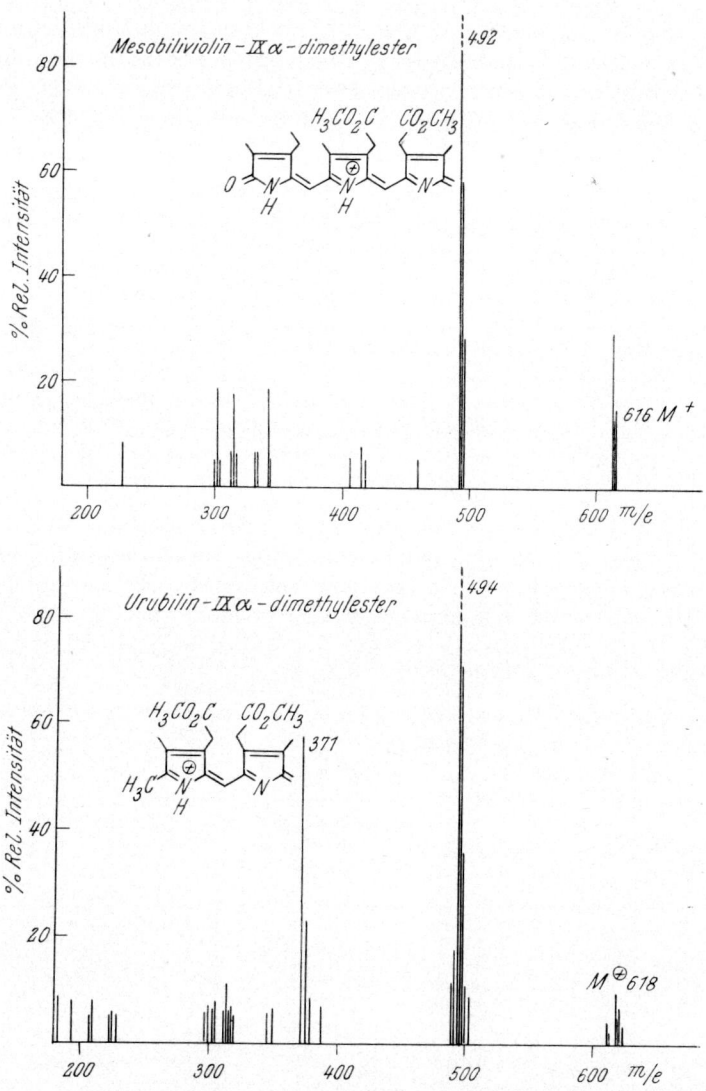

Abb. 6. Massenspektren von Mesobiladien-(a, b)-IXα-dimethylester (oben) und Mesobilen-(b)-IX α-dimethylester (unten) (nach (*214*))

bei der zeitabhängigen Intensitätsmessung der Massenverteilung (*118*). Die Redoxdisproportionierung ist wahrscheinlich auf thermische Einwirkung zurückzuführen; Erhitzen i. Vak. ergibt dieselbe Reaktion. (*118*, *233*).

Die Fragmentierung bei *Bilenen-(b)* verläuft ähnlich wie bei Biladienen-(a,b); der „base-peak" ist das Tripyrrol-Fragment mit der Methylen-Gruppe (Fragmentierungsreaktion 4). Die Fragmentierung an der zweiten Methylen-Brücke ist weniger ausgeprägt; das erwartete Fragment, das sich von den beiden mittleren Ringen ableitet und früher nicht gefunden wurde (*90*), wurde erst kürzlich beschrieben (*214*); seine Struktur wurde durch die genaue Massebestimmung gesichert (*38*).

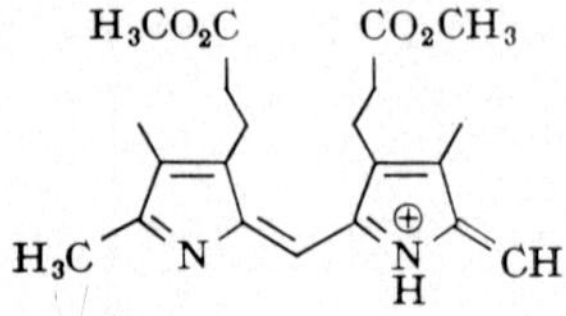

(38) *: m/e 371

* Hier als Immonium- statt Carbenium-Ion formuliert

4. NMR-Spektren

Mit Hilfe der NMR-Spektroskopie wurden die Strukturen der Phycobiline (unabhängig von der chemischen Strukturaufklärung) aufgeklärt. Bei einigen Gallenfarbstoffen erlaubte die NMR-Spektroskopie detaillierte Strukturaussagen, die mit anderen Methoden nicht zu erhalten waren. In jedem Fall erleichterte die Kenntnis der NMR-Spektren von Gallenfarbstoffen gesicherter Struktur die Zuordnung der Signale bei den unbekannten Pigmenten.

Tabelle 5. *NMR-Spektren von Gallenfarbstoffen als Dimethylester**

Phycobiliverdin (*32*)	Phycobiliverdin** (*43, 46*)		Phycobilivolin (*26*)	Zuordnung der Protonen
Deuteropyridin	Deuteropyridin	CF_3CO_2D	$CDCl_3$	
1,10	1,11	0,99		$-CH_3$ Äthyl
1,84—1,96	1,89—2,01	2,02—2,06	1,96—2,01	$-CH_3$ ungesättigter Ring
1,33	1,34	1,37	1,41	$-CH_3$ gesättigter Ring
1,59	1,58	1,84	1,89	$-CH_3$ Äthyliden
2,34	2,34	2,32	2,26	$-CH_2-$ Äthyl
2,51—2,91	2,70—2,97	2,65—3,04	2,51—2,84	$-CH_2-$ Propionsäure
3,19	3,2	3,43	3,09	$-C-H$ gesättigter Ring
			3,24	$-C-H$ angular (C-2', C-7')
3,46—3,48	3,17		3,61—3,63	$-OCH_3$ Ester
			4,32	$-CH_2-$ Methylen-Brücke
			5,33	$=CH_2$ Vinyl
			6,25	$=CH-$ Vinyl
6,18	6,17	6,62	6,40	$C=CH-$ Äthyliden
5,70—6,84	5,71—7,09	5,97—7,37	5,80—6,64	$=CH-$ Methin-Brücke
8,67	11,9			NH

Literaturverzeichnis: SS. 128—139

Tabelle 5 (Fortsetzung)

Mesobilatrien		Mesobiladien-(a,b)		Zuordnung der Protonen
(32) Deuteropyridin	(214) CDCl$_3$	(214) CDCl$_3$	(102) CDCl$_3$	
1,05—1,07	1,03—1,18	0,78—1,17	0,76—1,20	—CH$_3$ Äthyl
1,80—1,88	1,80—2,05	1,66—2,05	1,94—2,12	—CH$_3$ ungesättigter Ring
2,26			2,16—2,8	—CH$_2$— Äthyl
2,51—2,84	2,5—2,9	2,45—2,88	2,16—3,3	—CH$_2$— Propionsäure
			4,50	—C—H angular (C-7')
3,48	3,59—3,63	3,59—3,63	3,66	—OCH$_3$ Ester
			3,66	—CH$_2$— Methylen-Brücke
5,71—6,93	5,8—6,7	5,9—6,73	6,0—7,46	=CH— Methin-Brücke
7,80			7,5—12,7	NH

Protobilatrien	(135)	Protobiladien-(a,c)	(134)	Zuordnung der Protonen
CF$_3$CO$_2$D	CDCl$_3$	(CD$_3$)$_2$SO**	CDCl$_3$	
1,67—2,49	1,80—2,16	1,93—2,17	1,72—2,08	—CH$_3$ ungesättigter Ring
3,41—3,94	2,86—2,53	1,9—2,2	2,47—2,76	—CH$_2$— Propionsäure
3,88	3,67		3,65	—OCH$_3$ Ester
		4,02	4,13	—CH$_2$— Methylen-Brücke
		5,61—6,42	5,28—6,06	=CH$_2$ Vinyl
5,52—6,62***	5,75—7,12***	6,83	6,47	=CH— Vinyl
		5,57—6,11	5,54—5,89	=CH— Methin-Brücke
		9,88—10,01	10,14—11,11	NH

* δ-Werte (ppm) bezogen auf Tetramethylsilan bzw. Hexamethylsiloxan.
** Dicarbonsäure.
*** Nicht näher zugeordnet.

Bisher bekannte NMR-Daten von Gallenfarbstoffen sind in Tabelle 5 zusammengefaßt. Eine entsprechende graphische Übersicht, die überwiegend die Phycobiline berücksichtigt, findet sich bei (209). Einige Punkte, die für die Strukturfragen wichtig sind, seien hier diskutiert.

Die chemische Verschiebung von β-ständigen Methylgruppen am ungesättigten (Pyrrol- bzw. Pyrrolon-)Ring kann in zwei verschiedenen Bereichen liegen: In Nachbarschaft zum Pyrrolon-Sauerstoff („exo"-Stellung bei Gallenfarbstoffen, Formel (39a)) findet man δ = 1,7 bis 1,9 ppm, in anderen Positionen („endo"-Stellung bei Gallenfarbstoffen, Formel 39b und c) δ = 2,1—2,3 ppm (14, 15). Die entsprechenden Signale vom Dimethylpyrrolon liegen (in CCl$_4$) bei δ = 1,87 und 2,13 ppm (166). Isomere Gallenfarbstoffe mit unterschiedlicher Stellung der

Methyl-Gruppen an den äußeren Ringen können so voneinander unterschieden werden (*14, 15, 214*). Entsprechendes gilt auch für andere Signale, z. B. der Protonen von Vinylgruppen (*134*). Die Signale der mittleren (b-)Methin-Brücke liegen bei tieferem Feld ($\delta = 5{,}7-5{,}9$ ppm). Das gilt sowohl für Bilatriene als auch für Biladiene-(a,b) (*214*).

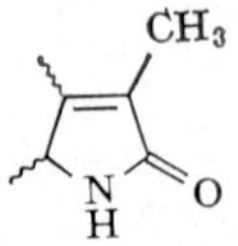

(39a) 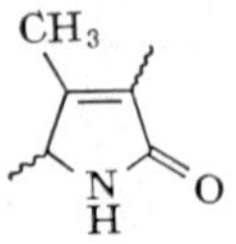   (39b) 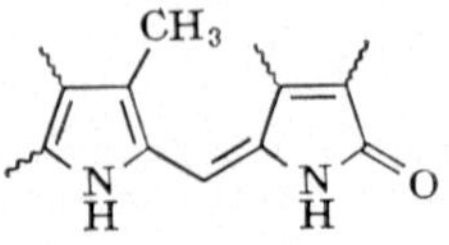   (39c)

Auf die Spin-Spin-Kopplung mit den Protonen an den benachbarten C-Atomen soll hier nicht eingegangen werden, auch wenn sich aus der Aufspaltung der Signale erst die eindeutige Zuordnung der Protonen und damit die Strukturen der Gallenfarbstoffe ergeben.

Auch über die Bindungsverhältnisse geben NMR-Messungen Auskunft: Austauschversuche (mit siedendem, neutralem Methanol und Äthanol) ergaben, daß zumindest in Phycocyanobilin C-gebundene Protonen leicht austauschbar sind; es sind dies die Protonen der c-Methin-Brücke, der Äthyliden-Methylgruppe und weniger ausgeprägt in Kern-Position 1 (*46*). Im Biliproteid Phycocyanin wurden Protein-Protein-Wechselwirkungen mit Hilfe der NMR-Spektroskopie untersucht. Dazu wurden Isotopen-Hybride benutzt, die durch Einbau von ^{1}H-Alanin, ^{1}H-Leucin bzw. ^{1}H-Methionin in das vollständig deuterierte Phycocyanin gewonnen wurden (*44*).

5. Chromatographie

Gallenfarbstoffe sind durch *Säulenchromatographie* an Talkum (*204*) oder desaktiviertem Aluminiumoxid (*220, 221, 222, 235*) gereinigt worden, jedoch werden sehr ähnliche Pigmente dadurch nicht getrennt. Auch durch *Papierchromatographie* lassen sich z. B. isomere Bilatriene nicht voneinander trennen (*135*), während das mit Hilfe der *Dünnschichtchromatographie* gelingt (*14, 141, 176*).

Die natürlich vorkommenden Gallenfarbstoffe besitzen zwei Propionsäure-Seitenketten, die frei oder verestert vorliegen können. Durch Dünnschichtchromatographie gelingt die Trennung der freien Dicarbonsäuren von den Mono- und Dimethylestern leicht (Tabelle 6). Das Auftreten dieser drei Produkte bei der partiellen Veresterung wurde zum Nachweis von zwei Säuregruppen bei Gallenfarbstoffen herangezogen (*212*).

Literaturverzeichnis: SS. 128—139

Tabelle 6. R_F-Werte der Dicarbonsäuren, Monomethylester und Dimethylester von Gallenfarbstoffen

Fließmittel	Isopropanol/ CHCl$_3$/NH$_3$ (4 : 2 : 1)	2,6-Lutidin/ Wasser (8 : 2) NH$_3$-Atmosphäre	Äthylacetat/Wasser (10 : 1)
Träger	Kieselgel G	Adsorbosil 1	
Literatur	(*177*)	(*212*)	(*212*)
Dicarbonsäuren	0,1—0,2	0,03	0,22
Monomethylester	0,4—0,6	0,25	0,50
Dimethylester	0,8—0,9	0,98	0,90

Tabelle 7. R_F-Werte von Gallenfarbstoffen bei der Dünnschichtchromatographie

	Dimethylester							Dicarbonsäuren	
Träger	Kieselgel G				Adsorbosil			Polyamid ITLC	
Fließmittel	A	B	C	D	E	F	G	H	I
Literatur	(*174, 177, 194*)			(*130*)	(*212*)			(*164*)	
Bilene-(b)									
Stercobilin				22				80	
Urobilin	11;	13	32	00	25			80	96
Biladiene-(a,b)									
Mesobiliviolin		19	33	08	35	42	24	05; 20	100
Mesobilirhodin		22	24	08					
Aplysioviolin		24	24	20					
Phycoerythrobilin		20	27	14					
Biladiene-(a,c)									
Bilirubin				36				00	06
Mesobilirubin				40				00	75
Bilatriene									
Biliverdin		30	12	43	70	37	54	50	10
Mesobiliverdin		28	20	40	75	48	49	65	40	98
Phycobiliverdin		35	15	38		52	35	67	
Phytochromobilin						61	42	72	
Biladien-(a,b)-one-(c)									
Bilipurpurin		40	60	55	62				
Mesobilipurpurin		36	56	52	60				

Fließmittel A: Benzol/Benzin/Methanol (9 : 5 : 1),
   B: CCl$_4$/Eisessig (1 : 1),
   C: CCl$_4$/Äthylacetat (1 : 1),
   D: Benzol/Äthanol (100 : 8),
   E: Benzol/Äthanol (100 : 2,5),
   F: CCl$_4$/Eisessig (100 : 25),
   G: CCl$_4$/Methylacetat (10 : 4),
   H: Methanol/Wasser (3 : 1),
   I: Methanol/10proz. NH$_3$/Wasser (9 : 1 : 2).

Tabelle 8. *Weitere Fließmittel für die Dünnschichtchromatographie von Gallenfarbstoffen als Dimethylester auf Kieselgel G*

Fließmittel	Trennung von	Lit.
Äthylendichlorid/Äthylacetat (7 : 3)	Phycobiline	*(26)*
CCl$_4$/Dichlormethan/Methylacetat/Methylpropionat (1 : 1 : 1 : 1)**	isomere Mesobiladiene und Mesobilatriene*	*(214)*
Äthylmethylketon/Äthylendichlorid (1 : 2)		
Benzol/Äthanol (25 : 2)***	isomere Bilene-(b)*	*(214)*
Benzol/Benzin/Methanol/Äthylacetat (48,5 : 40 : 10,5 : 9)	isomere Bilene-(b) und Biladiene-(a,b)*	*(156)*
Chloroform/Aceton (97 : 3) bzw. (95 : 5)		*(14, 15)*
n-Heptan/Äthylmethylketon/Eisessig (10 : 5 : 1)	isomere Bilatriene*	*(141)*
Benzol/Benzin/Methanol (60 : 5 : 3)		*(176)*
Benzol/Dioxan/Eisessig (12 : 2 : 1)		*(189)*

* Mehrfachentwicklung.
** Kieselgel D 5 (Camag).
*** Kieselgel/Kieselguhr (3 : 1).

Die Trennung verschiedener Gallenfarbstoffe voneinander in Form ihrer Dicarbonsäuren ist bisher nur für einige Beispiele beschrieben (Tabelle 7). Am häufigsten werden die Pigmente als Dimethylester getrennt (Tabelle 7 und Tabelle 8). Diäthyl- und Dipropylester, die in unpolaren Fließmitteln weiter wandern, bieten keine Vorteile *(214)*. In neutralen Fließmitteln (A, C, D, E, G in Tabelle 7) steigen die R_F-Werte mit der Oxydationsstufe, d. h. in der Reihenfolge Bilane — Bilene — Biladiene — Bilatriene — Biladienone. Die umgekehrte Reihenfolge (bis auf die Biladienone) gilt für säurehaltige Fließmittel (B, F). Sehr ähnliche (z. B. isomere) Gallenfarbstoffe werden am besten durch Mehrfachentwicklung *(213)* getrennt (in Tabelle 8 angegeben), wobei die optimale Zahl der „Läufe" aus den R_F-Werten beim ersten Lauf berechnet werden kann *(173)*. So wurden die durch Oxydation von Mesobilan-IXα hergestellten Bilene-(b) und Bilatriene in je drei Fraktionen *(156, 214)*, die Biladiene-(a,b) in sieben Fraktionen *(214)* getrennt.

Auch zur *präparativen Trennung* von Gallenfarbstoffen wird heute die Dünnschichtchromatographie häufig herangezogen *(14, 15, 26, 33, 188, 214)*. Zu beachten ist, daß die Trennwirkung abnimmt, wenn eine Trägerschicht größerer Dicke gewählt wird.

Der Nachweis der Gallenfarbstoffe ist leicht, da die Substanzen eine intensive Eigenfarbe besitzen. Zur Charakterisierung können die unter III,1 beschriebenen Farbreaktionen herangezogen werden. Auch der Chromat- oder Chromsäureabbau auf der Platte dient zur Identifizierung der Pigmente *(178)*.

Literaturverzeichnis: SS. 128—139

V. Bilirubin

Bilirubin ist als der am längsten bekannte Gallenfarbstoff recht genau untersucht worden. In neuerer Zeit sind einige interessante Gesichtspunkte zu Strukturproblemen von Bilirubin-Derivaten aufgetaucht.

Bei der ersten Totalsynthese des Bilirubins (59) wurden die Bausteine für die Endringe durch Oxydation des entsprechenden Pyrrols, der Opsopyrrolcarbonsäure (40) mit H_2O_2 gewonnen; die entstandenen Produkte wurden als Hydroxypyrrole angesehen (41a) und (42a). Erst als derartige Substanzen durch eine neue Ringsynthese leicht zugänglich wurden (167, 169), wurden sie als tautomere Pyrrolone (41b) und (42b) erkannt (vgl. 166); dementsprechend wurden auch die Endringe der Gallenfarbstoffe, die früher als Hydroxypyrrol-Ringe formuliert worden waren, jetzt als Pyrrolon-Ringe formuliert.

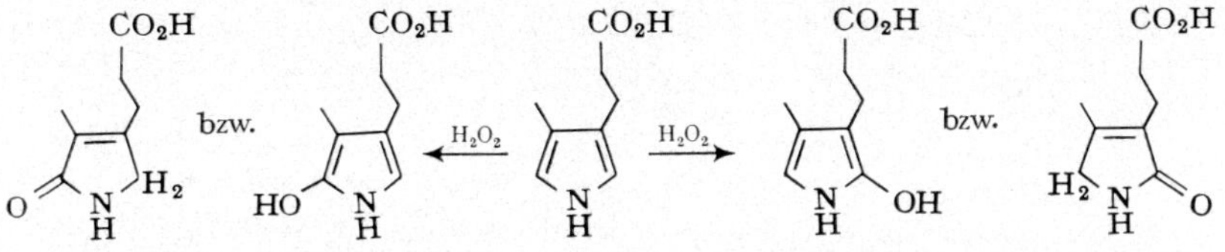

Die Pyrrolon-Struktur ist für die Gallenfarbstoffe inzwischen allgemein akzeptiert, wenn auch nicht in jedem Fall bewiesen worden. Nach NICHOL und MORELL (134) weisen die Elektronen-, IR- und NMR-Spektren von Bilirubin (im Gegensatz zu seinem Dimethylester sowie zu Biliverdin) auf eine Hydroxypyrrol-Struktur hin, obwohl die Meßdaten auch anders interpretiert werden können.

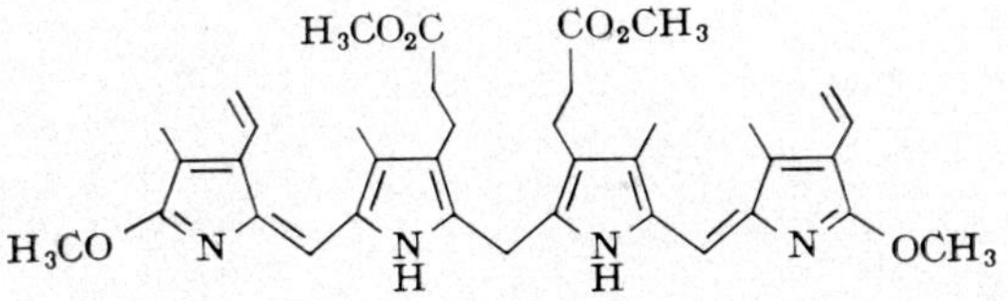

So wurde eine breite Bande bei 3230 cm^{-1} im IR-Spektrum des Bilirubins als NH-Schwingung angesehen, die auch beim Dimethylesterdimethyläther (43) und bei Dipyrromethenen und intramolekularen H-Brücken an derselben Stelle auftritt. Dagegen zeigt Bilirubindimethylester eine starke Absorption bei 3350 cm^{-1}, die dem Amid- und Pyrrol-NH zugeordnet wird. Eine zusätzliche scharfe Bande des Bilirubins bei 3405 cm^{-1} wird als OH-Bande der Wasserstoff-verbrückten Carboxylgruppen interpretiert, wohingegen diese Bande früher als Pyrrolon-NH-

Schwingung angesehen wurde (60); ihre (gegenüber normalen Pyrrolonen) verschobene Lage wurde als Nachweis einer Betain-Struktur des Bilirubins gewertet (48). Im NMR-Spektrum des Bilirubins (134) werden von den vier bei niedrigem Feld gefundenen Protonen zwei (bei $\delta =$ = 10,45 ppm) den OH-Gruppen der Propionsäure und zwei (bei $\delta =$ = 9,88 und 10,01 ppm) den NH-Gruppen zugeordnet; ein breites Signal bei $\delta = 5{,}25$ ppm soll den enolischen OH-Protonen (der Hydroxypyrrole) entsprechen. Beim Bilirubindimethylester werden dagegen die Signale bei $\delta = 10{,}14$ ppm den Pyrrol-, bei 10,45 und 11,11 ppm den Pyrrolon-NH-Protonen zugeordnet.

NICHOL und MORELL (134) schlagen eine Struktur für das Bilirubin vor, die ihren Meßdaten am besten gerecht wird (44). Eine andere Struktur mit Wasserstoffbrücken-Bindungen zeigt Formel (44a) (21, 60). In beiden Fällen muß noch berücksichtigt werden, daß zumindest Bilirubin-Anionen in Lösung als Dimere vorliegen (20).

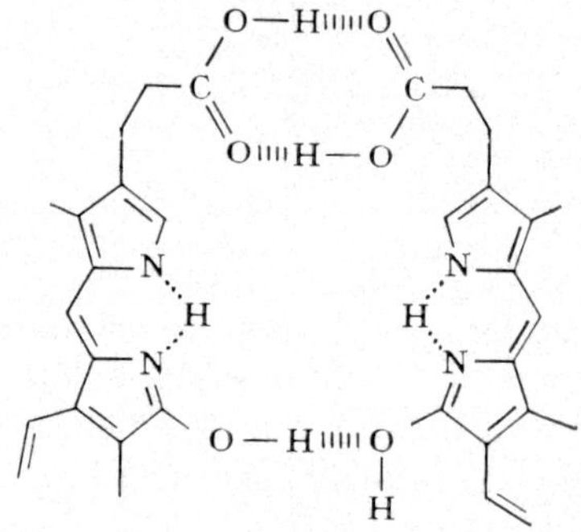

(44)

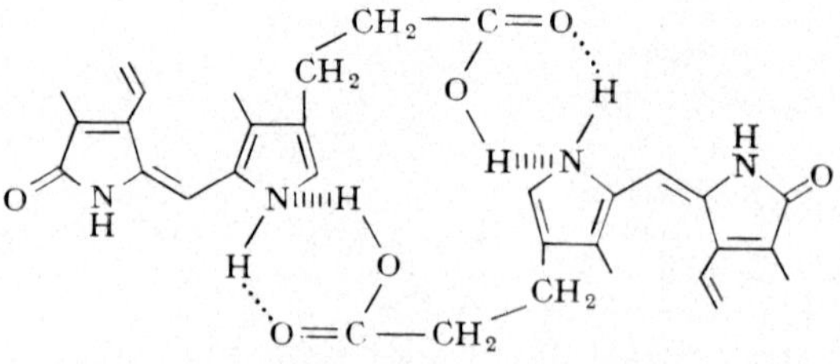

(44a)

1. Bilirubin-Konjugate

Bilirubin ist in neutralem, wäßrigem Medium schwerlöslich. Als Transportform (z. B. im Blut) kommen neben dem Protein-gebundenen Bilirubin (vgl. nächster Abschnitt) wasserlösliche Konjugate vor, die

Literaturverzeichnis: SS. 128—139

ihrerseits an Protein gebunden vorliegen können. COLE, LATHE und BILLING (*31*) isolierten zwei Konjugat-Fraktionen, die in der Folgezeit als Mono- und Diglucuronid des Bilirubins erkannt wurden (*9, 197, 215*) (*45*) und (*46*).

Abgesehen von der Analyse, die 1 Mol Glucuronsäure/Mol Bilirubin bei Fraktion I (Monoglucuronid) und 2 Mol Glucuronsäure/Mol Bilirubin bei Fraktion II (Diglucuronid) ergab, zeigte die Umsetzung mit Diazoniumsalzen, daß aus Fraktion II ausschließlich Azopigment B (*47*), aus Fraktion I daneben Azopigment A (*48*) entsteht. Azopigment A, welches auch durch Diazo-Reaktion aus freiem Bilirubin entsteht, bildet sich bei der Einwirkung von verdünntem Alkali oder β-Glucoronidase auf Azopigment B. Bei derselben Behandlung der Konjugate gehen diese in freies Bilirubin über. Die Labilität gegenüber Alkali und die Spaltbarkeit mit Hydroxylamin (*196*) deutet auf das Vorliegen von Acylglucuroniden hin. Die Zusammenhänge sind in Schema 4 veranschaulicht.

Das Monoglucuronid wird von einigen Autoren als 1:1-Komplex aus Diglucuronid und freiem Bilirubin (*25*), von anderen als ein chemisches Individuum (*45*) angesehen (*94*). Durch Dünnschichtchromatographie können die Azopigmente (mit diazotiertem Anthranilsäureester bzw. Anilin gewonnen) in eine Reihe von Fraktionen aufgetrennt werden. In der Gallenflüssigkeit des Hundes sind dabei u. a. Xylose- und Glucose-Konjugate (*52*), in menschlicher Galle eine Reihe von Disaccharid-Konjugaten (*106*) gefunden worden.

Im Jahre 1958 beschrieben ISSELBACHER und MCCARTHY (*88*) bei Ratten ein „Bilirubinsulfat". Intraperitoneal verabfolgtes $^{35}SO_4^{2-}$ wurde in einer Konjugat-Fraktion wiedergefunden, die bei Hydrolyse freies $^{35}SO_4^{2-}$ (und keine andere Schwefel-haltige Verbindung) ergab. Dieses Bilirubinsulfat verhält sich chromatographisch (*217*) wie ein aus Bilirubin und H_2SO_4/Acetanhydrid darstellbares, ebenfalls wasserlösliches „Bilirubinsulfat" (*227*). Während die Bindungsart des Sulfats beim natürlichen Konjugat unbekannt ist, müssen bei dem synthetisch erhaltenen Produkt die Vinylgruppen an der Bindung beteiligt sein: Mesobilirubin (und andere Gallenfarbstoffe, die Äthylgruppen an Stelle der Vinylgruppen besitzen) reagiert mit H_2SO_4/Acetanhydrid *nicht* zu einem wasserlöslichen Derivat (*181*).

2. Bilirubin-Proteide

Von den Bilirubin-Proteiden ist die Verbindung mit Serumalbumin am besten untersucht worden, da sie eine Transportform darstellt. Störungen bei der Bildung dieses Komplexes (wie andere Transportformen u. a. von Arzneimitteln beeinflußbar) besitzen daher klinisches Interesse (*4*).

Formelübersicht 4

Serumalbumin bindet bei Zugabe einer Bilirubin-Lösung das Pigment augenblicklich; es handelt sich um eine nichtkovalente Bindung, die z. B. durch Lösungsmittel wieder zerstört werden kann. Durch die Bindung an Albumin wird das Bilirubin gegen Oxydation (z. B. mit Peroxidase/ H_2O_2) geschützt. Das kann man zur Bestimmung von nicht-gebundenem neben gebundenem Bilirubin ausnutzen (93). So wurde eine feste Bindung von 1 Mol Pigment an 1 Mol menschliches Serumalbumin ($K_{Dissoz.}$ = = 7 · 10^{-9} M) und die schwächere Bindung von weiteren 2 Mol Pigment ($K_{Dissoz.}$ = 2 · 10^{-6} M) gefunden (93). Früher war mit verschiedenen Methoden die Bindung von 1—20 Mol Bilirubin pro Mol Serumalbumin gefunden worden (239); dabei war aber auf die Spezifität und die Stärke der Bindung nicht geachtet worden. Für die spezifische, feste Bindung von 1 Mol Bilirubin (und unspezifische, lockere Bindung von weiterem Pigment) spricht die Beobachtung, daß die durch freies Bilirubin hervorgerufene Entkopplung der oxydativen Phosphorylierung und die Schwellung von Mitochondrien durch Zusatz von Rinder- oder Menschen-Serumalbumin im Mol-Verhältnis 1 : 1 vollständig unterdrückt wird, während überschüssiges, locker an das Albumin gebundenes Bilirubin die Effekte noch zeigt (131).

Eine weitere Eigenschaft des spezifisch (Molekülverhältnis 1 : 1) an das Rinder-Serumalbumin gebundenen Bilirubins ist seine optische Aktivität (10, 11, 12). Die Amplitude im ORD-Spektrum, die der langwelligen Absorption im Elektronenspektrum des Bilirubins entspricht, ist die größte jemals im sichtbaren Bereich gemessene (1,5 · 10^6). Das ist insofern interessant, als Bilirubin selbst kein asymmetrisches Zentrum besitzt und dementsprechend keine Drehung zeigt.

Derart hohe Drehwerte wurden bisher auf inhärent dissymmetrische Chromophore (z. B. verdrillte) zurückgeführt (vgl. S. 78), und es ist sehr wahrscheinlich, daß die Dipyrromethen-Chromophore des Bilirubins durch die Bindung an das Serumalbumin ebenfalls eine Verdrillung erfahren. Das ist nur möglich, wenn jeder Chromophor (oder Molekülhälfte) an mindestens drei Stellen fixiert ist, d. h. das ganze Bilirubin-Molekül wenigstens vier Bindungen zum Albumin besitzt. Die Propionsäure-Seitenketten können viele Konformationen annehmen und sind daher zur Fixierung weniger geeignet. Am wahrscheinlichsten sind je zwei Wasserstoff-Brücken an den beiden Endringen (vgl. die Verhältnisse bei den Urobilinen), die sowohl in einer Pyrrolon- als auch einer Hydroxypyrrol-Form existieren können. BLAUER und KING diskutieren auch hydrophobe Wechselwirkungen zur Stabilisierung des Bilirubin-Albumin-Komplexes.

Wenn der Bilirubin-Spiegel erhöht ist (Hyperbilirubinämie), vermag das Serumalbumin nicht mehr das ganze Pigment zu binden. In einem solchen Fall wurde im Neugeborenen-Serum Bilirubin auch in einer β-Lipoproteid-Fraktion gefunden (*37*). Oberhalb eines Schwellenwertes tritt das Pigment in das Gewebe über und färbt dieses an (Ikterus); beschrieben wurde die Bindung von Bilirubin an Elastin und Deutin (*239*). Welche Aminosäuren mit dem Bilirubin in Wechselwirkung treten, ist bei keinem Protein bekannt.

VI. Umwandlungsprodukte des Bilirubins

1. Bilane und Bilene-(b) *(Urobilinoide)*

Bilirubin und seine Konjugate werden von den Bakterien des Dickdarms (besonders von Clostidien-Arten vgl. (*228*)) zu farblosen Bilanen hydriert. Obwohl die Glucoronide des Bilirubins leichter reduziert werden als der freie Farbstoff (*229*), sind bei den Reduktionsprodukten bisher keine Konjugate aufgefunden worden. Die Bilane gehen durch Autoxydation an der Luft leicht in die entsprechenden Bilene-(b) über. Die umgekehrte Reaktion läßt sich mit Na/Hg oder durch katalytische Reduktion erzielen. Eine Übersicht über die Entstehung der wichtigsten Verbindungen und ihre Trivialnamen gibt Schema 5, ältere wichtige Arbeiten s. bei (*136*) und (*228*).

Urobilin (von „Urin") und Stercobilin (von „stercus" = Kot) kommen vergesellschaftet vor; die Namen beziehen sich nur auf das erstmalige Auffinden dieser Pigmente. Die Strukturfragen beim optisch inaktiven i-Urobilin schienen 1936 mit der Totalsynthese (*205*) abgeschlossen zu sein, bis SCHWARTZ und WATSON (*199*) 1942 ein rechtsdrehendes d-Urobilin aus infizierter Fistelgalle isolierten, welches nach der Elementaranalyse (*231*) und der H_2-Aufnahme bei der katalytischen Hydrierung (*77*) 2 H weniger als das bis dahin allein bekannte Urobilin enthalten sollte. Auch im Massenspektrum wurde für das d-Urobilin (M = 588) ein Mindergehalt von 2 H gegenüber dem i-Urobilin (M = 590) gefunden (*90, 92*), jedoch sind gerade bei Urobilinoiden Redoxreaktionen im Massenspektrum so ausgeprägt, daß ein Mehr- oder Mindergehalt von 2 H nur schwierig zu erkennen ist (*118*). Für ein Urobilin mit dem Mol.-Gewicht 588 wurden Strukturen mit einem Azacyclopentadienon-Ring (54a, b) oder einer Vinylgruppe (55a, b) diskutiert (*77*), wobei sich letztere zwanglos durch partielle Reduktion aus dem Bilirubin ableiten lassen.

Die beiden Vinyl-substituierten Urobiline 55a und b wurden durch Totalsynthese gewonnen (*172*). (55a) besitzt im UV-Spektrum ein Maximum bei 230 nm, das bei (55b) und auch beim Naturprodukt fehlt. Wenn das d-Urobilin also eine Vinylgruppe besitzt, muß sie sich in Position 8 befinden. Jedoch sprechen neuere Ergebnisse, insbesondere das Fehlen von Methyl-vinyl-maleinimid beim Chromsäureabbau, gegen die Existenz einer Vinylgruppe beim d-Urobilin vom Mol.-Gewicht 588 (*140*). Neben dem aus d-Urobilin-588 durch katalytische Hydrierung und Re-

Literaturverzeichnis: SS. 128—139

oxydation darstellbaren d-Urobilin vom Mol.-Gewicht 590 (77) gibt es
ein natürlich vorkommendes, rechtsdrehendes Urobilin vom Mol.-Gewicht
590 (*118*); rechts- und linksdrehende Urobiline desselben Mol.-Gewichtes
wurden durch Totalsynthese erhalten (*168*).

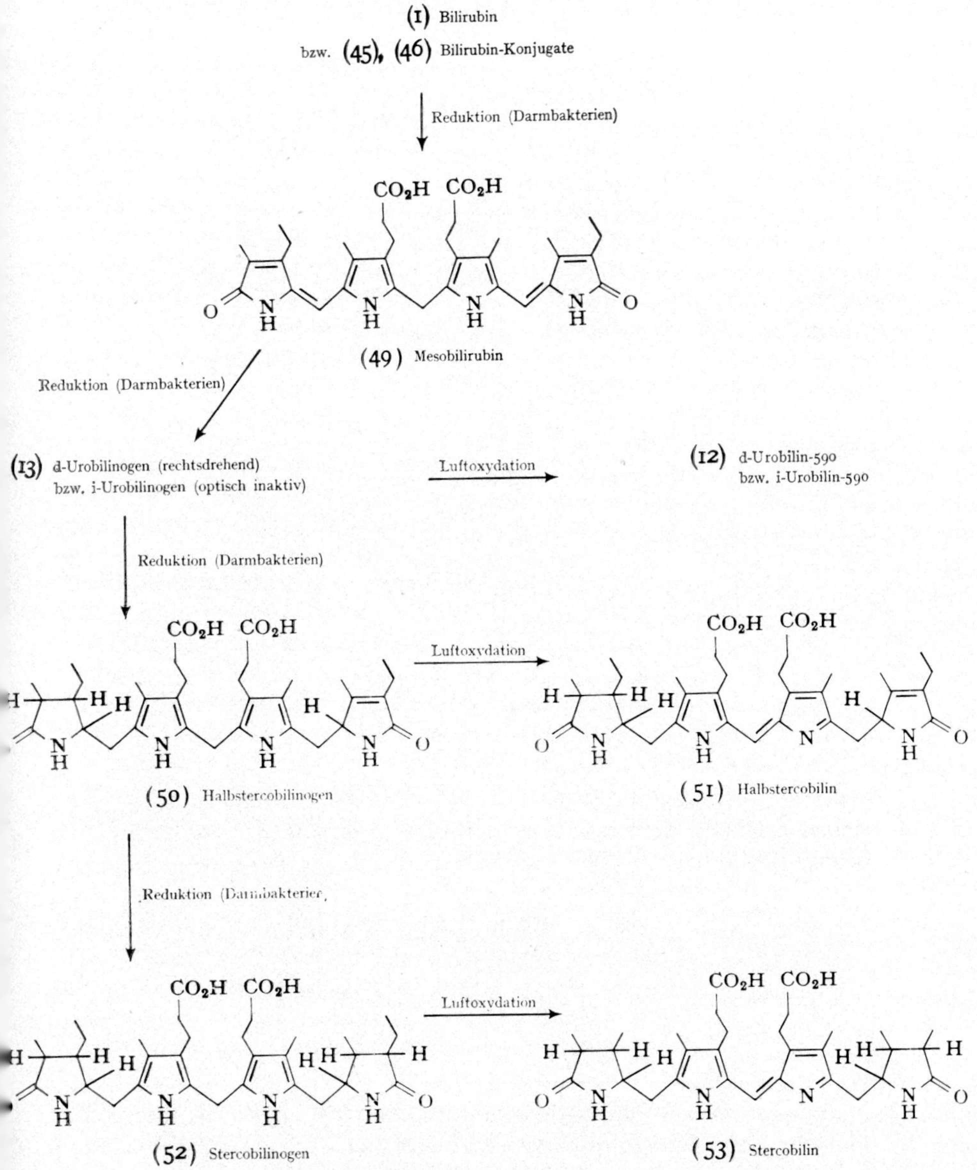

(**1**) Bilirubin
bzw. (**45**), (**46**) Bilirubin-Konjugate

↓ Reduktion (Darmbakterien)

(**49**) Mesobilirubin

Reduktion (Darmbakterien) ↙

(**13**) d-Urobilinogen (rechtsdrehend)
bzw. i-Urobilinogen (optisch inaktiv)

⟶ Luftoxydation ⟶

(**12**) d-Urobilin-590
bzw. i-Urobilin-590

↓ Reduktion (Darmbakterien)

(**50**) Halbstercobilinogen

⟶ Luftoxydation ⟶

(**51**) Halbstercobilin

↓ Reduktion (Darmbakterien)

(**52**) Stercobilinogen

⟶ Luftoxydation ⟶

(**53**) Stercobilin

Formelübersicht 5

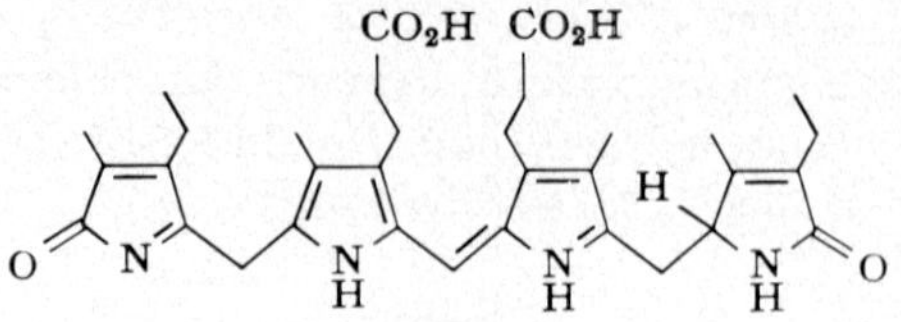

(54a)

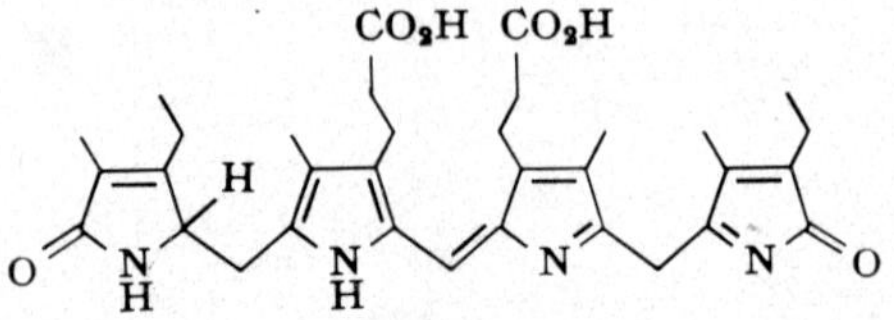

(54b)

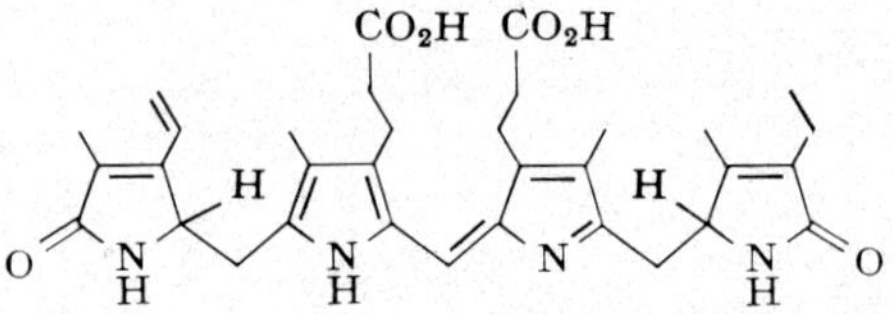

(55a)

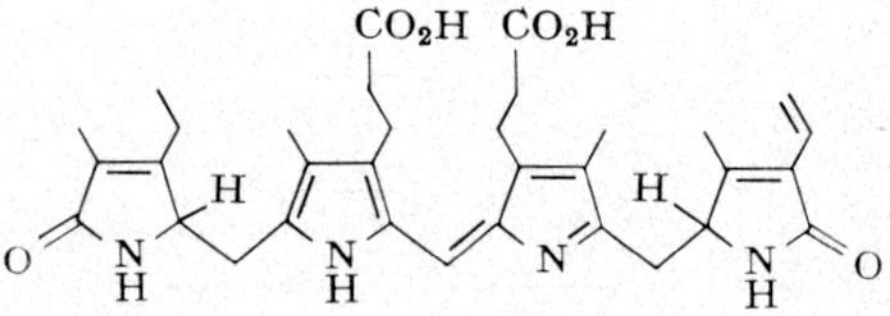

(55b)

Stercobilin besitzt 4 H mehr als i-Urobilin, deren Stellung in den Endringen durch die Isolierung von (optisch aktivem) Methyläthylsuccinimid beim CrO_3-Abbau gesichert wurde (77). Da das so gewonnene Imid die

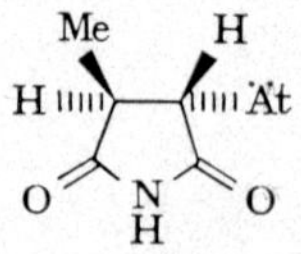

(56a) 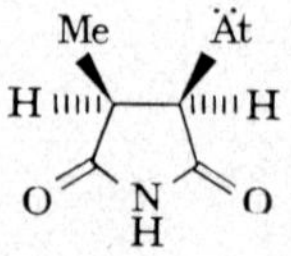   (56b)

trans- oder *threo*-Konfiguration besitzt (56a) (74), muß im natürlichen Pigment diese Konfiguration zweimal enthalten sein. Das durch katalytische Hydrierung von Bilirubin (und Reoxydation) gewonnene Pigment

Literaturverzeichnis: SS. 128—139

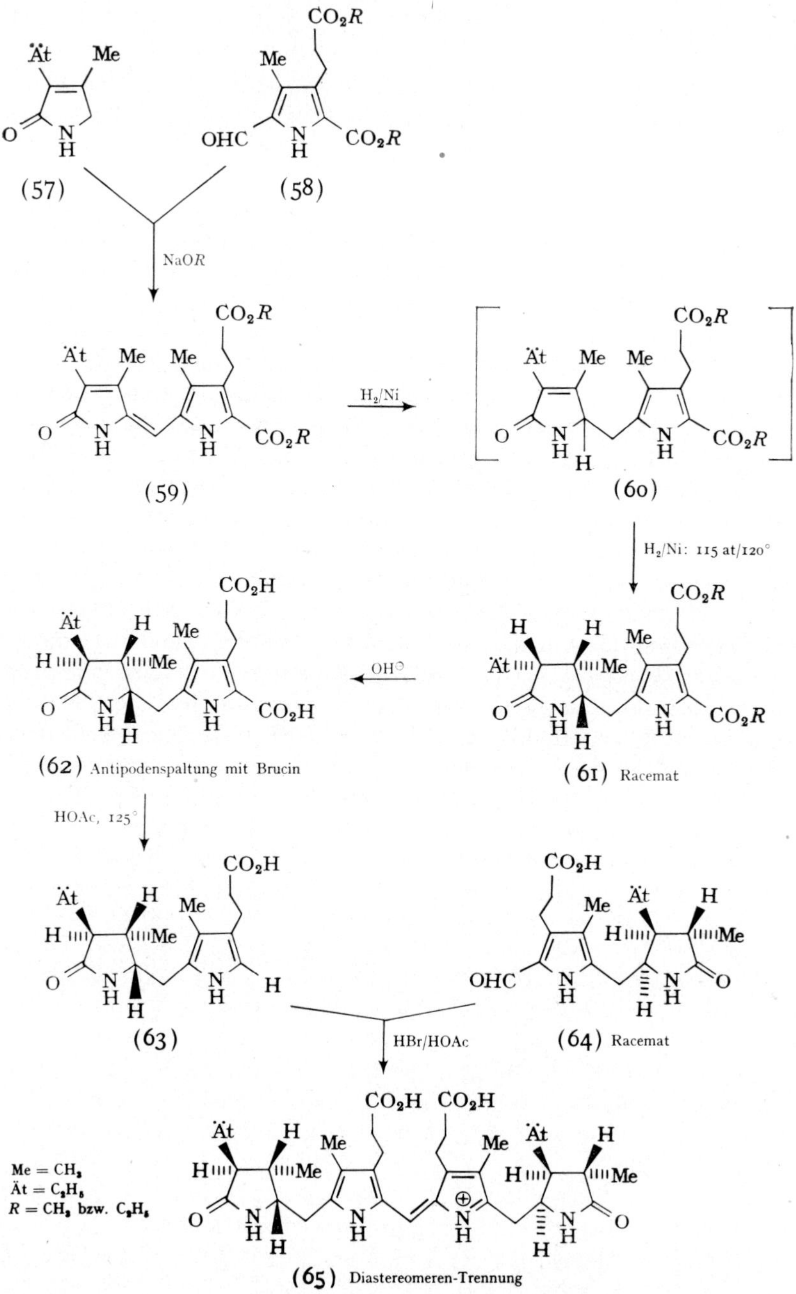

Gallenfarbstoffe und Biliproteide

(57) (58)

NaOR

(59)

H$_2$/Ni

(60)

H$_2$/Ni: 115 at/120°

(62) Antipodenspaltung mit Brucin   (61) Racemat

OH$^{\ominus}$

HOAc, 125°

(63)   (64) Racemat

HBr/HOAc

Me = CH$_3$
Ät = C$_2$H$_5$
R = CH$_3$ bzw. C$_2$H$_5$

(65) Diastereomeren-Trennung

Formelübersicht 6

(99) ist dagegen ein *cis-cis*-Stercobilin, da es beim CrO_3-Abbau das entsprechende *cis*- oder *erythro*-Imid (56b) liefert (163). Durch Totalsynthese wurden racemische (170) und optisch aktive (171) Stercobiline bekannter Konfiguration erhalten; die absolute Konfiguration des Stercobilins und damit des ersten Gallenfarbstoffs überhaupt ist also gesichert (165).

Die Totalsynthese des optischen aktiven Stercobilins (171) (Schema 6) geht von dem Pyrrolon (57) und dem Pyrrolaldehyd (58) aus, die zu dem Dipyrromethen (59) kondensiert werden. Bei der katalytischen Hydrierung entsteht zuerst das Dipyrromethan (60), welches nicht isoliert wird. Bei der weiteren Hydrierung entsteht aus sterischen Gründen überwiegend das all-cis-Produkt (61); die Zuordnung erfolgt durch NMR-Spektroskopie. Bei der alkalischen Verseifung tritt Inversion der den Carboxylgruppen benachbarten Substituenten ein. Die Dicarbonsäure (62) läßt sich als Brucin-Salz in die Antipoden spalten. Die linksdrehende Form wird zu (63) decarboxyliert und mit dem Dipyrromethenaldehyd (64), der als Racemat (mit bekannter relativer Konfiguration der drei Asymmetriezentren) eingesetzt wird, kondensiert. Bei der Diastereomeren-Trennung erhält man ein Produkt mit der Drehung $(\alpha)_{20}^{D} = -3750°$; dieses stimmt in allen Eigenschaften, auch im Debye-Scherrer-Diagramm, mit dem natürlichen Stercobilin überein. Damit ist die relative Konfiguration der sechs Asymmetriezentren zueinander gesichert. Die absolute Konfiguration ergibt sich aus dem Vergleich des durch CrO_3-Abbau erhaltenen, optisch aktiven Methyläthylsuccinimids mit dem Imid bekannter Konfiguration (19). Demnach besitzt das natürliche Stercobilin die 1S,2S,2'S,7S,7'S,8S-Konfiguration (65) (165).

Ein weiteres Bilen-(b) ist das Halbstercobilin, welches 2 H mehr als Urobilin, d. h. 2 H weniger als Stercobilin besitzt. Es wurde zunächst

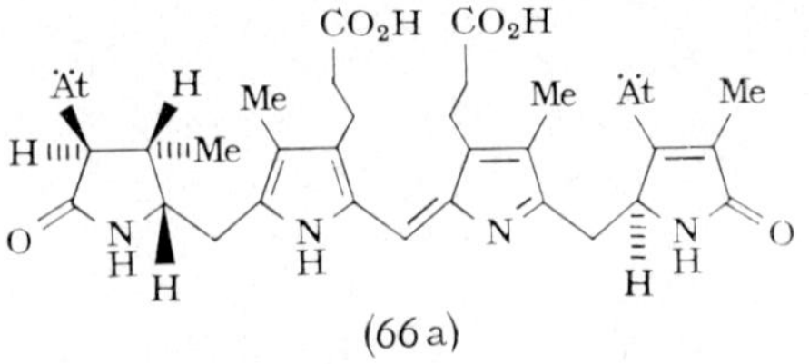

(66a)

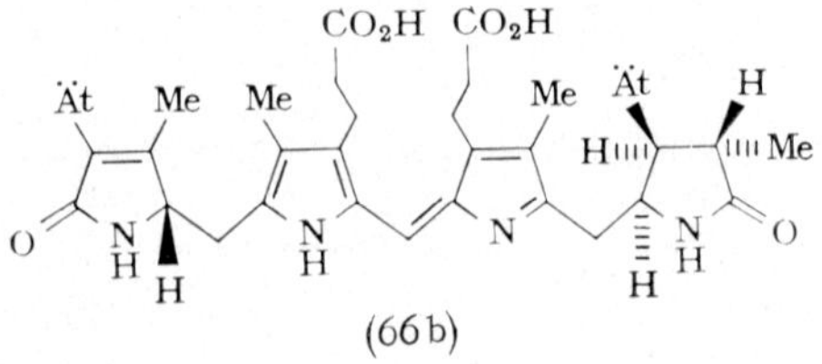

(66b)

Literaturverzeichnis: SS. 128—139

bei einem Fall mit hämolytischer Anämie isoliert (233), scheint aber weiter verbreitet zu sein (230). Die Struktur (66) folgt einmal aus dem Massenspektrum, welches (mit gewisser Unsicherheit vgl. S. 82) das Mol.-Gewicht 592 ergibt. Der Chromsäureabbau liefert Methyl-äthyl-maleinimid neben Methyl-äthyl-succinimid (und Hämatinsäureimid), d. h. einer der Endringe ist hydriert, der andere nicht. Ferner liefert die Oxydation mit $FeCl_3$ das sog. stabile Mesobiliviolin (vgl. S. 101). Welcher der beiden Endringe hydriert ist, ist bisher nicht bekannt.

Das dem Halbstercobilin entsprechende Bilan (Halbstercobilinogen) ist dem Bilan des Urobilins (Urobilinogen) in seiner Kristallform offensichtlich sehr ähnlich; jedenfalls wurden im natürlichen, kristallisierten Bilan beide Komponenten nebeneinander nachgewiesen (230).

2. Biladiene-(a, b)

Während Bilane bei Autoxydation an der Luft Bilene-(b) ergeben, entstehen bei der chemischen Dehydrierung mit Eisen(III)chlorid je nach den Bedingungen verschiedene Reaktionsprodukte (Schema 7). Bei tiefer Temperatur sind die Bilene-(b) Hauptprodukte (192), bei höherer Temperatur überwiegen je nach Einwirkungsdauer oder Acidität (CH_3OH bzw. CH_3CO_2H) der Oxydationslösung Biladiene-(a,b) oder Bilatriene.

Das aus Urobilinogen auf diese Weise dargestellte Biladien „Mesobiliviolin" (57) erwies sich später (200, 214) als uneinheitlich. Die Struktur (67) der violetten, bei der Chromatographie weiter wandernden Hauptkomponente wurde durch Synthese aus dem Dipyrromethenaldehyd (71) und dem Dipyrromethan (72) gesichert (200, 206); für diese wurde der Name *Mesobiliviolin* beibehalten. Die weniger weit wandernde Nebenkomponente, die auch synthetisch aus dem Dipyrromethanaldehyd (73) und dem Dipyrromethen (74) (allerdings in schlechter Ausbeute) erhalten wurde (200, 206), wurde zunächst „Mesobilrhodin" genannt (200), sie wird jedoch besser als *Isomesobiliviolin* bezeichnet (102, 165). Die reine Verbindung besitzt dasselbe Spektrum wie Mesobiliviolin; die hellrote Farbe des ursprünglichen Präparates dürfte auf gelbe oder braune Nebenprodukte zurückzuführen sein (165). Mesobiliviolin (67) entsteht auch aus dem Dipyrromethen (75) und dem Dipyrromethanaldehyd (76); die beiden prototropen Strukturen (67) und (67a) gehen also offensichtlich leicht ineinander über. Entsprechendes gilt für die prototropen Strukturen (68) und (68a) des Isomesobiliviolins; die beste Darstellung für dieses Pigment geht daher vom Dipyrromethan (77) und dem Dipyrromethenaldehyd (78) aus (102).

Halbstercobilin läßt sich unter entsprechenden Bedingungen zu einem violetten Farbstoff dehydrieren, der gegenüber der weiteren Einwirkung

Formelübersicht 7

von Eisen(III)chlorid beständig ist *(stabiles Mesobiliviolin)* *(164a)*. Die Struktur (69) ergibt sich aus folgenden Beobachtungen: Nach dem Massenspektrum enthält das Pigment 2 H-Atome mehr als (67). Beim

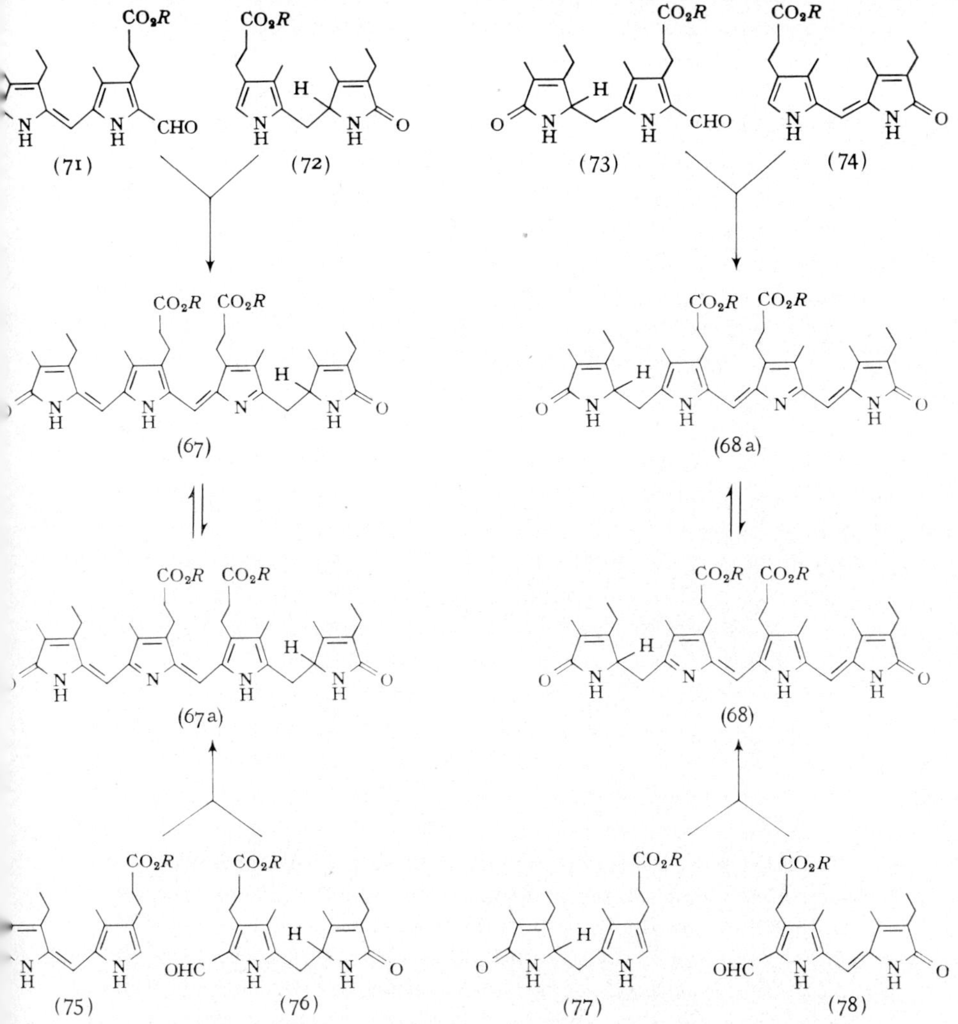

Formelübersicht 8

Chromsäureabbau entsteht neben Hämatinsäureimid und Methyl-äthylmaleinimid Methyl-äthyl-succinimid, wodurch die zusätzlichen H-Atome in den β-Stellungen eines Endringes lokalisiert werden. Welcher der Endringe hydriert ist, weiß man bisher nicht. Jedenfalls ist es derjenige, der durch die Methylen-Brücke mit den restlichen drei Ringen verknüpft ist, denn nur so kann das Pigment denselben Chromophor enthalten wie Mesobiliviolin (67). Tatsächlich besitzen beide Pigmente dieselbe Farbe.

Stercobilin ist im Gegensatz zu den anderen beiden erwähnten Bilenen-(b) gegenüber Oxydationsmitteln relativ beständig, jedenfalls geht es nicht in rote oder blaue Farbstoffe über. Die unterschiedliche Reaktion von Urobilin, Halbstercobilin und Stercobilin mit Eisen(III)chlorid kann man zur Bestimmung dieser Bilene-(b) im Gemisch ausnutzen (235). Die

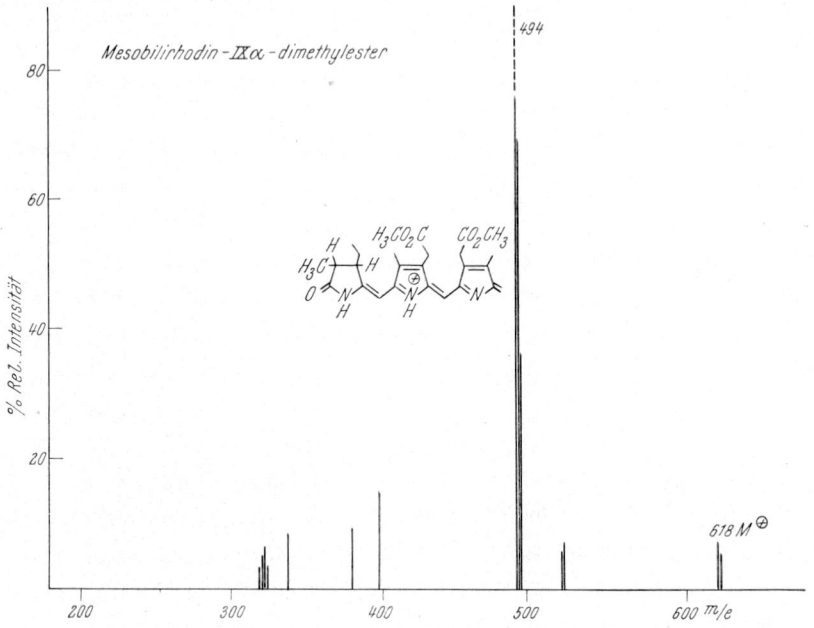

Abb. 7. Massenspektrum von Mesobilirhodin-IXα-dimethylester (nach (104))

in Schema 7 (S. 100) aufgezeigten strukturellen Zusammenhänge wurden durch Messung der Massenspektren der einzelnen Pigmente gesichert (118).

Neben diesen Biladienen der IXα-Struktur kommen im Oxydationsgemisch solche der XIIIα- und IIIα-Struktur vor, die durch Spaltung des Ausgangsmaterials und Kondensation der Bruchstücke entstehen (214). Diese Reaktion wird bei den Bilatrienen besprochen (S. 104).

Als Nebenkomponente tritt bei der Eisen(III)chlorid-Oxydation auch eine hellrote Komponente auf (174, 214), die besser durch Einwirkung von Alkali auf d-Urobilin (77, 145) oder i-Urobilin (143, 192) zu erhalten ist; für diese wird der (ursprünglich für ein Pigmentgemisch verwendete, vgl. oben) Name *Mesobilirhodin* beibehalten. Die Struktur wurde in unabhängigen Untersuchungen von zwei Arbeitskreisen aufgeklärt (143,

Literaturverzeichnis: SS. 128—139

192). Nach dem Massenspektrum (Abb. 7) ist das Mesobilirhodin ein Isomeres des i-Urobilins und des stabilen Mesobiliviolins, besitzt also 2 H-Atome mehr als das Mesobiliviolin. Die beiden H-Atome finden sich im Tripyrrol-Fragment wieder, welches sich von den durch Methin-Brücken verknüpften Ringen (vgl. Abschnitt Massenspektrometrie), d. h. vom Chromophor ableitet. Beim Chromsäureabbau entsteht auch aus diesem Pigment neben Methyl-äthyl-maleinimid und Hämatinsäure-imid Methyl-äthyl-succinimid, und zwar die *trans*-Form, wenn das Pigment durch Alkali-Einwirkung auf Urobilin (*143*, *192*), die *cis*-Form, wenn es durch Autoxydation von Urobilinogen in Eisessig (*140*) gewonnen wurde.

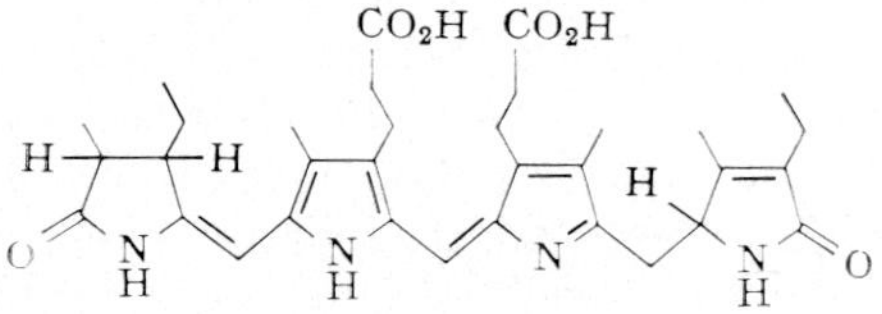

(79)

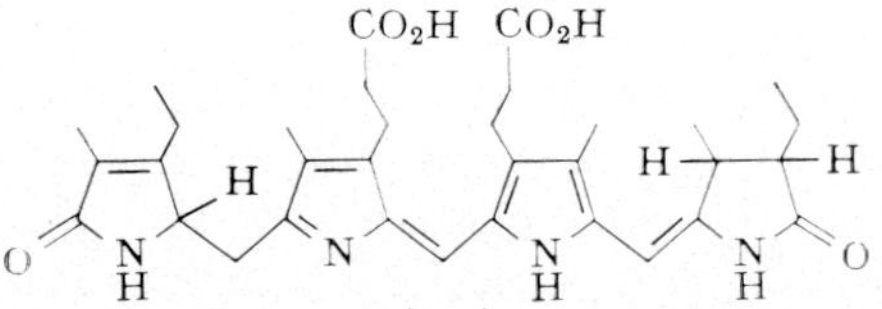

(79a)

Die Struktur (79) (bzw. (79a)) steht mit diesen Daten in Einklang, auch spricht das Elektronenspektrum für diese Struktur, da das langwellige Maximum (entsprechend der Ausdehnung des Chromophors) zwischen denen des Urobilins und des Mesobiliviolins liegt (Tabelle 9). Durch Einwirkung von Säure wird Mesobilirhodin leicht in Umkehrung der Alkali-Isomerisierung zu Urobilin umgesetzt.

Tabelle 9. *Langwellige Absorptionsmaxima von Gallenfarbstoffen in Methanol (a) bzw. Chloroform (b) (nm)*

Pigment	Freie Base	Hydrochlorid	Zink-Komplex	Lit.
Urobilin (a)	448	492	506—508	(*156*)
Urobilin (b)	454	499	510	(*143*)
Mesobilirhodin (a)	496	560—565	584	(*192*)
		557	578*	(*26*)
Mesobilirhodin (b)	504	576	583	(*143*)
Mesobiliviolin (a)		556	626*	(*26*)
Mesobiliviolin (b)	570	607	632*	(*102*)
	568	602	630	(*143*)

* In Äthanol.

VII. Bilatriene

1. Biliverdin und Mesobiliverdin

Biliverdin, welches *in vivo* die Vorstufe von Bilirubin (80) ist (*113, 219*), wird *in vitro* am besten durch Dehydrierung aus letzterem gewonnen. Die Bedingungen für diese Reaktion wurden zuerst bei der analogen Dehydrierung von Mesobilirubin (80a) zu Mesobiliverdin (Glaucobilin) ausgearbeitet (*53, 54*); als Oxydationsmittel diente Eisen(III)chlorid (*75, 111, 134*) oder Chinon (*221*). Die so erhaltenen Bilatriene erwiesen sich bei der Dünnschichtchromatographie als uneinheitlich (*33, 142, 156*); die Strukturen der einzelnen Fraktionen wurden erst kürzlich (*15, 214*) insbesondere mit Hilfe der NMR-Spektroskopie aufgeklärt.

Bei den Protobilatrienen (*15*) ergab sich die Strukturordnung aus der Lage der Signale für die Methylgruppe an den aromatischen Ringen (vgl. S. 85): Das Isomere IXα (81a) zeigt drei Singulets im Bereich der „*endo*"- und eines im Bereich der „*exo*"-Methylgruppen; dem Isomeren, das 1 Signal (jeweils entsprechend 6 H) in jedem der beiden Bereiche ergibt, wird die XIIIα-Struktur (82) zugeordnet. Das Isomere IIIα (83) besitzt nur *endo*-Methylgruppen; man findet hier zwei Signale (jeweils entsprechend 6 H) in diesem Bereich. Bei den Mesobilatrienen (81a)—(83a) (*214*) wurden auch die Signale der Äthylgruppen zur Interpretation herangezogen; zur Sicherung diente auch der chromatographische Vergleich mit den synthetisch erhaltenen Isomeren.

Als möglicher Mechanismus der Entstehung dieser isomeren Bilatriene wird die säurekatalysierte Spaltung der Biladiene-(a,c) und die Rekombination der erhaltenen Zweierbruchstücke im Verlauf der Oxydationsreaktion angenommen (*15*). Der Mechanismus dieser Spaltung ist analog derjenigen mit Diazoniumsalzen (vgl. S. 68 und vgl. *91*); die Rekombination dürfte auf ähnliche Weise erfolgen wie bei Porphyrinogenen, bei denen in einer analogen Isomerisierung markierter Formaldehyd eingebaut wird (*122*). Eine ähnliche Reaktionsfolge wurde bereits früher in zwei Schritten ausgeführt; aus (unsymmetrisch substituiertem) Mesobilirubin IXα (80a) entsteht bei der Resorcin-Spaltung ein Gemisch zweier Dipyrromethene (84a) und (85a), aus denen bei der Kondensation mit Ameisensäure/Bromwasserstoff das Gemisch der Mesobilatriene (81a), (82a) und (83a) entsteht (*53*).

2. Biliverdin und Biliverdin-Proteide bei Vertebraten

In der Gallenflüssigkeit der Wirbeltiere kommt neben dem Bilirubin stets Biliverdin in wechselnder Menge vor. Biliverdin ist das Hauptpigment in bestimmten Organen oder (insbesondere calcifizierten) Geweben einiger Wirbeltiere (Tabelle 10). In allen daraufhin untersuchten Fällen wurde als einziges Isomeres Biliverdin IXα gefunden.

Literaturverzeichnis: SS. *128—139*

Bei Fischen sind einige Biliverdin-Proteide gefunden worden. Beim Aal *(Anguilla japonica)* ist die toxische Komponente des Serums nicht mit diesem Chromoproteid identisch *(103)*; das Protein ist ein β_1-Lipo-

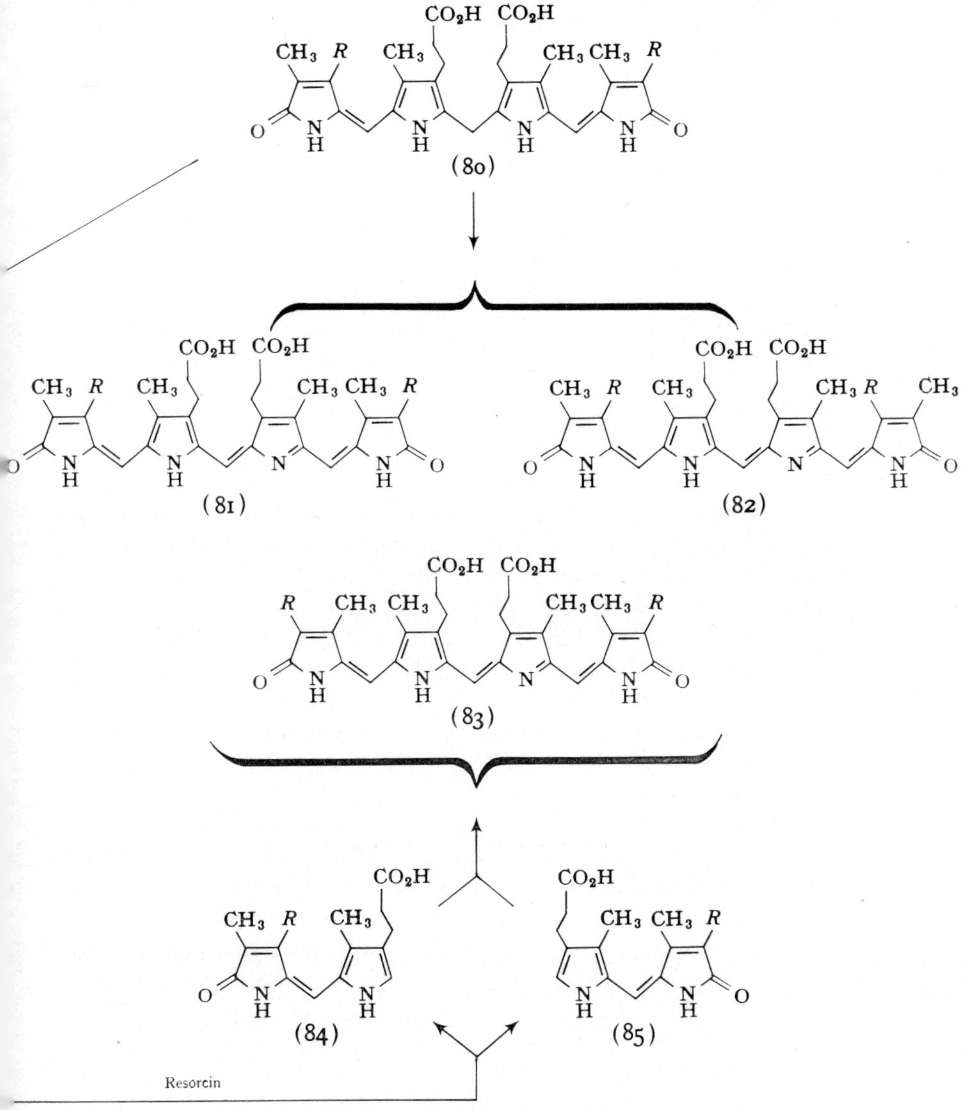

$R = CH=CH_2$
a: $R = C_2H_5$

Formelübersicht 9

Tabelle 10. *Biliverdin IX α und Biliverdin-Proteide bei Vertebraten*

Tier	Organ	Apoprotein (Mol.-Gew., pI)	Lit.
Hund	Placenta	—	*(112, 115)*
Möve	Eierschalen	—	*(110, 111, 112)*
Emu	Eierschalen	—	*(221)*
Fische			
Zoarces viviparus	Gräten	—	*(238)*
Katsuwonus pelamis L.	Gräten	—	*(61)*
Belone belone	Gräten	—	*(179)*
Cololabis saira Brevoort	Gräten	—	*(224)*
Anguilla japonica	Serum	Lipoproteid (89100; 4,7)	*(103, 240, 242)*
Myoxocephalus scorpioides	Serum, Haut	(46000; 3,1)	*(3)*
Cottus scorpius	Haut	nicht bestimmt	*(238)*
Crenilabrus pavo	Haut, Serum	$(0,5-1,2 \cdot 10^6)$*	*(1)*
Cheilinus undulatus	Muskel, Haut	(102000; 6,8)*	*(243)*

* Pigment kovalent gebunden.

proteid *(242)*, dessen Konzentration mit der Futteraufnahme jahreszeitlich schwankt *(241)*. Vermutlich kommt ihm eine Transportfunktion zu.

Besonderes Interesse verdienen die Lippfische (Labriden), da bei ihnen die bisher einzigen Biliproteide im Tierreich entdeckt wurden, bei denen der Gallenfarbstoff kovalent mit dem Protein verknüpft ist. Der Farbstoff wurde bei *Crenilabrus* *(1)* und *Cheilinus* *(243)* durch alkalische Hydrolyse freigesetzt und durch Vergleich der Spektren und des chromatographischen Verhaltens mit denen des authentischen Pigments als Biliverdin IX α identifiziert. Im Fall des *Crenilabrus*-Blaus wurden durch Chromsäureabbau darüberhinaus Aussagen über die Bindung des Biliverdins an das Protein gewonnen *(183)*.

Beim nicht-hydrolytischen Abbau werden Hämatinsäureimid (86) und Methyl-vinyl-maleinimid (87) in einer Menge freigesetzt, die jeweils einem Pyrrolkern entspricht. Demnach sind zwei Pyrrolkerne des Chromophors frei und zwei an das Protein gebunden (88). Beim Erhitzen in der sauren Oxydationslösung (hydrolytischer Abbau) erhält man zwar die doppelte Menge (86), jedoch fast kein (87) mehr. Offensichtlich wurde durch die Säurebehandlung die Vinylgruppe mit Thiolgruppen des Proteins verknüpft (89). Durch kurze Alkalibehandlung wird die ursprüngliche Bindung des äußeren Pyrrolkerns gespalten, während die des inneren intakt bleibt (90): Man erhält jetzt (87) aus zwei Pyrrolkernen, jedoch (86) nur aus einem Pyrrolkern. Beim Übergang von (88) zu (89) findet man eine Verschiebung des langwelligen Maximums von 640 nach 675 nm. Erst beim Erhitzen mit Alkali wird der Chromophor freigesetzt.

Literaturverzeichnis: SS. 128—139

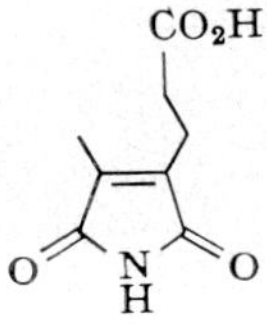

Gallenfarbstoffe und Biliproteide

(86) 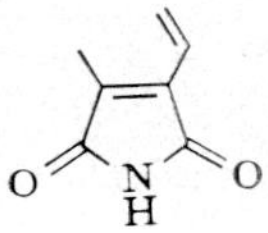 (87)

(88) $\lambda_{max} = 640$ nm

OCH_3^-, 20°C

$H^\oplus$, 100°C

(90) $\lambda_{max} = 675$ nm

(89)

OCH_3^-, 65°C

Biliverdin IXα

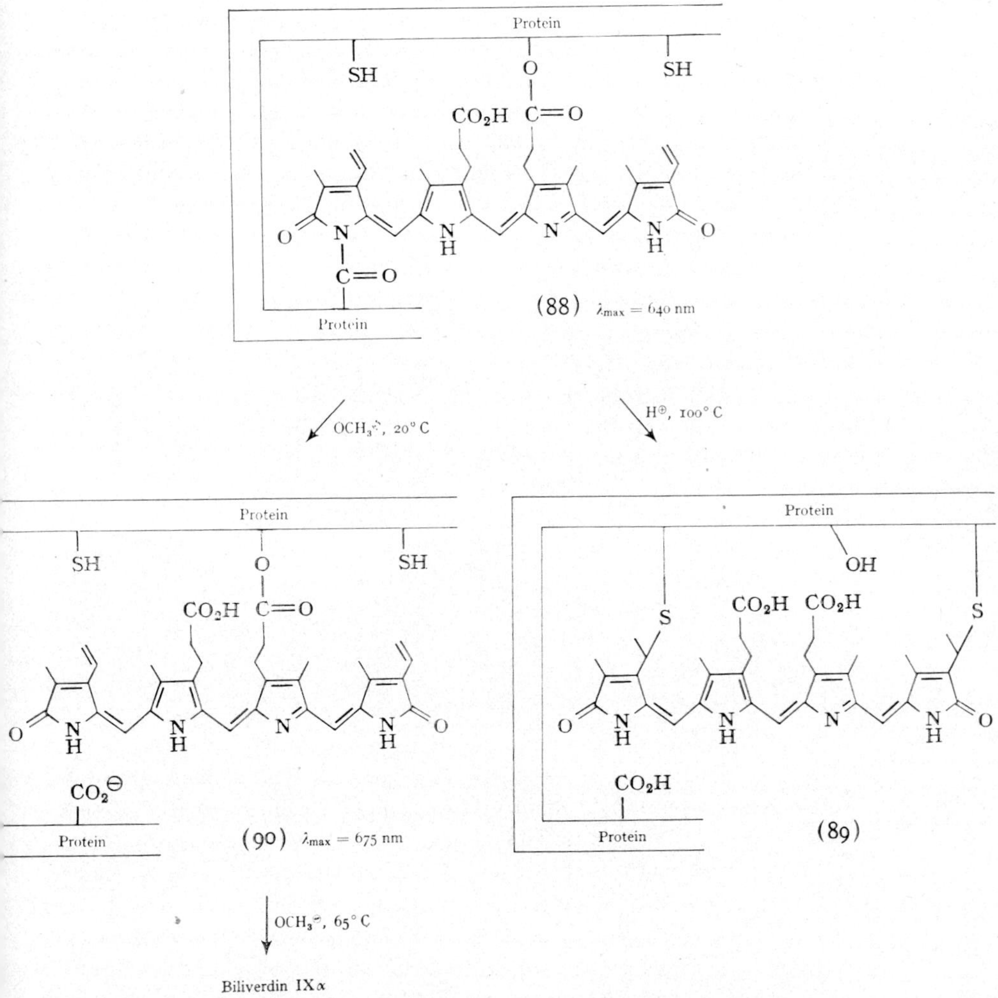

Formelübersicht 10

3. Bilatriene bei Invertebraten

Bei wirbellosen Tieren wurden zahlreiche Pigmente auf Grund von Spektren und Farbreaktionen als Gallenfarbstoffe angesehen, ohne daß ihre Struktur gesichert wäre (Literaturübersicht bei *180*). Gesichert sind die Strukturen von *Aplysia*-Farbstoffen (vgl. nächster Abschnitt) und von Bilatrienen.

Das Pigment im Kalkskelett der blauen Koralle *Heliopora coerulea* Pall. läßt sich — im Gegensatz zu dem der roten Koralle — nach Auflösen des Kalkgewebes in Säure in organische Lösungsmittel überführen. Für das so gewonnene „Helioporobilin" wurde zunächst eine von anderen Gallenfarbstoffen abweichende Struktur vermutet (*221*, dort auch ältere Literatur). Erst nach dünnschichtchromatographischer Reinigung konnte die Hauptfraktion kristallisiert und durch das Massenspektrum sowie Chromsäure- und Chromatabbau als Protobilatrien IXα (Biliverdin) charakterisiert werden (*188*). Eine blaue Nebenfraktion enthält wahrscheinlich ein Oxydations- oder Polymerisationsprodukt des Biliverdins mit verändertem Endring; möglicherweise bildet es sich aus dem Biliverdin erst nach dem Tod der Tiere (*188*).

Blaugrüne Biliproteide wurden im Tegument und der Hämolymphe zahlreicher Insekten gefunden (*180*). Der Farbstoff läßt sich jeweils mit neutralen oder sauren Lösungsmitteln vom Protein abspalten; es handelt sich also immer um (nicht näher untersuchte) Nebenvalenz-Bindungen. Zunächst wurden Biline mit besonderen Strukturen vermutet; die Chromoproteide wurden daher mit einem eigenen Namen („Insektoverdin") belegt (*97*). Das Bilin der Heuschrecken *Mantis religiosa* (*150*), *Locusta migratoria* (*153*) und *Oedipoda coerulescens* (*155*) wurde (als Dimethylester) durch Vergleich des Schmelzpunktes, des Elektronenspektrums und des chromatographischen Verhaltens, das von *Carausius morosus* (*177*, vgl. *237*), *Mantis religiosa* und *Clitumnus extradentatus* (*191*) darüber hinaus auf Grund des Chromsäure- und Chromatabbaus als Protobilatrien-IXα erkannt. Außer bei diesen Orthopteren kommt dasselbe Pigment auch bei der Neuroptere *Chrysopa carnea* (*187*) vor.

Das Biliverdin unterliegt einem dauernden Auf- und Abbau. Bei *Mantis und Locusta* wurde der Einbau von markiertem Glycin in das Pigment nachgewiesen (*154*). Die Umfärbung nach Braun wurde bei *Mantis*, *Sphodomantis* und *Locusta* auf einen Abbau des Biliverdins zu gelbbraunen Produkten zurückgeführt (*151, 152*); jedoch sind auch andere Mechanismen der Umfärbung möglich (*187*).

Bei Schmetterlingen (Lepidopteren) wurde ein anderes Bilatrien nachgewiesen. Das Pigment wurde zunächst aus den Flügeln von Pieriden isoliert und „Pterobilin" genannt (*236*); es kommt aber auch als Biliproteid im Tegument und in der Hämolymphe von Raupen und Puppen

Literaturverzeichnis: SS. 128—139

Formelübersicht 11

vor. Das aus dem Tegument der Raupen des Kohlweißlings *(Pieris brassicae)* isolierte, als Dimethylester dünnschichtchromatographisch gereinigte Pigment wurde auf Grund des Chromsäure- und Chromatabbaus sowie durch Vergleich mit authentischem Material als Protobilatrien-IXγ (95) erkannt *(189)*. Radioaktiv markierte Vorstufen normaler Biline (Glycin *(190)*, δ-Aminolävulinsäure *(182)*, Protoporphyrin IX) werden auch in das Pterobilin eingebaut. Die Aktivitätsverteilung entspricht der bei anderen Gallenfarbstoffen gefundenen *(190)*. Wahrscheinlich ist die Öffnung der γ-Methin-Brücke von Protohäm IX (statt der sonst gefundenen Oxydation der α-Methin-Brücke) die einzige Abweichung bei der Biogenese dieses Gallenfarbstoffs *(180)*. Den Weg der Biosynthese von Pterobilin zeigt Schema 11.

Die Aktivität des Glycins (91) findet sich im C-5 der Aminolävulinsäure (92), in zwei Positionen des Porphobilinogens (93) und in acht C-Atomen des Protoporphyrins (94) wieder; nach der oxydativen Eliminierung einer Methin-Brücke enthält das entstandene Pterobilin (95) noch sieben markierte C-Atome in den angegebenen Stellungen. Beim Chromatabbau des Dimethylesters (95a) erhält man neben Methylvinylmaleinimid, welches nicht näher untersucht wurde, den Pyrroldialdehyd (96), der noch drei markierte C-Atome, und das Imid (97), welches ein markiertes C-Atom trägt. Die gefundenen spezifischen Aktivitäten von (95) : (96) : (97) verhalten sich entsprechend der Erwartung wie $7:3:1$ *(190)*.

Tabelle 11. *Gallenfarbstoffe bei Lepidopteren (nach 225, 226)*

Familie	Species	Pigment
Pieridae	*Catopsilia florella, Mylothris* sp., *Nepheronia thalassina*	Pterobilin
Nymphalidae	*Charaxes eupale, Ch. dilitus, Ch. Khaldeni, Ch. subornatus, Morpho catenarius, Victorina steneles*	
Papilionidae	*Papilio graphium antheus, P. graphium tynderaeus, Graphium policenes, G. leonidas*	
	Graphium sarpedon, Papilio phorcas	Neopterobilin
Attacidae	*Actias artemis, A. selene, Antherea pernyi*	

Neben dem Pterobilin kommen bei Schmetterlingen neuartige Gallenfarbstoffe (Neopterobiline) *(225)* vor. Ihre Zugehörigkeit zur IXα- oder IXγ-Reihe konnte noch nicht geklärt werden, da sie beim Chromatabbau keinen Pyrroldialdehyd liefern *(182)*. Die Untersuchung des Vorkommens von Pterobilin und Neopterobilinen (vgl. Tabelle 11) wird möglicherweise zur Kenntnis der Phylogenie von Lepidopteren beitragen *(225, 226)*.

Literaturverzeichnis: SS. 128—139

VIII. Gallenfarbstoffe mit Äthylidengruppe

Alle bisher bekannten Gallenfarbstoffe mit Äthylidengruppe (vgl. Tabelle 12) sind pflanzlichen Ursprungs; die bei der Kiemenschnecke *Aplysia* gefundenen Biline stammen aus der pflanzlichen Nahrung der Tiere (*30*). Alle diese Pigmente kommen als Biliproteide vor. Die Chromophore der pflanzlichen Biliproteide sind kovalent an das Protein gebunden; sie werden bei Spaltung dieser Bindung leicht verändert. Dagegen lassen sich die nur locker mit Protein verknüpften Aplysia-Farbstoffe leicht ohne sekundäre Veränderung in freier Form gewinnen. Da ihre Strukturen dementsprechend zuerst aufgeklärt wurden, sollen sie hier zunächst behandelt werden.

Tabelle 12. *Übersicht über Gallenfarbstoffe mit Äthyliden-Gruppe*

Pflanze, Tier	Organ	Biliproteid	Chromophor (Bindung)
Rhodophyta Cyanophyta Cryptophyta	Phycobilisomen	Phycoerythrine, Phycocyanine	Phycoerythrobilin, Phycocyanobilin (kovalent)
höhere Pflanzen, einige Algen	Phycobilisomen? Lokalisation unbekannt	Phytochrom	Phytochromobilin
Aplysia sp.	Abwehr-Drüsen	Protein-Komplex	Aplysioviolin, Aplysioverdin, Aplysiourobilin (nichtkovalent)

1. Aplysia-Farbstoffe

Das Phänomen, daß die zu den Opisthobranchien gehörenden Seehasen (Aplysien) ein Sekret ausscheiden, wenn sie angegriffen werden, ist bereits lange bekannt (ältere Literatur vgl. *174*). Bei einigen, aber nicht allen Species ist dieses Abwehrsekret farbig.

Nach LEDERER und HUTTRER (*107*) liegen die Pigmente in der Form von Chromoproteiden vor, da sie die Dialyse-Membran nicht durchdringen und sich z. B. mit Ammoniumsulfat fällen lassen; bereits bei Einwirkung von Wärme, Lösungsmitteln oder Licht dissoziieren die Biliproteide. Diese Untersuchung wurde in ungepufferter Lösung (pH zirka 5,0—5,5) ausgeführt. In gepufferter Lösung (*177*) ergibt sich z. B. für die Fällbarkeit mit Ammoniumsulfat eine Abhängigkeit vom pH-Wert (Abb. 8): Bei gleicher Salzkonzentration wird am meisten Farbstoff im Bereich von pH 2 bis 5 gefällt; oberhalb pH 7 ist auch bei 100% Sättigung nicht mehr der gesamte Farbstoff mit dem Protein fällbar. Die Biline werden also bereits beim Neutralpunkt teilweise vom Protein abgespalten. Diese Dissoziation ist reversibel: Die Mischung von Lösungen der Protein-freien Farbstoffe und des isolierten Proteins verhält sich wie das ursprüngliche Sekret. Bei pH 7 kann man auf Grund dieser Dissoziation das Protein durch Chromatographie an Sephadex G-25 abtrennen; die Farbstoffe trennen sich dabei in eine violette Hauptzone (Aplysioviolin) sowie eine organgerote (Aplysiourobilin) und blaue (Aplysioverdin) Nebenzone (*177*).

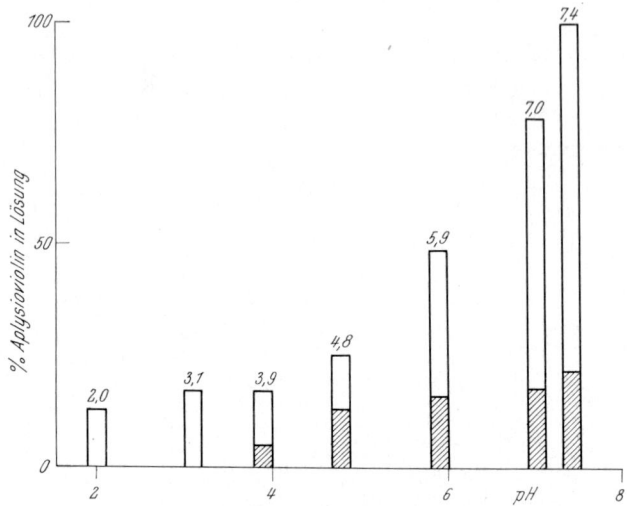

Abb. 8. Dissoziation der Chromoproteide aus dem Abwehrsekret von *Aplysia limacina*. In Lösung bleibendes Pigment bei Proteinfällung mit Ammoniumsulfat.
☐ 53% Sättigung; ▨ 87% Sättigung (nach (*177*))

Zur Gewinnung des Protein-freien Hauptpigments eignet sich am besten die Zerlegung des Chromoproteids mit neutralen Lösungsmitteln (*174*). Aus den analytischen Daten (Tabelle 13) ergibt sich, daß das native Aplysioviolin der Monomethylester einer Dicarbonsäure ist. Aus dem

Tabelle 13. *Analyse von Aplysioviolin (nach 174, 175)*

	Monomethylester (nativ)	Dimethylester
Zusammensetzung	$C_{34}H_{38-40}N_4O_6 \cdot H_2O$	$C_{35}H_{40-42}N_4O_6 \cdot H_2O$
Mol.-Gewicht ber.	598 bis 600 H_2O	612 bis 614 H_2O
Mol.-Gewicht gef.		
osmometrisch*	634	
massenspektrometr.		614
Äquiv.-Gewicht gef.	596	
Zeisel-Bestimmung	1 OCH_3	2 OCH_3

* In Tetrahydrofuran.

Literaturverzeichnis: SS. 128—139

Massenspektrum des Dimethylesters folgt, daß dieser ($M^+ = 614$) 2 H-Atome weniger besitzt als der des Mesobiliviolins ($M^+ = 616$); die zusätzliche Doppelbindung muß im isolierten Pyrrolkern lokalisiert sein, da das Tripyrrol-Fragment, welches dem Chromophor entspricht, in beiden Fällen dieselbe Massenzahl (m/e 492) besitzt (*177, 180*). Jedoch sind die beiden Chromophore voneinander verschieden (*174*). Beim Chromsäureabbau entstehen aus Aplysioviolin neben Methyl-vinyl-maleinimid, Hämatinsäureimid und dessen Methylester als neues Imid das Methyl-äthyliden-succinimid (98) (*174*), dessen Struktur durch Synthese gesichert wurde (*186*). Damit war zum ersten Mal eine Äthyliden-Seitenkette bei einem Gallenfarbstoff nachgewiesen. Aus dem Chromatabbau folgt die IXα-Struktur und die Stellung der veresterten Propionsäure-Seitenkette an C-5 (*177, 176*).

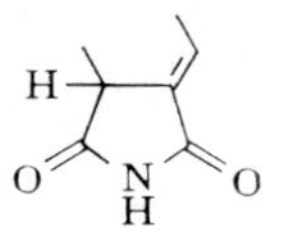

(98)

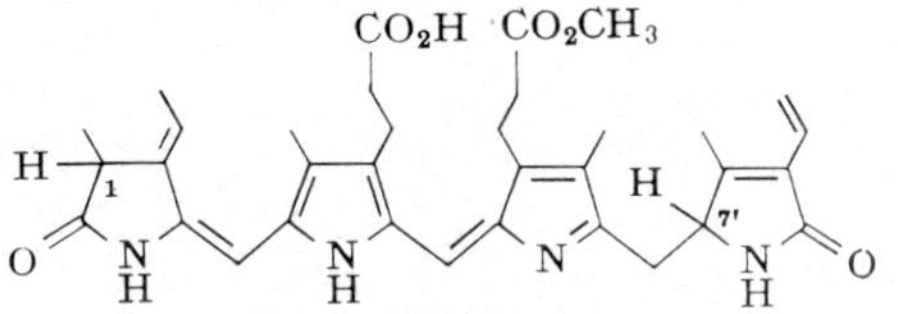

(99)

Die aus diesen Daten für das Aplysioviolin abgeleitete Struktur (99) besitzt einen weniger ausgedehnten Chromophor als das Mesobiliviolin (67). Für die Richtigkeit dieser Annahme spricht neben dem Elektronenspektrum (langwellige Bande für (99) 528 nm, für (67) 555 nm in Methanol) das Ergebnis der spektrophotometrischen pH-Titration: Der pK_a-Wert von 6,0 (*177*) spricht für eine schwächere Delokalisierung des bindenden Elektronenpaares, d. h. für ein weniger ausgedehntes mesomeres System bei (99) als bei Mesobiliviolin ($pK_a = 4,0$) (*177*).

Aplysioviolin besitzt nach der angegebenen Struktur 2 Asymmetriezentren (an C-1 und C-7'), tatsächlich erwies es sich als optisch aktiv (Abb. 9). Der Hauptabsorptionsbande des Pigments ($\lambda_{max}^{Dioxan} = 528$ nm; $\lambda_{max}^{CHCl_3/HCl} = 606$ nm) entspricht im ORD-Spektrum die positive Cotton-Effekt-Kurve mit Wendepunkten bei zirka 530 nm (Dioxan) bzw. 590 nm ($CHCl_3/HCl$). Aplysioviolin ist das erste natürlich vorkommende Biladien-(a,b), bei dem eine optische Aktivität gefunden wurde. Das aus dem Farbstoff gewonnene Methyl-äthyliden-succinimid zeigt keine meßbare Drehung im sichtbaren Spektralbereich, sondern eine solche erst im UV-Bereich in der Nähe des C=O-Chromophors (Abb. 9). Zur Interpretation der ORD-Spektren vgl. S. 79.

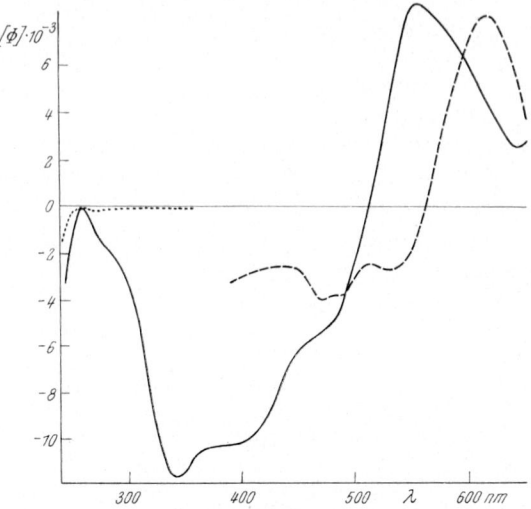

Abb. 9. ORD-Spektren von Aplysioviolin in Dioxan (————) bzw. CHCl$_3$/HCl (— — — —) und Methyl-äthyliden-succinimid in Methanol (·········) (nach (*177*))

Für die blaue Nebenkomponente (Aplysioverdin) wurde die Struktur (**100**) aus Spektraldaten und dem Ergebnis von Chromsäure- und Chromatabbau abgeleitet (*177, 180*). Die Struktur des orangeroten Pigments (Aplysiourobilin) ist noch unbekannt.

(**100**)

2. Phycobiliproteide

Biliproteide von roter (Phycoerythrine) und blauer Farbe (Phycocyanine) sind bei manchen Algen derart vorherrschend, daß sie zur Namengebung beigetragen haben (Rotalgen = Rhodophyten; Blaualgen = Cyanophyten); sie kommen außerdem bei Cryptophyten vor. Diese Biliproteide besitzen eine brillante Fluoreszenz. Nach den Elek-

Literaturverzeichnis: SS. 128—139

tronenspektren unterscheidet man sechs Phycoerythrine* und sechs Phycocyanine, deren Verbreitung und spektrale Eigenschaften in Tabelle 14 angegeben sind.

Tabelle 14. *Spektraldaten und Verbreitung von Phycobiliproteiden (22, 147)*

Biliproteid	Phycoerythrin			Phycocyanin		Allo-
	R-	B-	C-	R-	C-	
Maxima der Elektronenabsorption (nm)	499 540 568	546 565	557—560	553 615—620	615—620	(610) 650
Maxima der Fluoreszenzemission (nm)	578—580	578—580	575	565 637	647	660
Rhodophyta						
Porphyridiales		×		×	×	×
Bangiales	×	×		×	×	×
Nemalionales	×	×				
Gigartinales	×					
Cryptonemiales	×				×	×
Rhodymeniales	×			×		×
Ceramiales	×			×	×	×
Klassifizierung unbekannt:						
Cyanidium caldarium					×	×
Cyanophyta						
Chroococcales					×	×
Nostocales			×		×	×
Biliproteid	Cryptomonad-Phycoerythrin			Cryptomonad-Phycocyanin		
Maxima der Elektronenabsorption (nm)	544	555	565—568	580—585 583 (620—625) 643—650	625—630	585—588 615
Maxima der Fluoreszenzemission (nm)		578 bis 580			660	
Cryptophyta						
Cryptomonadales	×	×	×	×	×	×

* Zu einer möglichen weiteren Unterteilung und zur Phylogenie vgl. (*87*).

HAXO und BLINKS (*84*) zeigten, daß das Wirkungsspektrum der Photosynthese z. B. bei der Rotalge *Myriogramme spectabilis* dem Absorptionsspektrum von R-Phycoerythrin entspricht (Abb. 10). Die Banden der Fluoreszenzemission von Phycoerythrinen überlappen sich teilweise mit den Absorptionsbanden der Phycocyanine und deren Fluoreszenzemission entsprechend mit den Absorptionsbanden von Chlorophyll a des Photosynthese-Systems 2. Daher wird bei den Algen, die rote Biliproteide neben blauen enthalten, ein Übergang der Lichtenergie vom

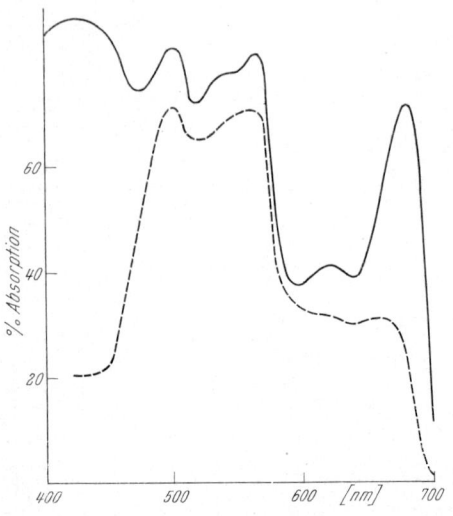

Abb. 10. Vergleich von Absorptionsspektrum des Thallus (————) und Wirkungsspektrum der Photosynthese (– – – –) bei *Myriogramme spectabilis* (nach (*84*))

Phycoerythrin zunächst zum Phycocyanin angenommen; nachgewiesen wurde die Energieübertragung von beiden Biliproteiden auf Chlorophyll (*2, 50*). Die Phycobiliproteide sind daher als accessorische Pigmente des Photosynthese-Systems 2 anzusehen (*51, 62*).

Biliproteide mit anderen Funktionen sind bei Algen nur gelegentlich beschrieben worden, z. B. Phytochrom (vgl. S. 124) (*49, 83, 216*), ferner ein dem Phycocyanin ähnliches Pigment, das bei der Photooxydation von Ascorbinsäure wirksam ist, bei einer Blaualge (*68*).

Die photosynthetisch aktiven Biliproteide sind in der Zelle an der Außenseite der Photosynthese-Lamellen in Form von „Phycobilisomen" lokalisiert (*69, 70, 71*). Bei *Porphyridium cruentum* konnte aus elektronenmikroskopischen Messungen berechnet werden, daß pro Phycobilisom (Durchmesser 350 Å) 1400 Chlorophyll- und zirka 35 Biliproteid-Moleküle (überwiegend Phycoerythrin) vorhanden sind (*69*).

Literaturverzeichnis: SS. 128—139

Die Biliproteide gehen aus den Algenzellen leicht in wäßrige Lösung, sobald die Zellwände zerstört sind. Die Absorptionsspektren der so gewonnenen, nativen Chromoproteide (Tabelle 14) werden verändert, sobald eine Dissoziation in Untereinheiten eintritt. So verliert das R-Phycoerythrin aus *Porphyridium cruentum* bei Einwirkung von Dodecylsulfat *(69)* oder p-Chloromercuribenzoat *(63, 65)* gleichzeitig mit seiner Fluoreszenz auch die Absorptionsbande bei 563 nm. Die Untereinheiten mit der Bande bei 500 und 545 nm konnten voneinander getrennt werden *(65)*; bei Zugabe von Mercaptoäthanol oder Glutathion (Entfernen der Quecksilberverbindung) assoziieren die Untereinheiten unter teilweiser Rückbildung der ursprünglichen Spektraleigenschaften. Die Verhältnisse sind bei anderen Phycoerythrinen *(158, 159)* und Phycocyaninen *(64)* ähnlich. Auch die Circulardichroismus-Spektren ändern sich bei Dissoziation und Reassoziation *(160)*.

Für die Phycobiliproteide wurden Mol.-Gewichte von 200 000 bis 300 000 gefunden *(148)*. Unterschiedliche Angaben für ein und dasselbe Biliproteid dürften auf eine reversible Dissoziation und Assoziation zurückzuführen sein, die vom pH-Wert, der Ionenstärke der Lösung, der Konzentration des Biliproteids und der Temperatur abhängig ist *(5, 133)*.

Am eingehendsten ist das C-Phycocyanin untersucht worden. Aus der Aminosäureanalyse *(6)* ergab sich ein Mindest-Mol.-Gew. von 15 200 *(82)*. Für die kleinsten bei der Dissoziation entstehenden Untereinheiten wurden Mol.-Gew. von 14 000 *(13)*, 20 000 *(139)*, 30 000 *(42, 98, 133)* oder 46 000 *(82)* angegeben. Am zuverlässigsten dürften die Meßwerte von BERNS und Mitarb. *(8, 98)* sein; mit drei unabhängigen Methoden (Sedimentation in 5 M Guanidin; Natriumdodecylsulfat-Polyacrylamidgel-Elektrophorese; Analyse der Peptide nach Spaltung mit Bromcyan) wurden durchweg Werte um 30 000 gefunden, die nur wenig von Species zu Species variieren (vgl. Tabelle 15). Hexamere dieser Untereinheit

Tabelle 15. *Mol.-Gewichte der Untereinheiten von C-Phycocyaninen. Bestimmung mit der Natriumdodecylsulfat-Polyacrylamidgel-Elektrophorese (nach 98)*

Isoliert aus	Mol.-Gewicht
Plectonema calothricoides	28 500
Phormidium luridum	28 000
Tolypothrix tenuis	28 500
Synechococcus lividus	27 000
Cyanidium caldarium*	25 500
Anabaena sp.*	30 000
Anacystis nidulans*	25 500
Porphyridium aerugenium*	25 500
Porphyridium cruentum*	28 700
Calothrix membranacea*	26 000

* Im Rohextrakt bestimmt.

(Mol.-Gew. 180000) (7, 42) liegen in wäßriger Lösung mit kleineren Einheiten im Gleichgewicht vor; diese wurden bei *Anacystis nidulans* als Dimere (Mol.-Gew. 60000) (133), bei anderen Blaualgen als Trimere (Mol.-Gew. 90000) (5) erkannt. Daneben wurden höhere Aggregate (Mol.-Gew. 360000) gefunden (108). Messungen an Gemischen dieser Assoziate ergeben abweichende Mol.-Gew. (133).

Für *R-Phycoerythrin* wurden Dissoziationsprodukte vom Mol.-Gew. 19500 (139) sowie 23000, 36000 und 85000 (123) gefunden; diese wurden unter Berücksichtigung des Mindest-Mol.-Gew. von 18000 als Monomere, Dimere und Polymere angesehen. Das native R-Phycoerythrin (Mol.-Gew. 300000) besitzt nach elektronenoptischen Aufnahmen — im Gegensatz zum ringförmigen hexameren C-Phycocyanin (7) — einen kompakten Aufbau (69). *R-Phycocyanin* läßt sich in rote Untereinheiten vom Mol.-Gew. 20500 und blaue vom Mol.-Gew. 18500 zerlegen (139).

Phycobiline

Der rote Chromophor der Phycobiliproteide wurde Phycoerythrobilin, der blaue entsprechend Phycocyanobilin genannt, lange ehe die Strukturen bekannt waren (109). Schwierigkeiten bei den Strukturuntersuchungen ergaben sich aus den sekundären Veränderungen, die die Pigmente bei der Spaltung der kovalenten Bindung zum Protein erfahren.

Mesobilatrien-IXα, der erste kristalline, bei der alkalischen Hydrolyse der Phycobiliproteide gewonnene Gallenfarbstoff, ist offensichtlich ein Sekundärprodukt (114). Saure Hydrolyse ergibt in der Hitze vorwiegend Bilipeptide (109), die wahrscheinlich erst im sauren Medium gebildete Thioäthergruppen enthalten (194). In der Kälte erhält man mit 12 N HCl aus Phycocyaninen drei Pigmente, die nach ihrem Absorptionsmaximum in saurem $CHCl_3$ als Phycobilin-608, -630 und -635 bezeichnet wurden (147); das zuerst gebildete Pigment -630 (Phycocyanobilin) geht zunächst in das (als Dimethylester allein stabile) Pigment -655 und dann in das Pigment -608 über (147). Auch mit neutralem, siedendem Methanol wird aus Phycocyaninen ein Protein-freies Bilin (blue pigment; ,,Phycocyanobilin" (32, 43, 66, 67, 144)) freigesetzt, was eher auf eine Eliminierung als auf eine Hydrolyse hindeutet. Um Verwechslungen mit dem Phycocyanobilin (= Phycobilin-630) zu vermeiden, schlug RÜDIGER (179) für dieses Pigment den Namen Phycobiliverdin vor. Auch bei der enzymatischen Hydrolyse mit Nagarse erhält man dieses Pigment (208).

Aus Phycoerythrinen erhält man mit 12 N HCl in der Kälte zuerst ein rotes Pigment (Phycoerythrobilin (145), Phycoerythrobilin-560 (209)), das dann in ein kurzwellig absorbierendes Pigment vom Urobilin-Typ übergeht (145). Ein längerwellig absorbierendes Pigment (,,purple pigment" (67, 144)), ,,Phycoerythrobilin" (26), ,,Phycoerythrobilin-590"

Literaturverzeichnis: SS. 128—139

(*209*)) wird mit siedendem Methanol erhalten; dieses wird zweckmäßigerweise als Phycobiliviolin bezeichnet (*179*). Phycobiliviolin und Phycoerythrobilin lassen sich reversibel ineinander überführen (*144, 179, 209*), entsprechendes gilt für Phycobiliverdin und Phycocyanobilin (*144, 179*).

Die Strukturen der mit siedendem Methanol erhältlichen Phycobiline wurden gleichzeitig in drei Arbeitskreisen aufgeklärt. SIEGELMAN und Mitarb. (*32, 33, 208, 209*) kristallisierten die Pigmente in Form der Dimethylester nach dünnschichtchromatographischer Reinigung. Im Massenspektrum fanden sie für Phycobiliverdin und Phycobiliviolin dasselbe Mol.-Gew. wie für Mesobiliverdin ($M^+ = 614$). Phycobiliverdin gibt im Gegensatz zu Mesobiliverdin ein intensives Fragment bei m/e 599, was sich auf den Verlust einer Methylgruppe von einem gesättigten C-Atom zurückführen läßt. Bei Phycobiliviolin entspricht der Peak bei m/e 492 dem bei Mesobiliviolin erhaltenen Tripyrrol-Fragment. Auf Grund der NMR-Spektren (vgl. Tabelle 5, S. 84) wurden die Strukturen (101) für Phycobiliverdin und (102) für Phycobiliviolin vorgeschlagen. Die Zuordnung der Protonen (z. B. des Methin-Protons der Äthyliden-Gruppe, das ein Quartett von Dubletts bei 6,18 ppm zeigt) wurde durch Entkopplungsversuche gesichert (*33*).

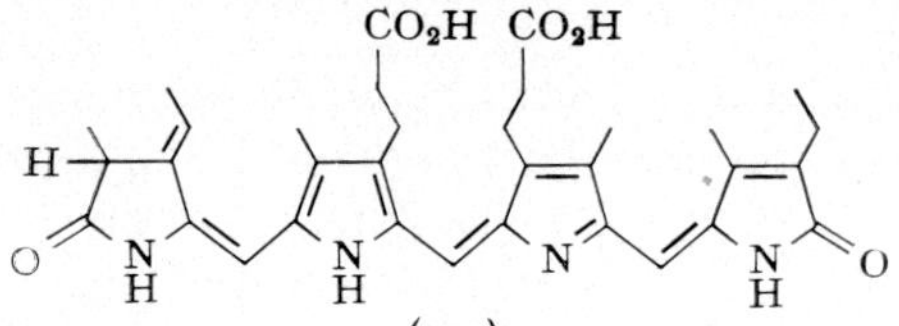

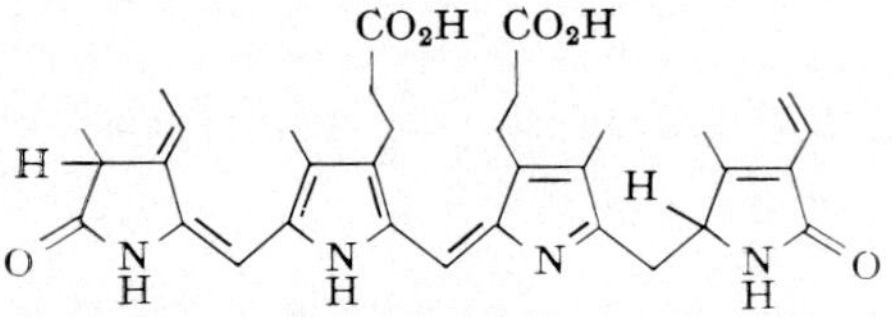

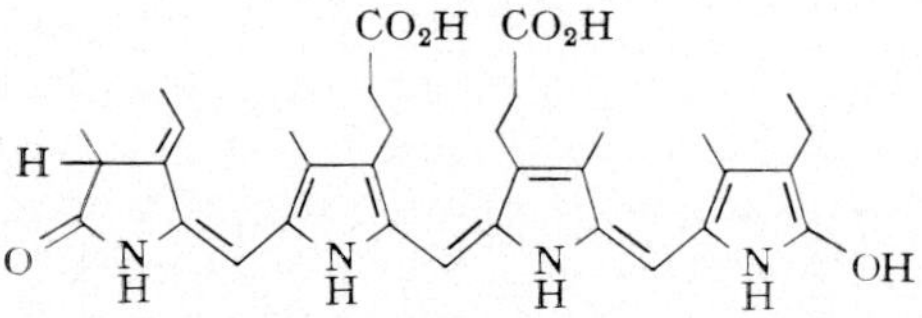

CRESPI, KATZ und Mitarb. (*43, 46*) untersuchten die Dicarbonsäure des Phycobiliverdins. Das NMR-Spektrum (Tabelle 5) entspricht dem des Dimethylesters. Der Mol.-Peak der Säure ($M^+ = 588$) besitzt jedoch zwei Masseneinheiten mehr als nach der Struktur des Dimethylesters zu erwarten wäre. Zunächst (*43*) wurde daher eine abweichende Struktur (**103**) für das Pigment vorgeschlagen, jedoch wurden im NMR-Spektrum von den sieben bei tiefem Feld zu erwartenden Protonen (4 NH, 3 OH) nur fünf gefunden. Später (*46*) wurde eine mögliche Disproportionierung des Farbstoffs im Massenspektrometer in Betracht gezogen und Struktur (**101**) (bzw. eine tautomere Struktur) für wahrscheinlich erachtet, jedoch ist die Diskussion über eine um 2 H-Atome reichere Struktur noch nicht beendet (vgl. *198*).

RÜDIGER wendete in Zusammenarbeit mit O'CARRA und O'HEOCHA (*194, 195*) den Chromsäureabbau auf die Phycobiline an. Unter Berücksichtigung von Elektronenspektren wurden dadurch dieselben Strukturen (**101**) und (**102**) für die Phycobiline abgeleitet. Der Dimethylester von (**102**) ist nach seinem spektralen und chromatographischen Verhalten identisch mit dem aus Aplysioviolin gewonnenen Dimethylester, was von anderen Autoren (*209*) bestätigt wurde. Zur Sicherung der Strukturen (**101**) und (**102**) diente auch die leichte Isomerisierung beider Pigmente mit Alkali zu Mesobiliverdin (*32, 33, 179, 194, 26*), das als reines IX α-Isomeres erkannt wurde; andere Isomere waren nicht nachzuweisen (*142*).

CHAPMAN, COLE und SIEGELMAN (*27*) untersuchten auch das aus Phycocyanin mit konz. Salzsäure erhaltene Pigment. Nach Veresterung erwies sich dieses als identisch mit dem Dimethylester von Phycobiliverdin. Die Autoren schlossen, daß sie die Identität von Phycobilin-630 mit Phycobiliverdin bewiesen hätten, obwohl O'HEOCHA (*147*) gezeigt hatte, daß der Ester-630 leicht in den Ester-655 übergeht. Die Spektraldaten (z. B. in $CHCl_3$/HCl 655 bzw. 660—680 mμ) sprechen eher dafür, daß Phycobilin-655 und Phycobiliverdin identisch sind (vgl. *198*). Der Chromsäureabbau lieferte aus den mit konz. Säure erhaltenen Pigmenten dieselben Abbauprodukte wie bei den mit Methanol abgespaltenen (*194, 195*); es wurde daher geschlossen, daß die beiden Gruppen von Pigmenten entweder dieselben Seitenketten besitzen oder daß diese (bei den Imiden) leicht ineinander umgewandelt werden, zumal auch die Pigmente ineinander übergehen können.

Die Ähnlichkeit zwischen Phycoerythrobilin und Mesobilirhodin (*79*) war bereits früher erkannt worden (*145*). Unter Berücksichtigung der heutigen Kenntnis über die Bindungen zwischen Chromophor und Protein im Phycoerythrin (vgl. S. 123) kann für das Phycoerythrobilin die Struktur (**104**) vorgeschlagen werden. Diese besitzt einen ähnlichen Chromophor wie das Mesobilirhodin; durch Eliminierung von H_2O sollte es in Phycobiliviolin übergehen, was auch beobachtet wird (Kochen mit Methanol). Die analoge Struktur (**105**) dürfte für Phycocyanobilin zutreffen.

Literaturverzeichnis: SS. 128—139

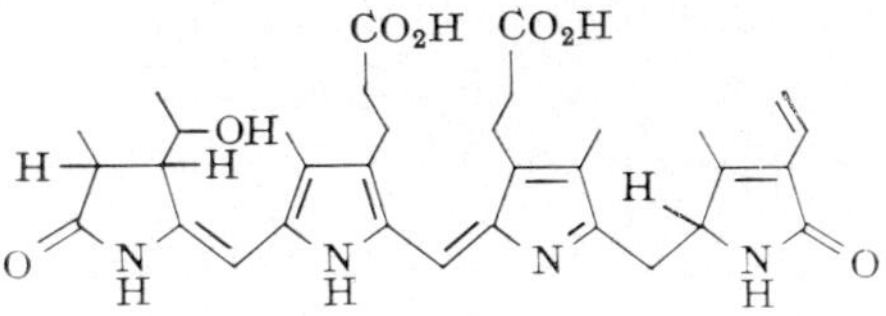

Gallenfarbstoffe und Biliproteide 121

(104)

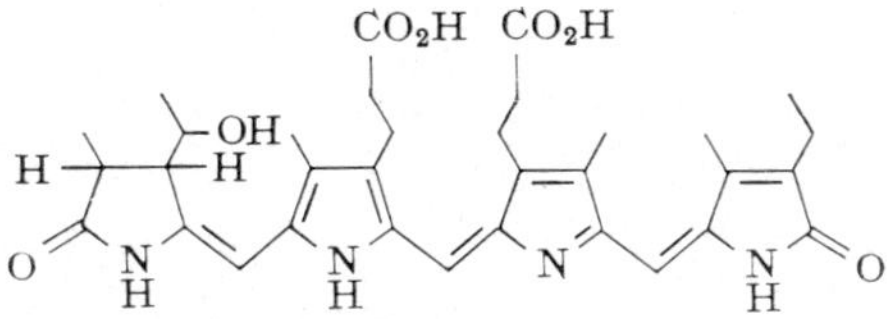

(105)

Alle Phycocyanine enthalten nur Phycocyanobilin als Chromophor (*148*) (bei C- und Allo-Phycocyanin als Phycobiliverdin isoliert (*28*)), R-Phycocyanin daneben Phycoerythrobilin (*28*). Das Vorkommen von Phycoerythrobilin in allen Phycoerythrinen ist gesichert (*29*, *145*, *157*); dagegen wird das Vorliegen eines Phycourobilins im R- und B-Phycoerythrin (*139*, *145*, *148*) bestritten (*29*); da das fragliche Maximum (zirka 500 nm) bei höheren pH-Werten (ab zirka 10) in allen Phycoerythrinen erscheint bzw. stärker ausgeprägt ist (*29*), könnte es auf die Absorption der freien Base von Phycoerythrobilin zurückzuführen sein (vgl. S. 123).

Für die *kovalenten Bindungen* der Phycobiline an das Protein wurden zunächst die Propionsäure-Seitenketten der Pigmente in Betracht gezogen; es wurden Amid- (*116*, dort S. 147) oder Ester-Bindungen (*145*, *147*) postuliert. Bei der Abspaltung von Phycobiliverdin aus einem vollständig deuterierten Phycocyanin mit siedendem ^{1}H-Methanol wurden, wie NMR-Messungen zeigten, u. a. Methyl-Protonen der Äthyliden-Gruppe ausgetauscht; daher wurde eine Bindung an dieser Stelle vermutet (*46*) (**106**). Ein anderer Strukturvorschlag (*45*) (**107**) beruht auf folgenden Versuchen: Nach enzymatischer Hydrolyse von Phycocyanin mit Nagarse wurden Chromopeptide isoliert, deren Zusammensetzung [neben Phycocyanobilin: Cystein, (Leucin, Cystein), (Leucin, Cystein, Alanin), (Asparaginsäure, Serin, Leucin, Cystein) und (Glycin, Asparaginsäure, Serin, Leucin, Cystein)] eine Bindung des Pigments an das Cystein vermuten ließen (*45*). Da Schwermetalle die Spaltung erleichtern, wird ein Thioäther (Analogie zum Cytochrom c) angenommen; eine zusätzliche „Ester"gruppe zwischen Asparaginsäure und dem Hydroxypyrrol soll durch das siedende Methanol gesprengt werden (Umesterung).

Beim Chromsäureabbau der Biliproteide werden unter hydrolytischen Bedingungen die Imide aus allen vier Pyrrolkernen erhalten, unter nicht-hydrolytischen Bedingungen jedoch nur die Imide aus den Kernen B und D; die aus den Kernen A und C abgeleiteten Imide können dann aus dem Protein-Niederschlag durch Hydrolyse freigesetzt werden (*176*, *179*,

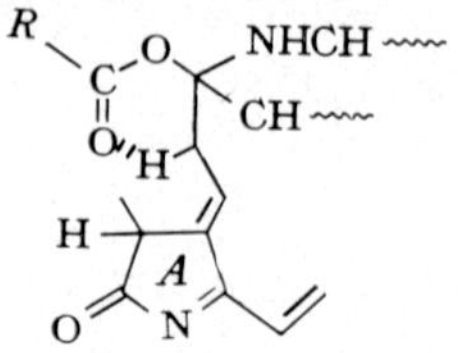

(106)

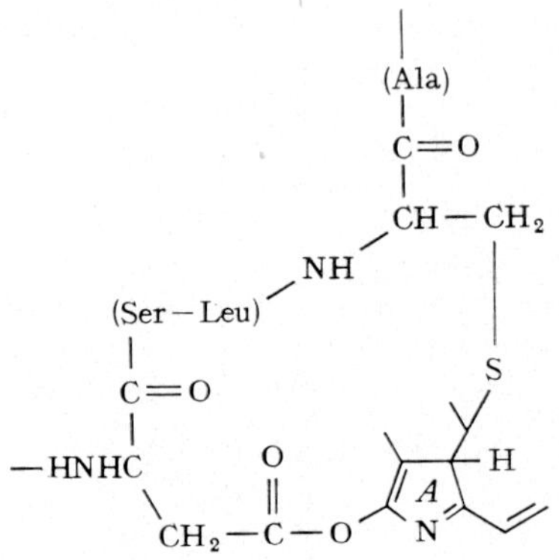

(107)

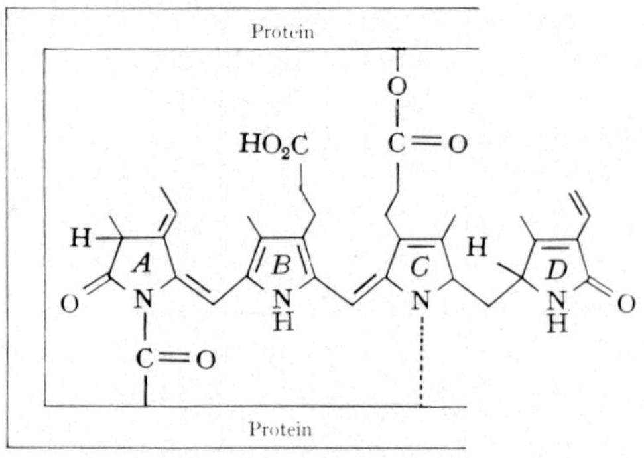

(108)

194). Aus diesem Ergebnis wurde auf Bindungen zwischen dem Protein und den Kernen A und C der Pigmente geschlossen; als mögliche Bindungen wurden eine Esterbindung der Propionsäure-Seitenkette und eine N-Acyl-Bindung des Pyrrolon-Stickstoffs angenommen (**108**) (*179*).

Literaturverzeichnis: SS. 128—139

Die aus R-, B- und C-Phycoerythrin durch Anwendung verschiedener proteolytischer Enzyme gewonnenen, kleinsten Chromopeptide enthielten stets Serin und Glutaminsäure, daneben in wechselnden Mengen Asparaginsäure, Threonin, Glycin und Alanin (*100*). Die Ergebnisse wurden im Sinne der Struktur (**108**) gedeutet, lassen sich aber auch, bei Berücksichtigung der im folgenden beschrieben Ergebnisse, anders deuten.

In einer eingehenden Untersuchung fand BROOKS (*22*), daß das aus dem Cryptomonad-Phycoerythrin von Rhodomonas durch enzymatische Hydrolyse (Pronase und Pankreatin) gewonnene, kleinste Chromopeptid Alanin, Cystein, Serin, Glutaminsäure und Phycoerythrobilin im Verhältnis 1 : 1 : 1 : 1 : 1 enthielt. Da das Alanin N-terminal, Serin und Glutaminsäure C-terminal stehen, ferner die Thiol-Gruppe des Cysteins frei, die Hydroxyl-Gruppe des Serins gebunden ist, ergeben sich zwei Bindungsstellen; die Hydroxyl-Gruppe des Serins und die Amino-Gruppe der Glutaminsäure. Die von BROOKS postulierten Ester- und Amid-Bindungen an beiden Propionsäure-Seitenketten stehen im Widerspruch zum Ergebnis des Chromsäureabbaus. Dagegen steht Struktur (**109**) mit allen Versuchsdaten in Einklang (vgl. auch Struktur (**104**) für Phycoerythrobilin). Weitere kovalente Bindungen zwischen Chromophor und Protein sind nach bisherigen Ergebnissen nicht ausgeschlossen, dürfen aber nur die Kerne A und C des Gallenfarbstoffs umfassen.

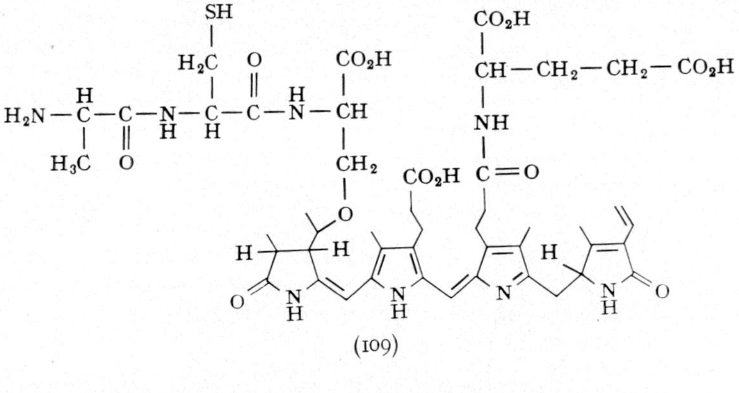

(**109**)

Dagegen sind *Nebenvalenz-Bindungen* auch zwischen anderen Kernen der Biline und dem Protein möglich. Solche Bindungen wurden durch Spektralmessungen wahrscheinlich gemacht (*4, 96, 145, 149*) und als Wasserstoff-Brückenbindung formuliert (*179*); möglicherweise spielen dabei auch Protonenübergänge eine Rolle, da die Hauptmaxima der Biliproteide durchweg eher denen der Bilin-Hydrochloride als denen der freien Bilin-Basen entsprechen. Diese Bindungen dürften auch bei der Aggregation der Biliproteide beteiligt sein (vgl. S. 117).

3. Phytochrom

Phytochrom ist als Lichtrezeptor für die photoreversible Kontrolle von Wachstum und Entwicklung höherer Pflanzen entdeckt worden (zur Geschichte der Entdeckung vgl. *207*). Es scheint im Pflanzenreich weit verbreitet zu sein und vielfältige Funktionen zu erfüllen. Bei der Photomorphogenese dürfte das Phytochrom eine differentielle Gen-Aktivierung bzw. -Repression bewirken; in eingehenden Versuchen mit niederenergetischer Bestrahlung wurden Enzyminduktion und Enzymrepression beim Senfkeimling (*Sinapis alba* L.) nachgewiesen (Zusammenfassung vgl. *125*, *126*). Auch unter Hochenergiebedingungen ist Phytochrom als Photorezeptor für die Photomorphogenese anzunehmen (*16*, *81*). Schnelle physiologische Effekte des Phytochroms, wie die Licht-abhängige Chloroplastendrehung bei der Alge *Mougeotia* (*83*), verlaufen wohl nach einem anderen Mechanismus, der bisher unbekannt ist. Mögliche Wirkungsweisen des Phytochroms werden von HENDRICHS und BORTHWICK diskutiert (*85*, *86*).

Charakteristisch für die durch das Phytochrom bewirkten Photomorphosen ist die Photoreversibilität: Die durch hellrotes Licht (660 nm, engl. ,,red light") hervorgerufenen Effekte werden durch nachfolgende Bestrahlung mit dunkelrotem Licht (730 nm; engl. ,,far-red light") rückgängig gemacht. Der Lichtrezeptor muß demnach in zwei photoreversiblen, bei diesen Wellenlängen absorbierenden Formen existieren (P_r und P_{fr}), von denen P_{fr} die physiologisch aktive ist (Effektor-Molekül (*125*)).

$$P_r \xrightleftharpoons[\text{Dunkelrot}]{\text{Hellrot}} P_{fr}$$

Diese Formulierung stellt eine Vereinfachung dar: Da die beiden Phytochrom-Formen in einem weiten Spektralbereich absorbieren, ist auch Licht anderer Wellenlänge wirksam; bei jeder Wellenlänge bildet sich ein charakteristisches, photostationäres Gleichgewicht aus (*81*, *125*). Möglicherweise beruhen komplexe Lichtreaktionen auch auf dem Vorliegen mehrerer Chromophore (*40*). Es gibt sowohl bei der Hin- als auch bei der Rückreaktion Zwischenstufen, die zum Teil nacheinander, zum Teil parallel geschaltet sind (*17*, *18*, *40*, *47*, *119*, *120*). Für viele Betrachtungen genügt jedoch das vereinfachte Schema.

Die der photoreversiblen Phytochrom-Umwandlung entsprechenden Absorptionsänderungen lassen sich *in situ* in Chlorophyll-freien Pflanzenteilen nachweisen (*23*); ein dem Absorptionsspektrum in vivo (Abb. 11) entsprechendes Spektrum findet man auch im Rohextrakt (*193*).

Phytochrom wurde zuerst von SIEGELMAN und FIRER (*210*) aus etiolierten Haferkeimlingen isoliert; die Isolierungsmethode wurde für Demonstrationszwecke vereinfacht (*124*) und für die präparative Ge-

Literaturverzeichnis: SS. 128—139

winnung (211) modifiziert. Wesentlicher Reinigungsschritt dieser Methode ist die wiederholte Chromatographie an Calciumphosphat („brushite") mit abwechselnder Gelfiltration. Für größere Ansätze eignet sich die Adsorption im „batch"-Verfahren an Hydroxylapatit und anschließende Chromatographie an DEAE-Sephadex (105). Eine weiterentwickelte

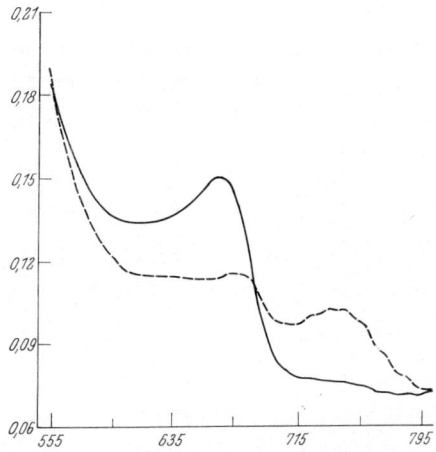

Abb. 11. Absorptionsspektrum von Koleoptilen etiolierter Roggenkeimlinge in vivo nach Bestrahlung bei 730 nm (————) und 660 nm (— — — —) (nach (41))

Isolierungsmethode lieferte unter Anwendung der trägerfreien Elektrophorese ein reines Produkt (130). Aus etiolierten Roggenkeimlingen wurde Phytochrom in weniger Reinigungsschritten (ohne Gelfiltration) isoliert (39). Die Ergebnisse der Isolierungsmethoden sind in Tabelle 16

Tabelle 16. *Isolierung von Phytochrom aus etiolierten Keimlingen*

Material	Anreicherung bez. auf den Rohextrakt = 1	Ausbeute Rohextrakt	E_{661}/E_{280} P_r-Form	Lit.
Hafer (*Avena sativa* L.)	61	0,06	0,11	(210)
	740	0,03	0,92	(130)
Roggen (*Secale cereale* L. cv. Balbo)	1200	0,25	0,57	(39, 41)

zusammengestellt. Als Reinheitskriterium ist der Anreicherungsfaktor weniger geeignet, da er sich auf den (eventuell unterschiedlichen) Gesamt-Gehalt an Proteinen im Rohextrakt bezieht. Zweckmäßigerweise vergleicht man das Verhältnis der Absorptionen (E_{664}/E_{280}) der P_r-Form, das umso höher ist, je weniger andere Proteine (mit $\lambda_{max} = 280$ nm) beigemengt sind (vgl. aber *41*). Das Absorptionsspektrum der entsprechend diesem Kriterium reinsten Präparation zeigt Abb. 12.

Das Molekulargewicht von isoliertem Hafer-Phytochrom wurde mit Hilfe der Gelfiltration zu 55000—62000 (*130*) und 55500 (*105*) bestimmt. Möglicherweise tritt während der Isolierung eine Dissoziation ein: Aus Sedimentationsversuchen (*41*) wurde beim Phytochrom aus Roggen auf das Vorliegen von 3 S-Tetrameren (Mol.-Gew. 150000—190000) und 14 S-Hexameren (Mol.-Gew. zirka 250000) geschlossen; 2 S-Monomere (Mol.-Gew. 42000 $\pm$ 3000) entstehen bei Carboxymethylierung oder Behandlung mit Dodecylsulfat bzw. Harnstoff sowie bei der Dialyse (Ab-

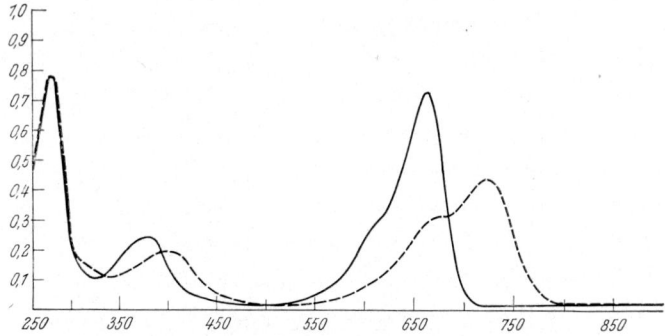

Abb. 12. Absorptionsspektrum von isoliertem Phytochrom aus Hafer nach Bestrahlung bei 730 nm (P_r-Form; ———) und 660 nm (P_{fr}-Form; — — — —) (nach (*130*))

sinken des pH-Wertes). Elektronenmikroskopische Aufnahmen bestätigten das Vorliegen von Tetrameren und Hexameren (*41*). Bei Einhalten eines pH-Wertes von 7,8 während der Isolierung wurden dieselben Aggregate auch beim Hafer-Phytochrom gefunden (*38*).

Für den Phytochrom-Chromophor (Phytochromobilin) wurde bereits in einem frühen Stadium der Untersuchungen die Struktur eines Gallenfarbstoffs postuliert, da das Wirkungsspektrum des Phytochroms Ähnlichkeit mit den Absorptionsspektren des Biliproteids Allo-Phycocyanin besitzt (vgl. *207*). Im Absorptionsspektrum des isolierten Chromoproteids (Abb. 12, P_r-Form) kommt diese Ähnlichkeit noch deutlicher zum Ausdruck (*210*). Die Abspaltung des Chromophors gelang (allerdings nur mit einer Ausbeute von 5—10%) durch Behandlung des denaturierten Chromoproteids mit siedendem Methanol (*212*). Die Bildung eines Mono- oder eines Dimethylesters wurde chromatographisch nachgewiesen; der Chromophor ist also eine Dicarbonsäure. Nach dem Absorptionsspektrum (λ_{max} = 380, 690 nm in 5% HCl/Methanol) handelt es sich um ein Bilatrien, was durch die rote Fluoreszenz des Zinkkomplexes nach Oxydation mit Jod (Bildung eines Biladienons) bestätigt wird.

Literaturverzeichnis: SS. 128—139

Durch den Chromsäure- und Chromatabbau wurde die Zugehörigkeit des Phytochrom-Chromophors zu den Gallenfarbstoffen bewiesen und gleichzeitig eine Struktur vorgeschlagen, die die Bindungen an das Protein berücksichtigt (*185*). Für diese Versuche wurde ein denaturiertes nicht mehr photoreversibles Phytochrom-Präparat aus Sommerroggen verwendet. Eine reversible Umfärbung ließ sich jedoch chemisch erreichen: Mit Säure färbte sich das Präparat blau (entsprechend P_r) mit Alkali grüngelb (entsprechend P_{fr}). Alle Gallenfarbstoffe mit Äthylidengruppe zeigen bei der Einwirkung von Alkali eine entsprechende Umfärbung, die durch Säure rückgängig gemacht werden kann (*179, 194*); die Verschiebung der Hauptmaxima ist sogar größer als bei der Umwandlung von P_r (λ_{max} 660 nm) in P_{fr} (λ_{max} 730 nm) (Tabelle 17). Für die bei längeren Wellen absorbierenden Formen, die Zwischenstufen bei der Wanderung der exocyclischen Äthyliden-Doppelbindung in die endocyclische Position sind, wurde die Struktur eines mesomeren Anions vorgeschlagen (*185*) (110).

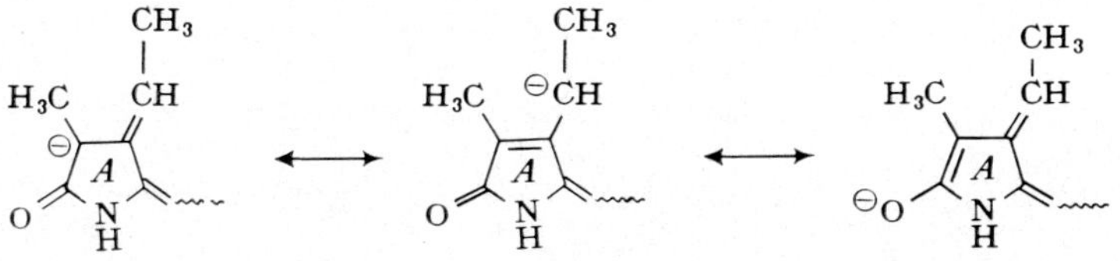

(110)

Tabelle 17. *Langwellige Absorptionsmaxima von Gallenfarbstoffen in neutraler und alkalischer Lösung (nach 194)*

Pigment	λ_{max} in Methanol (nm)	
	pH = 7	pH $\geq$ 11
Phycoerythrobilin (**104**)	500	600—605
Phycobiliviolin (**102**)	530	630—635
Phycocyanobilin (**105**)	575	770
Phycobiliverdin (**101**)	590	775
Mesobiliviolin (**67**)	580	580

Die Anwesenheit einer exocyclischen Doppelbindung in der blauen, denaturierten Phytochrom-Form gab sich durch das Auftreten von Methyl-äthyliden-succinimid beim Chromsäureabbau zu erkennen (*185*); der entsprechende Ring A des Chromophors liegt im denaturierten Präparat frei vor, dürfte aber im nativen Pigment Protein-gebunden sein, da das Imid dort unter entsprechenden Bedingungen nicht freigesetzt wird (*193*). Aus der denaturierten grüngelben Form erhält man beim Chromsäureabbau Methyl-äthyl-maleinimid statt Methyl-äthyliden-succinimid; hier liegt also eine endocyclische Doppelbindung vor. Bindung und Struktur der Ringe B, C und D entsprechen denen in den Phycocyaninen mit dem Unterschied, daß beim (nativen) Phytochrom Methyl-vinyl-maleinimid (statt Methyl-äthyl-maleinimid) aus Ring D entsteht (*193*); die früher (*185*) postulierte Bindung des Ringes D an das Protein ist offensichtlich ein Artefakt der Denaturierung. Aus diesen Ergebnissen wurden die Strukturen (111) für die blaue („P$_r$") und (112) für die grüngelbe („P$_{fr}$") Form des Phytochroms abgeleitet (modifiziert nach *185*). Welche Aminosäuren im Phytochrom mit dem Chromophor verknüpft sind, ist bisher unbekannt.

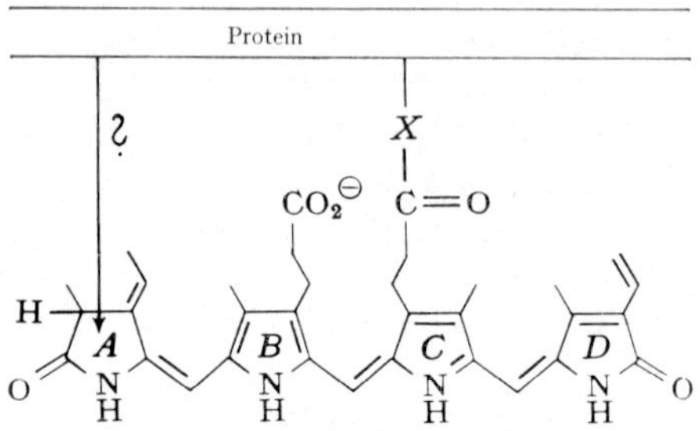

(111)

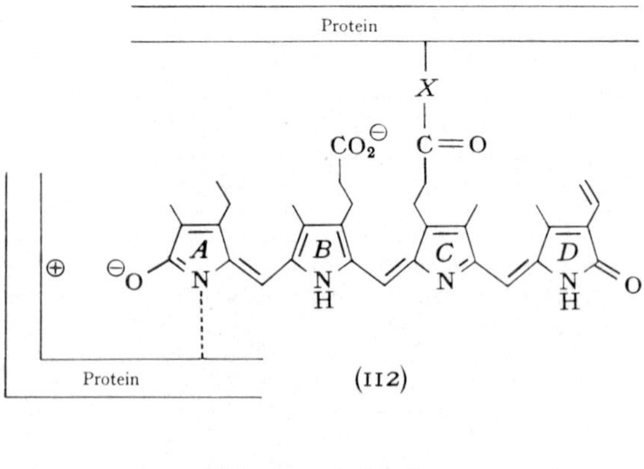

(112)

Literaturverzeichnis

1. ABOLINŠ, L., und W. RÜDIGER: Über die farbgebende Gruppe von Crenilabrus-Blau. Experientia **22**, 298 (1966).
2. ARNOLD, W., und J. R. OPPENHEIMER: Internal conversion in the photosynthetic mechanism of blue-green algae. J. gen. Physiol. **33**, 423 (1950).
3. BADA, J. L.: A blue-green pigment isolated from blood plasma of the Arctic sculpin *(Myoxocephalus scorpioides)*. Experientia **26**, 251 (1970).
4. BENNHOLD, H.: The transport of bilirubin in the circulating blood and its pathogenic importance. Acta Med. Scand. Suppl. **445**, S. 222 (1966).

5. BERNS, D. S.: Protein aggregation in phycocyanin. Osmotic pressure studies. Biochem. Biophys. Res. Comm. **38**, 65 (1970).
6. BERNS, D. S., H. L. CRESPI und J. J. KATZ: Isolation, amino acid composition and some physico-chemical properties of the protein deuterio-phycocyanin. J. Amer. chem. Soc. **85**, 8 (1963).
7. BERNS, D. S. und M. R. EDWARDS: Electron micrographic investigations of C-phycocyanin. Arch. Biochem. Biophys. **110**, 511 (1965).
8. BERNS, D. S., R. MACCOLL und J. J. LEE: The aggregation properties of C-phycocyanin. Biochem. J. **119**, 14p (1970).
9. BILLING, B. H., P. G. COLE und G. H. LATHE: The excretion of bilirubin as a diglucuronide giving the direct van den Bergh reaction. Biochem. J. **65**, 774 (1957).
10. BLAUER, G., D. HARMATZ und A. N. NEPARSTEK: Circular dichroism of bilirubin-human serum albumin complexes in aqueous solution. FEBS Letters **9**, 53 (1970).
11. BLAUER, G., und T. E. KING: Optical rotatory dispersion of bilirubin bound to bovine serum albumin. Biochem. Biophys. Res. Comm. **31**, 678 (1968).
12. — — Interactions of bilirubin with bovine serum albumin in aqueous solution. J. Biol. Chem. **245**, 372 (1970).
13. BLOOMFIELD, V. A., und B. R. JENNINGS: Molecular weight and shape of phycocyanin monomer and aggregates. Biopolymers **8**, 297 (1969).
14. BONNETT, R., und A. F. MCDONAGH: Oxidative cleavage of the haem system: The four isomeric biliverdins of the IX series. J. chem. Soc. (D) **1970**, 237.
15. — — The isomeric heterogeneity of biliverdin dimethyl ester derived from bilirubin. J. chem. Soc. (D) **1970**, 238.
16. BORTHWICK, H. A., S. B. HENDRICKS, M. J. SCHNEIDER, R. B. TAYLORSON und V. K. TOOTE: High-energy light action controlling plant responses and development. Proc. Nat. Acad. Sci. US **64**, 479 (1969).
17. BRIGGS, W. R., und D. C. FORK: Long-lived intermediates in phytochrome transformation. I. In vitro studies. Plant Physiol. **44**, 1081 (1969).
18. — — Long-lived intermediates in phytochrome transformation. II. In vitro and in vivo studies. Plant Physiol. **44**, 1089 (1969).
19. BROCKMANN, H. JR., und D. MÜLLER-ENOCH: Die absolute Konfiguration der (−)threo-3-Äthyl-2-methylbernsteinsäure. Angew. Chem. **80**, 562 (1968).
20. BRODERSEN, R.: Dimerisation of bilirubin anion in aqueous solution. Acta chem. Scand. **20**, 2895 (1966).
21. BRODERSEN, R., H. FLODGAARD und J. KROGH HANSEN: Intramolecular hydrogen bonding in bilirubin. Acta chem. Scand. **21**, 2284 (1967).
22. BROOKS, R. C.: Preparation, purification and analysis of peptides derived from Cryptomonad phycoerythrin. Dissertation Univ. Chicago, 1970.
23. BUTLER, W. L., K. H. NORRIS, H. W. SIEGELMAN und S. B. HENDRICKS: Detection, assay, and preliminary purification of the pigment controlling photoresponsive development in plants. Proc. Nat. Acad. Sci. US **45**, 1703 (1959).
24. CAHN, R. S., C. INGOLD und V. PRELOG: Spezifikation der molekularen Chiralität. Angew. Chem. **78**, 413 (1966).
25. CALLAHAN, E. W. JR., und R. SCHMID: Excretion of unconjugated bilirubin in the bile of Gunn rats. Gastroenterology **57**, 134 (1969).

26. CHAPMAN, D. J., W. J. COLE und H. W. SIEGELMAN: The structure of phycoerythrobilin. J. Amer. chem. Soc. **89**, 5976 (1967).
27. — — — Cleavage of phycocyanobilin from C-phycocyanin. Biochim. Biophys. Acta **153**, 692 (1968).
28. — — — Chromophores of Allophycocyanin and R-phycocyanin. Biochem. J. **105**, 903 (1967).
29. — — — A comparative study of the phycoerythrin chromophore. Phytochemistry **7**, 1831 (1968).
30. CHAPMAN, D. J., und D. L. FOX: Bile pigment metabolism in the sea-hare Aplysia. J. exper. Marine Biol. Ecology **4**, 71 (1969).
31. COLE, P. G., G. H. LATHE und B. H. BILLING: Separation of the bile pigments of serum, bile and urine. Biochem. J. **57**, 514 (1954).
32. COLE, W. J., D. J. CHAPMAN und H. W. SIEGELMAN: The structure of phycocyanobilin. J. Amer. chem. Soc. **89**, 3643 (1967).
33. — — — The structure and properties of phycocyanobilin and related bilatrienes. Biochemistry **7**, 2929 (1968).
34. COLE, W. J., C. H. GRAY und D. C. NICHOLSON: The chemistry of the bile pigments. V. The stereoisomerism of urobilins. J. chem. Soc. **1965**, 4085.
35. — — — The chemistry of the bile pigments. VI. The effect of metal complex formation upon optical activity and spectral absorption of urobilins. J. chem. Soc. (C) **1966**, 1321.
36. COLE, W. J., C. O'HEOCHA, A. MOSCOWITZ und W. R. KRUEGER: The optical activity of urobilins derived from phycoerythrobilin. Europ. J. Biochem. **3**, 202 (1967).
37. COOKE, J. R., und L. B. ROBERTS: Binding of bilirubin to serum proteins. Clin. Chim. Acta **26**, 425 (1969).
38. CORRELL, D. L., und J. L. EDWARDS: The aggregation state of phytochrome from etiolated rye and oat seedlings. Plant Physiol. **45**, 81 (1970).
39. CORRELL, D. L., J. L. EDWARDS, W. H. KLEIN und W. SHROPSHIRE JR.: Phytochrome in etiolated annual rye. III. Isolation of photoreversible phytochrome. Biochim. Biophys. Acta **168**, 36 (1968).
40. CORRELL, D. L., J. L. EDWARDS und W. SHROPSHIRE JR.: Multiple chromophore species in phytochrome. Photochem. Photobiol. **8**, 465 (1968).
41. CORRELL, D. L., E. STEERS JR., K. M. TOWE und W. SHROPSHIRE JR.: Phytochrome in etiolated annual rye. IV. Physical and chemical characterization of phytochrome. Biochim. Biophys. Acta **168**, 46 (1968).
42. CRAIG, I. W., und N. G. CARR: C-phycocyanin and Allophycocyanin in two species of blue-green algae. Biochem. J. **106**, 361 (1968).
43. CRESPI, H. L., L. J. BOUCHER, G. D. NORMAN, J. J. KATZ und R. C. DOUGHERTY: Structure of phycocyanobilin. J. Amer. chem. Soc. **89**, 3642 (1967).
44. CRESPI, H. L., und J. J. KATZ: High resolution proton magnetic resonance studies of fully deuterated and isotope hybrid proteins. Nature **224**, 560 (1969).
45. CRESPI, H. L., und U. H. SMITH: The chromophore-protein bounds in phycocyanin. Phytochemistry **9**, 205 (1970).
46. CRESPI, H. L., U. SMITH und J. J. KATZ: Phycocyanobilin. Structure and exchange studies by nuclear magnetic resonance and its mode of attachment in phycocyanin. A model for phytochrome. Biochemistry **7**, 2232 (1968).
47. CROSS, D. R., H. LINSCHITZ, V. KASCHE und J. TENENBAUM: Low-temperature studies on phytochrome: Light and dark reactions in the red to far-red transformation and new intermediate forms of phytochrome. Proc. Nat. Acad. Sci. US **61**, 1095 (1968).

48. DOBENECK, H. V., und E. BRUNNER: Über eine Ordnung der Dipyrromethene und über die Betainstruktur des Bilirubins. XI. Mitt. zur Stokvis-Reaktion. Hoppe-Seyler's Z. physiol. Chem. 341, 157 (1965).
49. DRING, M. J.: Phytochrome in red alga, *Porphyra tenera*. Nature 215, 1411 (1967).
50. DUYSENS, L. M. N.: Transfer of light energy within the pigment systems present in photosynthesizing cells. Nature 168, 548 (1951).
51. DUYSENS, L. M. N., J. AMESZ und D. M. KAMP: Two photochemical systems in photosynthesis. Nature 190, 510 (1961).
52. FEVERY, J., F. COMPERNOLLE und K. P. M. HEIRWEGH: Excretion of xylose and glucose conjugates of bilirubin in dog bile. Meeting on Bile Pigment Chemistry, Univ. Aarhus, August 1970.
53. FISCHER, H., H. BAUMGARTNER und R. HESS: Über Ferro- und Glaukobilin. Hoppe-Seyler's Z. physiol. Chem. 206, 201 (1932).
54. FISCHER, H., und H. W. HABERLAND: Über die Konstitution des Bilirubins sowie die seiner Azofarbstoffe und die Gmelinsche Reaktion. Hoppe-Seyler's Z. physiol. Chem. 232, 236 (1935).
55. FISCHER, H., und H. HALBACH: Über die Konstitution des Stercobilins. Hoppe-Seyler's Z. physiol. Chem. 238, 59 (1936).
56. FISCHER, H., H. HALBACH und A. STERN: Über Stercobilin und seine optische Aktivität. Liebigs Ann. Chem. 519, 254 (1935).
57. FISCHER, H., und G. NIEMANN: Zur Kenntnis des Gallenfarbstoffs. 8. Mitt. Mesobiliviolin, Mesobiliviolinogen und die Kondensation von Mesobilirubinogen mit Aldehyden unter Bildung von neuen Spaltprodukten. Diazofarbstoff des Mesobilirubins. Hoppe-Seyler's Z. physiol. Chem. 137, 293 (1924).
58. FISCHER, H., und H. ORTH: Die Chemie des Pyrrols. Akademische Verlagsgesellschaft Leipzig. Bd. II/1, S. 621 ff. (1937); Bd. II/2, S. 43 (1940).
59. FISCHER, H., und H. PLIENINGER: Synthese des Biliverdins (Uteroverdins) und Bilirubins, der Biliverdine XIIIα und IIIα, sowie der Vinylneoxanthosäure. Hoppe-Seyler's Z. physiol. Chem. 274, 231 (1942).
60. FOG, J., und E. JELLUM: Structure of bilirubin. Nature 198, 88 (1963).
61. FOX, D. L., und N. MILLOTT: A biliverdin-like pigment in the skull and vertebrae of the ocean skipjack, *Katsuwonus pelamis* (Linnaeus). Experientia 10, 185 (1954).
62. FRENCH, C. S., und V. K. YOUNG: The fluorescence spectra of red algae and the transfer of energy from phycoerythrin to phycocyanin and chlorophyll. J. gen. Physiol. 35, 873 (1952).
63. FUJIMORI, E.: Modified phycoerythrin from Porphyridium cruentum treated with p-chloro-mercuribenzoate. Nature 204, 1091 (1964).
64. FUJIMORI, E., und J. PECCI: Dissociation and association of phycocyanin. Biochemistry 5, 3500 (1966).
65. — — Distinct subunits of phycoerythrin from *Porphyridium cruentum* and their spectral characteristics. Arch. Biochem. Biophys. 118, 448 (1967).
66. FUJITA, Y., und A. HATTORI: Preliminary note on a new phycobilin pigment isolated from blue-green algae. J. Biochem. (Tokyo) 51, 89 (1962).
67. — — Occurrence of a purple bile pigment in phycoerythrin-rich cells of the blue-green algae *Tolypothrix tenuis*. J. gen. appl. Microbiol. (Tokyo) 9, 253 (1963).
68. FUJITA, Y., und T. TSUJI: Photochemically active chromoprotein isolated from the blue-green alga *Anabaena cylindrica*. Nature 219, 1270 (1968).
69. GANTT, E.: Properties and ultrastructure of phycoerythrin from *Porphyridium cruentum*. Plant Physiol. 44, 1629 (1969).

70. GANTT, E. und S. F. CONTI: Granules associated with the chloroplast lamellae of *Porphyridium cruentum*. J. Cell. Biol. **29**, 423 (1966).
71. — — Phycobiliprotein localization in algae. Brookhaven Symposia Biol. **19**, 393 (1966).
72. GRAY, C. H.: The Bile Pigments. London: Methuen. 1953.
73. GRAY, C. H., P. M. JONES, W. KLYNE und D. C. NICHOLSON: Optical activity of stercobilin and d-urobilin. Nature **184**, 41 (1959).
74. GRAY, C. H., G. A. LEMMON und D. C. NICHOLSON: The orientation of the alkyl groups of the end rings of natural and racemic stercobilin. J. chem. Soc. (C) **1967**, 178.
75. GRAY, C. H., A. LICHTAROWICZ-KULSZYCKA, D. C. NICHOLSON und Z. J. PETRYKA: The chemistry of the bile pigments. II. The preparation and spectral properties of biliverdin. J. chem. Soc. **1961**, 2264.
76. GRAY, C. H., A. KULCZYCKA und D. C. NICHOLSON: The chemistry of the bile pigments. IV. Spectrophotometric titration of the bile pigments. J. chem. Soc. **1961**, 2276.
77. GRAY, C. H., und D. C. NICHOLSON: The chemistry of the bile pigments. The structures of stercobilin and d-urobilin. J. chem. Soc. **1958**, 3085.
78. — — Recent developments in our knowledge of the urobilins. Medicine **46**, 83 (1967).
79. GRAY, C. H., D. C. NICHOLSON und R. A. NICOLAUS: The IX-α structure of the common bile pigments. Nature **181**, 183 (1958).
80. HARRIS, R. L. N., A. W. JOHNSON und I. T. KAY: A stepwise synthesis of unsymmetrical porphyrins. J. chem. Soc. (C) **1966**, 22.
81. HARTMANN, K. M.: Ein Wirkungsspektrum der Photomorphogenese unter Hochenergiebedingungen und seine Interpretation auf der Basis des Phytochroms (Hypokotylwachstumshemmung bei *Lactuca sativa* L.). Z. Naturforsch. **22 b**, 1172 (1967).
82. HATTORI, A., H. L. CRESPI und J. J. KATZ: Association and dissociation of phycocyanin and the effects of deuterium substitution on the processes. Biochemistry **4**, 1225 (1965).
83. HAUPT, W.: Die Chloroplastendrehung bei Mougeotia. I. Über den quantitativen und qualitativen Lichtbedarf bei Schwachlichtbewegung. Planta **53**, 484 (1959).
84. HAXO, F. T., und L. R. BLINKS: Photosynthetic action spectra of marine algae. J. gen. Physiol. **33**, 389 (1950).
85. HENDRICKS, S. B., und H. A. BORTHWICK: The physiological functions of phytochrome. In: T. W. GOODWIN (Ed.), Chemistry and Biochemistry of Plant Pigments, S. 405. London: Academic Press. 1965.
86. — — The function of phytochrome in regulation of plant growth. Proc. Nat. Acad. Sci. US **58**, 2125 (1967).
87. HIROSE, H., S. KUMANO und K. MADONO: Spectroscopic studies on phycoerythrins from Cyanophycean and Rhodophycean algae with special reference to their phylogenetical relations. Bot. Mag. Tokyo **82**, 197 (1969).
88. ISSELBACHER, K. J., und E. A. MCCARTHY: Identification of a sulfate conjugate of bilirubin in bile. Biochim. Biophys. Acta **29**, 658 (1958).
89. JACKSON, A. H., und G. W. KENNER: Recent developments in porphyrin chemistry. In: T. W. GOODWIN (Ed.), Porphyrins and Related Compounds, S. 3. London and New York: Academic Press. 1968.
90. JACKSON, A. H., G. W. KENNER, H. BUDZIKIEWICZ, C. DJERASSI und J. M. WILSON: Pyrroles and related compounds. X. Mass spectrometry in structural and stereochemical problems. XC. Mass spectra of linear di-, tri- and tetrapyrrolic compounds. Tetrahedron **23**, 603 (1967).

91. JACKSON, A. H., G. W. KENNER und G. S. SACH: Pyrroles and related compounds. XII. Stepwise synthesis of porphyrins through a-Oxo-bilanes. J. chem. Soc. (C) **1967**, 2045.
92. JACKSON, A. H., K. M. SMITH, C. H. GRAY und D. C. NICHOLSON: Molecular species of the urobilins. Nature **009**, 581 (1966)
93. JACOBSEN, J. G.: Binding of bilirubin to human serum albumin; determination of the dissociation constants. FEBS Letters **5**, 112 (1969).
94. JANSEN, H., und B. BILLING: Further evidence in support of the existance of bilirubin monoglucuronide as a chemical entity rather than as a complex. Meeting on Bile Pigment Chemistry, Univ. Aarhus, August 1970.
95. JIRSA, M., B. VEČERECK und M. LEDVINA: Di- and mono-taurobilirubin similar to a directly reacting form of bilirubin in serum. Nature **177**, 895 (1956).
96. JONES, R. F., und FUJIMORI, E.: Interactions between chromophore and protein in phycoerythrin from the red alga *Ceramium rubrum*. Physiol. Plantarum **14**, 253 (1961).
97. JUNGE, H.: Über grüne Insektenfarbstoffe. Hoppe-Seyler's Z. physiol. Chem. **268**, 179 (1941).
98. KAO, O., und D. S. BERNS: The monomer molecular weight of C-phycocyanin. Biochem. Biophys. Res. Comm. **33**, 457 (1968).
99. KAY, I. T., M. WEIMER und C. J. WATSON: The formation in vitro of ($\pm$)-stercobilin from bilirubin. J. Biol. Chem. **238**, 1122 (1963).
100. KILLILEA, S. D., und P. O'CARRA: Amino acids involved in chromophore-protein linkages. Biochem. J. **110**, 14p (1968).
101. — — Improvement of the starch-jodide method for detection of imides and other NH-containing compounds on thin-layer chromatograms. J. Chromatogr. **54**, 284 (1971).
102. KLINGA, K.: Synthese von Biladienen-(a,b) und deren spektroskopische Untersuchung. Dissertation Univ. Heidelberg, 1969.
103. KŌCHIYAMA, Y., K. YAMAGUCHI, K. HASHIMOTO und F. MATSUURA: Studies on a blue-green serum pigment of eel. I. Isolation and some physico-chemical properties. Bull. Japan. Soc. Sci. Fisheries **32**, 867 (1966).
104. KÖST, H.-P.: Die Struktur von Mesobilirhodin IXα. Diplomarbeit Univ. Saarbrücken, 1970.
105. KROES, H. H., A. v. ROVIJEN, J. M. GEERS und E. H. M. GREUELL: Large-scale isolation of phytochrome from oat seedlings. Biochim. Biophys. Acta **175**, 409 (1969).
106. KUENZLE, C. C.: The excretion of bilirubin in human bile as the ester glycosides of uronic acid-containing disaccharides. Meeting on Bile Pigment Chemistry, Univ. Aarhus, August 1970.
107. LEDERER, E., und C. HUTTRER: Quelques observations sur les pigments de la sécrétion des Aplysies *(Aplysia punctata)*. Trav. membres Soc. Chim. biol. **24**, 1055 (1942).
108. LEE, J. J., und D. S. BERNS: Protein aggregation. Studies of larger aggregates of C-phycocyanin. Biochem. J. **110**, 457 (1968).
109. LEMBERG, R.: Chromoproteide der Rotalgen. II. Spaltung mit Pepsin und Säuren. Isolierung eines Pyrrolfarbstoffs. Liebigs Ann. Chem. **477**, 195 (1930).
110. — Über Oocyan. I. Liebigs Ann. Chem. **488**, 74 (1931).
111. — Über Dehydro-bilirubin. Liebigs Ann. Chem. **499**, 25 (1932).
112. — Bile pigments. VI. Biliverdin, uteroverdin and oocyan. Biochem. J. **28**, 978 (1934).
113. — Transformation of haemin into bile pigments. Biochem. J. **29**, 1322 (1935).

114. LEMBERG, R., und G. BADER: Die Phycobiline der Rot-algen. Überführung in Mesobilirubin und Dehydro-mesobilirubin. Liebigs Ann. Chem. **505**, 151 (1933).
115. LEMBERG, R., und J. BARCROFT: Uteroverdin, the green pigment of the dog's placenta. Proc. Roy. Soc. Ser. B **110**, 362 (1932).
116. LEMBERG, R., und J. W. LEGGE: Hematin Compounds and Bile Pigments. New York: Interscience Publ. 1949.
117. LESTER, R., und R. F. TROXLER: Recent advances in bile pigment metabolism. Gastroenterology **56**, 143 (1969).
118. LIGHTNER, D. A., A. MOSCOWITZ, Z. J. PETRYKA, S. JONES, M. WEIMER, E. DAVIS, N. A. BEACH und C. J. WATSON: Mass spectrometry and ferric chloride oxidation applied to urobilinoid structures. Arch. Biochem. Biophys. **131**, 566 (1969).
119. LINSCHITZ, H., und V. KASCHE: The kinetics of phytochrome conversion. J. Biol. Chem. **241**, 3395 (1966).
120. — — Kinetics of phytochrome conversion: Multiple pathways in the P_r to P_{fr} reaction, as studied by double-flash technique. Proc. Nat. Acad. Sci. US **58**, 1059 (1967).
121. LOWRY, P. T., R. CARDINAL, S. COLLINS und C. J. WATSON: The isolation of crystalline d-urobilinogen. J. Biol. Chem. **218**, 641 (1956).
122. MAUZERALL, D.: The thermodynamic stability of porphyrinogens. J. Amer. chem. Soc. **82**, 2601 (1960).
123. MIERAS, G. A., und R. A. WALL: Sub-units of the algal biliprotein phycoerythrin. Biochem. J. **107**, 127. (1968).
124. MILLER, C. O., R. J. DOWNS und H. W. SIEGELMAN: A rapid procedure for the visible detection of phytochrome. BioScience **15**, 596 (1965).
125. MOHR, H.: Regulation der Enzymsynthese bei der höheren Pflanze. Naturwiss. Rundschau **23**, 187 (1970).
126. — Photomorphogenesis. In: M. B. WILKINS (Ed.), The Physiology of Plant Growth and Development, S. 507. New York: McGraw-Hill Publ. 1970.
127. MOSCOWITZ, A.: Some applications of the Kronig-Kramers theorem to optical activity. Tetrahedron **13**, 48 (1961).
128. MOSCOWITZ, A., W. C. KRUEGER, I. T. KAY, G. SKEWES und S. BRUCKENSTEIN: On the origin of the optical activity in the urobilins. Proc. Nat. Acad. Sci. US **52**, 1190 (1964).
129. MOSCOWITZ, A., K. MISLOW, M. A. W. GLASS und C. DJERASSI: Optical rotatory dispersion associated with dissymmetric non-conjugated chromophores. An extension of the octant rule. J. Amer. chem. Soc. **84**, 1945 (1962).
130. MUMFORD, F. E., und E. L. JENNER: Purification and characterization of phytochrome from oat seedlings. Biochemistry **5**, 3657 (1966).
131. MUSTAFA, M. G., M. L. COWGER und T. E. KING: Effects of bilirubin on mitochondrial reactions. J. Biol. Chem. **244**, 6403 (1969).
132. NAKAJIMA, O., und C. H. GRAY: Studies on haem-α-methenyl oxygenase. Isomeric structure of formylbiliverdin, a possible precursor of biliverdin. Biochem. J. **104**, 20 (1967).
133. NEUFELD, G. J., und A. F. RIGGS: Aggregation properties of C-phycocyanin from *Anacystis nidulans*. Biochim. Biophys. Acta **181**, 234 (1969).
134. NICHOL, A. W., und D. B. MORELL: Tautomerism and hydrogen bonding in bilirubin and biliverdin. Biochim. Biophys. Acta **177**, 599 (1969).
135. — — Studies on the isomeric composition of biliverdin and bilirubin by mass spectrometry. Biochim. Biophys. Acta **184**, 173 (1969).
136. NICHOLSON, D. C.: The urobilinoids. In: Formation and Breakdown of Haemoglobin. Symposium Leeds, 1960, S. 27. Amsterdam: Elsevier. 1961.

137. NICOLAUS, R. A., und L. CAGLIOTI: Ricerche di acidi pirrolici nelle miscele di ossidazione. Ricerca Sci. **27**, 113 (1957).
138. NICOLAUS, R. A., L. MANGONI und L. CAGLIOTI: Acidi pirrolici nella ossidazione delle porfirine. Ann. Chim. Appl. **46**, 793 (1956).
139. O'CARRA, P.: Algal biliproteins. Biochem. J. **119**, 2p (1970).
140. — Persönl. Mitteilung. Sept. 1970.
141. O'CARRA, P., und E. COLLERAN: Coupled oxidation of myoglobin with ascorbate as a model of haem breakdown in vivo. Biochem. J. **115**, 13p (1969).
142. — — Separation and identification of biliverdin isomers and isomer analysis of phycobilins and bilirubin. J. Chromatogr. **50**, 458 (1970).
143. O'CARRA, P., und S. D. KILLILEA: Mesobilirhodin, an isomeride of i-urobilin with the spectral characteristics of phycoerythrobilin. Tetrahedron Letters **1970**, 4211.
144. O'CARRA, P., und C. O'HEOCHA: Bilins released from algae and biliproteins by methanolic extraction. Phytochemistry **5**, 993 (1966).
145. O'CARRA, P., C. O'HEOCHA und D. M. CARROLL: Spectral properties of the phycobilins. II. phycoerythrobilin. Biochemistry **3**, 1343 (1964).
146. O'HEOCHA, C.: Comparative biochemical studies of the phycobilins. Arch. Biochem. Biophys. **73**, 207 (1958).
147. — Spectral studies of the phycobilins. I, Phycocyanobilin. Biochemistry **2**, 375 (1963).
148. — Biliproteins. In: T. W. GOODWIN (Ed.), Biochemistry of Chloroplasts, S. 407. London: Academic Press. 1966.
149. O'HEOCHA, C., und P. O'CARRA: Spectral studies of denatured phycoerythrins. J. Amer. chem. Soc. **83**, 1091 (1961).
150. PASSAMA-VUILLAUME, M.: Sur la pigmentation verte de *Mantis religiosa* (L.). C. R. Acad. Sc. Paris **258**, 6549 (1964).
151. — Étude de l'irradiation lumineuse, facteur essentiel du brunissement de *Mantis religiosa* (L.). C. R. Acad. Sc. Paris **261**, 3683 (1965).
152. — Sur la pigmentation brune de trois Orthoptères: *Mantis religiosa* (L.), *Sphodomantis viridis* (F.) et *Locusta migratoria* (L.). C. R. Acad. Sc. Paris **262**, 1597 (1966).
153. — Etude du pigment vert chez *Locusta migratoria* L. normal et albinos. Bull. Soc. Zool. France **90**, 485 (1965).
154. PASSAMA-VUILLAUME, M., und M. BARBIER: Sur la biosynthèse de la biliverdine IXα par la Mante *Mantis religiosa* et le Criquet *Locusta migratoria*. C. R. Acad. Sc. Paris **263**, 924 (1966).
155. PASSAMA-VUILLAUME, M., und B. LEVITA: Sur les pigments tétrapyrroliques d'*OEdipoda coerulescens*. C. R. Acad. Sc. Paris **263**, 1001 (1966).
156. PAULMANN, L.: Über die Fraktionierung der Dimethylester von Urobilin-IXα sowie Glaukobilin-IXα und die Untersuchung der Fraktionen. Diplomarbeit, Univ. Saarbrücken 1968.
157. PECCI, J., und E. FUJIMORI: The B-phycoerythrin chromophore. Phytochemistry **9**, 637 (1970).
158. — — Mercurial-induced dissociation of phycoerythrin from *Ceramium rubrum*. Biochim. Biophys. Acta **131**, 147 (1967).
159. — — Spectral change of phycoerythrin from Hydrocoleum species and its relationship to protein dissociation. Effect of mercurials on single- and double-peaked forms. Biochim. Biophys. Acta **154**, 332 (1968).
160. — — Mercurial-induced circular dichroism changes of phycoerythrin and phycocyanin. Biochim. Biophys. Acta **188**, 230 (1969).

161. PETRYKA, Z. J.: Identification of isomers differing from 9, α, in the early labelled bilirubin of the bile. Proc. Soc. Exper. Biol. Med. 123, 464 (1966).
162. PETRYKA, Z., D. C. NICHOLSON und C. H. GRAY: Isomeric bile pigments as products of the in vitro fission of hemin. Nature 194, 1047 (1962).
163. PETRYKA, Z. J., und C. J. WATSON: The stereospecific synthesis of *cis-cis*-stercobilinogen and *cis-cis*-stercobilin. Tetrahedron Letters 1967, 5223.
164. — — Separation of bile pigments by thin layer chromatography J. Chromatogr. 37, 76 (1968).
164a. PETRYKA, Z. J., C. J. WATSON, E. DAVIS, M. WEIMER, D. LIGHTNER und A. MOSCOWITZ: On the existence and structure of a stable mesobiliviolin. Tetrahedron Letters 1968, 5983.
165. PLIENINGER, H.: New results in the synthesis of bile pigments. Meeting on Bile Pigment Chemistry, Univ. Aarhus, August 1970.
166. PLIENINGER, H., H. BAUER, A. R. KATRITZKY und U. LERCH: Über Δ^3-Pyrrolone und Alkoxypyrrole. Liebigs Ann. Chem. 654, 165 (1962).
167. PLIENINGER, H., und M. DECKER: Eine neue Synthese für Pyrrolone, insbesondere für ,,Isooxyopsopyrrol" und ,,Isooxyopsopyrrol-carbonsäure". Liebigs Ann. Chem. 598, 198 (1956).
168. PLIENINGER, H., K. EHL und A. TAPIA: Die Synthese optisch aktiver Urobiline. Liebigs Ann. Chem. 736, 62 (1970).
169. PLIENINGER, H., und J. KURZE: Synthese der ,,Oxyopsopyrrolcarbonsäure" und weitere Untersuchungen in der Pyrrol-Reihe. Liebigs Ann. Chem. 680, 60 (1964).
170. PLIENINGER, H., und U. LERCH: Totalsynthese zweier racemischer Stercobiline-IX α. Liebigs Ann. Chem. 698, 196 (1966).
171. PLIENINGER, H., und J. RUPPERT: Synthese des (—)-Stercobilins IX α (,,nat." Stercobilin) und anderer optisch aktiver Stercobiline. Liebigs Ann. Chem. 736, 43 (1970).
172. PLIENINGER, H., und R. STEINSTRÄSSER: Gallenfarbstoffsynthesen, II. Synthese zweier Vinyl-substituierter Urobiline IX α. Liebigs Ann. Chem. 723, 149 (1969).
173. RÜDIGER, R., und H. RÜDIGER: Zur Berechnung optimaler Trennungsbedingungen für zwei Stoffe mit bekannten R_F-Werten bei mehrfach wiederholter Papierchromatographie. J. Chromatogr. 17, 186 (1965).
174. RÜDIGER, W.: Über die Abwehrfarbstoffe von *Aplysia*-Arten, I. Aplysioviolin, ein neuartiger Gallenfarbstoff. Hoppe-Seyler's Z. physiol. Chem. 348, 129 (1967).
175. — Über die Abwehrfarbstoffe von *Aplysia*-Arten, II. Die Struktur von Aplysioviolin. Hoppe-Seyler's Z. physiol. Chem. 348, 1554 (1967).
176. — Bile pigments: A new degradation technique and its application. In: T. W. GOODWIN (Ed.), Porphyrins and Related Compounds, S. 121. London: Academic Press. 1968.
177. — Vergleichende Biochemie der Gallenfarbstoffe. Habilitationsschrift, Univ. Saarbrücken, 1968.
178. — Chromsäure- und Chromatabbau von Gallenfarbstoffen. Hoppe-Seyler's Z. physiol. Chem. 350, 1291 (1969).
179. — Neues aus Chemie und Biochemie von Gallenfarbstoffen. Angew. Chemie 82, 527 (1970).
180. — Gallenfarbstoffe bei wirbellosen Tieren. Naturwiss. 57, 331 (1970).
181. — Partialsynthese von Konjugaten der Gallenfarbstoffe. Unveröffentlicht.
182. — Chromatabbau neuartiger Gallenfarbstoffe. Unveröffentlicht.

183. RÜDIGER, W., und L. ABOLINŠ: Zur Bindung von Biliverdin an das Protein im Crenilabrus-Blau. Experientia 25, 574 (1969).
184. RÜDIGER, W., und H. BUDZIKIEWICZ: Massenspektrometrie der Gallenfarbstoffe. Unveröffentlicht.
185. RÜDIGER, W., und D. L. CORRELL: Über die Struktur des Phytochrom-Chromophors und seine Protein-Bindung. Liebigs Ann. Chem. 723, 208 (1969).
186. RÜDIGER, W., und W. KLOSE: Synthese von Methyl-äthyliden-succinimid. Tetrahedron Letters 1967, 1177.
187. — — Über die Pigmente der Florfliege *Chrysopa carnea*. Experientia 26, 498 (1970).
188. RÜDIGER, W., W. KLOSE, B. TURSCH, N. HOUVENAGHEL-CREVECOEUR und H. BUDZIKIEWICZ: Zur Frage des Helioporobilins. Isolierung von Biliverdin-IX,α aus der blauen Koralle Heliopora coerulea Pall. Liebigs Ann. Chem. 713, 209 (1968).
189. RÜDIGER, W., W. KLOSE, M. VUILLAUME und M. BARBIER: On the structure of pterobilin, the blue pigment of *Pieris brassicae*. Experientia 24, 1000 (1968).
190. — — — — On the biosynthesis of biliverdin-IXγ in *Pieris brassicae*. Experientia 25, 487 (1969).
191. — — — — Sur la biliverdine IXα, pigment vert des insectes Orthoptères, *Clitumnus extradentatus* et *Mantis religiosa*. Bull. Soc. Chim. Biol. 51, 559 (1969).
192. RÜDIGER, W., H.-P. KÖST, H. BUDZIKIEWICZ und V. KRAMER: Die Struktur von Mesobilirhodin. Liebigs Ann. Chem. 738, 197 (1970).
193. RÜDIGER, W., und F. E. MUMFORD: Structural studies on native phytochrome. Unveröffentlicht.
194. RÜDIGER, W., und P. O'CARRA: Studies on the structures and apoprotein linkages of the phycobilins. Europ. J. Biochem. 7, 509 (1969).
195. RÜDIGER, W., P. O'CARRA und C. O'HEOCHA: Structure of phycoerythrobilin and phycocyanobilin. Nature 215, 5109 (1967).
196. SCHACHTER, D.: Nature of the glucuronide in direct-reacting bilirubin. Science 126, 507 (1957).
197. SCHMID, R.: The identification of „direct-reacting" bilirubin as bilirubin glucuronide. J. Biol. Chem. 229, 881 (1957).
198. SCHRAM, B. L.: Structure of phycocyanobilin. Biochem. J. 119, 15p (1970).
199. SCHWARTZ, S., und C. J. WATSON: Isolation of a dextrorotatory urobilin from human fistula bile. Proc. soc. Exper. Biol. Med. 49, 641 (1942).
200. SIEDEL, W.: Synthese des Glaukobilins sowie über Urobilin und Mesobiliviolin. Hoppe-Seyler's Z. physiol. Chem. 237, 8 (1935).
201. — Gallenfarbstoffe. Fortschr. Chem. Org. Naturst. 3, 81 (1939).
202. — Chemie und Physiologie des Blutfarbstoff-Abbaues. Ber. dtsch. chem. Ges. Abt. A 77, 21 (1944).
203. — Pyrrolfarbstoffe. In: Hoppe-Seyler/Thierfelder, Handbuch der physiologisch- und pathologisch-chemischen Analyse, 10. Aufl., Bd. 4/2, S. 845. Berlin-Göttingen-Heidelberg: Springer. 1960.
204. SIEDEL, W., und W. FRÖWIS: Über die Violettstufe der Gmelinschen Reaktion (Mesobilipurpurine). I. Mitt. über den Mechanismus der Gmelinschen Reaktion. Hoppe-Seyler's Z. physiol. Chem. 267, 37 (1941).
205. SIEDEL, W., und E. MEIER: Synthese des Urobilins (Urobilin-IX,α) sowie der isomeren Urobiline-III,α und -XIII,α. Hoppe-Seyler's Z. physiol. Chem. 242, 101 (1936).

206. SIEDEL, W., und H. MÖLLER: Über Mesobilivioline I. Konstitution des Mesobiliviolins, Synthesen des Mesobiliviolins-IX,α und Mesobiliviolins-XIII,α sowie über ψ-Mesobiliviolin und „Oxo"-urobilin. Hoppe-Seyler's Z. physiol. Chem. **264**, 64 (1940).
207. SIEGELMAN, H. W.: Phytochrome. In: M. B. WILKINS (Ed.), The Physiology of Plant Growth and Debelopment, S. 487. New York: McGraw-Hill Publ. 1970.
208. SIEGELMAN, H. W., D. J. CHAPMAN und W. J. COLE: Enzymatic cleavage of phycocyanobilin. Arch. Biochem. Biophys. **122**, 261 (1967).
209. — — — The bile pigments of plants. In: T. W. GOODWIN (Ed.), Porphyrins and Related Compounds, S. 107. London: Academic Press. 1968.
210. SIEGELMAN, H. W., und E. M. FIRER: Purification of phytochrome from oat seedlings. Biochemistry **3**, 418 (1964).
211. SIEGELMAN, H. W., und S. B. HENDRICKS: Purification and properties of phytochrome: A chromoprotein regulating plant growth. Fed. Proc. **24**, 863 (1965).
212. SIEGELMAN, H. W., B. C. TURNER und S. B. HENDRICKS: The chromophore of phytochrome. Plant Physiol. **41**, 1289 (1966).
213. STAHL, E.: Dünnschichtchromatographie, 2. Aufl., S. 87. Berlin-Heidelberg-New York: Springer. 1967.
214. STOLL, M. S., und C. H. GRAY: The oxidation products of crude mesobilirubinogen. Biochem. J. **117**, 271 (1970).
215. TALAFANT, E.: Properties and composition of the bile pigment giving a direct diazo reaction. Nature **178**, 312 (1956).
216. TAYLOR, A. O., und B. A. BONNER: Isolation of phytochrome from the algae *Mesotaenium* and liverwort *Sphaeocarpus*. Plant Physiol. **42**, 762 (1967).
217. TENHUNEN, R.: Thin-layer chromatography of bile pigments. Acta chem. Scand. **17**, 2127 (1963).
218. TENHUNEN, R., H. S. MARVER und R. SCHMID: Microsomal heme oxygenase. Characterization of the enzyme. J. Biol. Chem. **244**, 6388 (1969).
219. TENHUNEN, R., M. E. ROSS, H. S. MARVER und R. SCHMID: Reduced nicotinamide-adenine dinucleotide phosphate dependent biliverdin reductase: Partial purification and characterization. Biochemistry **9**, 298 (1970).
220. TIXIER, R.: Sur les pigments biliaires des coquilles de Mollusques du genre *Turbo*. C. R. Acad. Sc. Paris **225**, 508 (1947).
221. — Contribution a l'étude de quelques pigments pyrroliques naturels des coquilles de Mollusques, de l'oeuf d'Emeu et du squelette du corail bleu *(Heliopora caerulea)*. Ann. Inst. océanogr. Monaco **22**, 343 (1945).
222. TIXIER, R., und E. LEDERER: Sur l'haliotivioline, pigment principal des coquilles d'*Haliotis cracherodii*. C. R. Acad. Sc. Paris **228**, 1669 (1949).
223. TREIBS, A., und G. FRITZ: Die Substitutionsregeln des Pyrrols und der Mechanismus der Pyrrol-Austausch-Reaktionen. Liebigs Ann. Chem. **611**, 162 (1958).
224. TSUCHIYA, Y., und T. NOMURA: Ichthyoverdin, a pigment from *Cololabis saira*. C. R. Séances Soc. Biol. Fil. **155**, 34 (1961).
225. VUILLAUME, M., und M. BARBIER: Sur les pigments tétrapyrroliques des Lépidoptères. C. R. Acad. Sc. Paris **268**, 2286 (1969).
226. — — Mise au point sur les pigments bleus et verts des Lépidoptères. Recherche systèmatique de la Ptérobiline chez divers Rhopalocères. C. R. Séances Soc. Biol. Fil. **163**, 591 (1969).

227. WATSON, C. J.: Color reaction of bilirubin with sulfuric acid: A direct diazo-reacting bilirubin sulfate. Science 128, 142 (1958).
228. — Gold from dross: The first century of the urobilinoids. Ann. Internal Med. 70, 839 (1969).
229. WATSON, C. J., M. CAMPBELL und P. T. LOWRY: Preferential reduction of conjugated bilirubin to urobilinogen by normal fecal flora. Proc. Soc. Exper. Biol. Med. N. Y. 98, 707 (1958).
230. WATSON, C. J., D. A. LIGHTNER, A. MOSCOWITZ, E. DAVIS, Z. J. PETRYKA und M. WEIMER: A natural crystalline urobilinogen composed of d- and l-components of different molecular weight. Proc. Nat. Acad. Sci. US 61, 223 (1968).
231. WATSON, C. J., und P. T. LOWRY: A further study of crystalline d-urobilin. J. Biol. Chem. 218, 633 (1956).
232. WATSON, C. J., P. T. LOWRY, V. E. SBOROO, W. H. HOLLINSHEAD, S. KOHAN und H. O. MATTE: A simple method of isolation of crystalline stercobilin or urobilin from feces. J. Biol. Chem. 200, 697 (1953).
233. WATSON, C. J., A. MOSCOWITZ, D. A. LIGHTNER, Z. J. PETRYKA, E. DAVIS und M. WEIMER: On the existence and structure of a new urobilin of molecular weight 592. Proc. Nat. Acad. Sci. US 58, 1957 (1967).
234. WATSON, C. J., M. WEIMER und V. HAWKINSON: Differences in the formation of mesobiliviolin and glaucobilin from d- and i-urobilin. J. Biol. Chem. 235, 787 (1960).
235. WATSON, C. J., M. WEIMER, Z. J. PETRYKA, D. A. LIGHTNER, A. MOSCOWITZ, E. DAVIS und N. A. BEACH: A new method of interpretation of the ferric chloride oxidation patterns of the urobilinoids. Arch. Biochem. Biophys. 131, 414 (1969).
236. WIELAND, H., und A. TARTTER: Über die Flügelpigmente der Schmetterlinge, VIII. Pterobilin, der blaue Farbstoff der Pieridenflügel. Liebigs Ann. Chem. 545, 197 (1940).
237. WILLIG, A.: Die Carotinoide und der Gallenfarbstoff der Stabheuschrecke, *Carausius morosus* und ihre Beteiligung an der Entstehung der Farbmodifikationen. J. Insect Physiol. 15, 1907 (1969).
238. WILLSTAEDT, H.: Zur Kenntnis der grünen Farbstoffe von Seefischen. Enzymologia 9, 260 (1941).
239. WITH, T. K.: Biologie der Gallenfarbstoffe. Stuttgart: Thieme-Verlag. 1960.
240. YAMAGUCHI, K., K. HASHIMOTO und F. MATSUURA: Studies on a blue-green serum pigment of eel. III. Amino acid composition and constituent sugars. Bull. Japan. Soc. Sci. Fisheries 34, 214 (1968).
241. — — — Studies on a blue-green serum pigment of eel. IV. Seasonal variation of concentration of pigment in serum. Bull. Japan. Soc. Sci. Fisheries 34, 826 (1968).
242. YAMAGUCHI, K., Y. KŌCHIYAMA, K. HASHIMOTO und F. MATSUURA: Studies on a blue-green serum pigment of eel. II. Identification of prosthetic group. Bull. Japan. Soc. Sci. Fisheries 32, 873 (1966).
243. YAMAGUCHI, K., und F. MATSUURA: A blue pigment from the muscle of a marine Teleost, „Hirosa", *Cheilinus undulatus* Rüppell. Bull. Japan. Soc. Sci. Fisheries 35, 920 (1969).

(Eingelaufen am 20. Oktober 1970)

The Chemistry of Glutarimide Antibiotics

By F. JOHNSON, Wayland, Massachusetts, USA

Contents

	Page
I. Introduction	140
II. The Chemistry of the Glutarimide Antibiotics	141
1. Nomenclature	141
2. Isolation and Determination	145
3. The Structure of Cycloheximide and Its Isomers	146
a) The Gross Structures of Cycloheximide, Isocycloheximide and Naramycin-B	146
b) Absolute Configuration	151
c) Fine Structure	152
d) Miscellaneous Chemistry	167
4. The Streptovitacins and E-73	174
5. Inactone	179
6. Actiphenol (C-73)	180
7. Streptimidone and Protomycin	181
8. Fermicidin, Niromycin-A and Niromycin-B	186
III. Synthesis	186
1. Cycloheximide, Naramycin-B and Isocycloheximide	187
2. α-Epiisocycloheximide	191
3. Actiphenol	193
4. Homologs and Analogs of Cycloheximide and Other Related Substances	193
IV. Biosynthesis	197
References	202

I. Introduction

The glutarimide antibiotics constitute a group of near-neutral naturally-occurring substances whose discovery was heralded in 1946 by the isolation of a substance, called actidione (later changed to cycloheximide: *vide infra*) from a streptomycin-producing strain of *Streptomyces griseus* (*129*). Although this material displayed little or no activity against the usual pathogenic bacteria, it was found to be very effective in inhibiting the growth of many fungi. The characteristic feature of these antibiotics is the 3-ethylglutarimide group which is substituted at only the β-position of the ethyl group.

References, pp. 202—208

To date, seventeen distinct substances belonging to this class have been isolated (see Table 1) and they constitute a truly fascinating group of mold products. Their diverse spectrum of biological activity alone serves to place them in an almost unique category insofar as naturally-occurring organic compounds are concerned.

However, it is not the objective of this article to discuss in great detail the biological activity, the mechanism of biological action, or the structure-activity relationships that have been investigated for this group of antibiotics. Most of this information is already available in the excellent synopsis of these topics by SISLER and SIEGEL (*101*). Rather, this review is intended to summarize the degradative and synthetic chemistry that is known to date. Nevertheless, the major biological activities of interest will be cited in those sections where the compounds are treated individually. The only previous reviews that have appeared in this area are those due to OKUDA and his associates (*73*, *78*) which appeared in 1961. These dealt only with the chemistry of cycloheximide and its isomers, as it was known then and the subject has developed extensively since that time.

II. The Chemistry of the Glutarimide Antibiotics

1. Nomenclature

Based on systematic nomenclature all members of this group of antibiotics should be named as derivatives of glutarimide itself. Thus, the systematic name for cycloheximide is 3-[2-{3(S), 5(S)-dimethyl-2-oxocyclohexyl}-2(R)-hydroxyethyl]glutarimide. For the most part,

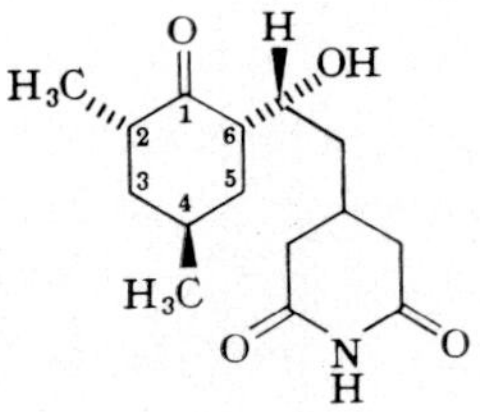

Cycloheximide (1)

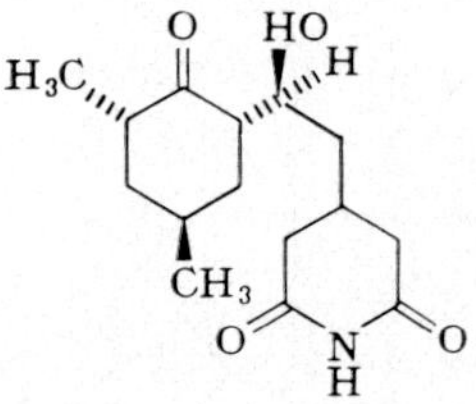

α-*epi*cycloheximide (2)

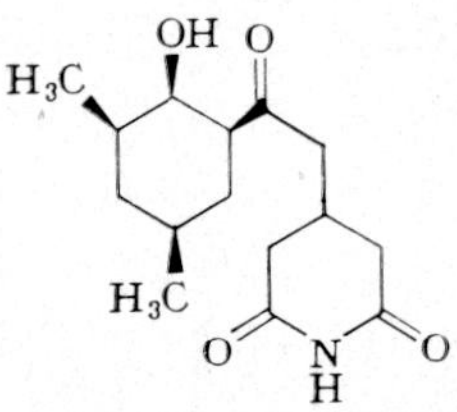

pseudo or ψ-Cycloisoheximide-I (3)

Table 1. *Naturally-Occurring Glutarimide Antibiotics*

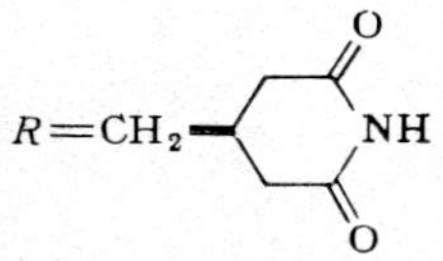

$R=CH_2-$ [glutarimide structure]

Name	Source	Structure	m. p.	$[\alpha]_D$ or ORD
Cycloheximide Naramycin-A (Actidione)	*Streptomyces griseus* *Streptomyces noursei* *Streptomyces naraensis*	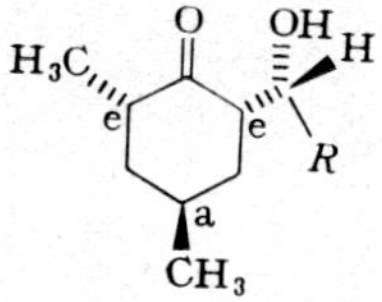 [structure]	115–117°	−33° (c 1.0 CHCl₃) −28° (c 9.6 MeOH) +6.8° (c 2.0, H₂O)
Naramycin-B	*Streptomyces naraensis*	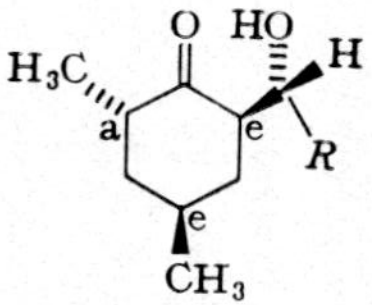 [structure]	112–113°	+55.8° (c 5.0, MeOH)
Isocycloheximide	*Streptomyces griseus*	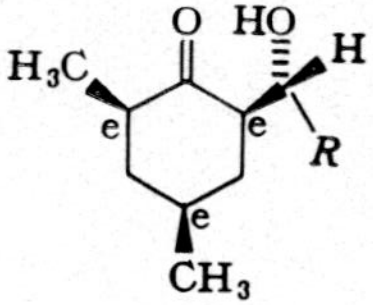 [structure]	101–103°	+32° (c 1.7, CHCl₃)
Streptovitacin-A	*Streptomyces griseus*	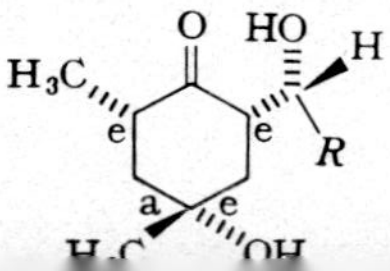 [structure]	156–159°	ORD: Negative Cotton Effect Curve

The Chemistry of Glutarimide Antibiotics

	Source	Formula/Structure	M.p.	$[\alpha]_D$	ORD
E-73	Streptomyces albulus	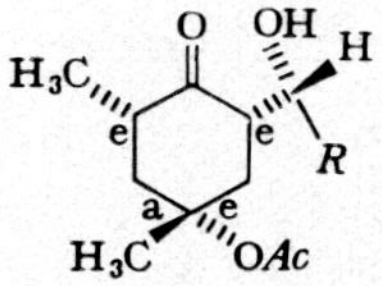 (structure with H₃C, O, OH, H, R, e, a, e, OAc, H₃C)	65–67° or 141–2°	−8.8° (c 1, MeOH)	
Streptovitacin-B	Streptomyces griseus	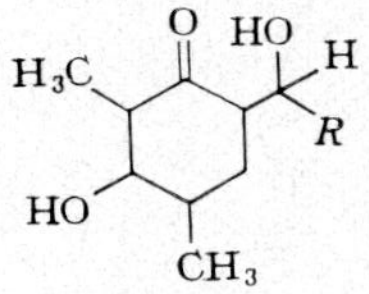 (structure: O HO H, R, H₃C, HO, CH₃)	124–128°		Negative Cotton Effect Curve
Streptovitacin-C₁	Streptomyces griseus	?	—		—
Streptovitacin-C₂	Streptomyces griseus	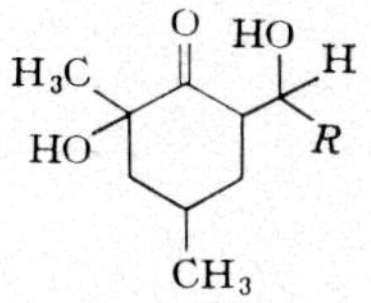 (structure: O HO H, R, H₃C, HO, CH₃)	91–96°		Positive Cotton Effect Curve
Streptovitacin-D	Streptomyces griseus	?	67–69°		—
Streptovitacin-E	Streptomyces griseus	?	—		—
Niromycin-A	Streptomyces albus	$C_{14-15}H_{21-23}O_{3-4}N$	98–105°	0° (c 1, EtOH)	
Niromycin-B	Streptomyces albus	$C_{14}H_{21}O_4N$	47–67°	0°	

Table 1 (continued)

Name	Source	Structure	m. p.	$[\alpha]_D$ or ORD
Inactone	*Streptomyces griseus*	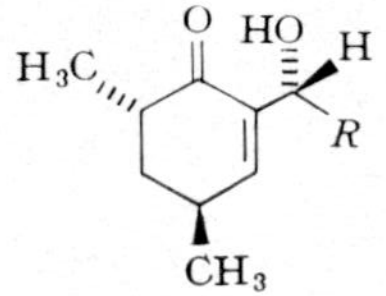 (cyclohexenone structure with H₃C, O, HO, H, R, CH₃ substituents)	116°	−55° (c 2, H₂O)
Actiphenol C-73	Actinomycetes Stammes ETH 7796 (Streptomyces)	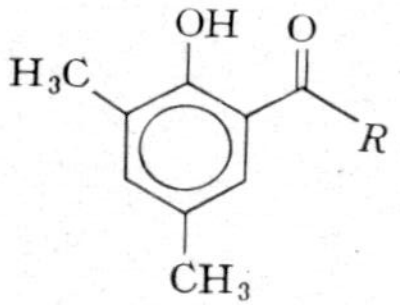 (aromatic phenol structure with OH, O, R, H₃C, CH₃)	198–199°	—
Fermicidin	*Streptomyces griseolus*	(Formula $C_{14}H_{21}O_4N$)	96–98°	+52.3° (c 0.62, H₂O)
Streptimidone	*Streptomyces rimosus* var. *paromomycicus*	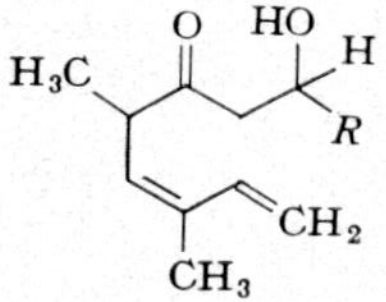 (open-chain structure with O, HO, H, R, CH₂, H₃C, CH₃)	72–73°	+245° (c 0.5, CHCl₃) +238° (c 0.5, H₂O)
Protomycin	*Streptomyces rimosus* var. *paromomycicus*	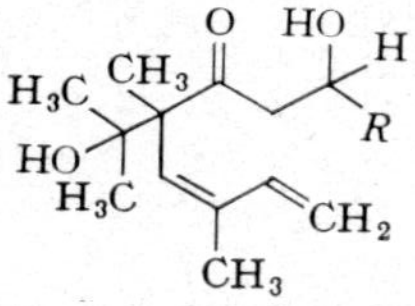 (open-chain structure with O, HO, H, R, CH₂, H₃C, CH₃, HO)	58–61°	+126° (c 1, CHCl₃)

however, because of convenience, the antibiotics and their derivatives are referred to by their trivial names. Apart from actiphenol, streptimidone and protomycin, chemists have by custom looked upon these antibiotics as substituted cyclohexanones and the usual numbering system has applied as shown above.

Two other types of trivial nomenclature have arisen. The first deals with the position in the side chain bearing the hydroxyl group. This is usually referred to as the α-position and those molecules which bear a hydroxyl group having the reverse configuration to that in cycloheximide at this position are usually referred to as α-epi derivatives. For instance, α-*epi*cycloheximide would have structure (2). Compounds in which the hydroxyl group and the ketone group have been interchanged as in (3) are usually referred to as *pseudo-* or ψ-compounds. Since these terms are well-established in the literature, they will be adhered to in this article.

2. Isolation and Determination

The isolation of these antibiotics presents no special difficulties. In some cases the material can be extracted directly from the broth with an organic solvent such as butanol, chloroform or ethyl acetate. Direct crystallization or more often than not chromatography on such substrates as silica gel or charcoal-diatomaceous earth mixtures is then used to obtain the crystalline compound. In other cases the antibiotic is adsorbed directly from the fermentation beer onto charcoal which is then treated with a solvent to remove the antibiotic. The latter is subsequently further purified by either of the methods mentioned above. In the case of the streptovitacins, extensive partition chromatography and countercurrent distribution operations were needed to separate the pure components (*15*).

For the commercial production (*9*) of cycloheximide *Streptomyces griseus* is cultured for four to five days at a temperature between 22 and 26° C in a medium containing soybean meal ($\sim$ 14 gm/l), glucose ($\sim$ 56 gm/l) and potassium dihydrogen phosphate (0.15–0.25 gm/ml), until the concentration of cycloheximide reaches at least 1000 mcg./ml. The fermentation beer is then acidified to pH 3.5–5.5 and clarified. The clarified beer is extracted with methylene chloride and the methylene chloride extract then concentrated and decolorized. Replacement of the methylene chloride by amyl acetate then affords crystalline cycloheximide.

The determination of the glutarimide antibiotics has been accomplished in three ways. If the antibiotic has antifungal activity, it can be assayed using the conventional disc-plate assay against a saccharomyces test organism. *Saccharomyces pastorianus* has been used for the estimation of cycloheximide (*127*) and the streptovitacins (*15*) whereas *Saccharomyces cerevisiae* and *saccharomyces sake* have been used (*93, 110*) for the estimation of E-73 and protomycin respectively. Two chemical assay methods are available. The first is due to OKUDA and his associates (*118*) and takes advantage of a color reaction that resorcinol gives with compounds capable of forming an α,β-unsaturated ketone when heated in hydrochloric acid. The second method is due to FORIST and THEAL (*27*) and is based on a reaction with alkaline hydroxylamine to produce a hydroxamic acid followed by conversion to the highly colored ferric hydroxamate which can be estimated colorimetrically. This latter method which was primarily evolved for the estimation of cycloheximide has recently been extended by GARRETT and NOTARI (*30*), so that it can now be used for the estimation of cycloheximide in the presence of its degradation products.

3. The Structure of Cycloheximide and Its Isomers

a) The Gross Structures of Cycloheximide, Isocycloheximide and Naramycin-B

Cycloheximide. The initial observation of WHIFFEN, BOHONOS, and EMERSON (*129*) of the production of an antifungal antibiotic by *streptomyces griseus* led to a report (*58*) a year later on the isolation of the active principle. The compound was initially assigned the formula $C_{27}H_{42}N_2O_7$ and named Actidione because it was thought to be a diketone. Subsequently, the formula was corrected (*26*) to $C_{15}H_{23}NO_4$ and on the basis of the analytical data from the oxime and semicarbazone derivatives, it was concluded that Actidione must be a mono-ketone. At a later date the official name of the compound was changed to cycloheximide and the term Actidione was retained by the Upjohn Company for use only as a trade name for the compound. This material was also isolated by OKUDA and his coworkers (*71*) from a culture of *Streptomyces naraensis*. They initially named it naramycin-A, but adopted the name cycloheximide once the identity of the two compounds was discovered.

Cycloheximide displays a broad spectrum of biological activity and is toxic to many types of organisms. In particular, it has potent antifungal activity (*129*) and is sold commercially to control certain fungal diseases of plants (*25, 128*), in particular cherry leaf spot (*8, 35, 63*), various turf diseases (*123, 126*) and white pine blister rust (*124*). In this connection the relationship of structure to fungitoxicity of cycloheximide and associated glutarimide derivatives has been studied recently (*100*). Cycloheximide is also sold as a rodent-repellant and is, in fact, the most potent repellant known so far (*125*). Rats refuse to drink water containing 1 ppm of cycloheximide and would rather die from thirst than drink water containing 4–5 ppm. The antibiotic has antitumor activity, but is apparently too toxic for practical use (*24*). Cycloheximide is also toxic to algae (*82*) and protozoa (*62*), but is is not effective against bacteria (*127*).

References, pp. 202—208

The mechanism of action of cycloheximide has been studied and its fungicidal effects appear to spring from its ability to inhibit protein synthesis within the living cell (6, 101). The structure-activity relationship of cycloheximide and its congeners with respect to this has been studied by ENNIS (21), and a theory has been outlined by GROLLMAN (34) to explain this inhibition at the molecular level. Finally, cycloheximide and its isomers have found a great deal of use recently in the study of memory and learning (3, 4, 10, 11, 32, 33).

The first serious work on the structure of cycloheximide was reported by KORNFELD and his associated in two papers (53, 54) in which they deduced the gross structure of the molecule as discussed below.

The empirical formula of cycloheximide was first determined and confirmed to be $C_{15}H_{23}NO_4$ as previously found by the Upjohn workers (58). In addition they found that all of the nitrogen of cycloheximide could be liberated as ammonia by simply boiling with sodium hydroxide solution which suggested the presence of an amide or imide grouping in the molecule. A methyl group determination suggested that two were present, and an acetate and a mono-p-nitrobenzoate were obtained easily indicating the presence of a secondary or primary hydroxyl group. That a ketone was also present in the molecule was evident from the formation of monoxime and semicarbazone derivatives.

Catalytic reduction of cycloheximide occurred with the absorption of one molecule of hydrogen to yield dihydroactidione (4) which was shown to be a diol by its conversion to a diacetate. However, (4) was not attacked by periodic acid and therefore could not be a 1,2-diol. Confirmation of these two functional groups was obtained by further physical studies. Cycloheximide showed a single maximum at 287 nm ($\varepsilon = 36.7$) characteristic of a simple ketone group and its infrared spectrum contained bands which could be assigned to hydroxyl ($3570\ cm^{-1}$), NH ($3330\ cm^{-1}$), carbonyl ($1724\ cm^{-1}$) and C-methyl ($1373\ cm^{-1}$). A semi-quantitative study of this spectrum in comparison with that of other compounds indicated that cycloheximide contained three carbonyl groups. Electrometric titration in water showed no titrable group below pH 10. However, a weakly acidic group could be titrated in the more alkaline region and this displayed a mid-point or pkα' of 11.2.

Degradative studies were next undertaken. Treatment of the molecule with 0.01 N-alkali at room temperature resulted in rapid cleavage to form, amongst other things, a fragrant volatile liquid. The latter material could be isolated either by ether extraction or by steam distillation and was found to be dextrorotatory and to have the empirical formula $C_8H_{14}O$. Physical and chemical data indicated that the material was a cyclic ketone and a comparison of these data with those of known ketones having this formula showed that the material was an optically active form of 2,4-dimethylcyclohexanone (5). Additional confirmation of the identity of the ketone was obtained by resolution of synthetic dl-2,4-dimethylcyclohexanone using L-menthydrazide. The least soluble L-menthydrazone was found to be identical with the L-menthydrazone of the naturally-derived ketone.

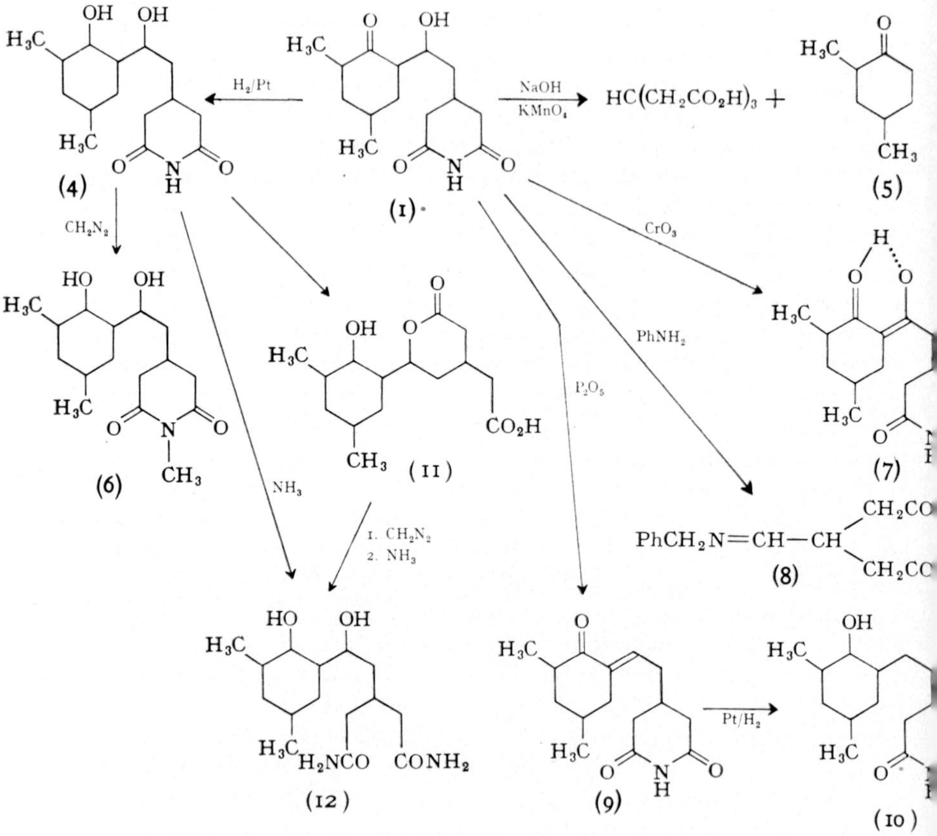

With the identification of the C_8 ketone, seven carbon atoms of cycloheximide remained unaccounted for. Nevertheless, the evidence accumulated to this point indicated that the remaining two carbonyl groups might be accounted for by a cyclic imide structure, a suggestion that was in accord with the very weakly acidic character of cycloheximide.

Support for this viewpoint was obtained by electrometric titration of glutarimide and succinimide. The former showed $pk\alpha'$ of 11.2 identical with that of cycloheximide whereas succinimide gave a value of 9.35. In order to determine the point of substitution of the glutarimide ring,

References, pp. 202—208

the infrared spectrum of cycloheximide was compared with those of α-ethyl- and β-ethylglutarimide. Both cycloheximide and β-ethylglutarimide had bands at 1142 cm^{-1} whereas this band was completely missing in the spectrum of the α-ethyl isomer. This was tentatively accepted as evidence that cycloheximide was a β-substituted glutarimide.

Chemical evidence for the β-substituted glutarimide structure was obtained in several ways. Methylation of dihydrocycloheximide (4) with diazomethane, produced an N-methyl derivative (6), which on alkaline degradation furnished methylamine. When cycloheximide was degraded with alkali and the nonvolatile residue oxidized with potassium permanganate, there was produced methanetriacetic acid. In yet another experiment cycloheximide was oxidized with chromic acid to produce dehydrocycloheximide (7) which was characterized a as 1,3-diketone by its ultraviolet absorption spectrum (λ_{max} 292 nm; ε, 9910) and by conversion to a copper complex. Degradation of (7) with alkali took place almost exclusively by fission of the carbon-carbon bond exocyclic to the cyclohexanone ring and furnished 2,4-dimethylcyclohexanone, ammonia and methanetriacetic acid. On the basis of this evidence structure (1) was proposed for cycloheximide. All subsequent transformations and reactions of the antibiotic have supported this proposal.

Although the propionaldehyde-2,2-diacetic acid formed on alkaline degradation of cycloheximide could not be obtained analytically pure, a derivative (8) could be obtained when (1) was heated with benzylamine. Evidence that the β-ethyl glutarimide was joined to the cyclohexanone ring at the 2-rather than the 6-position was obtained by dehydration of (1) by means of phosphorous pentoxide. The product of the reaction (9) exhibited absorption in the ultraviolet spectrum (λ_{max}, 241 nm; ε, 8250) typical of an α,β-unsaturated ketone. Catalytic reduction of (9) gave the tetrahydroderivative (10).

Several interesting transformations of dihydrocycloheximide (4) were also carried out. The carbon skeleton of this molecule was found to be stable to hydrolytic cleavage because of the absence of the aldol group. Alkaline hydrolysis merely opened the glutarimide ring with the elimination of ammonia, and subsequent acidification of the hydrolysis mixture yielded an acidic product (11). This acid formed a monomethyl ester when treated with diazomethane which was transformed by means of alcoholic ammonia into the diamide (12). This diamide was also produced directly by the reaction of (4) with ammonia and it could be reconverted by alkaline hydrolysis to the acid (11). All of these transformations are adequately accounted for on the basis of the formulations presented. All later chemical work on cycloheximide has, amongst other things, served to confirm structure (1) proposed by the ELI LILLY group.

Isocycloheximide. From the aged mother liquors after the recrystallization of cycloheximide a new material was isolated (*60*) which LEMIN and FORD (*61*) subsequently showed to be an isomer of cycloheximide. This they named isocycloheximide (13). Isocycloheximide was found to have about 30% of the activity of cycloheximide by *Saccharomyces pastorianus* bioassay and about 30% of the toxicity when determined intravenously in mice. Proof that isocycloheximide (13) was stereoisomeric with cycloheximide was obtained when it was found that cycloheximide was partially converted to isocycloheximide when shaken in

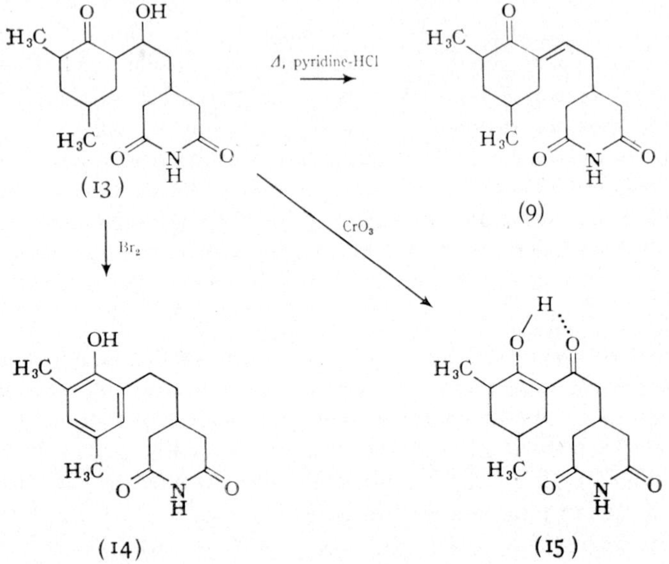

solution with acid-deactivated alumina. Further evidence for the identity of their skeletons was provided by bromination experiments. Both cycloheximide and isocycloheximide when reated with one mole equivalent of bromine gave the same phenol (14). That the secondary hydroxyl group was not involved in the isomerization of cycloheximide to isocycloheximide was shown by the fact that chromic acid oxidation of isocycloheximide gave a product dehydroisocycloheximide (15) different from (7) obtained from cycloheximide itself. Nevertheless, (7) could be isomerized to (15) by means of heating with pyridine hydrochloride. Isocycloheximide (13) however, did give the same dehydration product (9) as cycloheximide when heated with pyridine hydrochloride.

Naramycin-B. Naramycin-B (16) was discovered in a way similar to that described for isocycloheximide (13). OKUDA and his associates (*77*) isolated it from mother liquors left after the removal of cycloheximide derived from *Streptomyces naraensis*. Naramycin-B is active against microorganisms sensitive to cycloheximide, but with few exceptions its activity seems to be less than that of the latter compound. For example, it was found to have only 32% of the activity of (1) against *Saccharomyces sake*.

The infrared spectrum of naramycin-B in Nujol is different from that of cycloheximide especially in the 3000–4000 cm^{-1} region. By alkaline degradation (16) gives *cis-d*-dimethylcyclohexanone (5) whereas dehydration by phosphorus pentoxide or catalytic amounts of boron trifluoride etherate gives the product (9). Chromic acid oxidation of (16) leads to dehydrocycloheximide (7) identical with the compound obtained by the chromium trioxide oxidation of cycloheximide. From these results it was concluded that naramycin-B (16) was isomeric with cycloheximide (1), but the differences involved at the asymmetric centers giving rise to this isomerism were not understood. At a later date these same investigators found that cycloheximide (1) could be isomerized to naramycin-B (16) by means of base catalysis (79).

b) Absolute Configuration

The determination of the absolute configuration of any new type of natural product having a number of asymmetric centers usually follows a set procedure. One of the assymetric centers is excised intact as some simple derivative and this molecule in turn is related to a molecule of known absolute configuration. Provided that the asymmetric center chosen can be related stereochemically to the other asymmetric centers in the molecule the absolute configuration of the total molecule follows automatically. For cycloheximide the determination was done by DJERASSI and his coworkers (5, 20) at a stage before the full stereochemistry of cycloheximide was known. They determined the absolute configuration of the asymmetric carbon atom at the 4-position of the cyclohexanone ring in the following way.

The 2,4-dimethylcyclohexanone (5) from the base-induced degradation of (1), which exhibits a single negative COTTON effect curve in methanol solution (c, 0.097) with a trough at 297.5 nm (— 278°) and a peak at 275 nm (— 57°), was transformed into its enol acetate (17) which was ozonized to give (+)-4-methyl-6-oxoheptanoic acid (18). The latter compound was found to exhibit a single position COTTON effect curve opposite in sign to that of (+)-2-ethyl-4-pentanone (19). Since (19) had been synthesized previously from (—)-2-ethyl-1-propanol of known absolute configuration [(S)] it was deduced that 4-methyl group of (5) must have the (R)-configuration.

Rigorous confirmation of this argument was provided by hypobromide oxidation of (18) which led to (+)-3-methyladipic acid (20)

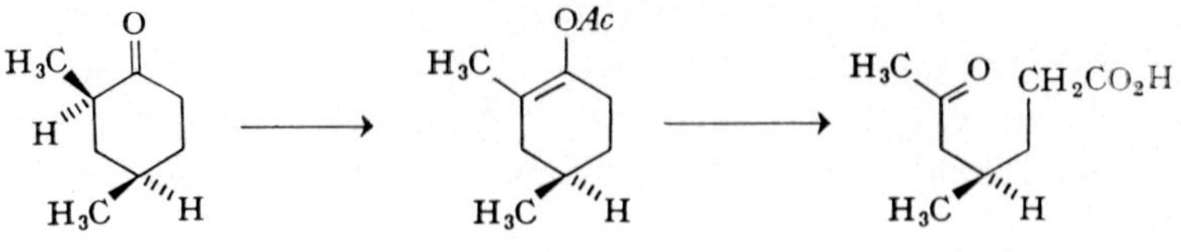

[Structures (5), (17), (18), (20), (19)]

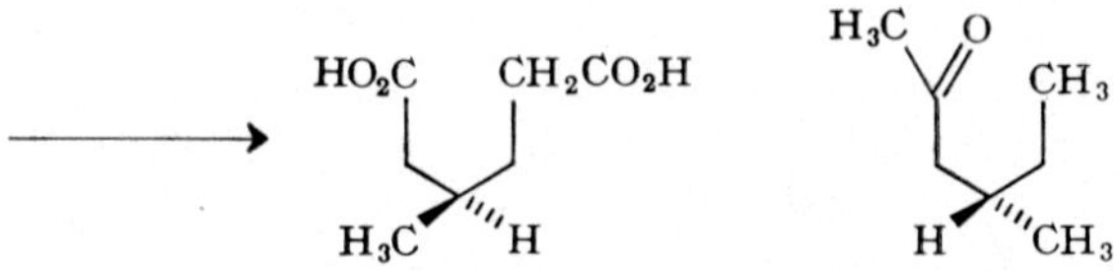

already related to (R)-glyceraldehyde. Since (5) was formed from cycloheximide under alkaline conditions, the two methyl groups logically were assumed to be *cis*-related from which it followed that the absolute configuration of the ketone (5) must be 2(R),4(R) as depicted. Perhaps it is needless to say that although at this point the absolute configuration of the asymmetric center at the 4-position of this cyclohexanone ring was known, no information was available that would determine whether the methyl group at this center was axially or equatorially oriented in cycloheximide itself. In addition the absolute configuration at the 4-position of (1), namely (S), must also be the same in its isomers (13) and (16) since this position is not involved in their base-induced isomerization or degradation (although the latter had not been rigorously proved at the time).

c) Fine Structure

The stereochemistry of cycloheximide and its isomers divides itself naturally into two areas, that of the cyclohexanone ring and that concerned with the hydroxyl group. The former problem, because of the biological activity of cycloheximide and of the then current high interest in the stereochemistry of 6-membered rings, occupied the attention of at least four research groups between the years 1958 and 1963. This problem and the variety of solutions offered to it must necessarily be discussed first, not only because of its greater intrinsic interest but because on the answer to this problem ultimately depended the determination of the orientation of α-hydroxyl group.

The Cyclohexanone Ring. From a theoretical point of view only four isomers are possible for the 2,4,6-trisubstituted cyclohexanone ring of these antibiotics. Based on the (S)-configuration already established as the absolute configuration at the 4-position, the most stable conformers of these isomers A, B, C, D are shown below.

References, pp. 202—208

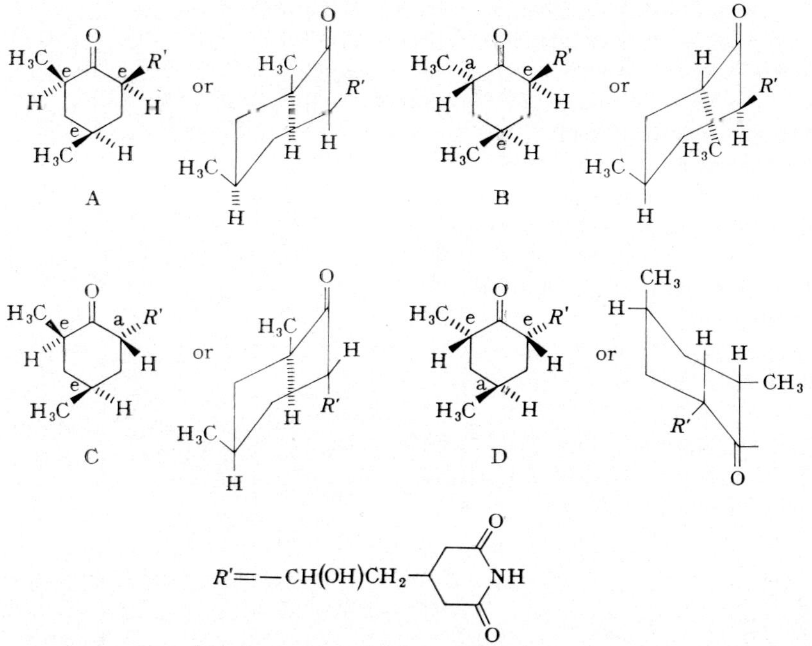

The Chemistry of Glutarimide Antibiotics 153

$R' = -CH(OH)CH_2-$ [glutarimide group]

If for the moment the assumption is made that the configuration of the hydroxyl group in cycloheximide and its isomers is the same and that this center is not involved in the base-catalyzed isomerization, then each of the three antibiotics must be represented by one of these formula.

The first workers to offer any opinion on this problem were LEMIN and FORD (*61*). They had shown that cycloheximide (*1*) and isocycloheximide (*13*) were related by some isomerization involving the substituents on the cyclohexanone ring. For this reason they suggested that the substituents at the 2- and 6-positions in cycloheximide were *trans*-related whereas in isocycloheximide they were *cis*-related. However, no further work was done by these investigators to support these contentions.

The first serious attempt to resolve this problem was carried out by OKUDA (*66, 68, 76*) who began with an infrared study of both (*1*) and (*16*) in dilute solution. In each case only bands due to intramolecular hydrogen-bonding were observed, and it was concluded that the glutarimide-containing substituent at the 6-position in both Naramycin-B and cycloheximide must be equatorially oriented. If it were to have been axially oriented, no intramolecular hydrogen bonding should have been observed. Infrared data from reasonably analogous systems supported this conclusion.

Okuda then undertook an analysis of the optical rotatory dispersion curves of cycloheximide and naramycin-B. Cycloheximide exibited a single negative Cotton effect curve similar to (+) cis-2,4-dimethylcyclohexanone (5) whereas Naramycin-B showed s single, but strong, positive Cotton effect curve. In the detailed analysis of these curves Okuda made a number of assumptions. The first was that a methyl group at the 2-position of a cyclohexanone ring whether it be equatorial or axial could not make as strong a contribution to the Cotton effect as another substituent bearing an asymmetric center. Therefore, while it was expected that 2-methyl group in both isomers might cause an amplitude variation, it was not thought that this group could cause a change in the sign of the Cotton effect. Inevitably then, it was concluded that the differences in signs of the Cotton effects in cycloheximide and Naramycin-B were the result of a configurational difference at C-6, a result in direct opposition to that obtained in the infrared studies.

The second assumption arose from a detailed investigation of the optical rotatory dispersion curves of cyclohexanones having an alkyl substituent at the 2-position. On the basis of the data obtained it was tentatively suggested that the 2-alkyl group behaved as if it were an axial halogen atom and that the sign of the Cotton effect could be interpreted according to the "axial halo-ketone dispersion rule". On the basis of this latter assumption, which Okuda seemed to feel took precedence over other considerations, he concluded that cycloheximide should be represented as C, but in the ring inverted form in which the two methyl groups were axially oriented and the glutarimide-bearing side-chain was equatorially-oriented. By the same token, he concluded that in naramycin-B all substituents were equatorially oriented and that it should be represented by structure A.

In two subsequent publications (72, 74) Okuda and his associates reported a more extensive examination of the optical rotatory dispersion curves including in the study those of isocycloheximide (13) and a synthetic optically active isomer α-epiisocycloheximide. From these studies they deduced that isocycloheximide, naramycin-B and cycloheximide should be represented respectively by structures A, B, and D, structures which are now accepted as correct.

Contemporaneously with the work of Okuda an attempt was made by Lawes (56, 57) to clarify the stereochemistry of the cyclohexanone ring of cycloheximide by chemical means. Lawes found, perhaps by following the slightly earlier work of Mole (64) on the thermal retro-aldol reaction, that cycloheximide when pyrolyzed gave rise to a 2,4-dimethylcyclohexanone (21) different from that obtained on alkaline degradation. Both ketones were clearly epimers because (21) could be isomerized in 79% yield to (5) (the ketone from the alkaline degradation). For this

References, pp. 202—208

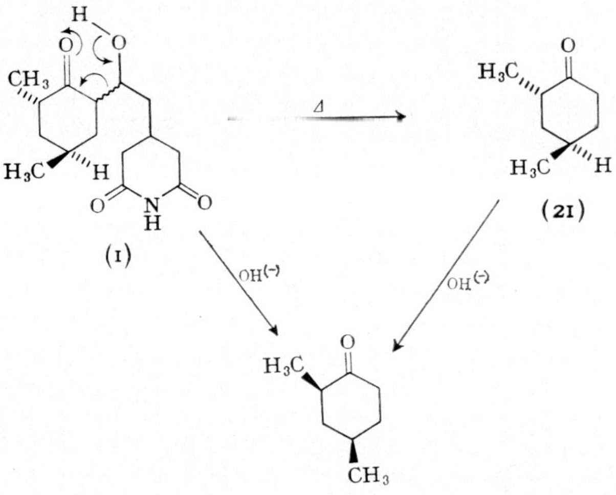

reason (21) was assigned *trans*-oriented methyl groups, a conclusion which also had to apply to cycloheximide itself.

Armed with this information LAWES then concluded that of the two possible structures, B and D, open for cycloheximide, B was the more likely probability. His arguments in support of this conclusion were based on a consideration of the ease of isomerization of cycloheximide to isocycloheximide which he felt was more consistent with an epimerization involving the labile sites at C-2 or C-6. It seemed to LAWES that of the two possible structures B and D an epimerization at C-2 in B should afford more readily, a stable product having all ring substituents in the equatorial configuration. On the other hand, it was considered that structure D should be resistant to epimerization for the two following reasons. (a) Any isomerization of D must involve an intermediate which is very unstable because it must necessarily have two axial groups situated 1,3 with respect to each other. (b) The epimerization must also at some point cause a disruption of whatever hydrogen bonding exists between the α-hydroxyl of the side chain and the carbonyl group of the ring, a factor which it was felt contributed to the ground state stability of D.

WOLINSKY and CHAN (*131*), however, contested that LAWES' conclusions were not necessarily justified by the reasoning involved. They pointed out that the epimerization of B and D might occur with almost equal ease, if D first underwent a transformation to B via ring inversion of the relatively strain free enolic form of D involving the 6-position as shown below.

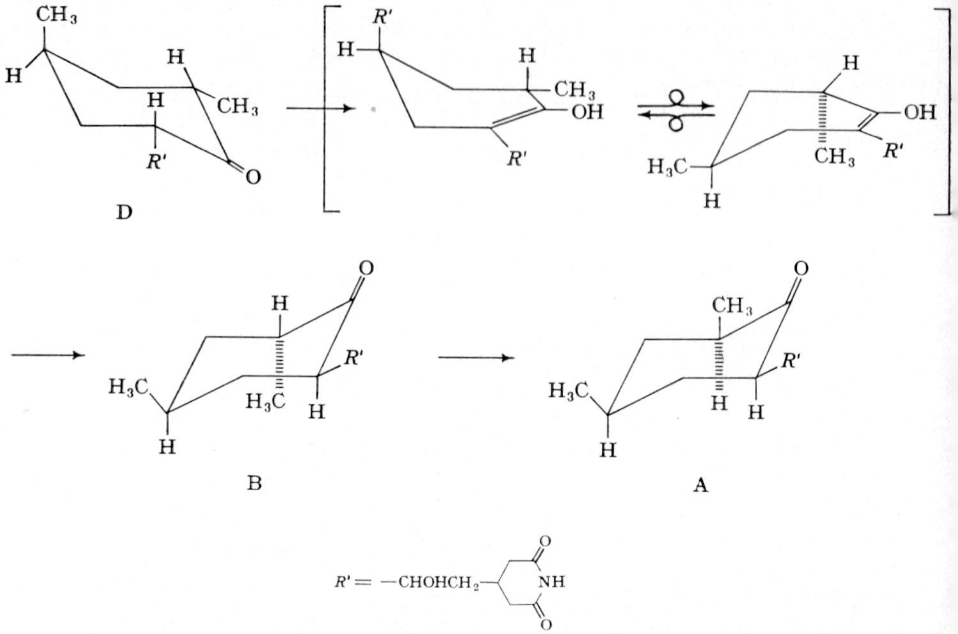

$R' = -CHOHCH_2-\underset{O}{\overset{O}{\big\langle}}NH$

The work of LAWES was pursued by OKUDA (79) who thermally degraded naramycin-B and isocycloheximide. The first afforded *trans*-2,4-dimethylcyclohexanone whereas the latter yielded the *cis*-isomer, results completely in accordance with the structures previously proposed by OKUDA for these two materials. OKUDA also carried out a careful study of the base-catalyzed isomerization of the three isomers. Cycloheximide could be isomerized completely to isocycloheximide, or to a mixture of naramycin-B and isocycloheximide depending upon the conditions used. Naramycin-B could be isomerized only to isocycloheximide and isocycloheximide underwent no epimerization under the conditions used. (1% sodium methoxide in boiling benzene solution.) Again, these results are entirely in accord with what could be expected for isomers having the ascribed stereochemistry.

JOHNSON and his associates (44, 45, 47) approched the problem of the cyclohexanone ring stereochemistry of cycloheximide and its isomers from an entirely different direction. Their studies arose from an attempt to synthesize cycloheximide based on structure C which had been tentatively assigned to this molecule by OKUDA in his early papers (66, 68, 76). Although they did succeed in synthesizing a molecule with the desired stereochemistry by the condensation of racemic *cis*-2,4-dimethylcyclohexanone (5) with the aldehyde (22), the product, 23), which was named

References, pp. 202—208

neocycloheximide, spectroscopically in no way corresponded to cycloheximide itself. (This and related Japanese synthetic work is described in Section III.)

The isolation of this new isomer did, however, aid materially in determining the orientation of the large side chain in both cycloheximide and isocycloheximide. Careful chromic acid oxidation of (23) and of the latter two compounds led to three different diones (24), (25), and (26) respectively. The rates of enolization of (25) and (26) to the previously known (7) and (15) respectively were rapid whether induced thermally or in solution in the presence of ferric chloride. On the other hand, (24) by comparison was relatively stable to enolization, but did eventually afford the racemic form of (15). On the basis of these qualitative rates of enolization (25) and (26) and therefore (1) and (13) were assigned equatorially oriented glutarimide-bearing side chains whereas the same side chain in (24), and therefore in (23) was assigned the axial orientation. These assignments were justified by making use of the finding by COREY and SNEEN (13) that in a rigid cyclic system a hydrogen atom adjacent to a ketone is removed at a faster rate when it is in an axial, as opposed to an equatorial position, in an enolization process.

Additional evidence for the axial nature of the large side chain in neocycloheximide came from a study of the hydroxylic infrared absorption bands of (1), (13) and neocycloheximide (23) itself. In dilute solution both of the former compounds showed a band at 3520 cm^{-1} characteristic of moderate intramolecular hydrogen bonding between the hydroxyl and ketone groups whereas (23) exhibited absorption only at 3585 cm^{-1} indicative of intermolecular hydrogen bonding.

Since neocycloheximide was derived from *cis*-2,4-dimethylcyclohexanone it was concluded that (23) also had *cis*-oriented methyl groups. This conclusion could also be applied to isocycloheximide (13) because both it and (23) are converted to the same enolic diketone (15) by processes which it was felt would not cause any isomerization of the methyl group at the 2-position. As a corollary to these deductions cycloheximide must have *trans*-methyl groups. Corroborative evidence for this was available from the work of LEMIN and FORD (61) who showed, as has already been mentioned, that of the two isomers (7) and (15) the latter is the stable one under equilibrating conditions. At this point it was concluded that neocycloheximide must be represented by structure C, isocycloheximide by structure A, and cycloheximide by either B or D. No comment was made on the stereochemistry of naramycin-B because sufficient material had not been available for chemical study.

In an attempt to bring the stereochemical studies on cycloheximide and its isomers to a more satisfactory conclusion JOHNSON and his co-workers undertook an extensive nuclear magnetic resonance study of (1), rac- (5) rac- (22) (23) (15) rac- (24)

(1) (25) (7)

(13) (26) (15)

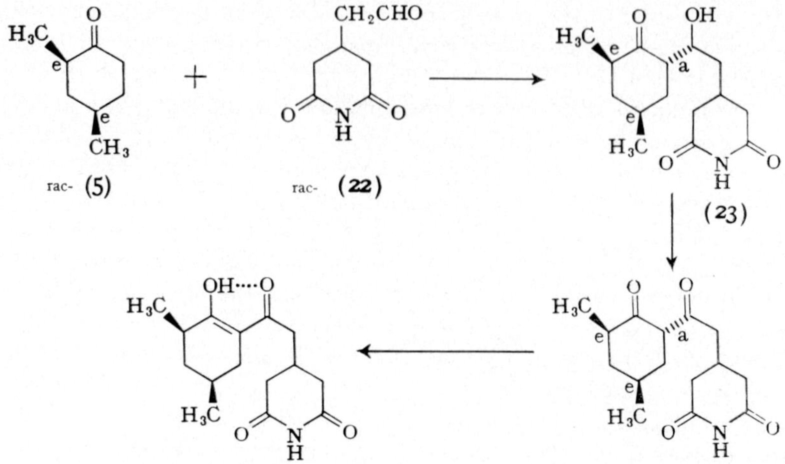

its isomers and some model systems. The models involved were the *cis*- and *trans*-isomers of both 4-*t*-butyl-2-methylcyclohexanone and 2-*t*-butyl-4-methylcyclohexanone, and these it was felt should be comparably as conformationally biased as cycloheximide and its isomers. A basic assumption was made that any special NMR spectral charac-

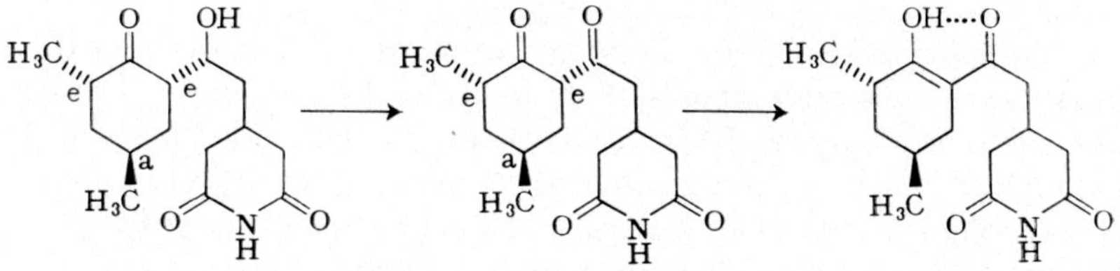

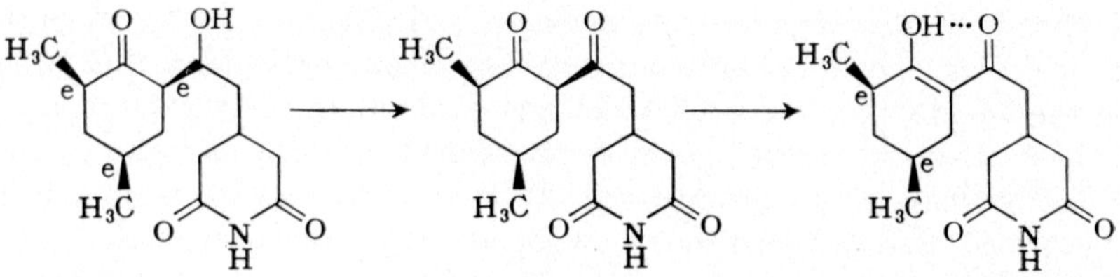

teristics of the methyl groups in the models also would be exhibited by the methyl groups in (1) and its isomers.

For the purposes of the study three parameters were measured: (a) the position of the center of the methyl group doublets in deuterochloroform, (b) the coupling constants J (in cps) of the methyl groups [and it should be recognized that differences between them are apparent rather than real (2)], (c) the direction and magnitude of the displacement of the methyl group doublets which occurred when pyridine was substituted for deuterochloroform as the solvent. The introduction of this third parameter marked for the very first time the use of the solvent-shift effect for the solution of a stereochemical problem. This technique was subsequently developed and applied extensively to other stereochemical problems by CONNELLY and MCCRINDLE (12) and more notably by WILLIAMS (7). The results for the model ketones are shown in Table 2.

Table 2. *N.M.R. Spectra of Model Ketones*

Cyclohexanone	Me group Stereochemistry	Positions of Me peaks (in c. p. s.)*	
		in $CDCL_3$	in pyridine
2-methyl	largely equat.	60.6 (6.2)	60.3 (6.3)
4-methyl	largely equat.	60.3 (5.5)	51.5 (5.4)
trans-4-tert-Bu-2-Me	axial	69.3 (7.2)	64.6 (7.2)
cis-4-tert-Bu-2-Me	equat.	61.1 (6.3)	61.8 (6.3)
trans-2-tert-Bu-4-Me	axial	66.8 (6.3)	60.8 (6.3)
cis-2-tert-Bu-4-Me	equat.	60.1 (5.6)	52.6 (5.6)

* J values are given in brackets

From this NMR data it was determined: (a) that axial methyl groups occur at lower field than equatorial groups irrespective of their being at the 2- or the 4-positions; (b) that the J value assumes the largest value for a 2ax-methyl group and the smallest value for a 4eq-methyl group; (c) that the change of solvent from deuterochloroform to pyridine led to a marked displacement to higher field of all methyl doublets except that of a 2eq-methyl group. The 4eq-methyl group doublet underwent the largest displacement upfield.

The NMR spectral data (Table 3) for cycloheximide, its isomers and a number of related compounds were then analyzed using these findings. As an example of the analysis the data for cycloheximide was assessed in the following way: This compound and its derivatives all display a low field methyl doublet indicative of an axial position which moves ($\sim$ 5 cps) to higher field in pyridine solution. On the other hand, the methyl doublet at higher field in deuterochloroform solution if anything moves to slightly lower fields when the medium is changed to pyridine. As judged in the light of Table 2 this can only be indicative of a molecule containing both 4ax- and 2eq-methyl groups. This deduction is confirmed by the fact that both of the doublets have intermediate J values which would be

Table 3. *N.M.R. Spectra of Cycloheximide and Related Compounds*

Compound	Position of Proton signals in cps.					
	In deuteriochloroform			in pyridine		
	methyl doublets*		CHOR*	methyl doublets		CHOR
	2-CH$_3$	4-CH$_3$		2-CH$_3$	4-CH$_3$	
Cycloheximide	58.9 (6.1)	73.6 (6.7)	244	59.1 (6.1)	68.7 (6.6)	265
Cycloheximide acetate	59.2 (6.0)	75.7 (6.6)	319	59.4 (6.4)	70.6 (6.6)	347
Cycloheximide chloroacetate	57.0 (6.4)	73.6 (7.0)	328	56.5 (6.3)	67.0 (6.7)	341
Cycloheximide tosylate	56.2 (6.3)	72.2 (6.9)	301	57.5 (6.2)	68.3 (6.7)	323
N-methylcyclohexi-mide acetate	58.3 (6.4)	74.8 (6.8)	317	59.3 (6.2)	71.0 (6.6)	346
Dehydrocyclohexi-mide (Diketo form)	59.0 (6.3)	72.5 (7.0)		—	—	
Isocycloheximide	59	59		58.3 (6.4)	52.5 (5.8)	254
Isocycloheximide acetate	60.6 (6.3)	58.2 (5.9)	317	58.4 (6.2)	52.1 (5.9)	
dl-α-*epi*isocyclo-heximide	60.2 (6.3)	60.7 (5.8)	252			
dl-α-*epi*isocyclo-heximide acetate	58.8	58.8	324	57.6 (6.5)	50.3 (5.8)	346
Dehydroisocyclohexi-mide (Diketo form)	60.6	60.6		—	—	
Naramycin-B	72.7 (7.4)	59.0 (5.0)	223	64.4 (7.1)	52.1 (5.8)	
Naramycin-B acetate	71.4 (7.2)	60.9 (6.0)	321	63.8 (7.2)	53.0 (5.8)	347
Neocycloheximide	—	—		63.8 (6.4)	50.3 (6.0)	261
Neocycloheximide acetate	57	57	319	61.4 (6.3)	48.4 (5.8)	
E-73	61.8 (5.8)	108.7 (Singlet)	253	61.4 (5.8)	111.6 (Singlet)	270
Streptovitacin A	—	—	—	65.6 (6.1)	105.0 (Singlet)	—

* In all the acetates compounds listed (except E-73), COCH$_3$ peaks occur within the range of 118—125 cps. in CDCl$_3$ *and* pyridine. *J* values in cps. given between brackets. In some vases overlap of peaks prevented accurate determination of splitting constants. Neocycloheximide was parctically insoluble in deuteriochloroform as was streptovitacin-A.

References, pp. 202—208

expected for the methyl groups at these positions having these orientations. Thus, in agreement with OKUDA's later optical rotatory dispersion data, cycloheximide was assigned structure D.

By similar analyses, naramycin-B was deduced to be structure B, isocycloheximide, structure A, and neocycloheximide, structure C. The latter isomer must, of course, be a 50 : 50 mixture of C with its mirror image since the substance is racemic. This, however, points out a major advantage of the NMR method since it can be used with racemic compounds whereas optical rotatory dispersion methods cannot.

An investigation into the stereochemistry of the methyl groups of cycloheximide was also reported by SCHAEFFER and JAIN (*94, 97*). These workers envisioned that if both the hydroxyl and ketone oxygen atoms could be removed under non-isomerizing conditions then the final dideoxy-derivative should be optically inactive if the methyl groups are *cis*-related because the substance would be a meso form, or optically active if the methyl groups were *trans*-related. Accordingly, they treated cycloheximide with phosphorus pentoxide and obtained the previously described dehydration product (9). Hydrogenation of (9) with a platinum catalyst in acetic acid afforded the alcohol (27). Treatment of (27) with thionyl chloride in dioxane solution then led in good yield to a crystalline monochloro compound which when reduced with zinc dust in acetic acid gave the dideoxy derivative (28). An optical rotation of — 19° was found for (28) and in view of this, the methyl groups in this compound and all of its precursors including cycloheximide were assigned a *trans*-relationship.

The same investigators also reported (*96*) that when cycloheximide tosylate (33) was boiled in dimethyl formamide a new type of dehydration product was formed. This was assigned structure (29). Hydrogenation of the latter with palladium in ethanol was then found to give a deoxy compound (30) which was claimed to be different from the deoxy compound (32) that could be obtained from (9) using the same conditions. When (30) was subjected to further hydrogenation over a platinum catalyst, a new alcohol (31) was obtained. Treatment of the latter compound with phosphorus tribromide followed by reduction of the intermediate bromo compound with zinc dust in acetic acid yielded a dideoxy compound (34) which was determined to be optically inactive. On this basis all of the compounds in this sequence which were given the designation *epi*, were assigned methyl groups that were *cis*-related. They explained this finding by postulating that isomerization took place at the 2-position under the vigorous conditions used to eliminate the toluene sulfonic acid from (33).

Superficially these reactions would appear to constitute a reasonable proof that the methyl groups in cycloheximide are *trans*-related. However, SCHAEFFER and JAIN had reported that the ketone (32) could not be isomerized under any conditions that were tried. This seemed anomalous to JOHNSON and his coworkers (*105*, *107*) who felt that if the compound had *trans*-methyl groups as was reported, then isomerization should have been easy. They therefore reinvestigated the problem and found that neither (32) nor (30) was affected by acid or base. Using a combination of physical methods including the NMR approach that they

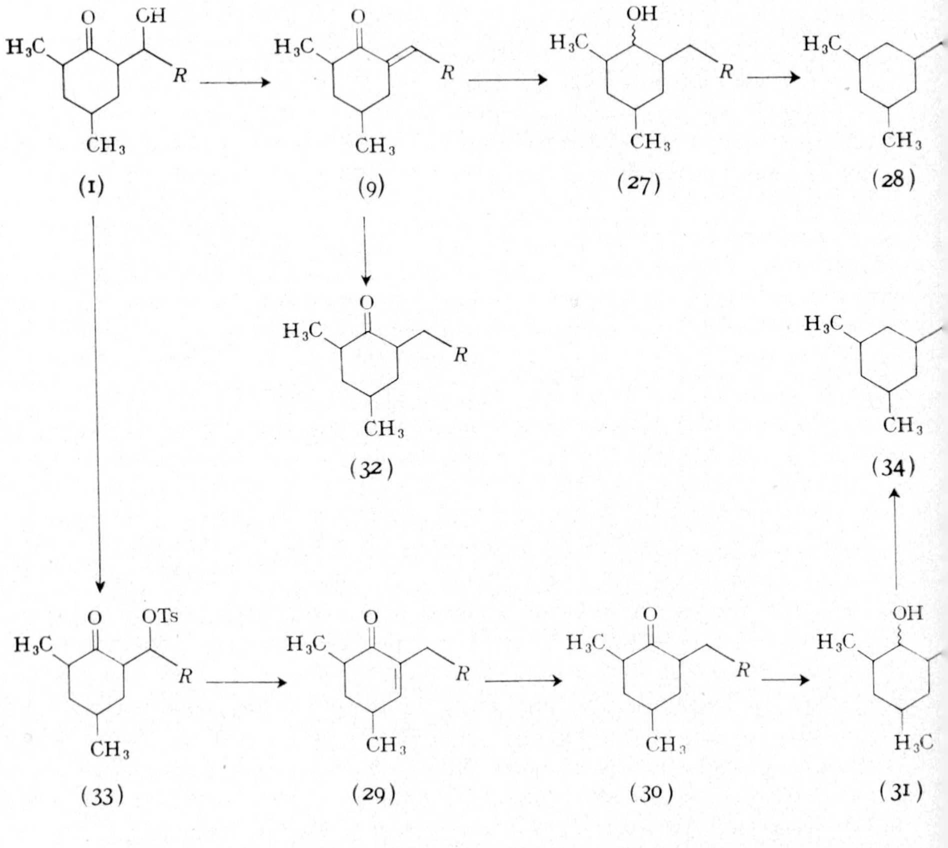

had previously developed, they not only showed that both of these ketones had *cis*-oriented methyl groups, but that in fact (30) was the racemic form of (32). This implied that all of the compounds of the so-called *epi*-series were racemic, a conclusion that was borne out by the fact that all of the compounds reported in this series by SCHAEFFER and JAIN (*96*) were virtually without optical activity.

JOHNSON and his coworkers (*105*, *107*) suggested, quite plausibly that racemization occurred during the conversion of (33) to (29) and involved a combination of an acid-catalyzed enolization processes and a

References, pp. 202—208

(29)

double bond migration in the ring as shown below. Although it was not re-examined, it appears likely that the optical rotation reported for the first dideoxy compound (28), is probably in error and is in fact zero. While the initial ideas in this particular approach to the stereochemistry of cycloheximide were good, fate was unkind.

The α-*hydroxyl group.* The determination of the configuration of the α-hydroxyl group in cycloheximide was carried out by relating it stereochemically to the ring-hydroxyl in the dihydrocycloheximides and their isomers. The problem was first tackled by OKUDA (*67, 70*). He assigned an equatorial orientation to the l-hydroxyl group of the dihydrocycloheximide (4) first obtained by KORNFELD and his coworkers (*54*) by the hydrogenation of (1) in acetic acid over a platinum catalyst. This assignment was based on the fact that hindered ketones tend to give equatorial alcohols under such conditions. OKUDA also examined the reduction of cycloheximide acetate (35) using the same catalytic conditions and also by sodium borohydride, or lithium tri(*t*-butoxy) aluminium hydride. The catalytic procedure led to a dihydrocycloheximide acetate (36) (m. p. 165°) as noted by KORNFELD (*54*), the hydroxyl group of which was thought to be on the ring and which was assigned an axial orientation. Complex metal hydride reduction afforded (37) (m. p. 177–8°), isomeric with (36), which was considered to have the ring hydroxyl group equatorially oriented.

Having made these assignments OKUDA then examined the optical activity of the borate ester of (4). This was found to differ little from that of the parent diol and logically the (S)-configuration was assigned to the side-chain asymmetric center of cycloheximide. Considerable support for this conclusion was adduced from an infrared study of the hydrogen bonding differences between the two supposedly stereoisomeric mono-acetates (36) and (37). The same conclusion was reached concerning the asymmetric center in the side chain of naramycin-B.

JOHNSON and his coworkers (*46, 106*) re-investigated the whole problem when they found that oxidation of (36) did not regenerate cycloheximide acetate (35) but instead afforded a new keto-acetate, (38), in excellent yield. They then re-examined the reduction chemistry of cycloheximide acetate and founds that the catalytic hydrogenation in reality gave a mixture of (36) and (37) in which the former predominated. Complex metal hydride reduction on the other hand afforded a similar mixture in which (37) was present in slight excess. Significantly, they showed that acetylation of (4), (36) and (37) with acetic anhydride in pyridine gave in each case the same diacetate, (39).

From these experiments they concluded that all of the reductions had the same stereochemical consequence and in the case of cycloheximide acetate a greater or lesser amount of acetyl group transfer from the side-

References, pp. 202—208

The Chemistry of Glutarimide Antibiotics

chain hydroxyl group to the ring hydroxyl occurred depending upon the reducing agent that was used. As expected, chromic acid oxidation of (37) regenerated cycloheximide acetate. Moreover they were able to show that when cycloheximide is reduced with diphenyltin dihydride a new diol, (40), could be isolated in addition to a large quantity of (4). Both (4) and (40) gave different triacetates, (41) and (42), respectively. Monoacetylation of (40) led to an hydroxyacetate (43) which on chromic acid oxidation afforded a new keto-acetate, (44), again belonging to the ψ-cycloheximide series.

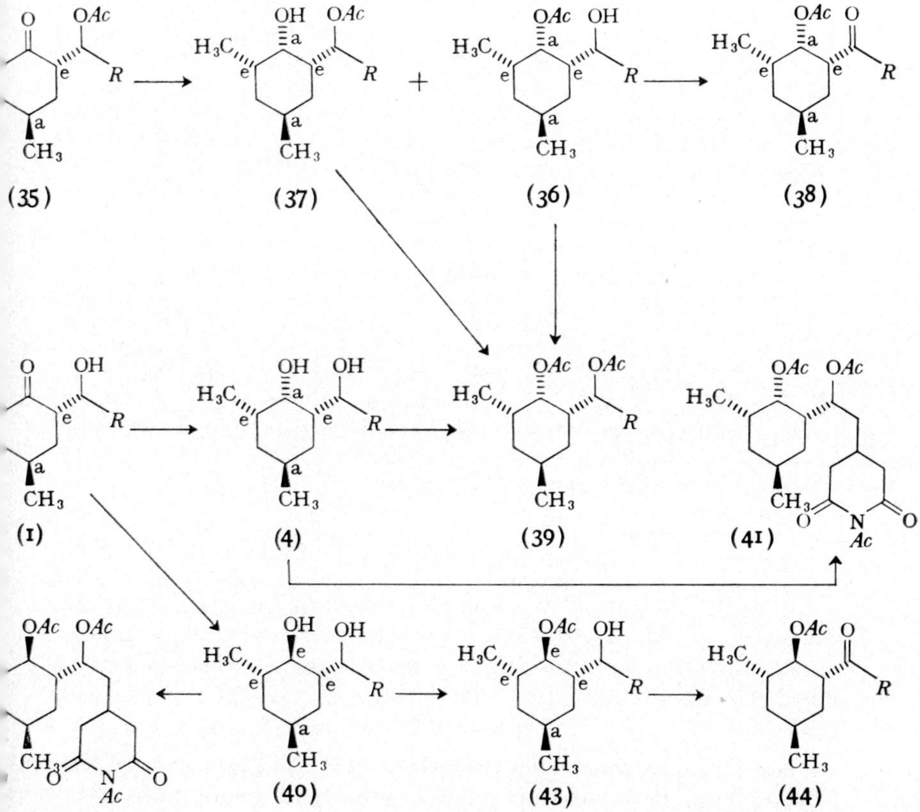

An extensive NMR study was carried out to prove the structures and stereochemistry depicted for the reduction products and their derivatives, not only in the cycloheximide series but also for the corresponding compounds from isocycloheximide and naramycin-B noted below. In this they used the results of LEMIEUXs work (59) to assign the orientations of the cyclohexane ring hydroxyl groups.

Catalytic reduction of isocycloheximide (13) gave a new diol (45) in good yield. On the other hand, reduction of (13) or of cycloheximide with aluminium isopropoxide in benzene solution led in moderate yield to a second diol (46). Only catalytic reduction of naramycin-B (16) was studied. This afforded diol (47).

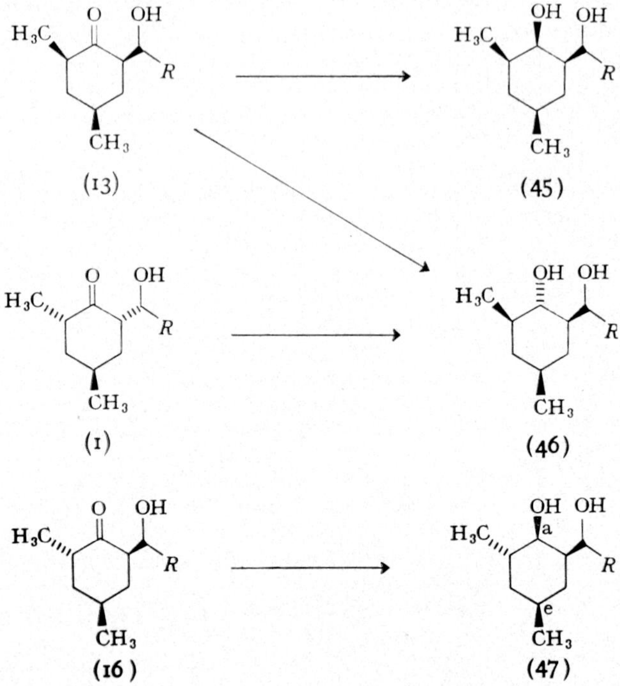

With the problem of the ring stereochemistry of the cyclohexanols completely solved it now became possible to determine the stereochemistry of the α-hydroxyl group. This was done by determining which of the dihydrocycloheximides, (4) or (40), would form an acetonide under mild conditions. The reasoning behind this statement was as follows.

Theoretically two possible acetonides, (48) and (49) could be derived from diol (4) depending upon the relative position of the side-chain hydroxyl group. If the latter were to have the (R)-configuration, then the formation of acetonide (48) should be possible because in that case the 3-glutarimidomethyl group (represented by R) necessarily would assume an equatorial orientation in the ketal ring. Thus, little steric resistance to ketal formation could be anticipated. On the other hand, f the hydroxyl group in question had the opposite configuration, namely (S), the ketal to be expected would be (49). In this latter molecule there

References, pp. 202—208

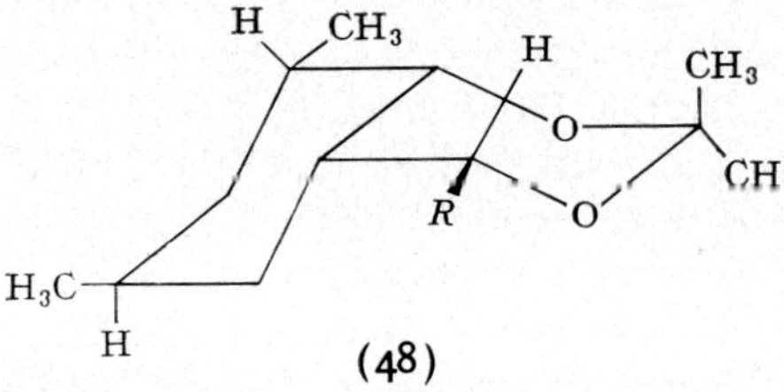

(48)

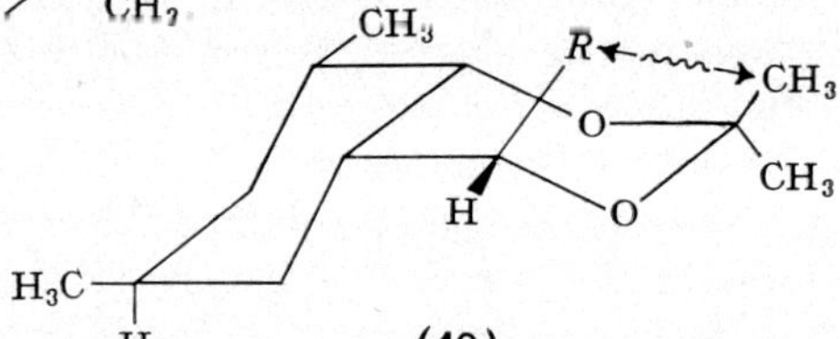

(49)

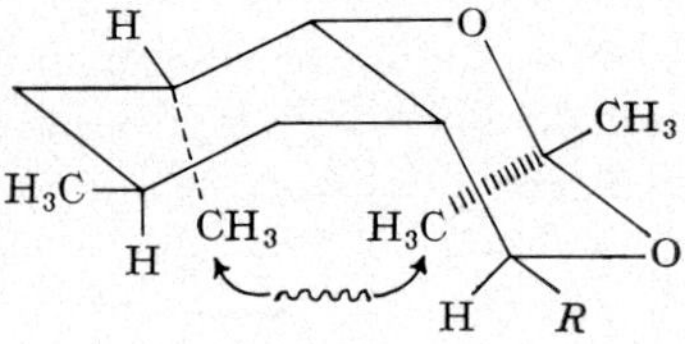

(50)

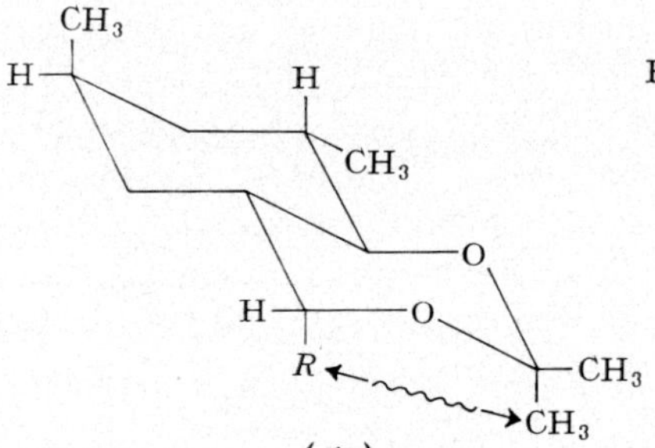

(51)

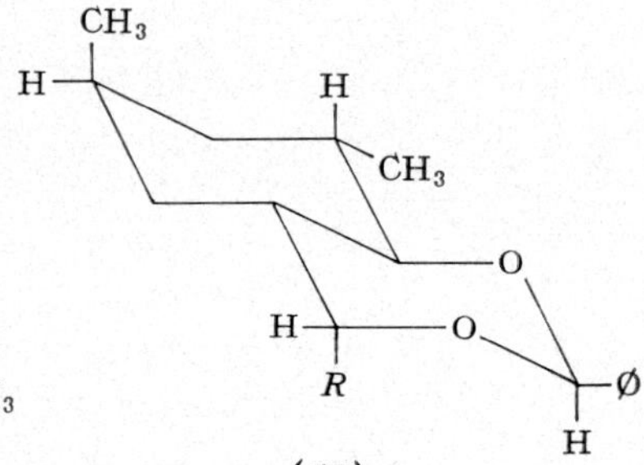

(52)

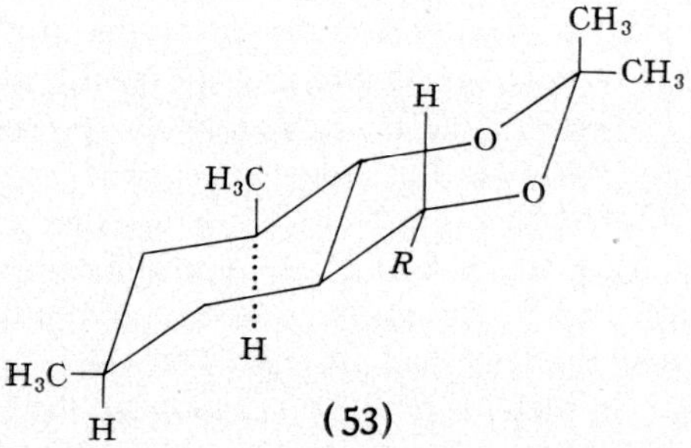

(53)

would be a large non-bonded 1,3-interaction between the R-group which would now necessarily be axial, and the axial methyl group of the isopropylidene function. The possibility that ketal formation could occur in this case via the cyclohexane ring-inverted form of diol (4) was considered remote since this conformer itself contains a 1,3-diaxial interaction involving the 2- and 6-groups of the cyclohexane ring. Even so, the ketal that would arise from such a conformer would have structure (50) and it was considered that the mutual interference of the methyl groups in the ketal and cyclohexane rings would be so great as to prohibit the formation of such a molecule under mild conditions. Similar considerations were thought to apply to the boat form of either ring. Thus, if (49) or (50) could be expected to form at all, their rate of formation should be very slow by comparison with that of (48).

On the basis of the above argument, the fact that, experimentally, diol (4) did give an acetonide under very mild conditions was considered compelling evidence that the side-chain asymmetric center of cycloheximide has the (R)-configuration. A logical deduction from this assignment was that the epimeric diol (40) should not give an acetonide or at best should form an acetonide only slowly. This followed because the expected acetonide should have structure (51) which once again would contain a severe 1,3-diaxial interaction. In practice it did not prove possible to prepare an acetonide from (40), although a benzylidene (52) could be prepared. These findings confirmed not only the argument but also the (R)-assignment made for the α-hydroxyl group of cycloheximide.

Using similar arguments JOHNSON and his associates were able to predict that, if as could be anticipated, the α-hydroxyl of isocycloheximide and naramycin-B had the same orientation as that in cycloheximide then diol (46) should form an acetonide (53) whereas diol (45) and (47) should not. Experimentally these predictions were fulfilled.

d) Miscellaneous Chemistry

Almost all of the chemistry of cycloheximide and its isomers has evolved because of attempts to determine their structure and stereochemistry. Very little of their chemistry has been studied *per se*. No intramolecular, free-radical or photochemical reactions have been reported. However, several points are dealt with in this section that were not clarified by the foregoing discussion.

Dehydration Products and their Reduction. Under acidic conditions of one type or another the dehydration of cycloheximide (54, 105) of cycloheximide tosylate (96), of naramycin-B (77), and of isocycloheximide (61) apparently gives the same anhydro product (9).

The kinetics of the dehydration of (1) and its hydrolysis products have been examined in great detail by GARRETT and NOTARI (29, 31) who showed that rehydration of (9) occurs as a competing reverse reaction in dilute acid solution. This is discussed later. The best yields (*105*) are obtained when cycloheximide is treated with acetic acid containing a trace of sulfuric acid for two hours at room temperature.

Catalytic reduction of (9) using a palladium catalyst afforded the saturated ketone (32) which, OKUDA (*69*) first suggested, had the all *cis*-configuration because of its optical rotatory dispersion curve. This suggestion was supported (*105*) by the NMR spectrum which was similar to that of isocycloheximide and suggested that (9) also had *cis*-oriented methyl groups. The conclusion was proved by STARKOVSKY, CARLSON and JOHNSON (*105*) in the following way. Reduction of (9) with diphenyltin dihydride gave the allylic alcohol (54) which when catalytically

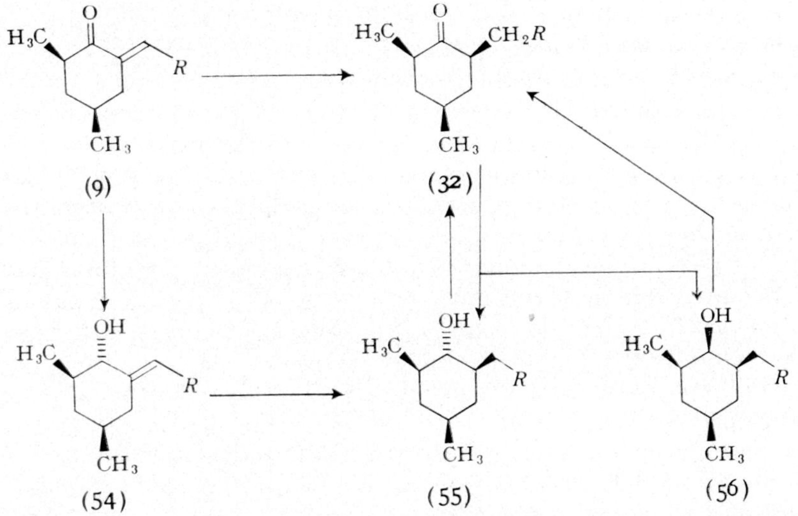

reduced in the presence of a platinum catalyst afforded a single alcohol, (55), m. p. 112–113° whose hydroxyl group was assigned an equatorial orientation on the basis of its NMR spectrum. Chromic acid oxidation of (55) then afforded (32). This sequence of reactions precluded the possibility that any isomerization of (9) could have occured during its palladium-catalyzed reduction to (32). Thus, (9) and (32) must be assigned to the isocycloheximide series and should be named anhydroisocycloheximide and deoxyisocycloheximide respectively.

So far, as has been mentioned earlier, no one appears to have been able to prepare an anhydro compound having *trans*-oriented methyl groups. Even when cycloheximide acetate was pyrolyzed at 180°, only (9) was produced. The same results were obtained with ethyl cycloheximide carbonate.

The reduction chemistry of (9) and (32) still remains somewhat enigmatic. OKUDA (69) found that when (32) was reduced with lithium aluminium tri-t-butoxyhydride it afforded a material m. p. 103–104° which on oxidation with chromic acid regenerated (32). The reduction product, however, was subsequently resolved (105) by chromatography into two components a major product, the alcohol (55) and a minor product, its epimer, a new alcohol (56) of m. p. 130–130.5°. When (32) was reduced catalytically using a platinum catalyst the same mixture of alcohols was obtained although (56) constituted the bulk of the product. Chromic acid oxidation of (56) afforded (32), thus providing a completely consistent picture.

The situation regarding the platinum-catalyzed reduction of (9) is less clear. OKUDA (69) examined this reaction and founds that it led to a mixture of alcohols that could be resolved by fractional crystallization from ether, into two substances having m. p. 117–118° and m. p. 130–131°. SCHAEFFER and JAIN (94) obtained essentially the same results. OKUDA found that oxidation of the lower melting alcohol gave exclusively (32) whereas the higher melting alcohol, to which he ascribed *trans*-oriented methyl groups, gave a new ketone isomeric with (32) of m. p. 123–124°. SCHAEFFER and JAIN on the other hand claimed just the reverse when they oxidized these two alcohols.

JOHNSON and his coworkers (105) have suggested that the source of these conflicting results may be due to the impurity of the anhydroisocycloheximide (9) used in the experimental work. The melting points quoted for (9) by these workers were approximately 20° and 10° low, respectively. Under the equilibrating conditions used for the preparation of (9), about 20% of the total crude material should have *trans*-methyl groups. There remains then the distinct possibility that the anomalous alcohols and ketones discussed above arise from this component of crude (9) and do have *trans*-methyl groups as suggested by OKUDA. No further work has been reported in this area and until a pure sample of the unknown anhydrocycloheximide is obtained, the differences seem likely to remain unresolved.

Cycloheximide Reduction Products and Their Reactions. The reduction of cycloheximide and in particular the reactions of the two alcohols that thus can be obtained were studied in great detail by SUZUKI (115). However, since the stereochemistry of cycloheximide was uncertain at the time the study was carried out, virtually no stereochemical assignments were given to the products. In the light of the currently accepted structure for cycloheximide the author, therefore, has made such stereochemical assignments as seem logical.

Reduction of (1) with platinum and acetic acid afforded the diol (4) that had been obtained previously by KORNFELD (54), whereas reduction with

References, pp. 202—208

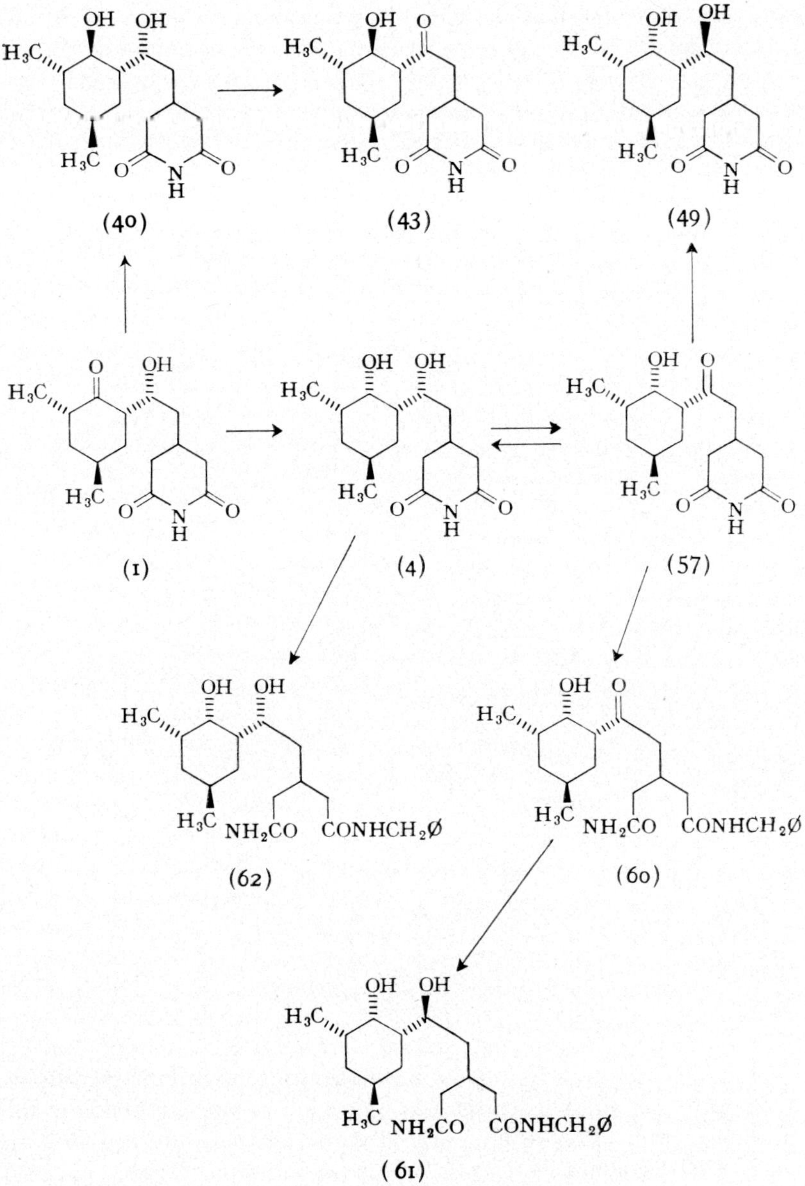

lithium aluminium hydride or lithium aluminum tri-*t*-butoxyhydride in tetrahydrofuran afforded the same diol, (40), that was later isolated by JOHNSON (*46*) from the reduction of (1) by means of diphenyltin dihydride. Selective oxidation of each of these diols by means of chromic

acid gave rise to position isomers of cycloheximide which were named
ψ-cycloheximide-I (57) and ψ-cycloheximide-II (58) respectively.

Only the chemistry of the former compound was investigated. Catalytic reduction of (57) over a platinum catalyst regenerated (4), but when ethanol was the solvent it afforded a new diol (59) whose α-hydroxyl

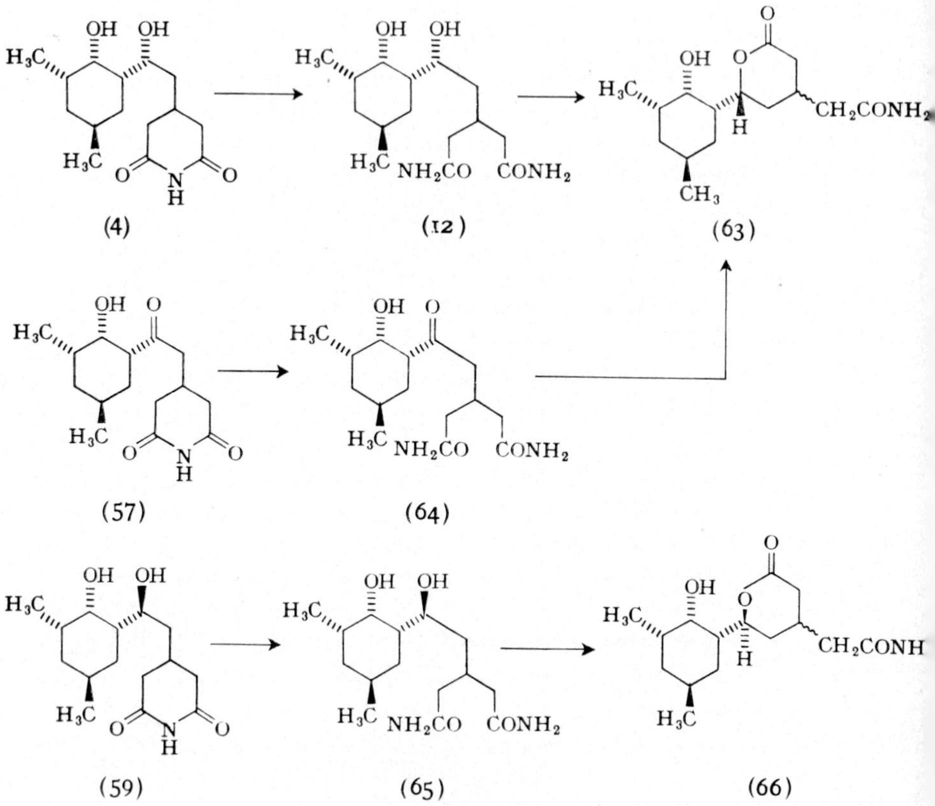

group must have the (S)-configuration. Treatment of (57) with benzylamine did not lead to a retro-aldol reaction with concomitant cleavage of cyclohexane ring as was anticipated but merely to (60), by fission of the glutarimide ring. The latter, on catalytic reduction, afforded (61) no matter whether ethanol or acetic acid were the solvent. This material was different from the substance (62) obtained by the reaction of (4) with benzylamine.

Suzuki (*116*) also studied the biological activity of (40), (4), (57) and (58) against *S. sake* and *P. orizae* but they showed only negligible effects compared with cycloheximide.

References, pp. 202—208

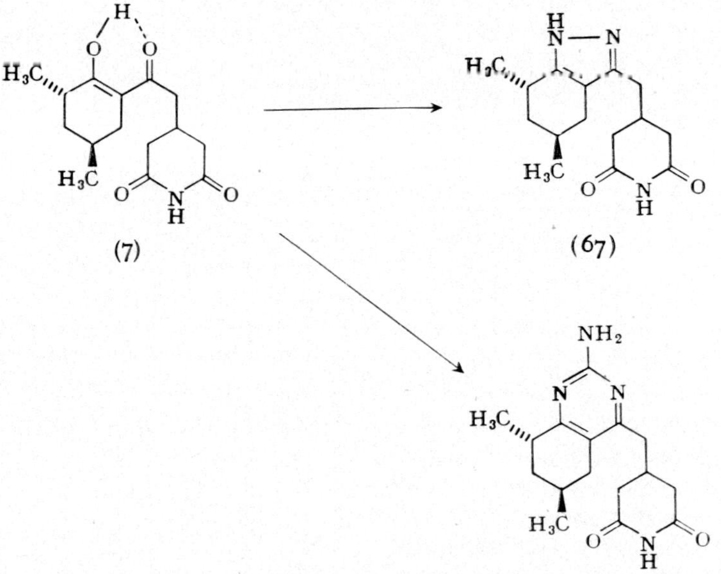

(7) → (67)

(68)

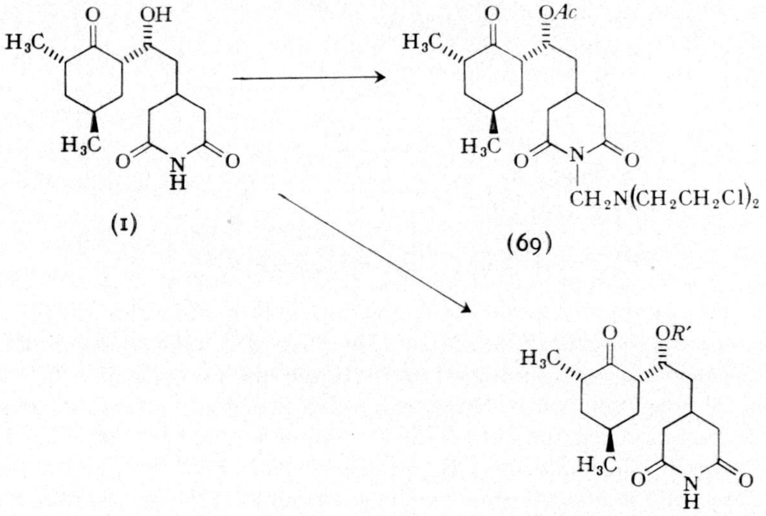

(1) → (69)

(70) (R' = acyl or aroyl)

The reactions of a number of these substances with ammonia in methanol were also examined (*115*). This reagent does not cause degradation of the molecule as does benzylamine (*54*), but simply opens the glurarimide ring. Diol (4) for example affords (*54*) the diamide (12) which when treated with acetic acid or one equivalent of nitrous acid yields the lactonic amide (63). The latter substance was also produced directly by the platinum-catalyzed reduction of (64) in acetic acid which itself was obtained by the treatment of ψ-cycloheximide-I (57) with ammonia. A lactone (67) isomeric with (63) was obtained by first reacting (59) with ammonia in methanol and then allowing the intermediate diamide (65) to react with acetic acid.

Much of the degradation chemistry of cycloheximide carried out in connection with studies on its biogenesis relies on the chemistry reported in this and the previous section. This is dealt with further in Section 4.

Other Reactions. In an attempt to develop new antineoplastic compounds several research groups have made derivatives of cycloheximide. PIATAK, YEN and KENNEDY (*86*) allowed dehydrocycloheximide (7) to react with hydrazine and obtained the cprresponding pyrazole (67). Similarly guanidine led to the corresponding 2-aminopyridine (68). Preliminary evaluation of (67) in the leukemia 1210 system showed the compound to be neither active nor toxic. Results for (68) are not available. PETTIT and DAS GUPTA (*84*) have prepared a nitrogen mustard derivative (69) of cycloheximide acetate and EGAWA, OSHIMA and UMEZAWA (*17*) have prepared a large number of aliphatic and aromatic esters (70) of cycloheximide. The latter were tested *in vivo* against *toxoplasma gondii* RH. The myristate, palmitate, and stearate showed far higher activity than cycloheximide at the dose rate of 12.5 mg/kg/day. The aromatic esters also showed significant prolongation of survival at the same dose at which the oxime, semicarbazone and thiosemicarbazone showed scarcely any effect.

4. The Streptovitacins and E-73

The streptovitacins are a group of closely related glutarimide antibiotics approximately six in number that were first isolated from a *Streptomyces griseus* fermentation by HERR and his associates (*16, 15*) at the Upjohn Company. Members of this group of compounds possess antitumor (*22, 23*) and antifungal (*104*) properties and are active in tissue culture by which they may be bioassayed (*102, 103*). A material closely related to this group of glutarimide antibiotics is E-73 an antitumor substance which was first isolated by RAO and CULLEN (*93*). E-73 is in fact the monoacetyl derivative of streptovitacin-A in which the acetyl group is on the hydroxyl at the 4-position. In common with cycloheximide, streptovitacin-A and E-73 inhibit protein synthesis (6). In antitumor tests the latter two compounds are respectively 100 and 200 times more potent than cycloheximide.

The streptovitacins because of their variation in hydrophilic properties can be separated by use of a modified gradient partition chromatogram (*15*). A solvent system containing ethyl acetate, cyclohexane and McILVAINES pH 5.0 buffer permitted the separation of streptovitacins-A and -B and a partial resolution of C and D. A further separation of streptovitacin-E from cycloheximide and of streptovitacins-C_1 and -C_2 was achieved by using less polar mixtures of these solvents.

The structures of the streptovitacins were determined by HERR (*37, 38*). Alkaline degradation of crude streptovitacin-A (**71**) afforded in addition to ammonia, a steam volatile product that could be resolved by chromatography into 2,4-dimethyl-2-cyclohexenone (**72**) and an unknown hydroxydimethylcyclohexanone (**73**). Since the latter compound was resistant to periodate and afforded 1,3-dimethylcyclohexanol on WOLFF-KISHNER reduction it was assigned the structure 4-hydroxy-2,4-dimethylcyclohexanone. By contrast alkaline degradation of streptovitacin-B gave predominantly (**72**). It had already been noted that both streptovitacin-A and -B were spectroscopically similar to cycloheximide and when it was found that acid-catalyzed dehydration of both substances gave the previously known (*61*) phenol (**14**), this together with the alkaline degradation evidence allowed their formulation as (**71**) and (**74**) respectively. Additional evidence for the tertiary and secondary nature of the extra hydroxyl groups (**71**) and (**74**) respectively was obtained when it was observed that under mild conditions the former gave a monoacetate whereas the latter afforded a diacetate.

Streptovitacin-C_2 was assigned structure (**75**) because on acid-catalyzed dehydration it yielded phenol (**14**), and when subjected to alkaline degradation provided a hydroxyketone which was spectroscopically similar to (**73**). Although no structure was assigned to it by HERR it is probably (**76**). Finally periodate oxidation of streptovitacin-C_2 took place with the consumption of one mole of oxident per mole of compound.

Acid degradation of streptovitacin-D again gave the phenolic derivative (**14**) showing that this compound also is a ring-hydroxylated cycloheximide. However, no further work was done to determine the exact location of the second hydroxyl group.

Sufficient quantities of streptovitacin-C_1 and -E were not available for degradative study and their structures remain unknown.

E-73 was isolated (*93*) as a colorless crystalline solid with antitumor activity from the culture filtrate of *Streptomyces albulus* and its structure was determined by RAO (*92*). In its spectral properties RAO found E-73 to resemble cycloheximide closely. The molecular formula was established as $C_{17}H_{25}O_6N$ and it gave rise to a monoacetate and a mono-p-benzoate, thus showing the presence of an hydroxyl group, and afforded the usual ketonic derivatives, a monosemicarbazone and monoxime. Preliminary alkaline hydrolysis gave rise to one equivalent of ammonia but no volatile ketone was isolated. Acid hydrolysis of E-73 under mild conditions afforded what was eventually recognized (91) as streptovitacin-A. Under more drastic conditions E-73 afforded the phenol (14) which on prolonged hydrolysis led to the diacid (78). At this point RAO realized that E-73 was an acetoxycycloheximide and then re-studied

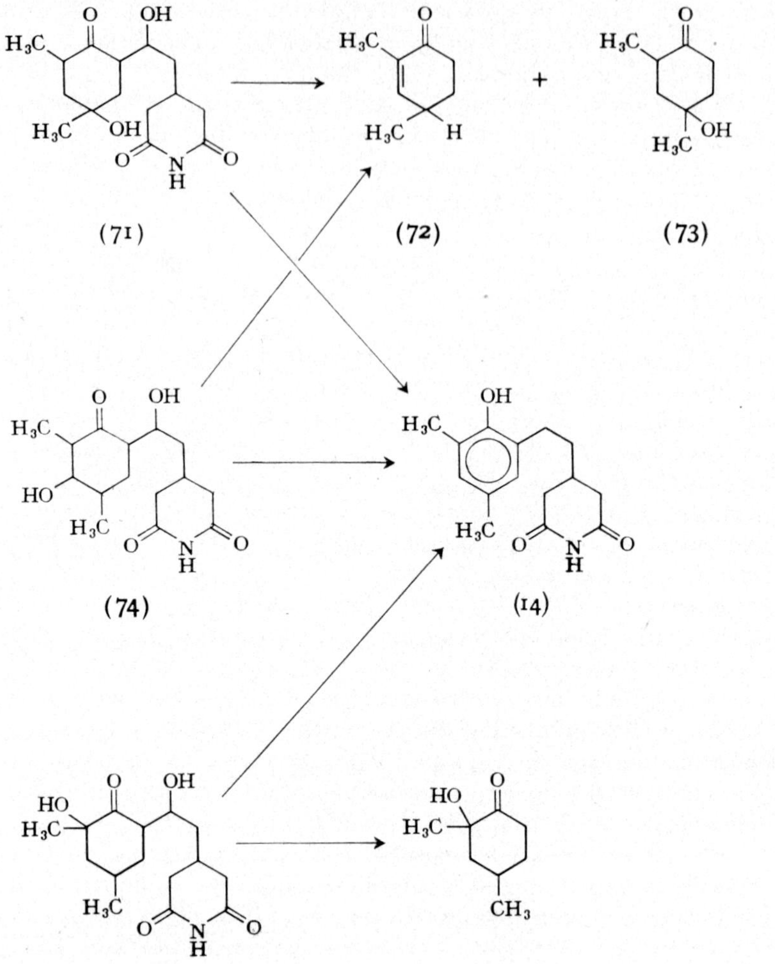

References, pp. 202—208

the alkaline hydrolysis to define the exact location of the acetoxy group. This was done in much the same way as had been done by HERR (37, 38) for streptovitacin-A, and on this basis it was possible to postulate structure (77) for E-73.

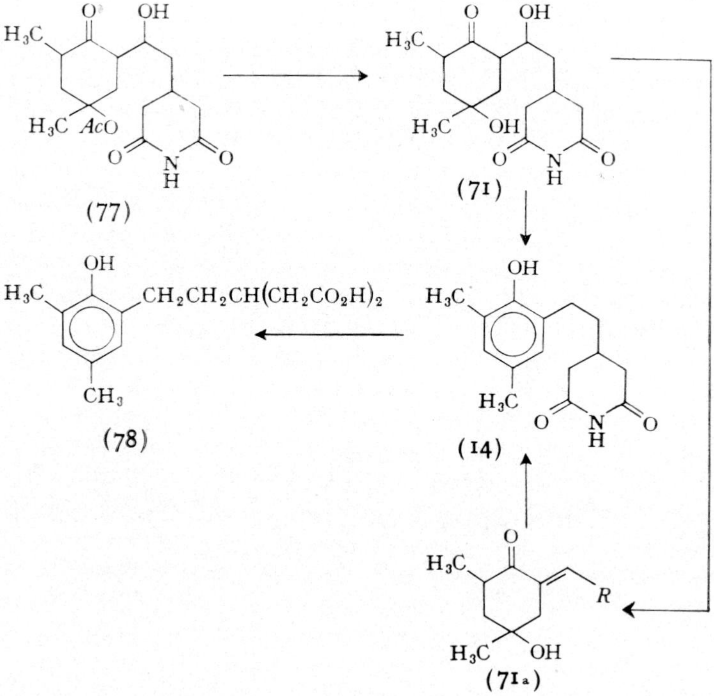

The only stereochemical work that has been done on the streptovitacin group is due to HENNIS, DUQUETTE and JOHNSON (36). They examined the optical rotatory dispersion curve of streptovitacin-A and found that it was virtually identical with that of cycloheximide [cf. HERR (38)], thus indicating that both molecules have the same absolute configuration. The stereochemistry at the 4-position was resolved by application of an observation by SHOPPEE and his associates (99). These investigators noted that in a pair of isomers such as (79) and (80), the NMR line width at half-height of the signal due to the C-1 axial methyl group in the former was approximately twice that of the latter. Since the character of the NMR absorption of the 4-methyl group of streptovitacin-A and E-73 approximated more closely that of (79) it was assigned an axial orientation in each case. Earlier NMR studies (47) had shown that the 2-methyl group was equatorially oriented.

The configuration of the α-hydroxyl group in streptovitacin-A was determined by the same method as had been used *(46)* for cycloheximide Catalytic reduction of (71) using a platinum catalyst afforded a dihydroderivative (81) whose NMR spectrum confirmed the presence of a new axial hydroxyl group. Acetonide formation with the latter compound occurred under very mild conditions and afforded (82) whose spectral data left little doubt that it was the compound depicted. The fact that such an acetonide will form proves that the side-chain asymmetric center of streptovitacin-A has the (R)-configuration as it does in cycloheximide. Thus the total stereochemistry of the molecule can then be represented as illustrated.

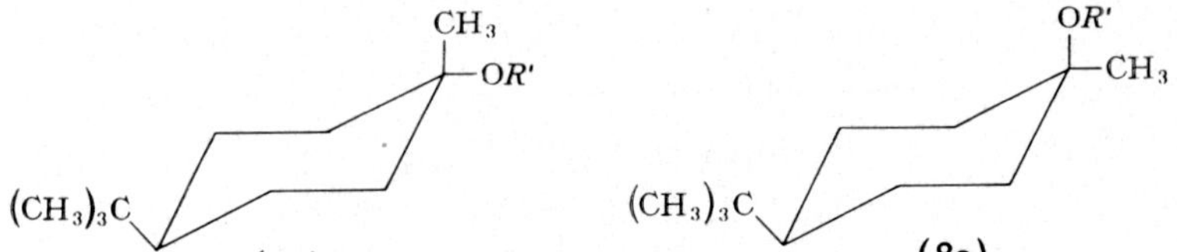

R' = H or COCH₃

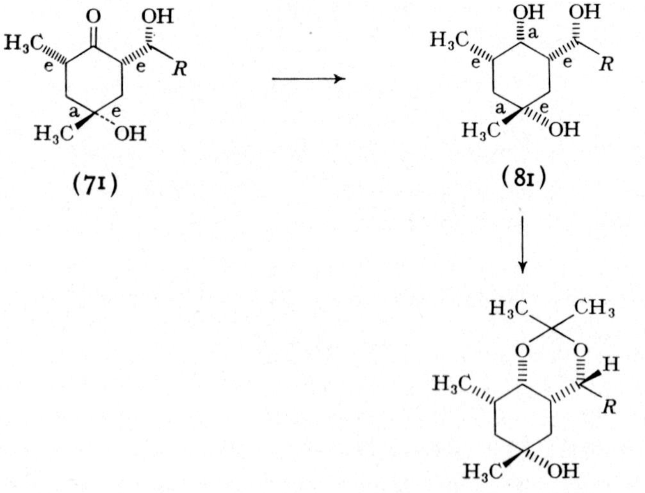

Apart from the reactions described above very little chemistry of the streptovitacin group has been published. Rao *(91)* examined a few reactions of E-73 and found that it behaved in much the same way as cycloheximide.

References, pp. 202—208

Oxidation with chromic acid, for instance, gives the enolic diketone (83) whereas catalytic reduction using a platinum catalyst yields the diol (84) which on alkaline hydrolysis leads to the lactonic acid (85). None of these derivatives had antitumor activity comparable with the parent substances E-73 or streptovitacin-A. NOTARI and CAIOLA (65) have studied the kinetics of dehydration of streptovitacin-A under both acidic and buffered basic conditions. The initial product in either case is (71a) but only in the acid medium does (71a) undergo further dehydration to (14).

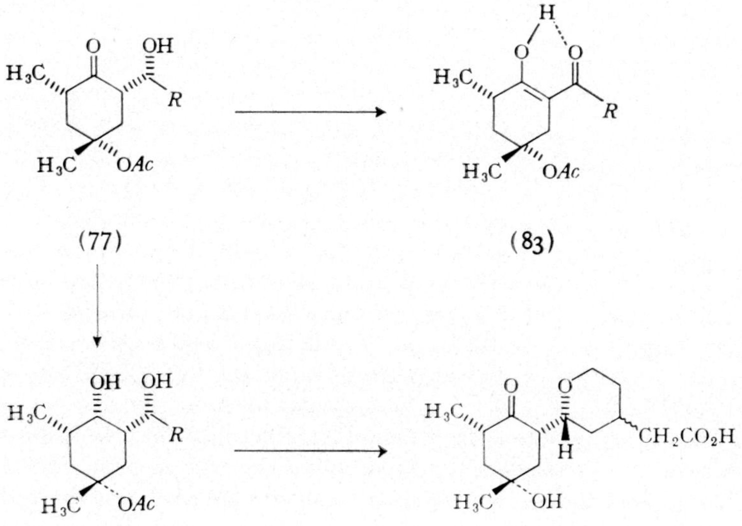

5. Inactone

Inactone was first isolated by PRUD'HOMME and DUBOST (87) from a culture of *Streptomyces griseus*. It is practically devoid of the antifungal activity characteristic of cycloheximide (1). However, because its formula, $C_{15}H_{21}O_4N$, was close to that of (1), PAUL and TCHELITCHEFF (83) presumed that it was related in structure. Its infrared spectrum was very similar to that of (1) except for two bands at 1670 and 828 cm^{-1} characteristic of a double bond. They also found that catalytic hydrogenation occurred with the absorption of two equivalents of hydrogen and from the reaction they were able to isolate two tetrahydro-derivatives. When the hydrogenation was halted after the absorption of one equivalent of hydrogen, three compounds isomeric with cycloheximide could be isolated, all having potent anti-fungal activity. The three isomers when subject to alkaline degradation afforded ammonia and 2,4-dimethylcyclohexanone as the volatile products. The non-volatile material was oxidized with potassium permanganate and afforded methanetriacetic acid. By contrast alkaline degradation of inactone did not yield any volatile ketone. Significantly, each of the dihydro-derivatives afforded dehydrocycloheximide (7) on oxidation and they concluded that inactone possessed the same skeleton as cycloheximide differing only in having a double bond in the cyclohexane ring. Since the compound did not undergo a normal fragmentation in alkali, they proposed that the ethylenic bond was at the 5-position.

The problem was re-investigated by JOHNSON, STARKOVSKY and GUROWITZ (47) because the hydrogenation products obtained by PAUL and TCHELITCHEFF did not appear to correspond to any known isomers of cycloheximide. The mass spectrum of inactone not only confirmed its molecular weight but also by comparison with the mass spectra of cycloheximide and streptimidone confirmed the presence of a double bond at the 5-position. In the spectra of the two latter compounds the dominant mode of decomposition took the anticipated retro-aldol course with major peaks at m/e 126 and 138 respectively. On the other hand the mass spectrum of inactone showed a major fragment at m/e 152 corresponding to allylic cleavage across the ethane bridge between the two rings, with hydrogen transfer, as would be expected for the cyclohexenone structure. Reduction of inactone over a rhodium catalyst afforded a product whose NMR spectrum showed it to be essentially a mixture of naramycin-B (16) and cycloheximide with the former predominating. These findings allowed the complete definition of the structure of inactone as (86).

(86)

6. Actiphenol (C-73)

The only glutarimide antibiotic having an aromatic ring was isolated almost simultaneously by two different groups. HIGHET and PRELOG (39) isolated it from an *Actinomycetes* culture (ETH 7796) and termed it actiphenol. RAO and CULLEN (93) obtained it from the broths of *Streptomyces albulus* together with cycloheximide and E-73 and gave it the designation C-73. Unlike cycloheximide or the streptovitacins, actiphenol has little or no antitumor activity in experimental animals.

Its structure was recognized quite quickly by both groups as (87). The key experiment on the part of HIGHET and PRELOG was the partial synthesis of (87) from cycloheximide (1) by treatment with N-bromosuccinimide whereas RAO identified it by means of CLEMMENSEN reduction which afforded the known phenol (14).

References, pp. 202—208

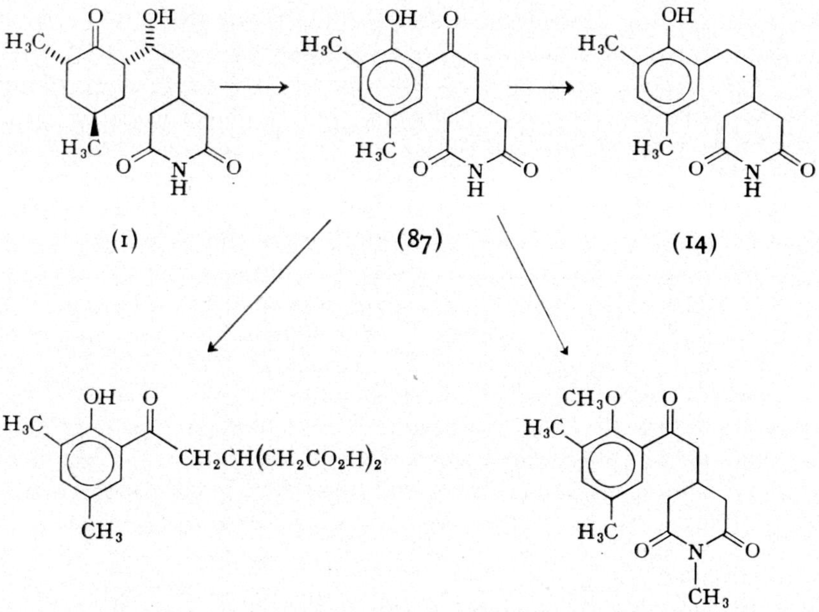

[Structures (1), (87), (14), (88), (89) shown with arrows from (1) → (87) → (14), and (87) leading down to (88) and (89).]

(1)

(87)

(14)

CH₂CH(CH₂CO₂H)₂ — attached to structure (88)

(88)

(89)

The physico-chemical data for actiphenol support this formulation. It is optically inactive and its ultraviolet spectrum shows maxima at 215 (ε, 23000), 264 (ε, 12900), 354 (ε, 4500) nm, whereas in alkaline solution the maxima occur at 229 (ε. 12600), 300 (ε, 2800), 375 (ε, 2100) nm. These absorptions are characteristic of an o-hydroxyacetophenone. Its infrared spectrum shows bands at 1724, 1694, 1639 and 1597 cm^{-1}. Actiphenol shows a bright yellow fluorescence under ultra-violet light, gives a dark color with alcoholic ferric chloride and is soluble in aqueous alkali to give a bright yellow solution, behavior typical of this type of phenol. It yields a colorless monoacetate and an orange-red 2,4-dinitrophenyl hydrazone. When boiled with aqueous alkali, actiphenol produces one molar equivalent of ammonia and the dicarboxylic acid (88). Treatment of (87) with dimethyl sulfate leads to the bis-methylated compound (89). All of these reactions are in accord with the proposed structure.

7. Streptimidone and Protomycin

Streptimidone and Protomycin are the only glutarimide antibiotics known that have an aliphatic chain in place of an alicyclic. ring Protomycin is potently antifungal ([110]) and streptimidone is active against a number of yeast strains ([28]). Both compounds are amebicides and streptimidone strongly inhibits protein synthesis *in vitro* ([6]). In contrast to cycloheximide, protomycin is not particularly toxic to rats. Streptimidone has also been found to have preemergent herbicide activity (52).

Streptimidone. Streptimidone was first isolated by FROHARDT and his associates (28) from the culture filtrates of *Streptomyces rimosus*. The crude antibiotic was initially precipitated by petroleum ether from the ethyl acetate extract of the culture filtrate. Subsequent chromatography of the impure material on activated carbon followed by crystallization from acetone-isopropyl ether afforded the pure antibiotic as fine colorless needles. The material is somewhat unstable and deteriorates rapidly when left in sunlight. It was determined to have the empirical formula $C_{16}H_{23}NO_4$ and it exhibited a characteristic ultraviolet spectrum in methanol with maxima at 232 (ε, 23100) and 291 (ε, 790) nm. Its infrared spectrum in the 3300 and 1700 cm^{-1} regions showed strong absorptions typical of the functionality of cycloheximide, but in the far infrared additional absorption bands were present at 870, 890, 900 and 990 cm^{-1} characteristic of olefinic unsaturation. The antibiotic gave a monoacetate and a monoxime and it quickly decolorized bromine and aqueous permanganate solution. When boiled with alkali all the nitrogen was liberated as ammonia.

Although from the spectral data it was obvious that the ketone was not conjugated with a double bond, an incorrect structure, (90), was initially deduced (28) for streptimidone. The salient features of the argument used for the deduction of this structure are as follows:

Catalytic reduction of streptimidone under various conditions gave sequentially a di-, tetra-, and hexahydro derivative. When tetrahydrostreptimidone was boiled with benzylamine, it afforded a crystalline product identical with (8); the compound obtaines by KORNFELD (54) from cycloheximide using the same procedure. Thus, the presence in streptimidone of a 3-(β-hydroxyethyl)glutarimide was established. Alkaline degradation of tetrahydrostreptimidone gave a volatile oil whose identity with 3,5-dimethyl-2-heptanone was established by synthesis thus identifying the other half of the carbon skeleton. Evidence for the environments of the double bonds in streptimidone itself were obtained from an ozonolysis experiment. Compounds identified as methyl ethyl ketone, formaldehyde, and acetaldehyde were isolated. This information together with the fact that when streptimidone was reduced with sodium borohydride only the weak ketonic absorption at 291 nm was lost, suggested that the conjugated diene unit (92) must be present and that streptimidone must be (90) or (93). That one end of the diene chromophore was a methylene group was also evident from the absence of the 900 cm^{-1} band from the spectrum of dihydrostreptimidone. A choice between (90) and (93) was made by studying the steam volatile oil obtained by the alkaline degradation of streptomidone itself. The ultraviolet spectrum of this material

References, pp. 202—208

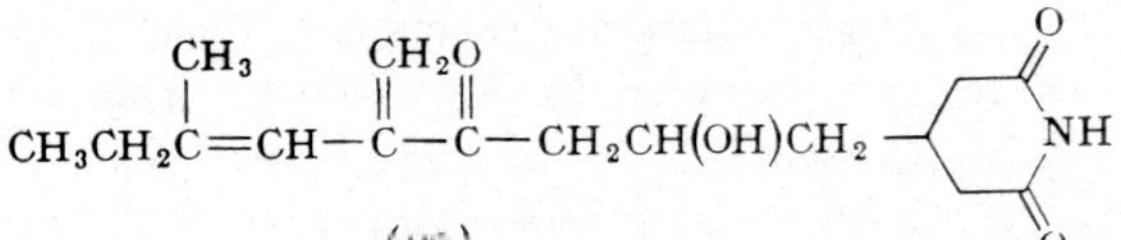

$$CH_3CH_2\underset{\underset{CH_3}{|}}{C}=CH-\underset{\underset{O}{||}}{C}-\underset{\underset{O}{||}}{C}-CH_2CH(OH)CH_2-\text{[glutarimide]}$$
$$\overset{CH_2O}{}$$

(90)

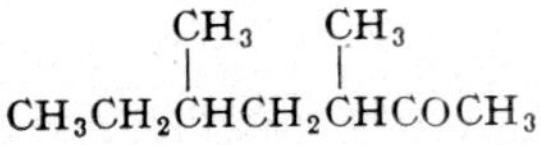

$$CH_3CH_2\underset{\underset{CH_3}{|}}{C}HCH_2\underset{\underset{CH_3}{|}}{C}HCOCH_3$$

(91)

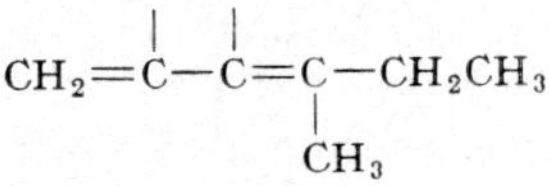

$$CH_2=\underset{|}{C}-\underset{|}{C}=\underset{\underset{CH_3}{|}}{C}-CH_2CH_3$$

(92)

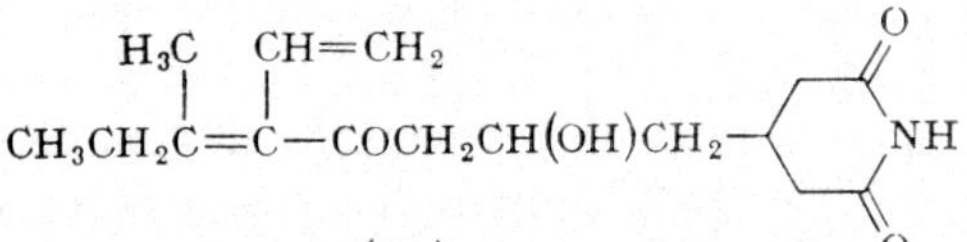

$$CH_3CH_2\underset{\underset{H_3C}{|}}{C}=\underset{\underset{CH=CH_2}{|}}{C}-COCH_2CH(OH)CH_2-\text{[glutarimide]}$$

(93)

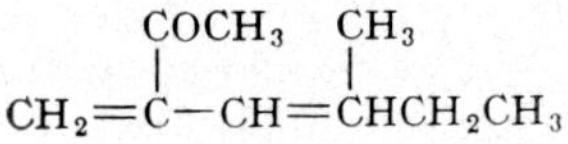

$$CH_2=\underset{\underset{COCH_3}{|}}{C}-CH=\underset{\underset{CH_3}{|}}{C}HCH_2CH_3 \qquad CH_2=CH-\underset{\underset{CH_3OC}{|}}{C}=\underset{\underset{CH_3}{|}}{C}-CH_2CH_3$$

(94) 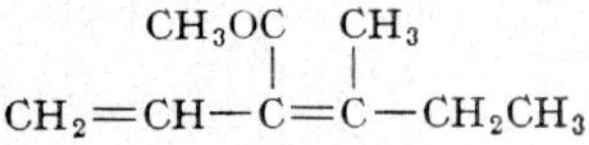   (95)

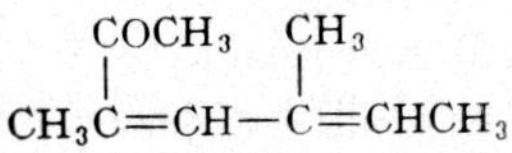

$$CH_3\underset{\underset{COCH_3}{|}}{C}=CH-\underset{\underset{CH_3}{|}}{C}=CHCH_3$$

(96)

$$CH_2=CH-\underset{\underset{CH_3}{|}}{C}=CH-\underset{\underset{CH_3}{|}}{C}HCOCH_2CH(OH)CH_2-\text{[glutarimide]}$$

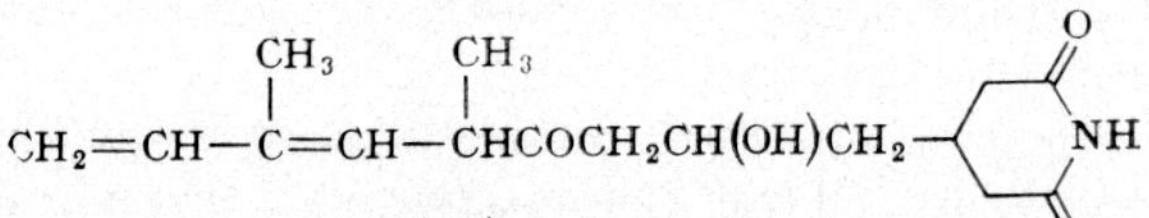

(97)

indicated that it was a dienone in which the diene was not fully conjugated with the ketone, but a material containing such a chromophore was obtained when the oil was treated under more drastic basic conditions. In the light of this and supportive evidence from similar reactions with-dihydrostreptimidone, it was concluded that the first dienone should have the structure (94) rather than the alternative (95) and that the product of further alkaline treatment was (96). Thus streptimidone was assigned structure (90).

However, VAN TAMELEN and HAARSTAD (*122*) took issue with this structure proposal. It seemed to them that certain of the recorded properties, in particular the base-catalyzed conversion of the existing chromophore to a 2,4-dienone system could be better interpreted in terms of the alternative structure (97). This was confirmed by the NMR spectrum of *o*-acetylstreptimidone which showed the presence of four olefinic hydrogens rather than three and whose pattern was more consistent with structure (97) than with (90). This new formula was later endorsed by WOO, DION and BARTZ (*132*). In particular, they re-examined the ozonolysis of streptimidone and found the product to consist of formaldehyde, pyruvaldehyde and acetaldehyde. In contrast with the earlier work, no methyl ethyl ketone was detected. The latter, in fact, was found to be an impurity in the reagent grade ethyl acetate used as a solvent in the isolation procedure. A reinterpretation of the chemistry of streptimidone is shown in Chart I.

No further chemical transformations of streptimidone have been reported. The stereochemistry around the trisubstituted double bond remains unknown as do the absolute configurations of the two asymmetric carbon atoms. It would not be surprising, however, if the latter turned out to have the same absolute configurations as are present at these centers in cycloheximide, *i.e.*, (R) for the carbon atom bearing the hydroxyl group and (S) for the tertiary carbon atom adjacent to the ketone.

Protomycin. This member of the series was first isolated from the culture filtrates of *Streptomyces reticuli var. protomycicus* (*110, 111*). It was purified in a series of steps involving chromatography on activated carbon followed by chromatography on acid-washed alumina, countercurrent distribution and finally molecular distillation. The pale yellow liquid that was obtained by these procedures slowly solidified over a period of two to four weeks, but was shown to be homogeneous by countercurrent distribution.

SUGAWARA (*109*) who first isolated the material quickly recognized from its physicochemical and biological properties that protomycin was a member of the glutarimide antibiotic series. Its empirical formula was found to be $C_{19}H_{29}NO_5$ and its ultraviolet spectrum was essentially identical with that of streptimidone (97) while its infrared spectrum showed many features of the latter material. Marked differences, however, were apparent in the 3300 and 1250 cm^{-1} regions. Like streptimidone, it gave a tetrahydro derivative (99) when hydrogenated in the presence of a palladium catalyst and alkali degradation of the latter

References, pp. 202—208

The Chemistry of Glutarimide Antibiotics 185

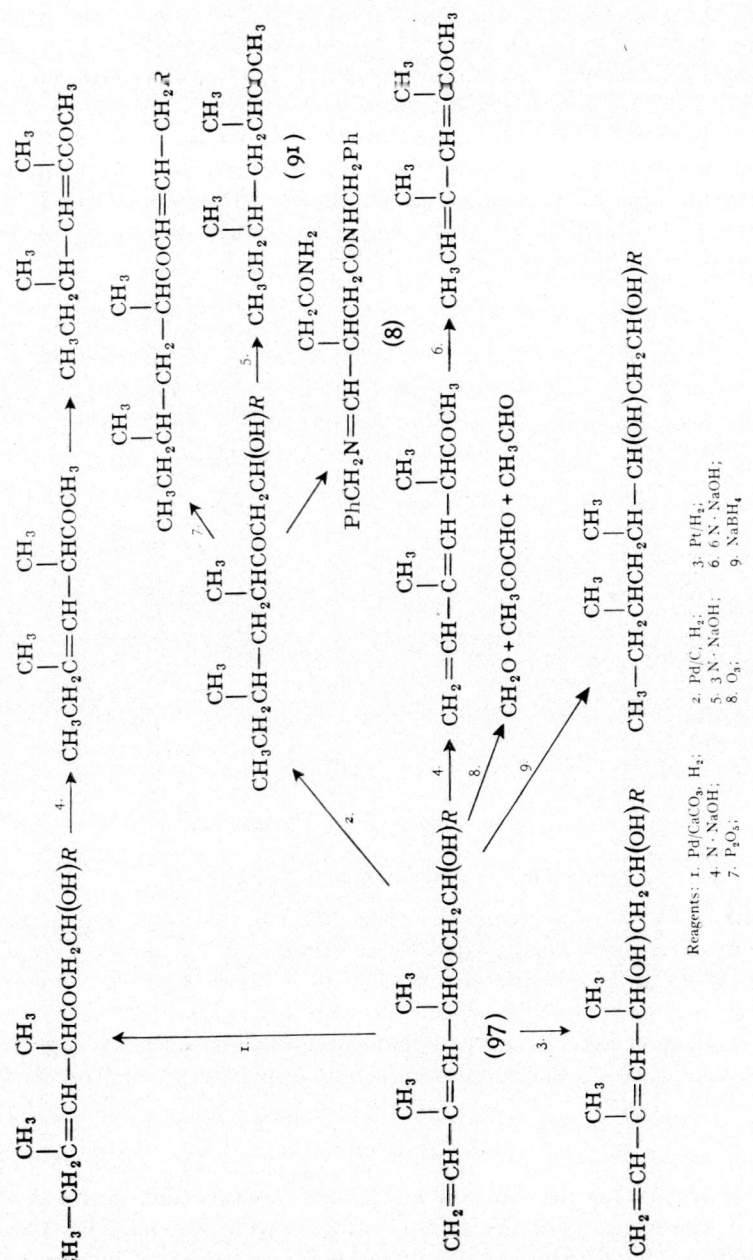

Chart I

material under steam distillation conditions produced 3,5-dimethyl-2-heptanone (91) and acetone. Protomycin itself under these conditions afforded acetone and 3,5-dimethyl-3,5-heptadiene-2-one identical with the material obtained from streptimidone by strong alkali treatment. Finally, when (99) was heated with benzylamine it gave the well-known product (8), thus establishing the presence of a β-ethylglutarimide moiety. On the basis of these data SUGAWARA (*109*) proposed structure (98) for protomycin. (For further comments on this structure see Section IV.)

As with streptimidone the double bond stereochemistry and absolute configuration remain unknown.

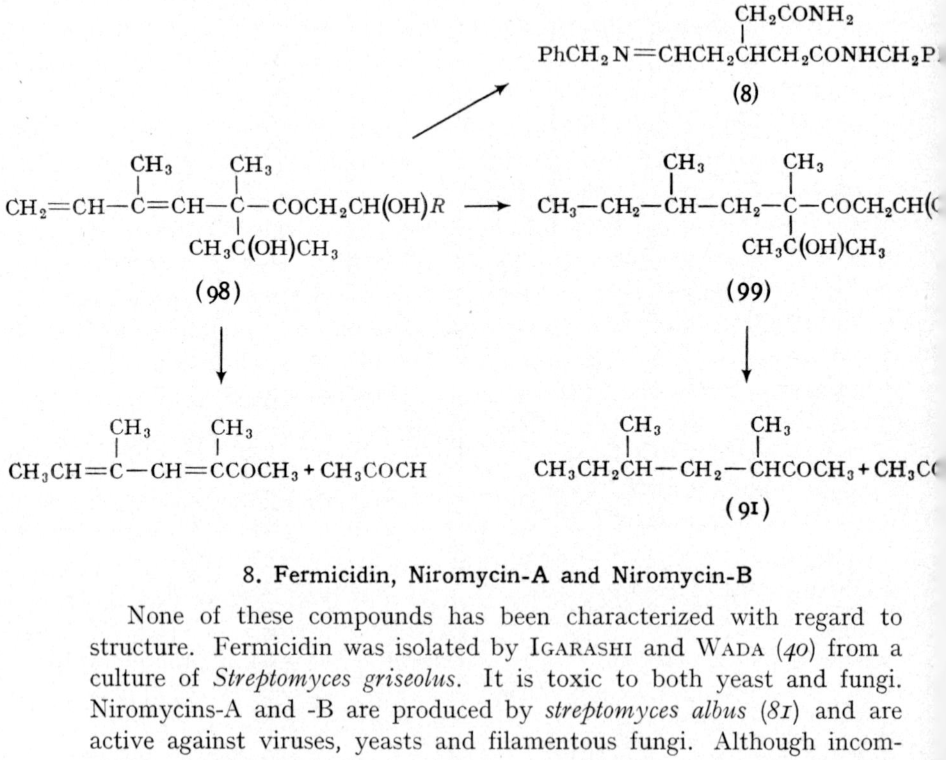

$$PhCH_2N=CHCH_2\overset{\overset{CH_2CONH_2}{|}}{C}HCH_2CONHCH_2P$$

(8)

$$CH_2=CH-\overset{\overset{CH_3}{|}}{C}=CH-\overset{\overset{CH_3}{|}}{\underset{\underset{CH_3C(OH)CH_3}{|}}{C}}-COCH_2CH(OH)R \longrightarrow CH_3-CH_2-\overset{\overset{CH_3}{|}}{C}H-CH_2-\overset{\overset{CH_3}{|}}{\underset{\underset{CH_3C(OH)CH_3}{|}}{C}}-COCH_2CH($$

(98)   (99)

$$CH_3CH=\overset{\overset{CH_3}{|}}{C}-CH=\overset{\overset{CH_3}{|}}{C}COCH_3 + CH_3COCH \qquad CH_3CH_2\overset{\overset{CH_3}{|}}{C}H-CH_2-\overset{\overset{CH_3}{|}}{C}HCOCH_3 + CH_3C($$

(91)

8. Fermicidin, Niromycin-A and Niromycin-B

None of these compounds has been characterized with regard to structure. Fermicidin was isolated by IGARASHI and WADA (*40*) from a culture of *Streptomyces griseolus*. It is toxic to both yeast and fungi. Niromycins-A and -B are produced by *streptomyces albus* (*81*) and are active against viruses, yeasts and filamentous fungi. Although incompletely characterized, they appear to resemble the streptovitacins (*80*).

III. Synthesis

Total synthesis in the glutarimide antibiotic area remains incomplete. Of all of the members of this group only cycloheximide, naramycin-B, isocycloheximide and actiphenol have yielded to synthesis. A number of isomers of cycloheximide which do not occur naturally have been pre-

References, pp. 202—208

The Chemistry of Glutarimide Antibiotics

pared, however, and, in addition, a substantial number of homologs and analogs of I and its derivatives have been prepared and tested for biological activity.

1. Cycloheximide, Naramycin-B and Isocycloheximide

All of the early attempts to synthesis cycloheximide were carried out at a time when the stereochemistry of the molecule was either unknown or, at best, imperfectly understood. None was accompanied by a large measure of good fortune and accordingly all failed. These approaches did have one thing in common however, the use of the intermediate 3-carboxymethylglutarimide (100).

This was first prepared by PHILLIPS, ACITELLI and MEINWALD (85). They carried out a Michael addition of methyl cyanoacetate to dimethyl glutaconate (101) and subjected the product (102) to a controlled acid hydrolysis which gave (100). The latter reaction proved capricious and consistently good yields could not be obtained. An improvement in the procedure was published by LAWES (55). He hydrolyzed (102) to methanetriacetic acid (103). Pyrolysis of the ammonium salt then gave a high yield of (100). A superior method was published by JOHNSON (41) [also independently developed by EGAWA (18)] which took advantage of a Cope condensation between dimethyl acetonedicarboxylate (104) and cyanoacetic acid. This afforded (105) in 54% yield, which could be hydrogenated almost quantitatively to the saturated nitrile (106). When the latter material was boiled with moderately concentrated

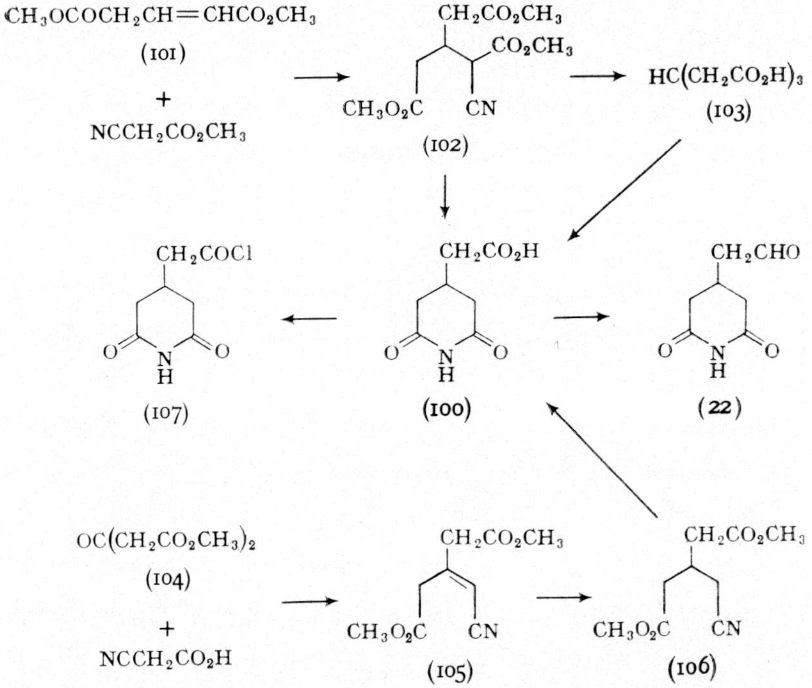

$CH_3OCOCH_2CH=CHCO_2CH_3$
(101)

+

$NCCH_2CO_2CH_3$

$\longrightarrow$

$CH_2CO_2CH_3$
CO_2CH_3
$CH_3O_2C \quad CN$
(102)

$\longrightarrow$ $HC(CH_2CO_2H)_3$
(103)

$\downarrow$

(107) CH$_2$COCl

$\longleftarrow$

(100) CH$_2$CO$_2$H

$\longrightarrow$

(22) CH$_2$CHO

$OC(CH_2CO_2CH_3)_2$
(104)

+

$NCCH_2CO_2H$

$\longrightarrow$

$CH_2CO_2CH_3$
$CH_3O_2C \quad CN$
(105)

$\longrightarrow$

$CH_2CO_2CH_3$
$CH_3O_2C \quad CN$
(106)

hydrochloric acid and the reaction mixture slowly raised to 235° and held there until gas evolution ceased, a 91% yield of (**100**) was obtained on cooling.

The two most useful derivatives of (**100**) namely (**22**) and (**107**) were obtained (*85*) by ROSENMUND reduction and by treatment with thionyl chloride respectively.

ACITELLI (*1*) was the first to attempt to couple (**22**) with *cis*-2,4-dimethylcyclohexanone. He used diethylamine as the catalyst, but was unable to isolate any tangible product. LAWES (*55*) investigated a different type of condensation reaction and was more fortunate. He allowed an aqueous solution of equimolar quantities of 2-hydroxymethylene-4,6-dimethylcyclohexanone (**108**), potassium carbonate and (**22**) to react at room temperature. This afforded in reasonable yield what must now be called anhydroisocycloheximide (**9**).

The NIELSEN condensation of the bromomagnesium salt of *cis*-2,4-dimethylcyclohexanone with (**22**) was examined by two groups of workers. JOHNSON, STARKOVSKY and GUROWITZ (*44, 47*) utilized the racemic form of *cis*-2,4-dimethylcyclohexanone in their work and obtained a mixture from which only one crystalline material was isolated. This was named neocycloheximide and its hydroxyethylglutarimide group was shown (*45, 47*) to be axially oriented. OKUDA, SUZUKI and EGAWA also examined this condensation, but used both the racemic and optically active forms of 2,4-dimethylcyclohexanone. In the latter case (*75, 117*), both the

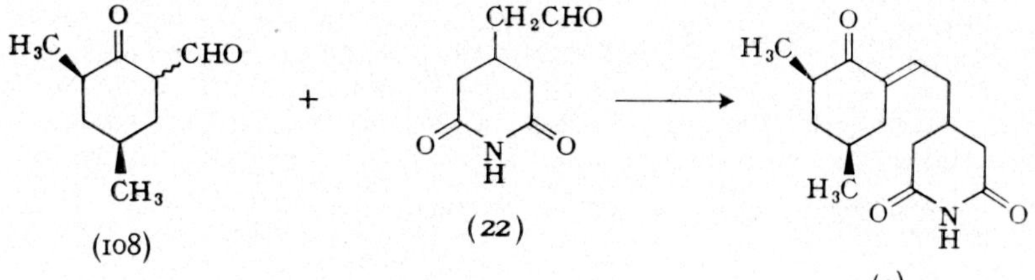

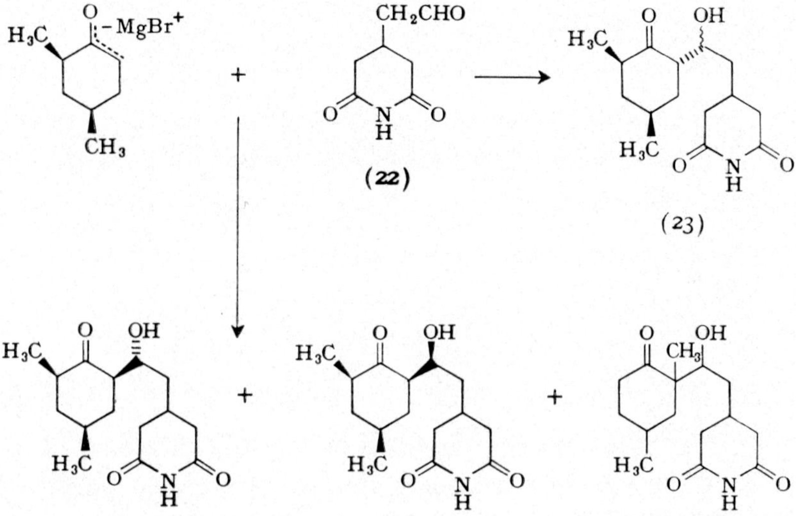

References, pp. 202—208

cis- and trans-form of the ketone were used, but both gave the same mixture of compounds, from which six crystalline substances were isolated by chromatography. Of these, one was identified as isocycloheximide (**13**) while another was identified as (**109**) since under dehydrating conditions it afforded anhydroisocycloheximide (**9**) and could not be isomerized by treatment with base. It was named α-epiisocycloheximide. Of the remaining four compounds, two were thought to have the normal cycloheximide structure, but insufficient work was done to characterize their stereochemistry. The final two compounds were characterized as gem-cycloheximides (**110**) because they did not give the resorcinol color reaction (*118*) characteristic of cycloheximide and its stereo-isomers. No attempt was made to determine the stereochemistry of either of these substances. When the Japanese workers (*18*) used racemic cis-2,4-dimethylcyclohexanone in this condensation, they were able to isolate two gem-cycloheximides (**110**) and after acetylation of the syrupy residue, a small quantity of racemic isocycloheximide acetate. Insofar as the published work is concerned the optically active gem-cycloheximides were not correlated with their racemic counterparts. In addition, no compound corresponding to neocycloheximide (**23**) was observed in these studies.

Recently the acid-catalyzed condensation of (**22**) with cis-2,4-dimethylcyclohexanone has been examined (*19*). This led to a mixture of two gem-cycloheximides (**110**) together with anhydroisocycloheximide (**9**) when the reaction was conducted using concentrated sulfuric acid in acetic acid. It was found also that (**9**) to a small extent could be rehydrated by means of 75% sulfuric acid in acetic acid. The product was found to be α-epiisocycloheximide (**109**) which is undoubtedly identical with a material, isomeric with cycloheximide, obtained by GARRETT and NOTARI (*29, 31*) during their studies of cycloheximide transformations in aqueous acid, but which was incompletely identified at the time.

When the stereochemistry of cycloheximide became known, a more rational approach to its synthesis was possible. This was accomplished by JOHNSON, STARKOVSKY, PATON and CARLSON (*48, 49*). In carrying out a stereoselective synthesis of cycloheximide three major problems had to be solved. The first of these concerned the 4-methyl group which had to be built into the molecule in the less stable axial orientation, but which is remote from all other functionality. A second difficulty was that of stereoselectively introducing the hydroxyl group of the large side-chain. Lastly, the final steps of the synthesis had to be accomplished under conditions which were neither too acidic nor too basic since cycloheximide itself is easily subject to fragmentation, dehydration and isomerization.

A method of establishing the methyl groups in a *trans*-relationship was suggested by the work of WILLIAMSON (*130*). He had proposed that enamines of 2-substituted cyclohexanones had the 2-substituent in the quasi-axial orientation and offered this as the explanation for the difficulty in alkylating such enamines that had been observed by other research workers. Any reagent approaching the double bond of the enamine would necessarily be involved in the non-bonded interaction with the quasi-axial substituent at the 2-position. Based on this proposition which turned out to be essentially true (*50, 98*), the synthesis of cycloheximide was accomplished in the following way.

Treatment of *cis*-2,4-dimethylcyclohexanone with morpholine led largely to the enamine (111) of *trans*-2,4-dimethylcyclohexanone. Acylation of (111) with 3-glutarimidylacetyl chloride (107) afforded, after decomposition with aqueous acetic acid, dehydrocycloheximide (7) in

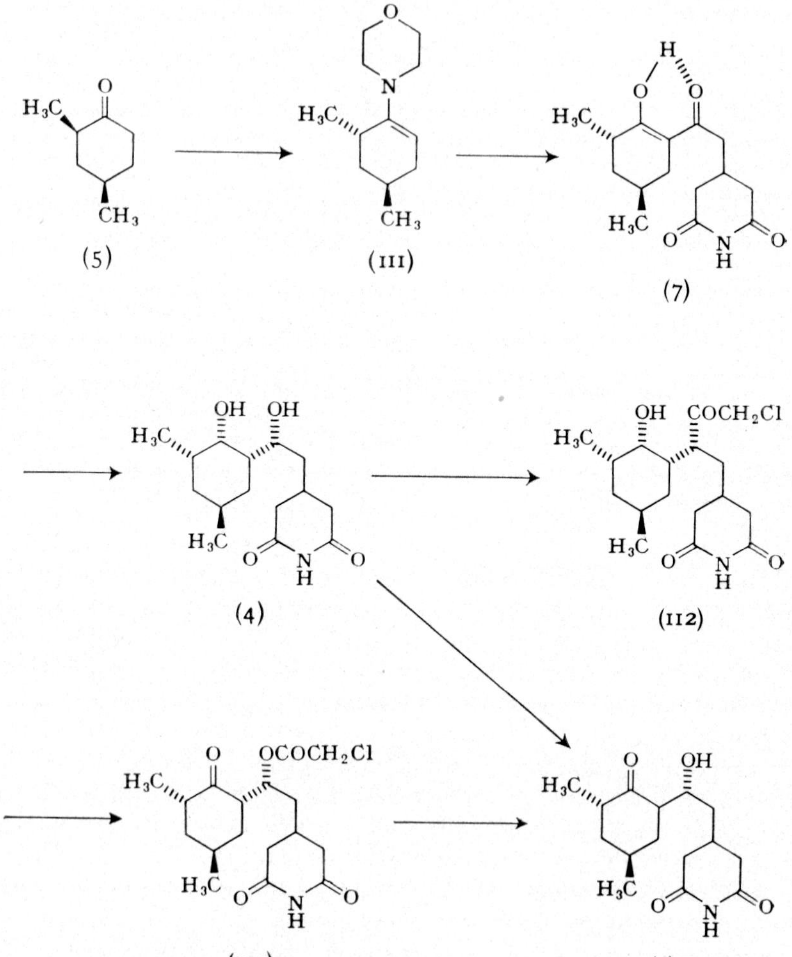

30% yield, a reaction also independently reported by SCHAEFFER and JAIN (*95, 98*). When the latter compound was hydrogenated over a platinum catalyst in acetic acid, dihydrocycloheximide (4) was obtained in 60% yield. Thus, in two steps the molecular framework of cycloheximide with the correct stereochemistry was completely established.

References, pp. 202—208

Acylation of (4) with chloroacetyl chloride in the presence of one equivalent of pyridine at 0° yielded the monochloroacetate (112) quite cleanly. Oxidation of the latter compound by means of chromic acid gave cyclohcximido chloroacetate (113) which on hydrolysis with potassium bicarbonate in aqueous methanol afforded cycloheximide (1) identical with the natural product. It also proved possible to oxidize (4) to cycloheximide directly by using one equivalent of chromic acid, but the yield was extremely poor. Using these procedures both dl- and l-cycloheximide were synthesized. Since cycloheximide can be isomerized under basic conditions to naramycin-B (16) and isocycloheximide (13), the total synthesis of (1) also constituted a formal total synthesis of these two isomers.

2. α-Epiisocycloheximide

A stereoselective synthesis of this substance was accomplished by JOHNSON, CARLSON and STARKOVSKY (42, 43) using a route similar to that described above for cycloheximide with the exception that a derivative (115) of 2,4-dimethylcyclohexanone with cis-related methyl groups was used as the starting point. Oxalylation of (5) afforded the ester (114) which easily underwent decarbonylation to the α-keto ester (115). The latter material by an ester exchange process was converted to the benzyl ester (116) whose bromomagnesium salt was treated with the acid chloride (107). The product of this reaction, a thick viscous oil which was characterized as (117) by infrared spectroscopy only, was subjected to hydrogenation over a palladium catalyst in ethyl acetate. The ethyl acetate solution was subsequently boiled to effect decarboxylation and dehydroisocycloheximide (15) was obtained in 50–70% yield.

Hydrogenation of (15) in acetic acid over a platinum catalyst led to a single crystalline diol, (118). Monochloroacetylation of (118) under the conditions used in the synthesis of cycloheximide afforded (119). Oxidation of the latter compound led to the keto-chloroacetate (120) which when hydrolyzed under mildly basic conditions yielded the racemic form of α-epiisocycloheximide (109).

The cyclohexanone ring stereochemistry of (109) was confirmed (47) by NMR spectroscopy (see Table 3) to be the same as that of isocycloheximide (13), namely all cis, thus confirming the cyclohexane ring stereochemistry assigned to it by EGAWA et al., (18). The relative orientation of the hydroxyl group in α-epiisocycloheximide was determined in a manner similar to that used for the hydroxyl group in cycloheximide. Reduction of (109) with lithium tri-t-butoxyaluminum hydride afforded a new diol (121) isomeric with (118). By NMR spectroscopy it was shown that these diols contained respectively an equatorial and an axial-hydroxyl group in their cyclohexane ring. When they were subjected to acetonide formation, (118), afforded an acetonide (122) whereas (121) was recovered unchanged.

From this it was concluded that the side-chain hydroxyl group of α-epiisocycloheximide must have the (S)-configuration an assignment opposite to that originally made by the Japanese workers (*117*), based on optical rotatory dispersion and hydrogen bonding studies.

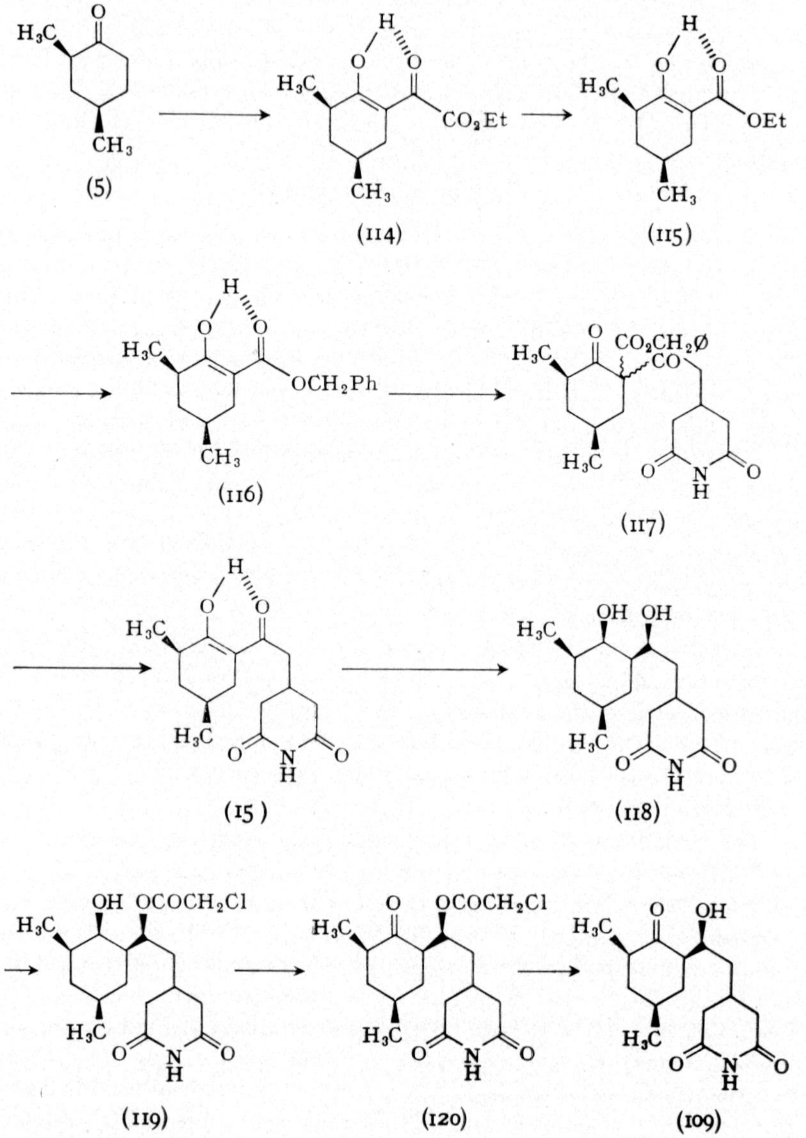

References, pp. 202—208

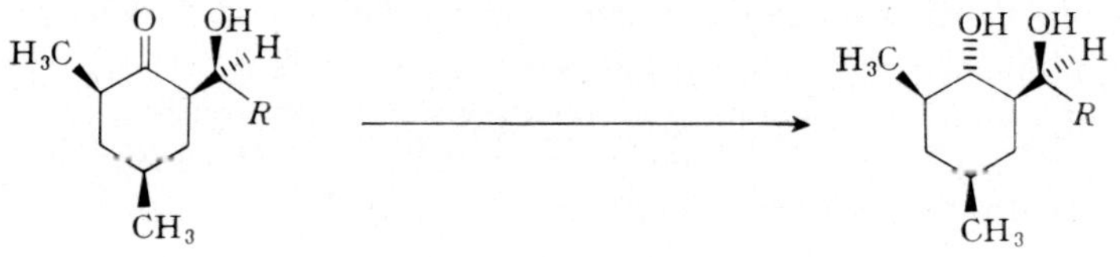

(109)   (121)

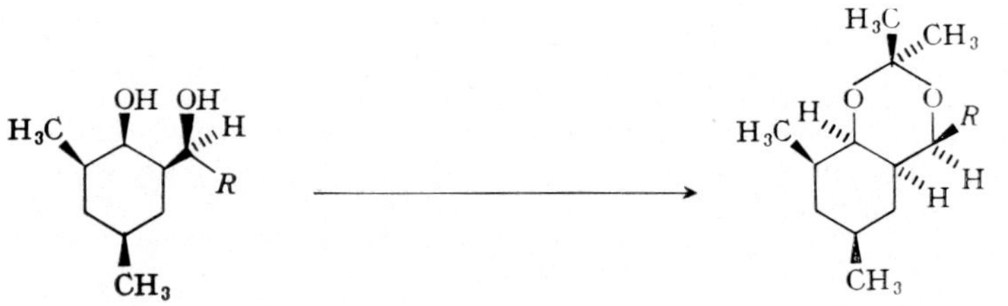

(118)   (122)

3. Actiphenol

Actiphenol was the first of the glutarimide antibiotics to be synthesized (*41*). This was accomplished quite simply by means of a FRIES rearrangement.

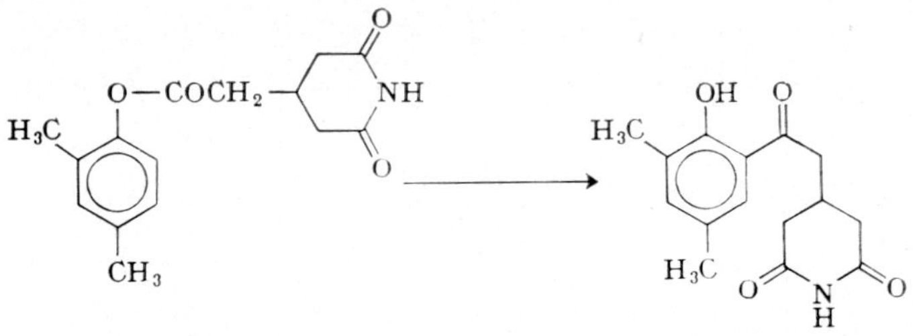

(123)   (87)

When the ester (**123**) was heated briefly with powdered aluminium chloride at 150–160°, a 55% yield of actiphenol (**87**) was obtained. Its identity with the natural product was confirmed by comparison with an authentic specimen.

4. Homologs and Analogs of Cycloheximide and Other Related Substances

In connection with studies on the synthesis of cycloheximide EGAWA, SUZUKI and OKUDA (*18*) studied the condensation of 3-glutarimidyl-acetaldehyde (**22**) with the bromomagnesium salts of a variety of ketones.

These included 2-octanone, α-tetralone, cyclohexanone, 4-methylcyclohexanone, and 2-methylcyclohexanone. In all cases β-hydroxyketones were obtained to the complete exclusion of any α,β-unsaturated ketones. In the case of 2-octanone, condensation took place at the methyl group and in the case of 2-methylcyclohexanone products derived from condensation at both 2- and 6-positions were isolated.

The condensation products showed no or only weak biological activity against *Saccharomyces cerevisiae*.

EGAWA and UMEZAWA (*19*) studied the acid-catalyzed condensation of 3-glutarimidylacetaldehyde (**23**) with acetone, methyl ethyl ketone and methyl isopropyl ketone in addition to 2,4-dimethylcyclohexanone which has been discussed previously in this article. The methods they employed involved either the addition of (**22**) to a saturated solution of hydrogen chloride in the requisite ketone or the slow addition of sulfuric acid to an acetic acid solution of (**22**) and the ketone under study.

$$CH_3COCH=CHR$$

(**124**)

$$CH_3CO\overset{\underset{\displaystyle CH_3}{|}}{C}=CHR$$

(**125**)

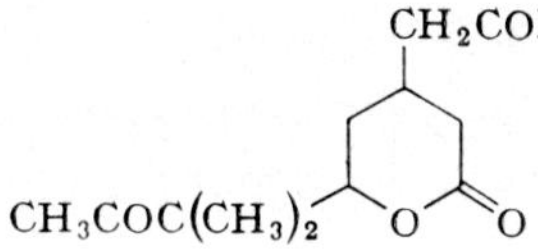

(**126**)

$$CH_3COC(CH_3)_2CH(OH)R$$

(**127**)

$$(CH_3)_2CHCOCH_2CH(OH)R$$

(**128**)

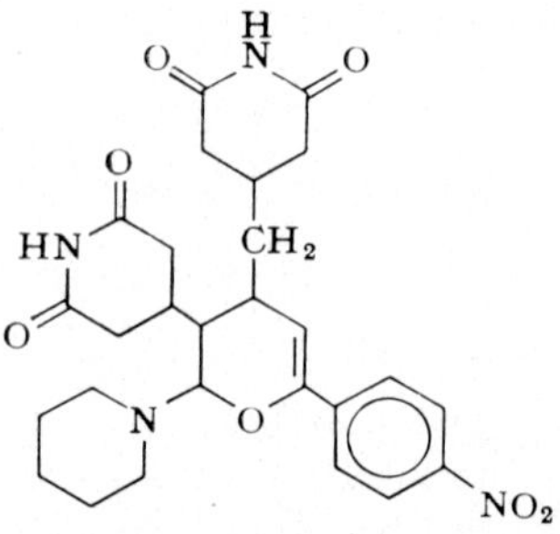

(**129**)

By the first method acetone and methyl ethyl ketone gave (**124**) and (**125**) respectively whereas methyl isopropyl ketone yielded (**126**) together with small amounts of (**127**). For comparison purposes the isomer (**128**) of the latter compound was synthesized by the condensation of (**22**) with the bromomagnesium salt of methyl isopropyl ketone.

References, pp. 202—208

UMEZAWA and EGAWA (*119*) also studied the amine-catalyzed reactions of 3-glutarimidylacetaldehyde (**22**) with some aromatic nitroketones which had failed to react in the presence of acid catalyst. When two equivalents of (**22**) were allowed to react with *p*-nitroacetophenone in the presence of piperidine, the cyclic product

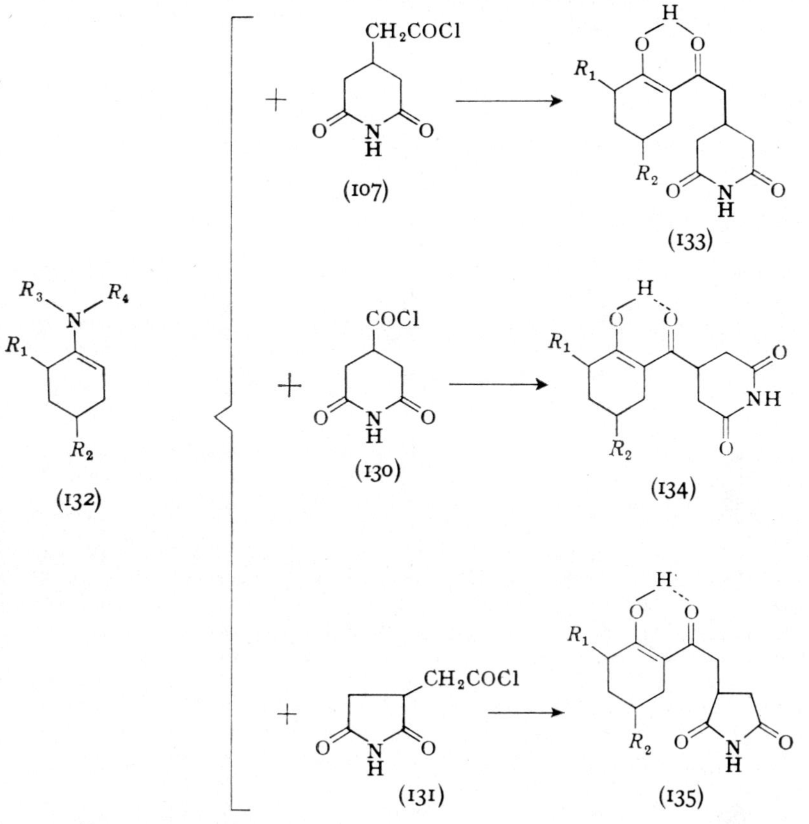

a) $R_1=R_2=H$;
b) $R_1=CH_3$, $R_2=H$;
c) $R_1=OCOCH_3$, $R_2=H$;
d) $R_1=H$, $R_2=CH_3$;

R_3, $R_4 = -(CH_2)_4-$; $-(CH_2)_5$; $O(CH_2CH_2-)_2$; CH_3, Ph.

(**129**) was obtained in 74% yield. Similar products were obtained when 2-acetyl-5-nitrofuran was employed in place of *p*-nitroacetophenone or when morpholine or pyrrolidine was substituted for piperidine. Attempts to condense the nitroketones with (**22**) using other organic amines such as N-methylaniline or triethylamine were unsuccessful. Acetone, acetophenone and *cis*-2,4-dimethylcyclohexanone failed to react with (**22**) even when piperidine was used as the catalyst.

STRUCK and his coworkers (*108*) have synthesized a number of homologs and analogs of dehydrocycloheximide. In this work three acid chlorides (**107**), (**130**) and (**131**) were used to acylate the enamines (**132**) of four different ketones. The latter comprised cyclohexanone, 2-methylcyclohexanone, 2-acetoxycyclohexanone and 4-methylcyclohexanone. Pyrrolidine, piperidine, morpholine and N-methylaniline, depending on the individual case, were used for the preparation of the enamines (**132**). The products obtained from these acylation reactions, after hydrolysis, were (**133**), (**134**) and (**135**) respectively. Of the twelve possible compounds only (**135b**) displayed slight toxic activity at a concentration of 100 µg/ml. against EAGLES KKB cells. Cell growth in these tests was less than 50% of the growth of the controls. Most of the compounds were also screened in the Sarcoma 180 system, the Adenocarcinoma 755 system, and in the Leukemia L-1210 system, but did not display any significant reproducible activity.

Finally, it is worth mentioning that SUZUKI (*112, 113, 114*) examined the condensation of several *p*-substituted benzaldehydes with *cis-d*-2,4-dimethylcyclohexanone in connection with stereochemical studies on cycloheximide. Two sets of reaction conditions were examined. In the first, the aldehydes were allowed to react with the bromomagnesium salt of the ketone whereas in the second tetramethylammonium hydroxide was used to catalyze the condensation of the two components. The major products from these differing reaction conditions, (**136**) and (**137**) respectively, were isomers. SUZUKI found, however, that (**136**) (X = NO$_2$) could be isomerized to (**137**) (X = NO$_2$) when percolated through a column of acid-washed alumina. Thus, he concluded that the products of NIELSEN condensation were the result of a kinetically-controlled reaction whereas those from the tetramethylammonium hydroxide-catalyzed reaction were products of thermodynamic control. Although it is not clear from SUZUKIS analysis of the optical rotatory dispersion data, it seems certain that the products derived from the former reaction have the benzyl alcohol substituent axially oriented because they all give intense negative COTTON effect curves. On the other hand, the products from the simple base-catalyzed condensation must have all substituents on the cyclohexanone ring equatorially oriented both from a consideration of their mode of preparation and because they all display rather weak positive COTTON-effect curves. In both series the relative orientation of the hydroxyl group is the same and on the basis of similar evidence to that used for cycloheximide and its derivatives the side-chain asymmetric center of these

(X = NO$_2$, NHAc, Br or OH)

(**136**)
(X = NO$_2$)

(X = NHAc, NO$_2$)
(**137**)

References, pp. 202—208

compounds was assigned the (R)-configuration. However, since the Japanese assignment of this center in cycloheximide was shown to be incorrect (*46, 106*), it seems likely that the assignment for this center in the simpler derivatives (**136**) and (**137**) must be reversed and should be designated (S).

These compounds have been tested against several yeasts, but show little biocidal activity compared with cycloheximide itself (*116*).

IV. Biosynthesis

Almost all of the work concerned with the biogenesis of cycloheximide and related compounds has been carried out by Czech workers. Recently, they have summarized their results (*121*). KHARATYAN and his associates (*51*) found that when *Streptomyces noursei* was cultured in the presence of [methyl-^{14}C]-methionine, labelled cycloheximide was obtained corresponding to 15% incorporation of the radioactivity. The incorporated isotope was found to be localized in the methyl groups and equally distributed between them. Incorporation (4%) of sodium [1-^{14}C]acetate proceeded with little randomization to give an alternate-carbon labelling pattern, typical of a polyketide system. They found also that sodium [1-^{14}C]propionate was ineffectively utilized (0.15% incorporation) by the system, thus confirming that the biosynthesis of cycloheximide and its close relatives involves *trans*-methylation from a C_1-donor pool rather than the mixed condensation of acetate and propionate units which is known also to occur with Streptomycetes. No incorporation was observed with [2-^{14}C]mevalonic acid nor with generally-labelled glutamic acid or lysine. These investigators adumbrated two main ways, represented diagramatically by (**138**) and (**139**), by which the polyketide skeleton of the glutarimide antibiotics could be derived.

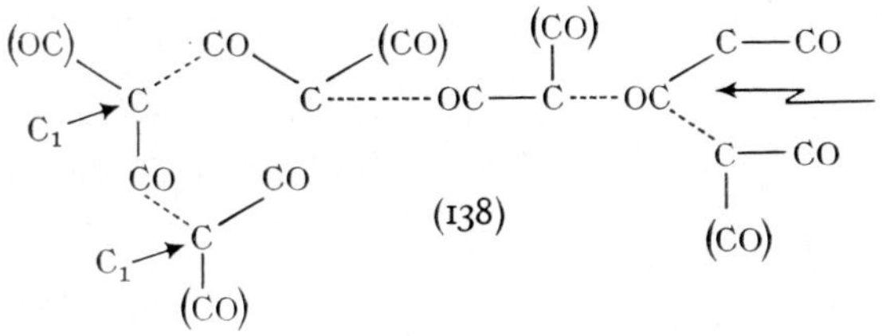

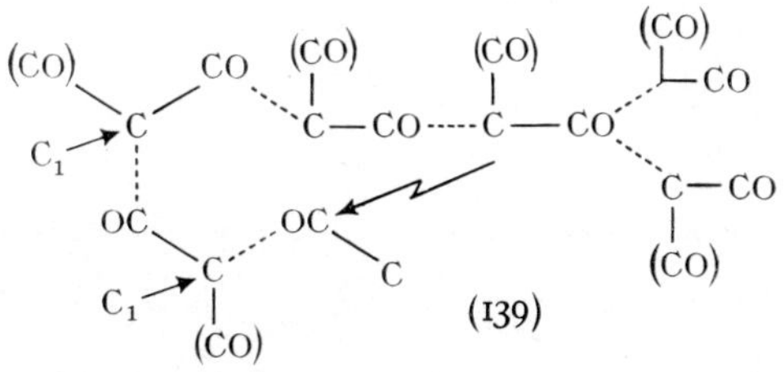

References, pp. 202—208

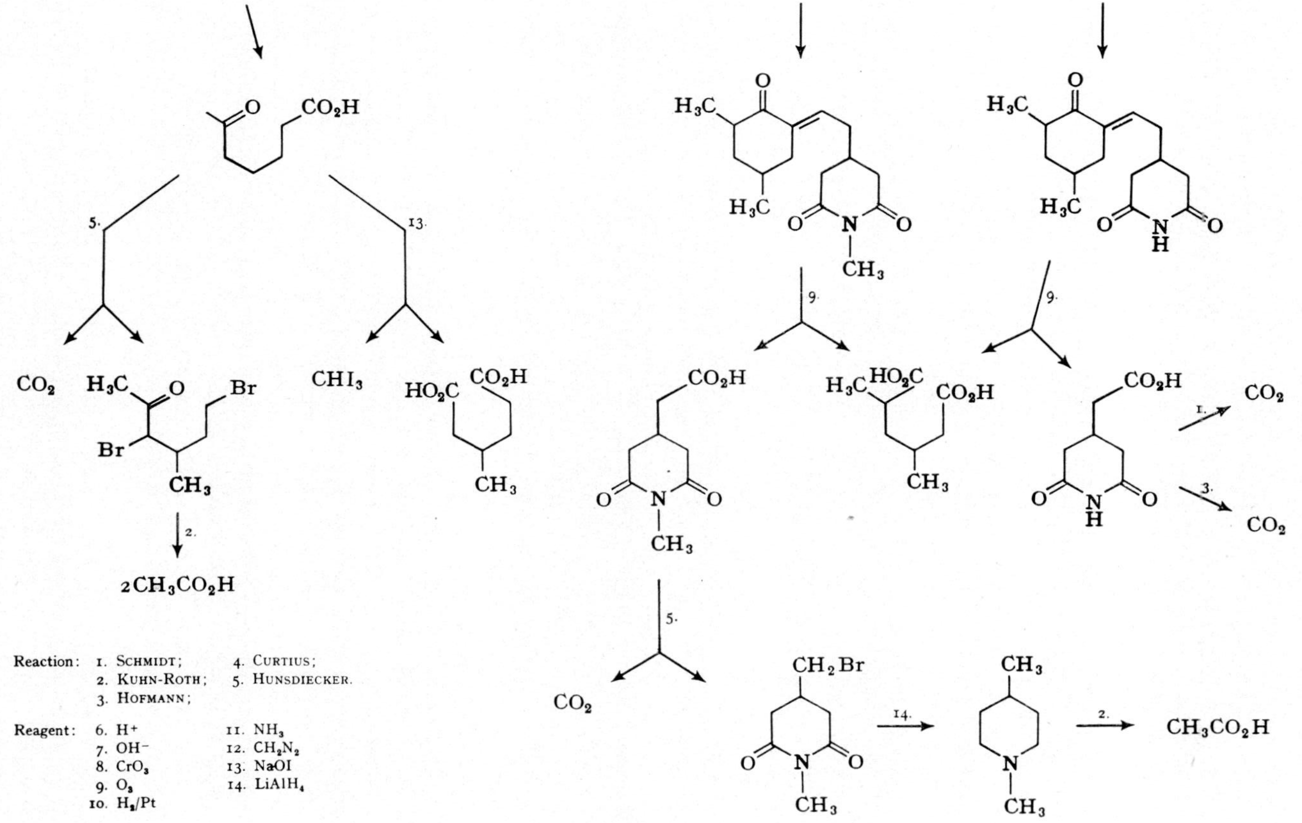

Chart II

Reaction:
1. Schmidt;
2. Kuhn-Roth;
3. Hofmann;
4. Curtius;
5. Hunsdiecker.

Reagent:
6. H^+
7. OH^-
8. CrO_3
9. O_3
10. H_2/Pt
11. NH_3
12. CH_2N_2
13. $NaOI$
14. $LiAlH_4$

In (138) the chain would be initiated by the malonate unit (as malonyl or even malonamyl coenzyme-A) indicated and extended by the addition of five more units, with decarboxylation of these latter units at some stage. The biosynthesis of streptimidone by this route would involve the addition of a seventh malonate unit which subsequently would have to suffer double decarboxylation. In route (139) the chain would be initiated by the acetate unit marked and extended by six malonate units. The terminal methyl group of this C_{16} precursor would then have to be oxidatively lost to yield the C_{15} bicyclic group of the glutarimide antibiotics and retained only in the case of streptimidone (and protomycin).

In order to differentiate between these two pathways the Czech workers compared the levels of radioactivity at positions labelled by [1-^{14}C]acetate in the degradation products 2,4-dimethylcyclohexanone and methanetriacetic acid. Considerably more activity was found in the former compound at these positions, a finding which was better explained by route (138) since some of the atoms of the glutarimide unit come from the malonyl coenzyme-A unit which initiates the synthesis. The latter probably differs in history (i. e., comes from a nonradioactive source) from those involved in chain extension. In agreement with this it was also found that sodium [1,3-^{14}C]malonate caused labelling of methanetriacetic acid and 2,4-dimethylcyclohexanone in precisely a 4:3 ratio. Furthermore, it was found that incorporation of radioactive carbon from [1-^{14}C]propionic and [1,4-^{14}C]succinic acid occurred preferentially (85–90%) in the tri-acid fragment. These two substances are not utilized as such, but on enzymic degradation yield labelled carbon dioxide which would be generally utilized for the carboxylation of acetyl coenzyme-A, but retained after condensation only in the case of the "primer" malonyl unit.

Further proof (*88, 120*) that the biosynthesis of cycloheximide commenced with the glutarimide ring was obtained by adding radioactive sodium carbonate to the fermentation medium. Degradation experiments then showed that the cyclohexanone ring of cycloheximide contained three radioactive carbon atoms whereas the β-ethylglutarimide unit contained four radioactive carbon atoms. The larger part of the radioactivity (80.5%) was found in methanetriacetic acid while the 2,4-dimethylcyclohexanone contained only 17.1%.

The degradation (*121*) pathways utilized to determine the radioactivity in individual carbon atoms of the cycloheximide molecule are shown in Chart II.

The stereospecific biogenesis of the carbon atom at the 3-position in the glutarimide ring of cycloheximide also has been elucidated (*14, 121*). Biogenetically, the glutarimide ring involves one malonate unit which undergoes decarboxylation to form an acetate unit and another unchanged

References, pp. 202—208

malonate unit, thereby destroying the symmetry and forming a new center of asymmetry at the 3-position of the glutarimide ring. Since the differences between these units is evident only during the biogenesis of (1), labelling experiments were undertaken to determine whether (1) is generated from a pure biodiasteromer of the (R)-form (140), or the (S)-form (141), or from a mixture of both.

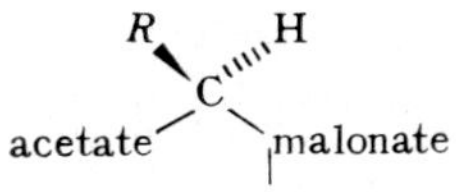

$$
\begin{array}{cc}
\underset{\text{acetate}}{R}\diagdown\underset{|}{\overset{H}{C}}\diagup\underset{\text{malonate}}{} & \underset{\text{acetate}}{H}\diagdown\underset{|}{\overset{R}{C}}\diagup\underset{\text{malonate}}{} \\
(140) & (141)
\end{array}
$$

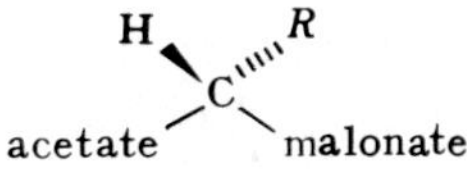

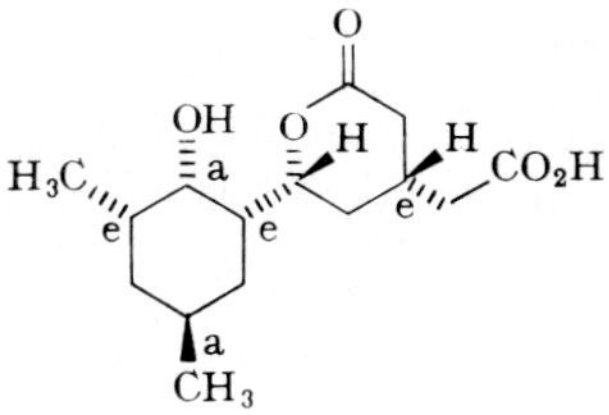

(II)

The diversity of the units and the existence of more or less than one of the units was established by stereospecific degradation. The carbonyl group of the acetate unit was labelled using [1-^{14}C]acetic acid as the precursor, so that the carbonyl group of the malonate unit eventually destined for the 2-position would arise by unlabelled carbon dioxide fixation. Separate labelling of the carbonyl group of the malonate unit was accomplished by means of radioactive sodium carbonate. Selective degradation of the glutarimide ring was accomplished by decarboxylating the dihydrocycloheximide acid lactone (11) since only that stereoisomer posessing an equatorial actetic acid group on the lactone ring is produced. The assay results indicated that during the biogenesis of cycloheximide both biodiastereomers are formed in the ratio of 1 : 2.2, (141) being the dominant form.

The biogenetic origin of the terminal methylene group of streptimidone has also been studied (89). Radioactive assay showed that this group arises from the C-2 carbon atom of a malonate unit which has undergone double decarboxylation. So far, no biogenesis studies on protomycin (98) appear to have been carried out. On the basis of the established biogenetic pattern for streptimidone, it would appear somewhat difficult to account for the incorporation of the isopropyl group. The distinct possibility exists that protomycin is not a naturally-occurring antibiotic. It may simply be an artifact generated by a base-catalyzed condensation of streptimidone with the acetone solvent used during its isolation (*110*). From a chemical point of view the preferred sight of aldol condensation in streptimidone would be precisely that at which the isopropanol group is found in protomycin.

Finally, although they have not been studied specifically, there can be little doubt that the streptovitacins arise biogenetically in exactly the same way as cycloheximide. Further oxygenation would appear to be necessary, at least to produce streptovitacins-A and -C_2. However, no attempt to convert cycloheximide to the streptovitacins, by means of a biological process, has been reported so far.

References

1. ACITELLI, M. A.: Actidione. The Synthesis of the Glutarimide Moiety. Ph. D. Thesis, Cornell University, 1957.
2. ANET, F. A. L.: Nuclear Magnetic Resonance Spectra of Compounds Containing C-Methyl Groups. Canad. J. Chem. **39**, 2262 (1961).
3. BARONDES, S. H.: Cerebral Protein Synthesis Inhibitors Block Long-Term Memory. Int. Rev. Neurobiol. **12**, 177 (1970).
4. — In *Protein Metabolism of the Nervous System*. Ed., A. Lajtha Plenum Press, New York, 545 (1970).
5. BEARD, C., C. DJERASSI, J. SICHER, F. ŠIPOŠ, and M. TICHY: Optical Rotatory Dispersion Studies. LXXXI. Stereochemical Studies. XXVII. Conformational Distortion in 2-Methylcyclohexanones. Tetrahedron **19**, 919 (1963).
6. BENNETT, L. L., V. L. WARD and R. W. BROCKMAN: Inhibition of Protein Synthesis *in vitro* by Cycloheximide and Related Glutarimide Antibiotics. Biochim. Biophys. Acta **103**, 478 (1965).
7. BHACCA, N. S., and D. H. WILLIAMS: Solvent Effects in NMR Spectroscopy III. Chemical Shifts Induced by Benzene in Ketones. Tetrahedron **21**, 2021 (1965).
8. CATION, D.: Actidione Stops Cherry Leaf Spot Fungus. Amer. Fruit Grower **74**, 29 (1954).
9. CHURCHILL, B. W.: Production of Cycloheximide. United States Patent 2885326 (1959).
10. COHEN, H. D., and S. H. BARONDES: Puromycin and Cycloheximide. Different Effects on Hippocampal Electrical Activity. Science **154**, 1557 (1966).
11. — — Puromycin Effect on Memory may be due to Occult Seizures. Science **157**, 333 (1967).
12. CONNOLLY, J. D., and R. MC CRINDLE: Nuclear Magnetic Resonance Solvent Shifts of Methyl Groups in Alicyclic Ketones: Pyrrole and Other Aromatic Solvents. J. Chem. Soc. (London) **1966**, 1613.
13. COREY, E. J., and R. A. SNEEN: Stereoelectronic Control in Enolization-Ketonization Reactions. J. Amer. Chem. Soc. **78**, 6269 (1956).
14. CUDLÍN, J., M. PŮŽA, M. VONDRÁČEK, Z. VANĚK and R. W. RICKARDS: Metabolites of *Streptomyces noursei*. VIII. Biogenesis of Cycloheximide from the Point of View of the Glutarimide Ring Symmetry. Folia Microbiol. **12**, 376 (1967).
15. EBLE, T. E., M. E. BERGY, R. R. HERR and J. A. FOX: The Separation and Properties of the Streptovitacins. Antibiotics and Chemotherapy **10**, 479 (1960).
16. EBLE, T. R., M. E. BERGY, C. M. LARGE, R. R. HERR and W. G. JACKSON: Isolation Purification and Properties of Streptovitacins-A and -B. Antibiotics Annual 1958–1959, New York, Medical Encyclopedia Inc. **1959**, 555.

17. EGAWA, Y., S. OSHIMA and S. UMEZAWA: Studies on Cycloheximide-Related Compounds. I. Esters of Cycloheximide and Their Antitoxoplasmic Activity. J. Antibiotics (Japan) Ser. A **18**, 171 (1965).
18. EGAWA, Y., M. SUZUKI and T. OKUDA: Studies on Streptomyces Antibiotic, Cycloheximide. XVI. Synthesis of Cycloheximide Analogous Compounds. Chem. Pharm. Bull. (Tokyo) **11**, 589 (1963).
19. EGAWA, Y., and S. UMEZAWA: Studies of Cycloheximide-Related Compounds II. The Acid-Catalyzed Condensation of Glutarimide-β-Acetaldehyde with Ketones. Bull. Chem. Soc. Japan **38**, 2169 (1965).
20. EISENBRAUN, E. J., J. OSIECKI and C. DJERASSI: On the Absolute Configuration of the Antibiotic Actidione. J. Amer. Chem. Soc. **80**, 1261 (1958).
21. ENNIS, H. L.: Structure Activity Studies with Cycloheximide and Congeners. Biochem. Pharm. **17**, 1197 (1968).
22. EVANS, J. S., G. D. MENGEL, J. CERU and R. L. JOHNSTON: Biological Studies on Streptovitacin-A New Antitumor Agent. Antibiotics Annual 1958–1959, New York, Medical Encyclopedia Inc. **1959**, 565.
23. FIELD, J. B., A. MIRELES, H. R. PACHL, L. BASCOY, L. CANO and W. K. BULLOCK: Experimental Evaluation of a New Antitumor Agent, Streptovitacin-A. Antibiotics Annual 1958–1959, New York, Medical Encyclopedia Inc. **1959**, 572.
24. FORD, J. H., and W. KLOMPARENS: Cycloheximide (Actidione) and its Non-agricultural Uses. Antibiotics and Chemotherapy **10**, 682 (1960).
25. FORD, J. H., W. KLOMPARENS and C. L. HAMNER: Cycloheximide (Actidione) and its Agricultural Uses. Plant Disease Reptr. **42**, 680 (1958).
26. FORD, J. H., and B. E. LEACH: Actidione an Antibiotic from *Streptomyces griseus*. J. Amer. Chem. Soc. **70**, 1223 (1948).
27. FORIST, A. A., and S. THEAL: Spectrophotometric Determination of Cycloheximide. Analyt. Chem. **31**, 1042 (1959).
28. FROHARDT, R. P., H. W. DION, Z. L. JAKUBOWSKI, A. RYDER, J. C. FRENCH and Q. R. BARTZ: Chemistry of Streptimidone, A New Antibiotic. J. Amer. Chem. Soc. **81**, 5500 (1959).
29. GARRETT, E. R., and R. E. NOTARI: Cycloheximide Transformations. II. Kinetics and Stability in a Pharmaceutically Useful pH Range. J. Pharm. Sci. **54**, 209 (1965).
30. — — Determination of Cycloheximide and its Degradation Products Alone and in Mixtures. J. Pharm. Sci. **54**, 561 (1965).
31. — — Cycloheximide Transformations. I. Kinetics and Mechanism in Aqueous Acid. J. Organ. Chem. (U. S. A.) **31**, 425 (1966).
32. GELLER, A., F. ROBUSTELLI, S. H. BARONDES, H. D. COHEN and M. E. JARVIK: Impaired Performance by Post-Trial Injections of Cycloheximide in a Passive Avoidance Task. Psychopharmacologia (Berl.) **14**, 371 (1969).
33. GELLER, A., F. ROBUSTELLI and M. E. JARVIK: A Parallel Study of the Amnesic Effects of Cycloheximide and ECS Under Different Strengths of Conditioning. Psychopharmacologia (Berl.) **16**, 281 (1970).
34. GROLLMAN, A. P.: Structural Basis for Inhibition of Protein Synthesis by Emetine and Cycloheximide Based on an Analogy Between Ipecac Alkaloids and Glutarimide Antibiotics. Proc. Nat. Acad. Sci. U. S. A. **56**, 1867 (1966).
35. HAMILTON, J. M., M. SZKOLNIK and E. SONDHEIMER: Systemic Control of Cherry Leaf Spot Fungus by Foliar Sprays of Actidione Derivatives. Science **123**, 1175 (1956).
36. HENNIS, H. E., L. G. DUQUETTE and F. JOHNSON: Glutarimide Antibiotics. XIII. Comment on the Stereochemistry of Streptovitacin-A and E-73. J. Organ. Chem. (U.S.A.) **33**, 904 (1968).

37. HERR, R. R.: Structures of the Streptovitacins. J. Amer. Chem. Soc. 81, 2595 (1959).
38. — Structure Studies on Streptovitacins-A and -B. Antibiotics Annual 1958–1959, New York, Medical Encyclopedia Inc. 1959, 560.
39. HIGHET, R. J. and V. PRELOG: Stoffwechselprodukte von Actinomycetin. Actiphenol. Helv. Chim. Acta 42, 1523 (1959).
40. IGARASHI, S., and S. WADA: Fermicidin, A New Antibiotic Active Against Yeasts and Trichomonas. J. Antibiotics (Japan), Ser. B 7, 221 (1954).
41. JOHNSON, F.: Glutarimide Antibiotics. I. The Synthesis of Actiphenol. J. Organ. Chem. (U. S. A.) 27, 3658 (1962).
42. JOHNSON, F., and A. A. CARLSON: Glutarimide Antibiotics. Part VIII. A Stereoelective Synthesis of dl-α-Epiisocycloheximide. Tetrahedron Letters 1965, 885.
43. JOHNSON, F., A. A. CARLSON and N. A. STARKOVSKY: Glutarimide Antibiotics. XI. A Total Synthesis of dl-α-Epiisocycloheximide. J. Organ. Chem. (U. S. A.) 31, 1327 (1966).
44. JOHNSON, F., W. D. GUROWITZ and N. A. STARKOVSKY: Glutarimide Antibiotics. Part II. The Synthesis and Stereochemistry of dl-Neocycloheximide, A New Isomer of Cycloheximide. Tetrahedron Letters 1962, 1167.
45. JOHNSON, F., and N. A. STARKOVSKY: Glutarimide Antibiotics. Part III. The Determination of the Stereochemistry of the Methyl Groups of Cycloheximide Isomers by Nuclear Magnetic Resonance Spectroscopy. Tetrahedron Letters 1962, 1173.
46. JOHNSON, F., N. A. STARKOVSKY and A. A. CARLSON: Glutarimide Antibiotics. IX. The Stereochemistry of the Dihydrocycloheximides and the Configuration of the Hydroxyl Group of Cycloheximide. J. Amer. Chem. Soc. 87, 4612 (1965).
47. JOHNSON, F., N. A. STARKOVSKY and W. D. GUROWITZ: Glutarimide Antibiotics. Part VII. The Synthesis of dl-Neocycloheximide and the Determination of the Cyclohexanone Ring Stereochemistry of Cycloheximide, its Isomers, and Inactone. J. Amer. Chem. Soc. 87, 3492 (1965).
48. JOHNSON, F., N. A. STARKOVSKY, A. C. PATON and A. A. CARLSON: Glutarimide Antibiotics. Part VI. The Total Synthesis of dl- and l-Cycloheximide. J. Amer. Chem. Soc. 86, 118 (1964).
49. — — — — The Total Synthesis of Cycloheximide. J. Amer. Chem. Soc. 88, 149 (1966).
50. JOHNSON, F., and A. WHITEHEAD: The Stereochemistry of 2-Substituted Cyclohexanone Enamines and the Corresponding Schiff's Bases. Tetrahedron Letters 1964, 3825.
51. KHARATYAN, S., M. PŮŽA, T. SPÍŽEK, L. DOLEŽIKOVÁ and Z. VANĚK: Biogenesis of Cycloheximide and Related Compounds. Chem. and Ind. 1963, 1038.
52. KOHBERGER, D. L., M. W. FISHER, M. M. GALBRAITH, A. B. HILLEGAS, P. E. THOMPSON and J. ERLICH: Biological Studies of Streptimidone a New Antibiotic. Antibiotics and Chemotherapy 10, 9 (1960).
53. KORNFELD, E. C., and R. G. JONES: The Structure of Actidione, An Antibiotic from *Streptomyces griseus*. Science 108, 437 (1948).
54. KORNFELD, E. C., R. G. JONES and T. V. PARKE: The Structure and Chemistry of Actidione, An Antibiotic from *Streptomyces griseus*. J. Amer. Chem. Soc. 71, 150 (1949).
55. LAWES, B. C.: The Synthesis of Anhydroactidione. J. Amer. Chem. Soc. 82, 6413 (1960).
56. — Absolute Configuration of Cycloheximide from Thermal Degradation. Abstracts of the 139th Meeting American Chemical Society, St. Louis, Mo., March 1961, p. 33N.

57. LAWES, B. C.: Absolute Configuration of Cycloheximide from Thermal Degradation. J. Amer. Chem. Soc. **84**, 239 (1962).
58. LEACH, B. E., J. H. FORD and A. J. WHIFFEN: Actidione an Antibiotic from *Streptomyces griseus*. J. Amer. Chem. Soc. **69**, 474 (1947).
59. LEMIEUX, R. A., R. K. KULLIG, H J BERNSTEIN and W. G. SCHNEIDER: Configurational Effects on the Proton Magnetic Resonance Spectra of Six-Membered Ring Compounds. J. Amer. Chem. Soc. **80**, 6098 (1958).
60. LEMIN, A. J.: Chemical Process. United States Patent 2,903,458 (1959).
61. LEMIN, A. J., and J. H. FORD: Isocycloheximide. J. Organ. Chem. (U.S.A.) **25**, 344 (1960).
62. LOEFER, J. B., and T. S. MATNEY: Growth Inhibition of Free-Living Protozoa by Actidione. Physiol. Zoöl. **25**, 272 (1952).
63. Mc LURE, T. T.: Experiences with Cherry Sprays in 1951. Phytopathology **42**, 14 (1952).
64. MOLE, T.: Thermal Retro-aldol Reaction. Chem. and Ind. **1960**, 1164.
65. NOTARI, R. E., and S. M. CAIOLA: Catalysis of Streptovitacin-A Dehydration. Kinetics and Mechanisms. J. Pharm. Sci. **58**, 1203 (1969).
66. OKUDA, T.: Some Configurational Aspects of Naramycin-A (Cycloheximide) and its Stereoisomeric Antibiotic, Naramycin-B. Chem. Pharm. Bull. (Tokyo) **7**, 137 (1959).
67. — Further Studies on the Absolute Configuration of Naramycin-A (Cycloheximide) and its Isomeric Naramycin-B. Chem. Pharm. Bull. (Tokyo) **7**, 259 (1959).
68. — Studies on Streptomyces Antibiotic Cycloheximide. IV. Some Observations on Stereochemical Configurations of Naramycin-A (Cycloheximide) and Naramycin-B. Chem. Pharm. Bull. (Tokyo) **7**, 659 (1959).
69. — Studies on Streptomyces Antibiotic Cycloheximide. V. Synthesis and Stereochemistry of Deoxycycloheximide. Chem. Pharm. Bull. (Tokyo) **7**, 666 (1959).
70. — Studies on Streptomyces Antibiotic, Cycloheximide. VI. The Absolute Configuration of Naramycin-A (Cycloheximide) and its Isomeric Naramycin-B. Chem. Pharm. Bull. (Tokyo) **7**, 671 (1959).
71. OKUDA, T., K. ASHINO, Y. EGAWA and M. SUZUKI: The Streptomyces Antibiotic Cycloheximide I. Isolation of Naramycin-A and its Identity with Cycloheximide. Chem. Pharm. Bull. (Tokyo) **6**, 711 (1958).
72. OKUDA, T., and M. SUZUKI: Absolute Configuration of Cycloheximide. Chem. Pharm. Bull. (Tokyo) **9**, 1014 (1961).
73. — — Cycloheximide (Naramycin) Part I-1. Yakugaku Kenkyu **33**, 371 (1961).
74. OKUDA, T., M. SUZUKI and Y. EGAWA: Studies on Streptomyces Antibiotic, Cycloheximide. VII. On the Configuration of Naramycin-B and Isocycloheximide. Chem. Pharm. Bull. (Tokyo) **8**, 336 (1960).
75. — — — Synthesis of Cycloheximide Isomers Including Isocycloheximide. J. Antibiotics (Japan), Ser. A **14**, 158 (1961).
76. OKUDA, T., M. SUZUKI, Y. EGAWA and K. ASHINO: Studies on Cycloheximide and its New Stereoisomeric Antibiotic. Chem. Pharm. Bull. (Tokyo) **6**, 328 (1958).
77. — — — — Studies on Streptomyces Antibiotic Cycloheximide. II. Naramycin-B an Isomer of Cycloheximide. Chem. Pharm. Bull. (Tokyo) **7**, 27 (1959).
78. OKUDA, T., M. SUZUKI, Y. EGAWA and K. KOTERA: Studies on Streptomyces Antibiotic, Cycloheximide XIV. Supplemental Physicochemical Data on Cycloheximide and its Related Compounds. Yakugaku Kenkyu **33**, 530 (1961).

79. OKUDA, T., M. SUZUKI, T. FURUMAI and H. TAKAHASHI: Studies on Streptomyces Antibiotic Cycloheximide. XVIII. Isomerization Study of Cycloheximides and Thermal Degradation of Naramycin-B. Chemical Support of the Proposed Absolute Configuration of Cycloheximides. Chem. Pharm. Bull. (Tokyo) 11, 730 (1963).
80. OSATO, T., Y. MORIKUBO and H. UMEZAWA: Production and Extraction of Niromycins, Antiviral Antibiotics. J. Antibiotics (Japan), Ser. A 13, 110 (1960).
81. OSATO, T., Y. MORIKUBO, S. YAMAZAKI, T. HIKIJI, K. YANO, M. KANAO, T. OSONO and H. UMEZAWA: Screening Studies of Antiviral Substances Produced by Actinomycetes and New Antiviral Substances, Niromycins. J. Antibiotics (Japan), Ser. A 13, 97 (1960).
82. PALMER, C., and T. E. MALONEY: Preliminary Screening for Potential Algicides. Ohio. J. Sci. 55, 1 (1955).
83. PAUL, R., and S. TCHELITCHEFF: Constitution Chimique de l'Inactone. Synthèse Partielle de trois Isomères de l'Actidione. Bull. Soc. Chim. France 1955, 1366.
84. PETTIT, G. R., and A. K. DAS GUPTA: Actidione Acetate Nitrogen Mustard. Chem. and Ind. 1962, 1016.
85. PHILLIPS, D. D., M. A. ACITELLI and J. MEINWALD: Actidione. I. The Synthesis of the Glutarimide Moiety. J. Amer. Chem. Soc. 79, 3517 (1957).
86. PIATAK, D. M., C. C. YEN and R. V. KENNEDY, Jr.: Antineoplastic Research. I. Pyrazole and Pyrimidine Derivatives of Dehydrocycloheximide Analogs. J. Med. Chem. 13, 770 (1970).
87. PREUD'HOMME, J., and M. DUBOST: L'Inactone, Nouvelle Substance Voisine de l'Actidione Estraite des Cultures de *Streptomyces griseus*. Handbook 14th Int. Cong. Pure and Applied Chem. Zurich 1955, 382.
88. PŮŽA, M., J. CUDLÍN, L. DOLEŽILOVÁ, M. VONDRÁČEK, J. SPÍŽEK, Z. VANĚK and J. BENES: Origin of Terminal Group in Cycloheximide. Antibiotic Advan. Res., Prod. Clin. Use, Proc. Congr. Prague 1964, 582.
89. PŮŽA, M., J. CUDLÍN and Z. VANĚK: Diversity of Streptimidone and Cycloheximide Biosynthesis. Folia Microbiol. 13, 533 (1968).
90. RAO, K. V.: C-73: A Metabolic Product of *Streptomyces Albulus*. J. Organ. Chem. (U.S.A.) 25, 661 (1960).
91. — E-73: An Antitumor Substance. Part III. Some Derivatives. Antibiotics and Chemotherapy 12, 123 (1962).
92. — E-73: An Antitumor Substance. Part II. Structure. J. Organ. Chem. (U.S.A.) 82, 1129 (1960).
93. RAO, K. V., and W. P. CULLEN: E-73: An Antitumor Substance. Part I. Isolation and Characterization. J. Amer. Chem. Soc. 82, 1127 (1960).
94. SCHAEFFER, H. J., and V. K. JAIN: Determination of the Absolute Configuration of Cycloheximide. J. Pharm. Sci. 50, 1048 (1961).
95. SCHAEFFER, H. J., and V. K. JAIN: Synthesis of Dehydrocycloheximide. J. Pharm. Sci. 52, 509 (1963).
96. — — Investigation of the Stereochemistry of Cycloheximide and its Degradation Products. J. Pharm. Sci. 52, 639 (1963).
97. — — Investigation of the Stereochemistry of Cycloheximide. J. Pharm. Sci. 53, 144 (1964).
98. — — The Synthesis of Dehydrocycloheximide and the Conversion of *cis*-2,4-Dimethylcyclohexanone to its *trans*-Isomer. J. Organ. Chem. (U.S.A.) 29, 2595 (1964).

99. SHOPPEE, C. W., F. P. JOHNSON, R. E. LACK and S. STERNHELL: Line Widths of Nuclear Magnetic Resonance Signals Due to Tertiary Methyl Groups. Chem. Comm. **1965**, 347.
100. SIEGEL, M. R., H. D. SISLER and F. JOHNSON: The Relationship of Structure to Fungitoxicity of Cycloheximide and Related Glutarimide Derivatives. Biochem. Pharm. **15**, 1213 (1966).
101. SISLER, H. D., and M. R. SIEGEL: Cycloheximide and Other Glutarimide Antibiotics. Antibiotics (Mechanism of Action) I, 283 (1967).
102. SMITH, C. G.: Tissue Culture. Bioassay Methods for Streptovitacin-A. Proc. Soc. Exper. Biol. and Med. **100**, 747 (1959).
103. SMITH, C. F., W. L. LUMMIS and J. E. GRADY: An Improved Tissue Culture Assay. I. Methodology Studies: Cytotoxicities of Streptovitacins and Cycloheximide. Cancer Research, Sept. 1959.
104. SOKOLSKI, W. T., N. J. EILERS and G. M. SAVAGE: Paper Chromatography and Microbiological Assay of the Streptovitacins, Antibiotics Annual 1958–1959, New York, Medical Encyclopedia Inc. **1959**, 951.
105. STARKOVSKY, N. A., A. A. CARLSON and F. JOHNSON: Glutarimide Antibiotics. XII. A Stereochemical Investigation of the Dehydration Products of Cycloheximide. J. Organ. Chem. (U.S.A.) **31**, 2516 (1966).
106. STARKOVSKY, N. A., and F. JOHNSON: Glutarimide Antibiotics. V. The Absolute Configuration of the Side-Chain Asymmetric Center of Cycloheximide. Tetrahedron Letters **1964**, 919.
107. STARKOVSKY, N. A., F. JOHNSON and A. A. CARLSON: Glutarimide Antibiotics. VI. The Nature of Deoxycycloheximide and Epideoxycycloheximide. Tetrahedron Letters **1964**, 1015.
108. STRUCK, R. F., H. J. SCHAEFFER, C. A. KRAUTH, R. J. KEMP, Y. F. SHEALY and J. A. MONTGOMERY: Synthesis of Potential Antineoplastic Agents. XXXIII. β-Diketone Analogs of the Glutarimide Antibiotics. J. Med. Chem. **7**, 646 (1964).
109. SUGAWARA, R.: Protomycin, A New Antibiotic of Cycloheximide Group. III. Degradation Products. J. Antibiotics (Japan) Ser. A **16**, 167 (1963).
110. — Protomycin, A New Antibiotic of the Cycloheximide Group. II. Production and Properties. Antibiotics (Japan) Ser. A **16**, 115 (1963).
111. SUGAWARA, R., A. MATSUMAE and T. HATA: Protomycin, A New Antibiotic of Cycloheximide Group. I. J. Antibiotics (Japan) Ser. A **16**, 111 (1963).
112. SUZUKI, M.: Studies on Streptomyces Antibiotic Cycloheximide. VIII. Hydroxycarbonylation of Optocally Active 2,4-Dimethylcyclohexanone and Chemical Correlation of the Products (1). Chem. Pharm. Bull. (Tokyo) **8**, 706 (1960).
113. — Studies on Streptomyces Antibiotic, Cycloheximide. VIII. Hydroxycarbonylation of Optically Active 2,4-Dimethylcyclohexanone and Chemical Correlation of the Products (2). Chem. Pharm. Bull. (Tokyo) **8**, 713 (1960).
114. — Studies of Streptomyces Antibiotic, Cycloheximide. IX. Absolute Configuration of Optically Active 2,4-Dimethyl-6-[α-hydroxy(p-substituted-)benzyl]cyclohexanones. Chem. Pharm. Bull. (Tokyo) **8**, 717 (1960).
115. — Studies on Streptomyces Antibiotic Cycloheximide. XI. Preparation and Chemical Structure of the Oxidation Products (ψ-Cycloheximides) from Dihydrocycloheximides. Chem. Pharm. Bull. (Tokyo) **9**, 778 (1960).
116. — Studies on Streptomyces Antibiotic Cycloheximide. X. Structure-Antimicrobiol Activity Relationship of Cycloheximide and Related Compounds. Yakugaku Zasshi **80**, 1217 (1960).

117. SUZUKI, M., Y. EGAWA and T. OKUDA: Studies on Streptomyces Antibiotic Cycloheximide. XV. Hydroxycarbonylation of Optically Active 2,4-Dimethylcyclohexanones with Glutarimide-β-acetaldehyde (Synthesis of Isocycloheximide and its Isomers). Chem. Pharm. Bull. (Tokyo) **11**, 582 (1963).
118. TAKASHITA, M., H. TAKAHASHI and T. OKUDA: Studies on Streptomyces Antibiotic, Cycloheximide. XIII. New Spectrophotometric Determination of Cycloheximide. Chem. Pharm. Bull. (Tokyo) **10**, 304 (1962).
119. UMEZAWA, S., and Y. EGAWA: Studies of Cycloheximide-Related Compounds. III. A Novel Reaction of Aromatic Nitro Ketones with Aldehydes and Secondary Amines. Bull. Chem. Soc. Japan **40**, 908 (1967).
120. VANĚK, Z., M. PŮŽA, J. CUDLÍN, L. DOLEŽILOVÁ and M. VONDRÁČEK: Metabolites of *Streptomyces noursei*. III. Incorporation of ^{14}C-Carbon Dioxide into Cycloheximide. Biochem. Biophys. Res. Commun. **17**, 532 (1964).
121. VANĚK, Z., M. PŮŽA, J. CUDLÍN, M. VONDRÁČEK and R. W. RICKARDS: Metabolites of *Streptomyces noursei*. X. Biogenesis of Cycloheximide. Folia Microbiol. **14**, 388 (1969).
122. VAN TAMELEN, E. E., and V. HAARSTAD: Structure of the Antibiotic Streptimidone. J. Amer. Chem. Soc. **82**, 2974 (1960).
123. VAUGHN, J. R.: Cycloheximide, An Antibiotic Effective Against Turf Disease. Phytopathology **41**, 36 (1951).
124. VELAIRE, C. D.: Improved Formulation of Cycloheximide Composition and Process for the Control of Blister Rust and Like Fungal Infections in Trees, United States Patent 3,014,840 (1961).
125. WELCH, J. J.: Rodent Control. A Review of Chemical Repellents for Rodents. J. Agr. Food Chem. **2**, 142 (1954).
126. WELLS, H. D., and B. P. ROBINSON: Cottony Blight of Rye Grass Caused by *Pythium aphanidermatum*. Phytopathology **44**, 509 (1954).
127. WHIFFEN, A. J.: The Production, Assay and Antibiotic Activity of Actidione an Antibiotic from *Streptomyces griseus*. J. Bacteriol. **56**, 283 (1948).
128. — The Activity *in vitro* of Cycloheximide (Actidione) Against Fungi Pathogenic to Plants. Mycologia **42**, 253 (1950).
129. WHIFFEN, A. J., N. BOHONAS and R. L. EMERSON: The Production of an Antifungal Antibiotic by *Streptomyces griseus*. J. Bact. **52**, 610 (1946).
130. WILLIAMSON, W. R. N.: The Alkylation of Pyrrolidine Enamines. Tetrahedron **3**, 314 (1958).
131. WOLINSKY, J., and D. CHAN: Preparation of (−)-*cis*-2,4-Dimethylcyclohexanone from (+)-Pulegone. J. Amer. Chem. Soc. **85**, 937 (1963).
132. WOO, P. W. K., H. W. DION and Q. R. BARTZ: The Structure of Streptimidone. J. Amer. Chem. Soc. **83**, 3085 (1961).

(Received, March 1, 1971)

Chemie und Biosynthese der Flechtenstoffe

Von S. HUNECK, Halle (Saale)

Mit 4 Abbildungen

Inhaltsübersicht

	Seite
I. Einleitung	209
II. Methoden zum Nachweis und zur Strukturaufklärung der Flechtenstoffe	210
A. Dünnschichtchromatographie	210
B. Papierchromatographie	211
C. Gaschromatographie	211
D. Infrarotspektroskopie	211
E. Ultraviolettspektroskopie	211
F. NMR-Spektroskopie	212
G. Massenspektrometrie	213
H. Röntgenstrukturanalyse	215
I. Chemische Methoden	215
III. Einteilung der Flechtenstoffe	216
IV. Strukturaufklärung und Synthese der Flechtenstoffe	216
1. Produkte des Primärstoffwechsels	216
2. Acetogenine	220
3. Phenylalanin-Derivate	269
4. Vitamine	273
5. Enzyme	273
V. Biosynthese der Flechtenstoffe	273
VI. Aus Mycobionten isolierte Verbindungen	285
VII. Chemotaxonomie der Flechten	285
VIII. Antibiotische und weitere biologische Wirkungen der Flechtenstoffe	287
Literaturverzeichnis	288

I. Einleitung

Seit der letzten Zusammenfassung von ASAHINA (24) über „Neuere Entwicklungen auf dem Gebiete der Flechtenstoffe" in dieser Serie, die Literatur bis etwa 1950 umfassend, hat die Chemie, Taxonomie und Physiologie der Flechten einen neuen Aufschwung genommen und große Fortschritte gemacht. Anknüpfend an diesen Artikel soll im folgenden über die seit 1950 erzielten neuen Ergebnisse auf dem Gebiet der Chemie und Biosynthese der Flechtenstoffe sowie der Chemotaxonomie der Flechten berichtet werden, unter Berücksichtigung der Literatur bis 1970.

Der insbesondere durch die Anwendung der modernen physikalischen Methoden erzielte Fortschritt bei der Strukturaufklärung von Flechtenstoffen wird durch folgende Gegenüberstellung demonstriert: 1900 war lediglich die Struktur der Lecanorsäure bekannt (*136*), während 1954 in der Monographie von ASAHINA und SHIBATA (*26*) die Struktur von etwa 74 charakteristischen Flechtenstoffen beschrieben wird. In den 16 Jahren seit 1954 wurden dagegen etwa 65 neue Flechtenstoffe bekannt (*80, 82a*).

Untersuchungen zur Biosynthese der Flechtenstoffe datieren erst aus jüngster Zeit; vor 1965 existierten nur spekulative Vorstellungen über dieses Gebiet, das wegen der Doppelnatur der Flechten schwierig, aber gleichzeitig besonders interessant ist.

Nach der Einführung der Dünnschichtchromatographie, welche die Analyse kleiner Herbarproben in kürzester Zeit gestattet, sind zahlreiche Arbeiten über die Chemotaxonomie der Flechten erschienen, aus denen wertvolle Schlüsse über die systematische Stellung einzelner Arten und verschiedener Gattungen bzw. Familien gezogen wurden.

Ganz neu sind Untersuchungen über die Stoffwechselprodukte der Flechtenpartner (Mycobiont und Phycobiont), die zur Aufklärung der Symbiose zwischen beiden Partnern wichtig sind und neue Einsichten in die Natur der Flechten versprechen.

II. Methoden zum Nachweis und zur Strukturaufklärung der Flechtenstoffe

A. Dünnschichtchromatographie

Die Dünnschichtchromatographie ist heute die Methode der Wahl zur Analyse der Flechtenstoffe und hat wegen ihrer besseren Trennfähigkeit, höheren Empfindlichkeit und einfachen Handhabung die vordem angewandte Papierchromatographie völlig verdrängt, zumal seit einiger Zeit fertig beschichtete Träger im Handel sind. Besonders bewährt hat sich die Kodak Chromagramfolie K 301 R 2, mit der SANTESSON (*260*) die R_F-Werte von 90 Depsiden, Depsidonen, Depsonen, Dibenzofuranen, Chinonen, Pulvinsäurederivaten, Xanthonen, Chromonen und aliphatischen Flechtensäuren bestimmte. Xanthone geben im Dünnschichtchromatogramm nach Behandlung mit Ammoniak im UV fluoreszierende Flecke (*265*). CULBERSON und KRISTINSSON (*84*) beschreiben eine Routinemethode zur dünnschichtchromatographischen Bestimmung von 104 Flechtensubstanzen. BENDZ et al. (*37*) gelang die dünnschichtchromatographische Trennung des Komplexes von Brucin und racemischer Usninsäure in die beiden Antipoden. Eine Zusammen-

Literaturverzeichnis: SS. 288—306

stellung der R_F-Werte weiterer Flechtenstoffe geben RAMAUT (*242*, *243*) und HUNECK (*143*). Tabelle 1 enthält die Nachweismethoden für die wichtigsten Flechtenstoffe bei der Dünnschichtchromatographie.

Tabelle 1. *Dünnschichtchromatographischer Nachweis verschiedener Flechtenstoffe*

Substanzgruppe	Nachweis
Aliphatische Säuren	Alkalische Lösung von Bromkresolgrün
Di- und Triterpene, Steroide	Konz. Schwefelsäure, Verkohlung
Aromatische Aldehyde	p-Phenylendiamin, o-Dianisidin
Nichtaldehydische Depside und Depsidone	Bis-diazotiertes Benzidin plus Natronlauge
Dibenzofurane, Xanthone, Chromone	Fluoreszenz im UV nach Behandlung mit Ammoniak
Pulvinsäurederivate, Usninsäure	Fluoreszenz im UV bei 365 nm

B. Papierchromatographie

Da die Papierchromatographie in der Analytik der Flechtenstoffe weitgehend durch die Dünnschichtchromatographie ersetzt wurde, sollen hier nur zusammenfassende Arbeiten über diese Methode zitiert werden: (*26*, *281*, *282*, *241*, *143*).

C. Gaschromatographie

Die Gaschromatographie hat in der Flechtenchemie bisher nur wenig Anwendung gefunden. IKEKAWA et al. (*168*) und SHIBATA et al. (*287*) bestimmten die Retentionszeiten von Zeorin, Zeorinin, Isozeorinin, Zeorinon, Zeorininon und Isozeorininon. Von WILSON (*336*) stammt eine quantitative Analyse von Depsiden und von FURUYA et al. (*121*) eine gaschromatographische Bestimmung von Parietin.

D. Infrarotspektroskopie

Die CO- und OH-Banden von zahlreichen Flechtenstoffen wurden von HUNECK (*143*) tabelliert. Insbesondere bei Phenolcarbonsäuren liefern die Lage der CO- und OH-Banden wertvolle Aussagen über die Stellung der freien OH-Gruppen. Auf Grund von intramolekularen Wasserstoffbrückenbindungen treten bei o-Hydroxycarbonsäuren Verschiebungen der CO-Bande von etwa 1750 cm^{-1} auf 1650 cm^{-1} auf. SANTESSON (*271*) diskutiert die IR-Spektren von 25 Xanthonen im Bereich von 1500—1700 cm^{-1} und 2000—3100 cm^{-1}.

E. Ultraviolettspektroskopie

Wie die Spektren in Abb. 1*a—i* zeigen, weisen die aromatischen Flechtenstoffe im UV charakteristische Unterschiede auf und können daher sehr gut zur Zuordnung neuer Verbindungen herangezogen werden. 1-Hydroxylierte Anthrachinone und Xanthone sowie 5-hydroxylierte Chromone geben auf Zusatz von Aluminiumchlorid typische bathochrome Bandenverschiebungen (siehe Spektrum der Abb. 1i) (*135*). Eine Zusammenstellung der UV-Maxima weiterer Flechtenstoffe findet sich bei HUNECK (*143*).

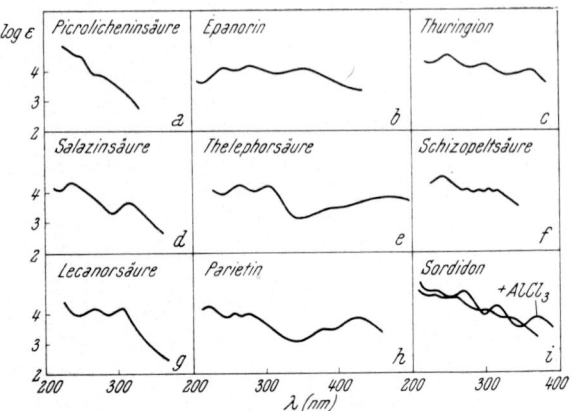

Abb. 1 a—i. UV-Spektren einiger aromatischer Flechtenstoffe

F. NMR-Spektroskopie

Bei der Strukturaufklärung der in den letzten Jahren neu aufgefundenen Flechtenstoffe leistete die NMR-Spektroskopie ausgezeichnete Dienste (*143*). Tabelle 2 zeigt die Zuordnung der Signale bei Depsiden und Depsidonen (*159*).

Tabelle 2. *Zuordnung der Protonensignale bei Depsiden und Depsidonen*

Gruppierung	Lage der Signale (in ppm)
aliphat. $-CH_3$	0,71 — 1,1
aliphat. $-CH_2-$	1,33 — 1,65
$CH_3-CO-O-$	1,93 — 2,40
aromat. $-CH_3$	2,06 — 2,81
aliphat. $-CH_2-CO-$	2,31 — 2,71
benzyl. $-CH_2-$	2,50 — 3,20
CH_3-O-	3,45 — 4,65
[pyrone structure with R, H]	3,90

Literaturverzeichnis: SS. 288—306

Tabelle 2 (Fortsetzung)

Gruppierung	Lage der Signale (in ppm)
benzyl. $-CH_2-CO-CH_2-$ aromat. $-H$	3,90— 4,10 6,20— 7,20
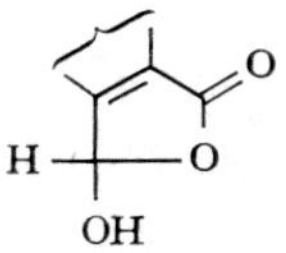 [Struktur mit OH, H, O]	6,80— 7,05
aromat. $-CH(OAc)_2$ aromat. $-CHO$ phenol. $-OH$	7,85— 8,20 10,3—10,6 10,9 —12,5

Die NMR-Spektren von 37 Xanthonen diskutiert SANTESSON (*269*). STEGLICH und LÖSEL (*307*) ziehen die NMR-Spektroskopie zur Bestimmung der Stellung von O-Substituenten bei 1,8-Dihydroxyanthrachinonen heran.

G. Massenspektrometrie

Sowohl Elektronenstoß- als auch Elektronenanlagerungsmassenspektrometrie haben bei der Strukturaufklärung von Flechtenstoffen breite Anwendung gefunden. Beide Methoden liefern nicht nur die Molmasse, sondern auch aus den Fragmentionen wertvolle strukturelle Einzelheiten. Nach HUNECK et al. (*145*) sowie LETCHER et al. (*181*) werden Depside bevorzugt an der Esterbrücke fragmentiert: Schema 1.

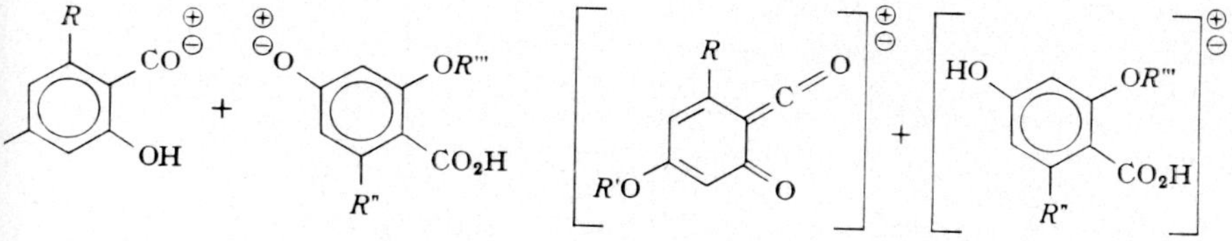

Schema 1. Fragmentierung von Depsiden bei Elektronenstoß und Elektronenanlagerung

Bei Elektronenstoß ist oft der Peak des Fragmentions A besonders intensiv, während bei Elektronenanlagerung der Peak des Fragmentions S intensiver ist, was zur Zuordnung der Peaks bei Depsiden unbekannter Struktur herangezogen werden kann (vgl. Abb. 2).

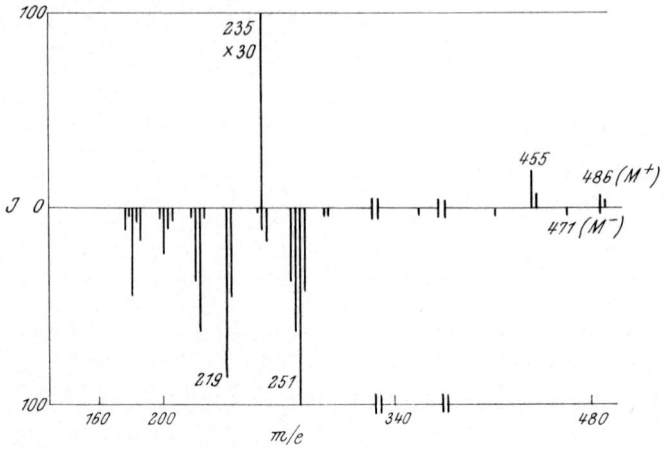

Abb. 2. Massenspektren von Planasäuremethylester

Depsidone spalten sowohl an der Ester- als auch an der Ätherbrücke (vgl. Abb. 3).

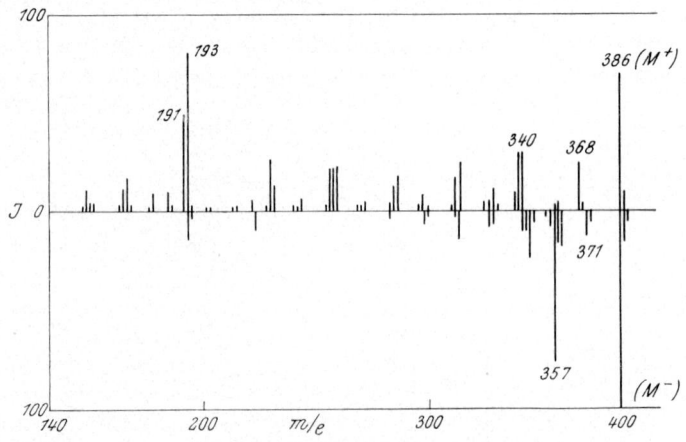

Abb. 3. Massenspektren von Stictinsäure

Bei chlorhaltigen Verbindungen kann aus der Intensitätsverteilung der Molmassenpeaks sofort die Anzahl der Chloratome pro Molekül abgeleitet werden: Abb. 4.

Literaturverzeichnis: SS. 288—306

Die Massenspektren von Pulvinsäurederivaten wurden von LETCHER (*182*, *180*) und die von Xanthonen von SANTESSON (*271*) diskutiert.

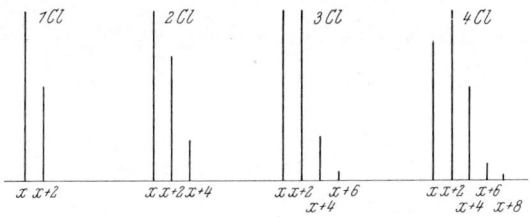

Abb. 4. Muster der Isotopen-Peaks von Verbindungen mit 1 bis 4 Atomen Chlor im Molekül

Bei der „Flechten-Massenspektrometrie" nach SANTESSON (*267*) werden kleine Thallusstücke von Flechten direkt im Massenspektrometer erhitzt, wobei die niedermolekularen Inhaltsstoffe unzersetzt verdampfen und charakteristische Spektren geben, aus denen auf die vorliegenden Verbindungen rückgeschlossen werden kann. Diese Methode scheint besonders zur Identifizierung von chlorhaltigen Xanthonen und Anthrachinonen geeignet zu sein.

H. Röntgenstrukturanalyse

Auch die Röntgenstrukturanalyse wurde zur Aufklärung einer Reihe von Flechtenstoffen herangezogen, und zwar von Acetyldihydro-p-brombenzoylportentol (*103a*), 4-Bromnephroarctin (*225*), Jodacetylvicanicin (*98a*), (+)-Dibromdehydrotetrahydrorugulosin (*174*, *174a*), 6α-p-Brombenzoylzeorin (*217a*), 16β-p-Brombenzoyl-6-ketoleucotylin (*217*, *346*) und 3β,12β-Di-p-brombenzoylpyxinol (*337*).

I. Chemische Methoden

Einige funktionelle Gruppen lassen sich leicht durch Farbreaktionen erkennen: Carbonsäuren mit blauem Lackmuspapier, phenolische OH-Gruppen mit Eisen(III)-chlorid, meta-ständige OH-Gruppen mit Natriumhypochlorit und Aldehyde mit p-Phenylendiamin. Esterbrücken in Depsiden und Depsidonen können mit konz. Schwefelsäure bei 0°, wäßriger oder methanolischer Kalilauge, wäßriger Trinatriumphosphatlösung bei 40° oder mit methanolischer Kaliumcarbonatlösung bei 40° gespalten werden, wobei entweder die freien Säuren, deren Decarboxylierungsprodukte oder die Methylester erhalten werden. Die Ätherbrücke in Depsidonen wird beim Schmelzen mit Kaliumhydroxid oder durch Oxydation mit Chromsäure gespalten. Die Alkalischmelze wird auch zur Fragmentierung von Dibenzofuranen, Chromonen und Xanthonen angewandt (*143*). 1-Hydroxyxanthone geben nach Überführung in die 1,4-Dihydroxyverbindungen bei milder Oxydation Salizylsäuren (*250*).

III. Einteilung der Flechtenstoffe

Die bisher bekannten Flechtenstoffe können nach Tabelle 3 eingeteilt werden.

Tabelle 3. *Einteilung der Flechtenstoffe*

1.	Produkte des Primärstoffwechsels
1.1.	Aminosäuren, Amine und Proteine
1.2.	Polyole
1.3.	Mono-, Oligo- und Polysaccharide
1.4.	Schwefelhaltige Verbindungen
2.	Acetogenine
2.1.	Acetat- und Malonat-Derivate
2.1. 1.	Mono-, di- und tribasische Fettsäuren
2.1. 2.	γ-Lactonsäure-Derivate
2.1. 3.	Cycloaliphatische Verbindungen
2.2.	Aromatische Derivate
2.2. 1.	Orcin-Derivate
2.2. 2.	Naphthochinone
2.2. 3.	Anthrachinone und Anthrone
2.2. 4.	Chromone
2.2. 5.	Xanthone
2.2. 6.	Dibenzofurane
2.2. 7.	Usnin- und Isousninsäure
2.2. 8.	Depside
2.2. 9.	Benzylester-Derivate
2.2.10.	Depsidone
2.2.11.	Depsone
2.2.12.	Bisanthrachinone und Bisanthronyle
2.2.13.	Bisxanthone
2.3.	Mevalonat-Derivate
2.3. 1.	Diterpene
2.3. 2.	Triterpene
2.3. 3.	Steroide
2.3. 4.	Carotinoide
3.	Phenylalanin-Derivate
3.1.	Picroroccellin
3.2.	Polyporsäure
3.3.	Pyxiferin
3.4.	Thelephorsäure
3.5.	Pulvinsäure-Derivate
3.6.	Epanorin und Rhizocarpsäure
4.	Vitamine
5.	Enzyme

IV. Strukturaufklärung und Synthese der Flechtenstoffe

1. Produkte des Primärstoffwechsels

1.1. Aminosäuren, Amine und Proteine

Wie zu erwarten, enthalten Flechten die meisten der auch in höheren Pflanzen nachgewiesenen Aminosäuren (*237, 255, 238, 239, 240, 305, 120b*).

Literaturverzeichnis: SS. 288—306

Für das von HUNECK et al. aus *Roccella vicentina* (Wain.) Wain. (*153a*) und *Roccella canariensis* Darb. (*155a*) isolierte hochschmelzende Neutralprodukt wurden von BOHMAN (*49a*) die beiden folgenden alternativen Cyclotetrapeptidstrukturen vorgeschlagen:

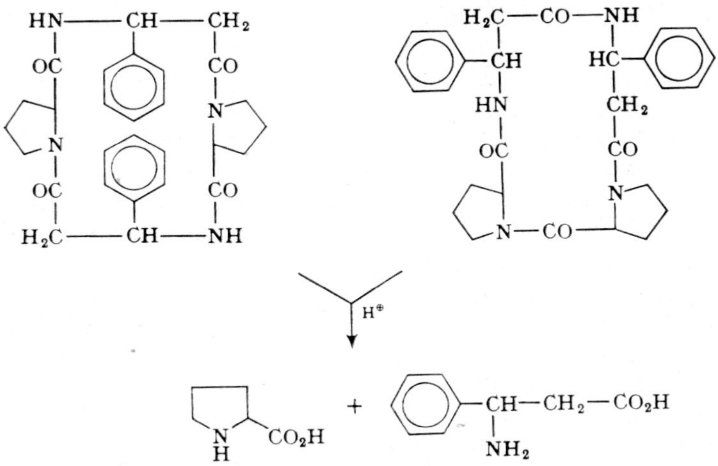

Saure Hydrolyse des Cyclopeptides liefert L-Prolin und D-β-Amino-β-phenylpropionsäure.

BERNARD (*40a*) untersuchte verschiedene Stictaceen auf Amine und fand Mono-, Di- und Trimethylamin in Konzentrationen zwischen 0,02 und 1,2% Trockengewicht in *Lobaria laetevirens* (Lightf.) Zahlbr., *Sticta sylvatica* (Huds.) Ach., *Sticta fuliginosa* (Dicks.) Ach. und *Sticta limbata* (Sm.) Ach. Dagegen enthält *Lobaria pulmonaria* (L.) Hoffm. nur Spuren von Trimethylamin.

SOLBERG (*303*) untersuchte 44 verschiedene Flechtenarten und fand in 40 einen durchschnittlichen Proteingehalt zwischen 1,6 und 11,4% vom Trockenmaterial; dagegen liegt der Proteingehalt von *Xanthoria parietina* (L.) Th. Fr., *Anaptychia fusca* (Huds.) Vain., *Lobaria pulmonaria* (L.) Hoffm. und *Lobaria scrobiculata* (Scop.) DC. zwischen 16,0 und 20,3% der Trockensubstanz. Auf den Nährwert von Flechten hat SANKARA SUBRAMANIAN (*253*) hingewiesen.

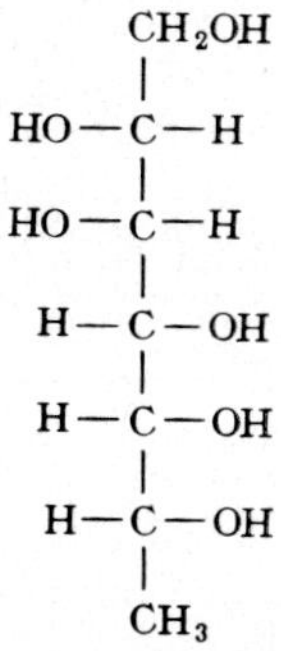

1.2. Polyole

Aus Flechten sind folgende Polyole bekannt: Glycerin, Ribit, D-Arabit, meso-Erythrit, D-Mannit, myo-Inosit, D-Volemit und D-Siphulit (*80*). D-Siphulit (= 1-Desoxy-D-glycero-D-taloheptit) (I) aus *Siphula ceratites* (Fr.)

$$\begin{array}{c} CH_2OH \\ | \\ HO-C-H \\ | \\ HO-C-H \\ | \\ H-C-OH \\ | \\ H-C-OH \\ | \\ H-C-OH \\ | \\ CH_3 \end{array}$$

(I)

Th. Fr. wurde aus D-Mannose über D-Glycero-D-taloheptose und 1-Desoxy-1-nitro-D-glycero-D-taloheptit synthetisiert (*185*).

1.3.] *Mono-, Oligo- und Polysaccharide*

Die Mono- und Disaccharide D-Glucose, D-Fructose, D-Galactose, D-Xylose, D-Tagatose, Arabinose, Saccharose und Trehalose (*80*) dürften primär der Photosynthese der Mycobionten entstammen. Darüber hinaus wurden einige Zuckeralkohol-Glykoside in Flechten gefunden, nämlich 3-O-β-D-Glucopyranosyl-D-mannit (2) (*187, 188*), dessen Struktur durch Synthese aus 4-O-β-D-Glucopyranosyl-D-mannose bewiesen wurde, 3-O-β-D-Galaktofuranosyl-D-mannit (= Peltigerosid) (3) (*188, 235*) und Umbilicin (4) (*189, 190, 186, 191, 192*). (4) wurde aus Aldit auf folgendem

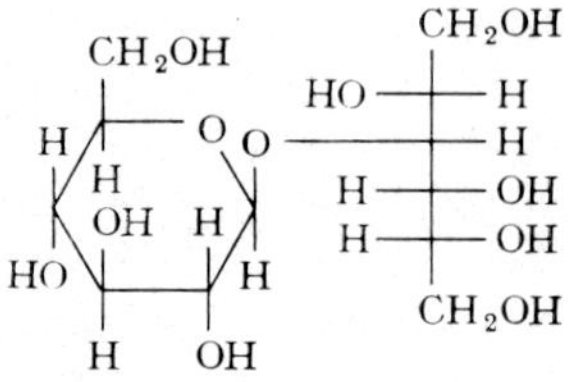

Wege synthetisiert: Benzyl-3,4-O-isopropyliden-β-arabinopyranosid (5) liefert mit Allylbromid und nachfolgender Behandlung mit Säure Benzyl-2-O-allyl-β-D-arabinopyranosid (6), das bei weiterer saurer Hydrolyse und Reduktion mit Natriumborhydrid 2-O-Allyl-D-arabinit (7) gibt. Benzylierung von (7), Isomerisierung zu (8) und Hydrolyse führt zu 1,3,4,5-Tetra-O-benzyl-D-arabinit (9), das mit 3,5,6-Tri-O-acetyl-1,2-O-methyl-orthoacetyl-α-D-galaktofuranose (10) zum Galaktofuranosid (11) umgesetzt wird. (11) wird entacetyliert und zu 2-O-β-D-Galaktofuranosyl-D-arabinit (4) hydriert (*44*) (Schema 2).

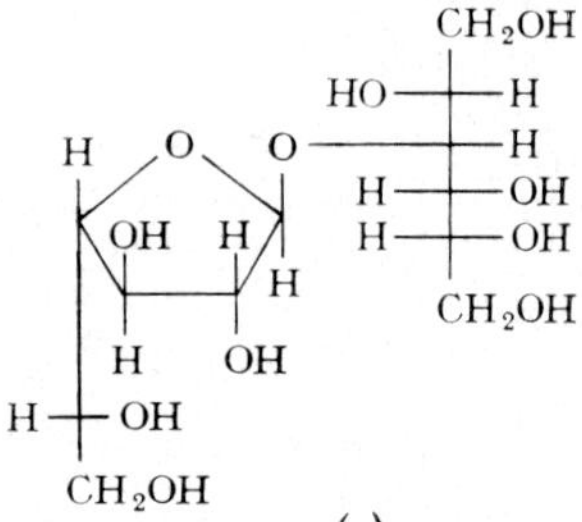

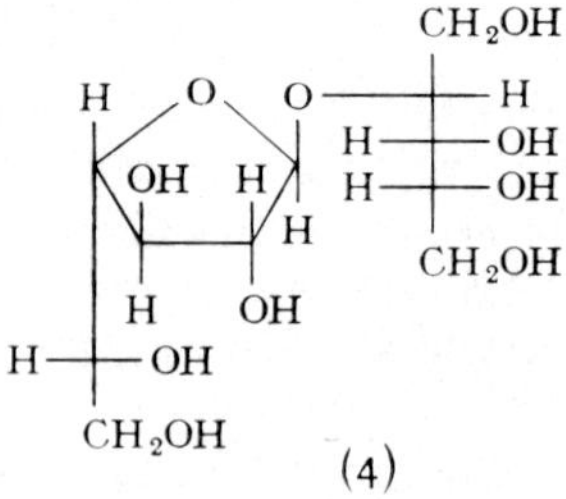

Isolichenin, das in verschiedenen Flechten vorkommt, ist ein lineares Polysaccharid, das nur aus Glucose-Einheiten in α-(1 → 3) und α-(1 → 4)-Verknüpfung im Verhältnis 45:55 aufgebaut ist (*68, 227, 104*).

Literaturverzeichnis: SS. 288—306

Lichenin ist Poly-β-D-glucopyranose mit β-(1 → 3) und β-(1 → 4)-glykosidischen Bindungen im Verhältnis 27 : 73 *(80)*. Wie der enzymatische Abbau zeigt, unterscheidet es sich in der Feinstruktur von Poly-β-D-glucosan aus Getreidesamen *(229)*.

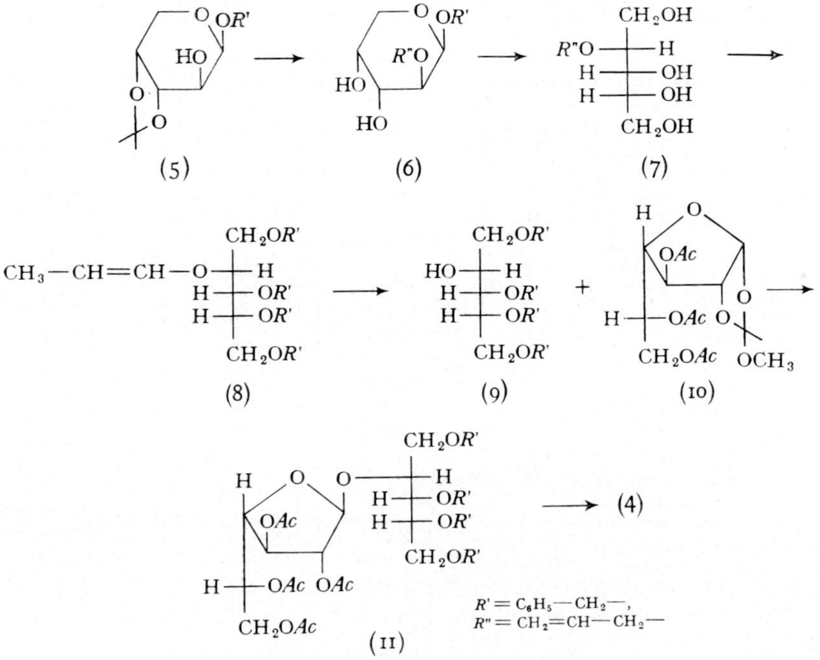

Schema 2. Synthese von Umbilicin

Pustulan, ein Polysaccharid aus *Lasallia pustulata* (L.) Mérat und *Umbilicaria hirsuta* (Sw.) Ach. em. Frey *(96)* liefert bei Hydrolyse Glucose, Gentiobiose, Gentiotriose und Gentiotetraose und ist daher ein lineares Glucan mit β-(1 → 6)-Bindungen *(184)*. Mit der partiellen Methylierung von Pustulan hat sich NORRMAN *(224b)* beschäftigt.

1968 berichteten SHIBATA et al. *(289)* über die Isolierung von zwei neuen Glucanen aus *Gyrophora esculenta* Miyoshi und *Lasallia papulosa* (Ach.) Llano und deren Inhibitorwirkung auf das Wachstum des Sarkoms 180 bei Mäusen. Weitere Untersuchungen zeigten, daß beide Glucane identisch sind und ein zu etwa 2% O-acetyliertes Poly-β-(1 → 6)-glucan vorliegt, dessen Molgewicht mittels Ultrazentrifuge zu 20000 bestimmt wurde *(288, 224)*. Dieses antitumorwirksame Glucan kommt auch in *Umbilicaria angulata* Tuck., *Umbilicaria caroliniana* Tuck. und *Umbilicaria polyphylla* (L.) Baumg. vor *(224a)*.

Parmelia caperata (L.) Ach. enthält mehrere Polysaccharide, von denen das eine, PC-3, elektrophoretisch rein isoliert wurde. Es zeigt eine positive einfache ORD-Kurve und keine Färbung mit Jod. Methanolyse des Methyläthers von PC-3 gibt gleiche Mengen 2,3,6-Tri-O-methyl- und Methyl-2,4,6-tri-O-methyl-D-glucopyranosid neben wenig Methyl-2,3,4,6-tetra-O-methyl-D-glucopyranosid. Daraus, aus dem IR-Spektrum, der Endgruppenbestimmung und der Perjodatoxydation von PC-3 folgt ein lineares α-(1 → 3)- und α-(1 → 4)-Glucan der Struktur (12).

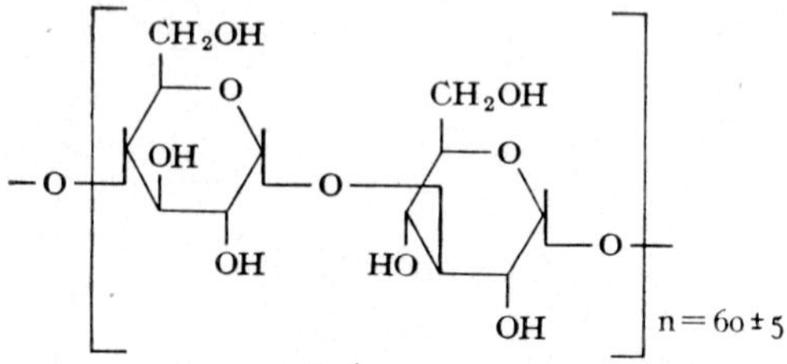

(12)

PC-3 ist ebenfalls antitumorwirksam (*324*).

1.4. Schwefelhaltige Verbindungen

Bis jetzt wurden nur drei schwefelhaltige organische Verbindungen aus Flechten isoliert: Dimethylsulfon aus *Cladonia deformis* Hoffm. (*63*), das interne Salz von Cholinschwefelsäure aus verschiedenen *Roccella*-Arten (*183*), *Dermatiscum thunbergii* (Ach.) Nyl. (*129*) und *Cladonia convoluta* (Lam.) P. Cout. (*103*) sowie Thioäthanolamin aus *Cladonia convoluta* (Lam.) P. Cout. (*103*).

2. Acetogenine

2.1. Acetat- und Malonat-Derivate

2.1.1. Mono-, di- und tribasische Fettsäuren

SCHADE (*276a*) sowie SCHADE und SEITZ (*276b*) fanden weitere Vorkommen von Calciumoxalat in Flechten.

(±)-Norrangiformsäure wurde von ÅKERMARK (*14*) nach Schema 3 synthetisiert: Isopren und Fumarsäuredimethylester liefern in einer DIELS-ALDER-Reaktion 4-Methyl-4-cyclohexen-1,2-dicarbonsäuredimethylester (13), dessen trans-Form ozonisiert wird. Das Ozonid wird mit aktivem Zinkstaub zersetzt, die gebildete threo-3,4-Dimethoxycarbonyl-6-oxoheptansäure (14) elektrolytisch mit Myristinsäure gekuppelt, das Reaktionsgemisch hydrolysiert und die entstandene threo-2-Oxo-4,5-nonadecandicarbonsäure (15) mit Hypobromit zur (±)-Norrangiformsäure (16) oxydiert.

Literaturverzeichnis: SS. 288—306

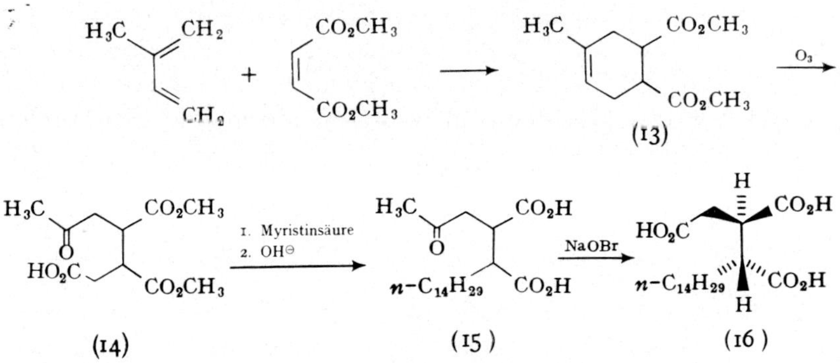

Schema 3. Synthese der (±)-Norrangiformsäure

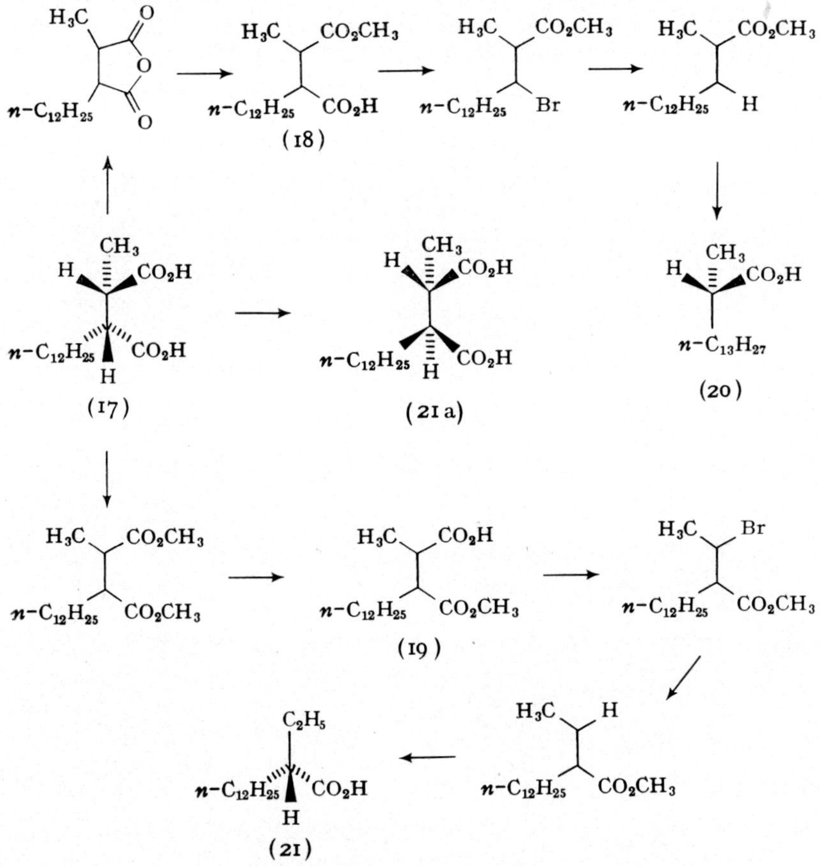

Schema 4. Bestimmung der Absolutkonfiguration von (+)-Roccellsäure

Auf dem eben dargelegten Wege erhielt ÅKERMARK (*15*) aus (—)-4-Methyl-4-cyclohexen-(1 R : 2 R)-dicarbonsäure (—)-Norrangiformsäure (= (—)-1.(2 R : 3 R)-Heptadecantricarbonsäure), der damit die in (16) gezeigte Absolutkonfiguration zukommt. Die Stellung der Methylestergruppe in Rangiformsäure (= Monomethylester von (16)) ist immer noch unbekannt.

Die absolute Konfiguration von (+)-Roccellsäure (17) bestimmte ÅKERMARK (*13*) durch Abbau der beiden isomeren Methylester (18) und (19) zu (+)-2 S-Methylpentadecansäure (20) bzw. (+)-2 S-Äthyltetradecansäure (21). (+)-Roccellsäure ist daher (+)-2 S-Methyl-3 R-dodecylbernsteinsäure und besitzt im Gegensatz zur Rangiformsäure erythro-Konfiguration (Schema 4). Bei der Isomerisierung mit konzentrierter Schwefelsäure wird Roccellsäure teilweise racemisiert und teilweise in threo-2-Methyl-3-dodecylbernsteinsäure (21a) umgewandelt, deren IR-Spektrum von dem der Roccellsäure deutlich verschieden ist: ein weiterer Hinweis für die erythro-Konfiguration (17) der Roccellsäure (*15a*).

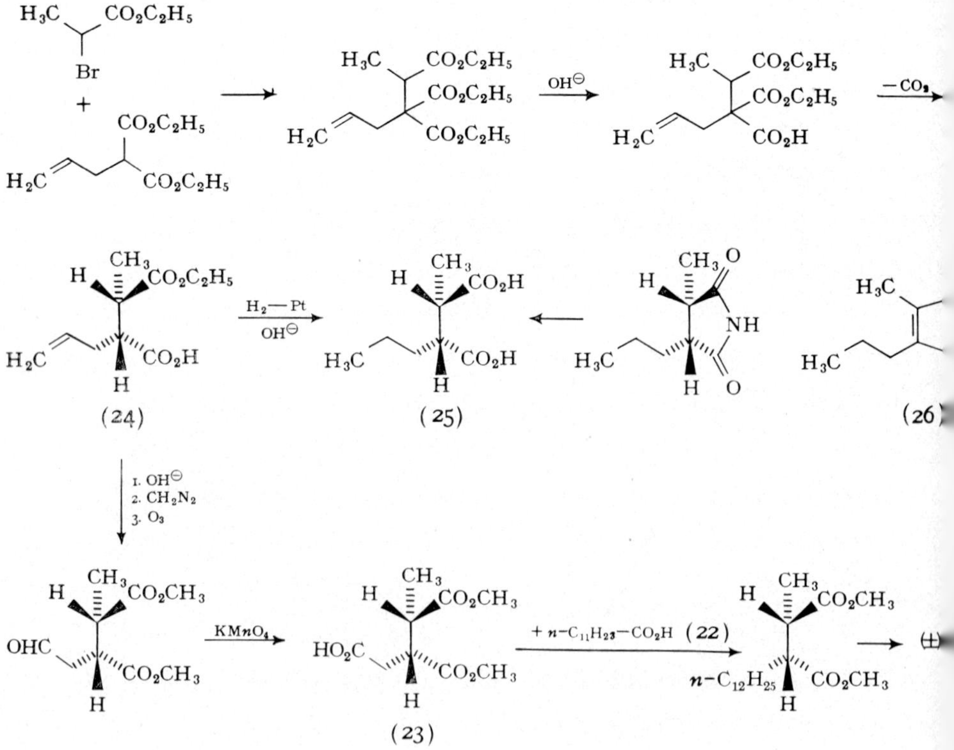

Schema 5. Synthese von (±)-Roccellsäure

Literaturverzeichnis: SS. 288—306

ÅKERMARK und JOHANSSON (*17*) gelang dann auch die Synthese von (±)-Roccellsäure durch anodische Kupplung von Laurinsäure (22) mit erythro-3,4-Dimethoxycarbonylpentansäure (23), die aus erythro-2-Allyl-3-methylbernsteinsäure (24) durch Ozonolyse des Dimethylesters gewonnen wurde. Die Konfiguration von (24) wurde durch Hydrierung zu erythro-2-Methyl-3-propylbernsteinsäure (25) bewiesen, die auf stereoselektivem Wege durch katalytische Reduktion und anschließende Hydrolyse aus Methylpropylmaleimid (26) zugänglich ist (*18*) (Schema 5).

Bei den aus zahlreichen Flechten isolierten Hydroxyfettsäuren handelt es sich nach SOLBERG (*301*, *302*) um ein Gemisch aus 9,10,12,13-Tetrahydroxy-heneicosansäure und 9,10,12,13-Tetrahydroxy-docosansäure. *Parmelia centrifuga* (L.) Ach. enthält wahrscheinlich Tetrahydroxytricosansäure, $C_{23}H_{46}O_6$. WAGNER und FRIEDRICH (*334*) wiesen in *Hypogymnia physodes* (L.) Nyl., *Usnea barbata* (L.) Wigg. und *Cetraria islandica* (L.) Ach. folgende ungesättigten Fettsäuren nach: C_{16} (1 Δ), C_{16} (3 Δ), C_{18} (1 Δ), C_{18} (2 Δ), C_{18} (4 Δ), C_{20} (1 Δ) und C_{22} (1 Δ).

2.1.2. γ-Lactonsäurederivate

(±)-Protolichesterinsäure [(±)-(27)] wurde von VAN TAMELEN und BACH (*327*) nach Schema 6 synthetisiert: trans-Hexadec-2-ensäuremethylester (28) wurde über das Epoxid mit Malonsäuredimethylester zum α,β-Dimethoxycarbonyl-γ-tridecyl-γ-butyrolacton (29) kondensiert und das Monokaliumsalz der Lactondisäure mit Formaldehyd und Diäthylamin behandelt.

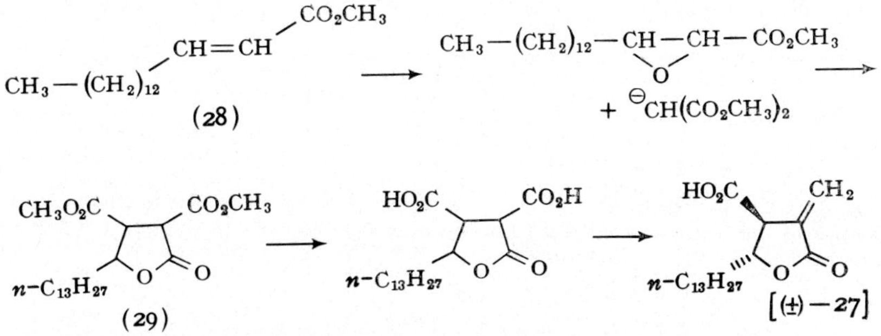

$$CH_3-(CH_2)_{12}-CH=CH-CO_2CH_3 \quad (28)$$

$$\longrightarrow CH_3-(CH_2)_{12}-CH-CH-CO_2CH_3 \quad + \quad {}^{\ominus}CH(CO_2CH_3)_2$$

$$CH_3O_2C \quad CO_2CH_3 \qquad HO_2C \quad CO_2H \qquad HO_2C \quad CH_2$$
$$n\text{-}C_{13}H_{27} \quad O \quad O \quad \longrightarrow \quad n\text{-}C_{13}H_{27} \quad O \quad O \quad \longrightarrow \quad n\text{-}C_{13}H_{27} \quad O \quad O$$
$$(29) \qquad\qquad\qquad\qquad\qquad\qquad\qquad\qquad [(\pm)-27]$$

Schema 6. Synthese von (±)-Protolichesterinsäure

Die Struktur von (+)-Roccellarsäure aus *Roccellaria mollis* (Hampe) Zahlbr. wurde im Sinne von (30) aufgeklärt und durch Partialsynthese bewiesen (*151*): (+)-Protolichesterinsäure wird zu (+)-Dihydroprotolichesterinsäure hydriert und deren Methylester mit Natriummethylat zu (+)-Neodihydroprotolichesterinsäuremethylester isomerisiert, der bei gleichzeitiger Verseifung (+)-Roccellarsäure liefert (Schema 7).

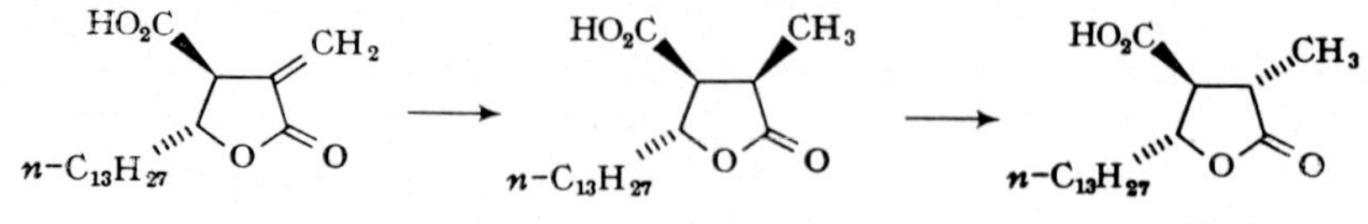

Schema 7. Partialsynthese der (+)-Roccellarsäure

Die in (30) angegebene Absolutkonfiguration folgt nach der Regel von HUDSON und KLYNE aus dem positiven COTTON-Effekt des (+)-Roccellarsäuremethylesters. Da Neodihydroprotolichesterinsäure mit Protolichesterinsäure und diese mit allo-Protolichesterinsäure und allo-Dihydroprotolichesterinsäure sowie Nephromopsinsäure verknüpft ist (327), kommen den rechtsdrehenden Antipoden der genannten Verbindungen folgende Absolutkonfigurationen zu:

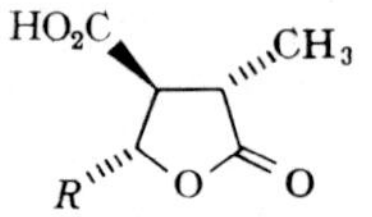

(+)-Roccellarsäure

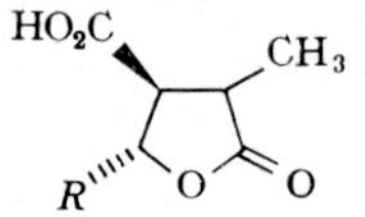

(+)-Dihydroprotolichesterinsäure

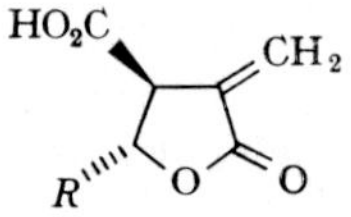

(+)-Protolichesterinsäure

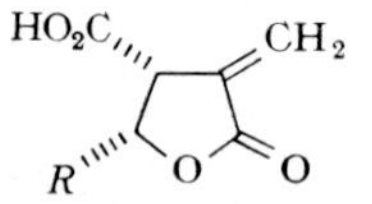

(+)-allo-Protolichesterinsäure

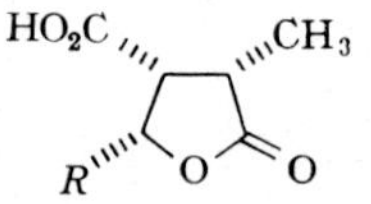

(+)-allo-Dihydroprotolichesterinsäure

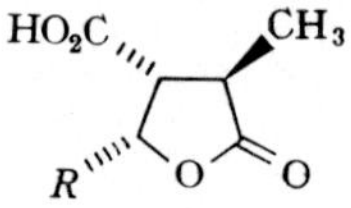

(+)-Nephromopsinsäure

$R = n\text{-}C_{13}H_{27}-$

Aus dem negativen COTTON-Effekt von (−)-Lichesterinsäure leitete BOLL (50) für diese Lactonsäure die absolute Konfiguration (31) ab.

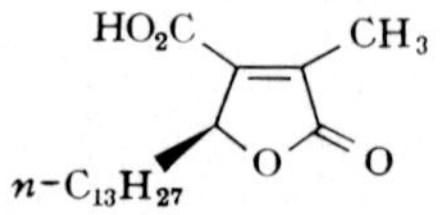

(31)

Die bereits von ZOPF (354) aus *Acarospora chlorophana* (Wg.) Mass. isolierte Pleopsidsäure erwies sich als Gemisch zweier Lactonsäuren, die von SANTESSON (261) strukturell aufgeklärt und Acaranosäure (32) und Acarenosäure (33) genannt wurden.

Literaturverzeichnis: SS. 288—306

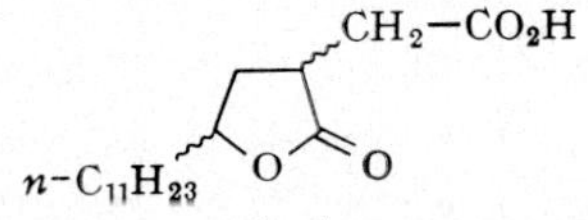

Ein Gemisch von zwei diastereomeren Racematen von (32) synthetisierte SANTESSON (*261*) nach Schema 8.

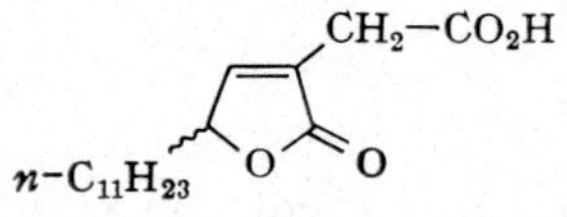

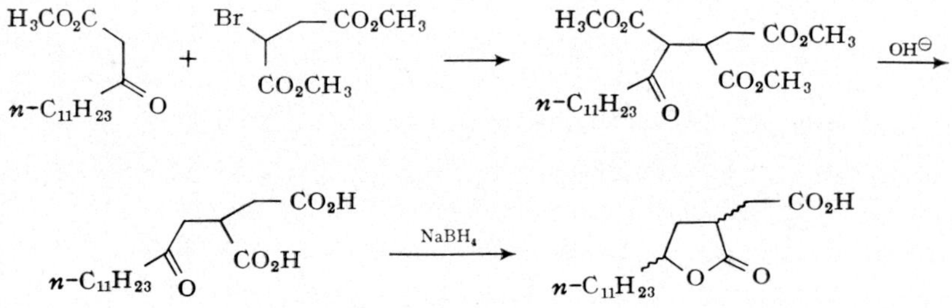

Schema 8. Synthese von zwei diastereomeren Racematen der Acaranosäure

2.1.3. Cycloaliphatische Verbindungen

Portentol (34) aus *Roccella galapagoensis* Follm. (vorher *Roccella portentosa* (Mont.) Darb.) (*156*), *Dirina repanda* (Nyl.) Fr. (*166*) und *Roccellina condensata* Darb. (*153*) ist das erste Polyketid bekannter Struktur aus Flechten (*2*, *3*). Portentol (34) und Acetylportentol (35), das in *Roccella fuciformis* DC. (*160*) und *Roccella maderensis* (Stein.) Stein. (*108*) vorkommt, gehen beim Erhitzen mit äthanolischer Kalilauge in Decarboxyportentol (36) über, das sich zu Decarboxyportenton (37) oxydieren läßt. (37) liefert beim Erhitzen mit verdünnter Schwefelsäure über die Zwischenstufe (38) das Mesitolderivat (39), das nach Methylierung zu (40) und dreistufigem oxydativem Abbau (OsO_4, $NaJO_4$ und $H_2O_2/OH^{\ominus}$) das Keton (41) gibt. (41) ist mit aus Mesitol zugänglichem Acetylmesitolmethyläther identisch. Nach der von FERGUSON und MACKAY (*103a*) durchgeführten Röntgenstrukturanalyse hat Acetyldihydro-p-brombenzoylportentol die in (42) gezeigte Absolutkonfiguration. Damit wird die mittels chemischer Methoden für Portentol abgeleitete Stereostruktur bestätigt (Schema 9).

Vermutlich handelt es sich bei dem von ZOPF (*354*) aus *Arthonia impolita* (Ehrh.) Borr. (syn. *Leprantha impolita* Ehrh.) isolierten Lepranthin ebenfalls um ein Polyketid (*161*).

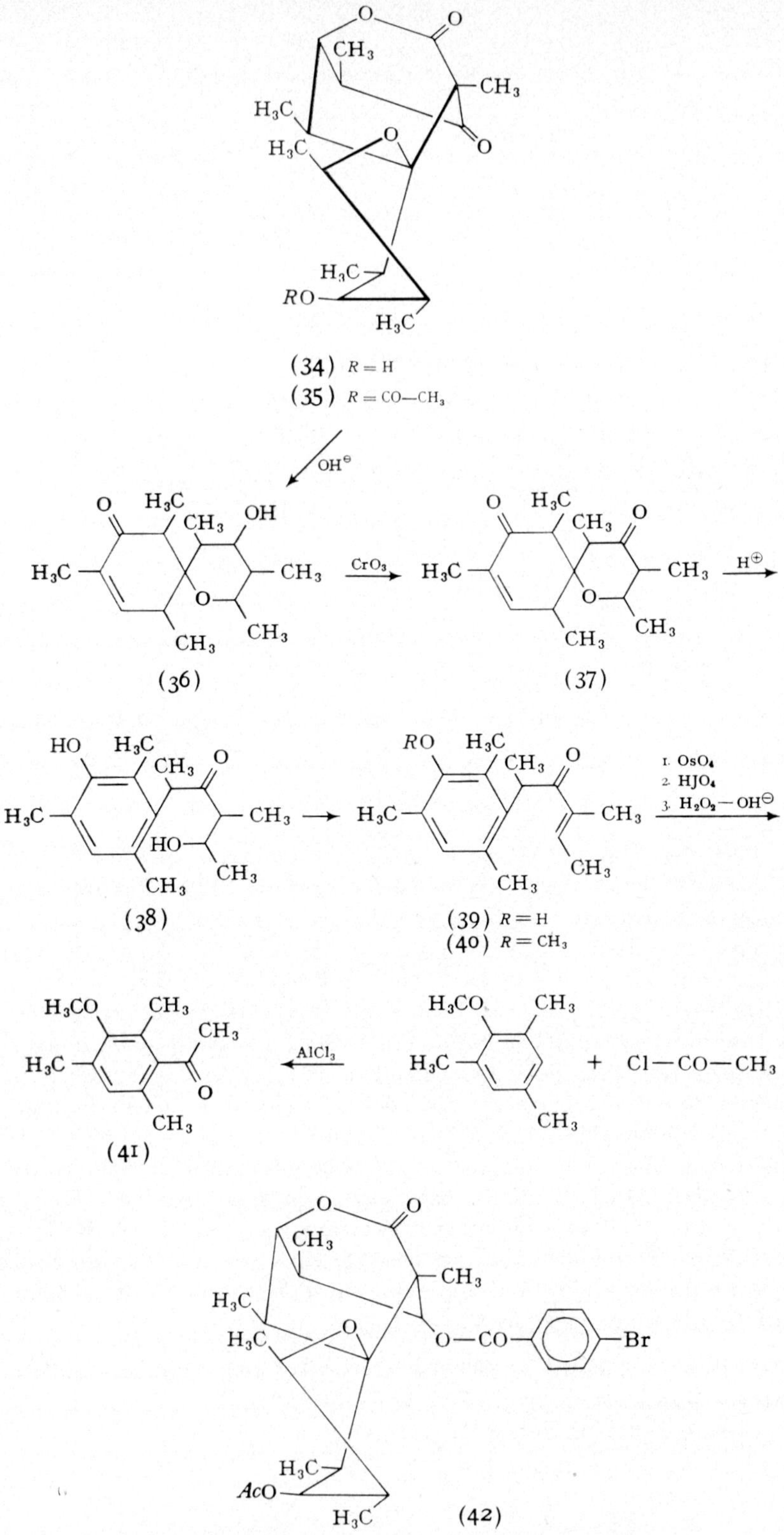

Schema 9. Abbau von Portentol

2.2. Aromatische Derivate

2.2.1. Orcin Derivate

Obwohl sich zahlreiche aromatische Flechtenstoffe von Phenolcarbonsäuren ableiten, ist es bisher nur in wenigen Fällen gelungen, die zu Grunde liegenden Phenolcarbonsäuren bzw. deren Derivate aus Flechten zu isolieren. Möglicherweise kommen diese nicht frei vor, sondern sind in einem Enzymkomplex gebunden. Folgende Orcinderivate wurden in Flechten gefunden: Orcin (257), Orsellinsäureäthylester (1), β-Orcincarbonsäuremethylester (215) und (+)-Montagnetol (257).

Die bereits von SONN (305a) durchgeführte Synthese der Orsellinsäure wurde von KLOSS und CLAYTON (173) durch Hydrolyse des Dibromorsellinsäureäthylesters mit Schwefelsäure zu Dibromorsellinsäure und von SANTESSON (272b) durch Enthalogenierung von Dibromorsellinsäure mit Raney-Nickel verbessert. Nach SANTESSON (272b) entsteht bei der Bromierung von Dihydroorsellinsäureäthylester Dibromorsellinsäureäthylester und nicht, wie KLOSS und CLAYTON angeben, Dibromdihydroorsellinsäureäthylester. Nach SANTESSON (272b) wird Orsellinsäure in 30% Ausbeute neben 10% p-Orsellinsäure und 10% 2,4-Dihydroxy-6-methylisophthalsäure bei der Umsetzung von Orcin mit Magnesiummethylcarbonat in Dimethylformamid erhalten. Chlorierung von Orsellinsäure mit Sulfurylchlorid gibt nur 5-Chlororsellinsäure, Jodierung mit Jod und Quecksilberoxid 5-Jodorsellinsäure (272b).

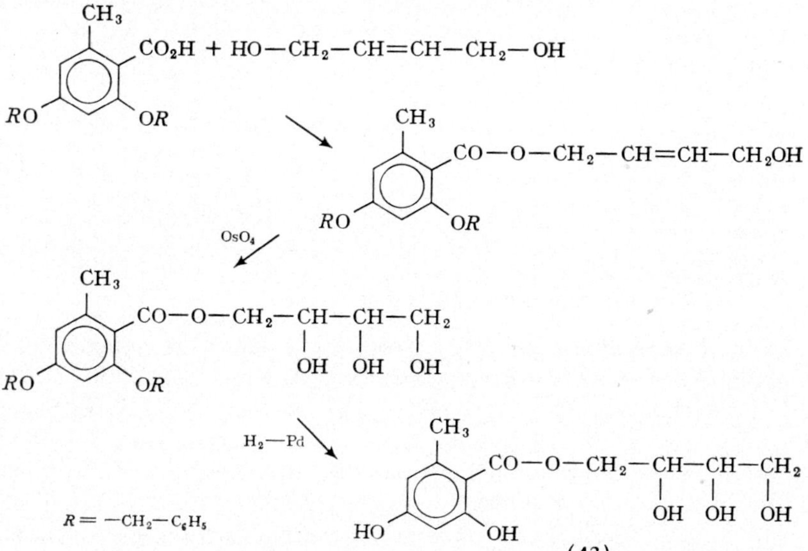

(43)

$R = -CH_2-C_6H_5$

Schema 10. Synthese von (±)-Montagnetol

Eine einfache Synthese von Olivetol stammt von BAECKSTRÖM und SUNDSTRÖM (*27b*).

MANAKTALA *et al.* (*198*) synthetisierten (±)-Montagnetol (**43**) aus dem Ester von Di-O-benzylorsellinsäure und 2-Buten-1,4-diol durch Hydroxylierung mit Osmiumtetroxid und nachfolgende Entbenzylierung (Schema 10).

Der aus verschiedenen *Roccella-, Lecanora-* und *Variolaria-*Arten durch Behandeln mit Ammoniak und Luft gewonnene Lackmus-Farbstoff ist nach Untersuchungen von MUSSO (*36*) ein Gemisch aus komplexen polymeren Komponenten mit 7-Oxo-phenoxazon-2-Chromophoren. Diese Chromophoren bilden im sauren Bereich ein rotes Kation und im alkalischen Bereich ein blaues Anion (Schema 11).

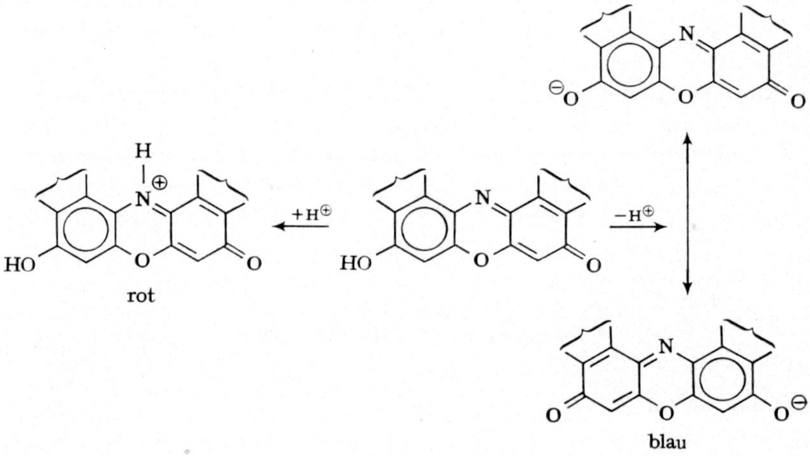

Schema 11. Lackmus-Chromophore bei verschiedenem pH

2.2.2. Naphthochinone

Für Rhodocladonsäure, den Apothecienfarbstoff der rotfrüchtigen *Cladonia-*Arten (Sektion *Cocciferae*) war von ASAHINA und SHIBATA (*26*) die Anthrachinonstruktur (**44**) vorgeschlagen worden. Nach BAKER und BULLOCK (*28*) hat Rhodocladonsäure jedoch die massenspektrometrisch ermittelte Summenformel $C_{15}H_{10}O_8$ und die Naphthochinonstruktur (**45**), wofür auch die beim Erhitzen von (**45**) mit Kaliumcarbonat erhaltenen Spaltprodukte (**46**) und (**47**) sprechen. Neuerdings wird jedoch auch die Chromonstruktur (**47a**) für Rhodocladonsäure in Betracht gezogen (*324a*).

Literaturverzeichnis: SS. 288—306

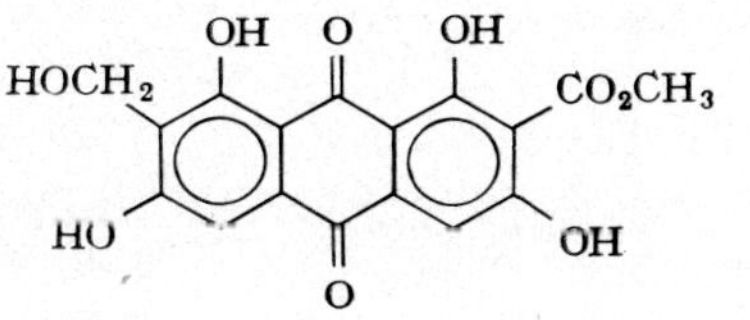

(44) 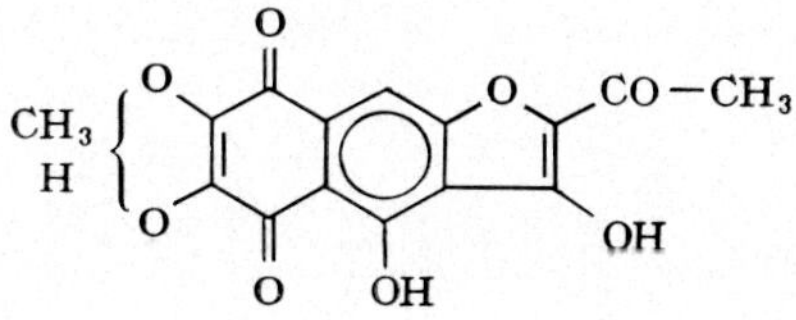 (45) 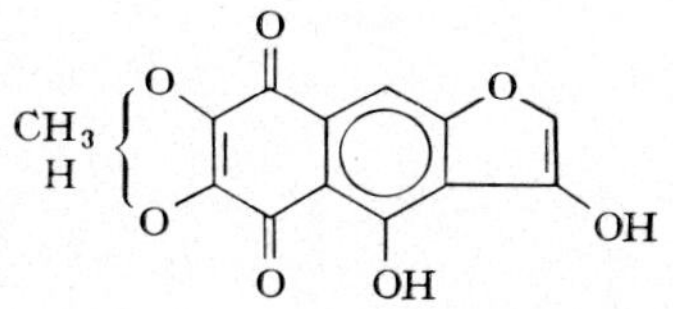 (46) 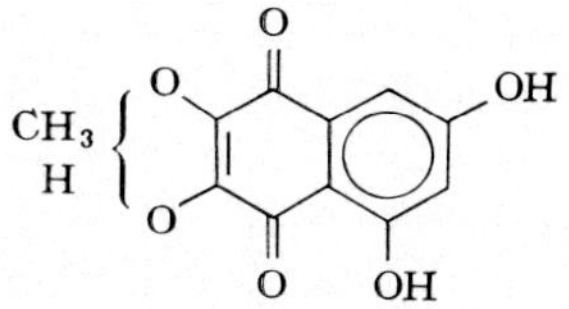 (47) (47a)

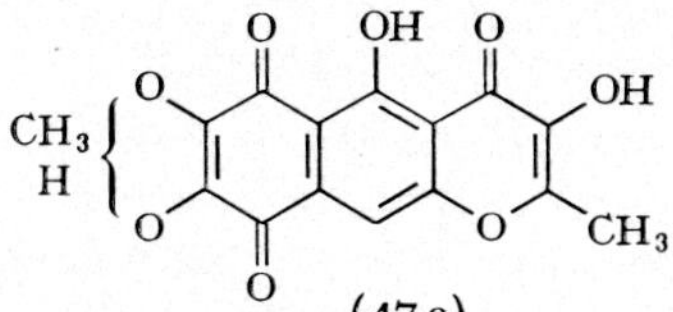

2.2.3. Anthrachinone und Anthrone

Die Struktur des von ZOPF (*354*) aus *Sphaerophorus fragilis* Pers. und *Sphaerophorus globosus* Vain. isolierten Fragilins wurde von BRUUN et al. (*62*) im Sinne von (48) aufgeklärt und von SARGENT et al. (*274*) durch Synthese bewiesen. Chlorierung von Parietin (49) mit 1 Mol. Chlor gibt 5-Chlorparietin (50), mit 2 Mol. Chlor 4,5-Dichlorparietin (51) und mit überschüssigem Chlor 4,5,7-Trichlorparietin (52), das mit Hydrazinhydrat und Palladiumkohle und nachfolgender Methylierung Fragilindimethyläther (53) liefert. (53) geht bei partieller Entmethylierung in Fragilin (48) und bei vollständiger Entmethylierung in 7-Chloremodin über. (52) wird mit überschüssigem Natriumdithionit wieder zu Parietin enthalogeniert (Schema 12).

Inzwischen wurden weitere chlorhaltige Anthrachinone aus Flechten isoliert: 7-Chloremodin (54), 7-Chloremodin-1-methyläther (55) und 7-Chloremodin-1,6-dimethyläther (56) aus *Nephroma laevigatum* Ach. (*38*), 5,7-Dichloremodin (57) und (54) aus *Anaptychia obscurata* (Nyl.) Vain. (*352*), Papulosin (= 7-Chlor-5-hydroxyemodin) (58) und (54) aus *Lasallia papulosa* (Ach.) Llano (*118*), 7-Chlor-4-hydroxyemodin (59) und (58) aus *Lasallia papulosa* (Ach.) Llano var. *rubiginosa* Pers. (*48*) sowie (54) aus *Caloplaca arenaria* (Pers.) Müll.-Arg. und *Caloplaca percrocata* (Arn.) Stein. (*47*).

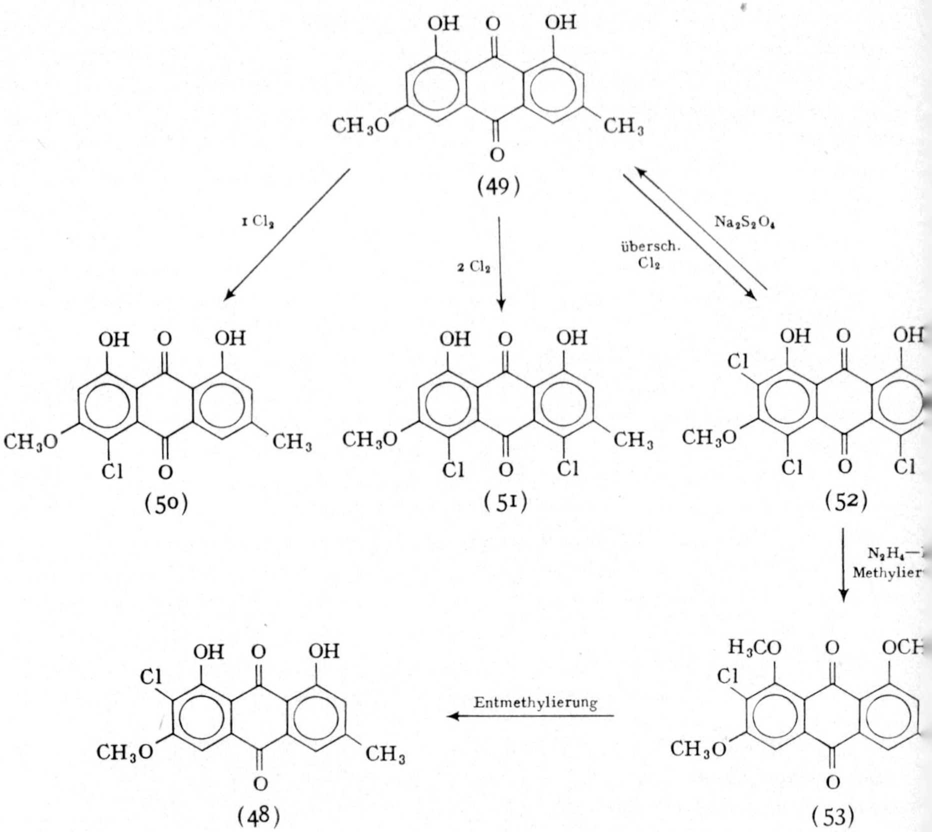

Schema 12. Synthese von Fragilin

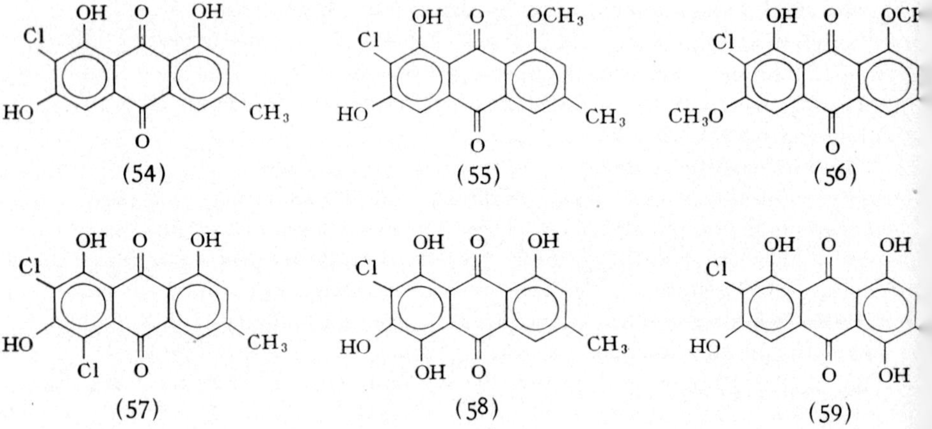

Emodin (60) selbst kommt ebenfalls in *Nephroma laevigatum* Ach. (*38*) und in *Xanthoria parietina* (L.) Beltram (*232*) sowie *Fulgensia fulgida* (Nyl.) Szat. (*131*) vor, Chrysophanol (61) in *Acroscyphus sphaerophoroides* Lév. (*293*), Erythroglaucin (61a) in zahlreichen *Xanthoria-*, *Teloschistes-*, *Caloplaca-* und *Protoblastenia-*Arten (*308a*, *131a*) und Emodinaldehyd (61b) in *Xanthoria aureola* (Ach.) Erichs., *Caloplaca bryochrysion* Poelt, *Caloplaca cinnamomea* (Th. Fr.) Oliv., *Caloplaca leucoraea* (Ach.) Deichm. und *Caloplaca tetraspora* (Nyl.) Oliv.; (61b) ist synthetisch durch Oxydation von Citreoroseintriacetat mit Dimethylsulfoxid in Acetanhydrid zugänglich (*131a*).

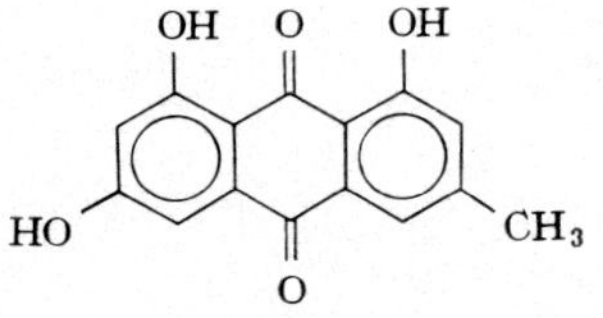

(60) 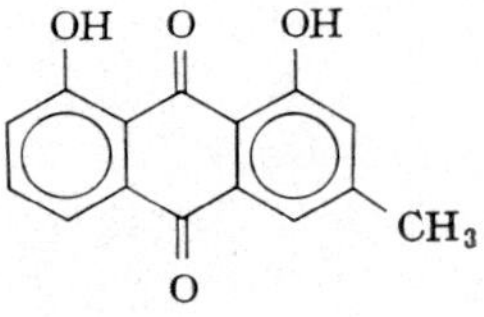 (61)

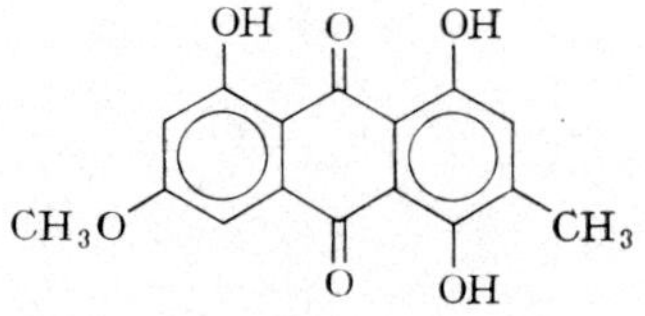

(61a) (61b)

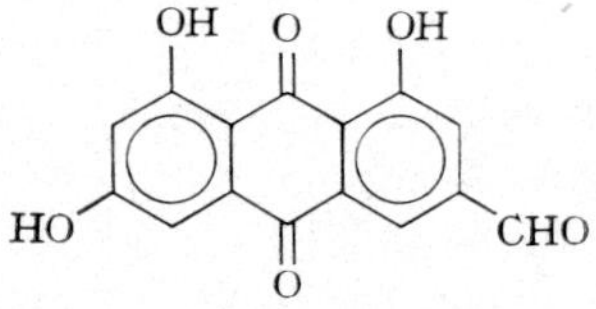

Xanthoria parietina (L.) Beltram enthält außer Emodin und Parietin noch Parietinsäure (62) (*100*, *232*), Fallacinal (63), Fallacinol (= Teloschistin) (64), Citreorosein (65) und Emodinsäure (66) (*232*).

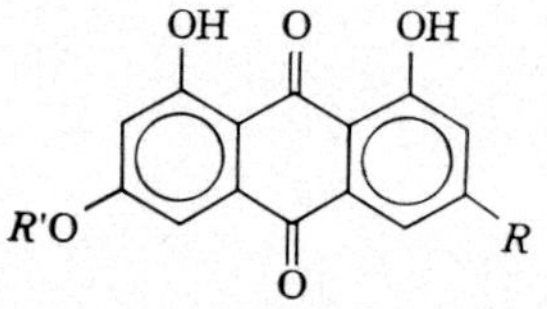

	R	R'
(62)	CO_2H	CH_3
(63)	CHO	CH_3
(64)	CH_2OH	CH_3
(65)	CH_2OH	H
(66)	CO_2H	H

Die Struktur von Fallacinal und Fallacinol wurde von MURAKAMI (*214*) sowie SESHADRI et al. (*220*, *221*, *222*, *236*) durch Synthese bewiesen. Nach NEELAKANTAN et al. (*222*) wird Diacetylparietin (67) mit N-Bromsuccinimid bromiert, das Bromderivat (68) mit Acetanhydrid-Silberacetat umgesetzt und das resultierende Triacetat (69) mit Schwefelsäure zu (64) hydrolysiert (Schema 13). Die von ASAHINA und FUZIKAWA (*25*)

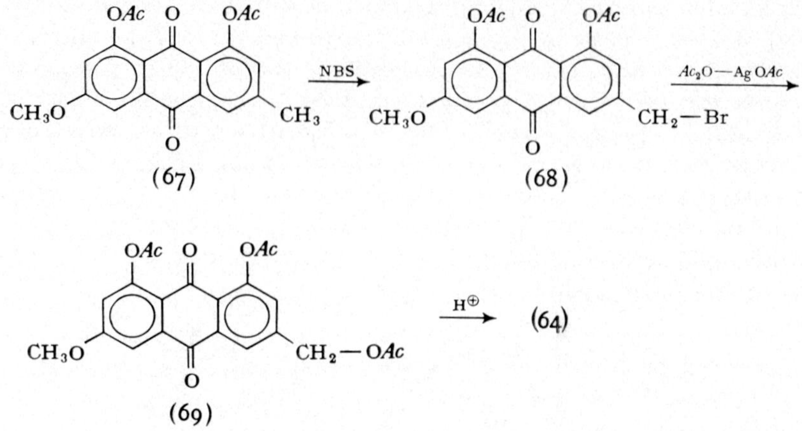

Schema 13. Synthese von Fallacinol

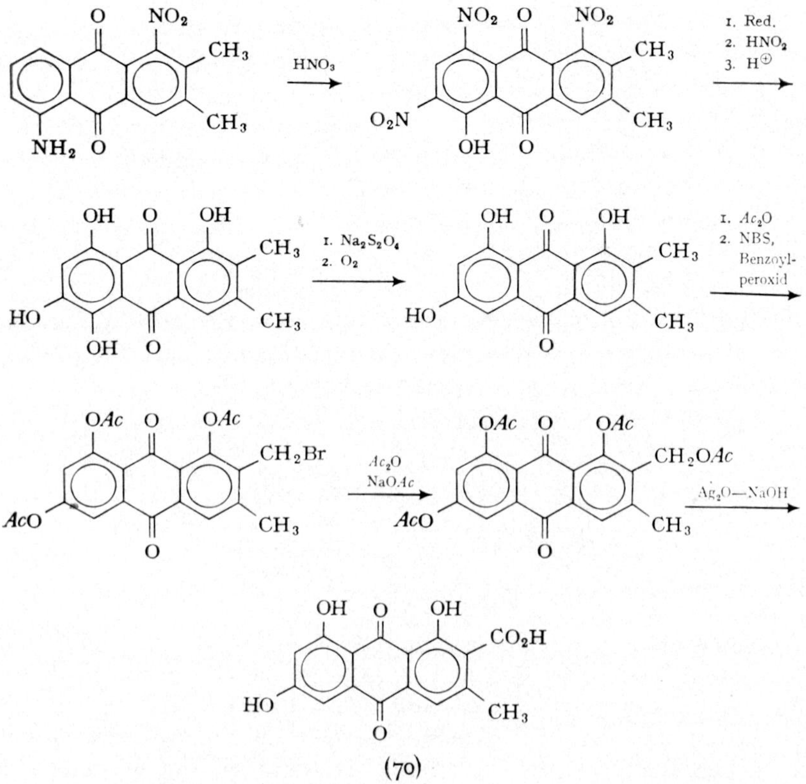

Schema 14. Synthese von Endocrocin nach JOSHI *et al.* (171)

Literaturverzeichnis: SS. 288—306

für Endocrocin aufgestellte Struktur (70) wurde durch die beiden Synthesen nach Schema 14 (*171*) und Schema 15 (*309*) bewiesen. FRANCK et al. (*120a*) synthetisierten radioaktiv markiertes [10-^{14}C]-Endocrocin auf folgendem Wege: FRIEDEL-CRAFTS-Kondensation von [1-^{14}C]-3,5-Dimethoxyphthalsäureanhydrid mit 2,3-Dimethylphenol zur Benzoylbenzoesäure, anschließende Cyclisierung in borsäurehaltigem Oleum und Entmethylierung mit Kaliumjodid-Phosphorsäure gab [10-^{14}C]-2-Methylemodin, das nach Acetylierung einer Benzylbromierung mit N-Bromsuccinimid unterworfen wurde. Das Benzylbromid wurde mit Natrium-

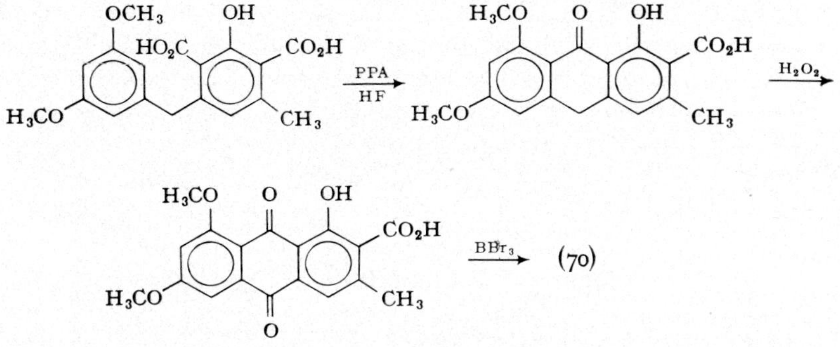

Schema 15. Synthese von Endocrocin nach STEGLICH und REININGER (*309*).
PPA = Polyphosphorsäure

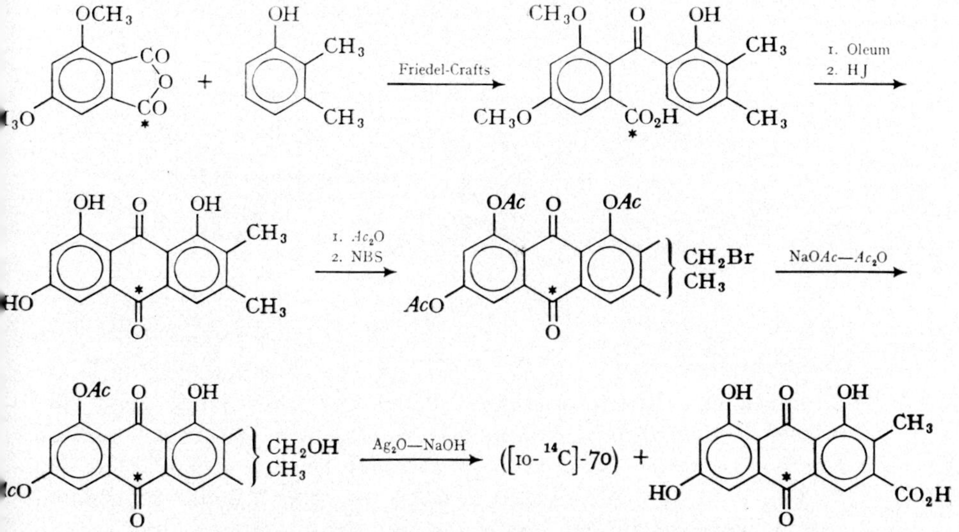

Schema 15a. Synthese von [10-^{14}C]-Endocrocin nach FRANCK et al. (*120a*)

acetat-Acetanhydrid hydrolysiert und dann mit Silberoxid-Natronlauge zu einem Gemisch aus Endocrocin und der isomeren 3-Carbonsäure oxydiert, das chromatographisch auf weinsaurem Kieselgel G getrennt wurde (Schema 15a).

SANTESSON (*272c*) untersuchte 230 Arten der Gattung *Caloplaca* mittels Flechtenmassenspektrometrie und Dünnschichtchromatographie auf Anthrachinone und fand Emodin, Parietin, Fallacinol, Fallacinal, Parietinsäure, Xanthorin, 2-Chloremodin, Fragilin und 1-O-Methylfragilin.

Aus *Solorina crocea* (L.) Ach. isolierten ANDERSON *et al.* (*19*) neben Solorinsäure (71) auch Norsolorinsäure (72), deren Struktur durch Synthese aus 1,3,6,8-Tetrahydroxyanthrachinon-2-carbonsäure bewiesen wurde (Schema 16). In der gleichen Flechte fanden EBIZUKA *et al.* (*98b*) neben Gyrophorsäuremethylester Averythrin-6-monomethyläther (72a).

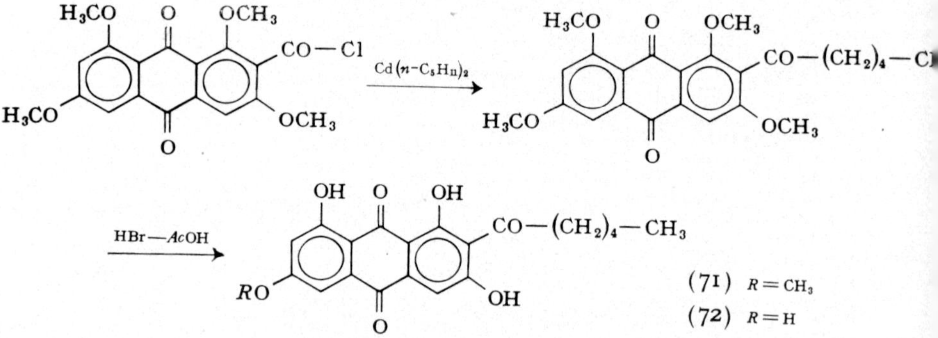

Schema 16. Synthese von Norsolorinsäure

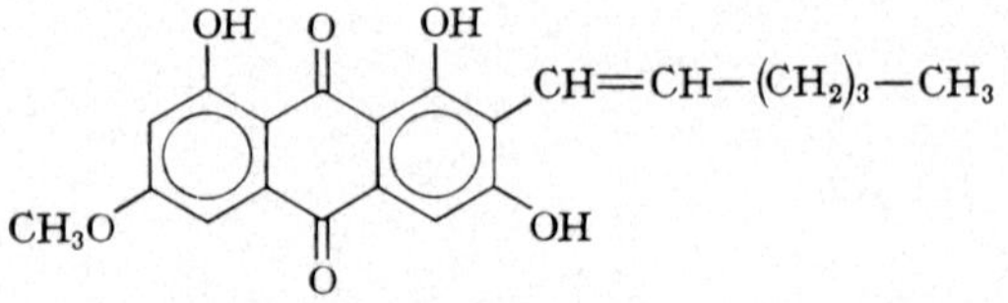

(72a)

(72) kommt auch in *Lecidea piperis* (Spreng.) Nyl. vor (*272*).

Neben dem bereits erwähnten Papulosin (*118*) wurden noch zwei weitere 5,8-Dihydroxyanthrachinone in Flechten gefunden, nämlich Xanthorin (73) in *Xanthoria elegans* (Link) Th. Fr. (*308*) und *Laurera purpurina* (Nyl.) Zahlbr. (*311*) sowie 1,2,5,6,7,8-Hexahydroxy-3-methylanthrachinon (74) in den Apothecien von *Mycoblastus sanguinarius* (L.) Norm. (*49*). Xanthorin wurde aus Parietin durch ELBS-Oxydation mit

Literaturverzeichnis: SS. 288—306

Kaliumperoxydisulfat (*311*) und aus Emodin (60) durch Erhitzen mit Kalilauge und anschließende Methylierung des entstandenen 1,5,6,8-Tetrahydroxy-3-methylanthrachinons mit Dimethylsulfat (*308*) synthetisiert.

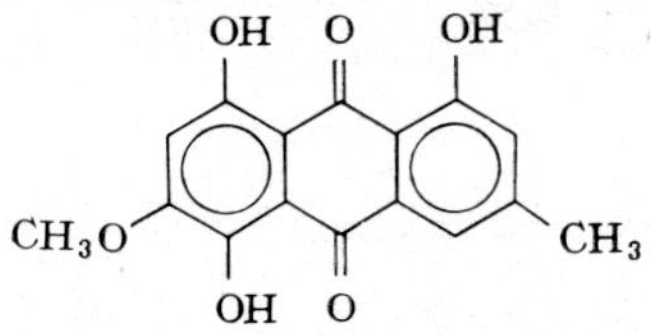

(73) 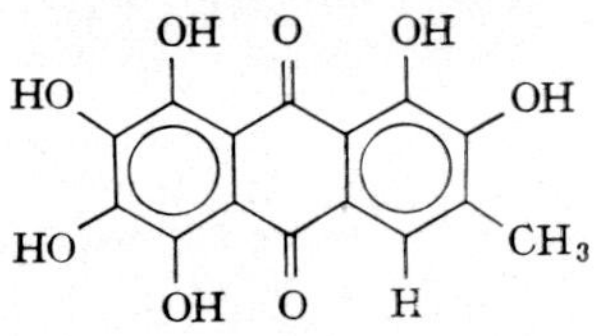 (74)

YOSIOKA et al. (*351a*) isolierten aus *Anaptychia obscurata* Vain. die beiden Anthrone (74a) und (74b).

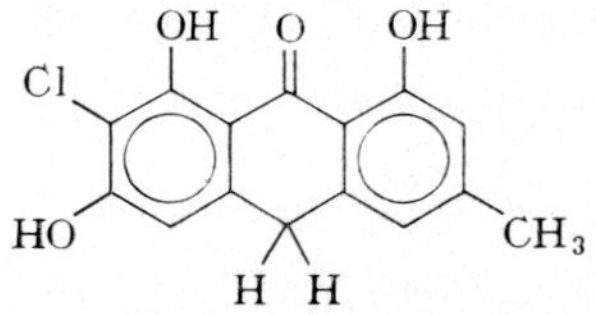

(74a) 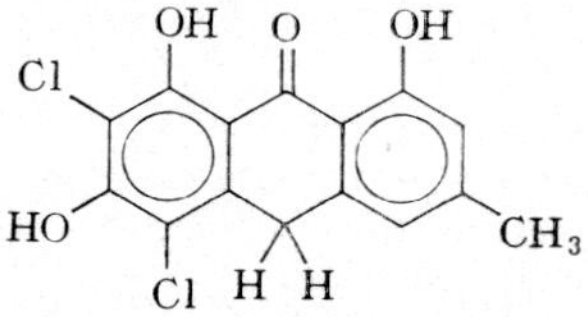 (74b)

2.2.4. Chromone

Bisher kennt man nur wenige Chromone aus Flechten. In *Siphula ceratites* (Fr.) Th. Fr. fand BRUUN (*58*) Siphulin, das bei Alkalischmelze in 3,5-Dihydroxy-n-heptylbenzol, 3,5-Dihydroxyphenylessigsäure und Essigsäure zerfällt (Schema 17). Daraus und aus spektroskopischen Daten ließ sich Struktur (75) ableiten (*59*).

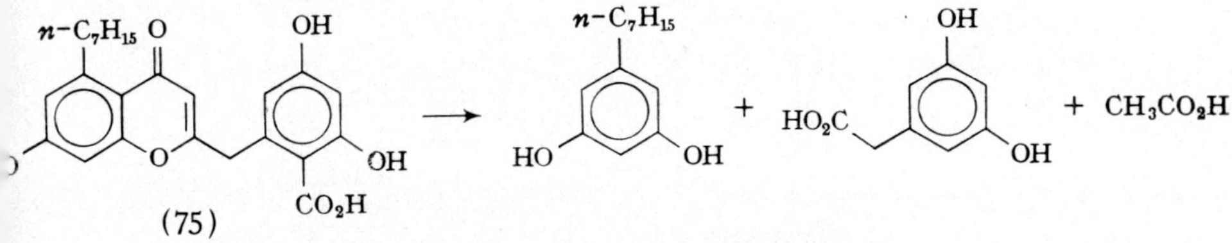

(75)

Schema 17. Abbau von Siphulin bei Alkalischmelze

Ein weiteres Chromon, 8-Chlor-5,7-dihydroxy-2,6-dimethylchromon (= Sordidon = Rupicolon) (76) kommt in *Lecanora rupicola* (L.) Zahlbr. und *Lecanora carpinea* (L.) Ach. em. Vain. vor und wurde von ARSHAD et al. (*22*) sowie HUNECK und SANTESSON (*162*) durch Chlorierung von Eugenitol mit Sulfurylchlorid in Tetrahydrofuran synthetisiert (Schema 18).

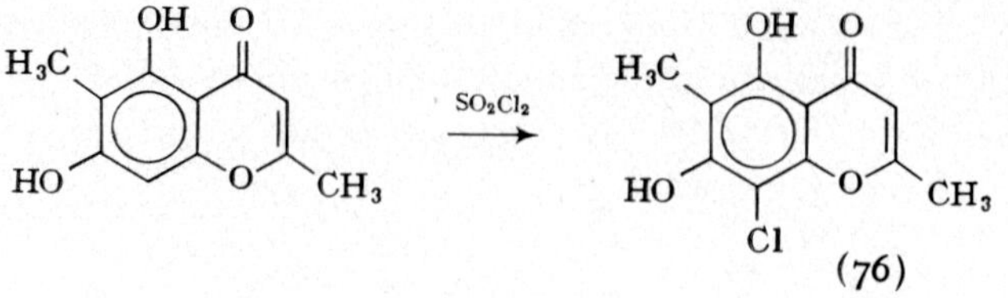

Schema 18. Synthese von Sordidon

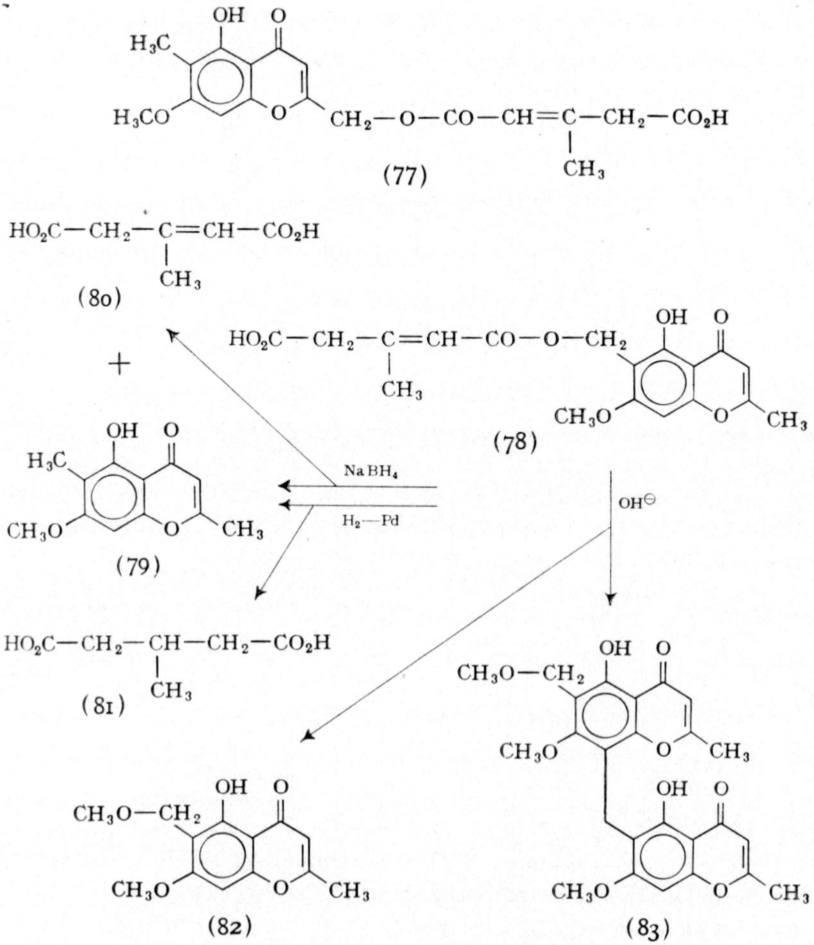

Schema 19. Abbaureaktionen von Leprarsäure

Literaturverzeichnis: SS. 288—306

Für die von Zopf (*354*) unter der Bezeichnung Leprarin aus *Lepraria latebrarum* Ach. isolierte Leprarsäure wurde von Soviar et al. (*306*) die Struktur (77) aufgestellt, die von Aberhart et al. (*1a*) auf Grund NMR-spektroskopischer Daten zu (78) korrigiert wurde. (78) läßt sich mit Natriumborhydrid in Tetrahydrofuran in Eugenitin (79) und β-Methylglutaconsäure (80) spalten, während Hydrierung über Palladium (79) und β-Methylglutarsäure (81) liefert. Alkalihydrolyse von (78) in Methanol gibt das Chromon (82) neben wenig Bis-chromon (83) (Schema 19).

Roccella fuciformis DC. enthält neben (78) noch das Chromon (84) (*1a*).

Chiodectonsäure aus der tiefroten tropischen Rindenflechte *Chiodecton sanguineum* (Sw.) Vain. ist nach Steglich und Huneck (*306a*) ein Chromon der Summenformel $C_{15}H_{10}O_9$, hat Struktur (84a) und steht somit in enger Beziehung zur Rhodocladonsäure.

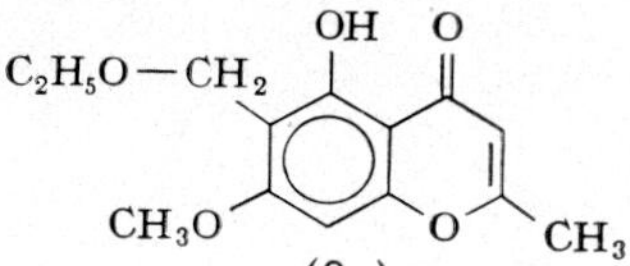

(84)

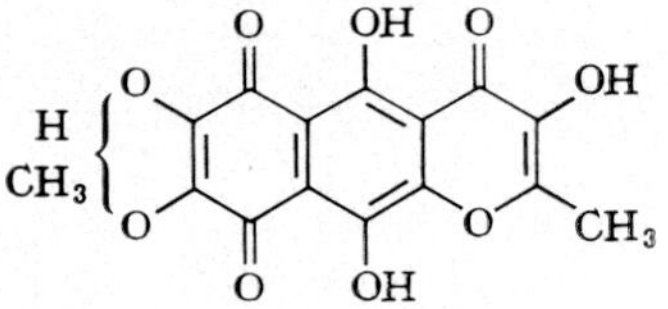

(84 a)

2.2.5. Xanthone

Bis 1966 war Lichexanthon (85) das einzige bekannte Flechtenxanthon, für das von Aghoramurty und Seshadri (*10*) und Grover et al. (*127*) neue Synthesen entwickelt wurden. 1966 wurde die Struktur der Thiophansäure im Sinne von (86) aufgeklärt (*142, 162*) und später durch Synthese bewiesen (*169, 273, 23*).

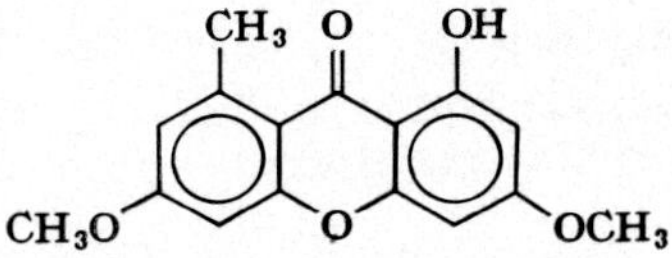

(85)

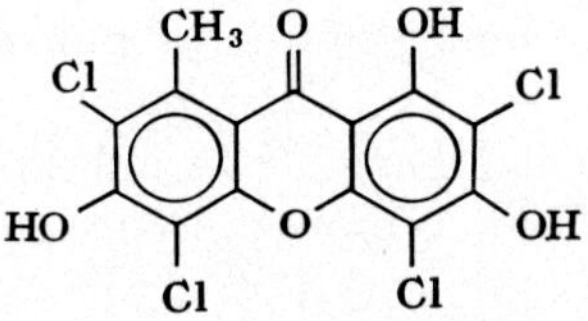

(86)

Die Synthese von Jayalakshmi et al. (*169*) geht von Lecanorsäure und Phloroglucin aus über Norlichexanthontrimethyläther, der mit 4 Mol. Chlor in Tetrachlorkohlenstoff chloriert wird. Der resultierende Tetrachlormethyläther wird mit Aluminiumchlorid in Benzol zu (86) entmethyliert. Santesson und Sundholm (*273*) synthetisierten Norlichexanthon (87) auf drei verschiedenen Wegen: a) durch Kondensation von Orsellinsäure mit Phloroglucin in Gegenwart von Zinkchlorid und

Phosphoroxychlorid, b) durch Pyrolyse von Phloroglucinylorsellinat und c) durch Kondensation von Orsellinsäure mit Phloroglucin in Gegenwart von Trifluoracetanhydrid. Chlorierung von (87) mit Chlor in Eisessig liefert in 31% Ausbeute Thiophansäure (Schema 20).

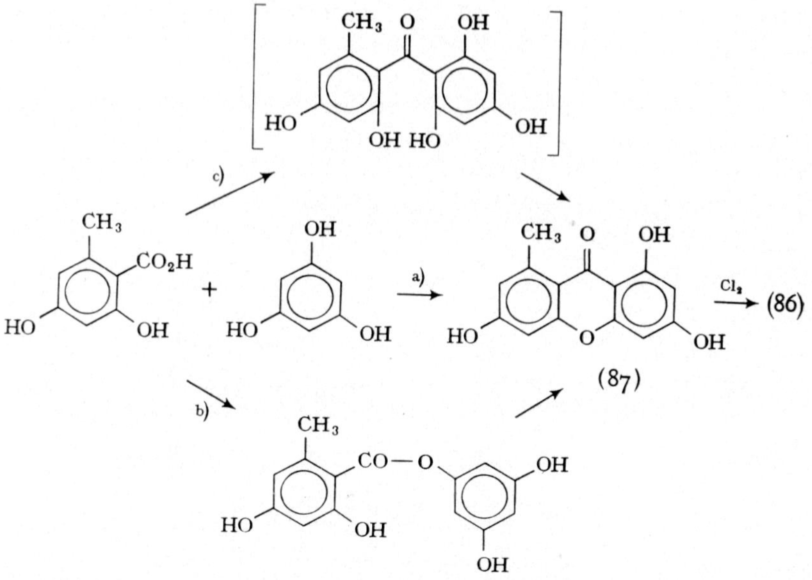

Schema 20. Synthese von Thiophansäure nach SANTESSON und SUNDHOLM *(273)*

ARSHAD und OLLIS *(23)* kondensieren 3,5-Dichloreverninsäure mit Phloroglucin in Gegenwart von Zinkchlorid und Phosphoroxychlorid zu 5,7-Dichlor-1,3-dihydroxy-6-methoxy-8-methylxanthon, das mit Sulfurylchlorid in Tetrahydrofuran Thiophansäure-6-methyläther gibt; Entmethylierung dieses Äthers mit Pyridinhydrochlorid bei 190° führt schließlich ebenfalls zu (86).

Insbesondere SANTESSON *(264)* hat zahlreiche neue Xanthone aus Krustenflechten der Gattungen *Pertusaria, Lecanora, Buellia* und *Lecidea* isoliert und strukturell aufgeklärt; Arthothelin (88) aus *Arthothelium pacificum* Follm. *(143)* und *Lecanora straminea* (Wahlbg.) Ach. *(268)*, 2,7-Dichlornorlichexanthon (89), 2-Chlornorlichexanthon (90) und 2,4-Dichlornorlichexanthon (91) ebenfalls aus *Lecanora straminea* (Wahlbg.) Ach. *(269, 270)*, Norlichexanthon (87) aus *Lecanora reuteri* Schaer. *(262)*, 2,7-Dichlorlichexanthon (92), 2,5,7-Trichlornorlichexanthon (93), 3-O-Methyl-2,5,7-trichlornorlichexanthon (94), 3-O-Methyl-2,5-dichlornorlichexanthon (95) und 2,5-Dichlorlichexanthon (96) *(265, 266)*.

Literaturverzeichnis: SS. 288—306

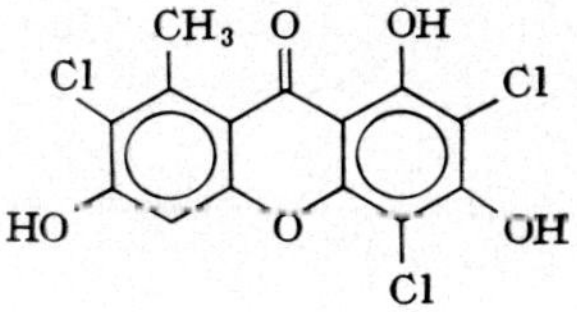

Huneck und Santesson (*163*) fanden in *Lecidea carpathica* (Koerb.) Szat. Thuringion (97) und Poelt und Huneck (*234*) in *Lecanora vinetorum* Poelt et Hun. das chlorhaltige Vinetorin, dem wahrscheinlich Struktur (98) zukommt. Bei der bereits von Zopf (*354*) aus *Pertusaria wulfenii* DC. und *Pertusaria lutescens* (Hoffm.) Lamy isolierten Thiophaninsäure handelt es sich um 6-O-Methyl-2,4-dichlornorlichexanthon (99) (*163*).

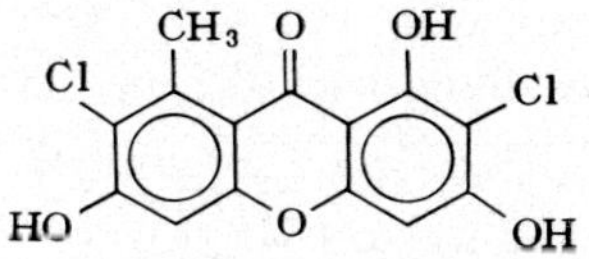

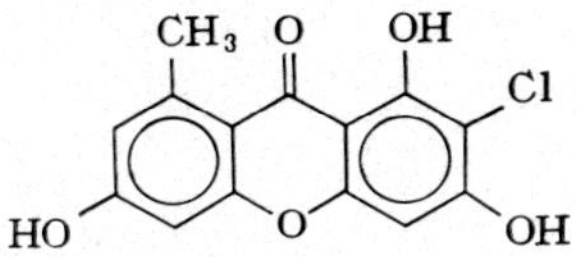

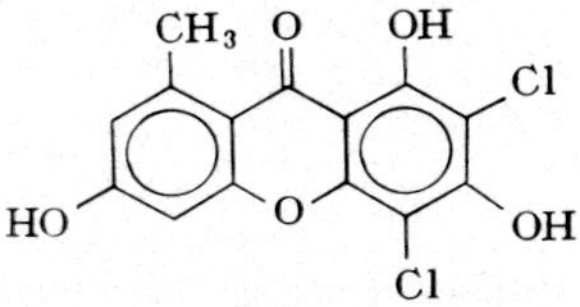

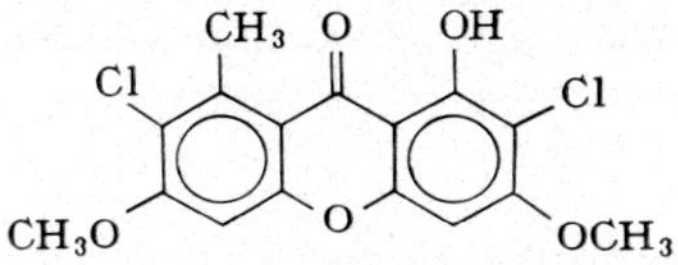

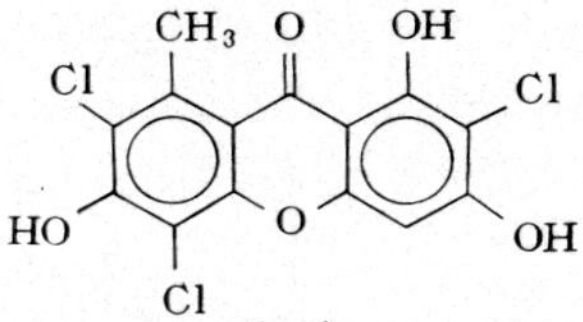

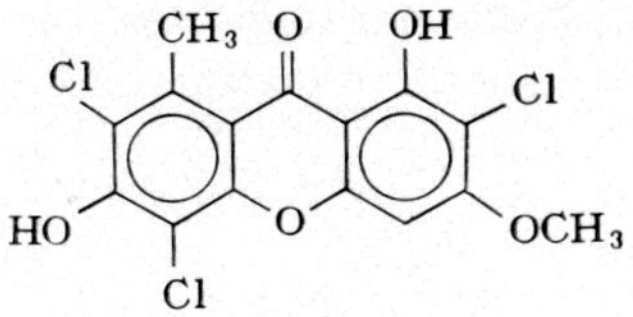

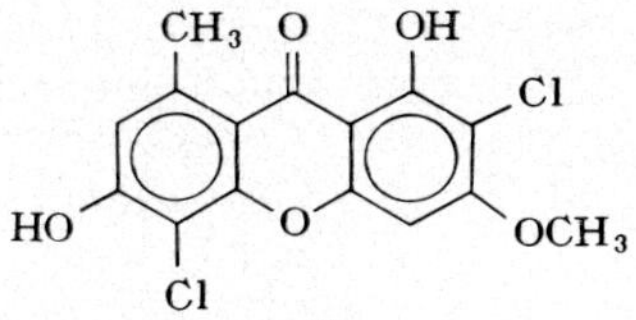

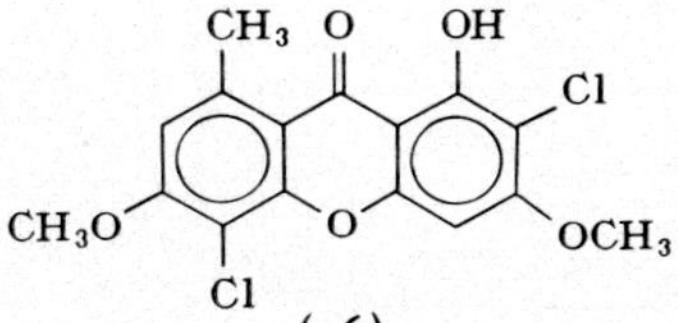

Über die Dünnschichtchromatographie, NMR- und IR-Spektroskopie von Flechtenxanthonen liegen ebenfalls von Santesson Untersuchungen vor (*260, 265, 267, 269, 271*), der auch einen empfindlichen Tüpfeltest mittels DIMROTHs-Reagens (Tetraanhydrid von Essigsäure mit Borsäure) für 1-Hydroxyxanthone angibt (*263*).

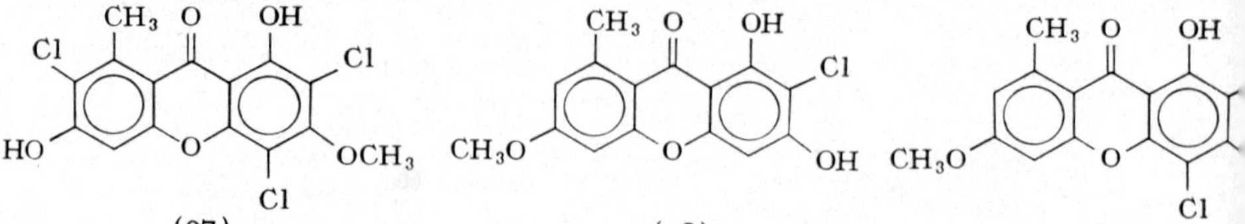

(97) (98) (99)

Bemerkenswert ist das gehäufte Vorkommen von chlorsubstituierten Xanthonen, die sich alle von einem Grundtyp, dem Norlichexanthon, ableiten.

Kürzlich wurden einige biogenese-artige Synthesen von Xanthonen beschrieben (*99, 233, 27, 193*), die sich auch auf Flechtenxanthone anwenden lassen dürften.

2.2.6. Dibenzofurane

Porphyrilsäure aus *Haematomma coccineum* (Dicks.) Koerb., *Haematomma porphyrium* (Pers.) Zopf. und *Lecidea silacea* (Ach.) Ach. (*80*) hat Struktur (100) (*331, 332*). Porphyrilsäuredimethyläther (101) gibt bei Oxydation mit Kaliumpermanganat die Tetracarbonsäure (102),

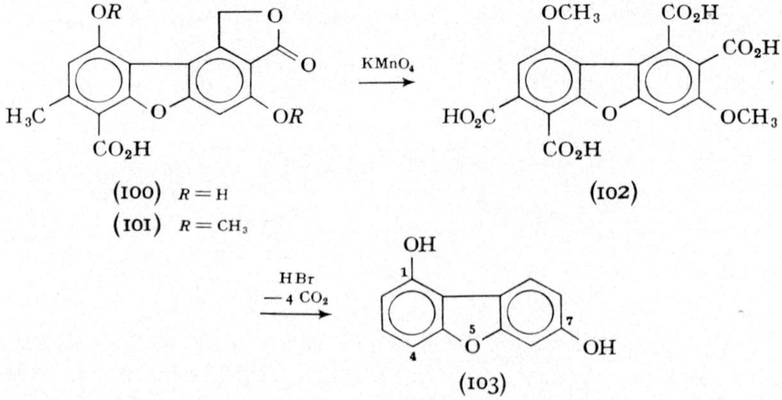

(100) R = H
(101) R = CH$_3$

(102)

(103)

Schema 21. Abbau von Porphyrilsäure zu 1,7-Dihydroxydibenzofuran

die nach Entmethylierung und Decarboxylierung in 1,7-Dihydroxydibenzofuran (103) übergeht, das synthetisch aus 2,6,2′,4′-Tetramethoxybiphenyl zugänglich ist (Schema 21).

Literaturverzeichnis: SS. 288—306

Lepraria membranacea (Dicks.) Vain. enthält ein weiteres Dibenzofuran, Pannarsäure (104), deren Struktur aus Decarboxylierung zu Pannarol (105) folgt, das aus 3,5-Dimethoxy-4-jodtoluol (106) und 3,5-Dimethoxy-2-jodtoluol (107) synthetisiert werden kann. Ferner ist das Kaliumpermanganat-Oxydationsprodukt von Pannarsäuredimethyläther (108) mit 1,2,6,7-Tetracarboxy-3,9-dimethoxydibenzofuran (102) aus Porphyrilsäure identisch (16) (Schema 22).

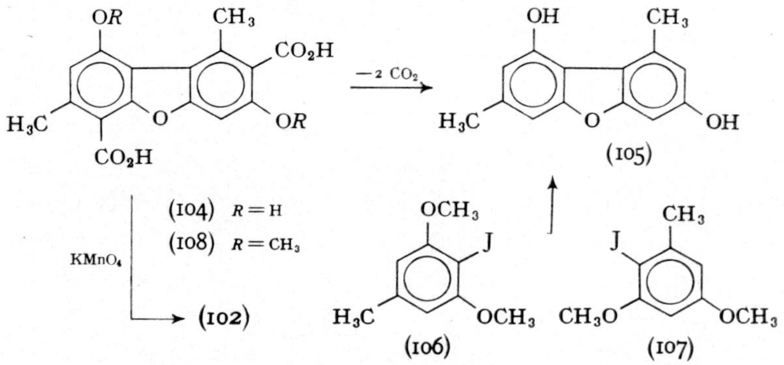

Schema 22. Abbau von Pannarsäure zu Pannarol

Schizopeltsäure (109) aus *Schizopelte californica* Th. Fr., *Reinkella parishii* Hasse (259) und *Roccellina luteola* Follm. (153) geht mit Diazomethan in Dimethylätherpannarsäuremethylester über und liefert beim Erhitzen Decarboxyschizopeltsäure (110), die sich mit Bortribromid zu Desmethyldecarboxyschizopeltsäure (111) entmethylieren läßt; da (111) eine positive GIBBssche Reaktion zeigt, kommt Schizopeltsäure Struktur (109) zu (165) (Schema 23).

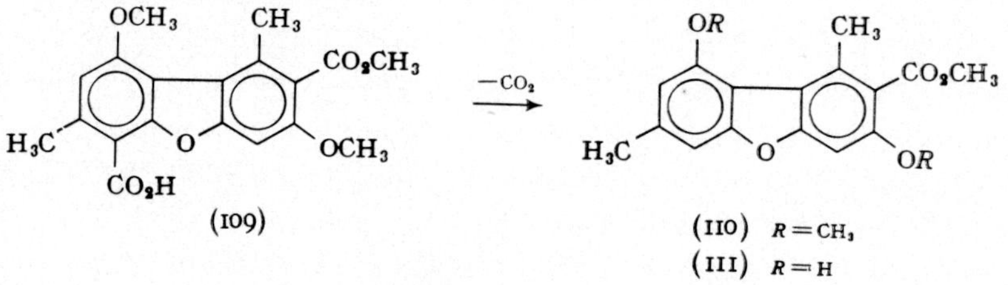

Schema 23. Decarboxylierung von Schizopeltsäure

BREWER und ELIX (53) synthetisierten Strepsilinmethyläther auf folgendem Wege (Schema 24): Everninaldehyd (112) wird mit Bromacetaldehyd zu 6-Methoxy-4-methylbenzofuran-2-aldehyd (113) umgesetzt, der nach WITTIG-Reaktion mit Methoxymethylentriphenylphosphoran ein Gemisch aus cis- und trans-6-Methoxy-4-methyl-2-(β-methoxyvinyl)-benzofuran (114a) und (114b) gibt. Dieses Gemisch liefert mit Acetylendicarbonsäuredimethylester 3,4-Dihydro-3,7-dimethoxy-9-methyldibenzofuran-1,2-dicarbonsäuredimethylester (115), der mit N-Bromsuccinimid und Dibenzoylperoxid zum 3,7-Dimethoxy-9-methyldibenzofuran-1,2-dicarbonsäuredimethylester (116) dehydriert wird. Lithiumaluminiumhydrid-Reduktion von (116) zum Diol (117) und dessen Oxydation mit Bichromat-Schwefelsäure liefert schließlich Strepsilinmethyläther (118).

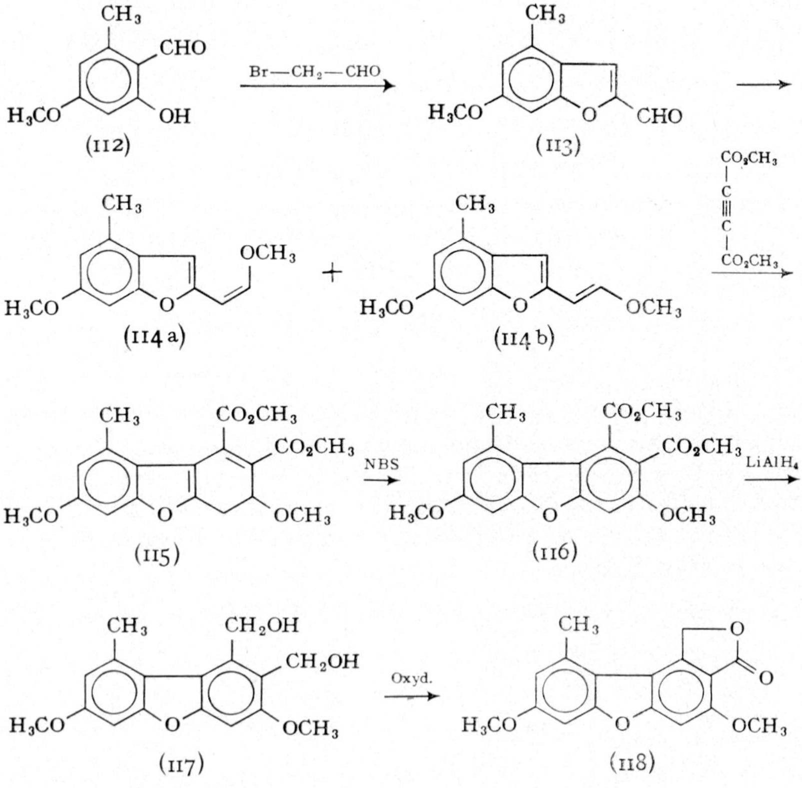

Schema 24. Synthese von Strepsilinmethyläther

2.2.7. Usnin- und Isousninsäure

IR- und NMR-Spektren zeigen, daß Usninsäure (119) drei intramolekulare Wasserstoffbrücken bildet: a, b und c (*114*). In wasserhaltigen Lösungsmitteln verhält sich (119) einbasisch, in nichtwäßrigen dagegen zweibasisch (*280*). Die Racemisierung von Usninsäure, die durch Erhitzen oder UV-Bestrahlung bewirkt wird (*35*), wurde von STORK (*312*), MAC KENZIE (*197*) sowie BERTILSSON und WACHTMEISTER (*42*) untersucht und verläuft über das Diradikal (120) bzw. das Ketenanalogon (121).

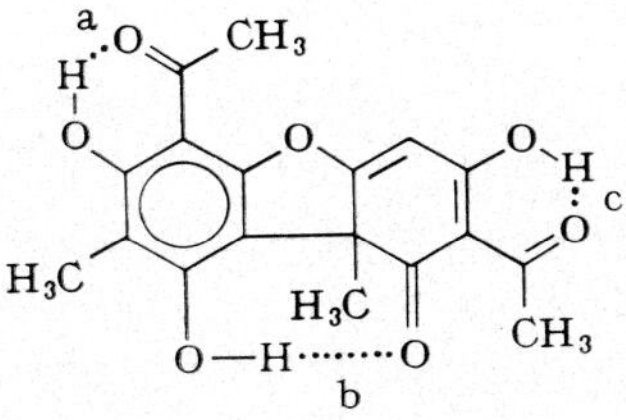

(119)

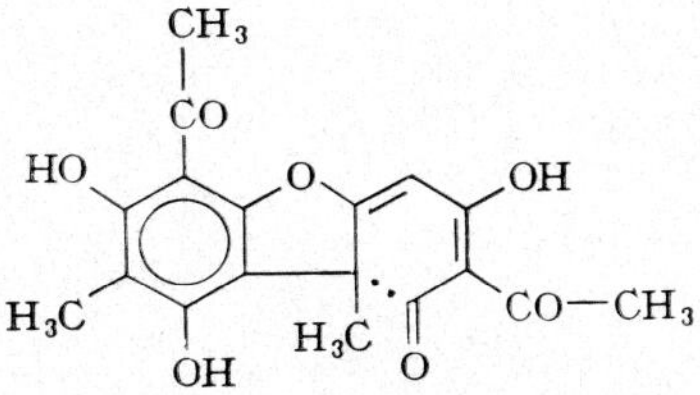

(120) 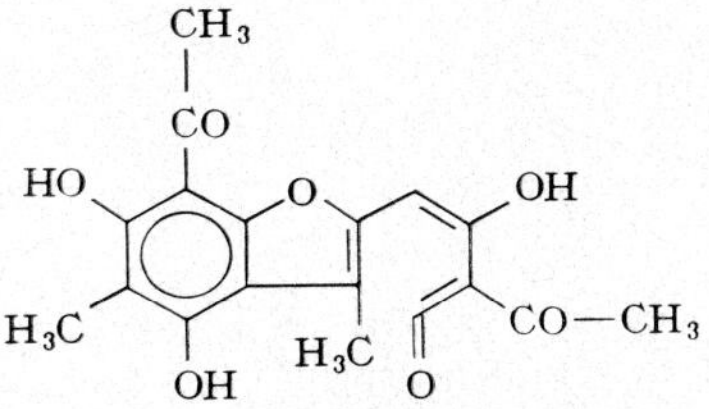   (121)

BERTILSSON und WACHTMEISTER (*42*) haben die Racemisierungsrate von (+)-Usninsäure und deren 7-O-Methyl-, 8-C-Methyl-, 7-O-Acetyl-, 9-O-Acetyl- und 7,9-Di-O-acetyl-Derivate gemessen und gefunden, daß die Stabilität der genannten Verbindungen vom Vorliegen der intramolekularen Wasserstoffbrückenbindung zwischen der Hydroxylgruppe am C-Atom 9 und der C-1-CO-Gruppe abhängt. Die Methylierung von (119) mit Dimethylsulfat-Natronlauge führt über eine C-Methylierung zu 8-Methylusninsäure (122), während Methylierung mit Methyljodid-Kaliumcarbonat 7-O-Methylusninsäure (123) gibt (*42*).

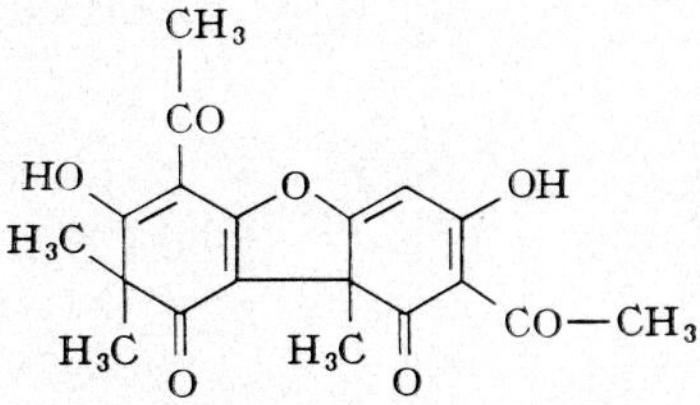

(122) 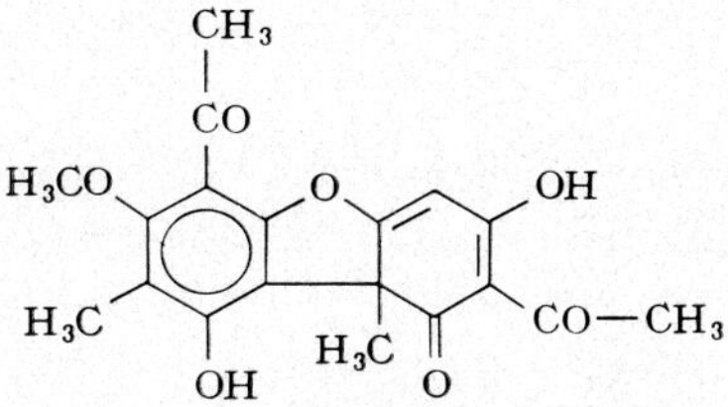   (123)

Diazomethan bildet mit Usninsäure das tetracyclische Furanderivat (124) (*43*).

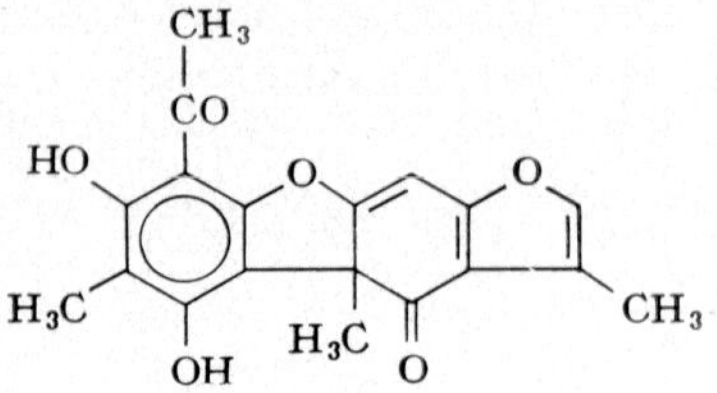

(124)

(+)-Usninsäure läßt sich in Tetrahydrofuran mit Palladiumschwarz zu (—)-Dihydrousninsäure (125) hydrieren (292), die beim Erhitzen unter Wasserstoff zu Isodihydrousninsäure (126) isomerisiert (294).

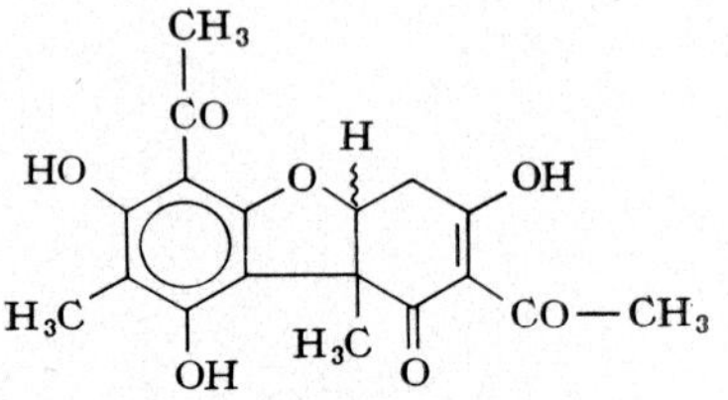

(125) 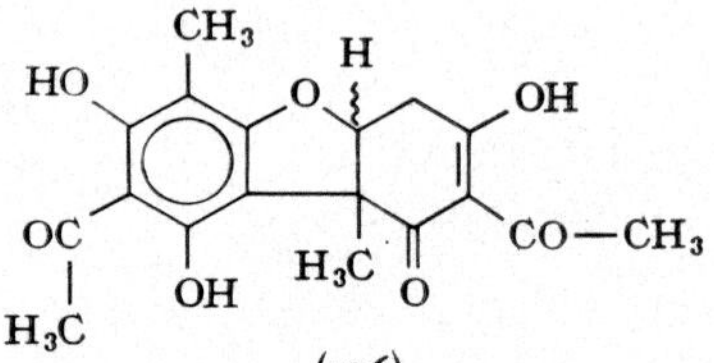   (126)

Ozonolyse von O,O-Diacetylusninsäure (127) gibt (128) und bestätigt damit die Struktur (119) von Usninsäure (290).

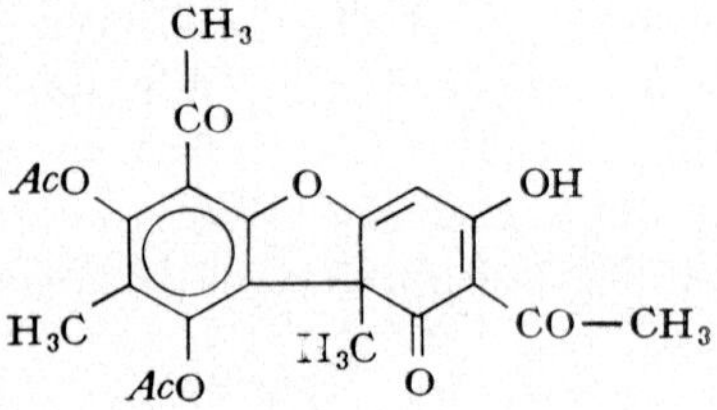

(127) 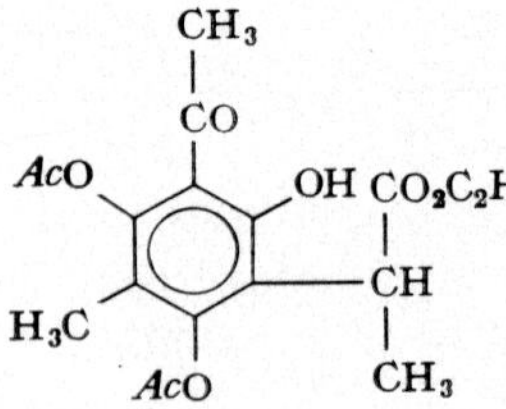   (128)

Die Synthese von (±)-Usninsäure gelang BARTON et al. (33) durch Oxydation von Methylphloracetophenon mit Kaliumferricyanid (Schema 25).

Die Racematspaltung von (±)-Usninsäure war bereits von DEAN et al. (93) über das Brucinsalz durchgeführt worden.

Über weitere chemische Umsetzungen mit Usninsäure und deren Derivate vergleiche folgende Literatur: *26, 32, 94, 21, 91, 93, 92, 95, 319, 317, 320, 322, 322a, 318, 321, 249, 330* und *199*.

Literaturverzeichnis: SS. 288—306

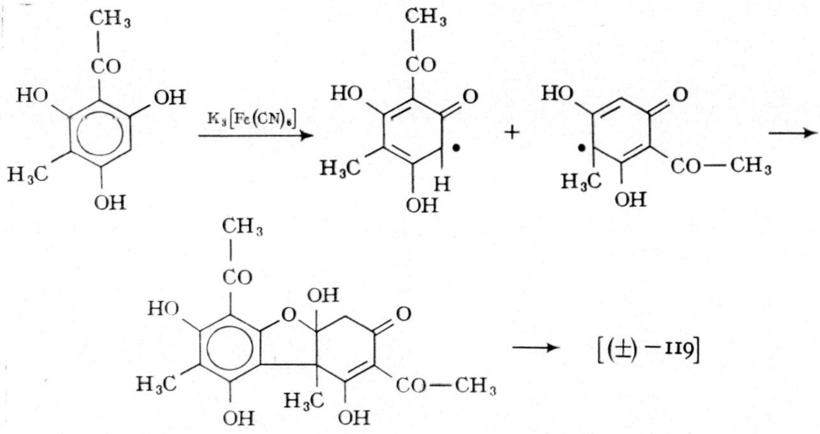

Schema 25. Synthese von (±)-Usninsäure

Einige Analoga der Usninsäure synthetisierten DAVIS und ELIX (*90*).

Aus *Cladonia mitis* Sandst., *Cladonia submitis* Evans, *Cladonia pleurota* Schaer. und *Cladonia sylvatica* Harm. isolierten SHIBATA und TAGUCHI (*291*) ein Isomeres der Usninsäure, Isousninsäure (**129**), die ebenso wie Usninsäure durch oxydative Kupplung von Methylphloracetophenon entsteht. Diacetylisousninsäure (**130**) liefert bei Ozonoloyse das Cumaronderivat (**131**), das mit konz. Schwefelsäure unter Hydrolyse in das isomere Cumaron (**132**) umgewandelt wird; (**132**) ist auf analogem Wege aus Diacetylusninsäure zugänglich. Isousninsäure geht bei katalytischer Hydrierung mit Palladiumschwarz in Tetrahydrofuran in Isodihydrousninsäure (**133**) über, die bereits beim Umkristallisieren in Dihydrousninsäure (**125**) umgewandelt wird (*316*) (Schema 26).

Isousninsäure wurde auch in *Lecanora saligna* (Schrad.) Zahlbr., *Lecanora sarcopis* (Wahlb.) Roehl. (*107*) und in einer noch unbestimmten *Sphaerophorus*-Art (*144*) gefunden.

2.2.8. Depside

Ein neues Depsid aus verschiedenen *Ramalina*-Arten (*39, 146, 147, 148, 149, 150*) erwies sich als 3,5-Dichlorlecanorsäuremethylester (**134**) (*40, 141*). Die Struktur wurde durch Hydrolyse zu 3,5-Dichlororsellinsäure und Orsellinsäuremethylester und Synthese durch Chlorierung von Lecanorsäuremethylester mit Sulfurylchlorid bewiesen (Schema 27).

In *Stereocaulon ramulosum* (Sw.) Räusch. aus Neuseeland fand CAMBIE (*66*) neben Atranorin und Anziasäure den bisher unbekannten Anziasäuredimethyläther (**135**).

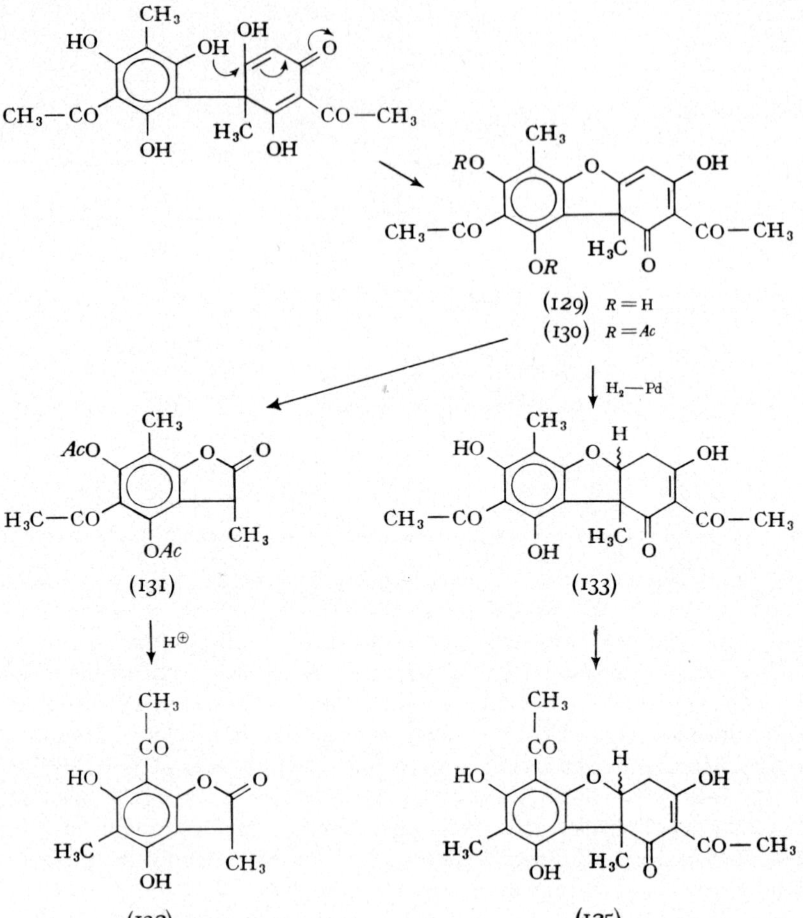

Schema 26. Abbau und Hydrierung von Isousninsäure

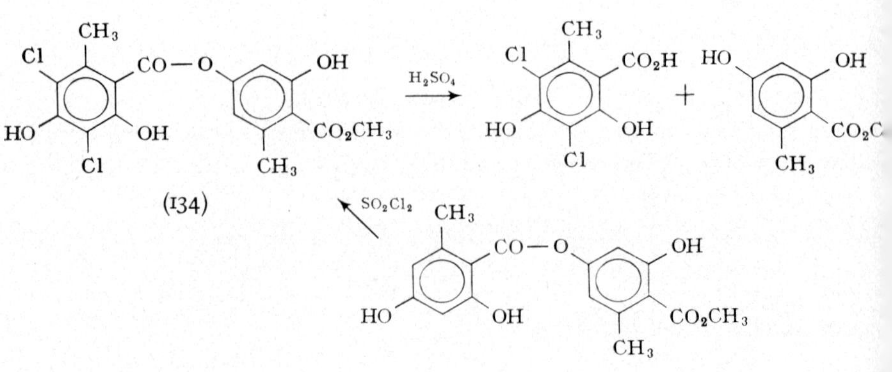

Schema 27. Hydrolyse und Synthese von 3,5-Dichlorlecanorsäuremethylester

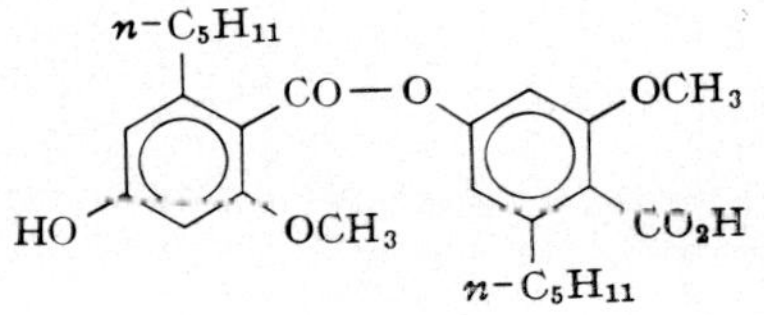

(135)

Planasäure aus *Lecidea plana* (Lahm ex Koerb.) Nyl. (*139*) und *Lecidea lithophila* (Ach.) Ach. em. Th. Fr. (*140*) liefert bei Hydrolyse Dimethylätherolivetolcarbonsäure und o-Methylätherolivetolcarbonsäure und ist daher mit Dimethylätherperlatolinsäure (136) identisch (Schema 28).

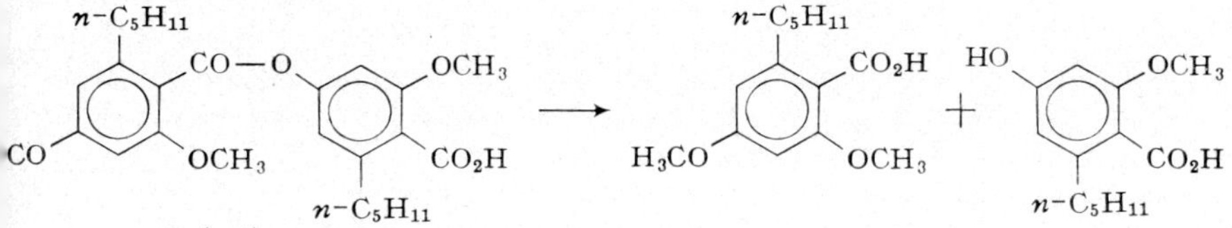

(136)

Schema 28. Hydrolyse von Planasäure

Die Struktur der Stenosporsäure (137) aus *Ramalina stenospora* Müll.-Arg. wurde von CULBERSON (*82*) durch Synthese des Methylesters (138) aus Divaricatinsäure (139) und Olivetolcarbonsäuremethylester (140) mit Trifluoracetanhydrid als Kondensationsmittel bewiesen (Schema 29).

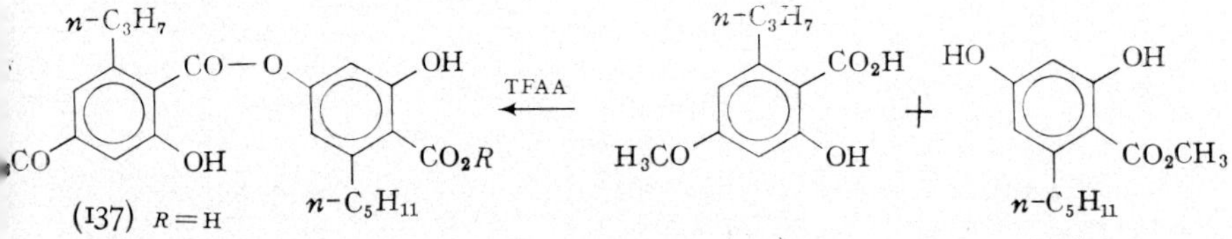

(137) R=H
(138) R=CH$_3$
(139)
(140)

Schema 29. Synthese von Stenosporsäuremethylester

Confluentinsäure (141) spaltet mit Ameisensäure in p-Methylätherolivetonid (142) und Olivetolmonomethyläther (143), ein Beweis für ihre Konstitution (*138*) (Schema 30).

Arthonia impolita (Ehrh.) Borr. enthält Arthoniasäure, deren Struktur aus den UV-, NMR- und Massen-Spektren und der Hydrolyse zu Olivetonid abgeleitet wurde; es handelt sich um 4-O-Desmethyl-2'-O-methylmicrophyllinsäure (144) (*164*).

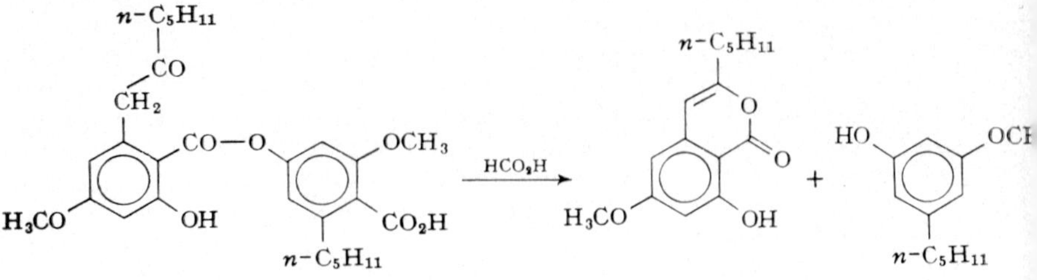

Schema 30. Spaltung von Confluentinsäure mit Ameisensäure

Miriquidisäure (144a) ist ein neues Depsid aus *Lecidea lilienstroemii* D. R. und *Lecidea leucophaea* (Flk.) Nyl. und weicht in bezug auf die Stellung der Seitenkettencarbonylgruppe von der Acetatregel ab; ihre Struktur folgt aus der Hydrolyse zu 2-Hydroxy-6-(3-oxo-n-pentyl)-anissäure und Olivetolcarbonsäure sowie insbesondere aus NMR-Doppelresonanzexperimenten (*164a*).

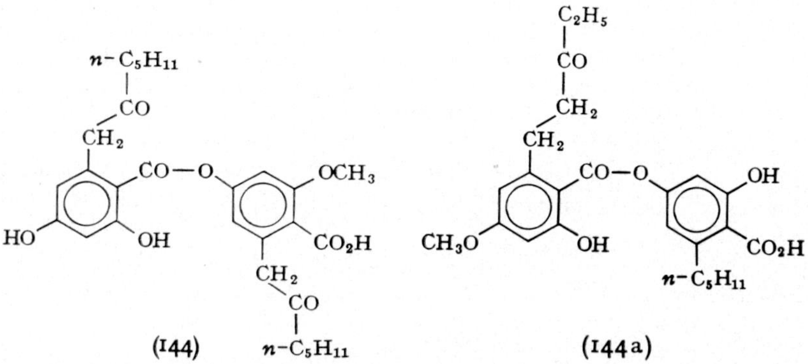

In *Lobaria* cfr. *dissecta* (Sw.) Räusch. und einigen anderen Arten der Sektion *Lobaria* fand CULBERSON (*81*) 4-O-Methylgyrophorsäure (145), die bei Hydrolyse Orsellinsäure und Everninsäure und bei kurzer Methylierung mit Diazomethan Tenuiorin (146) lieferte (Schema 31).

Das von ZOPF (*355*) aus verschiedenen *Peltigera*-Arten isolierte Peltigerin ist mit Tenuiorin (146) identisch (*167*).

BACHELOR und KING (*27a*) gelang die Isolierung des ersten Tetradepsides, Aphthosin (146a), aus *Peltigera aphthosa* (L.) Willd.; es gibt bei Hydrolyse Everninsäure, Orsellinsäure und Orsellinsäuremethylester.

Zur Synthese von Depsiden entwickelten BROWN et al. (*55*) und NEELAKANTAN et al. (*219*) zwei neue Methoden. Nach der einen wird die Säure mit dem Phenol in Gegenwart von Trifluoressigsäureanhydrid

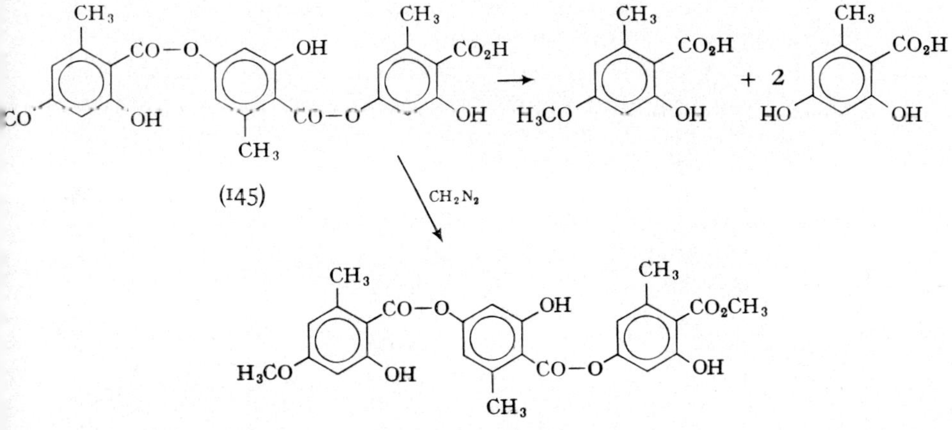

(145)

Schema 31. Hydrolyse und Methylierung von 4-O-Methylgyrophorsäure

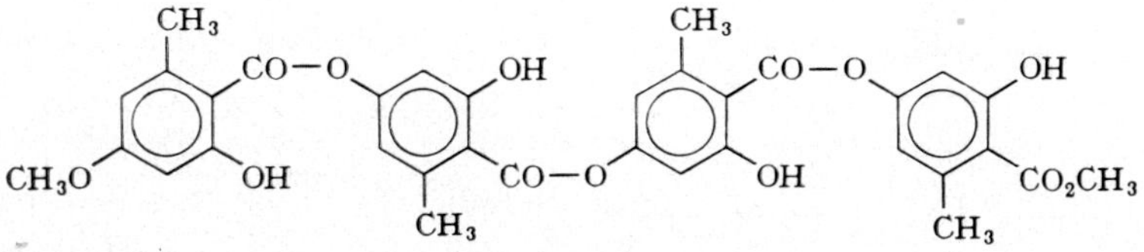

(146 a)

kondensiert, nach der anderen wird N,N'-Dicyclohexylcarbodiimid (DCC) als Kondensationsmittel benutzt.

Nach der bereits bei Montagnetol erwähnten Methode wurde auch (±)-Erythrin synthetisiert: Tricarbäthoxylecanorsäure wird mit cis-2-Buten-1,4-diol in Gegenwart von DCC verestert, der Ester mit MILAS'-Reagens hydroxyliert und danach decarbäthoxyliert.

Diploschistessäure (147) läßt sich aus 3-Formyllecanorsäure durch Oxydation mit alkalischem Wasserstoffperoxid gewinnen (278) (Schema 32).

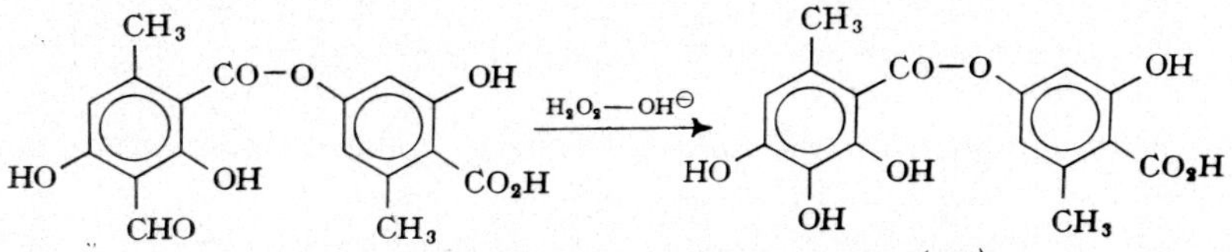

Schema 32. Synthese von Diploschistessäure

Die Synthese von Homodiploschistessäuremethylester beschreiben SESHADRI und VENKATASUBRAMANIAN (279).
Die Struktur von Cryptochlorophaesäure (148) aus *Cladonia cryptochlorophaea* Asah. und Merochlorophaesäure (149) aus *Cladonia merochlorophaea* Asah. wurde von SHIBATA und CHIANG (285) aufgeklärt. Dimethyläthercryptochlorophaesäuremethylester (150) liefert bei Hydrolyse Dimethylätherolivetolcarbonsäure (151) und 2,4-Dimethyläther-6-n-pentyl-pyrogallolcarbonsäuremethylester (152). Die Struktur von (149) folgt aus der Hydrolyse zu Divaricatinsäuremethyläther (153) und 6-n-Pentylpyrogallolcarbonsäure (154) (Schema 33).

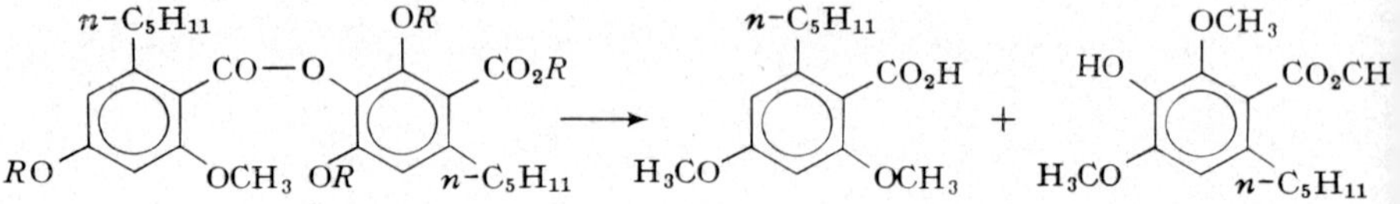

(148) R = H
(150) R = CH₃

Schema 33. Hydrolyse von Crypto- und Merochlorophaesäure

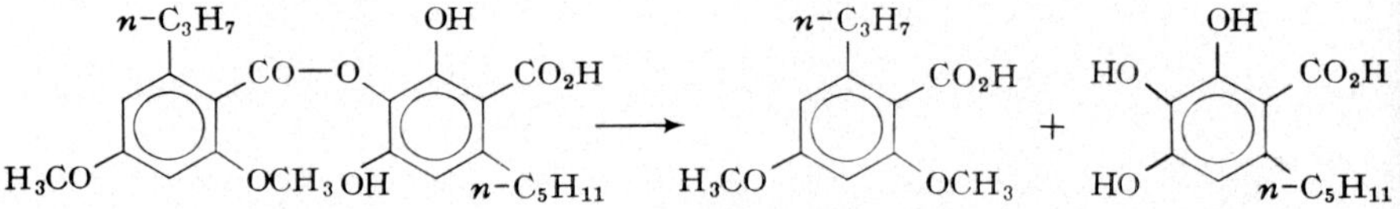

4-O-Methylcryptochlorophaesäure (155) kommt in *Cladonia perlomera* Krist. und *Cladonia merochlorophaea* Asah. vor (83).

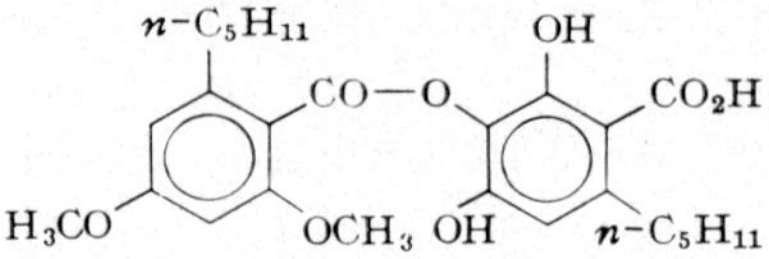

(155)

Novochlorophaesäure ist nach CULBERSON und KRISTINSSON (83) ein Gemisch aus Sekika- und Homosekikasäure.

Scrobiculin, ein neues Depsid aus *Lobaria verrucosa* Hoffm. (syn. *Lobaria scrobiculata* (Scop.) DC.) und *Lobaria amplissima* (Scop.) Forss. gibt bei Hydrolyse Divaricatinsäure und 2,3,4-Trihydroxy-6-n-propyl-

Literaturverzeichnis: SS. 288—306

benzoesäure und bei Methylierung Sekikasäuremethylester; daraus folgt Struktur (156) (79).

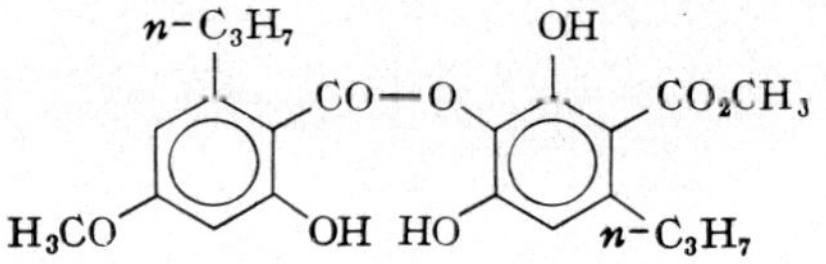

(156)

In *Ramalina paludosa* Moore kommt neben Atranorin, Usninsäure und Cryptochlorophaesäure ein niederes Homologes letzterer Verbindung vor: Paludossäure, die als 4-O-Desmethylmerochlorophaesäure (157) identifiziert wurde (78).

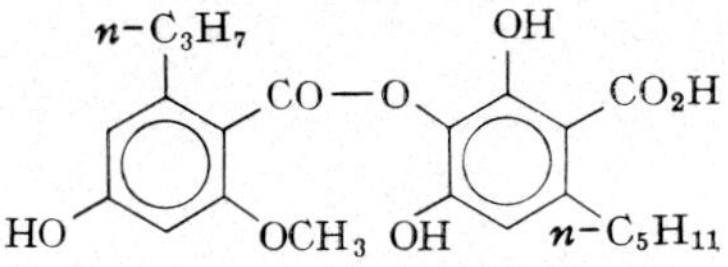

(157) 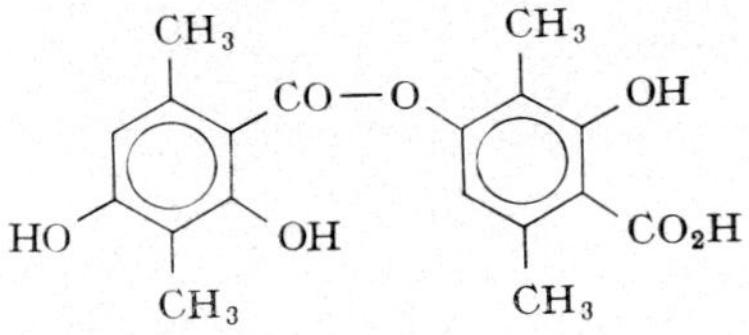   (158)

Ramalina subdecipiens Stein. enthält neben (+)-Usninsäure und Salazinsäure 4-O-Desmethylbarbatinsäure (158), deren Struktur aus der Hydrolyse zu β-Orcincarbonsäure sowie den NMR- und Massenspektren folgt (*155*).

Aus *Nephroma arcticum* (L.) Torss. isolierten NUNO et al. (*225*) das Depsid Nephroarctin (159), dessen Konstitution durch Röntgenstrukturanalyse des 4-Bromderivates aufgeklärt wurde. *Nephroma arcticum* (L.) Torss. enthält ferner das vollständig substituierte Depsid Phenarctin (160), dessen Struktur von BRUUN (*61*) aus NMR-spektroskopischen Daten abgeleitet wurde. Offenbar entsteht (159) aus (160) durch Decarboxylierung und Methylierung.

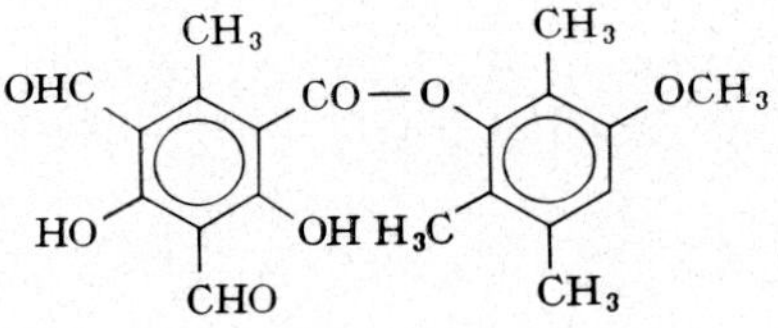

(159) 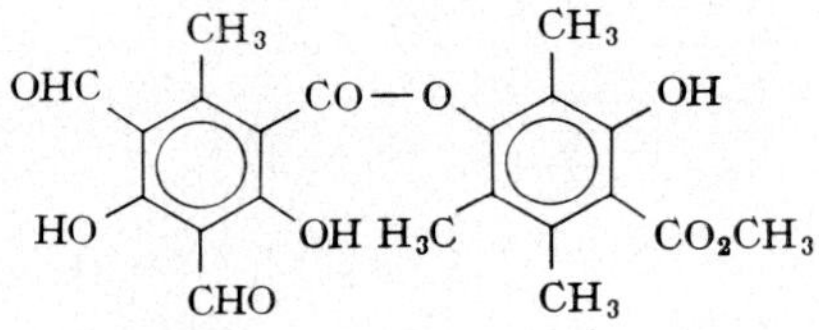   (160)

Die Halogenierung von Lecanorsäure und Atranorin untersuchten NEELAKANTAN et al. (*218*) mit den im Schema 34 gezeigten Ergebnissen.

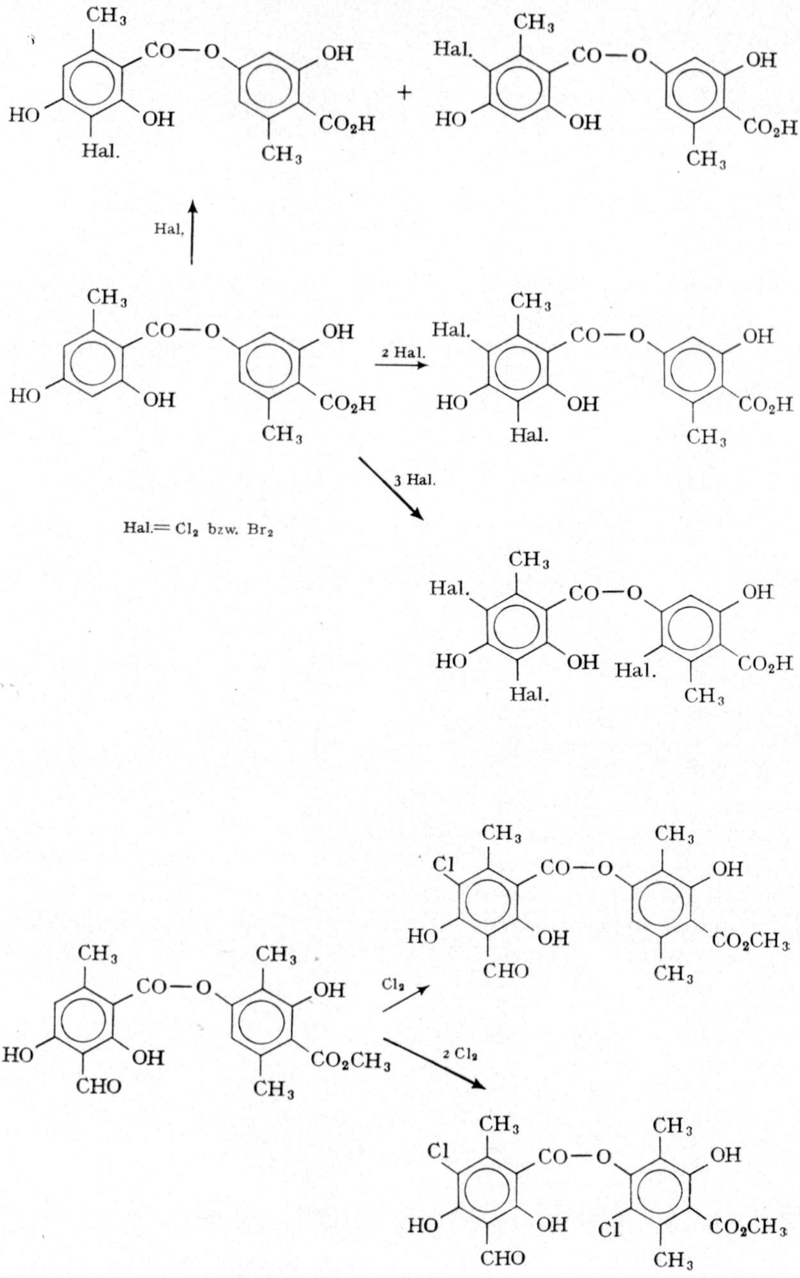

Hal.= Cl₂ bzw. Br₂

Schema 34. Halogenierung von Lecanorsäure und Atranorin

Literaturverzeichnis: SS. 288—306

Haemathamnolsäure (**161**) ist ein gelbes Depsid aus der Krustenflechte *Pertusaria rhodesiaca* Vain. (*130*). Kurze Einwirkung von Kalilauge auf (**161**) gibt Haemathamnol (**162**), längere Einwirkung Haematommsäure-4-methyläther (**163**), während mit Ameisensäure Thamnol (**164**) entsteht (Schema 35).

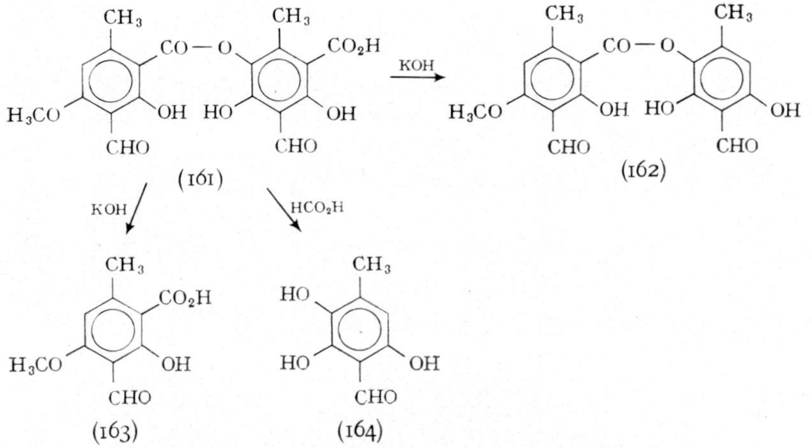

Schema 35. Hydrolyse von Haemathamnolsäure

2.2.9. Benzylester-Derivate

Bei der bereits von ZOPF (*354*) aus *Alectoria nigricans* (Ach.) Nyl. isolierten Alectorialsäure handelt es sich nach SOLBERG (*304*) und PERSSON und SANTESSON (*231*) um das Benzylesterderivat (**165**), das bei

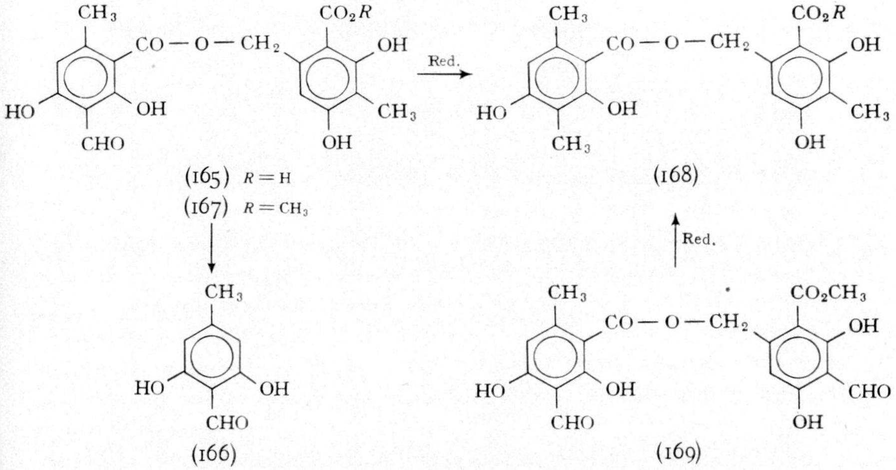

Schema 36. Hydrolyse und Reduktion von Alectorialsäure

Hydrolyse Atranol (166) liefert. Der Methylester (167) gibt bei Reduktion mit Jodwasserstoffsäure und Zink (168), das mit dem aus Barbatolsäuremethylester (169) durch Reduktion erhaltenen Produkt identisch ist (Schema 36).

2.2.10. Depsidone

Norlobaridon (170) aus *Parmelia conspersa* (Ehrh.) Ach. weicht insofern von der Acetat-Regel ab, als sich die Ketogruppe der Seitenkette direkt am aromatischen Ring befindet. Bei längerer Alkalihydrolyse des Monomethyläthers (171) wird Lobariol (172) erhalten; dagegen wird (170) von 1 n Natronlauge zu Isonorlobaridon (173) isomerisiert (*122, 123*) (Schema 37).

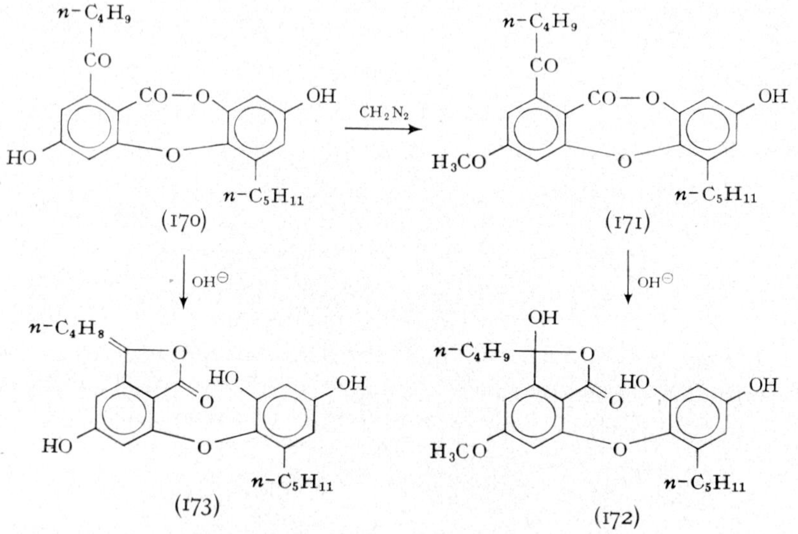

Schema 37. Hydrolyse von Norlobaridon

KOMIYA und KUROKAWA (*174b*) isolierten aus *Parmelia flavescentireagens* Gyeln., *Parmelia abessinica* Krempelh., *Parmelia australiensis* Cromb., *Parmelia dichotoma* Müll.-Arg., *Parmelia furcata* Müll.-Arg., *Parmelia hababiana* Gyeln., *Parmelia metamorphosa* Gyeln., *Parmelia paulensis* Zahlbr., *Parmelia recipienda* Nyl., *Parmelia scabrosa* Tayl., *Parmelia subdistorta* Kurok., *Parmelia subtinctoria* Zahlbr., *Parmelia thamnoides* Kurok. und *Parmelia tortula* Kurok. das neue Depsidon Loxodin und bewiesen dessen Identität mit Norlobarsäuremethylester (172a).

Constictinsäure aus *Usnea aciculifera* Vain. und anderen Flechten ist nach YOSIOKA *et al.* (*341a*) 4-O-Methylsalazinsäure (172b).

Literaturverzeichnis: SS. 288—306

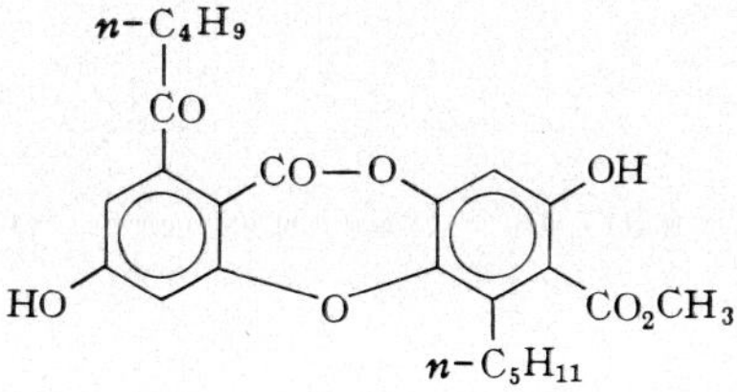

Chemie und Biosynthese der Flechtenstoffe

(172 a)

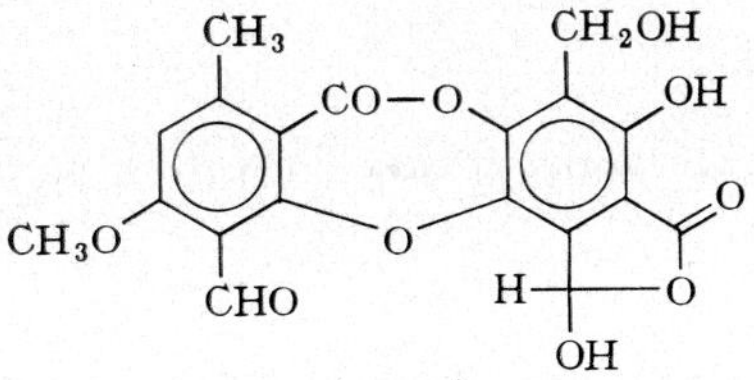

(172 b)

Die erste Synthese eines Depsidones, Diploicin, gelang BROWN *et al.* (55) auf zwei Wegen. Der eine geht von Dichlororsellinsäuredimethyläther (174) und 2,4-Dichlororcin-3-methyläther (175) aus, die in Gegenwart von Trifluoracetanhydrid zu (176) kondensiert werden. Entmethy-

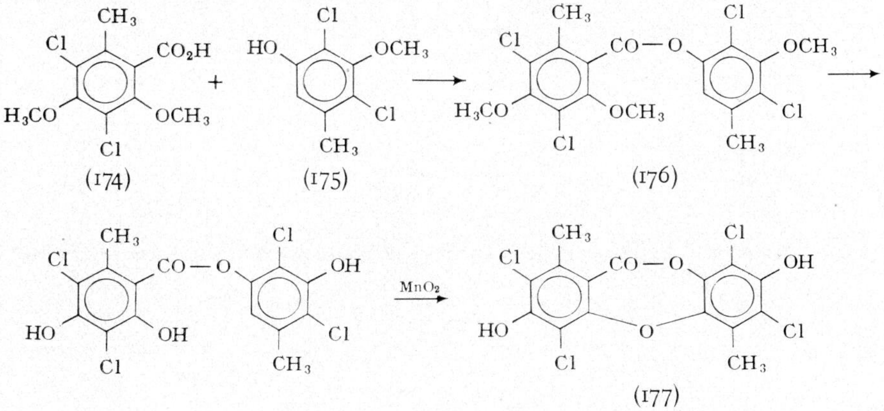

Schema 38. Synthese von Nordiploicin

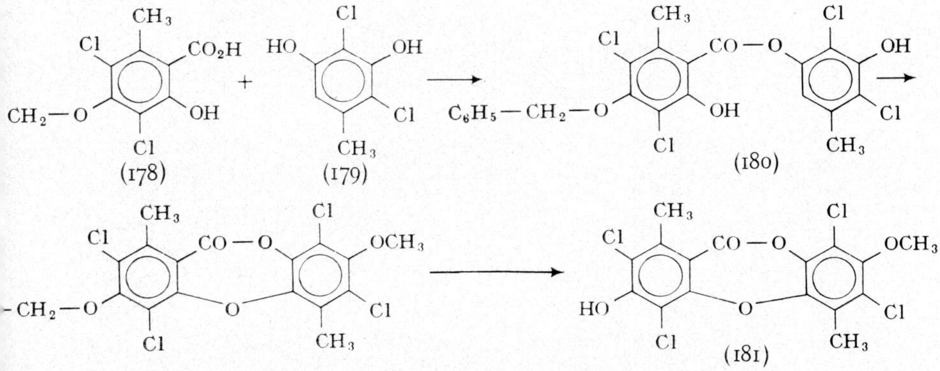

Schema 39. Synthese von Diploicin nach BROWN *et al.* (55)

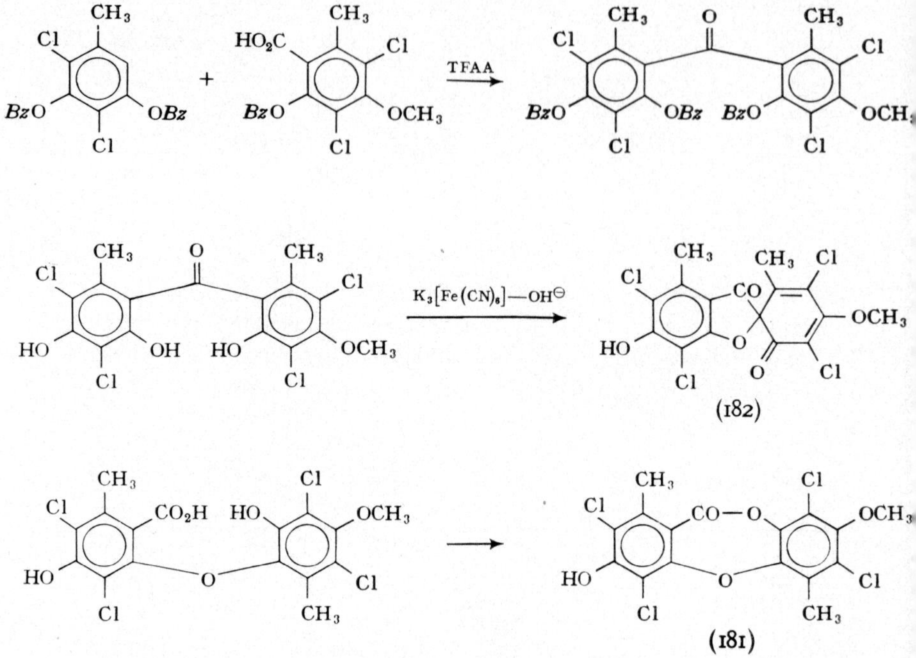

Schema 40. Synthese von Diploicin nach HENDRICKSON und RAMSAY (132)

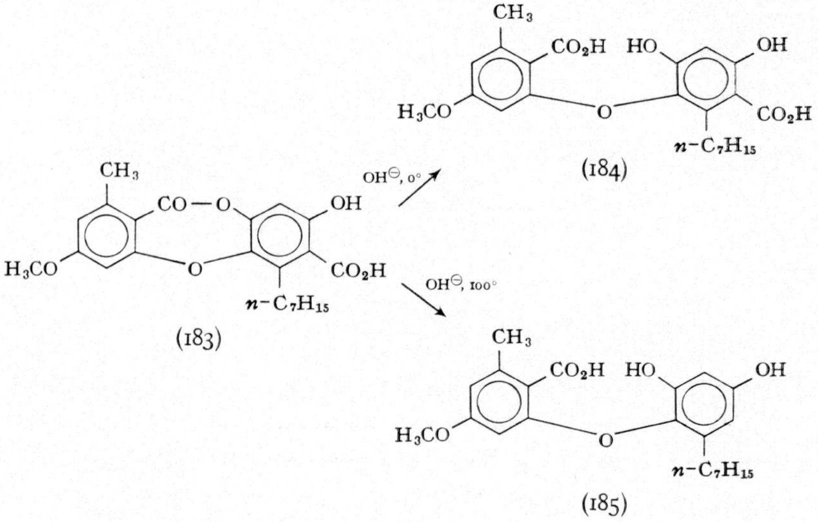

Schema 41. Hydrolyse von Grayansäure

Literaturverzeichnis: SS. 288—306

lierung von (176) und Oxydation mit aktivem Mangandioxid gibt Nordiploicin (177) (Schema 38). Der zweite Weg geht von 3,4-Dichlor-4-O-benzylorsellinsäure (178) aus, die mit 2,4-Dichlororcin (179) in Trifluoracetanhydrid zu (180) verestert wird; dann wird (180) mit Mangandioxid oxydiert, methyliert und katalytisch zu Diploicin (181) hydriert (Schema 39).

Eine weitere Diploicin-Synthese schließt die hydrolytische Spaltung des Grisans (182) ein (*132*) (Schema 40).

Die Struktur der Grayansäure (183) aus *Cladonia grayi* Merrill wurde vorwiegend aus spektroskopischen Daten abgeleitet (*284*); sie liefert bei vorsichtiger alkalischer Hydrolyse bei 0° Grayanoldicarbonsäure (184), bei 100° unter Decarboxylierung Grayanolsäure (185) (Schema 41).

Ein Depsid aus *Parmelia livida* Tayl. erwies sich mit 4-O-Methylphysodsäure (186) identisch (77).

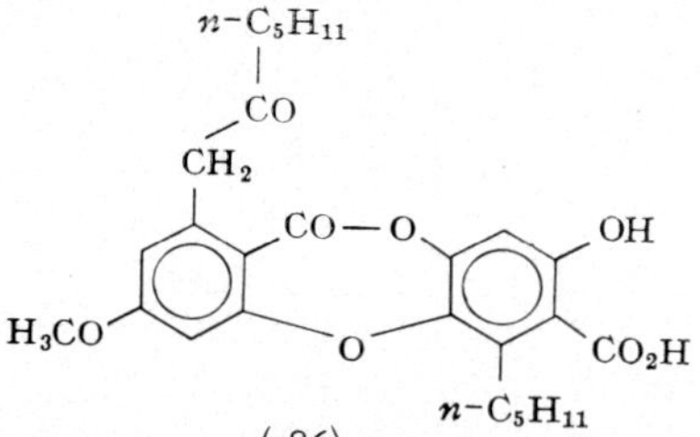

(186)

Das Depsidanalogon zur Perlatolinsäure, die in zahlreichen *Stereocaulon*-Arten vorkommt, Colensoinsäure (187), wurde von Fox et al. (*117*) in *Stereocaulon colensoi* Bab. gefunden.

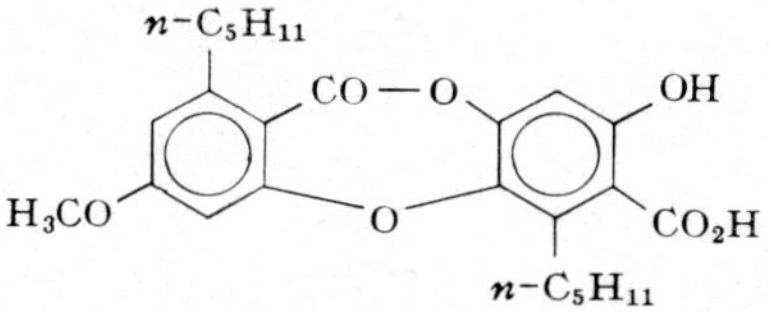

(187)

Die Struktur der Virenssäure (188) aus *Alectoria virens* Tayl. wurde durch katalytische Reduktion zu Hypoprotocetrarsäure (189) bewiesen

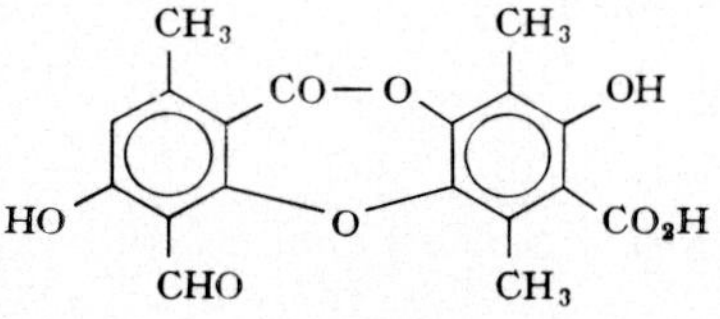

(188) 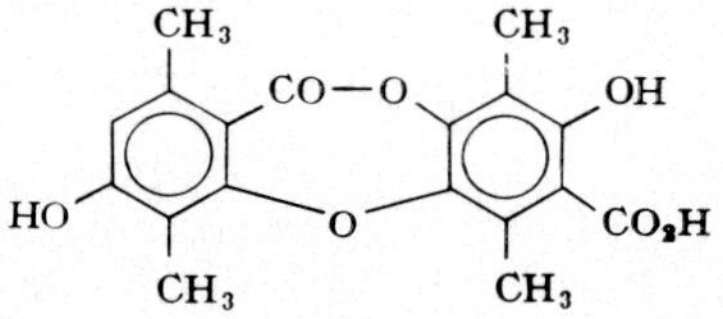   (189)

(9). Hypoprotocetrarsäure wurde später auch in verschiedenen *Ramalina*-Arten gefunden (*76, 146, 158*).

Die von NEELAKANTAN et al. (*223*) für das aus *Teloschistes flavicans* Sw. isolierte Chlordepsidon Vicanicin vorgeschlagene Konstitution (190) wurde von DYER et al. (*98a*) auf Grund der Röntgenstrukturanalyse der Jodacetylverbindung (191) zu (192) korrigiert. Vicanicin liefert nach Methylierung und Methanolyse (193) (*223*) (Schema 42).

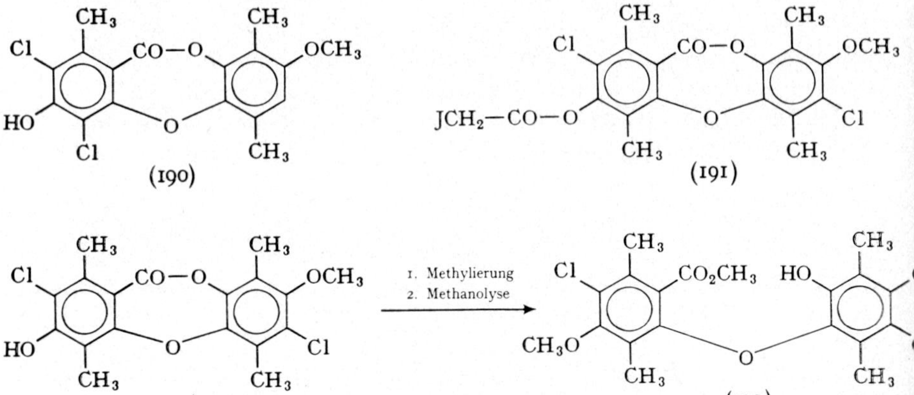

Schema 42. Methylierung und Methanolyse von Vicanicin

2.2.11. Depsone

Bisher wurde nur ein einziges Depson, Picrolicheninsäure (194), das bittere Prinzip von *Pertusaria amara* (Ach.) Nyl. aus Flechten isoliert. Die einzigartige Struktur von (194) wurde von WACHTMEISTER (*333*)

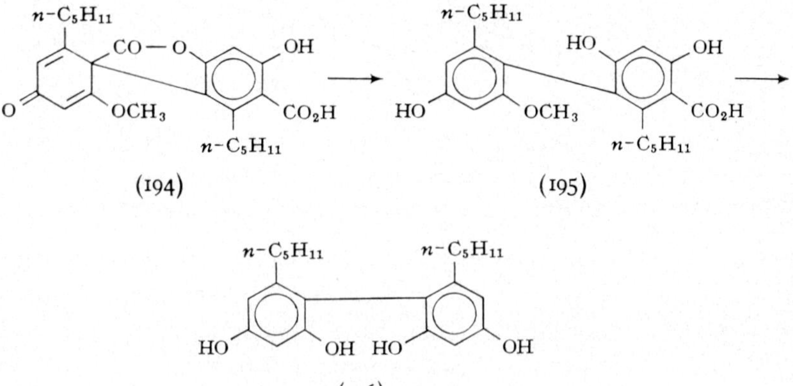

Schema 43. Abbau von Picrolicheninsäure

Literaturverzeichnis: SS. 288—306

aufgeklärt. Bei milder alkalischer Hydrolyse und nachfolgender Säureeinwirkung wird die Esterbindung unter Bildung von (195) gespalten, dessen Decarboxylierung und Entmethylierung 2,2′-Di-n-pentyl-4,6,4′,6′-tetrahydroxydiphenyl (196) gibt (Schema 43).

Die Struktur (194) von Picrolicheninsäure wurde durch eine biogeneseähnliche Synthese von DAVIDSON und SCOTT (*89*) bewiesen: Oxydation von Anziasäure-2-methyläther (197) mit aktivem Mangandioxid führt über das Diradikal (198) direkt zu (194) (Schema 44).

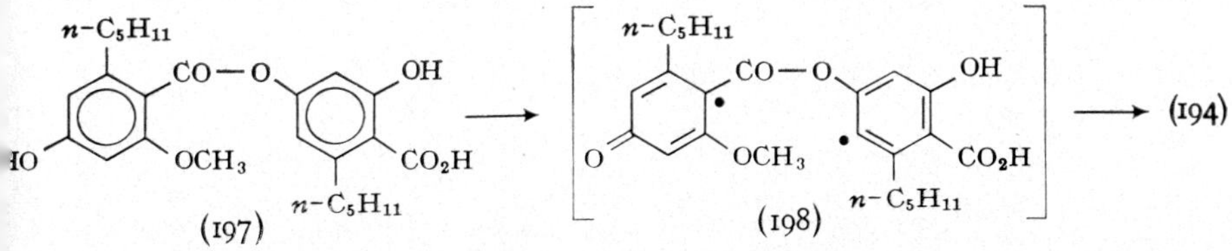

Schema 44. Synthese von Picrolicheninsäure

2.2.12. Bisanthrachinone und Bisanthronyle

In einer Serie von ausgezeichneten Arbeiten haben SHIBATA und Mitarbeiter die Struktur der unter anderem auch in der Strauchflechte *Acroscyphus sphaerophoroides* Lév. (*293*) vorkommenden dimeren Anthrachinone Rugulosin und Skyrin aufgeklärt (*283, 258, 226*). Aus der Röntgenstrukturanalyse von (+)-Dibromdehydrotetrahydrorugulosin (199) folgt für Rugulosin die Struktur und absolute Konfiguration (200) (*174, 174a*).

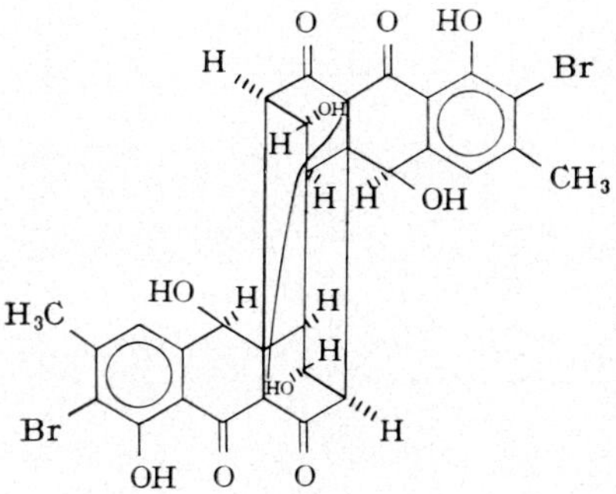

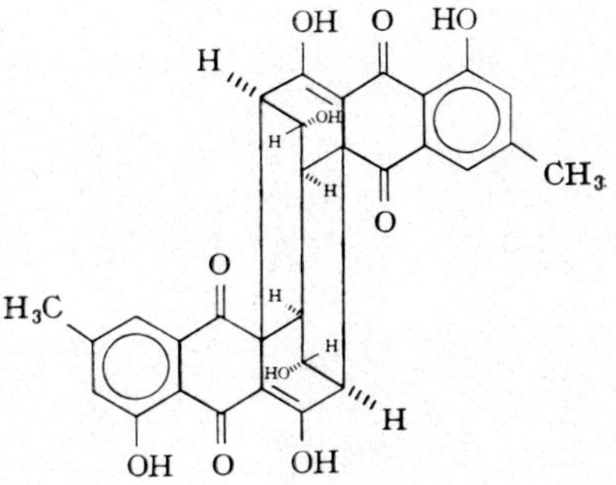

Skyrin (201, $R = R' = H$) ist wegen der eingeschränkten Rotation um die C-5—C-5′-Achse chiral und zeigt einen positiven COTTON-Effekt (*226*).

SANTESSON (272a) wies in *Trypetheliopsis boninensis* Asah. neben
(+)-Skyrin (201, $R'_4 = R' = H$) noch Oxyskyrin (201, $R = H$, $R' = OH$)
und Skyrinol (201, $R = R' = OH$) nach.

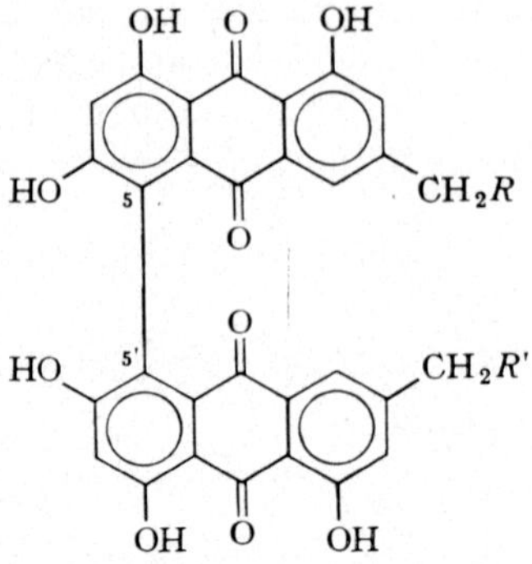

(201)

Anaptychia obscurata Vain. enthält neben den bereits erwähnten
Anthrachinonen (54) und (57) (352) die Bisanthronyle Flavoobscurin A
(202), Flavoobscurin B$_1$ und Flavoobscurin B$_2$ (353). Flavoobscurin B$_1$

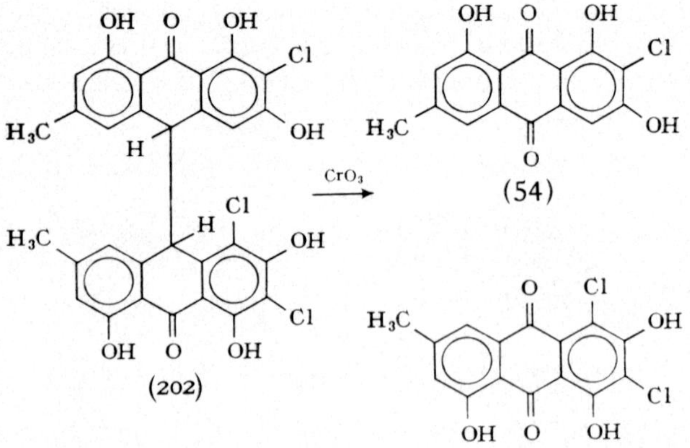

und B$_2$ sind wahrscheinlich Rotationsisomere (wegen der beschränkten
Drehbarkeit um die C-10—C-10'-Achse) der Struktur (203) und gehen
bereits bei längerem Kochen in Aceton in ein Gleichgewichtsgemisch aus
B$_1$ und B$_2$ über. Bei der Oxydation mit Chromtrioxid in Eisessig liefert
(202) ein Gemisch aus (54) und (57), (203) dagegen nur (57) (Schema 45).

Literaturverzeichnis: SS. 288—306

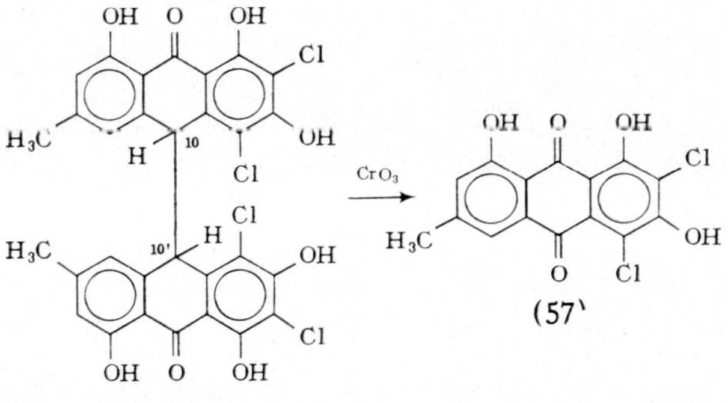

(203)

Schema 45. Oxydative Spaltung der Flavoobscurine B_1 und B_2

2.2.13. Bisxanthone

Das von YOSIOKA et al. aus *Parmelia entotheiochroa* Hue (*349*), *Parmelia perisidians* Nyl., *Parmelia aurulenta* Tuck. und *Parmelia subaurulenta* Nyl. (*342*) isolierte gelbe Entothein erwies sich identisch mit dem Mutterkornfarbstoff Ergochrom AA (= Secalonsäure A) (204) (*120*). *Parmelia entotheiochroa* Hue enthält ferner Ergochrom AB (= Secalonsäure C) (205) (*120*).

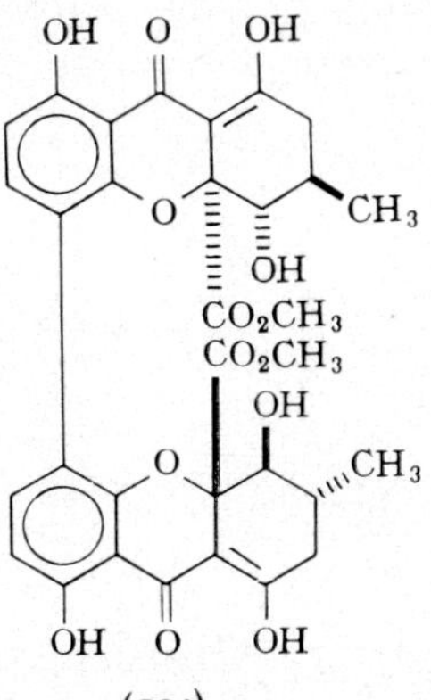

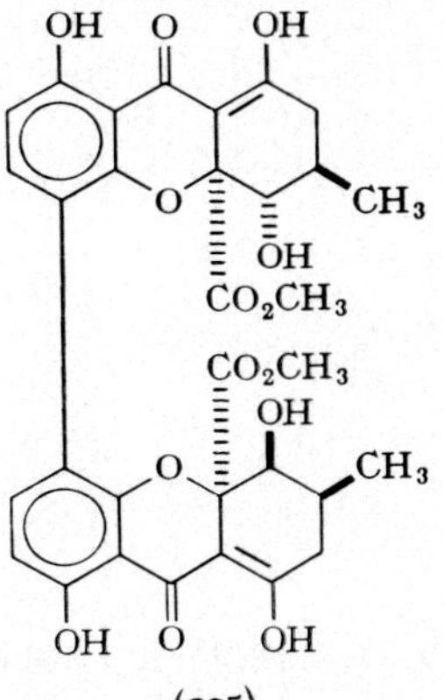

2.3. Mevalonat-Derivate

2.3.1. Diterpene

Das einzige bisher in seiner Struktur bekannte Diterpen aus Flechten ist (−)-16α-Hydroxykauran (206), das in verschiedenen *Ramalina*-Arten vorkommt (*146, 179, 149, 39*).

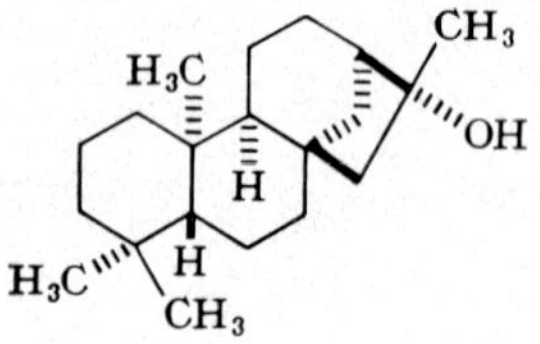

(206)

2.3.2. Triterpene

BARTON et al. (*31, 34*) klärten in zwei grundlegenden Arbeiten die Konstitution von Zeorin, das in zahlreichen Flechten vorkommt, bis auf die Stereochemie der Verknüpfung der Ringe D und E und die Stellung (α oder β) der C-21-Hydroxyisopropylgruppe auf. Die postulierte α-Stellung der Seitenkette (*137, 157, 344*) wurde zwar von YOSIOKA et al. (*345, 347*) angezweifelt, dann aber von der gleichen Arbeitsgruppe bestätigt (*348a*). Den endgültigen Beweis der α-Seitenkette im Zeorin (207) erbrachte die Röntgenstrukturanalyse von 6α-p-Brombenzoylzeorin (*217a*). Zum Unterschied von Leucotylin hat der Ring E im Zeorin „envelop"-Konformation.

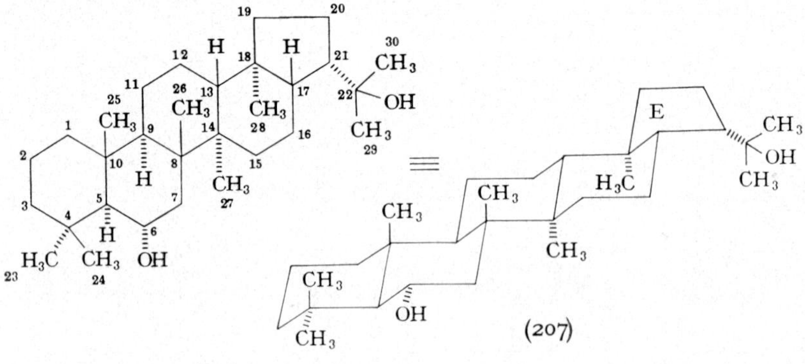

(207)

Bei der Behandlung von Zeorinon (208) mit Phosphoroxychlorid-Pyridin entstehen die beiden 22-Desoxy-Verbindungen (209) und (210). (209) gibt nach katalytischer Hydrierung im Neutralen (211), das sich weiter zu Hopan (211a) reduzieren läßt, während (210) nach katalytischer Hydrierung im Sauren (212) liefert, das weiter zu Isohopan (212a) reduziert werden kann (*348a*). (212) ist mit dem aus Leucotylin über Diacetylleucotylin, Diacetylanhydroleucotylin, 6α,16β-Dihydroxyisohopan und 6,16-Dioxo-isohopan hergestellten 6-Oxo-isohopan identisch (*348a, 347*) (Schema 46).

Literaturverzeichnis: SS. 288—306

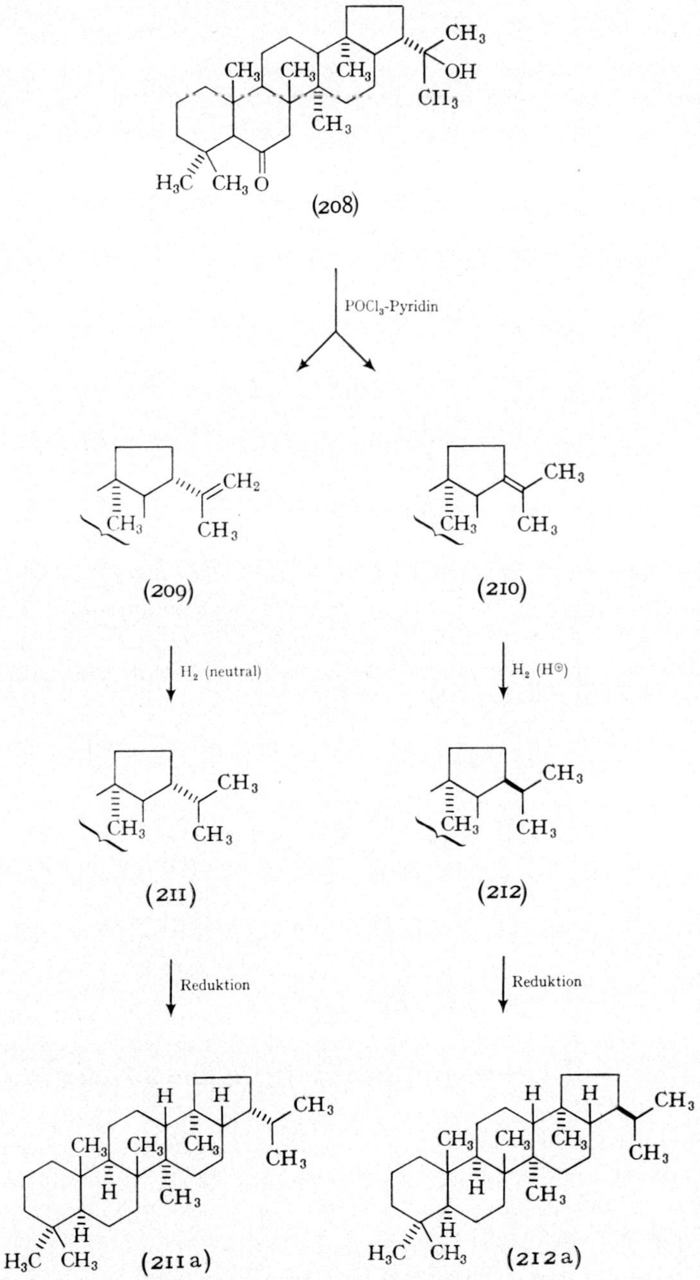

Schema 46. Dehydratisierung von Zeorinon

Nach TSUDA et al. (*326*) entsteht bei der katalytischen Hydrogenolyse von 22-Hydroxyhopan vorwiegend Isohopan (**212a**) neben Hopan (**211a**).

Leucotylin aus verschiedenen *Parmelia*-Arten und *Lecanora muralis* (Schreb.) Rabenh. (*80*) kommt nach YOSIOKA et al. (*340, 345, 346*) und der Röntgenstrukturanalyse des 6-Keto-16-β-p-brombenzoyl-leucotylins (*217*) die Stereoformel (**213**) mit einer Halbsesselform des Ringes E zu.

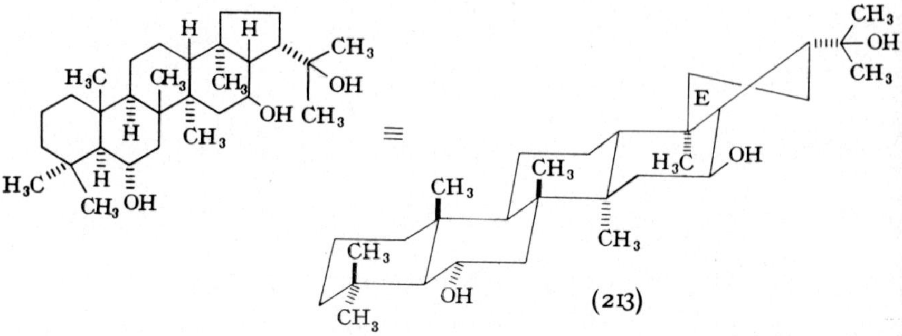

YOSIOKA et al. (*349*) isolierten aus *Parmelia entotheiochroa* Hue neben Atranorin, Entothein, Zeorin, Leucotylin, Leucotylsäure und 16β-Acetylleucotylsäure folgende Leucotylinderivate: 6α-Acetylleucotylin (**214**), 6-Desoxy-16β-acetylleucotylin (**215**), 6-Desoxyleucotylin (**216**) und 6α,16β-Diacetylleucotylin (**217**).

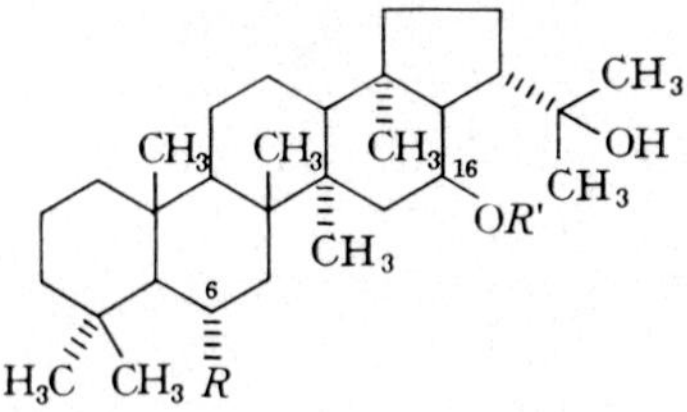

	R	R'
(**214**)	OAc	H
(**215**)	H	Ac
(**216**)	H	H
(**217**)	OAc	Ac

Leucotylsäure aus *Parmelia leucotyliza* Nyl. (*348*) und *Parmelia entotheiochroa* Hue (*349*) erwies sich als 16β,22-Dihydroxyhopan-23-säure (**218**) und konnte nach Schema 47 in Hopen-I (**219**) überführt werden.

Wie bereits erwähnt, kommt auch 16β-Acetylleucotylsäure (**220**) in *Parmelia entotheiochroa* Hue vor (*349*). Leucotylsäuremethylester (**221**) wird mit äthanolischer Salzsäure zu Isoleucotylsäuremethylester (**222**) isomerisiert (*350*).

Die katalytische Hydrierung von Anhydroleucotylin und Anhydroleucotylsäuremethylester untersuchten YOSIOKA et al. (*343*).

Literaturverzeichnis: SS. 288—306

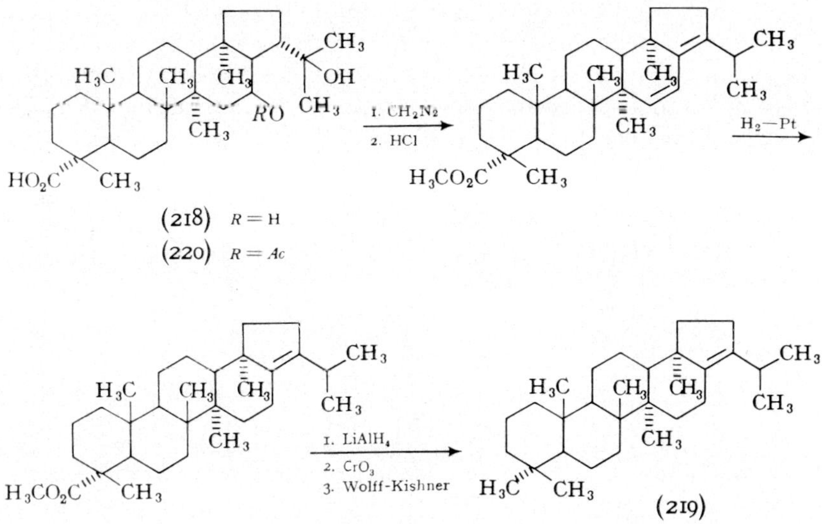

(218) R = H
(220) R = Ac

Schema 47. Umwandlung von Leucotylsäure in Hopen-I

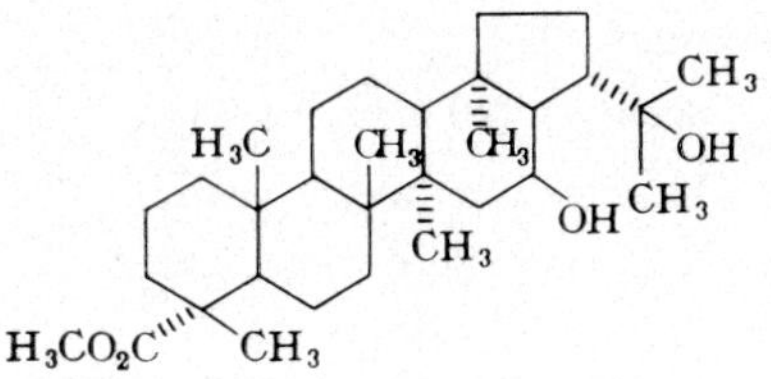

(221) 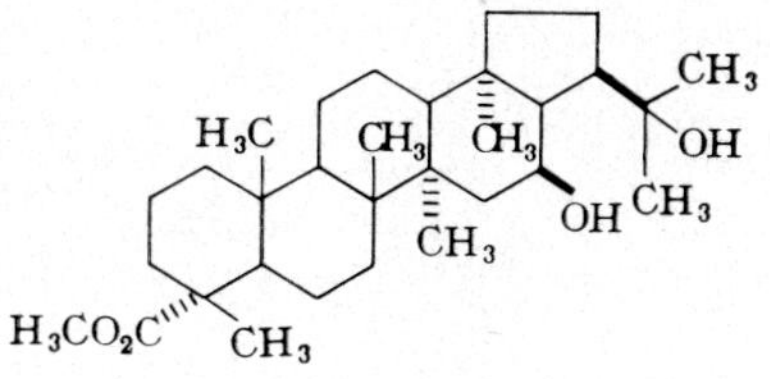   (222)

Eine weitere Triterpensäure, Pyxinsäure (223), isolierten YOSIOKA et al. (*341*) aus *Pyxine endochrysina* Nyl.; Reduktion des Methylesters von (223) mit Lithiumaluminiumhydrid und nachfolgende Spaltung des Diols mit Bleitetraacetat führt zum Keton (224), identisch mit einem aus Hydroxyhopanon gewonnenen Präparat (Schema 48).

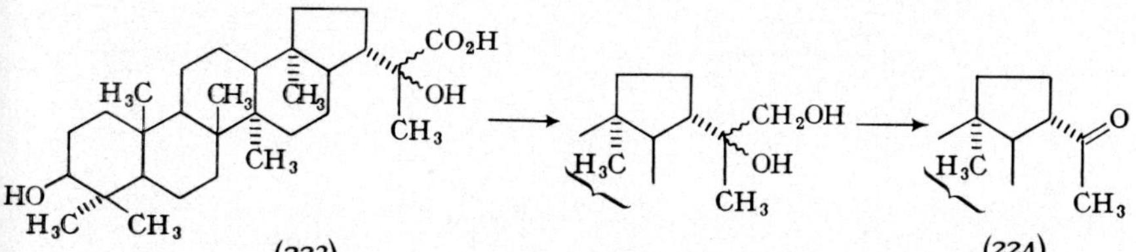

Schema 48. Umwandlung von Pyxinsäure in 3β-Hydroxy-norphopan-22-on

Die Struktur der beiden Triterpene aus *Pseudocyphellaria billardierii* (Del.) Raes. (syn. *Sticta billardierii* Del.) wurde von CORBETT und YOUNG (*73, 74*) aufgeklärt. Es handelt sich um 7β-Acetoxy-22-hydroxyhopan (**225**) und 15α,22-Dihydroxyhopan (**226**), die später auch in *Pseudocyphellaria intricata* (Del.) Wain. gefunden wurden (*152*).

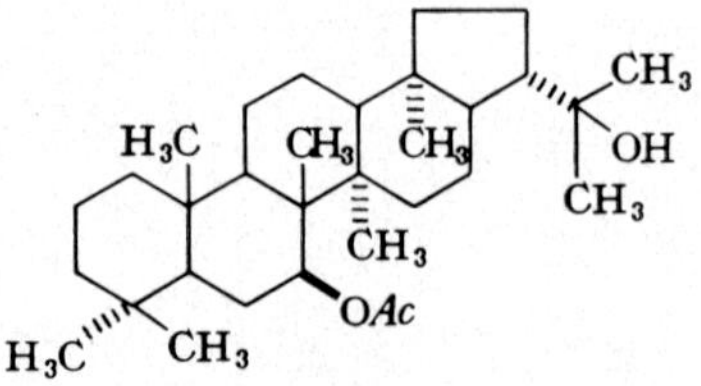

(**225**)

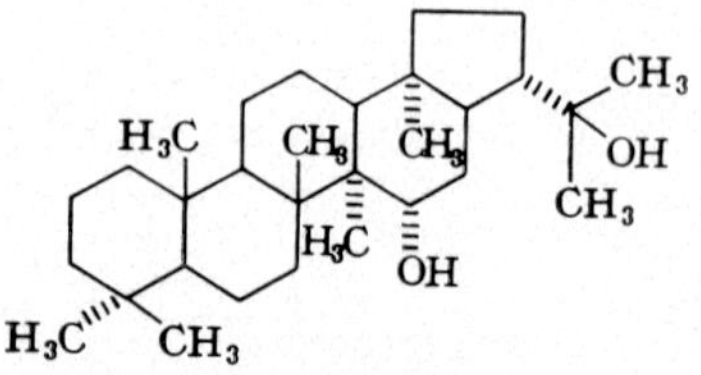

(**226**)

CORBETT und Mitarb. untersuchten ferner zahlreiche Umlagerungsreaktionen von (**225**) und (**226**) (*70, 71, 72*).

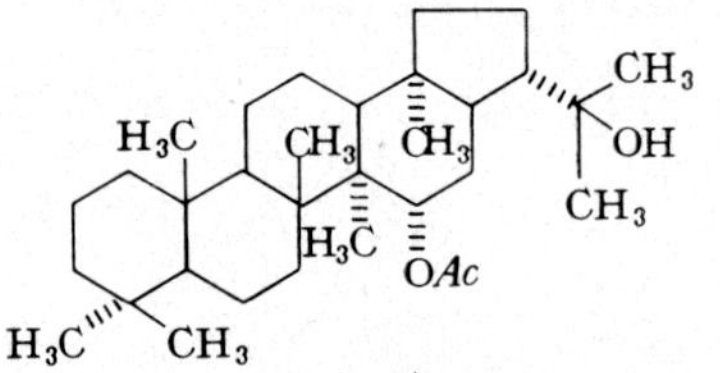

(**227**)

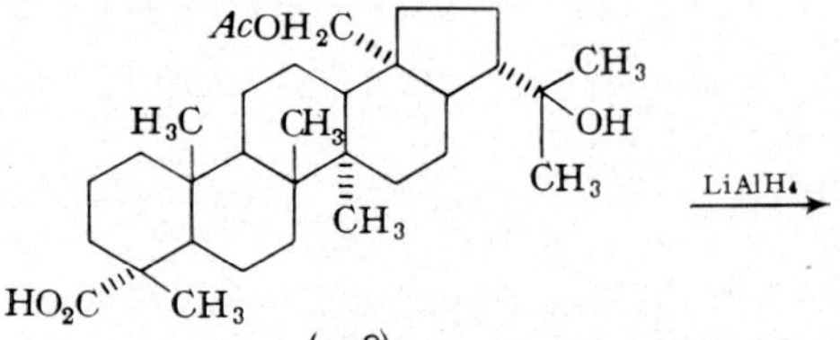

(**228**)

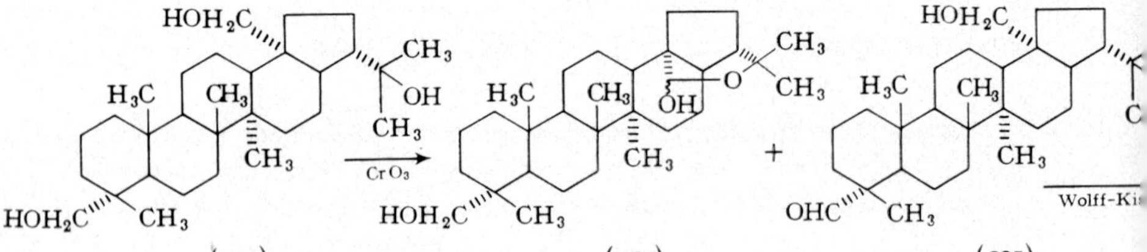

(**229**) (**230**) (**231**)

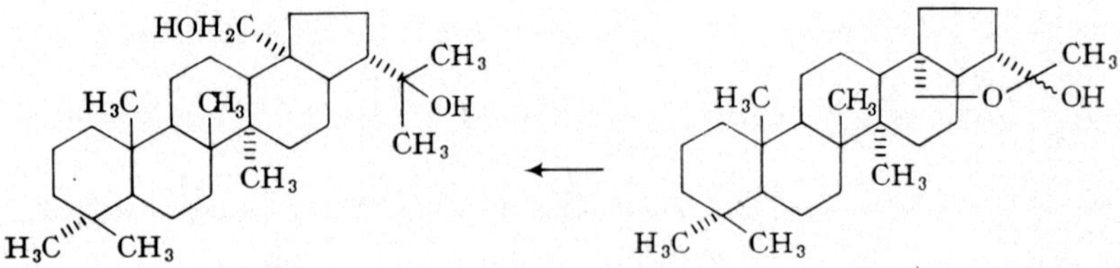

(**232**) (**233**)

Schema 49. Korrelation von Phlebsäure A mit Adipedatol

Literaturverzeichnis: SS. 288—306

In *Peltigera aphthosa* (L.) Willd. kommen neben Tenuiorin und Zeorin 15α-Acetoxy-22-hydroxyhopan (227) und Phlebsäure A vor, die von TAKAHASHI et al. (*323*) als 28-Acetoxy-22-hydroxyhopan-23-säure (228) erkannt wurde. (228) läßt sich mit Lithiumalumininmhydrid zum Triol (229) reduzieren, das bei Oxydation mit Chromtrioxid in Pyridin neben (230) den Aldehyd (231) gibt. WOLFF-KISHNER-Reduktion des Aldehydes liefert das Diol (232), das mit einem aus Adipedatol (233) gewonnenen Produkt identisch ist (Schema 49).

Phlebsäure B, ebenfalls aus *Peltigera aphthosa* (L.) Willd., ist 22-Hydroxyhopan-23-säure (233a, $R = H$); ihr Methylester (233a, $R = CH_3$) liefert bei Reduktion mit Lithiumaluminiumhydrid das Diol (233b), das mit Chromtrioxid zum Aldehyd (233c) oxydiert wird; dieser Aldehyd wird schließlich nach HUANG-MINLON zum bekannten 22-Hydroxyhopan (233d) reduziert (*323a*) (Schema 49a).

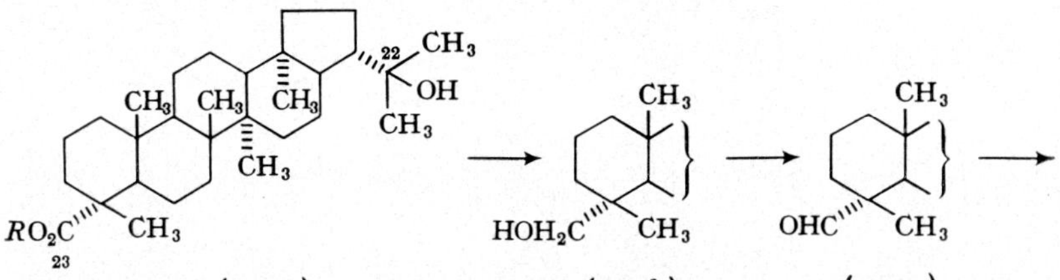

Schema 49a. Korrelation von Phlebsäure B mit 22-Hydroxyhopan

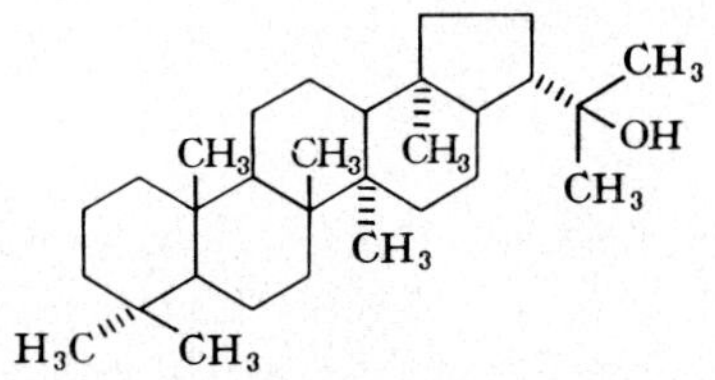

Weitere Hopanderivate, 6α,7α,22-Trihydroxyhopan (233e), 6α-Acetoxy-7α,22-dihydroxyhopan (233f) und 11α,22-Dihydroxyhopan (233g) fanden CORBETT und CUMMING (*69a*) in *Sticta mougeotiana* var. *dissecta* Del.

Diacetylpyxinol (234) aus *Pyxine endochrysina* Nyl. ist das erste aus Flechten isolierte Triterpen mit Dammaran-Gerüst und wurde durch Röntgenstrukturanalyse des 3β,12β-Di-p-brombenzoylpyxinols aufgeklärt (*351, 337*).

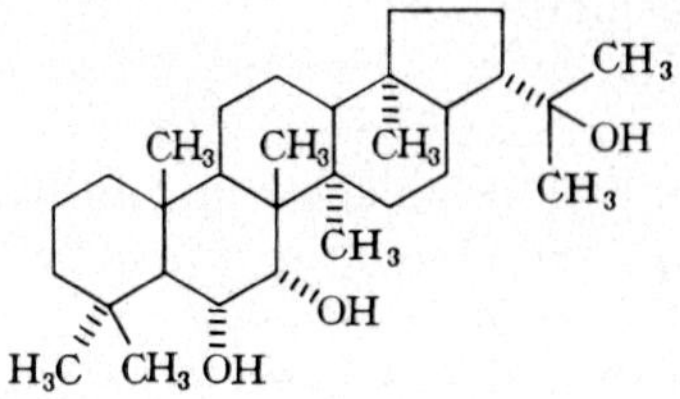

(233e)

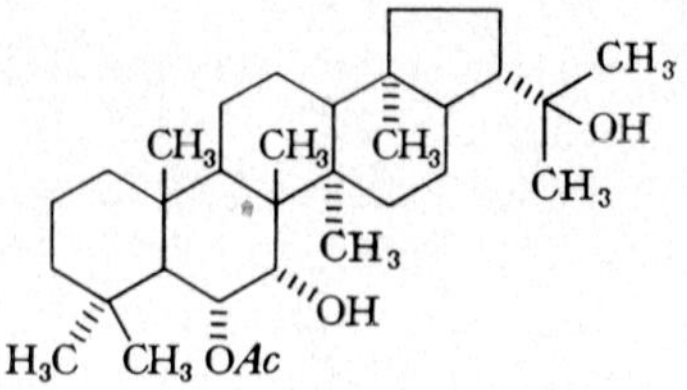

(233f)

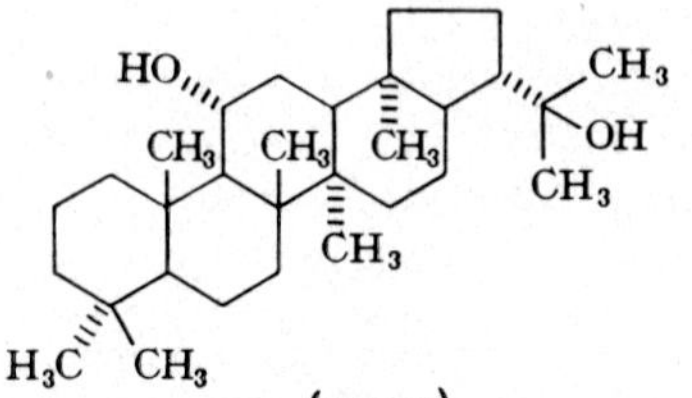

(233g)

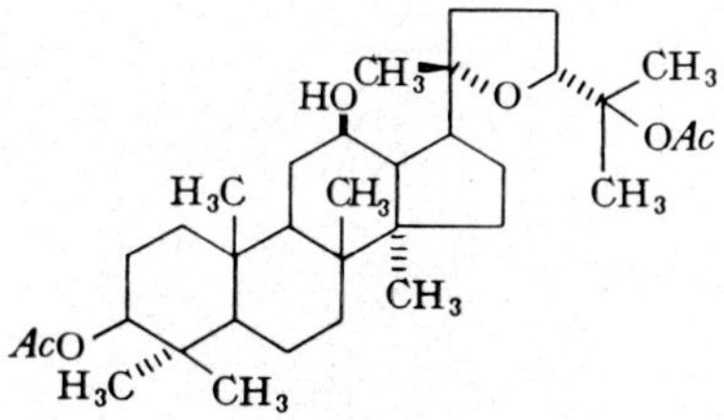

(234)

Bruun isolierte aus *Cetraria nivalis* (L.) Ach. Friedelin, Friedelan-3β-ol (56), α-Amyrin, Lupeol, Ursolsäure, Cerin und einen unbekannten Triterpenalkohol der Summenformel $C_{30}H_{50}O_2$ (60) sowie aus *Cladonia deformis* (L.) Hoffm. Taraxeren (57).

Weitere Triterpene aus Flechten harren noch der Strukturaufklärung (80).

2.3.3. Steroide

Ergosterin, Fungisterin und β-Sitosterin sind die einzigen bekannten Flechtensterine; sie liegen nur in sehr geringen Konzentrationen vor (80).

2.3.4. Carotinoide

Aus Flechten wurden β- und γ-Carotin und aus dem Phycobiont von *Xanthoria parietina* (L.) Th. Fr. Violaxanthin und Xanthophyll isoliert (80).

Literaturverzeichnis: SS. 288—306

3. Phenylalanin-Derivate

3.1. Picroroccellin

Trotz intensiver Suche fanden wir dieses Dipeptid in keiner der von uns analysierten Proben von *Roccella fuciformis* DC. verschiedenster Herkunft.

3.2. Polyporsäure

Polyporsäure (235) ist ein Terphenylchinon aus *Sticta colensoi* Bab. und *Sticta coronata* Müll.-Arg. (*80*).

3.3. Pyxiferin

Für das einzige bisher aus Flechten isolierte Dibenzochinon, Pyxiferin (aus *Pyxine coccifera* Nyl.) wurde Struktur (236) vorgeschlagen (*69*).

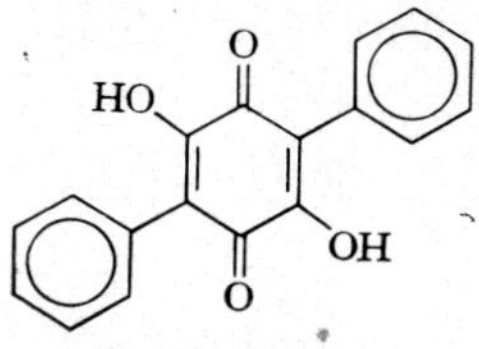

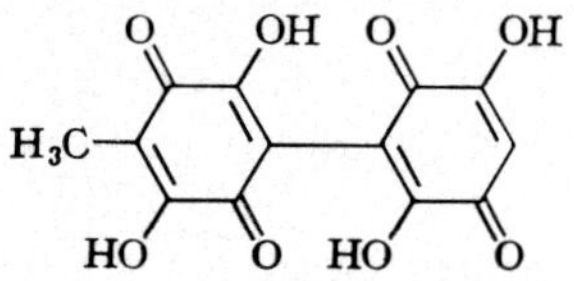

3.4. Thelephorsäure

Thelephorsäure wurde von verschiedenen Arbeitsgruppen untersucht (*244, 8*), jedoch erst von GRIPENBERG (*124*) durch Synthese strukturell endgültig aufgeklärt: Kondensation von 2 Mol. 3,4-Dimethoxyphenol mit Chloranil und nachfolgende Entmethylierung liefert Thelephorsäure (237) (Schema 50).

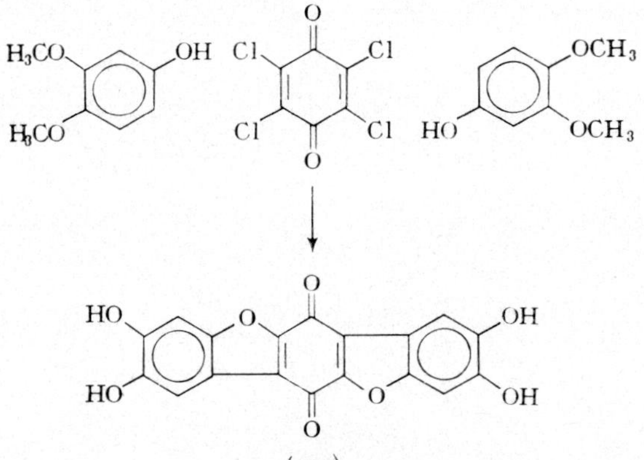

Schema 50. Synthese von Thelephorsäure nach GRIPENBERG (*124*)

In einer neueren Synthese wird 5-Jod-2-hydroxyhydrochinontrimethyläther (238) mit 2,5-Dijodhydrochinondimethyläther (239) in Gegenwart von Kupferbronze zu 2,4,5,2',5',2'',4'',5''-Octamethoxy-p-terphenyl (240) umgesetzt, das über (241) Tetraacetylthelephorsäure (242) liefert (*194*) (Schema 51).

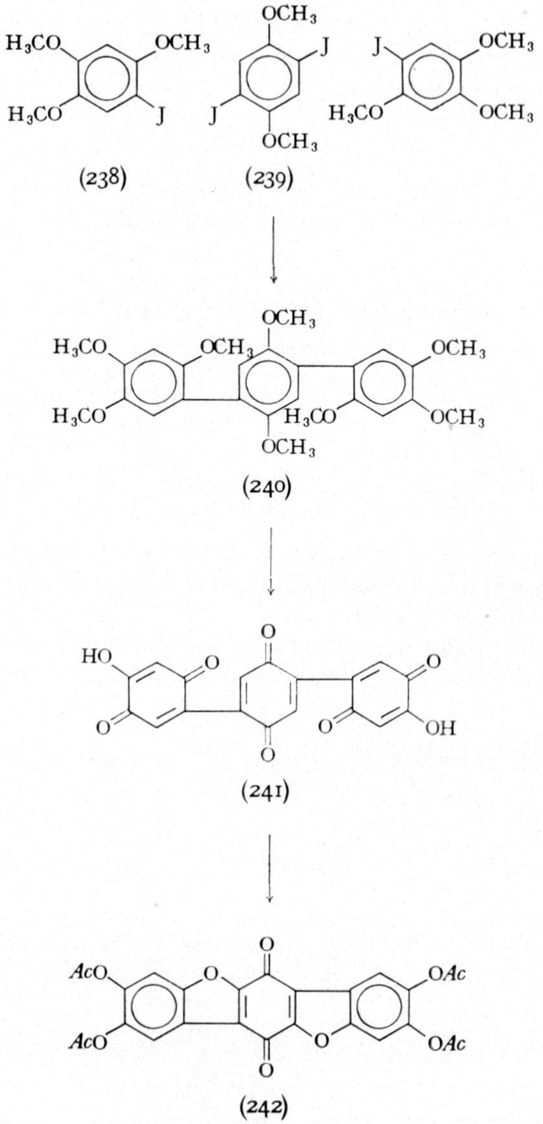

Schema 51. Synthese von Tetraacetylthelephorsäure nach LOUNASMAA (*194*)
Literaturverzeichnis: SS. 288—306

3.5. Pulvinsäure-Derivate

MAASS (*194b*) fand in *Pseudocyphellaria crocata* (L.) Vain. das bereits synthetisch hergestellte Pulvinsäureamid (242a), eine mögliche biosynthetische Vorstufe der Pulvinsäure.

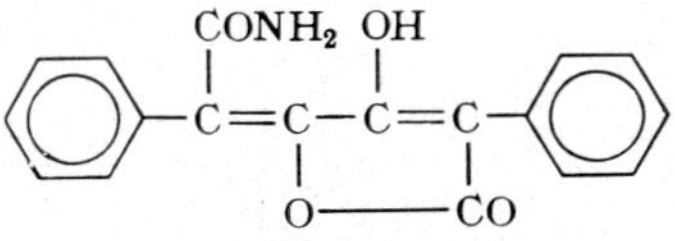

(242a)

Pulvinsäurelacton erhielten RUNGE und KOCH (*252a*) bei der Kondensation von Oxalylchlorid mit Phenacetylchlorid in Gegenwart von Pyridin oder Triäthylamin in 20% Ausbeute.

ÅKERMARK (*12*) synthetisierte Calycin nach Schema 52: das Kondensationsprodukt aus o-Methoxybenzylcyanid und Oxalsäurediäthylester (243) wird mit Benzylcyanid zu (244) kondensiert, das bei Hydrolyse mit Schwefelsäure Calycin (245) liefert.

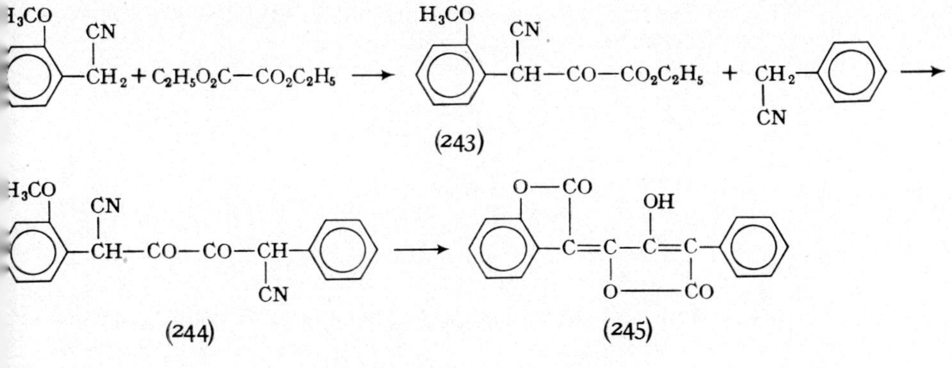

Schema 52. Synthese von Calycin

Die ursprünglich für Pinastrinsäure aufgestellte Struktur (246) (*26, 125*) kommt nach AGARWAL und SESHADRI (*4*) der Isopinastrinsäure zu, die bei Ozonolyse und hydrierender Spaltung des Ozonids mit Palladium in Essigsäureäthylester p-Methoxybenzoylameisensäuremethylester, Benzoylameisensäure und Benzoesäure liefert. Dagegen gibt Pinastrinsäure (247) bei analoger Spaltung Benzoylameisensäuremethylester, p-Methoxybenzoylameisensäure und Anissäure (Schema 53).

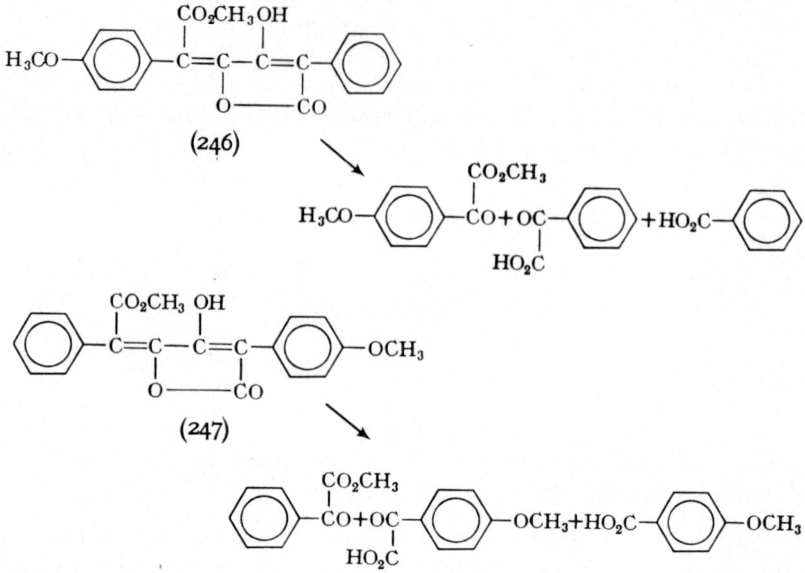

Schema 53. Ozonolytischer Abbau von Isopinastrin- und Pinastrinsäure

Die Synthese von Pinastrinsäure gelingt durch Hydrolyse von 2-(p-Methoxyphenyl-)-5-phenyl-3,4-dioxoadipodinitril (248) und anschließende Methylierung mit Diazomethan (6) (Schema 54).

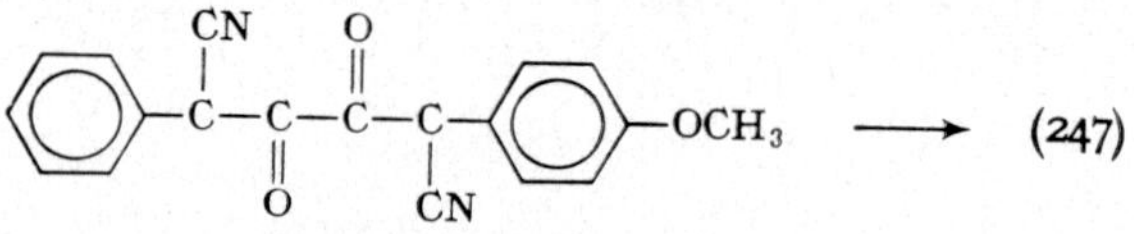

Schema 54. Synthese von Pinastrinsäure

Leprapinsäure (249) und Leprapinsäuremethyläther (250) kommen in verschiedenen *Lepraria*-Arten vor (*204, 7, 275*) und leiten sich möglicherweise von Calycin ab. Über die Synthese von Leprapinsäure berichten MITTAL und SESHADRI (*205*).

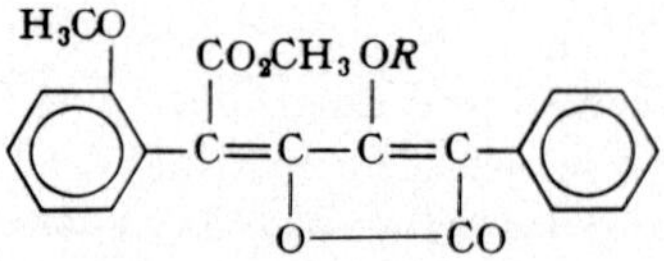

(249) R = H
(250) R = CH$_3$

Literaturverzeichnis: SS. 288—306

Die Umsetzung von Pulvinsäure mit Brom und o-Phenylendiamin wurde von SESHADRI und Mitarb. untersucht (126, 44a, 5).

3.6. Epanorin und Rhizocarpsäure

Für Epanorin und Rhizocarpsäure werden eine Reihe neuer Funde in Flechten beschrieben (80).

4. Vitamine

In Flechten wurden bisher folgende Vitamine gefunden: Ascorbinsäure, Biotin, Folsäure, Folinsäure, Nicotinsäure, Panthothensäure, Riboflavin und Vitamin B_1 (80).

5. Enzyme

Folgende Enzyme wurden in Flechten nachgewiesen: Amylase, Asparaginase, Katalase, Cellulase, eine Depsid-hydrolysierende Esterase, Invertase, Lichenase, Lipase, Orsellinsäuredecarboxylase, Oxidase, Peroxidase, Phenolase, Protease, Ribonuclease, Tannase, Tyrosinase, Urease und Zymase (80).

Um die mögliche Beteiligung des Phycobionten beim Katabolismus der Depside festzustellen, verglichen MOSBACH und EHRENSVÄRD (213) die durch teilweise gereinigte zellfreie Extrakte aus *Lasallia pustulata* (L.) Mérat (syn. *Umbilicaria pustulata* (L.) Hoffm.) bewirkte enzymatische Hydrolyse und Decarboxylierung verschiedener Depside und der zugrundeliegenden Phenolcarbonsäuren mit der Zersetzung, die durch zellfreie Extrakte des entsprechenden Phycobionten *(Trebouxia)* hervorgerufen wird. Sie fanden, daß die Substratspezifität beider Enzympräparate gleich ist: Gyrophorsäure, Umbilicarsäure und Evernsäure werden gespalten, während bei Benzoesäurephenylester und m-Digallussäure keine Hydrolyse eintritt. Orsellinsäuredecarboxylase, die nur aus dem Mycobionten isoliert werden konnte, zeigt eine ziemlich hohe Substratspezifität: Orsellinsäure, Homoorsellinsäure und Evernsäure werden decarboxyliert, 6-Methylsalizylsäure und β-Resorcylsäure dagegen nicht.

Glutamat-glyoxylat-, Glutamat-hydroxypyruvat- und Glutamatoxalacetat-Transaminase wurden in *Lobaria laetevirens* (Lightf.) Zahlbr., *Lobaria pulmonaria* (L.) Hoffm. und *Sticta sylvatica* (Huds.) Ach. nachgewiesen (41).

V. Biosynthese der Flechtenstoffe

Untersuchungen zur Biosynthese der Flechtenstoffe sind erst in den letzten Jahren durchgeführt worden. Zusammenfassend wurde darüber von HUNECK (143), FOX (115), MOSBACH (212) und CULBERSON (80) berichtet. Diese Untersuchungen werden durch die Doppelnatur, nicht ganz einfache Züchtung und das langsame Wachstum der Flechten erschwert.

Nach POKORNY et al. (*234a*) ist der Metabolismus von L- und D-Methionin bei den von ihnen verwendeten Flechten qualitativ gleich: beide Antipoden werden desaminiert, acetyliert, malonyliert und in α-Hydroxy- bzw. α-Keto-γ-methylthiobuttersäure umgewandelt.

Nach SMITH (*296*) kann *Peltigera polydactyla* (Neck.) Hoffm. relativ große Mengen von Asparagin, Glutamin, Glutaminsäure und Asparaginsäure akkumulieren, wobei jedoch die Amide erst nach Entamidierung in den Stoffwechsel einbezogen werden. Respirationsfähigkeit und Stickstoffgehalt der Algenschicht sind größer als die des Markes (*297*).

Über den Kohlenhydratstoffwechsel der Flechten liegen zahlreiche Untersuchungen von SMITH und FEIGE vor. Glucose wird von *Peltigera polydactyla* (Neck.) Hoffm. innerhalb 24 Stunden zu 20% zu Kohlendioxid veratmet, zu 45—50% in D-Mannit und zu 20% in 3-O-β-D-Glucopyranosyl-D-mannit verwandelt, wobei letzteres nur in der Algenzone synthetisiert wird (*298*). Interessanterweise scheiden frisch isolierte Phycobionten *(Nostoc)* von *Peltigera polydactyla* (Neck.) Hoffm. die Hauptmenge des photosynthetisch fixierten Kohlendioxids als Glucose ins Nährmedium aus, verlieren aber diese Eigenschaft bereits nach 48 Stunden (*97*). Nach Inkubation von *Peltigera polydactyla* (Neck.) Hoffm. mit $NaH^{14}CO_3$ im Licht ist nach 1 Min. ^{14}C-Glucose und nach 2 Min. ^{14}C-Mannit nachweisbar; nach Zusatz von höheren Konzentrationen an ^{12}C-Glucose wird kein ^{14}C-Mannit gebildet. Dies deutet darauf hin, daß Glucose vor der Alge zum Pilz wandert und ^{14}C-Glucose bei kompetitiver Hemmung durch ^{12}C-Glucose im Medium den Pilz nicht erreicht (*98*). Bei *Xanthoria aureola* (Ach.) Erichs. sind Ribit, D-Mannit, D-Arabit und Saccharose die Hauptkohlenhydrate, wie durch Fütterung mit $NaH^{14}CO_3$ ermittelt wurde; davon ist Ribit der erste Zucker, der in größeren Mengen angereichert und wahrscheinlich von der Alge synthetisiert wird (*246, 247*). In Flechten mit Grünalgen sind Polyole (bei *Trebouxia*, *Myrmecia* und *Coccomyxa* Ribit, bei *Trentepohlia* Erythrit und bei *Hyalococcus* Sorbit) primäre Kohlenhydrate, während Flechten mit Blaualgen *(Nostoc, Calothrix, Scytonema)* Glucose synthetisieren (*248, 245*). Nach FEIGE (*101, 295*) lassen sich auf Grund von ^{14}C-Markierungsversuchen folgende Markierungstypen aufstellen:

1. Mannit-Typen (Blaualgenflechten), in denen selbst nach 10 Stunden nur Mannit als markierter Zuckeralkohol nachweisbar ist. Hierher gehören wahrscheinlich alle *Peltigera*-Arten mit Blaualgen.

2. Pentit-Typen, in denen die Aktivität vorwiegend in D-Arabit, D-Mannit und Erythrit lokalisiert ist. Hierher gehören u. a. *Cladonia endiviaefolia* (Dicks.) Fr., *Lobaria pulmonaria* (L.) Hoffm., *Cetraria*

Literaturverzeichnis: SS. 288—306

islandica (L.) Ach., *Hypogymnia physodes* (L.) Nyl., *Ramalina maciformis* (Del.) Nyl. und *Ramalina fastigiata* (Pers.) Ach.

3. Kombinationstypen zwischen 1. und 2., zu denen z. B. *Solorina crocea* (L.) Ach. zählt und

4. Typen, die von 1., 2. und 3. abweichen. Bei dieser Gruppe ist D-Arabit das am stärksten markierte Produkt; im Gegensatz zu Typ 2 wird jedoch eine stärkere Markierung des Erythrits gegenüber D-Mannit beobachtet. Zu diesem Typ zählen *Letharia vulpina* Hue und *Evernia prunastri* (L.) Ach.

Wird an die Grünalgenflechte *Cladonia convoluta* (Lam.) P. Cout. uniform markierte ^{14}C-Glucose oder uniform markiertes ^{14}C-Glycerin verfüttert, so ähnelt das Verteilungsmuster der Markierungsprodukte (Mannit, Mannitglykosid und wahrscheinlich Threit) dem einer Blaualgenflechte nach photosynthetischer ^{14}C-Fixierung (*102a*).

Kürzlich fand FEIGE (*102*), daß die Basidiolichene *Cora pavonia* (Sw.) Fr. im Gegensatz zu allen bisher untersuchten Blaualgenflechten einen Pentit (Ribit oder Arabit) synthetisiert und in Form eines Pentitgalaktosids speichert.

BLOOMER et al. (*45, 45a, 46*) untersuchten die Biosynthese der (+)-Protolichesterinsäure mit *Cetraria islandica* (L.) Ach. in Hydroponkultur. Nach Fütterung mit 1-^{14}C-Acetat wurde die radioaktiv markierte (+)-Protolichesterinsäure (Einbaurate: 0,01—0,08%) isoliert und nach Schema 55 abgebaut.

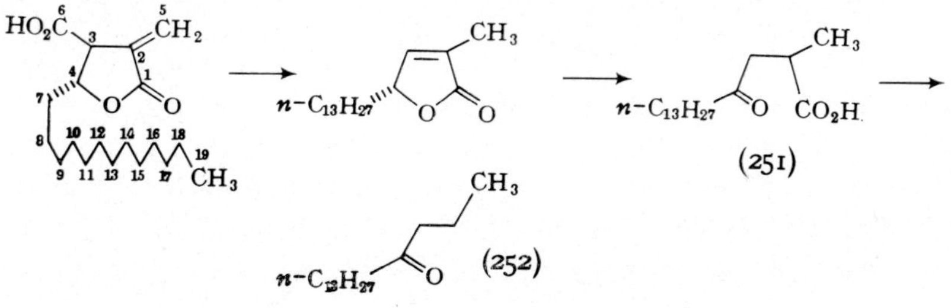

Schema 55. Abbau von (+)-Protolichesterinsäure

Aus der in Tabelle 4 gezeigten Verteilung der Radioaktivität kann geschlossen werden, daß (+)-Protolichesterinsäure durch Kondensation einer Fettsäure (C-6,3,4,7—19) mit einer C_3- oder C_4-Einheit aus dem Acetatstoffwechsel gebildet wird.

Tatsächlich wird von *Cetraria islandica* (L.) Ach. im Sommer 1,4-^{14}C-Bernsteinsäure spezifisch in die Stellungen 1, 2 und 5 eingebaut, wonach sich folgendes Biosyntheseschema für (+)-Protolichesterinsäure ergibt (Schema 56).

Tabelle 4. *Verteilung der Radioaktivität in (+)-Protolichesterinsäure*

Kohlenstoffatom	Gemessen als	% der totalen Radioaktivität
C-1	CO_2	3
C-2	CO_2	2
C-5	CO_2	2
C-6	CO_2	12
C-4, C-7—19	$C_{13}H_{27}CO_2CH_3$	80
C-7—19	$C_{12}H_{25}CO_2CH_3$	67
C-8—19	$C_{11}H_{23}CO_2CH_3$	67
C-1—5, C-7—19	(251)	85
C-2, C-5	CH_3CO_2Na	4
C-2—5, C-7—19	(252)	81

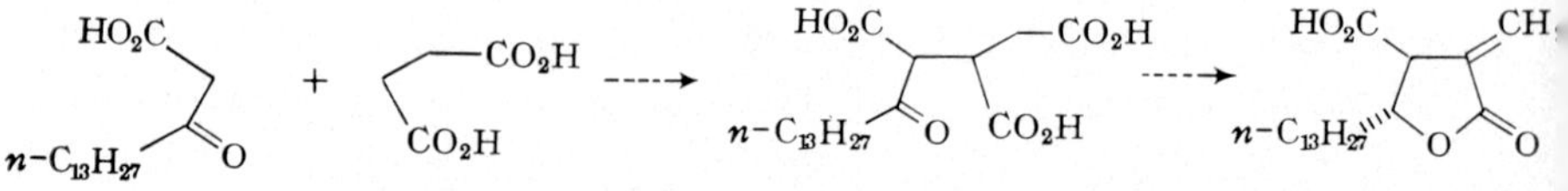

Schema 56. Biosynthese der (+)-Protolichesterinsäure

ABERHART et al. (*1*) haben 1- und 2-^{14}C-Acetat, Methyl-^{14}C-Methionin, 1-^{14}C-Malonat, 1- und 2-^{14}C-Propionat sowie radioaktives Mevalonat an *Roccella fuciformis* DC. verfüttert und gefunden, daß die Biosynthese des Acetylportentols (35) offenbar über eine Polyketidzwischenstufe verläuft. Acetat und Malonat werden in (35) inkorporiert, Propionat und Mevalonat dagegen nicht. Während die terminale sekundäre Methylgruppe vom C-2 des Acetats stammt, leiten sich die restlichen 5 Methylgruppen vom Methionin ab (Schema 56a).

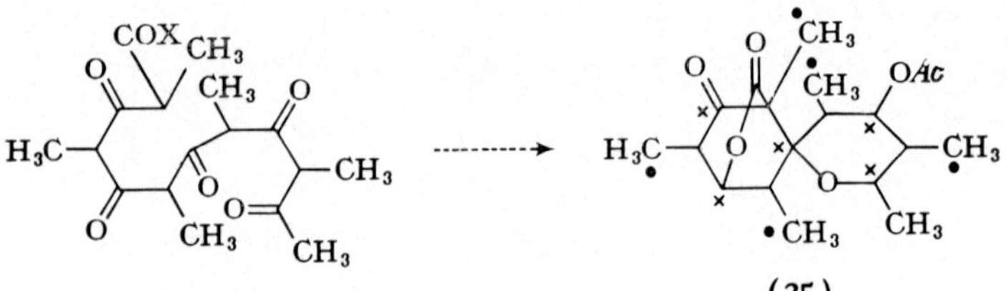

● ~ Methyl-^{14}C-Methionin
× ~ 1-^{14}C-Acetat oder 1-^{14}C-Malonat

Schema 56a. Biosynthese von Acetylportentol

Mit der Biosynthese aromatischer Flechtenstoffe haben sich MOSBACH, SHIBATA und MAASS beschäftigt. MOSBACH (*208, 207*) inkubierte die beiden Pilze *Chaetomium cochliodes* Pall. und *Penicillium baarnense* v.

Literaturverzeichnis: SS. 288—306

Beyma mit 1- und 2-^{14}C-Malonsäure und beobachtete eine dem Acetatschema (*194a*) entsprechende Verteilung der Radioaktivität in Orsellinsäure (Schema 57).

$$HO_2C-CH_2-\overset{*}{C}O_2H \longrightarrow \quad\ldots\quad \longrightarrow \text{(Orsellinsäure)}$$

$$HO_2\overset{*}{C}-\overset{*}{C}H_2-CO_2H \longrightarrow \quad\ldots\quad \longrightarrow \text{(Orsellinsäure)}$$

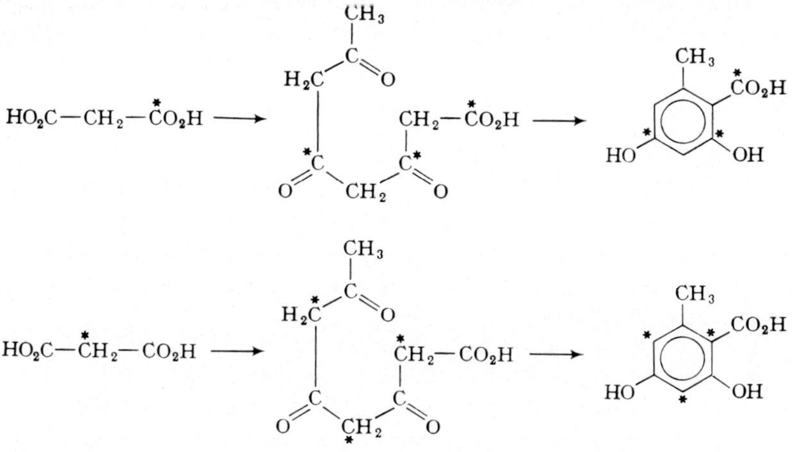

Schema 57. Biosynthese der Orsellinsäure

Unter Photosynthesebedingungen wird $^{14}CO_2$ von *Lasallia pustulata* (L.) Mérat uniform in alle C-Atome von Gyrophorsäure eingebaut (*119*).

Aus weiteren Versuchen mit radioaktiv markierter Essigsäure geht hervor, daß Orsellinsäure aus einer Acetyl-Co-A-Einheit und 3-Malonyl-Co-A-Einheiten aufgebaut wird. Dieser Befund wird durch Fütterung von 1-^{14}C-Malonsäurediäthylester an *Lasallia pustulata* (L.) Mérat und Abbau der radioaktiven Gyrophorsäure bestätigt (*209*); es ergab sich die im Schema 58 gezeigte Verteilung der radioaktiven ^{14}C-Atome.

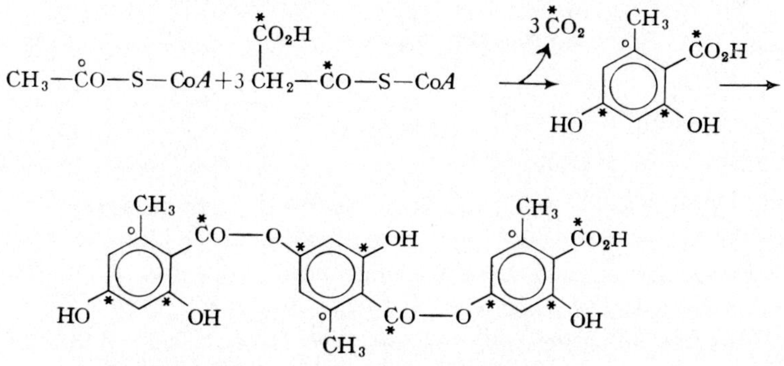

Schema 58. Biosynthese der Gyrophorsäure

Nach Fütterung von 1-^{14}C-Essigsäure an *Parmelia tinctorum* Despr. konnten sowohl radioaktive Lecanorsäure als auch radioaktives Atranorin und Chloratranorin isoliert werden; dagegen waren nach Applikation von ^{14}C-Ameisensäure nur Atranorin und Chloratranorin aktiv. Abbau der Lecanorsäure lieferte das zu erwartende Aktivitätsmuster und zeigte, daß auch Lecanorsäure nach dem Acetatschema biosynthetisiert wird. Die Aldehydgruppe am C-Atom 3 und die Methylgruppen an den C-Atomen 3' und 6' in Atranorin und Chloratranorin leiten sich dagegen von einem C_1-Fragment ab *(338)*. Wird tritiummarkierte Orsellinsäure und β-Orcincarbonsäure *Parmelia tinctorum* Despr. verabreicht, so wird zwar Orsellinsäure in Lecanorsäure und β-Orcincarbonsäure in Atranorin, nicht aber Orsellinsäure in Atranorin und β-Orcincarbonsäure in Lecanorsäure eingebaut, ein Beweis dafür, daß die C_1-Einheit vor der Aromatisierung zur Orsellinsäure angeknüpft wird *(339)*.

Die von SCHÖPF und ROSS *(277)* postulierte und von BARTON et al. *(33)* in vitro durchgeführte Biosynthese der Usninsäure aus zwei Molekülen Methylphloracetophenon wurde kürzlich von SHIBATA und Mitarb. bestätigt *(313, 314)*. Sie fanden, daß Natriumacetat-2-^{14}C, Natriumformiat-^{14}C, Malonsäure-2-^{14}C und Methylphloracetophenon-CO-^{14}CH$_3$ von *Usnea longissima* Ach., *Usnea diffracta* Wain., *Evernia mesomorpha* Müll.-Arg., *Parmelia caperata* (L.) Ach., *Cladonia mitis* Sandst. und *Cladonia alpestris* (L.) Rabenh. in Usninsäure eingebaut werden, Phloracetophenon-CO-^{14}CH$_3$ jedoch nicht. Aus Abbauversuchen ergab sich folgende Verteilung der radioaktiven C-Atome (Schema 59).

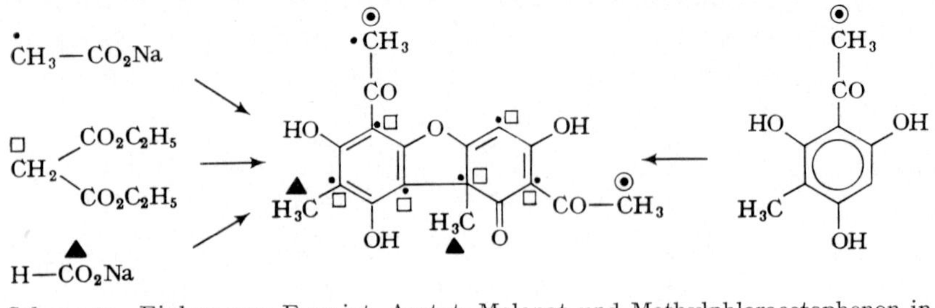

Schema 59. Einbau von Formiat, Acetat, Malonat und Methylphloracetophenon in Usninsäure

Das Muster zeigt, daß Usninsäure aus zwei C_8-Polyketidketten synthetisiert wird, die beiden mit ▲ markierten Methylgruppen von C_1-Einheiten stammen und diese Methylgruppen in die C_8-Polyketidketten vor der Cyclisierung eingebaut werden. Außerdem wurde durch die Isotopenverdünnungsmethode in *Cladonia mitis* Sandst. Methylphloracetophenon nachgewiesen und die letzte Stufe der Usninsäurebiosynthese durch Fütterung und Umwandlung von tritiummarkiertem Usninsäurehydrat in Usninsäure bewiesen. Damit nimmt die Biosynthese der Usninsäure den im Schema 60 gezeigten Weg.

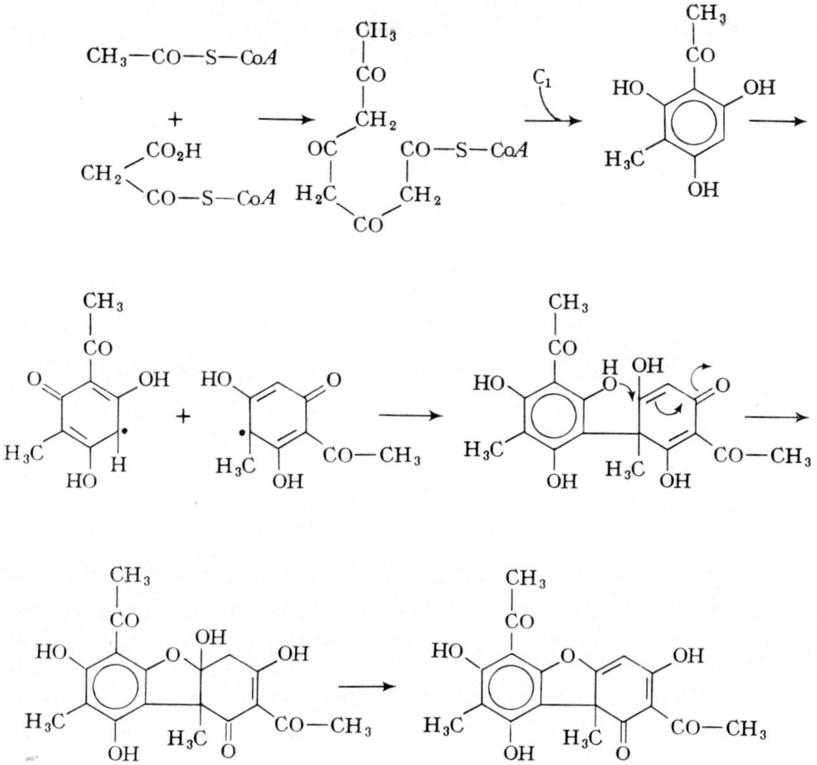

Schema 60. Biosynthese der Usninsäure

Die enzymatische Umwandlung von Methylphloracetophenon in Usninsäurehydrat mittels Meerrettichperoxidase gelang PENTTILA und FALES (*228*).

Wie bereits erwähnt, synthetisiert *Cetraria islandica* (L.) Ach. nur in den Sommermonaten (+)-Protolichesterinsäure; ähnliche Verhältnisse fanden TAGUCHI et al. (*315*) bei *Parmelia caperata* (L.) Ach. und *Usnea diffracta* Wain. in Japan, mit einem ausgeprägten Maximum der Syntheseaktivität für Usninsäure im Februar; dagegen unterliegt die Syntheseaktivität für Diffractasäure und Protocetrarsäure kaum jahreszeitlichen Schwankungen.

Durch Fütterung von DL-Phenylalanin-1-^{14}C an *Letharia vulpina* (L.) Hue und Abbau der radioaktiven Vulpinsäure fand MOSBACH (*210*) das im Schema 61 gezeigte Verteilungsmuster der radioaktiven C-Atome und folgerte daraus die Biosynthese der Vulpinsäure aus 2 Molekülen Phenylalanin über Polyporsäure.

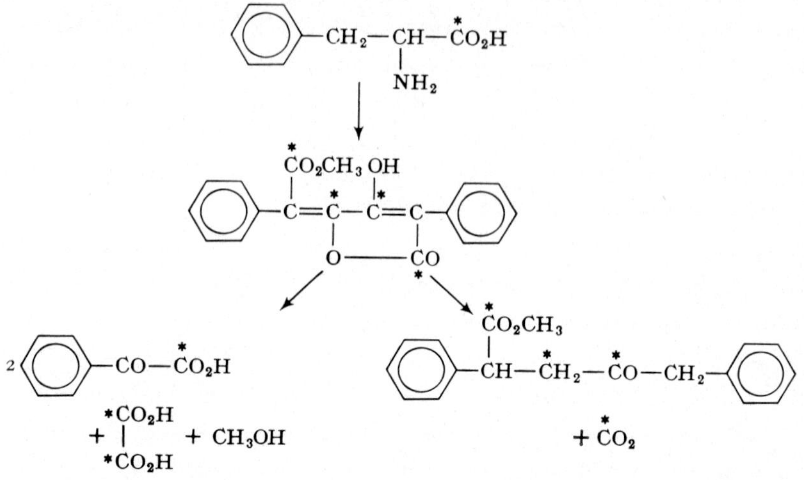

Schema 61. Einbau von Phenylalanin in Vulpinsäure

Zu einem ähnlichen Ergebnis kamen MAASS et al. (*196*), die bei *Pseudocyphellaria crocata* (L.) Vain. Phenylalanin als besten Vorläufer von Pulvinsäurelacton beobachteten. Zimtsäure-2-^{14}C, Glucose-U-^{14}C und Natriumacetat-2-^{14}C werden praktisch nicht inkorporiert. Der Abbau des aus L-Phenylalanin-1-^{14}C, DL-Phenylalanin-2-^{14}C und ringmarkiertem DL-Phenylalanin-^{14}C biosynthetisierten Pulvinsäurelactons über Vulpinsäure zu Benzoylameisensäure, Oxalsäure und Benzoesäure zeigte, daß die Kohlenstoffatome 1 und 2 des Phenylalanins den Carbonyl- und Enolkohlenstoffatomen des Pulvinsäurelactons entsprechen (Schema 62).

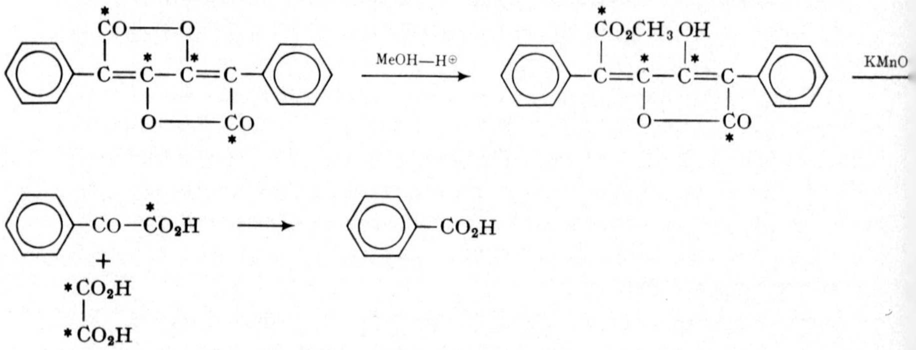

Schema 62. Abbau von radioaktiv markiertem Pulvinsäurelacton

Literaturverzeichnis: SS. 288—306

In weiteren Experimenten stellten MAASS und NEISH (*195*) fest, daß neben ^{14}C-markiertem Phenylalanin auch ^{14}C-markierte Phenylmilchsäure und Polyporsäure in guter Ausbeute von *Pseudocyphellaria crocata* (L.) Vain. in Pulvinsäurelacton und Calycin eingebaut werden. Bei Pulsmarkierung wird Pulvinsäurelacton schneller als Calycin markiert. Daraus ergibt sich Schema 63 für die Biosynthese beider Verbindungen.

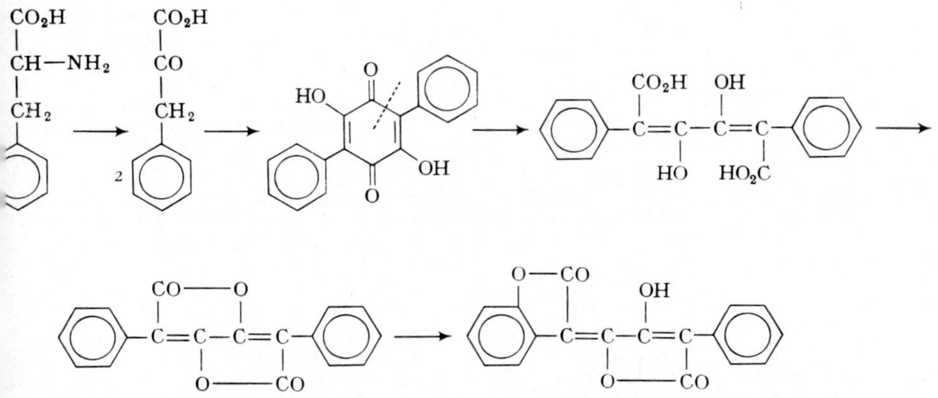

Schema 63. Biosynthese von Pulvinsäurelacton und Calycin

Eine Alternative zur Hypothese, daß Polyporsäure aus 2 Molekülen Phenylbrenztraubensäure gebildet wird, postuliert MAASS (*194b*). Danach soll Phenylalanin nach Bindung an Pyridoxalphosphat eine aktive Methylengruppe ausbilden, die nach Kondensation mit Phenylbrenztraubensäure ein Chinoniminderivat der Polyporsäure liefert. Oxydative Spaltung des Chinonimins führt zum Pyridoxal-gebundenen Pulvinamid, von dem sich durch Hydrolyse, Methanolyse oder Transamidierung Pulvinamid, Pulvinsäure, Vulpinsäure bzw. die Aminosäurekonjugate der Pulvinsäure ableiten (Schema 63a).

Die Biosynthese der Sterine sowie Di- und Triterpene verläuft auch in den Flechten zweifellos über Acetyl-Co-A und Mevalonsäure. Beim (—)-16α-Hydroxykauran (206) wird als Zwischenstufe der Cyclisierung des Geranylgeranylpyrophosphates eine enantio-Pimaradien-Zwischenstufe angenommen. Bei den Sterinen und Triterpenen werden 2 Moleküle Farnesylpyrophosphat Schwanz-an-Schwanz zum Squalen verknüpft und dieses zum 2,3-Oxidosqualen oxydiert. Das Epoxid wird dann je nach den vorliegenden Faltungen enzymatisch zu verschiedenen Endprodukten cyclisiert. Die meisten Flechtentriterpene tragen am C-Atom 3 keinen Sauerstoff mehr: entweder wird die Cyclisierung des Squalens

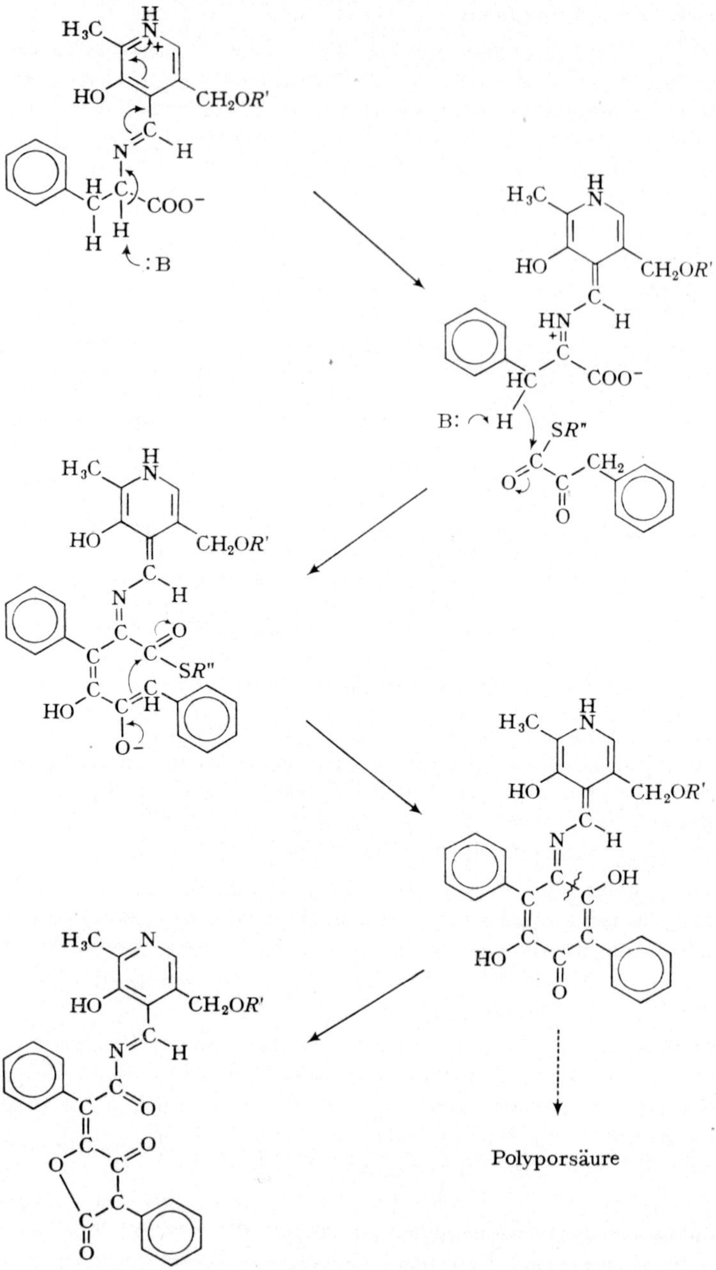

R' = an Protein gebundener Phosphatrest, R'' = Coenzym A oder Äquivalent

Schema 63a. Biosynthese von Pulvinsäurederivaten nach Maass (*194a*)

Literaturverzeichnis: SS. 288—306

direkt durch ein Proton ausgelöst oder der Sauerstoff wird nachträglich eliminiert (53a). Im Falle der Hopanderivate (z. B. 233g) liegt das Squalen bzw. dessen 2,3-Epoxid in einer durchgehend sesselgefalteten Kette vor (Schema 63b).

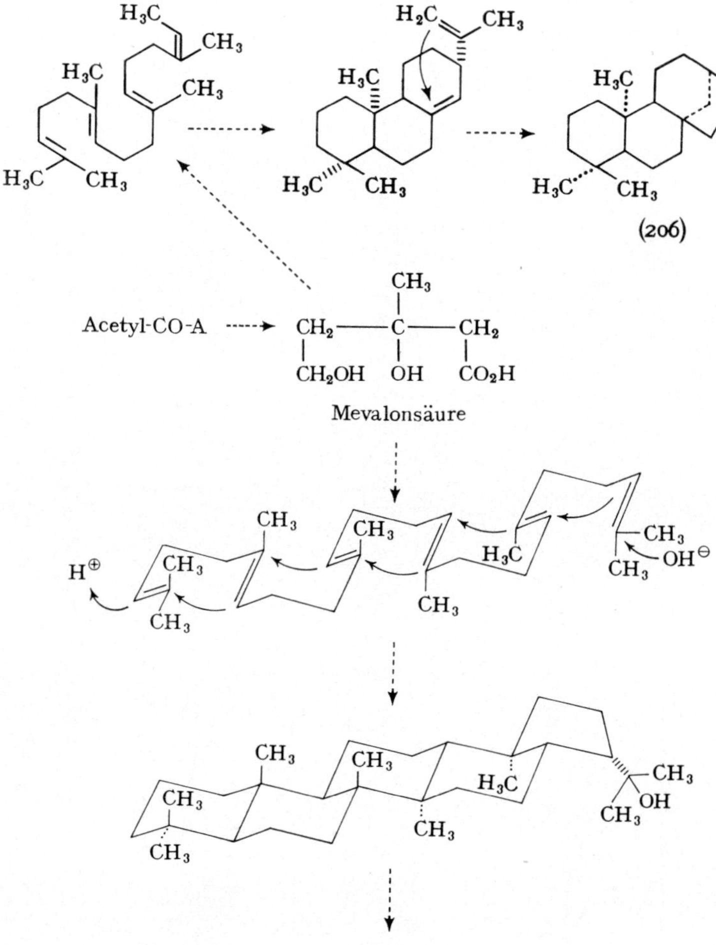

Schema 63b. Biosynthese von Di- und Triterpenen

Schema 64 stellt den Versuch dar, alle Flechtenstoffe in biogenetischer Sicht in einem „Stammbaum" zusammenzufassen.

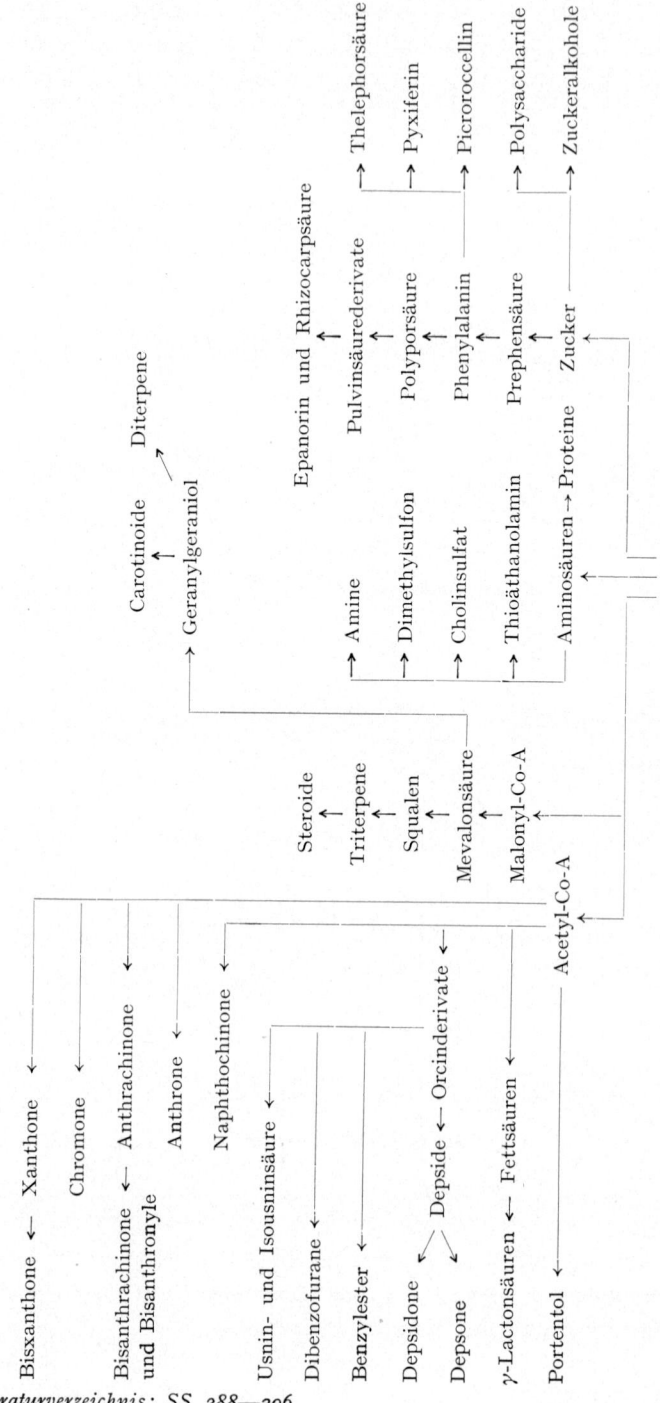

Schema 64. Stammbaum der Flechtenstoffe

Literaturverzeichnis: SS. 288—306

VI. Aus Mycobionten isolierte Verbindungen

CASTLE und KUBSCH (67) berichteten 1949 über die Isolierung von Usninsäure, Didymsäure und Rhodocladonsäure aus dem Mycobionten von *Cladonia cristatella* Tuck., in der die gleichen Verbindungen vorkommen. Diese Ergebnisse konnten jedoch von AHMADJIAN (11) und FOX (115) nicht bestätigt werden. Immerhin fand FOX (115) in 23 von 100 Pilzkulturen aus 73 verschiedenen Flechtenarten Substanzen, die mit Eisen(III)-chlorid eine Färbung gaben, jedoch in keinem Fall mit den von der intakten Flechte synthetisierten Verbindungen identisch waren.

Nach TOMASELLI (325) synthetisiert der Mycobiont von *Xanthoria parietina* (L.) Th. Fr. Parietin, das Hauptanthrachinon der Flechte.

MOSBACH (211) isolierte aus dem auf festem Malzextrakt-Agar gezüchteten Pilz von *Candelariella vitellina* (Ehr.) Müll.-Arg. neben Pulvinsäure, Pulvinsäurelacton und Calycin überraschenderweise auch Vulpinsäure, die von der Flechte selbst nicht gebildet wird.

In Extrakten des Pilzpartners einer Probe von *Lecanora rupicola* (L.) Zahlbr. aus Schweden fanden FOX und HUNECK (116) Roccellsäure, Eugenitol, Eugenitin und 8-Chlor-5,7-dihydroxy-2,6-dimethylchromon; in der Flechte konnten Eugenitin und Eugenitol nicht nachgewiesen werden.

Schließlich isolierten KOMIYA und SHIBATA (175) aus dem Mycobionten von *Ramalina crassa* (Del.) Mot. (+)-Usninsäure und Salazinsäure und aus dem Mycobionten von *Ramalina yasudae* Räs. ebenfalls (+)-Usninsäure und bewiesen damit, daß die Pilzpartner alleine unter geeigneten Kulturbedingungen in der Lage sind, typische Flechtenstoffe zu synthetisieren.

In den Mycobionten von *Collema tenax* (Sw.) Ach. em. Degel., *Baeomyces roseus* Pers., *Lecidea coarctata* (Turn. ex Sm. et Sow.) Nyl., *Lecidea plana* (Lahm. ex Koerb.) Nyl. und *Cladonia cristatella* Tuck. wiesen HENRIKSSON und PEARSON (133) verschiedene nicht näher identifizierte Carotinoide nach.

VII. Chemotaxonomie der Flechten

Die Chemotaxonomie der Flechten liefert Beiträge zur systematischen Abgrenzung von Taxa und hat in den letzten Jahren große Fortschritte gemacht. Da das Gebiet hier nicht im Detail besprochen werden kann, sollen lediglich einige Aspekte und zusammenfassende Arbeiten erwähnt werden.

Insbesondere ASAHINA und SHIBATA (286) und CULBERSON (83) haben zahlreiche Chemovarietäten in den Rang von Arten erhoben, wie z. B. *Cladonia chlorophaea* (Flk.) Spreng., die auf Grund ihrer unterschiedlichen Sekundärstoffausstattung in folgende Arten aufgespalten wurde (Tabelle 5).

Tabelle 5. *Die Cladonia chlorophaea-Gruppe und ihre Inhaltsstoffe*

Art	Inhaltsstoffe
Cladonia chlorophaea (Flk.) Spreng	Fumarprotocetrarsäure
C. cyatomorpha W. Wats.	Fumarprotocetrarsäure
C. conistea (Del.) Asah.	Atranorin
C. subconistea Asah.	Fumarprotocetrarsäure ($\pm$), Psoromsäure, Atranorin
C. cryptochlorophaea Asah.	Fumarprotocetrarsäure ($\pm$), Cryptochlorophaesäure
C. merochlorophaea Asah.	Fumarprotocetrarsäure ($\pm$), Merochlorophaesäure, 4-O-Methylcryptochlorophaesäure
C. perlomera Krist.	Perlatolsäure, Merochlorophaesäure, 4-O-Methylcryptochlorophaesäure
C. grayi Merr.	Fumarprotocetrarsäure, Grayansäure
C. conistea (Ach.) Robb. sensu Evans	Fumarprotocetrarsäure, unbekannte Substanz
nicht benannt	Fumarprotocetrarsäure, Homosekikasäure
nicht benannt	Fumarprotocetrarsäure ($\pm$), Sekikasäure, Homosekikasäure
nicht benannt	Imbricarsäure
nicht benannt	Fumarprotocetrarsäure, Rangiformsäure
nicht benannt	Protolichesterinsäure

Auf interessante Beziehungen zwischen Chemie und Evolution der Flechten weist HALE (*128*) hin: O-methylierte Depside und Depsidone kommen vorzugsweise in morphologisch höher strukturierten Arten vor, fehlen dagegen den auf niedriger Stufe stehenden Gruppen. Unter Berücksichtigung dieses Kriteriums und bisher Bekanntem wäre die Familie der *Ramalinaceae* vorläufig als höchstentwickelte Flechtengruppe anzusehen (*106*).

Auch zwischen Aminosäureausstattung und systematischer Zugehörigkeit von Flechten-Familien und -Gattungen wurden Zusammenhänge beobachtet (*256*).

Ramalina siliquosa (Huds.) A. L. Sm. bildet je nach Standort an der europäischen Atlantikküste auf engstem Raume 6 verschiedene Rassen aus; eine derart starke ökologische und chemische Differenzierung wurde bei anderen Pflanzen bisher nicht festgestellt (*86, 87*).

Zusammenfassend wird von CH. F. CULBERSON (*80*), W. L. CULBERSON (*85*), CULBERSON und CULBERSON (*88*), HUNECK und FOLLMANN (*154*) sowie FOLLMANN und HUNECK (*109*) über die Chemotaxonomie der Flechten berichtet.

Literaturverzeichnis: SS. 288—306

VIII. Antibiotische und weitere biologische Wirkungen der Flechtenstoffe

Die antibiotische Wirkung der Usninsäure wurde von BURKHOLDER und EVANS (64) entdeckt und von MARSHAK (200), VARTIA (328), KROG (177), HESS (134), ARK et al. (20) und PERLMAN (230) bestätigt.

Zwei Usninsäurederivate sind als Antibiotica im Handel: Usno [Benzyldimethyl-{2-[2-(p-1,1,3,3-tetramethylbutylphenoxy-)-äthoxy-]-äthyl-}-ammoniumusneat] (178, 329, 176) und Binan (Mononatriumsalz der Usninsäure) (276).

Usninsäure scheint in die oxydative Phosphorylierung einzugreifen (202, 170, 335), die Synthese gewisser Bakterienproteine zu blockieren (75, 54) und den Nucleinsäurestoffwechsel zu stören (201, 206, 51, 52, 310).

Nach BANDONI und TOWERS (29) sind zahlreiche Bodenpilze zum Abbau von (+)- und (—)-Usninsäure befähigt.

Über die Wirkung von Flechtenstoffen auf Algen und höhere Pflanzen liegen nur wenige Beobachtungen vor. FOX (115) sowie KINRAIDE und AHMADJIAN (172) untersuchten den Einfluß von Flechtensäuren auf Flechtenmycobionten.

Die Keimung und Entwicklung von Gartenkresse (*Lepidium sativum* L.) wird sowohl durch Usninsäure (30) als auch durch andere Flechtenstoffe, insbesondere Pinastrinsäure (144) stark gehemmt.

In Extrakten von *Lasallia papulosa* (Ach.) Nyl. fanden MILLER et al. (203) zwei Typen von Wachstumsinhibitoren. Keim- und Wachstumshemmung weiterer Flechtenextrakte bei höheren Pflanzen werden von FOLLMANN und NAKAGAVA (110), FOLLMANN und PETERS (111) und RONDON (251, 252) beschrieben. Ferner wirken Flechtenstoffe als Virusinhibitoren (112), als Hemmer bei der Stecklingsbewurzelung (105) und erhöhen die Zellpermeabilität (113).

Polyporsäure (65) und gewisse Flechtenglucane (289, 324, 224a) besitzen Antitumorwirkung und Pulvinsäurelacton cardiotonische Aktivität (216).

Nach SÖDERBERG (299, 300) wirkt Vulpinsäure auf die Blutzuckerregulation.

Eine zusammenfassende Darstellung über Arzneimittel und Antibiotica aus Flechten geben SUBRAMANIAN (254) sowie BLANK und STRUHAL (44b).

Literaturverzeichnis

1. ABERHART, D. J., A. CORBELLA and K. H. OVERTON: The Biosynthesis of Portentol: Assembly of a Linear Pentapropionate from Acetate and Methionine. Chem. Commun. **1970**, 664.
1a. ABERHART, D. J., K. H. OVERTON and S. HUNECK: Studies on Lichen Substances. Part LXII. Aromatic Constituents of the Lichen *Roccella fuciformis* DC. A Revised Structure for Lepraric Acid. J. Chem. Soc. (London) C **1969**, 704.
2. ABERHART, D. J., K. H. OVERTON and S. HUNECK: Portentol: A Novel Polypropionate from the Lichen *Roccella portentosa*. Chem. Commun. **1969**, 162.
3. ABERHART, D. J., K. H. OVERTON and S. HUNECK: Portentol: An Unusual Polypropionate from the Lichen *Roccella portentosa*. J. Chem. Soc. (London) C **1970**, 1612.
4. AGARWAL, S. C., and T. R. SESHADRI: Application of Ozonolysis to the Study of Substituted Derivatives of Vulpinic Acid. Tetrahedron **19**, 1965 (1963).
5. AGARWAL, S. C., and T. R. SESHADRI: Condensation of o-Phenylenediamine and Pulvinic Acid Derivatives. Tetrahedron **20**, 17 (1964).
6. AGARWAL, S. C., and T. R. SESHADRI: A Reinvestigation of the Structure of Pinastric Acid and Isopinastric Acid. Indian J. Chem. **2**, 17 (1964).
7. AGARWAL, S. C., and T. R. SESHADRI: Constitution of Leprapinic Acid. Tetrahedron **21**, 3205 (1965).
8. AGHORAMURTHY, K., K. G. SARMA and T. R. SESHADRI: The Structure of Thelephoric Acid. Tetrahedron Letters **1960**, 5.
9. AGHORAMURTHY, K., K. G. SARMA and T. R. SESHADRI: Chemical Investigation of Indian Lichens. XXIV. The Chemical Components of *Alectoria virens* Tayl. Constitution of a New Depsidone, Virensic Acid. Tetrahedron **12**, 173 (1961).
10. AGHORAMURTHY, K., and T. R. SESHADRI: An Improved Synthesis of Lichexanthone. J. Sci. Ind. Res. (India) **12 B**, 350 (1953).
11. AHMADJIAN, V.: Further Studies on Lichenized Fungi. Bryologist **67**, 87 (1964).
12. ÅKERMARK, B.: Studies on the Chemistry of Lichens. 14. The Structure of Calycin. Acta Chem. Scand. **15**, 1695 (1961).
13. ÅKERMARK, B.: Studies on the Chemistry of Lichens. 16. The Absolute Configuration of Roccellic Acid. Acta Chem. Scand. **16**, 599 (1962).
14. ÅKERMARK, B.: Studies on the Chemistry of Lichens. 26. A Stereospecific Synthesis of ($\pm$)-Norrangiformic Acid. Arkiv Kemi **27**, 11 (1967).
15. ÅKERMARK, B.: Studies on the Chemistry of Lichens. 27. The Absolute Configuration of Rangiformic Acid. Acta Chem. Scand. **21**, 589 (1967).
15a. ÅKERMARK, B.: Studies on the Chemistry of Lichens. 28. Additional Evidence for the erythro Configuration of Roccellic Acid. Acta Chem. Scand. **24**, 1456 (1970).
16. ÅKERMARK, B., H. ERDTMAN and C. A. WACHTMEISTER: Studies on the Chemistry of Lichens. XIII. The Structure of Pannaric Acid. Acta Chem. Scand. **13**, 1855 (1959).
17. ÅKERMARK, B., and N.-G. JOHANSSON: Studies on the Chemistry of Lichens. 25. A Stereospecific Synthesis of ($\pm$)-Roccellic Acid. Arkiv Kemi **27**, 1 (1967).
18. ÅKERMARK, B., and N.-G. JOHANSSON: Some Methods for Stereospecific or Stereoselective Preparation of erythro-2-Methyl-3-propylsuccinic Acid. Acta Chem. Scand. **21**, 583 (1967).
19. ANDERSON, H. A., R. H. THOMSON and J. W. WELLS: Naturally Occurring Quinones. Part VIII. Solorinic Acid and Norsolorinic Acid. J. Chem. Soc. (London) C **1966**, 1727.

20. ARK, P. A., A. T. BOTTINI and J. P. THOMPSON: Sodium Usnate as an Antibiotic for Plant Diseases. Pl. Dis. Reptr. **44**, 200 (1960).
21. ARKLEY, V., F. M. DEAN, A. ROBERTSON and P. SIDISUNTHORN: Usnic Acid. XII. Pummerer's Ketone. J. Chem. Soc. (London) **1956**, 603.
22. ARSHAD, M., J. P. DEVLIN, W. D. OLLIS and R. E. WHEELER: The Constitution of Sordidone and its Relation to Thiophanic Acid. Chem. Commun. **1968**, 154.
23. ARSHAD, M., and W. B. OLLIS: Privatmitteilung.
24. ASAHINA, Y.: Neuere Entwicklungen auf dem Gebiete der Flechtenstoffe. Fortschr. Chem. organ. Naturstoffe **8**, 207 (1951).
25. ASAHINA, Y., und F. FUZIKAWA: Untersuchungen über Flechtenstoffe. LV. Mitteilung. Über Endocrocin, ein neues Oxy-anthrachinon-Derivat. Ber. dtsch. chem. Ges. **68**, 1558 (1935).
26. ASAHINA, Y., and S. SHIBATA: Chemistry of Lichen Substances. Tokyo: Japan Society for the Promotion of Science. 1954.
27. ATKINSON, J. E., and J. R. LEWIS: Oxidative Coupling. Part VII. Biogenetic Type Synthesis of Naturally Occurring Xanthones. J. Chem. Soc. (London) C **1969**, 281.
27a. BACHELOR, F. W., and G. G. KING: Chemical Constituents of Lichens: Aphthosin, a Homologue of Peltigerin. Phytochem. **9**, 2587 (1970).
27b. BAECKSTRÖM, P., and G. SUNDSTRÖM: A Simple Synthesis of Olivetol. Acta Chem. Scand. **24**, 716 (1970).
28. BAKER, P. M., and E. BULLOCK: Structure of Rhodocladonic Acid. Canad. J. Chem. **47**, 2733 (1969).
29. BANDONI, R. J., and G. H. N. TOWERS: Degradation of Usnic Acid by Microorganisms. Canad. J. Biochem. **45**, 1197 (1967).
30. BARBALIĆ, L.: Beitrag zur Kenntnis der Einwirkung von L-Usninsäure auf höhere Pflanzen. Qual. Plant. Mater. Veg. **9**, 286 (1963).
31. BARTON, D. H. R., and T. BRUUN: Triterpenoids. Part VI. Some Observations on the Constitution of Zeorin. J. Chem. Soc. (London) **1952**, 1683.
32. BARTON, D. H. R., and T. BRUUN: Constitution of Usnic Acid. J. Chem. Soc. (London) **1953**, 603.
33. BARTON, D. H. R., A. M. DEFLORIN and O. E. EDWARDS: The Synthesis of Usnic Acid. J. Chem. Soc. (London) **1956**, 530.
34. BARTON, D. H. R., P. DE MAYO and J. C. ORR: Triterpenoids. Part XXIV. Further Investigations on the Constitution of Zeorin. J. Chem. Soc. (London) **1958**, 2239.
35. BARTON, D. H. R., and G. QUINKERT: Photochemical Transformations. VI. Photochemical Cleavage of Cyclohexadienones. J. Chem. Soc. (London) **1960**, 1.
36. BEECKEN, H., E.-H. GOTTSCHALK, U. V. GIZYCKI, H. KRÄMER, D. MAASSEN, H.-G. MATTHIES, H. MUSSO, C. RATHJEN und U. I. ZAHORSZKY: Orcein und Lackmus. Angew. Chem. **73**, 665 (1961).
37. BENDZ, G., G. BOHMAN and J. SANTESSON: Chemical Studies on Lichens. 5. Separation and Identification of the Antipodes of Usnic Acid by Thin Layer Chromatography. Acta Chem. Scand. **21**, 1376 (1967).
38. BENDZ, G., G. BOHMAN and J. SANTESSON: Chemical Studies on Lichens. 9. Chlorinated Anthraquinones from *Nephroma laevigatum*. Acta Chem. Scand. **21**, 2889 (1967).
39. BENDZ, G., J. SANTESSON and C. A. WACHTMEISTER: Studies on the Chemistry of Lichens. 20. The Chemistry of the *Ramalina ceruchis* Group. Acta Chem. Scand. **19**, 1185 (1965).

40. BENDZ, G., J. SANTESSON and C. A. WACHTMEISTER: Studies on the Chemistry of Lichens. 21. The Isolation and Synthesis of Methyl 3,5-Dichlorolecanorate, a New Depside from *Ramalina* sp. Acta Chem. Scand. **19**, 1188 (1965).

40 a. BERNARD, T.: Notiz in: International Lichenological Newsletter **4**, No. 1, 2 (1970).

41. BERNARD, T. et G. GOAS: Contribution à l étude du métabolism azoté des lichens. Mise en évidence de quelques transaminases; activité de la glutamate-oxaloacetate transaminase dans cinq especès de la famille des Stictacées. C. R. Acad. Ser. D **269**, 1657 (1969).

42. BERTILSSON, L., and C. A. WACHTMEISTER: Methylation and Racemisation Studies on Usnic Acid. Acta Chem. Scand. **22**, 1791 (1968).

43. BERTILSSON, L., and C. A. WACHTMEISTER: Formation of a Tetracyclic Furan Derivative from Usnic Acid and Diazomethane. Acta Chem. Scand. **22**, 3081 (1968).

44. BEVING, H. F. G., H. B. BORÉN and P. J. GAREGG: Synthesis of Umbilicin (2-O-β-D-Galactofuranosyl-D-arabinitol). Acta Chem. Scand. **22**, 193 (1968).

44 a. BHUTANI, S. P., S. S. CHIBBER and T. R. SESHADRI: A Study of Bromopulvinic Acid Derivatives. Indian J. Chem. **8**, 406 (1970).

44 b. BLANK, H., und H. STRUHAL: Die kleinen Antibiotica. In: R. BRUNNER und G. MACHEK, Die Antibiotica, Bd. III. Nürnberg: Verlag Hans Carl. 1970.

45. BLOOMER, J. L., W. R. EDER and W. F. HOFFMAN: The Biosynthesis of (+)-Protolichesterinic Acid. Chem. Commun. **1968**, 354.

45 a. BLOOMER, J. L., W. R. EDER and W. F. HOFFMAN: Biosynthesis of (+)-Protolichesterinic Acid in *Cetraria islandica*. J. Chem. Soc. (London) C **1970**, 1848.

46. BLOOMER, J. L., and W. F. HOFFMAN: On the Origin of the C_3-Unit in (+)-Protolichesterinic Acid. Tetrahedron Letters **1969**, 4339.

47. BOHMAN, G.: Anthraquinones from the Genus *Caloplaca*. Phytochem. **8**, 1829 (1969).

48. BOHMAN, G.: Chemical Studies on Lichens. 22. Anthraquinones from the Lichen *Lasallia papulosa* var. *rubiginosa* and the Fungus *Valsaria rubricosa*. Acta Chem. Scand. **23**, 2241 (1969).

49. BOHMAN, G.: Chemical Studies on Lichens. 25. A New Anthraquinone from *Mycoblastus sanguinarius*. Tetrahedron Letters **1970**, 445.

49 a. BOHMAN, G.: Chemical Studies on Lichens. 31. A Cyclic Tetrapeptide from *Roccella canariensis*. Tetrahedron Letters **1970**, 3065.

50. BOLL, P. M.: Naturally Occurring Lactones and Lactames. I. The Absolute Configuration of Ranunculin, Lichesterinic Acid, and some Lactones Related to Lichesterinic Acid. Acta Chem. Scand. **22**, 3245 (1968).

51. BRACHET, J.: Quelques effets cytologiques et cytochimiques des inhibiteurs des phosphorylations oxydatives. Experientia **7**, 344 (1951).

52. BRACHET, J.: Quelques effets des inhibiteurs des phosphorylations oxydatives sur des fragments nucléés et énucléés d'organismes unicellulaires. Experientia **8**, 347 (1952).

53. BREWER, J. D. and J. A. ELIX: The Synthesis of Di-O-methylstrepsilin. Tetrahedron Letters **1969**, 4139.

53 a. BRIGGS, M. H., and J. BROTHERTON: Steroid Biochemistry and Pharmacology. London, New York: Academic Press. 1970.

54. BROCK, T. D.: Effect of Antibiotics and Inhibitors on M Protein Synthesis. J. Bact. **85**, 527 (1963).

55. BROWN, C. J., D. E. CLARK, W. D. OLLIS and P. L. VEAL: Synthesis of Diploicin. Proc. Chem. Soc. (London) **1960**, 393.

56. BRUUN, T.: Triterpenoids in Lichens. I. The Occurrence of Friedelin and Epifriedelanol. Acta Chem. Scand. **8**, 71 (1954).
57. BRUUN, T.: Triterpenoids in Lichens. II. Taraxerene, a Naturally Occurring Triterpene. Acta Chem. Scand. **8**, 1291 (1954).
58. BRUUN, T.: Slphulin, a Chromanone Type Lichen Acid. Tetrahedron Letters **1960**, No. 4, 1.
59. BRUUN, T.: Siphulin, a Chromenone Type Lichen Acid. Acta Chem. Scand. **19**, 1677 (1965).
60. BRUUN, T.: Triterpenoids in *Cetraria nivalis* (L.) Ach. Acta Chem. Scand. **23**, 3038 (1969).
61. BRUUN, T.: Phenarctin, a Fully Substituted Depside from *Nephroma arcticum*. Acta Chem. Scand. **23**, 3601 (1969).
62. BRUUN, T., D. P. HOLLIS and R. RYHAGE: The Constitution of Fragilin. Acta Chem. Scand. **19**, 839 (1965).
63. BRUUN, T., and N. A. SÖRENSEN: A Note on the Occurrence of Dimethyl Sulphone in *Cladonia deformis* Hoffm. Acta Chem. Scand. **8**, 703 (1954).
64. BURKHOLDER, P. R. and A. W. EVANS: Further Studies on the Antibiotic Activity of Lichens. Bull. Torrey bot. Club **72**, 157 (1945).
65. CAIN, B. F.: Potential Anti-Tumor-Agents. Part I. Polyporic Acid Series. J. Chem. Soc. (London) **1961**, 936.
66. CAMBIE, R. C. The Depsides from *Stereocaulon ramulosum* (Sw.) Räusch. New Zealand J. Sci. **11**, 48 (1968).
67. CASTLE, H., and F. KUBSCH: The Production of Usnic, Didymic and Rhodocladonic Acids by the Fungal Component of the Lichen *Cladonia cristatella*. Arch. Biochem. **23**, 158 (1949).
68. CHANDA, N. B., E. L. HIRST and D. J. MANNERS: A Comparision of Isolichenin and Lichenin from Iceland Moss *(Cetraria islandica)* J. Chem. Soc. (London) **1957**, 1951.
69. CHANDRASENAN, K., S. NEELAKANTAN and T. R. SESHADRI: Naturally Occurring Dibenzoquinones. Bull. Natl. Inst. Sci. India No. **28**, 92 (1965).
69a. CORBETT, R. E., and S. D. CUMMING: Lichens and Fungi. Part VII. Extractives from the Lichen *Sticta mougeotiana* var. *dissecta* Del. J. Chem. Soc. (London) **C 1971**, 955.
70. CORBETT, R. E., and R. A. J. SMITH: Lichens and Fungi. Part IV. Rearrangements at C-21 in the Hopane Series. J. Chem. Soc. (London) C **1967**, 1622.
71. CORBETT, R. E., R. A. J. SMITH and H. YOUNG: Lichens and Fungi. Part V. Dehydration Rearrangements of 7-Hydroxyhopanes. J. Chem. Soc. (London) C **1968**, 1823.
72. CORBETT, R. E., and R. A. J. SMITH: Lichens and Fungi. Part VI. Dehydration Rearrangements of 15-Hydroxyhopanes. J. Chem. Soc. (London) C **1969**, 44.
73. CORBETT, R. E., and H. YOUNG: Lichens and Fungi. Part II. Isolation and Structural Elucidation of 7β-Acetoxy-22-hydroxyhopane from *Sticta billardierii* Del. J. Chem. Soc. (London) C **1966**, 1556.
74. CORBETT, R., and H. YOUNG: Lichens and Fungi. Part III. Structural Elucidation of $15\alpha,22$-Dihydroxyhopane from *Sticta billardierii* Del. J. Chem. Soc. (London) C 1966, 1564.
75. CREASER, E. H.: The Induced (Adaptive) Biosynthesis of Galactosidase in *Staphylococcus aureus*. J. gen. Microbiol. **12**, 288 (1955).
76. CULBERSON, CH. F.: Some Constituents of the Lichen *Ramalina siliquosa*. Phytochem. **4**, 951 (1965).

77. CULBERSON, CH. F.: The Structure of a New Depsidone from the Lichen *Parmelia livida*. Phytochem. **5**, 815 (1966).
78. CULBERSON, CH. F.: The Chemical Constituents of *Ramalina paludosa*. The Bryologist **70**, 397 (1967).
79. CULBERSON, CH. F.: The Structure of Scrobiculin, a New Lichen Depside in *Lobaria scrobiculata* and *Lobaria amplissima*. Phytochem. **6**, 719 (1967).
80. CULBERSON, CH. F.: Chemical and Botanical Guide to Lichen Products. Chapel Hill: The University of North Carolina Press. 1969.
81. CULBERSON, CH. F.: Chemical Studies in the Genus *Lobaria* and the Occurrence of a New Tridepside, 4-O-Methylgyrophoric Acid. The Bryologist **72**, 19 (1969).
82. CULBERSON, CH. F.: Stenosporic Acid, a New Depside in *Ramalina stenospora*. Phytochem. **9**, 841 (1970).
82a. CULBERSON, CH. F.: Supplement to "Chemical and Botanical Guide to Lichen Products". The Bryologist **73**, 177 (1970).
83. CULBERSON, CH. F., and H. KRISTINSSON: Studies on the *Cladonia chlorophaea* Group: A New Species, a New meta-Depside, and the Identity of "Novochlorophaeic Acid". The Bryologist **72**, 431 (1969).
84. CULBERSON, CH. F., and H.-D. KRISTINSSON: A Standardized Method for the Identification of Lichen Products. J. Chromatog. **46**, 85 (1970).
85. CULBERSON, W. L.: The Use of Chemistry in the Systematics of the Lichens. Taxon **18**, 152 (1969).
86. CULBERSON, W. L. The Behavior of the Species of *Ramalina siliquosa* Group in Portugal. Österr. Bot. Z. **116**, 85 (1969).
87. CULBERSON, W. L., and CH. F. CULBERSON: Habitat Selection by Chemically Differentiated Races of Lichens. Science **158**, 1195 (1967).
88. CULBERSON, W. L., and CH. F. CULBERSON: A Phylogenetic View of Chemical Evolution in the Lichens. The Bryologist **73**, 1 (1970).
89. DAVIDSON, T. A., and A. I. SCOTT: Oxidative Pairing of Phenolic Radicals. Part II. The Synthesis of Picrolichenic Acid. J. Chem. Soc. (London) **1961**, 4075.
90. DAVIS, D., and J. A. ELIX: Synthetic Analogues of Usnic Acid. Tetrahedron Letters **1969**, 2901.
91. DEAN, F. M., C. A. EVANS, T. FRANCIS and A. ROBERTSON: Usnic Acid. XIII. The Orientation and Synthesis of Usnolic Acid. J. Chem. Soc. (London) **1957**, 1577.
92. DEAN, F. M., E. EVANS and A. ROBERTSON: Usnic Acid. X. The Exploration of a Route to 4,6-Dimethoxy-3,5-dimethyl Coumarilic Acid. J. Chem. Soc. (London) **1954**, 4565.
93. DEAN, F. M., P. HALEWOOD, S. MONGKOLSUK, A. ROBERTSON and W. B. WHALLEY: Usnic Acid. IX. A Revised Structure for Usnolic Acid and the Resolution of ($\pm$)-Usnic Acid. J. Chem. Soc. (London) **1953**, 1250.
94. DEAN, F. M., and A. ROBERTSON: Usnic Acid. VIII. C-Diacetyl Derivatives of Phloroglucinol and C-Methyl Phloroglucinol. J. Chem. Soc. (London) **1953**, 1241.
95. DEAN, F. M., and A. ROBERTSON: Usnic Acid. XI. Synthesis of 7-Acetyl-4,6-dihydroxy-3,5-dimethylcoumaran-2-one. J. Chem. Soc. (London) **1955**, 2166.
96. DRAKE, B.: Untersuchungen über einige Polysaccharide der Flechten, vornehmlich das Lichenin und das neuentdeckte Pustulin. Biochem. Z. **313**, 388 (1943).
97. DREW, E. A., and D. C. SMITH: Studies in the Physiology of Lichens. VII. The Physiology of the *Nostoc* Symbiont of *Peltigera polydactyla* Compared with Cultured and Free-living Forms. New Phytol. **66**, 379 (1967).

98. DREW, E. A., and D. C. SMITH: Studies in the Physiology of Lichens. VIII. Movement of Glucose from Alga to Fungus during Photosynthesis in the Thallus of *Peltigera polydactyla*. New Phytol. **66**, 389 (1967).
98a. DYER, J. R., A. C. BAILLIE, V. M. BALTHIS and J. A. BERTRAND: Abstracts of Papers Presented at the Southeastern Regional Meeting of the American Chemical Society, Atlanta, Georgia. Nov. 1—3, 1967.
98b. EBIZUKA, Y., U. SANKAWA and S. SHIBATA: The Constituents of *Solorina crocea*: Averythrin 6-Monomethyl Ether and Methyl Gyrophorate. Phytochem. **9**, 2061 (1970).
99. ELLIS, R. C., W. B. WHALLEY and (in part) K. BALL: Biogenetic-type Synthesis of Xanthones. Chem. Commun. **1967**, 803.
100. ESCHRICH, W.: Über Parietinsäure, einen neuen Inhaltsstoff der gelben Wandflechte *Xanthoria parietina* (L.) Th. Fr. Biochem. Z. **330**, 73 (1958).
101. FEIGE, B.: Untersuchungen zum Kohlenstoff- und Phosphatstoffwechsel der Flechten unter Verwendung radioaktiver Isotope. Inaugural-Dissertation. Würzburg: Nat. Fakultät. 1967.
102. FEIGE, B.: Stoffwechselphysiologische Untersuchungen an der tropischen Basidiolichene *Cora pavonia* (Sw.) Fr. Flora, Abt. A, **160**, 169 (1969).
102a. FEIGE, B.: Zur Verwertung uniform ^{14}C-markierter Glukose und uniform ^{14}C-markierten Glycerins durch die Flechte *Cladonia convoluta* (Lam.) P. Cout. Z. Pflanzenphysiol. **63**, 211 (1970).
103. FEIGE, B., und W. SIMONIS: Cholinsulfat in der Flechte *Cladonia convoluta* (Lam.) P. Cout. Planta (Berl.) **86**, 202 (1969).
103a. FERGUSON, G., and I. R. MACKEY: The Structure of Portentol: X-Ray Analysis of a Heavy-atom Derivative. Chem. Commun. **1970**, 665.
104. FLEMING, M., and D. J. MANNERS: The Fine Structure of Isolichenin. Biochem. J. **100**, 24 P (1966).
105. FOLLMANN, G.: Flechtenstoffe und Stecklingsbewurzelung. Naturwiss. **52**, 266 (1965).
106. FOLLMANN, G., und S. HUNECK: Mitteilungen über Flechteninhaltsstoffe. LXI. Zur Chemotaxonomie der Flechtenfamilie *Ramalinaceae*. Willdenowia **5**, 181 (1969).
107. FOLLMANN, G., und S. HUNECK: Mitteilungen über Flechteninhaltsstoffe. LXVIII. Zur Phytochemie und Chemotaxonomie der Sammelgattung *Lecanora*. Willdenowia **5**, 351 (1969).
108. FOLLMANN, G., und S. HUNECK: Mitteilungen über Flechteninhaltsstoffe. LIX. Zur Chemotaxonomie einiger Roccellaceen. J. Hattori Bot. Lab. (Tokyo) **1969**, 35.
109. FOLLMANN, G., und S. HUNECK: Chemotaxonomie der Flechten. II. In Vorbereitung.
110. FOLLMANN, G., und M. NAKAGAVA: Keimhemmung von Angiospermen durch Flechtenstoffe. Naturwiss. **50**, 696 (1963).
111. FOLLMANN, G., und R. PETERS: Flechtenstoffe und Bodenbildung. Z. Naturforsch. **21 b**, 386 (1966).
112. FOLLMANN, G., und V. VILLAGRÁN: Flechtenstoffe als Virusinhibitoren. Naturwiss. **51**, 543 (1964).
113. FOLLMANN, G., und V. VILLAGRÁN: Flechtenstoffe und Zellpermeabilität. Z. Naturforsch. **20 b**, 723 (1965).
114. FORSEN, S., M. NILSSON and C. A. WACHTMEISTER: Spectroscopic Studies on Enols. IV. Hydrogen Bonding in Usnic Acid. Acta Chem. Scand. **16**, 583 (1962).
115. FOX, C. H.: Studies on the Physiology and Biochemistry of Lichens and their Isolated Symbionts. Dissertation. Worcester: Clark University. 1968.

116. Fox, C. H., and S. Huneck: The Formation of Roccellic Acid, Eugenitol, Eugenitin, and Rupicolin by the Mycobiont of *Lecanora rupicola*. Phytochem. **8**, 1301 (1969).
117. Fox, C. H., E. Klein und S. Huneck: Colensoinsäure, ein neues Depsidon aus *Stereocaulon colensoi*. Phytochem. **9**, 2567 (1970).
118. Fox, C. H., W. S. G. Maass and T. P. Forrest: Papulosin, a Novel Chlorinated Anthraquinone from *Lasallia papulosa* (Ach.) Llano. Tetrahedron Letters **1969**, 919.
119. Fox, C. H., and K. Mosbach: On the Biosynthesis of Lichen Substances. Part 3. Lichen Acids as Products of a Symbiosis. Acta Chem. Scand. **21**, 2327 (1967).
120. Franck, B.: Struktur und Biosynthese der Mutterkorn-Farbstoffe. Angew. Chem. **81**, 269 (1969).
120a. Franck, B., U. Ohnsorge und H. Flasch: Einfache Totalsynthese von [10-^{14}C]-Endocrocin. Tetrahedron Letters **1970**, 3773.
120b. Fujikawa, F., K. Hirai, T. Hirayama, T. Toyota, T. Nakamura, T. Nishimaki, T. Yoshikawa, S. Yasuda, S. Nishio, K. Kojitani, T. Nakai, T. Ando, Y. Tsuji, K. Tomisaki, M. Watanabe, M. Fujisawa, M. Nagai, M. Koyama, N. Matsuami, M. Urasaki and M. Takagawa: On the Free Amino Acids in Lichens of Japan. I. Yakugaku Zasshi (J. Pharmaceut. Soc. Japan) **90**, 1267 (1970).
121. Furuya, T., S. Shibata and H. Iizuka: Gas-liquid Chromatography of Anthraquinones. J. Chromatog. **21**, 116 (1966).
122. Gream, G. E., and N. V. Riggs: Chemistry of Australian Lichens. II. A New Depsidone from *Parmelia conspersa* (Ehrh.) Ach. Austr. J. Chem. **13**, 285 (1960).
123. Gream, G. E., and N. V. Riggs: Reaction Mechanism of Certain 2,6-Disubstituted Benzoic Acid Derivatives. Austr. J. Chem. **13**, 314 (1960).
124. Gripenberg, J.: Fungus Pigments. XII. The Structure and Synthesis of Thelephoric Acid. Tetrahedron **10**, 135 (1960).
125. Grover, P. K., and T. R. Seshadri: Constitution of Pinastric Acid. Tetrahedron **6**, 312 (1959).
126. Grover, P. K., and T. R. Seshadri: Bromo-Derivatives of Pulvinic Acid. J. Chem. Soc. (London) **1960**, 2134.
127. Grover, P. K., G. D. Shah and R. C. Shah: Xanthones. Part V. A New Synthesis of Lichexanthone. J. Sci. Ind. Res. (India) **15B**, 629 (1956).
128. Hale, M. E., Jr.: Chemistry and Evolution in Lichens. Israel J. Bot. **15**, 150 (1966).
129. Harper, S. H., and R. M. Letcher: Isolation of the Internal Salt of Choline Sulphuric Acid from *Dermatiscum thunbergii*. Chem. and Ind. **1966**, 419.
130. Harper, S. H., and R. M. Letcher: Chemistry of Lichen Constituents. Part III. Haemathamnolic Acid: A New β-Orcinol Depside from *Pertusaria rhodesiaca* Vainio. J. Chem. Soc. (London) C **1967**, 1603.
131. Hauschild, G., M. Steiner und K. W. Glombitza: Emodin in Flechten. Naturwiss. **55**, 346 (1968).
131a. Hauschild, G., M. Steiner und K.-W. Glombitza: Emodinaldehyd und Erythroglaucin in Flechten. Planta Medica **19**, 363 (1971).
132. Hendrickson, J. B., and M. V. J. Ramsay: A New Synthesis of Depsidones; Diploicin. Chem. Commun. **1968**, 1101.
133. Henriksson, E., and L. C. Pearson: Carotenoids Extracted from Mycobionts of *Collema tenax*, *Baeomyces roseus*, and Some Other Lichens. Sv. Bot. Tidskr. **62**, 441 (1968).

134. HESS, D.: Untersuchungen über die hemmende Wirkung von Extrakten aus Flechtenpilzen auf das Wachstum von *Neurospora crassa.* Z. Bot. **48**, 136 (1959).
135. HESSE, M., und H. SCHMID: Natürliche Chromone. In: K. PAECH and M. V. TRACEY (ed.), Modern Methods of Plant Analysis, Vol. 6, p. 108. Berlin- Göttingen-Heidelberg: Springer. 1963.
136. HESSE, O.: Beitrag zur Kenntnis der Flechten und ihrer charakteristischen Bestandteile. 5. Mitteilung. J. prakt. Chem. **62**, 430 (1900).
137. HUNECK, S.: Zur Struktur von Zeorin und Leucotylin. Chem. Ber. **94**, 614 (1961).
138. HUNECK, S.: Über Flechteninhaltsstoffe. I. Konstitution der Confluentinsäure. Chem. Ber. **95**, 328 (1962).
139. HUNECK, S.: Über Flechteninhaltsstoffe. XVIII. Konstitution der Planasäure, eines neuen Depsides aus *Lecidea plana* (Lahm ex Koerb.) Nyl. Z. Naturforsch. **20 b**, 1119 (1965).
140. HUNECK, S.: XXI. Mitteilung über Flechteninhaltsstoffe. Über die Inhaltsstoffe von *Lecidea lithophila* (Ach.) Ach. emend. Th. Fr., *Lecidea macrocarpa* (DC.) Steud. und *Lecidea fuscoatra* (L.) Ach. Z. Naturforsch. **20 b**, 1137 (1965).
141. HUNECK, S.: Flechteninhaltsstoffe. XXIV. Die Struktur von Tumidulin, einem neuen chlorhaltigen Depsid. Chem. Ber. **99**, 1106 (1966).
142. HUNECK, S.: Flechteninhaltsstoffe. XXXII. Thiophansäure, ein neues chlorhaltiges Xanthon aus *Lecanora rupicola* (L.) Zahlbr. Tetrahedron Letters **1966**, 3547.
143. HUNECK, S.: Lichen Substances. In: L. REINHOLD and Y. LIWSCHITZ (ed.), Progress in Phytochemistry, Vol. 1, p. 223. London-New York-Sydney: Interscience Publ. 1968.
144. HUNECK, S.: unveröffentlichte Ergebnisse.
145. HUNECK, S., C. DJERASSI, D. BECHER, M. BARBER, M. V. ARDENNE, K. STEINFELDER und R. TÜMMLER: Flechteninhaltsstoffe. XXXI. Massenspektrometrie und ihre Anwendung auf strukturelle und stereochemische Probleme. CXXIII. Massenspektrometrie von Depsiden, Depsidonen, Depsonen, Dibenzofuranen und Diphenylbutadienen mit positiven und negativen Ionen. Tetrahedron **24**, 2707 (1968).
146. HUNECK, S., und G. FOLLMANN: Zur Chemie chilenischer Flechten. V. Über die Inhaltsstoffe von *Ramalina ceruchis* (Ach.) De Not. var. *tumidula* (Tayl.) Nyl. Z. Naturforsch. **20 b**, 611 (1965).
147. HUNECK, S., und G. FOLLMANN: Acerca de la composicion quimica de los liquenes chilenos. VI. La presenca de tumidulina en *Ramalina peruviana* Ach. Bol. Univ. Chile **7**, 56 (1965).
148. HUNECK, S., und G. FOLLMANN: Zur Chemie chilenischer Flechten. VIII. Über die Inhaltsstoffe von *Ramalina chilensis* Bert. Z. Naturforsch. **21 b**, 90 (1966).
149. HUNECK, S., und G. FOLLMANN: Zur Chemie chilenischer Flechten. XI. Über die Inhaltsstoffe von *Ramalina tigrina* Follm. und *Ramalina inanis* Mont. Z. Naturforsch. **21 b**, 713 (1966).
150. HUNECK, S., und G. FOLLMANN: Zur Chemie chilenischer Flechten. XV. Über die Inhaltsstoffe von *Ramalina cactacearum* Follm., *Ramalina ecklonii* (Spreng.) Mey. et Flot. var. *ambigua* Mont. und *Medusulina chilena* Dodge. Z. Naturforsch. **22 b**, 110 (1967).
151. HUNECK, S., und G. FOLLMANN: Zur Chemie chilenischer Flechten. XIV. Über die Inhaltsstoffe von *Roccellaria mollis* (Hampe) Zahlbr. und die Struktur sowie absolute Konfiguration der Roccellarsäure. Z. Naturforsch. **22 b**, 666 (1967).

152. HUNECK, S., und G. FOLLMANN: Zur Chemie chilenischer Flechten. XVIII. Über die Inhaltsstoffe einiger Stictaceen. Z. Naturforsch. **22 b**, 1182 (1967).
153. HUNECK, S., und G. FOLLMANN: Zur Chemie chilenischer Flechten. XIX. Über die Inhaltsstoffe einiger Roccellaceen und die Struktur der Schizopeltsäure, eines neuen Dibenzofuran-Derivates aus *Roccellina luteola* Follm. Z. Naturforsch. **22 b**, 1185 (1967).
153a. HUNECK, S., und G. FOLLMANN: 48. Mitteilung über Flechteninhaltsstoffe. Über die Inhaltsstoffe von *Combea mollusca* (Ach.) De Not., *Roccella vicentina* (Wain.) Wain., *Roccella gayana* Mont. und *Roccella fucoides* (Neck.) Wain. Z. Naturforsch. **22 b**, 1369 (1967).
154. HUNECK, S., und G. FOLLMANN: Chemotaxonomie der Flechten. I. In Vorbereitung.
155. HUNECK, S., G. FOLLMANN und J. SANTESSON: 49. Mitteilung über Flechteninhaltsstoffe. 4-O-Desmethylbarbatinsäure, ein neues Depsid aus *Ramalina subdecipiens* Stein. Z. Naturforsch. **23 b**, 856 (1968).
155a. HUNECK, S., G. FOLLMANN und H. ULLRICH: 50. Mitteilung über Flechteninhaltsstoffe. Über die Inhaltsstoffe einiger Roccellaceen von den Kanarischen Inseln. Z. Naturforsch. **23 b**, 292 (1968).
156. HUNECK, S., G. FOLLMANN, W. A. WEBER und G. TROTET: 37. Mitteilung über Flechteninhaltsstoffe. Über die Inhaltsstoffe einiger *Roccella*-Arten. Z. Naturforsch. **22 b**, 671 (1967).
157. HUNECK, S., et J.-M. LEHN: Résonance magnétique nucléaire de produits naturels — V. Triterpènes — VII. Triterpènes de la série du hopane. Structure et stéréochimie de la zéorine. Bull. Soc. Chim. France **1963**, 1702.
158. HUNECK, S., und J.-M. LEHN: 27. Mitteilung über Flechteninhaltsstoffe. Die Identität von Coquimbosäure und Hypoprotocetrarsäure. Z. Naturforsch. **21 b**, 299 (1966).
159. HUNECK, S., und P. LINSCHEID: NMR-Spektroskopie einiger Depside und Depsidone. Z. Naturforsch. **23 b**, 717 (1968).
160. HUNECK, S., A. MATHEY und G. TROTET: 46. Mitteilung über Flechteninhaltsstoffe. Über die Inhaltsstoffe von *Roccella fuciformis* DC. Z. Naturforsch. **22 b**, 1367 (1967).
161. HUNECK, S., und K. H. OVERTON: unveröffentlichte Ergebnisse.
162. HUNECK, S., und J. SANTESSON: 64. Mitteilung über Flechteninhaltsstoffe. Über die Inhaltsstoffe von *Lecanora rupicola* (L.) Zahlbr. und *Lecanora carpinea* (L.) Ach. em. Vain. und die Strukturaufklärung sowie Synthese von 8-Chlor-5,7-dihydroxy-2,6-dimethylchromon. Z. Naturforsch. **24 b**, 750 (1969).
163. HUNECK, S., und J. SANTESSON: 65. Mitteilung über Flechteninhaltsstoffe. Die Inhaltsstoffe von *Lecidea carpathica* (Koerb.) Szat. und die Struktur des Thuringions, eines neuen Xanthons. Z. Naturforsch. **24 b**, 757 (1969).
164. HUNECK, S., K. SCHREIBER, G. SNATZKE und H.-W. FEHLHABER: 70. Mitteilung über Flechteninhaltsstoffe. Arthoniasäure, ein neues Depsid aus *Arthonia impolita* (Ehrh.) Borr. Z. Naturforsch. **25 b**, 49 (1970).
164a. HUNECK, S., K. SCHREIBER, G. SNATZKE und H.-W. FEHLHABER: 85. Mitteilung über Flechteninhaltsstoffe. Miriquidisäure, ein neues Depsid aus *Lecidea lilienstroemii* und *Lecidea leucophaea*. Z. Naturforsch., im Druck.
165. HUNECK, S., K. SCHREIBER, G. SNATZKE und P. TRŠKA: 72. Mitteilung über Flechteninhaltsstoffe. Struktur der Schizopeltsäure. Z. Naturforsch. **25 b**, 265 (1970).
166. HUNECK, S., und G. TROTET: 40. Mitteilung über Flechteninhaltsstoffe. Über die Inhaltsstoffe von *Dirina repanda* (Nyl.) Fr. Z. Naturforsch. **22 b**, 363 (1967).

167. HUNECK, S., und R. TÜMMLER: Flechteninhaltsstoffe. XII. Die Struktur von Peltigerin. Liebigs Ann. Chem. **685**, 128 (1965).
168. IKEKAWA, N., S. NATORI, H. AGETA, K. IWATA and M. MATSUI: Gas Chromatography of Triterpenes. X. Hopane-Zeorinane and Onocerane Groups. Chem. Pharm. Bull. (Japan) **13**, 320 (1965).
169. JAYALAKSHMI, V., S. NEELAKANTAN and T. R. SESHADRI: A Synthesis of Thiophanic Acid. Current Sci. **37**, 196 (1968).
170. JOHNSON, R. B., G. FELDOTT and H. A. LARDY: The Mode of the Antibiotic, Usnic Acid. Arch. Biochem. **28**, 317 (1950).
171. JOSHI, B. S., S. RAMANANTHAN and K. VENKATARAMAN: Constitution and Synthesis of Endocrocin. Tetrahedron Letters **1962**, 951.
172. KINRAIDE, W. T. B., and V. AHMADJIAN: The Effects of Usnic Acid on the Physiology of Two Cultured Species of the Lichen Alga *Trebouxia* Puym. Lichenologist **4**, 234 (1970).
173. KLOSS, R. A., and D. A. CLAYTON: A Synthesis of Orsellinic Acid. J. Organ. Chem. (USA) **30**, 3566 (1965).
174. KOBAYASHI, N., Y. IITAKA, U. SANKAWA, Y. OGIHARA and S. SHIBATA: The Crystal and Molecular Structure of a Bromination Product of (+)-Tetrahydrorugulosin. Tetrahedron Letters **1968**, 6135.
174a. KOBAYASHI, N., Y. IITAKA and S. SHIBATA: X-ray Structure Determination of (+)-Dibromodehydrotetrahydrorugulosin, a Heavy Atom Derivative of (+)-Rugulosin. Acta Cryst. **B 26**, 188 (1970).
174b. KOMIYA, T., and S. KUROKAWA: Loxodin, a Depsidone of Lichens of *Parmelia* Species. Phytochem. **9**, 1139 (1970).
175. KOMIYA, T., and S. SHIBATA: Formation of Lichen Substances by Mycobionts of Lichens. Isolation of (+)-Usnic Acid and Salazinic Acid from Mycobionts of *Ramalina* spp. Chem. Pharm. Bull. (Japan) **17**, 1305 (1969).
176. KORTEKANGAS, A. E., and O. E. VIRTANEN: The Antibiotic Activity of Some Amino Compound Derivatives of Usnic Acid. Suomen Kemistilehti **B 29**, 2 (1956).
177. KROG, H.: Determination of the Antibiotic Effect of Lichen Acids. Det Kongel. Norske Vidensk. Selsk. Forh. **27**, 1 (1954).
178. LÄÄKE OY: Turku, Finnland, nach Angaben dieser Firma.
179. LEHN, J.-M., und S. HUNECK: Über Flechteninhaltsstoffe. XVIII. Die erstmalige Isolierung des Diterpens (—)-16α-Hydroxykauran aus einer Flechte. Z. Naturforsch. **20 b**, 1013 (1965).
180. LETCHER, R. M.: Chemistry of Lichen Constituents. VII. Mass Spectra of Some Pulvic Acid Derivatives. Org. Mass Spectrometry **1**, 805 (1968).
181. LETCHER, R. M., P. A. ALSOP and S. H. HARPER: Chemistry of Lichen Constituents. Part 5. Proceed. a. Transact. of the Rhodesia Scientific Assoc. **53**, 70 (1969).
182. LETCHER, R. M., and S. H. EGGERS: Chemistry of Lichen Constituents. Part IV. Tetrahedron Letters **1967**, 3541.
183. LINDBERG, B.: Studies on the Chemistry of Lichens. VIII. Investigation of a *Dermatocarpon* and Some *Roccella* Species. Acta Chem. Scand. **9**, 917 (1955).
184. LINDBERG, B., and J. MCPHERSON: Studies on the Chemistry of Lichens. VI. The Structure of Pustulan. Acta Chem. Scand. **8**, 985 (1954).
185. LINDBERG, B., and H. MEIER: Studies on the Chemistry of Lichens. 15. Siphulitol, a New Polyol from *Siphula ceratites*. Acta Chem. Scand. **16**, 543 (1962).
186. LINDBERG, B., A. MISIORNY and C. A. WACHTMEISTER: Studies on the Chemistry of Lichens. IV. Investigation of the Low-molecular Carbohydrate Constituents of Different Lichens. Acta Chem. Scand. **7**, 591 (1953).

187. LINDBERG, B., B.-G. SILVANDER and C. A. WACHTMEISTER: Studies on the Chemistry of Lichens. 18. 3-O-β-D-Glucopyranosyl-D-mannit from *Peltigera aphthosa* (L.) Willd. Acta Chem. Scand. **17**, 1348 (1963).
188. LINDBERG, B., B.-G. SILVANDER and C. A. WACHTMEISTER: Studies on the Chemistry of Lichens. 19. Mannitol Glycosides in *Peltigera* Species. Acta Chem. Scand. **18**, 213 (1964).
189. LINDBERG, B., C. A. WACHTMEISTER and B. WICKBERG: Studies on the Chemistry of Lichens. II. Umbilicin, an Arabitol Galactoside from *Umbilicaria pustulata* (L.) Hoffm. Acta Chem. Scand. **6**, 1052 (1952).
190. LINDBERG, B., and B. WICKBERG: Studies on the Chemistry of Lichens. III. Disaccharides from *Umbilicaria pustulata* (L.) Hoffm. Acta Chem. Scand. **7**, 140 (1953).
191. LINDBERG, B., and B. WICKBERG: Studies on the Chemistry of Lichens. V. The Furanoside Structure of Umbilicin. Acta Chem. Scand. **8**, 821 (1954).
192. LINDBERG, B., and G. WICKBERG: Studies on the Chemistry of Lichens. 17. The Structure of Umbilicin. Acta Chem. Scand. **16**, 2240 (1962).
193. LOCKSLEY, H. D., and I. G. MURRAY: Extractives from *Guttiferae*. Part XVI. Biogenetic-type Synthesis of Xanthones from their Benzophenone Precursors. J. Chem. Soc. (London) C **1970**, 392.
194. LOUNASMAA, M.: Neue Synthese von Thelephorsäure. Acta Chem. Scand. **19**, 540 (1965).
194a. LYNEN, F.: Biosynthetic Pathways from Acetate to Natural Products. Pure Appl. Chem. **14**, 137 (1967).
194b. MAASS, W. S. G.: Pulvinamide and Possible Biosynthetic Relationships with Pulvinic Acid. Phytochem. **9**, 2477 (1970).
195. MAASS, W. S. G., and A. C. NEISH: Lichen Substances. II. Biosynthesis of Calycin and Pulvinic Dilactone by the Lichen *Pseudocyphellaria crocata*. Canad. J. Bot. **45**, 59 (1967).
196. MAASS, W. S. G., G. H. N. TOWERS and A. C. NEISH: Flechtenstoffe. I. Untersuchungen zur Biogenese des Pulvinsäureanhydrids. Ber. Dtsch. Bot. Ges. **77**, 157 (1964).
197. MAC KENZIE, S.: The Racemization of Usnic Acid. J. Amer. Chem. Soc. **77**, 2214 (1955).
198. MANAKTALA, S. K., S. NEELAKANTAN and T. R. SESHADRI: Synthesis of ($\pm$)-Montagnetol and ($\pm$)-Erythrin. Tetrahedron **22**, 2373 (1966).
199. MANAKTALA, S. K., S. NEELAKANTAN and R. T. SESHADRI: A Study of the Condensation Products of Usnic Acid with Amino Compounds. Indian J. Chem. **5**, 29 (1967).
200. MARSHAK, A.: A Crystalline Antibacterial Substance from the Lichen *Ramalina reticulata*. Publ. Hlth. Rep. (Wash.) **62**, 3 (1947).
201. MARSHAK, A., and J. FAGER: Prevention of Nuclear Fusion and Mitosis and Inhibition of Desoxyribonuclease by D-Usnic Acid. J. cell. comp. Physiol. **35**, 317 (1950).
202. MARSHAK, A., and J. HARTING: Inhibition of Cleavage and P^{32} Uptake in *Arbacia* by d-Usnic Acid. J. cell. comp. Physiol. **31**, 321 (1948).
203. MILLER, E. V., C. E. GRIFFIN, T. SCHAEFERS and M. GORDON: Two Types of Growth Inhibitors in Extracts of *Umbilicaria papulosa*. Bot. Gaz. **126**, 100 (1965).
204. MITTAL, P. P., and T. R. SESHADRI: Chemical Investigation of Indian Lichens. XIX. *Lepraria*: Constitution of Leprapinic Acid. J. Chem. Soc. (London) **1955**, 3053.

205. MITTAL, P. P., and T. R. SESHADRI: Synthesis of Leprapinic Acid and Constitution of Pinastric Acid. J. Chem. Soc. (London) **1956**, 1734.
206. MIURA, Y., Y. NAKAMURA et H. MATSUDAIRA: Le mode d'action de l'acide usnique sur les bacteries. C. R. **232**, 1710 (1951).
207. MOSBACH, K.: Die Rolle der Malonsäure in der Biosynthese der Orsellinsaure. Naturwiss. **48**, 525 (1961).
208. MOSBACH, K.: Studies on the Biosynthesis of Aromatic Compounds in Fungi and Lichens. Inaugural Dissertation. Lund: Fac. of Science. 1964.
209. MOSBACH, K.: On the Biosynthesis of Lichen Substances. I. The Depside Gyrophoric Acid. Acta Chem. Scand. **18**, 329 (1964).
210. MOSBACH, K.: On the Biosynthesis of Lichen Substances. Part 2. The Pulvic Acid Derivative Vulpinic Acid. Biochem. Biophys. Res. Comm. **17**, 363 (1964).
211. MOSBACH, K.: On the Biosynthesis of Lichen Substances. Part 4. The Formation of Pulvic Acid Derivatives by Isolated Lichen Fungi. Acta Chem. Scand. **21**, 2331 (1967).
212. MOSBACH, K.: Zur Biosynthese von Flechtenstoffen, Produkten einer Symbiontischen Lebensgemeinschaft. Angew. Chem. **81**, 233 (1969).
213. MOSBACH, K., and U. EHRENSVÄRD: Studies on Lichen Enzymes. Part I. Preparation and Properties of a Depside Hydrolysing Esterase and of Orsellinic Acid Decarboxylase. Biochem. Biophys. Res. Comm. **22**, 145 (1966).
214. MURAKAMI, T.: The Coloring Matters of *Xanthoria fallax* (Hepp.) Arn. Pharm. Bull. (Japan) **4**, 298 (1956).
215. MURTY, T. K.: Isolation of Methyl β-Orcinolcarboxylate from *Parmelia tinctorum* Despr. J. Sci. Ind. Res. (India) **19 B**, 508 (1960).
216. NADOR, K., J. SZEGI and M. MARKÓ: Preparation of α,β-Unsaturated Lactones with Digitaloid Properties. Vegyipari Kutato Intézetek Közleményei **4**, 85 (1954).
217. NAKANISHI, T., T. FUJIWARA and K. TOMITA: The Crystal Structure of 16β-O-p-Bromobenzoate of 6-Keto-leucotylin. Tetrahedron Letters **1968**, 1491.
217a. NAKANISHI, T., H. YAMAUCHI, T. FUJIWARA and K. TOMITA: The Crystal Structure of 6-O-p-Bromobenzoyl Zeorin. Tetrahedron Letters **1971**, 1157.
218. NEELAKANTAN, S., R. PADMASANI and T. R. SESHADRI: Halogenation of the Depsides Lecanoric Acid and Atranorin. Indian J. Chem. **2**, 478 (1964).
219. NEELAKANTAN, S., R. PADMASANI and T. R. SESHADRI: New Reagents for the Synthesis of Depsides. Methyl Evernate, Methyl Lecanorate, Evernic Acid and Atranorin. Tetrahedron **21**, 3531 (1965).
220. NEELAKANTAN, S., S. RANGASWAMI, T. R. SESHADRI and S. SANKARA SUBRAMANIAN: Chemical Investigation of Indian Lichens. Part XI. Constitution of Teloschistin- the Position of the Methoxyl Group. Proc. Indian Acad. Sci. **33A**, 142 (1951).
221. NEELAKANTAN, S., and T. R. SESHADRI: A New Synthesis of Teloschistin. J. Sci. Ind. Res. (India) **13B**, 884 (1954).
222. NEELAKANTAN, S., T. R. SESHADRI and S. SANKARA SUBRAMANIAN: Chemical Investigation of Indian Lichens. Part XX. A New Synthesis of Teloschistin. Proc. Indian Acad. Sci. **44A**, 42 (1956).
223. NEELAKANTAN, S., T. R. SESHADRI and S. SANKARA SUBRAMANIAN: Chemical Investigation of Indian Lichens. XXVI. Constitution of Vicanicin from *Teloschistes flavicans*. Tetrahedron **13**, 597 (1962).
224. NISHIKAWA, Y., T. TAKEDA, S. SHIBATA and F. FUKUOKA: Polysaccharides in Lichens and Fungi. III. Further Investigation on the Structures and the Antitumor Activity of the Polysaccharides from *Gyrophora esculenta* Miyoshi and *Lasallia papulosa* (Ach.) Llano. Chem. Pharm. Bull. (Japan) **17**, 1910 (1969).

224a. NISHIKAWA, Y., M. TANAKA, S. SHIBATA and F. FUKUOKA: Polysaccharides of Lichens and Fungi. IV. Antitumour Active O-Acetylated Pustulan-Type Glucans from the Lichens of *Umbilicaria* Species. Chem. Pharm. Bull. (Japan) **18**, 1431 (1970).
224b. NORRMAN, B.: Methylation Studies on Pustulan, Methyl α- and β-D-Glucopyranoside and Some Derivatives. Acta Chem. Scand. **22**, 1623 (1968).
225. NUNO, M., Y. KUWADA and K. KAMIYA: The Structure of Nephroarctin. Chem. Commun. **1969**, 78.
226. OGIHARA, Y., N. KOBAYASHI and S. SHIBATA: Further Studies on the Bianthraquinones of *Penicillium islandicum* Sopp. Tetrahedron Letters **1968**, 1881.
227. PEAT, S., W. J. WHELAN, J. R. TURVEY and K. MORGAN: The Structure of Isolichenin. J. Chem. Soc. (London) **1961**, 623.
228. PENTTILA, A., and H. M. FALES: On the Biosynthesis In Vitro of Usnic Acid. Chem. Commun. **1966**, 656.
229. PERLIN, A. S., and S. SUZUKI: The Structure of Lichenin: Selective Enzymolysis Studies. Canad. J. Chem. **40**, 50 (1962).
230. PERLMAN, D.: Antibiotic Inhibition of Algal Growth. Antimicrob. Ag. Chemother. **1964**, 114.
231. PERSSON, B., and J. SANTESSON: Chemical Studies on Lichens. 27. The Structure of the Depside Alectorialic Acid. Acta Chem. Scand. **24**, 345 (1970).
232. PIATTELLI, M., and M. GIUDICI DE NICOLA: Anthraquinone Pigments from *Xanthoria parietina*. Phytochem. **7**, 1183 (1968).
233. PIKE, D. G., J. J. RYAN and A. I. SCOTT: Synthesis and Aromatisation of the Linear Hepta-β-carbonyl System. Chem. Commun. **1968**, 629.
234. POELT, J., und S. HUNECK: *Lecanora vinetorum* nova spec., ihre Vergesellschaftung, ihre Ökologie und ihre Chemie. Österr. Bot. Z. **115**, 411 (1968).
234a. POKORNY, M., E. MARČENKO and D. KEGLEVIĆ: Comparative Studies of L- and D-Methionine Metabolism in Lower and Higher Plants. Phytochem. **9**, 2175 (1970).
235. PUEYO, G.: Présence de mannitol et d'arabitol dans de nouvelles espèces de lichens. Un hétéroside nouveau (peltigéroside) dans *Peltigera horizontalis* Hoffm. Rev. Bryol. Lichenol. **29**, 124 (1960).
236. RAJAGOPALAN, T. R., and T. R. SESHADRI: Chemical Investigation of Indian Lichens. Part XXI. Occurrence of Fallacinal in *Teloschistes flavicans*. Proc. Indian Acad. Sci. **49A**, 1 (1959).
237. RAMAKRISHNAN, S., and S. SANKARA SUBRAMANIAN: Amino Acids of *Roccella montagnei* and *Parmelia tinctorum*. Indian J. Chem. **2**, 467 (1964).
238. RAMAKRISHNAN, S., and S. SANKARA SUBRAMANIAN: Amino Acid Composition of *Cladonia rangiferina*, *Cladonia gracilis* and *Lobaria isidiosa*. Current Sci. **34**, 345 (1965).
239. RAMAKRISHNAN, S., and S. SANKARA SUBRAMANIAN: Amino Acids of *Lobaria subisidiosa*, *Umbilicaria pustulata*, *Parmelia nepalensis* and *Ramalina sinensis*. Current Sci. **35**, 124 (1966).
240. RAMAKRISHNAN, S., and S. SANKARA SUBRAMANIAN: Amino Acids of *Dermatocarpon moulinsii*. Current Sci. **35**, 284 (1966).
241. RAMAUT, J. L.: Reactions thallines, microcristallisations et chromatographie de partage sur papier en lichenologie. Les Naturalistes belges **43**, 359 (1962).
242. RAMAUT, J. L.: Chromatographie en couche mince des depsidones du β orcinol. Bull. Soc. Chim. Belg. **72**, 97 (1963).
243. RAMAUT, J. L. Chromatographie sur couche mince des depsides et des depsidones. Bull. Soc. Chim. Belg. **72**, 316 (1963).

244. READ, G., and L. C. VINING: Thelephoric Acid. Canad. J. Chem. **37**, 1442 (1959).
245. RICHARDSON, D. H. S., D. JACKSONHILL and D. C. SMITH: Lichen Physiology. XI. The Role of the Alga in Determining the Pattern of Carbohydrate Movement between Lichen Symbionts. New Phytol. **67**, 469 (1968).
246. RICHARDSON, D. H. S., and D. C. SMITH: Lichen Physiology. IX. Carbohydrate Movement from the *Trebouxia* Symbiont of *Xanthoria aureola* to the Fungus. New Phytol. **67**, 61 (1968).
247. RICHARDSON, D. H. S., and D. C. SMITH: Lichen Physiology. X. The Isolated Algal and Fungal Symbionts of *Xanthoria aureola*. New Phytol. **67**, 69 (1968).
248. RICHARDSON, D. H. S., D. C. SMITH and D. H. LEWIS: Carbohydrate Movement between the Symbionts of Lichens. Nature **214**, 879 (1967).
249. RIEDL, W.: Über Usninsäure. I. Liebigs Ann. Chem. **597**, 148 (1955).
250. ROBERTS, J. C.: Studies in Mycological Chemistry. Part VI. A Novel Method for the Degradation of 1-Hydroxy-xanthones. J. Chem. Soc. (London) **1960**, 785.
251. RONDON, Y.: Action inhibitrice de l'extrait du lichen *Roccella fucoides* (Dicks.) Vain. sur la germination. Bull. Soc. Bot. France **113**, 1 (1966).
252. RONDON, Y.: Action sur la germination de l'extrait du lichen *Parmelia furfuracea* (L.) Ach. var. *olivetorina* (Zopf) Zahlbr. Bull. Soc. Bot. France **115**, 121 (1968).
252a. RUNGE, F., und U. KOCH: Darstellung von Polycarbonsäuren, III. Die Reaktion von Oxalylchlorid mit Acetylchlorid und einigen phenylsubstituierten Carbonsäurechloriden. Chem. Ber. **91**, 1217 (1958).
253. SANKARA SUBRAMANIAN, S.: Lichens and their Food Value. J. Nutr. Dietet. **2**, 217 (1965).
254. SANKARA SUBRAMANIAN, S.: Drugs from Lichens. J. Indian Pharm. Manufact. **4**, 9 (1966).
255. SANKARA SUBRAMANIAN, S., and S. RAMAKRISHNAN: Amino Acids of *Peltigera canina*. Current Sci. **33**, 522 (1964).
256. SANKARA SUBRAMANIAN, S., and S. RAMAKRISHNAN: Significance of Amino Acids in Lichen Chemotaxonomy. Bull. Natl. Inst. Sci. India No. **34**, 375 (1967).
257. SANKARA SUBRAMANIAN, S., and M. N. SWAMY: A Note on *Roccella* found in Pondicherry. J. Sci. Ind. Res. (India) **20C**, 275 (1961).
258. SANKAWA, U., S. SEO, N. KOBAYASHI, Y. OGIHARA and S. SHIBATA: Further Studies on the Structure of Luteoskyrin, Rubroskyrin and Rugulosin. Tetrahedron Letters **1968**, 5557.
259. SANTESSON, J.: Chemical Studies on Lichens. 8. Schizopeltic Acid, a Novel Lichen Dibenzofuran. Acta Chem. Scand. **21**, 1111 (1967).
260. SANTESSON, J.: Chemical Studies on Lichens. 4. Thin Layer Chromatography of Lichen Substances. Acta Chem. Scand. **21**, 1162 (1967).
261. SANTESSON, J.: Chemical Studies on Lichens. 7. Acaranoic and Acarenoic Acid, Two New Aliphatic Lichen Acids. Acta Chem. Scand. **21**, 1993 (1967).
262. SANTESSON, J.: Chemical Studies on Lichens. 12. A New Lichen Xanthone from *Lecanora reuteri*. Acta Chem. Scand. **22**, 1698 (1968).
263. SANTESSON, J.: Chemical Studies on Lichens. 13. A Spot Test for Lichen Xanthones. Acta Chem. Scand. **22**, 2393 (1968).
264. SANTESSON, J.: Chemical Studies on Lichens. Inaugural Dissertation. Uppsala: Almqvist & Wiksell. 1969.
265. SANTESSON, J.: Chemical Studies on Lichens. 20. The Xanthones of Some Crustaceous Lichens. Arkiv Kemi **31**, 57 (1969).

266. SANTESSON, J.: Chemical Studies on Lichens. 21. Two Novel Chlorinated Lichen Xanthones. Arkiv Kemi **31**, 121 (1969).
267. SANTESSON, J.: Chemical Studies on Lichens. 10. Mass Spectrometry of Lichens. Arkiv Kemi **30**, 363 (1969).
268. SANTESSON, J.: Chemical Studies on Lichens. 16. The Xanthones of *Lecanora straminea*. I. Arthothelin and Thiophanic Acid. Arkiv Kemi **30**, 449 (1969).
269. SANTESSON, J.: Chemical Studies on Lichens. 17. The Xanthones of *Lecanora straminea*. II. 2,7-Dichloronorlichexanthone. Arkiv Kemi **30**, 455 (1969).
270. SANTESSON, J.: Chemical Studies on Lichens. 18. The Xanthones of *Lecanora straminea*. III. Norlichexanthone, 2-Chloronorlichexanthone, and 2,4-Dichloronorlichexanthone. Arkiv Kemi **30**, 461 (1969).
271. SANTESSON, J.: Chemical Studies on Lichens. 19. Infrared and Mass Spectra of Some Lichen Xanthones. Arkiv Kemi **30**, 479 (1969).
272. SANTESSON, J.: Chemical Studies on Lichens. 24. Norsolorinic Acid in *Lecidea piperis*. Acta Chem. Scand. **23**, 3270 (1969).
272a. SANTESSON, J.: Chemical Studies on Lichens. 30. Anthraquinonoid Pigments of *Trypetheliopsis boninensis* and *Ocellularia domingensis*. Acta Chem. Scand. **24**, 3331 (1970).
272b. SANTESSON, J.: Syntheses of Orsellinic Acid and Related Compounds. Acta Chem. Scand. **24**, 3373 (1970).
272c. SANTESSON, J.: Anthraquinones in *Caloplaca*. Phytochem. **9**, 2149 (1970).
273. SANTESSON, J., and G. SUNDHOLM: Chemical Studies on Lichens. 14. Syntheses and Chlorinations of Norlichexanthone. Arkiv Kemi **30**, 427 (1969).
274. SARGENT, M. V., D. O'N. SMITH and J. A. ELIX: The Minor Anthraquinones of *Xanthoria parietina* (L.) Beltram, the Chlorination of Parietin, and the Synthesis of Fragilin and 7-Chloroemodin (AO-1). J. Chem. Soc. (London) C **1970**, 307.
275. SARMA, K. G., und S. HUNECK: Über Flechteninhaltsstoffe. 52. Mitteilung. Über die Inhaltsstoffe einiger Flechten aus dem Himalaja. Pharmazie **23**, 583 (1968).
276. SAVICZ, V. P., M. A. LITVINOV und E. N. MOISSEJEVA: Ein Antibiotikum aus Flechten als Arzneimittel. Planta Medica **8**, 191 (1960).
276a. SCHADE, A.: Über Herkunft und Vorkommen der Ca-oxalat Exkrete in kortizikolen Parmeliaceen. Nova Hedwigia, im Druck.
276b. SCHADE, A., und W. SEITZ: Extremes Auftreten von Calciumoxalat-Exkreten bei einer Art der Gattung *Usnea* (Lichenes). Ber. Dtsch. Bot. Ges. **83**, 121 (1970).
277. SCHÖPF, C., und F. ROSS: Die Konstitution der Usninsäure. II. Liebigs Ann. Chem. **546**, 1 (1941).
278. SESHADRI, T. R., and G. B. VENKATASUBRAMANIAN: A New Synthesis of Diploschistesic Acid. J. Chem. Soc. (London) **1959**, 1658.
279. SESHADRI, T. R., and G. B. VENKATASUBRAMANIAN: Synthesis of Homodiploschistesic Acid Methyl Ester. J. Indian Chem. Soc. **40**, 7 (1963).
280. SHARMA, R. K., and P. J. JANNKE: Acidity of Usnic Acid. Indian J. Chem. **4**, 16 (1966).
281. SHIBATA, S.: Especial Compounds of Lichens. In: W. RUHLAND (ed.), Handbuch der Pflanzenphysiologie, Bd. 10, Der Stoffwechsel sekundärer Pflanzenstoffe, S. 560. Berlin-Göttingen-Heidelberg: Springer. 1958.
282. SHIBATA, S.: Lichen Substances. In: K. PAECH and M. V. TRACEY (ed.), Modern Methods of Plant Analysis, Vol. 6, p. 155. Berlin-Göttingen-Heidelberg: Springer. 1963.

283. SHIBATA, S.: Chemistry and Biosynthesis of Some Fungal Metabolites. Chem. in Britain 3, 110 (1967).
284. SHIBATA, S., and H.-C. CHIANG: Grayanic Acid, a New Lichen Depsidone. Chem. Pharm. Bull. (Japan) 11, 926 (1963).
285. SHIBATA, S., and H.-C. CHIANG: The Structure of Cryptochlorophaeic Acid and Merochlorophaeic Acid. Phytochem. 4, 133 (1965).
286. SHIBATA, S., and H.-C. CHIANG: Some Chemotaxonomical Aspects of Phenolic Compounds in Lichens. Bull. Natl. Inst. Sci. India No. 31, 151 (1965).
287. SHIBATA, S., T. FURUYA and H. IIZUKA: Gas-liquid Chromatography of Lichen Substances. I. Studies on Zeorin. Chem. Pharm. Bull. (Japan) 13, 1254 (1965).
288. SHIBATA, S., Y. NISHIKAWA, T. TAKEDA and M. TANAKA: Polysaccharides in Lichens and Fungi. I. Antitumour Active Polysaccharides of *Gyrophora esculenta* Miyoshi and *Lasallia papulosa* (Ach.) Llano. Chem. Pharm. Bull. (Japan) 16, 2362 (1968).
289. SHIBATA, S., Y. NISHIKAWA, T. TAKEDA, M. TANAKA, F. FUKUOKA and M. NAKANISHI: Studies on the Chemical Structures of the New Glucans Isolated from *Gyrophora esculenta* Miyoshi and *Lasallia papulosa* (Ach.) Llano and their Inhibiting Effect on Implanted Sarcoma 180 in Mice. Chem. Pharm. Bull. (Japan) 16, 1639 (1968).
290. SHIBATA, S., J. SHOJI, N. TOKUTAKE, Y. KANEKO, H. SHIMIZU and H.-C. CHIANG: Decomposition of Usnic Acid. VI. The Ozonolytic Products of O,O-Diacetylusnic Acid. Chem. Pharm. Bull. (Japan) 10, 477 (1962).
291. SHIBATA, S., and H. TAGUCHI: Occurrence of Isousnic Acid in Lichens with Reference to "Isodihydrousnic Acid" Derived from Dihydrousnic Acid. Tetrahedron Letters 1967, 4867.
292. SHIBATA, S., K. TAKAHASHI and Y. TANAKA: Decomposition of Usnic Acid. V. Pyrolysis of Dihydrousnic Acid. II. Some Observations on Dihydrousnic Acid. Pharm. Bull. (Japan) 4, 65 (1956).
293. SHIBATA, S., O. TANAKA, U. SANKAWA, Y. OGIHARA, R. TAKAHASHI, S. SEO, D.-M. YANG and Y. IDA: The Constituents of *Acroscyphus sphaerophoroides* Lév. J. Jap. Bot. 43, 335 (1968).
294. SHOJI, J.: Decomposition of Usnic Acid. VII. Pyrolysis of Dihydrousnic Acid. III. Isodihydrousnic Acid. Chem. Pharm. Bull. (Japan) 58, 483 (1963).
295. SIMONIS, W., und B. FEIGE: Untersuchungen über den Intermediärstoffwechsel zwischen Alge und Pilz bei *Peltigera aphthosa* (L.) Willd. Flora, Abt. A, 158, 599 (1967).
296. SMITH, D. C.: Studies in the Physiology of Lichens. 2. Absorption and Utilization of Some Simple Organic Nitrogen Compounds by *Peltigera polydactyla*. Annals of Botany, N. S. 24, 172 (1960).
297. SMITH, D. C.: Studies in the Physiology of Lichens. 3. Experiments with Dissected Discs of *Peltigera polydactyla*. Annals of Botany, N. S. 24, 186 (1960).
298. SMITH, D. C.: Studies in the Physiology of Lichens. IV. Carbohydrates in *Peltigera polydactyla* and the Utilization of Absorbed Glucose. New Phytol. 62, 205 (1963).
299. SÖDERBERG, U.: Note on the Action of Vulpinic Acid. Acta Physiol. Scand. 27, 97 (1952).
300. SÖDERBERG, U.: On the Action of Lichen Acids on Blood Sugar Regulation and Carbohydrate Metabolism. Acta Neurovegetativa (Wien) 9, 168 (1954).
301. SOLBERG, Y. J.: Studies on the Chemistry of Lichens. II. Chemical Components of *Haematomma ventosum* (L.) Mass. var. *lapponicum* Räs. Acta Chem. Scand. 11, 1477 (1957).

302. SOLBERG, Y. J.: Studies on the Chemistry of Lichens. III. Long-chain Tetrahydroxy Fatty Acids from Some Norwegian Lichens. Acta Chem. Scand. **14**, 2152 (1960).
303. SOLBERG, Y. J.: Studies on the Chemistry of Lichens. IV. The Chemical Composition of Some Norwegian Lichen Species. Ann. Bot. Fenn. **4**, 29 (1967).
304. SOLBERG, Y. J.: Studies on the Chemistry of Lichens. VI. Chemical Investigations of the Lichen Species *Alectoria nigricans* (Ach.) Nyl. and *Parmelia alpicola* Th. Fr. Z. Naturforsch. **22 b**, 777 (1967).
305. SOLBERG, Y. J.: Studies on the Chemistry of Lichens. VII. Chemical Investigations of the Lichen Species *Lecanora (Aspicilia) Myrinii* (Fr.) Nyl. Z. Naturforsch. **24 b**, 447 (1969).
305 a. SONN, A.: Eine neue Synthese der Orsellinsäure (5. Mitteilung über Flechtenstoffe). Ber. dtsch. chem. Ges. **61**, 926 (1928).
306. SOVIAR, K., O. MOTL, Z. SAMEK and J. SMOLIKOVÁ: The Structure of Lepraric Acid, a Lichen Chromone. Tetrahedron Letters **1967**, 2277.
306 a. STEGLICH, W., und S. HUNECK: unveröffentlichte Ergebnisse.
307. STEGLICH, W., und W. LÖSEL: Bestimmung der Stellung von O-Substituenten bei 1,8-Dihydroxy-Anthrachinon-Derivaten mit Hilfe der NMR-Spektroskopie. Tetrahedron **25**, 4391 (1969).
308. STEGLICH, W., W. LÖSEL und W. REININGER: Xanthorin, ein Anthrachinonpigment aus *Xanthoria elegans* (Link) Th. Fr. Tetrahedron Letters **1967**, 4719.
308 a. STEGLICH, W., und W. REININGER: Erythroglaucin und weitere Anthrachinonpigmente aus *Xanthoria elegans* (Link) Th. Fr. Z. Naturforsch. **24 b**, 1196 (1969).
309. STEGLICH, W., and W. REININGER: A Synthesis of Endocrocin, Endocrocin-9-anthrone, and Related Compounds. Chem. Commun. **1970**, 178.
310. STEINERT, M.: Metabolism de l'acide ribonucleique dans l'oeuf d'amphibien traite and dinitrophenol. Biochem. biophys. Acta **10**, 427 (1953).
311. STENSIÖ, K.-E., and C. A. WACHTMEISTER: 1,5,8-Trihydroxy-6-methoxy-3-methylanthraquinone from *Laurera purpurina* (Nyl.) Zahlbr. Acta Chem. Scand. **23**, 144 (1969).
312. STORK, G.: The Racemization of Usnic Acid. Chem. and Ind. **1955**, 915.
313. TAGUCHI, H., U. SANKAWA and S. SHIBATA: Biosynthesis of Usnic Acid in Lichens. Tetrahedron Letters **1966**, 5211.
314. TAGUCHI, H., U. SANKAWA and S. SHIBATA: Biosynthesis of Natural Products. VI. Biosynthesis of Usnic Acid in Lichens (1). A General Scheme of Biosynthesis of Usnic Acid. Chem. Pharm. Bull. (Japan) **17**, 2054 (1969).
315. TAGUCHI, H., U. SANKAWA and S. SHIBATA: Biosynthesis of Natural Products. VII. Biosynthesis of Usnic Acid in Lichens. Seasonal Variation Observed in Usnic Acid Biosynthesis. Chem. Pharm. Bull. (Japan) **17**, 2061 (1969).
316. TAGUCHI, H., and S. SHIBATA: The Structure of Isousnic Acid with Reference to "Isodihydrousnic Acid" Derived from Dihydrousnic Acid. Chem. Pharm. Bull. (Japan) **18**, 374 (1970).
317. TAKAHASHI, K., A. ARAI, K. OSHIMA, Y. UEDA and S. MIYASHITA: Usnic Acid. II. Methylusnic Acid. Chem. Pharm. Bull. (Japan) **10**, 607 (1962).
318. TAKAHASHI, K., Y. HONDA and S. MIYASHITA: Usnic Acid. V. Some Decomposition Reactions of Methyl- and Methyldihydrousnic Acids. The Revised Structure for Usnic Acid Isomethoxide. Chem. Pharm. Bull. (Japan) **11**, 1229 (1963).
319. TAKAHASHI, K., and S. MIYASHITA: Usnic Acid. I. Methyldihydrousnic Acid. Chem. Pharm. Bull. (Japan) **10**, 603 (1962).

320. TAKAHASHI, K., and S. MIYASHITA: Usnic Acid. III. Anhydromethyldihydrousnic Acid. Chem. Pharm. Bull. (Japan) **11**, 209 (1963).
321. TAKAHASHI, K., and S. MIYASHITA: Usnic Acid. VI. The Ozonolysis of Anhydromethyldihydrousnic Acid Monoacetate. Chem. Pharm. Bull. (Japan) **16**, 988 (1968).
322. TAKAHASHI, K., S. MIYASHITA and Y. UEDA: Usnic Acid. IV. Isoanhydromethyldihydrousnic Acid. Chem. Pharm. Bull. (Japan) **11**, 473 (1963).
322a. TAKAHASHI, K., and M. TAKANI: Usnic Acid. VII. The Pyrolysis of Methyldihydrousnic Acid. Chem. Pharm. Bull. (Japan) **18**, 1831 (1970).
323. TAKAHASHI, R., O. TANAKA and S. SHIBATA: Occurrence of 15α-Acetoxy-22-hydroxyhopane and Phlebic Acid A in the Lichen *Peltigera aphthosa*. Phytochem. **8**, 2345 (1969).
323a. TAKAHASHI, R., O. TANAKA and S. SHIBATA: The Structure of Phlebic Acid B, a Constituent of the Lichen *Peltigera aphthosa*, and the Occurrence of 15α-Acetoxy- and 7β-Acetoxy-22-hydroxyhopane in *P. dolichorrhiza*. Phytochem. **9**, 2037 (1970).
324. TAKEDA, T., Y. NISHIKAWA and S. SHIBATA: A New α-Glucan from the Lichen *Parmelia caperata* (L.) Ach. Chem. Pharm. Bull. (Japan) **18**, 1074 (1970).
324a. THOMSON, R. H.: Naturally Occurring Quinones. 2. Ed. London, New York: Academic Press. 1971.
325. TOMASELLI, R.: Nuovo contributo alle ricerche sulla presenza di "fiscione" in culture pure di *Xanthoriomyces*. Atti dell' Ist. Bot. et Lab. Crittogamico dell' Univ. di Pavia **14**, 128 (1957).
326. TSUDA, Y., K. ISOBE, S. FUKUSHIMA, H. AGETA and K. IWATA: Final Clarification of the Saturated Hydrocarbons Derived from Hydroxyhopanone, Diploptene, Zeorin, and Dustanin. Tetrahedron Letters **1967**, 23.
327. VAN TAMELEN, E. E., and S. R. BACH: The Synthesis of d,l-Protolichesterinic Acid. J. Amer. Chem. Soc. **80**, 3079 (1958).
328. VARTIA, K. C.: On Antibiotic Effects of Lichens and Lichen Substances. Ann. Med. exp. Biol. Fenn., Suppl. No. 7, **28**, 5 (1950).
329. VIRTANEN, O. E.: The Antibiotic Activity of Some Amino Compound Derivatives of L-Usnic Acid. II. Soumen Kemistilehti **B27**, 67 (1954).
330. VIRTANEN, O. E., and A. E. KORTEKANGAS: Usnic Acid Derivatives of 4-Amino-3-isoxazolidone. Suomen Kemistilehti **29B**, 30 (1956).
331. WACHTMEISTER, C. A.: Studies on the Chemistry of Lichens. VII. Structure of Porphyrilic Acid. Acta Chem. Scand. **8**, 1433 (1954).
332. WACHTMEISTER, C. A.: Studies on the Chemistry of Lichens. X. The Structure of Porphyrilic Acid. Acta Chem. Scand. **10**, 1404 (1956).
333. WACHTMEISTER, C. A.: Studies on the Chemistry of Lichens. XI. Structure of Picrolichenic Acid. Acta Chem. Scand. **12**, 147 (1958).
334. WAGNER, H., und H. FRIEDRICH: Über die ungesättigten Fettsäuren von Moosen, Bärlappgewächsen und Flechten. Naturwiss. **52**, 305 (1965).
335. WHITEHOUSE, M. W., and P. D. G. DEAN: Biochemical Properties of Antiinflammatory Drugs. V. Uncoupling of Oxidative Phosphorylation by Some γ-Resorcyl and Other Dihydroxybenzol Compounds. Biochem. Pharmac. **1965**, 557.
336. WILSON, J. L.: Quantitative Analysis of Depsides and Related Plant Phenolics by Gas Chromatography. Diss. Abstr. **B28**, (6), 2314 (1967).
337. YAMAUCHI, H., T. FUJIWARA and K. TOMITA: The Crystal and Molecular Structure of 3β,12β-O-Di-p-bromobenzoylpyxinol. Tetrahedron Letters **1969**, 4245.
338. YAMAZAKI, M., M. MATSUO and S. SHIBATA: Biosynthesis of Lichen Depsides, Lecanoric Acid and Atranorin. Chem. Pharm. Bull. (Japan) **13**, 1015 (1965).

339. YAMAZAKI, M., and S. SHIBATA: Biosynthesis of Lichen Substances. II. Participation of C_1-Unit to the Formation of β-Orcinol Type Lichen Depsides. Chem. Pharm. Bull. (Japan) **14**, 96 (1966).
340. YOSIOKA, I.: Structure of Leucotylin. Chem. Pharm. Bull. (Japan) **11**, 1468 (1963).
341. YOSIOKA, I., A. MATSUDA and I. KITAGAWA: Pyxinic Acid, a Novel Lichen Triterpene with 3β-Hydroxyl Function. Tetrahedron Letters **1966**, 613.
341a. YOSIOKA, I., Y. MORITA and K. EBIHARA: The Structure of Constictic Acid. Chem. Pharm. Bull. (Japan) **18**, 2364 (1970).
342. YOSIOKA, I., T. NAKANISHI, S. IZUMI and I. KITAGAWA: Structure of a Lichen Pigment Entothein and its Identity with Secalonic Acid A, a Major Ergot Pigment. Chem. Pharm. Bull. (Japan) **16**, 2090 (1968).
343. YOSIOKA, I., T. NAKANISHI and I. KITAGAWA: On Catalytic Hydrogenation of Anhydroleucotylin and Methyl Anhydroleucotylate. Tetrahedron Letters **1966**, 5185.
344. YOSIOKA, I., T. NAKANISHI and I. KITAGAWA: The Chemical Proof of Hopane Skeleton of Zeorin. Chem. Pharm. Bull. (Japan) **15**, 353 (1967).
345. YOSIOKA, I., T. NAKANISHI and I. KITAGAWA: On the Stereostructures of Zeorin and Leucotylin. Tetrahedron Letters **1968**, 1485.
346. YOSIOKA, I., T. NAKANISHI and I. KITAGAWA: Lichen Triterpenoids. I. The Structure of Leucotylin. Chem. Pharm. Bull. (Japan) **17**, 279 (1969).
347. YOSIOKA, I., T. NAKANISHI and I. KITAGAWA: Lichen Triterpenoids. II. The Stereostructure of Zeorin. Chem. Pharm. Bull. (Japan) **17**, 291 (1969).
348. YOSIOKA, I., T. NAKANISHI and E. TSUDA: The Structure of Leucotylic Acid, a New Triterpenic Acid from a Lichen. Tetrahedron Letters **1966**, 607.
348a. YOSIOKA, I., T. NAKANISHI, H. YAMAUCHI and I. KITAGAWA: Revised Structure of Zeorin and its Correlation with Leucotylin. Tetrahedron Letters **1971**, 1161.
349. YOSIOKA, I., M. YAMAKI and I. KITAGAWA: On the Triterpenic Constituents of a Lichen, *Parmelia entotheiochroa* Hue; Zeorin, Leucotylin, Leucotylic Acid, and Five New Related Triterpenoids. Chem. Pharm. Bull. (Japan) **14**, 804 (1966).
350. YOSIOKA, I., M. YAMAKI, T. NAKANISHI and I. KITAGAWA: The Structure of Methyl Isoleucotylate, an Acid Isomerized Product of Methyl Leucotylate. Tetrahedron Letters **1966**, 2227.
351. YOSIOKA, I., H. YAMAUCHI and I. KITAGAWA: Diacetylpyxinol, a Triterpene Alcohol from a Lichen: *Pyxine endochrysina* Nyl. Tetrahedron Letters **1969**, 4241.
351a. YOSIOKA, I., H. YAMAUCHI, K. MORIMOTO and I. KITAGAWA: The Pigment Constituents of Some *Anaptychia* Species. J. Jap. Bot. **43**, 343 (1968).
352. YOSIOKA, I., H. YAMAUCHI, K. MORIMOTO and I. KITAGAWA: Two New Chlorine Containing Anthraquinones from a Lichen, *Anaptychia obscurata* (Nyl.) Vain. Tetrahedron Letters **1968**, 1149.
353. YOSIOKA, I., H. YAMAUCHI, K. MORIMOTO and I. KITAGAWA: Three New Chlorine Containing Bisanthronyls from a Lichen, *Anaptychia obscurata* Vain. Tetrahedron Letters **1968**, 3749.
354. ZOPF, W.: Die Flechtenstoffe in chemischer, botanischer, pharmakologischer und technischer Beziehung. Jena: G. Fischer-Verlag. 1907.
355. ZOPF, W.: Zur Kenntnis der Flechtenstoffe. 17. Mitteilung. Über die in den Lappenflechten (Peltigeraceen) vorkommenden Stoffe. Liebigs Ann. Chem. **364**, 273 (1909).

(Eingelaufen am 15. September 1970)

The Cucurbitanes, a Group of Tetracyclic Triterpenes

By D. LAVIE and E. GLOTTER, Rehovot, Israel

With 1 Figure

Contents

	Page
I. Introduction	308
II. The Carbon Skeleton up to 1960	309
III. Nomenclature	310
IV. Structure Determination and Chemistry of the Cucurbitacins	311
1. Cucurbitacins B, D, E and I	311
1.1. Interrelationship Between Cucurbitacins B (**5**), D (Elatericin A) (**6**), E (Elaterin) (**2**), and I (Elatericin B) (**7**)	311
1.2. The Skeleton	312
1.3. The Side Chain	312
1.4. The Ring A Substituents; the α-Hydroxyketones (**5**), (**6**) and the Diosphenols (Enolized α-Diketones) (**2**), (**7**)	314
1.5. The 19-Methyl Group and Ring C Carbonyl	316
1.6. The Ring B Double Bond	317
1.7. The 16-Hydroxy Group	322
1.8. Alkaline Treatment of Elaterin	323
2. Cucurbitacins A (**61**) and C (**69**)	325
2.1. Structure Determination	325
2.2. Interrelationship Between Cucurbitacins A, B and C	329
3. Stereochemistry of the Cucurbitacins	329
4. Stereochemistry of Ring A Ketols	332
5. Interrelationship Between the Cucurbitane and Lanostane Series	334
6. Synthesis of a 32-nor-Cucurbitane Skeleton	336
V. Cucurbitacins G, H, L, J, K, Dihydrocucurbitacin B and 22-Deoxocucurbitacin D	337
1. Cucurbitacins G and H (**108**)	337
2. Cucurbitacins J, K (**109**), and L (**110**)	337
3. Dihydrocucurbitacin B (**111**)	338
4. 22-Deoxocucurbitacin D (**112**)	339
5. "β-Elaterin"	339

VI. Isocucurbitacin B (**119**), 22-Deoxoisocucurbitacin D (**120**) and Tetrahydrocucurbitacin I (**121**) 341
    1. Isocucurbitacin B (**119**) 341
    2. 22-Deoxoisocucurbitacin D (**120**) 341
    3. Tetrahydrocucurbitacin I (**121**) 342

VII. Bryodulcosigenin (**123**), Bryosigenin (**124**), Bryogenin (**125**), Gratiogenin (**126**), and 16-Hydroxygratiogenin (**127**)............................ 343

VIII. Cucurbitacin F (**134a**), O (**135a**), P (**136a**), and Q (**137a**)........... 346

IX. Cucurbitacins of Unknown Structure 348

X. Biogenetic Aspects 348

XI. Physical Methods Used in the Structure Elucidation of the Cucurbitacins 350

XII. Biological Properties of the Cucurbitacins 351

XIII. Tables.. 352
    1. Occurrence of the Cucurbitacins in Nature 352
    2. Physical Constants of the Cucurbitacins 356

References.. 357

I. Introduction

The potent physiological activity of plants belonging to the Cucurbitaceae family has been known since antiquity. They were feared on account of their high toxicity (Elisha's Miracle)*, and yet valued because of the medicinal properties ascribed to them (*40*). Greeks and Romans used them, the doctors of the Middle Ages praised their virtues and some were still described in the British Pharmacopoeia of 1914.

In present times the interest in cucurbitaceous plants still persists for different reasons. The sporadic occurrence of toxic bitter principles in cultivated species constitutes a health hazard and a commercial nuisance chiefly noted in South Africa (*107, 122*); cattle mortality due to grazing on toxic species of Cucurbitaceae; a search for substances with tumor necrotizing capacity in experimental tumors (*10, 43, 46*); genetic selection and taxonomic studies of species (*115*); specific attraction of beetles (*20, 21*).

All these and other aspects (*25*), as well as the challenge of the chemical problems have sparked research programs in different parts of the world and brought about studies directed towards the elucidation of the chemical constituents of the Cucurbitaceae and their biological evaluation.

From our point of view the Cucurbitaceae can be subdivided into two groups; one containing the genera possessing a β-glycosidase, elaterase

---

* The Bible, Kings II, Chap. 4. Vers. 38–41.

References, pp. 357—362

(*32*, *103*), which cleaves the glycosides releasing the free aglycones, and a second group in which such an enzyme seems to be absent and therefore the aglycones can be obtained only following enzymatic hydrolysis in the laboratory.

The aglycones which have been isolated from various species of the Cucurbitaceae have been given the general name of *cucurbitacins*; this is following by various letters according to the chronology of isolation, although in certain cases other names have been given. The first isolation of a crystalline substance, elaterin, was reported as early as 1831 (Cf. *11*). Attempts to solve its structure were reported already at the beginning of this century (*11*, *12*, *52*, *95*, *101*, *103*, *135*), but only during the last decade has the structural problem found its solution. Following the identification of their structures, cucurbitacins have been found to occur as well in certain Cruciferae (*18*, *47*, *49*) and Scrophulariaceae (*50*); and thus they seem to have a wider distribution in the plant kingdom.

II. The Carbon Skeleton up to 1960

The literature up to 1960 contains extensive studies dealing with the chemistry of the cucurbitacins. The first tetracyclic triterpenoid structure proposed for them was based on the formation of 1,2,8-trimethylphenanthrene during selenium dehydrogenation of cucurbitacin A (*38*) and subsequently of elaterin (*88*) and of other compounds (*74*). Because of this, the nature of the side chain, and the observation that the compounds dealt with have a thirty carbon skeleton, a lanostane-type tetracyclic triterpene structure was proposed (*76*); elaterin for example was assigned structure (1).

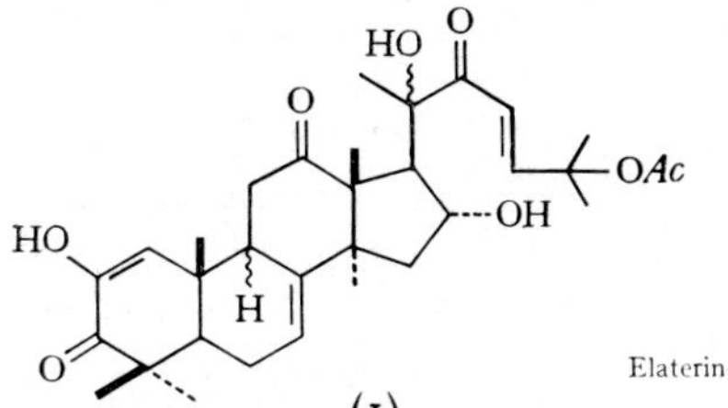

(1) Old structure — Elaterin

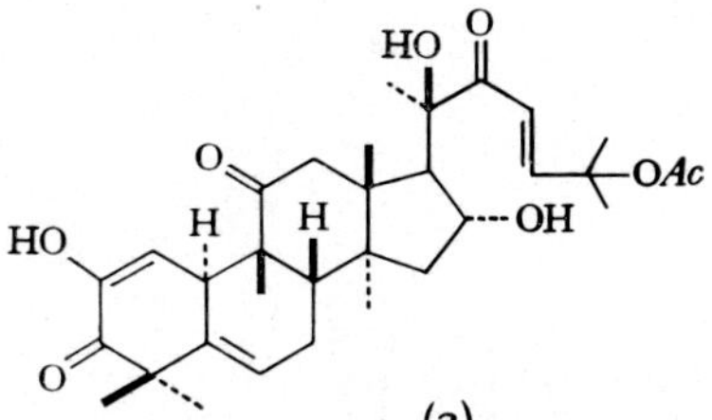

(2) Actual structure

In the same year (1960), in a communication dealing with the NMR spectrum of a compound containing a diosphenol-chromophore, NOLLER and coworkers (*28*) observed that the olefinic proton involved in the diosphenol system appeared as a doublet δ 5.97 (J ~ 3 Hz), an observation which required the presence of an adjacent proton at C-10 or a different structure for ring A. Objections to structure (1) were also based (23) on a comparison of molecular rotation differences between compounds (5) and (2), as well as between (6) and (7) with those of appropriate 4,4-dimethyl-cholestane derivatives.

One year later, 1961, the Rehovot group presented the first publication (79) containing a modified skeletal structure of type 2, thus confirming the NMR observations, and provided chemical and spectroscopical evidence for the location of a methyl group at C-9 instead of C-10. The cucurbitacins were thus the first representatives of tetracyclic triterpenoids with a C-9 methyl group. Final assignment of the ring C carbonyl to C-11 was given in 1962 by the same group (119), and also in a communication presented through the combined effort of three groups from Pretoria, London and Pittsburgh (59). The problem of the structure of the cucurbitacins was heading towards its solution.

Much of the earlier observations concerning the substituents and other parts of the skeleton remained valid and could be applied to the new structure (2) of elaterin. In this review, structural work on the cucurbitacins, even that published before 1960, will be summarized and discussed in terms of the correct structure.

III. Nomenclature

The name *cucurbitane* has been proposed for the hydrocarbon skeleton of the cucurbitacins. Originally (66) it referred to a tetracyclic triterpene with 10α and 8β hydrogens, and 9β, 13β, 14α methyl groups. According to the nomenclature proposed in and known as the "IUPAC-IUB 1967 Revised Tentative Rules for Steroid Nomenclature" the name 5α-cucurbitane or 19 (10 → 9β)-*abeo*-5α-lanostane is proposed for the hydrocarbon skeleton shown in (3).

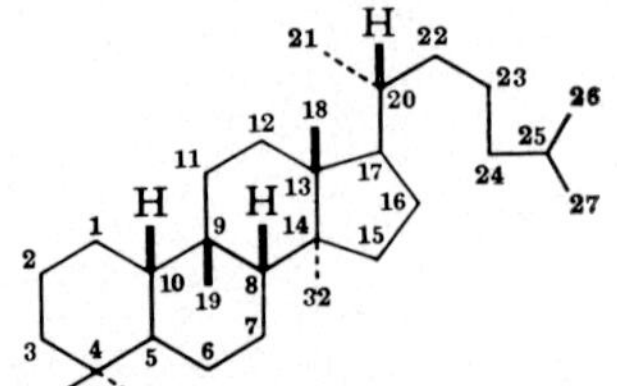

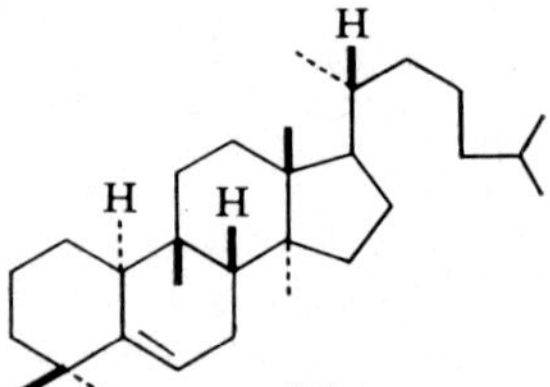

Accordingly, a hydrocarbon as shown in (4) with 10-H α-oriented and a double bond between C-5 and C-6 (as in all naturally occurring cucurbitacins) should be named 10α-cucurbit-5-ene or 19 (10 → 9β)-*abeo*-10α-lanost-5-ene.

References, pp. 357—362

IV. Structure Determination and Chemistry of the Cucurbitacins

1. Cucurbitacins B, D, E, and I

1.1. Interrelationship Between Cucurbitacins B (5), D (Elatericin A) (6), E (Elaterin) (2), and I (Elatericin B) (7)

The above four cucurbitacins will be dealt together since they were interrelated (Chart 1) at an early stage of the structural work (*84*); thus any description of the relevant structural arguments for each compound will contribute to structure elucidation of the others, as well to the whole field.

Cucurbitacin B (5) $\xrightarrow{Bi_2O_3}$ Cucurbitacin E (2)
$C_{32}H_{46}O_8$ (elaterin)
$C_{32}H_{44}O_8$

$\downarrow$ enzymatic hydrolysis

Cucurbitacin D (6) $\xrightarrow{Bi_2O_3}$ Cucurbitacin I (7)
(elatericin A) (elatericin B)
$C_{30}H_{44}O_7$ $C_{30}H_{42}O_7$

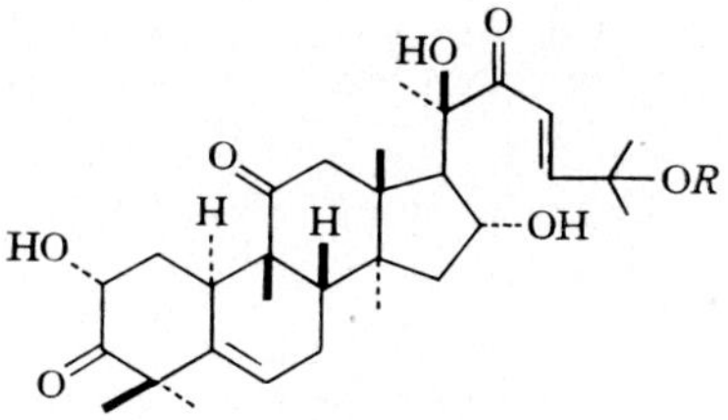

Elatericin A (6) $R = H$
Cucurbitacin B (5) $R = Ac$

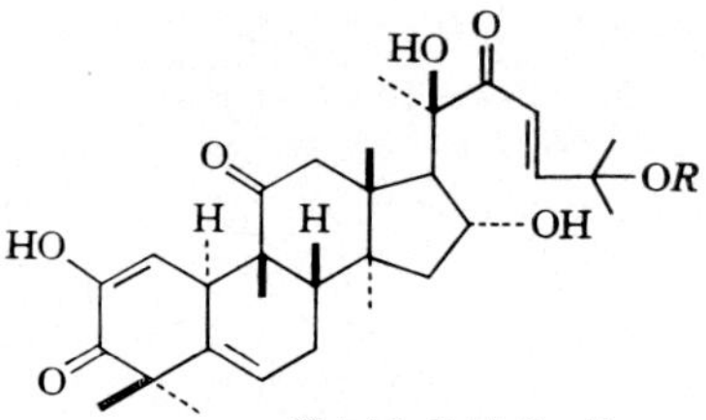

Elatericin B (7) $R = H$
Elaterin (2) $R = Ac$

Chart 1. Interrelation of Cucurbitacins B, D, E and I

The empirical formulae of elaterin (2) and cucurbitacin B (5) as well as of A (61) and C (69) (see later), have been determined using X-ray unit cell and density measurements (*110*).

The difference between compounds (5) and (6) resides in the presence of an extra acetate group in (5); also (5) has two hydrogen atoms more than (2). The same relationship holds for the pairs (2), (7) and (6), (7) respectively. The conversion of (2) to (7) could be performed only enzymatically, using the fresh juice of the Golden Hubbard Squash (*36*); with alkali deep seated changes took place in the molecule. The interconversion between the pairs, (5) → (2) and (6) → (7) was done with bismuth oxide, a reagent known to oxidize ketols to diosphenols.

1.2. The Skeleton

As mentioned in Section II the isolation of 1,2,8-trimethylphenanthrene, a well established degradation product of tetracyclic triterpenes, from the mixture obtained from the selenium dehydrogenation of cucurbitacin A (*38*), elatericin A (*74*) and elaterin (*88*) was the original impetus for believing that the cucurbitacins possessed a tetracyclic triterpenoid skeleton. However, in a subsequent set of experiments performed on elatericin A (*80*), 1,2,5-trimethylnaphthalene was identified along with the trimethylphenanthrene. The concurrent formation of these two degradation products was unique and implied an unusual distribution of the methyl groups in the skeleton, a methyl being present between rings B and C. Evidence for its location at C-9 will be described subsequently.

The previously reported isolation of 1,4-dimethyl naphthalene (*95*, *106*) seems to have been erroneous (Cf. *80*).

1.3. The Side Chain

The identification of the eight-carbon atom side chain of the cucurbitacins and its substitution pattern as present in compounds (2), (5), (6) and (7) is based on the reactions displayed in Chart 2. Following various cleavage reactions, the bulk of the molecule afforded different compounds depending on the starting material and the reaction conditions; these compounds are indicated by partial structures (8), (14), (17), (18) and (20) and will be discussed separately.

Treatment of elatericin A (6) with periodic acid induced cleavage of the C-20, C-22 bond with formation of a methyl ketone (partial structure (8)), and of *trans*-4-hydroxy-4-methyl-pent-2-enoic acid (9a), λ_{max} 208 nm (ε 9500) and ν_{max} 1703, 1651 and 980 cm^{-1} (the last value is indicative of the *trans* configuration). Hydrogenation of (9a) afforded isocaprolactone (10) which was obtained in turn by hydrogenation of the acetylenic acid (11) (*71*, *83*). A similar oxidation performed on cucurbitacin B (5) afforded the corresponding acetate (9b) identified by hydrolysis to the original acid (9a) (*33*). Similar results were obtained with cucurbitacin A (61), C (69), E (2) and their acetates, thus confirming that the tertiary hydroxyl (or acetate) is at the end of the side chain (*29*, *33*).

Further information on the nature of the side chain was obtained by treatment of elaterin (2) with alkali. This reaction which resulted in the isolation of ecballic acid (21a) (partial structure (14)) and of acetoin (15), (*109*, *110*) results in cleavage of the C-23, C-24 bond, and can be rationalized by the sequence (12) → (13) → (15). The first step of this reaction consisted in the hydration of the Δ^{23} bond to a mixture of 24-OH isomers (12) which no longer contained the characteristic enone absorption at 230 nm (*82*, *83*).

References, pp. 357—362

Such compounds have also been found to occur in nature and are known as cucurbitacins J and K (**109**) (*34, 48*). The mixture obtained by hydration of the double bond (Δ^{23}) of (**2**) has been called elateridin (*85*).

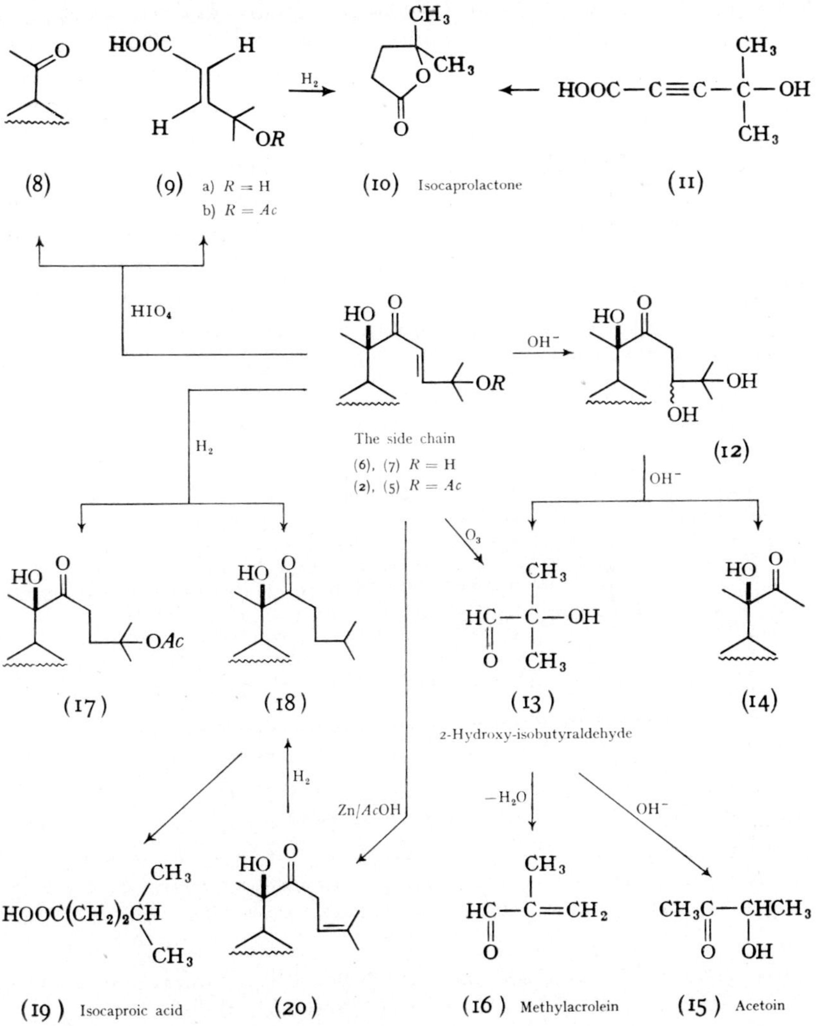

Chart 2. Degradation of the Side Chain

The retro-aldol cleavage (*110*) of the 24-hydroxy-22-one moiety present in (**12**) should have given 2-hydroxy-isobutyraldehyde (**13**) which has not been, however, isolated, since under the alkaline conditions of the reaction it underwent an acyloin rearrangement leading to acetoin

(15); the validity of this postulate was confirmed by alkaline treatment of a synthetic sample of (13) (*83*).

Corroboration of the structure of the side chain as deduced from the two sequences described above was reached by ozonolysis of elatericin A (6), as well as of cucurbitacins A (61), C (69) and E (2). This led to the isolation of (13) and of its dehydration product, 2-methyl acrolein (16), both being identified as the corresponding 2,4-dinitrophenylhydrazones (*83, 109*).

A sequence of reactions complementing those already described were performed on cucurbitacin B (5), (*93, 111*) and are summarized in partial formulae (17)–(20). Catalytic hydrogenation of (5) (λ_{max} 230 nm) led to the dihydro derivative (17) and the dihydrodeacetoxy compound (18) (no major UV absorption).

Dihydrocucurbitacin B (17) has also been found to occur in nature in *Marah oreganus* (*61*) (see section V.3).

Compound (18) when submitted to cleavage with sodium periodate afforded isocaproic acid (19) which was identified as its methyl ester. Upon reduction with zinc in acetic acid of (5) the deacetoxy derivative (20) was obtained, the removal of the C-25 acetoxy group being accompanied by the migration of the double bond ($\Delta^{23} \to \Delta^{24}$). Catalytic hydrogenation of (20) gave the dihydrodeacetoxy derivative (18), (*93, 111*).

The identification of the side chain was at the time a decisive factor in the assignment of a tetracyclic skeleton to the cucurbitacins.

1.4. The Ring A Substituents; the α-Hydroxyketones (5), (6) and the Diosphenols (Enolized α-Diketones) (2), (7)

As already described, the interrelation between the α-hydroxyketones (5), (6) and the diosphenols (2), (7) could be performed by oxidation with bismuth oxide, a reagent which did not affect the other functional groups of the molecule (*74*). The α-hydroxyketone function was characterized by the formation of a formazan with triphenyltetrazolium chloride, whereas the diosphenol gave a deep green color with ferric chloride. Furthermore, the previous system could be converted into the latter by autoxidation in alkaline medium (see later).

A series of reactions which resulted in cleavage of the side chain and the concomitant contraction of ring A, leading to a common degradation product (23) is compiled in Chart 3. Treatment of elaterin (2) or elatericin B (7) with hot alkali induced cleavage of the C-23, C-24 bond and a benzilic acid type rearrangement of the ring A diosphenol thus resulting in the formation of ecballic acid. This compound which was obtained in the early studies of elaterin (*17*) was assigned structure (21a) (*69; 80*), a structure which was confirmed through lithium aluminium hydride

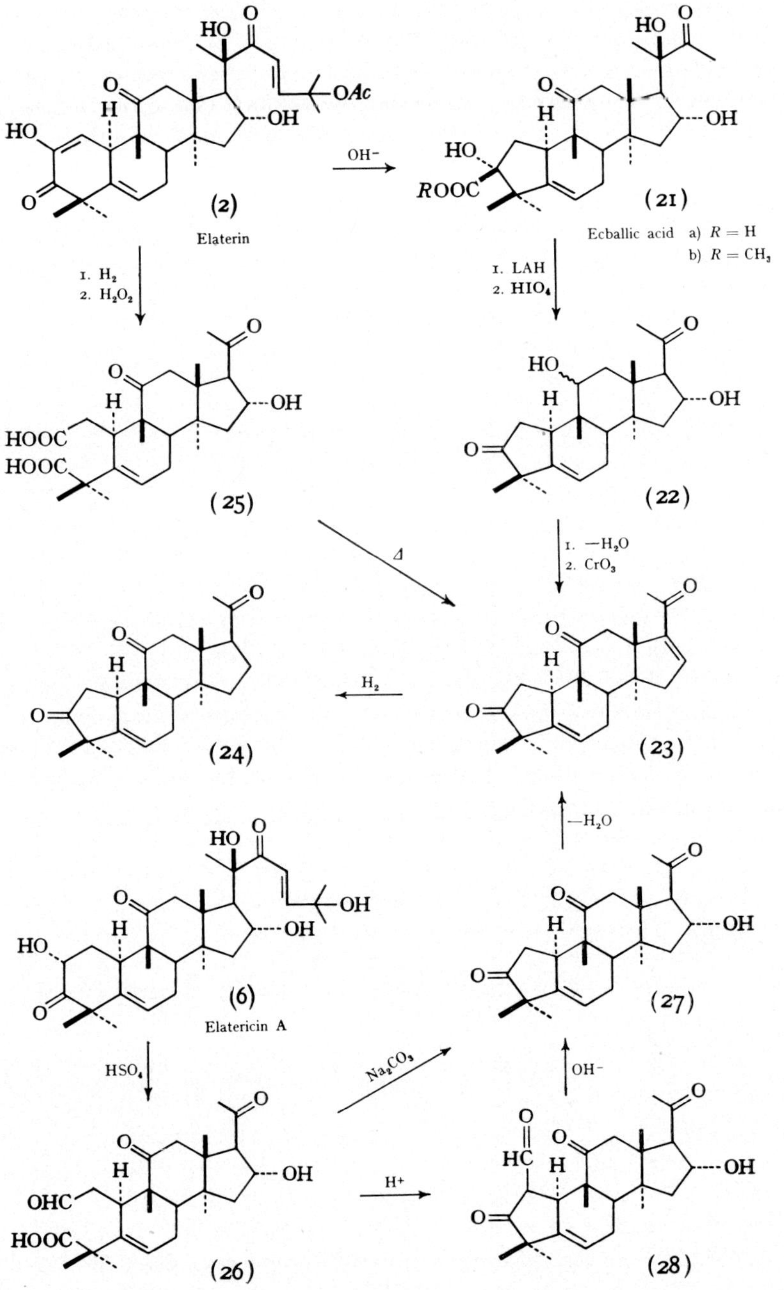

The Cucurbitanes, a Group of Tetracyclic Triterpenes 315

(2) Elaterin

(21) Ecballic acid a) $R = H$ b) $R = CH_3$

1. H_2
2. H_2O_2

1. LAH
2. HIO_4

(25)

(22)

(24)

(23)

(6) Elatericin A

(27)

HSO_4

Na_2CO_3

OH^-

(26)

H^+

(28)

Chart 3. Reactions of Ring A

reduction of the corresponding methyl ester (**21 b**) followed by periodic acid oxidation to the dihydroxy-diketone (**22**). The relation between the 16-hydroxyl and the 20-one was disclosed by β-elimination to the corresponding Δ^{16}-20-one (λ_{max} 240 nm); reoxidation of the ring C hydroxyl afforded finally the unsaturated triketone (**23**) which was then reduced to (**24**).

A second route to the triketone (**23**) consisted in alkaline hydrogen peroxide oxidation of 23,24-dihydroelaterin (dihydro-**2**) to the seco-dicarboxylic acid (**25**) (it is noteworthy that under these conditions the C-20, C-22 bond was also cleaved). Pyrolysis of (**25**) afforded now (**23**) directly (*88*).

Oxidation of elatericin A (**6**) with two moles of periodic acid resulted in cleavage of the two ketol groups present in the molecule and furnished the seco-aldehydoacid (**26**). Upon heating in aqueous sodium carbonate or bicarbonate solution, the latter cyclized to form the hydroxytriketone (**27**) accompanied by small quantities of (**23**). The complete conversion of (**27**) into (**23**) was carried out in the presence of p-toluenesulfonic acid (*74, 80*). A similar cleavage with periodic acid performed on cucurbitacin B (**5**) afforded the same seco-aldehydoacid (**26**) which could be cyclized with dilute mineral acid to the β-aldehydoketone (**28**) (enolic form λ_{max} 282 nm, ε 2740, shifted with base to 309 nm, ε 8450). Sodium carbonate treatment of (**28**) gave a mixture of (**23**) and (**27**) (*59, 60*).

In the presence of dilute alcoholic alkali the absorption band of the side chain enone (λ_{max} 230 nm) in elatericin A (**6**) gradually disappeared due to hydration of the double bond (Δ^{22}). At the same time however, a new absorption developed at 310 nm (ε 5100) (enolate ion) which was shifted following acidification to 268 nm (ε 6900); the reaction involves consequently the oxidation of the ring A ketol to the enolized diketone (*72*). This spectroscopic test has been widely used for the characterization of ring A functionality in several cucurbitacins.

1.5. *The 19-Methyl Group and Ring C Carbonyl*

The locations of the 19-methyl at C-9 and of the ring C carbonyl at C-11 are based on a series of independent experiments performed on derivatives of elatericin A (**6**) (*80, 81*) and cucurbitacin A (**61**) (*60*), as well as on the conversion of the latter (**7**) into a derivative (**100**) of eburicoic acid. All these determinations converged to the same conclusions. The first group of experiments will be described in the sequel (Chart 4), the latter in the appropriate subsequent sections (Charts 11, 12 and 17).

The triketone (**24**) yielded only a bisethylene-ketal (**24 b**) thus revealing the hindered nature of one of the ketones. Lithium aluminium hydride reduction to (**24 c**) followed by dehydration produced a compound with

References, pp. 357—362

a double bond in ring C. The NMR spectrum of the diketone (29) obtained after hydrolysis of the ketal groups exhibited an AB pattern for the two vinylic cis protons (11-H and 12-H) (δ 5.28 and 6.0, J = 10 Hz), thus indicating that no vicinal proton is present at C-9 to induce additional coupling of the 11-H (*80*).

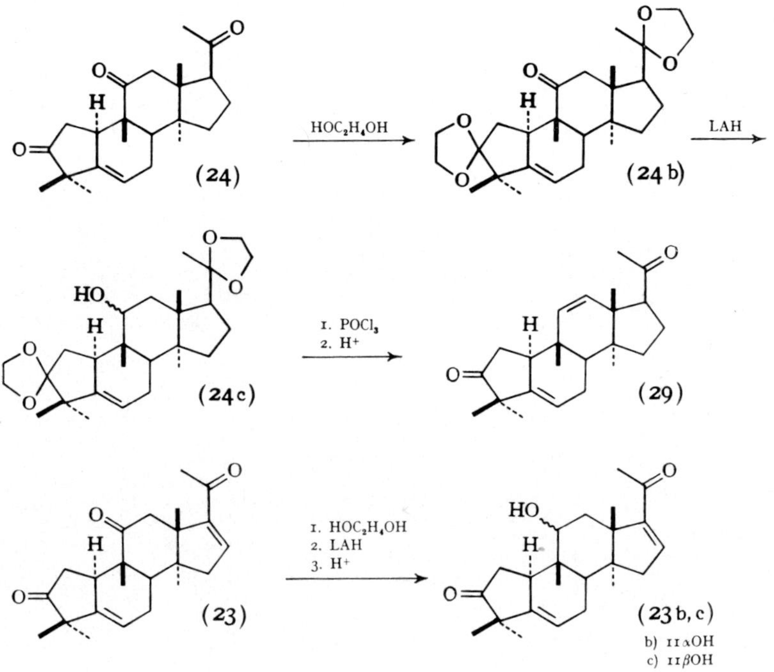

Chart 4. Reactions of Ring C Carbonyl

The formation of a C-11, C-12 double bond in (29) left two possibilities for the ring C ketone, either at C-11 or at C-12. At this stage, its assignment to C-11 rather than to C-12 was based on the UV and IR spectra of (23) (λ_{max} 240 nm, and ν_{max} 1666 cm^{-1} for the unsaturated 20-one), thus eliminating the C-12 alternative. Furthermore, neither of the isomeric C-11 alcohols (23 b, c), obtained by reduction with lithium aluminium hydride of the bisethylene ketal of (*23*) form a hydrogen bond with the 20-ketone; this again excludes the 12-position for the ketone.

1.6. The Ring B Double Bond

Reactions which permitted assignment of the ring B double bond to the 5,6-position also established the absence of a methyl group from C-10 which became a characteristic feature of the cucurbitane skeleton.

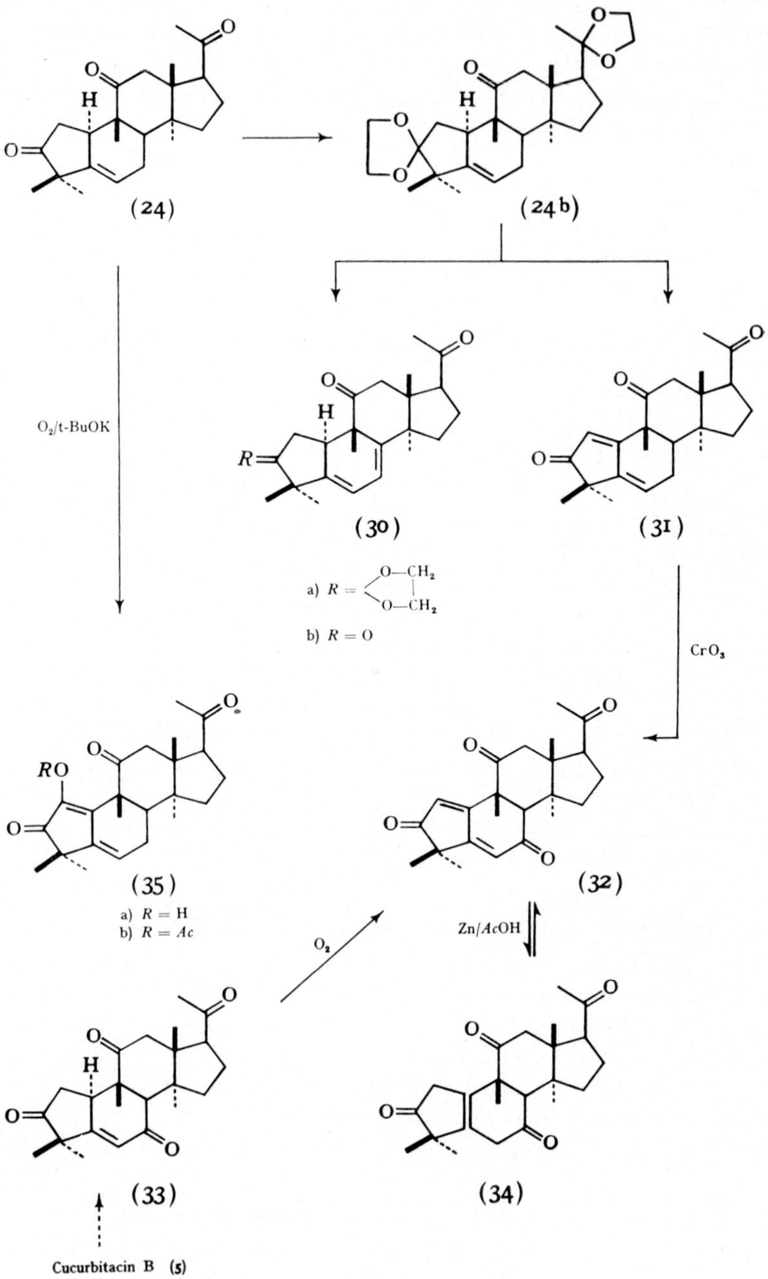

Chart 5. Reactions of Ring B

Reaction of N-bromosuccinimide with the bis-ethyleneketal (24b) (Chart 5) induced allylic bromination. This was followed by spontaneous dehydrobromination to the homoannular diene (30a) (λ_{max} 275 nm, ε 7800) and the heteroannular diene (31) (λ_{max} 289 nm ε 9000). The NMR spectrum of (30b) confirmed the presence of an allylic proton at C-10 (X part of an ABX system, $J_{AB} = 4$ Hz, $J_{AX} = 2$ Hz, $J_{BX} = 0$). The other two allylic positions are occupied by methyl groups.

The IR spectrum of the second compound (31) indicated conjugation of the ring A carbonyl: 1742 cm^{-1} in (24) against 1707 cm^{-1} in (31).

Oxidation of (31) with chromium trioxide in acetic acid yielded the yellow conjugated dienedione (32) (λ_{max} 282 nm (ε 18,700); ν_{max} 1717, 1709, 1666 and 1569 cm^{1-}), the NMR spectrum of which showed two doublets at δ 6.21 and 6.08 (J = 1 Hz) for the two vinylic protons which couple their spins through the conjugated system. The same dienedione (32) was obtained by oxidative degradation of cucurbitacin B (5) (60); such a reaction can be explained by analogy to the base catalysed autoxidation of 1,4-diketones to enediones, the 2-ene-1,6-dione bearing a vinylogous relationship to 1,4-diketones. Subsequent reduction with zinc in acetic acid afforded the β,γ-unsaturated ketone (34) which was reconverted into (32) by oxidation under alkaline conditions (60). Compound (24) could also be autoxidized in benzene solution with potassium-t-butoxide in t-butanol to the diosphenol system shown in (35a) and was characterised as the crystalline acetate (35b): λ_{max} 291 nm (ε 10,900), ν_{max} 1773 cm^{-1} (enol acetate) (80).

During the prolonged treatment with acid of the unsaturated triketone (24), a migration of the double bond took place and produced the conjugated unsaturated triketone (40) (λ_{max} 236 nm, ε 16,300). Actually, during the dehydration process of the 16-OH in (27), the $\Delta^{1(10),16}$-diene (41) was formed as a by-product together with the $\Delta^{5,16}$-compound (23), which was then isomerised to (41) by the acid treatment (Chart 6) (66).

An alternative sequence of reactions (80) was based on the hydroxylation with osmium tetroxide of the carbocyclic double bond (Δ^5) in (24) to the secondary-tertiary glycol (36a), subsequently oxidized to the tetraketone (37). The latter was then dehydrated under acidic conditions to the ene-1,5-dione (38), and extracted with sodium carbonate from the reaction mixture as the corresponding enol (39a) [λ_{max} 335 (ε 10,300) and 405 nm (ε 3100); enolate ion $\lambda_{max}^{1\% KOH}$ 405 nm (ε 21,700)]; the compound was completely characterized as the enol acetate (39b), λ_{max} 290 nm (ε 25,800). Additional work on the hydroxylation of (24) showed that the reaction is non-stereospecific, about equal quantities of both α and β cis-5,6-glycols (36) being formed. The two glycols underwent oxidation to (37) (5α- and 5β-isomers) and dehydration to the same enolized ene-1,5-dione (39a), the reaction proceeding at a faster rate with the 5β-isomer*.

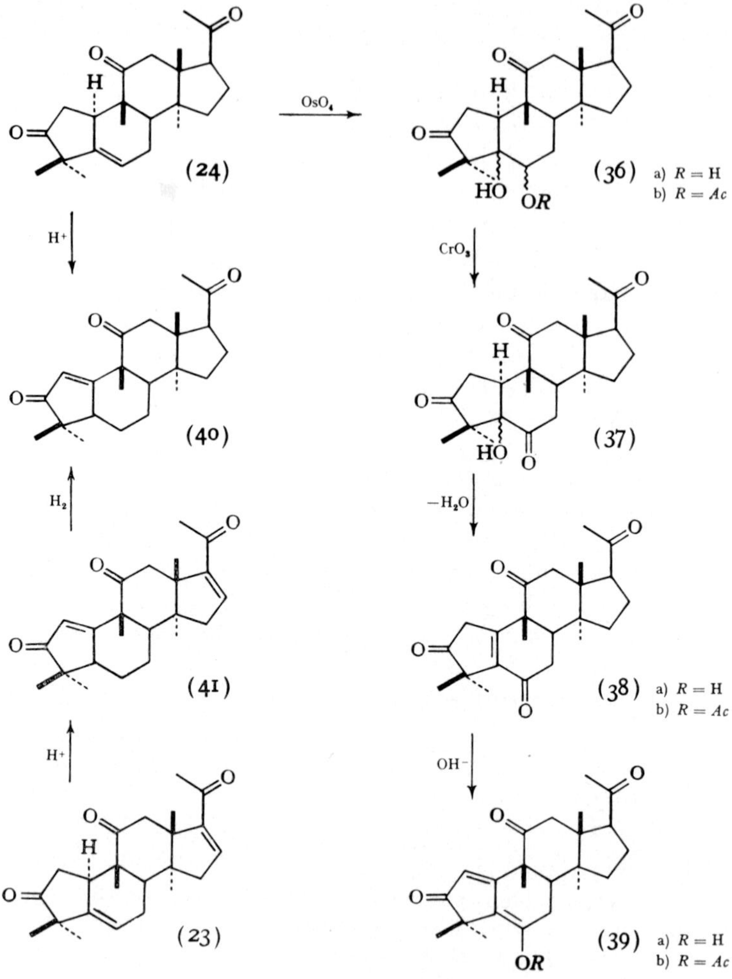

Chart 6. Reactions of Ring B

Reactions confirming the position of the double bond (Δ^5) and the presence of a hydrogen at C-10 were further performed on series of compounds which had the six-membered ring A with the original functionality present in cucurbitacin B (5) and elatericin A (6) (Chart 7). Oxidation of cucurbitacin B acetate with chromium trioxide afforded "cucur-

---

* M. COHEN, Ph. D. thesis, Weizmann Institute, Rehovot, Israel, 1969.

References, pp. 357—362

bitone B" (42) (91) which rearranged with methanolic hydrochloric acid to the conjugated dienedione (44), λ_{max} 238 (ε 12,800) and 288 nm (ε 19,800), in which the 1-H and 6-H showed the expected NMR doublets at δ 5.84 and 6.27. The mechanism proposed for the formation of the latter is based on the intermediate enol (43) (see arrows). This dienedione (44) behaved with zinc in acetic acid in the same way as (32), being reduced to the $\beta\gamma$-unsaturated compound (45) (60).

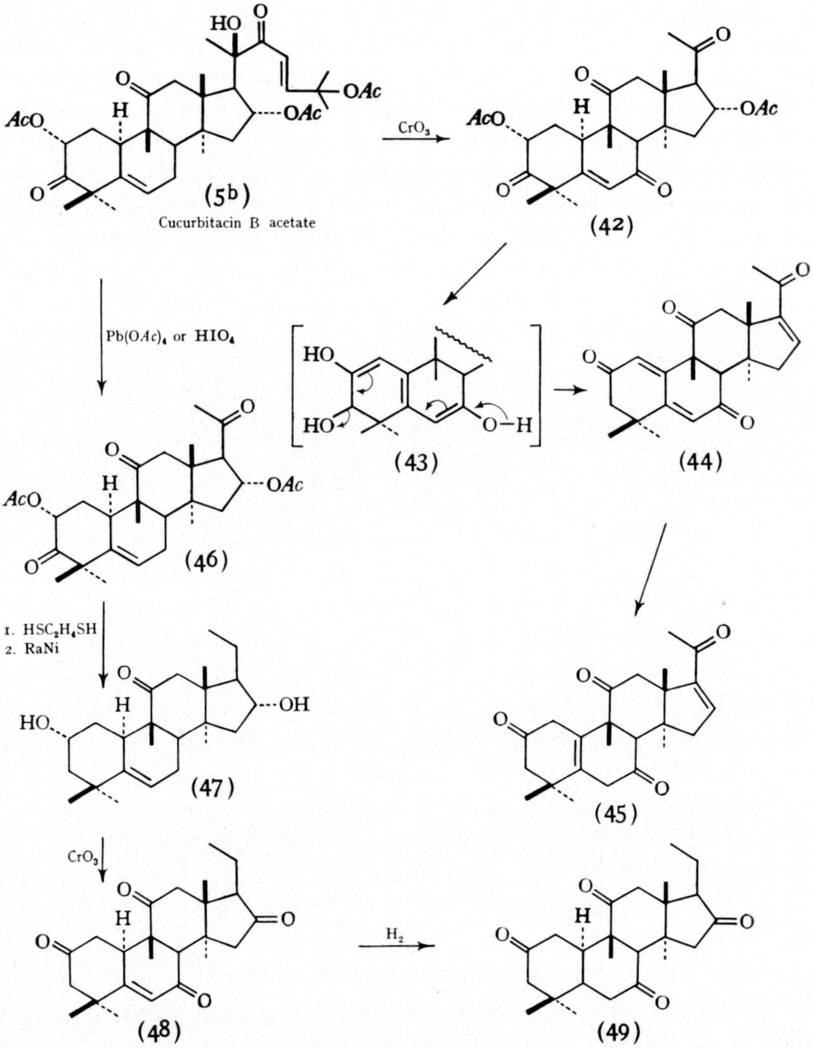

Chart 7. Reactions of Rings A and B

In a more recent paper (*19*) an efficient method for the cleavage of the C-20, C-22 bond is described: treatment with lead tetraacetate in refluxing benzene of cucurbitacin B (5) acetate afforded in excellent yield the triketone (46). The reaction was also applied to the acetates of cucurbitacin A (61) and E (2).

A rather different sequence (*51*) of reactions leading to the tetraketone (49) was performed on (46), obtained at the time by periodic acid treatment of elatericin A (6) acetate. Thioketalization of (46) followed by Raney nickel reduction afforded the 2,16-diol (47) which was oxidized with chromium trioxide to the unsaturated tetraketone (48). Catalytic hydrogenation of the latter led to (49) (ν_{max} 1740 and 1700 cm^{-1}); however, the stereochemistry at C-5 in this compound has not yet been determined.

The relationship between the 5,6-double bond and the *gem*-dimethyl group in ring A could be demonstrated by means of the nitrile (52), ν_{max} 2243 cm^{-1}, which is formed as a result of the BECKMANN rearrangement of the 3,20-bisoxime (50) (*80*). Whereas the 20-oxime afforded in presence of p-toluene-sulfonyl chloride the expected eneamine leading to the 17-ketone, the 3-oxime underwent an abnormal rearrangement (51) (see arrows) which resulted in cleavage of ring A to (52). The UV band of the latter (229 nm, ε 9000) is characteristic of a conjugated diene; the nature of this chromophore was confirmed by the appropriate NMR signals for the C-30 terminal methylene (δ 4.70 and 4.90) (Chart 8).

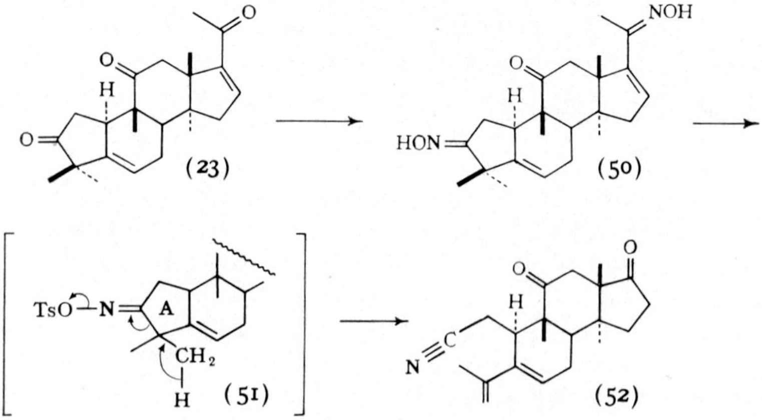

Chart 8. Beckmann Degradation of (23)

1.7. The 16-Hydroxy Group

The location of the hydroxy group at C-16 is based on reactions performed on different degradation products obtained from several of the cucurbitacins analyzed in the present chapter (Chart 9). Treatment with hot alkali of elaterin (2) followed by esterification afforded methyl ecballate (21b) which was then oxidized with chromium trioxide in acetone to the 16-one derivative (53) (ν_{max} 1743 cm^{-1} for this carbonyl). Under

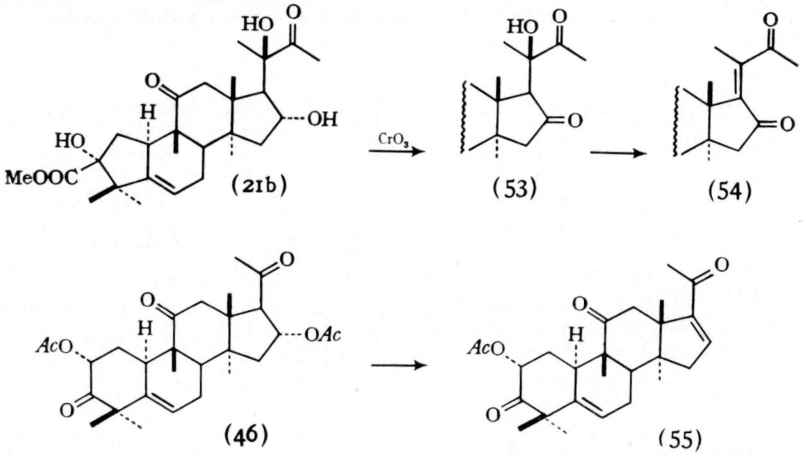

Chart 9. The 16-Hydroxyl

alkaline conditions, elimination of the 20-hydroxyl and formation of the enedione (54), λ_{max} 254 nm (ε 10,500), took place. Since the location of the 20-hydroxyl is well documented by means of previously described reactions, the conversion of (53) into (54) constitutes a good proof for assignment of the secondary hydroxyl group of elaterin to the 16-position (88). Compound (46) obtained by cleavage of the C-20 C-22 bond in the acetate of elatericin A (6) (74) or cucurbitacin B (5) (19) could be converted into the Δ^{16}-20-one (55) λ_{max} 240 nm (ε 10,000), ν_{max} 1660 and 1590 cm^{1-}. Similar eliminations which are compatible with the 1,3-relationship between the 20-one and the 16-hydroxyl have already been mentioned in connection with other reactions of the cucurbitacins *(vide supra)*.

1.8. Alkaline Treatment of Elaterin

A series of complicated reactions takes place when elaterin, and other cucurbitacins containing a diosphenol system are treated with aqueous alkali. Ring A undergoes a benzilic-type rearrangement to give an α-hydroxyacid. This is accompanied by a hydrolytic cleavage in the side chain, which arises through hydration and subsequent cleavage of the double bond conjugated to the 22-carbonyl group, the product being primarily ecballic acid (21a). The latter, when submitted to prolonged treatment with alkali yields elateric (56) and isoelateric (57) acid which are characterized by IR bands indicative of the newly formed five membered ring ketone (1751 cm^{-1}) and cyclic ether (1156 cm^{-1}). Aluminum amalgam reduction of (56) afforded the 20β-OH derivative (60) which was used for ORD measurements (69).

The formation of ring E in (56) and (57) can be explained by a sequence of reactions involving an acyloin rearrangement in the α-hydroxyketone side chain of (21a) to form (58), followed by β-elimination of the 16-hydroxyl (59) and cyclisation to either (56) or (57) (Chart 10).

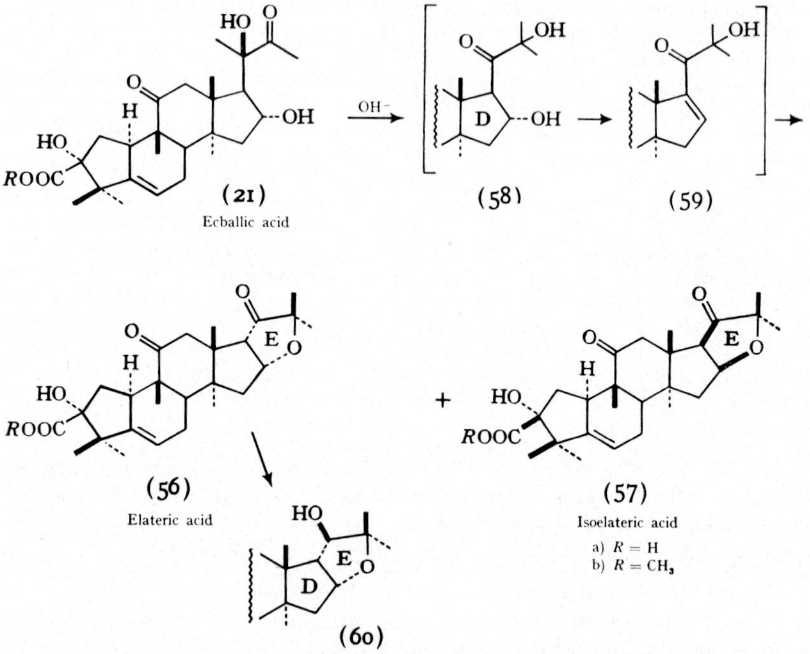

Chart 10. Formation of Elateric and Isoelateric Acids

Assignment of the two possible orientations of ring E with respect to the plane of ring D to methyl elaterate and methyl isolaterate was made possible by NMR and ORD measurements. When ring E is α-oriented, the 12α-H assumes a position equidistant from the deshielding cones of the carbonyl groups at C-11 and C-20, and should therefore appear at lower field than when ring E is β-oriented. Since this is true for methyl elaterate, structure (56) was assigned to this compound, and methyl isolaterate with the 12α-H signal at higher field is therefore (57). Similar conclusions were reached on the basis of ORD measurements.

A detailed analysis of the benzilic acid rearrangement taking place in ring A provided a description of the complete stereochemistry of compounds (21), (56) and (57). The rearrangement has to be unidirectional since logically, in alkaline solution, the diosphenol exists largely as the $C_2\text{-}O^-$ anion, and the 3-carbonyl group offers the preferred acceptor site for attack by the hydroxyl-ion (69).

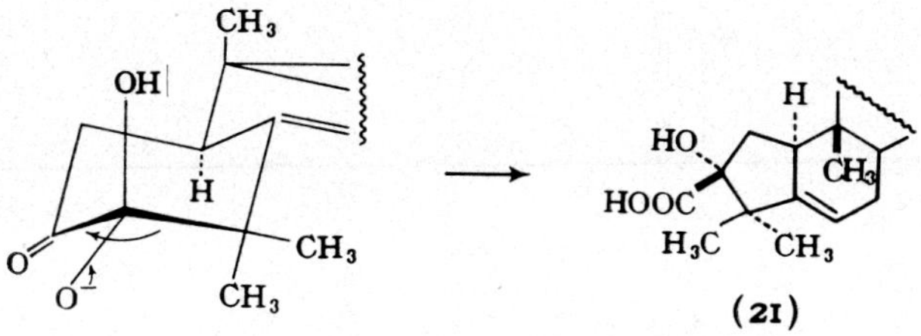

2. Cucurbitacins A (61) and C (69)

2.1. Structure Determination

The characteristic feature of cucurbitacins A and C is that one of the five angular methyl groups is replaced by a hydroxymethyl group. The difference between A and C is in the substitution pattern of ring A, a 2-hydroxy-3-one in cucurbitacin A (61) and a 3-hydroxyl in cucurbitacin C (69) (60). The presence of the ring A ketol in cucurbitacin A, and that of the usual side chain, the C-16 hydroxyl and the trisubstituted double bond in both compounds, has been established independently (29, 30) by methods similar to those used for the cucurbitacins described in the previous section.

Chromic acid oxidation of (61 b) leads to "cucurbitone A" (62) (29), the NMR spectrum of which disclosed only four quaternary C-methyl groups, and the pattern of an AB system centered at δ 3.98 and 4.41 (J = 11 Hz) which can be assigned to a $-\overset{|}{\underset{|}{C}}-CH_2OAc$ group replacing the fifth angular methyl (60). Enol-acetylation of (62) afforded a yellow crystalline enol-acetate (64) (λ_{max} 238 and 357 nm; ε 14400 and 9200 respectively), its formation being interpreted as outlined in (63) (see arrows). Mild acid hydrolysis (0.1% aqueous perchloric acid in acetic acid) afforded the dienedione (65) which differs from the corresponding compound (44) obtained from cucurbitacin B (5) in having one quaternary methyl replaced by a $-CH_2OAc$ group. To establish the position of the latter, the 5,6-dihydrocucurbitone A (66) was subjected to the action of aqueous ethanolic alkali. This resulted in loss of formaldehyde and formation of the norcompound (67). Since the loss of formaldehyde must represent a reverse aldol reaction, the $-CH_2OH$ group had to be α to a ketone, position 9 being the only one which could fit with all the experi-

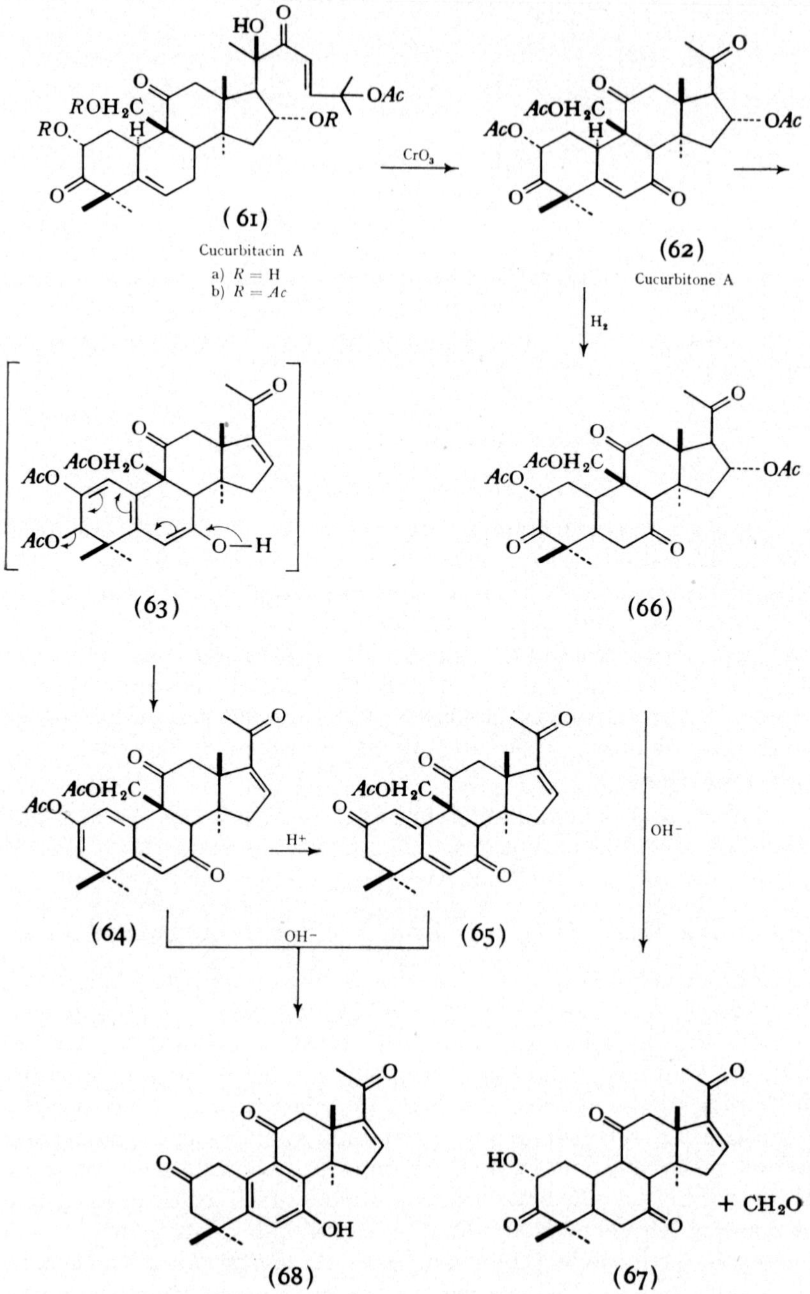

(61) Cucurbitacin A
a) R = H
b) R = Ac

(62) Cucurbitone A

(63)

(66)

(64)

(65)

(68)

(67) + CH₂O

Chart 11. Reactions of Cucurbitacin A

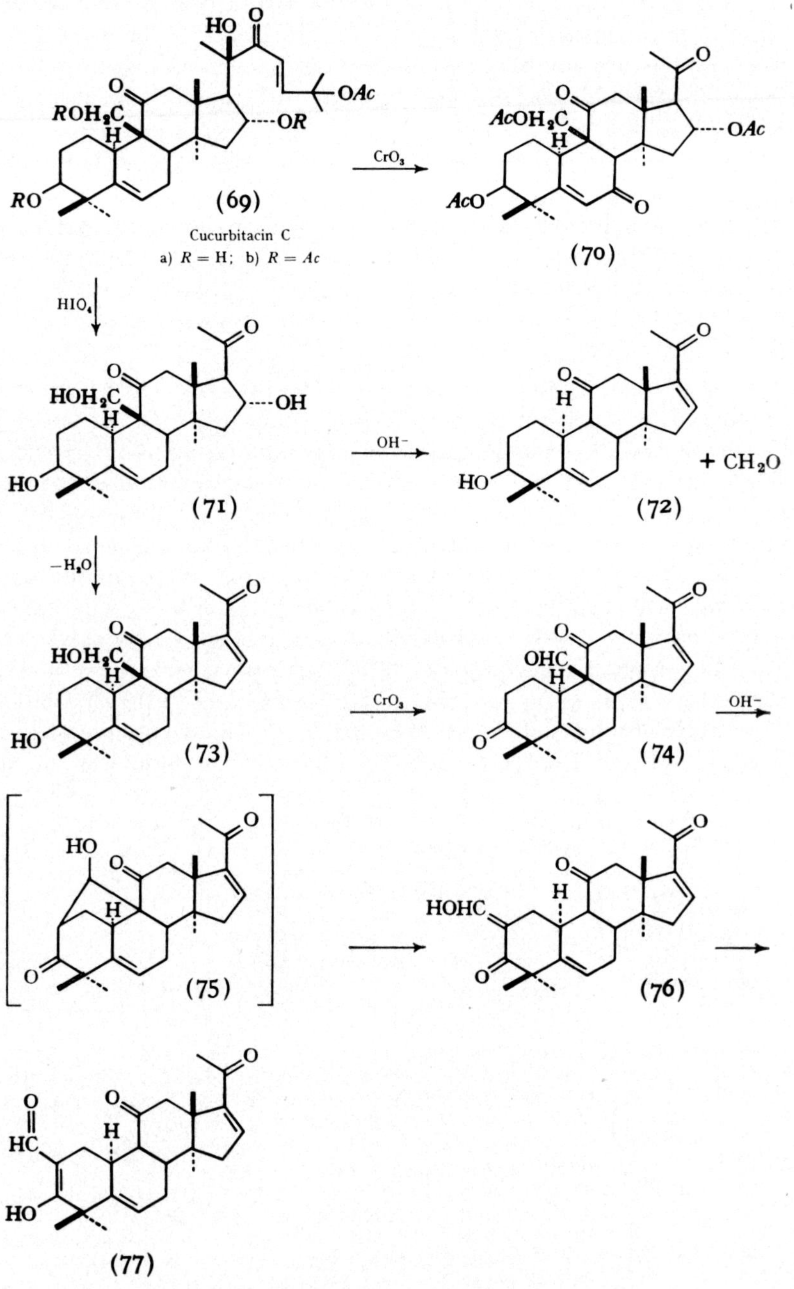

The Cucurbitanes, a Group of Tetracyclic Triterpenes

Chart 12. Reactions of Cucurbitacin C

mental details; this was also in agreement with a similar reaction performed with a derivative (71) of cucurbitacin C (69). Furthermore, when either the enol acetate (64) or the dienedione (65) was treated with dilute alkali, formaldehyde was obtained together with the phenolic norketone (68), (60) (Chart 11).

Cucurbitacin C acetate (69b) on oxidation with chromium trioxide afforded "cucurbitone C" (70) (30). It was also possible to cleave the side chain with periodic acid without concomitant allylic oxidation at C-7. The resulting hexanorderivate (71) behaved towards alkali like (66), one mole of formaldehyde being lost and the heptanor-derivative (72) being obtained, hence this reaction confirmed that the —CH$_2$OH group had to be α to the 11-ketone (60).

A remarkable reaction takes place with the triketoaldehyde (74) that is produced by oxidation of the 16-desoxy compound (73). In presence of ethanolic alkali (room temperature) (74) affords two isomeric enolic compounds (76) and (77). This conversion which is formulated as an intramolecular transfer of a β formyl group implies the presence of two keto-groups; the readily enolisable 3-ketone enables the anionic 2-carbon atom to attack the formyl group with formation of the intermediate aldol (75). By cleavage of the C-9 C-19 bond in the latter, the formyl group is transferred to C-2. As expected, the reaction did not take place with the 3-hydroxy analogue of (74) (Chart 12). The transfer of the formyl group [(74) → (75) → (76)] greatly contributes to the elucidation of the stereochemistry of rings A/B and B/C. This reaction can take place only if the B/C ring fusion is *cis* with the formyl group at

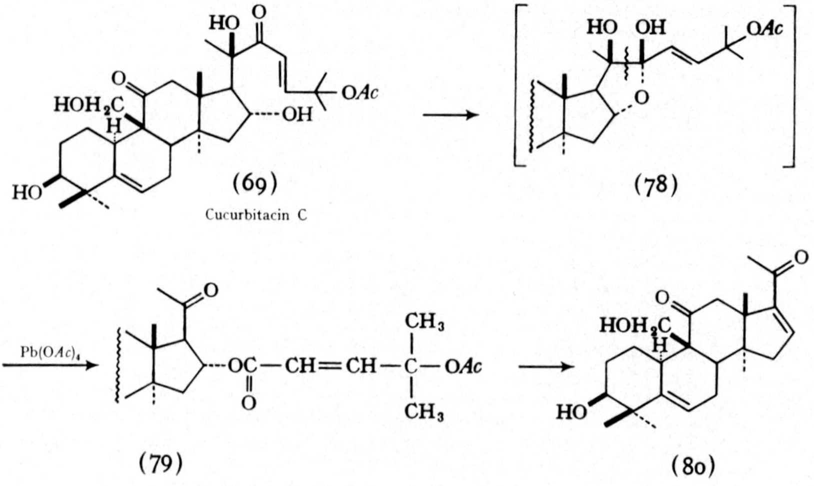

Chart 13. A Cleavage of the Side Chain

position 9 and consequently the hindered ketone at C-11. Furthermore, the configuration of the 10-H has to be opposite to the configuration of the 9-formyl group (*60*).

An interesting aspect of the chemistry of the cucurbitacins is the opportunity for intramolecular attack on the 22-one by the 16α-OH with formation of a "pseudo-glycol" (78). The formation of the 16-ester (79) following lead tetraacetate cleavage of (69) has been interpreted (*19*), as taking place through cleavage of C-20 C-22 bond in (78). Under alkaline conditions the 16-ester (79) was dehydrated to the corresponding Δ^{16}-20 one (80). Such an ester (79) has been previously encountered as a by-product in the periodic acid cleavage of (69); however, it could not be characterized at the time (*30*) (Chart 13).

2.2. Interrelationship Between Cucurbitacins A, B and C

The relation between cucurbitacins A (61) and C (69) on the one hand and cucurbitacin B (5) on the other is based on the degradation of all three compounds to the same monoketone (83) (*60*) (Chart 14). Reduction of the acetate (5b) with calcium in liquid ammonia followed by cleavage with periodate and alkaline hydrolysis afforded the hydroxydiketone (81) (Cf. *91*). Oxidation of the 11-ol and reduction of the 16,17-double bond in the latter gave the triketone (82) which was converted into (83) via thioketalization and Raney nickel desulfurization. Similarly, cucurbitacin A acetate (61b) was tranformed into the hydroxytriketone (86) (obtained along with two other compounds (84) and (85)) which upon oxidation yielded the aldehydo-triketone (87); thioketalization followed by desulfurization gave the same monoketone (83) (*60*).

Conversion of cucurbitacin C (69) to the aldehydotriketone (87) connected this compound (69) with A (61) and B (5). Since the latter has been previously related to cucurbitacins D, E and I (*84*) all compounds described heretofore have been interrelated.

3. Stereochemistry of the Cucurbitacins

The work which has been reviewed in the previous chapter has unequivocally demonstrated that the cucurbitacins possess the 19-methyl at position 9 and a hydrogen at position 10; moreover, these two substituents had to be in a *trans* relationship, which is the only arrangement in agreement with the sequence (74) → (75) → (76) (Chart 12) i. e. 10α-H, 9β-CH$_3$ and 8β-H.

An independent demonstration for the α orientation of the 10-H is based on ORD measurements (*81*). Subtraction of the ORD curve of compound (88) from that of (24) gave the contribution of the ring A ketone which was comparable but opposite in sign with that exhibited by 4,4-dimethyl-A(2)-norcholest-5-en-3-one (89).

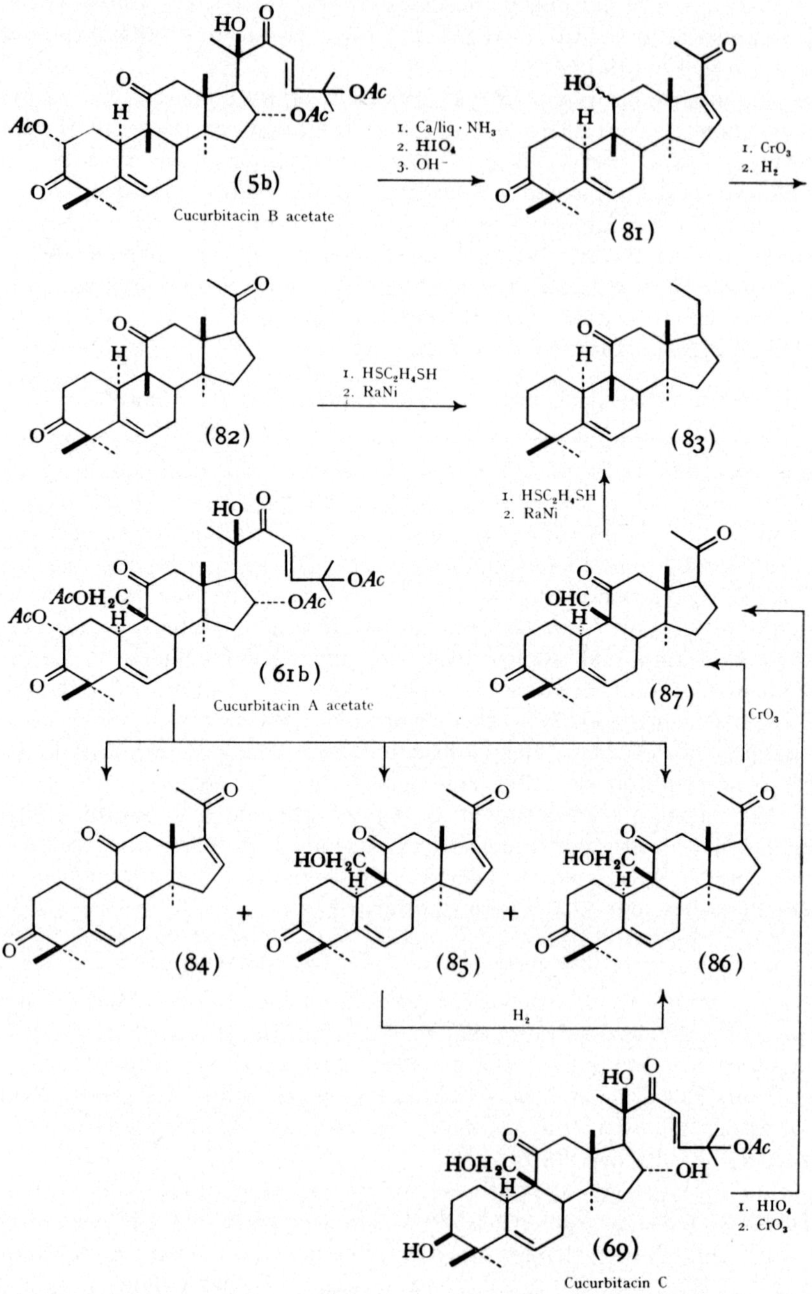

Chart 14. Interrelation Between Cucurbitacins A, B and C

References, pp. 357—362

The B/C rings fusion was studied using the ORD curve of the monoketone (90) which gave a large positive Cotton effect in agreement with data recorded for a steroidal B/C-*cis*-fused system (Cf. ref. 28b in *81*). Indeed, models of compounds with 8β,9β orientation, such as (90), when studied in the light of the octant rule, were found to be entirely in positive octants (*81*).

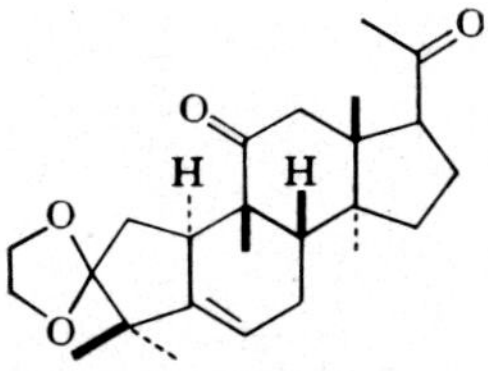

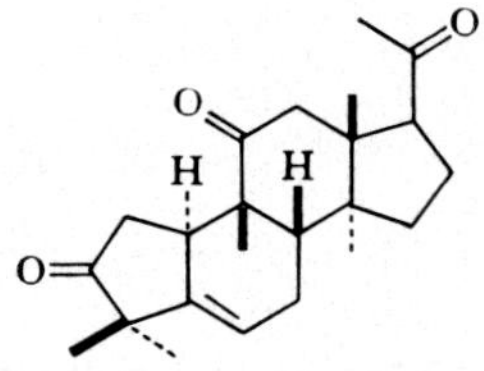

Chart 15. Compounds Used for ORD Measurements

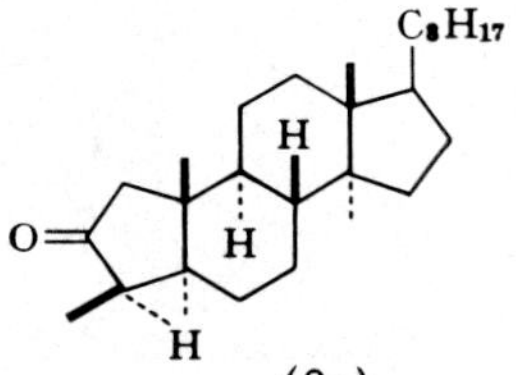

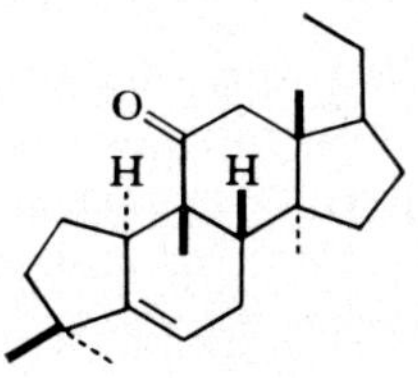

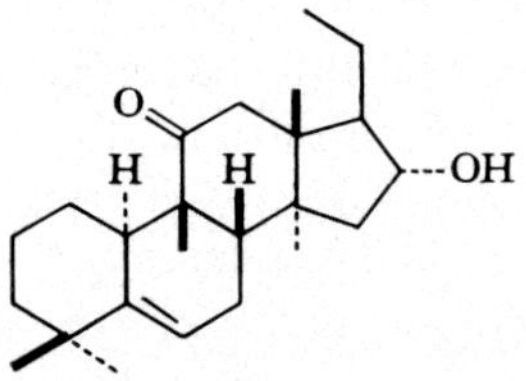

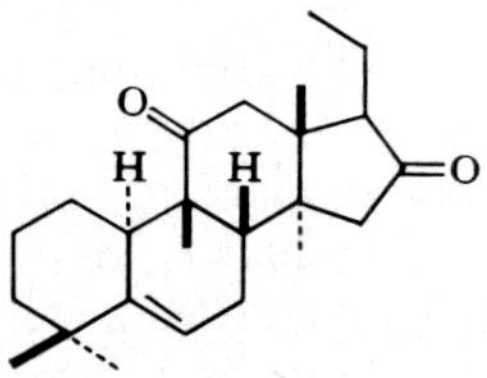

The stereochemistry of the C/D ring fusion was determined by ORD measurements on the 16-ketone derivative (92). Subtraction of the ORD curve of the monoketone (91) from the curve of the diketone (92) eliminated the contribution of the C-11 carbonyl and gave a curve for a 16-ketone which showed a strong negative Cotton effect characteristic of 16-keto-13β, 14α steroids and tetracyclic triterpenes. Hence a lanostane-type fusion was indicated (*76*).

The orientation of the side chain at C-17 was studied using the Cotton effect associated with the 20-ketone group. The compounds selected for these measurements were (88) and (90), the curve of the latter being subtracted from that of the former, eliminating thereby the effect of the 11-ketone. The resultant was a positive curve of weak amplitude, indeed smaller than found in 17β-20-one steroids. On the other hand whenever such a side chain is α-oriented, a negative effect is observed. The orientation of the side chain is therefore β (*81*).

To establish the stereochemistry of the 16-hydroxyl, the molecular rotation shift resulting from its acetylation was compared with that observed for similar groups in steroids. For this calculation elatericin B diacetate (7b) and elatericin B (7a) were selected, and the difference was found to be in agreement with an α-oriented hydroxyl group. The same orientation was deduced by NMR spectroscopy using the observed

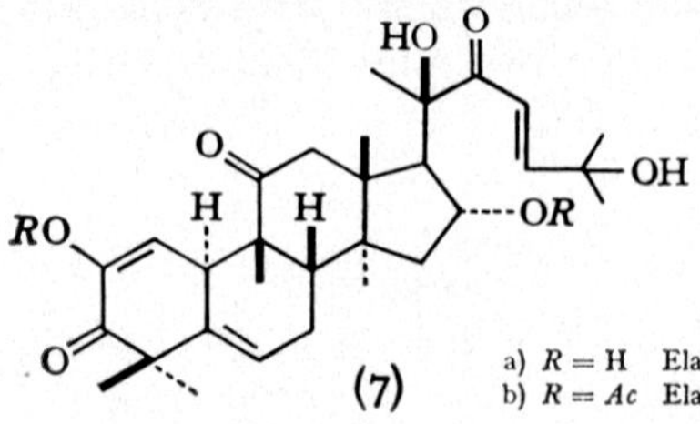

(7) a) R = H Elatericin B
    b) R = Ac Elatericin B diacetate

coupling constants between the 16β-H and the adjacent protons and the known relationship between the dihedral angles and coupling constants (*81*).

4. Stereochemistry of Ring A Ketols

While the study of ring A ketol stereochemistry has been carried out (*121*) on anhydro-22-deoxocucurbitacin D (93a), anhydro-22-deoxoisocucurbitacin D (94a) and anhydro-22-deoxo-3-epi-isocucurbitacin D (95a), the conclusions are valid for all ring A ketols in the cucurbitacin

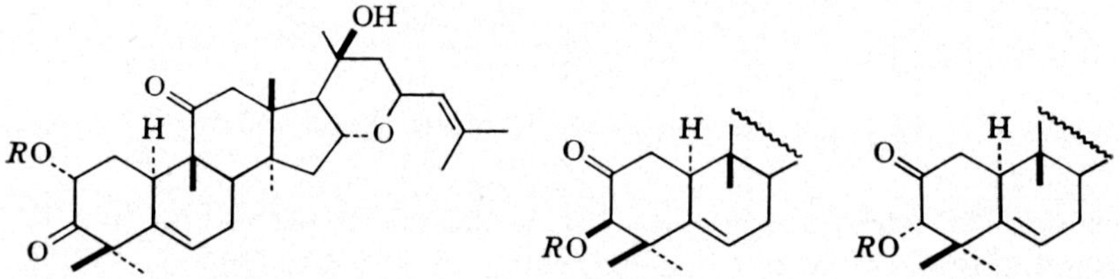

(93)   (94)   (95) a) R = H
                      b) R = Ac

Chart 16. Stereochemistry of Ring A Ketols

series. The NMR signal of 2-H in (93b) (double doublet, δ 5.45, J = 13 and 6 Hz) indicates that this proton is axial; consequently the equatorial 2-OAc should be β oriented when ring A is chair, or α when twisted. The equatorial orientation of the latter is in agreement with the UV band (290 nm) of the 3-ketone and its IR absorption (1736 cm^{-1} in CS_2). The α configuration in a twisted ring A which has been assigned to the 2-OAc rests on CD measurements (Fig. 1) and on comparison with gratiogenin (126) derivatives*.

---

* Professor S. M. KUPCHAN has kindly informed the authors that on the basis of an X-ray analysis performed on the di-p-iodobenzoate ester of a new glycoside of Cucurbitacin D, the 2-OH was found to be β-oriented (KUPCHAN, S. M., C. W. SIGEL, J. L. GUTTMAN, R. J. RESTIVO and R. F. BRYAN: Datiscoside, a Novel Antileukemic Cucurbitacin Glycoside from *Datisca glomerata*, J. Amer. Chem. Soc., in press).

In compounds (94b) and (95b) the NMR signals of the 3-H are singlets at δ 5.02 and 4.95, respectively. The UV absorption band of both compounds is at 291 nm (288–289 nm in the free ketols), and both compounds display the IR frequencies of the 2-one at 1736–1737 cm^{1-} (in CS$_2$).

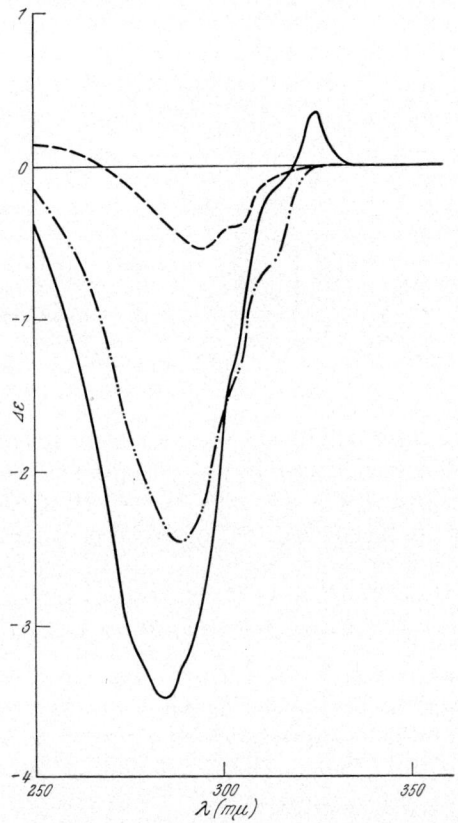

Fig. 1. Cotton effects of the ketol chromophores in compounds (93a) (—··—··—), (94a) (————), and (95a) (— — — —). These data have been computed by subtracting the CD curves of these compounds from that of the corresponding 2,3-diacetate (*121*)

This suggests that the 3-acetate is equatorial. The deduction is supported by the CD curves of (94b) and (95b), the difference in the amplitudes being attributed to different conformations of ring A, namely a twist form with an equatorial 3β-OAc in (95b) and a chair form with an equatorial 3α-OAc in (95b). Isomerization of (93a) and (94a) in concentrated alkaline solution leads to the formation, in high yield, of (95a); this fact does not prove that (95a) is more stable, since it crystallizes out during the reaction (*28, 121*).

It is worthwhile to refer here to the catalytic reduction of the diosphenol system of the cucurbitacins (67). Hydrogenation of elatericin B (7a) proceeded with the absorption of two moles of hydrogen to yield a tetrahydroderivative (96a) different from dihydroelatericin A (dihydro-6a); in the NMR spectrum of (96a) a singlet was observed at δ 3.98 [δ 5.0 in (96b)]. Compound (96a) was therefore assigned the isostructure (2-keto-3-hydroxy) with the OH group equatorial oriented, its formation being interpreted through 1,4-addition of the reagent. Hydrogenation of the diosphenol acetate system present in (7b) proceeded,

(7a) ⟶ (96) a) R = H b) R = Ac

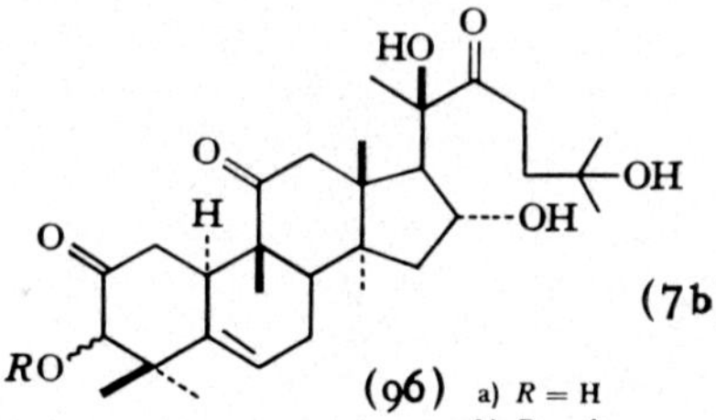

(7b) ⟶ 23,24-dihydro (6b)

however, by normal 1,2-addition to yield a tetrahydroderivative identical with dihydroelatericin A diacetate (dihydro-(6b)) in which the 2α-H is axial (double doublet, δ 5.6, J = 13.5 and 5.1 Hz). Since no CD data are available for compound (96) the conformation of ring A in this compound is as yet unknown.

5. Interrelationship Between the Cucurbitane and Lanostane Series

This interrelation which has been carried out by BARTON and co-workers (7) is based on the transformation of cucurbitacin A (61) into the tetraketone (100), which in turn has been obtained (8) by degradation of eburicoic acid, a triterpenoid of the lanostane group with a structure that has been established unequivocally (Chart 17).

Cucurbitacin A triacetate (61b) was converted into the dienedione (65a) *via* the enol-acetate (64) as already described (59, 60). Since chromous chloride reduction of (65a) yielded the βγ-unsaturated dihydroderivative (101a), it was expected that a similar reduction of the p-toluenesulphonate (65b) would result either directly, or *via* the intermediate (101b) and spontaneous intramolecular solvolysis, in the cyclopropane derivative (102). Actually, the sequence led to a ring B-homoderivative (103), the undesired opening of the intermediate cyclopropane ring being due to the relationship of the C-2 and C-11 ketones in (102). To overcome these difficulties, (65b) was submitted to controlled sodium borohydride reduction which gave a mixture of the 2α and 2β-hydroxy derivatives (97), a third compound, namely the 7β-OH derivative of (65b) being

References, pp. 357—362

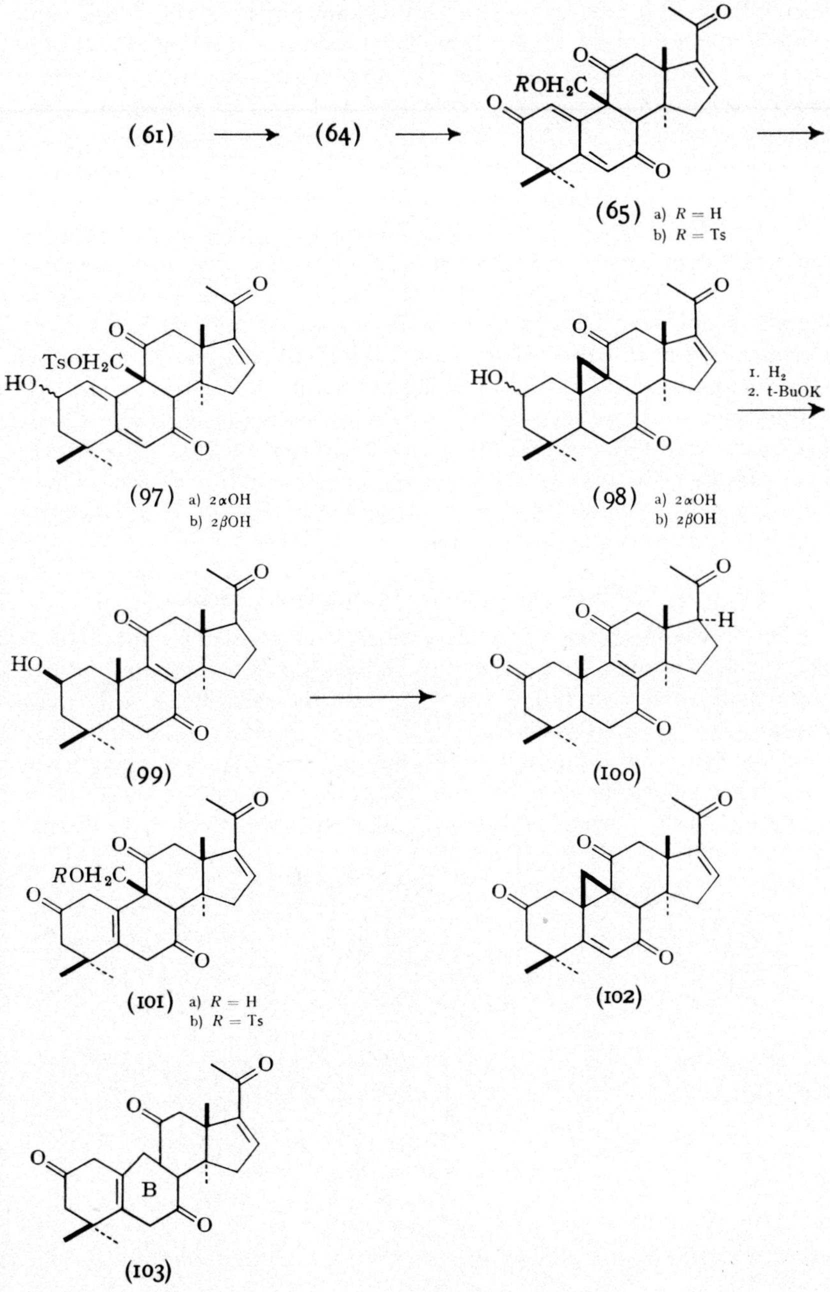

(61) ⟶ (64) ⟶

(65) a) R = H
b) R = Ts

(97) a) 2αOH
b) 2βOH

(98) a) 2αOH
b) 2βOH

1. H$_2$
2. t-BuOK

(99)

(100)

(101) a) R = H
b) R = Ts

(102)

(103)

Chart 17. Interrelation Between the Cucurbitane and Lanostane Skeletons

also formed. Further reduction with sodium borohydride of (97a) and (97b) led to formation of a cyclopropane ring; the reaction was accompanied by reduction of the 2-one which on reoxidation with manganese dioxide gave (98a) and (98b) respectively. Following reduction of the 16,17-double bond of (98b), the saturated cyclopropane-trione (16,17-dihydro-(98b)) was obtained and submitted to the key step of the sequence, cyclopropane ring opening under anionic conditions (potassium-t-butoxide in t-butanol and dimethyl sulfoxide). This resulted in the desired enedione (99) which upon oxidation afforded 4,4,14α-trimethylpregn-8-ene-2,7,11,20 tetraone (100). It is of interest that the 16,17-dihydroderivative of (98a) remained unchanged under the conditions which induced the smooth isomerization of dihydro (98b). This remarkable difference is attributed to participation of the axial 2β-OH in protonating the incipient 19-methyl carbanion at the same time as the C-9, C-19 cyclopropane bond is breaking.

This interrelation confirms the stereochemistry at C-9, C-13 and C-14 in cucurbitacin A (61), as well as in the other cucurbitacins which have been previously correlated with it. The configurational assignment at C-20 is based only on biogenetic arguments.

6. Synthesis of a 32-nor-Cucurbitane Skeleton

This synthesis is based on a migration of the 19-methyl group (10 → 9β) which was induced by cleavage of the 9α,11α epoxide in a 4,4-dimethylandrostane derivative (104) to yield the dienedione (105) (λ_{max} 235 nm, ε 6150) (4) (Chart 18).

Structure (105) follows from its rearrangement in basic conditions into the homoannular dienone (106) (λ_{max} 300 nm, ε 6150) as well as by formation of the Michael addition product (107) when milder basic con-

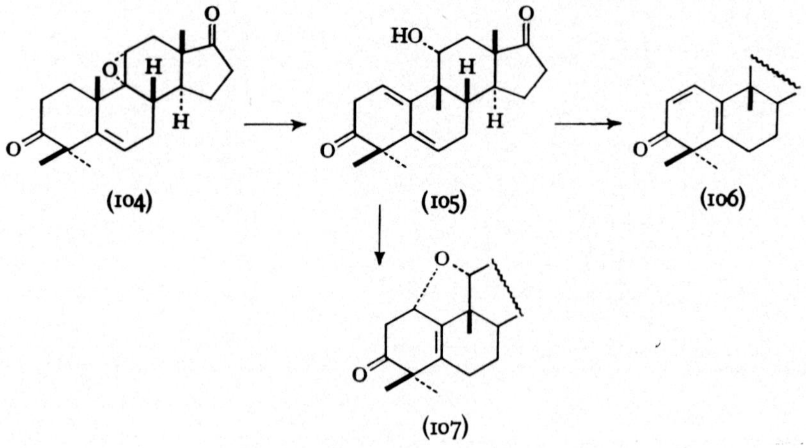

(104) (105) (106)

(107)

Chart 18. Synthesis of a Cucurbitane Skeleton

References, pp. 357—362

ditions were applied. The last reaction constitutes also a good proof for the *cis* B/C ring juncture. Spectroscopic properties of (105) are also compatible with the proposed formula.

Mechanistically, one of the most interesting aspects of the conversion of (104) into (105) is that the stabilisation of the developing cationic center by the double bond (Δ^5), directs the migration of the 19-methyl group. Such a reaction does not occur in 9α,11α-epoxylanostan-3β-ol, i. e. in a compound devoid of the C-5 C-6 double bond (*4*).

V. Cucurbitacins G, H, L, J, K, Dihydrocucurbitacin B and 22-Deoxocucurbitacin D

1. Cucurbitacins G and H (108)

These two compounds have been isolated from roots of *Cucumis hirsutus* in which they occur together with cucurbitacin B and D (elatericin A) (*37*). Cucurbitacin G is crystalline, m. p. 150–152°, whereas cucurbitacin H could not be induced to crystallize. They are both α-hydroxy-ketones (positive triphenyltetrazolium chloride test) and do not display the characteristic 230 nm band for the αβ-unsaturated carbonyl of the side chain. Treatment of either of these two compounds with dilute hydrochloric acid or with deoxygenated dilute alcoholic sodium hydroxide, produces a mixture of cucurbitacin D (6) and the two isomers G and H (108).

Structure (108) assigned to these isomeric compounds is based on their spectral behavior (lack of major UV absorption) and the ready loss of the elements of water under acetylating conditions which leads to cucurbitacin D diacetate (6b). No attempts have been reported so far to determine the stereochemistry at C-24 for each of the two isomers (*53*).

2. Cucurbitacins J, K (109), and L (110)

Cucurbitacins J, K and L which possess a diosphenol grouping in ring A occur as bitter glycosides in the roots of *Citrullus ecirrhosus* Cogn. and have been isolated by enzymatic hydrolysis (*34*). They have been also obtained directly as the free aglycones together with other cucurbitacins (see Table 1) from the roots of *Bryonia dioica* Jack. (*48*), in which they occur probably as glycosides. The glycosidic bond is cleaved by the enzyme elaterase existing in this plant. Cucurbitacin L (110) occurs in *Citrullus colocynthis* L. Schrad. together with elaterin (2) and cucurbitacin I (7) as the corresponding glycosides, and has been isolated following enzymatic hydrolysis (*90*). Cucurbitacin J and K (109) have also been isolated from *Iberis amara* L. (Cruciferae) (*47*).

Cucurbitacins J and K (109) which are isomeric at C-24 do not show the UV absorption of the Δ^{23}-22-one chromophore, and their empirical

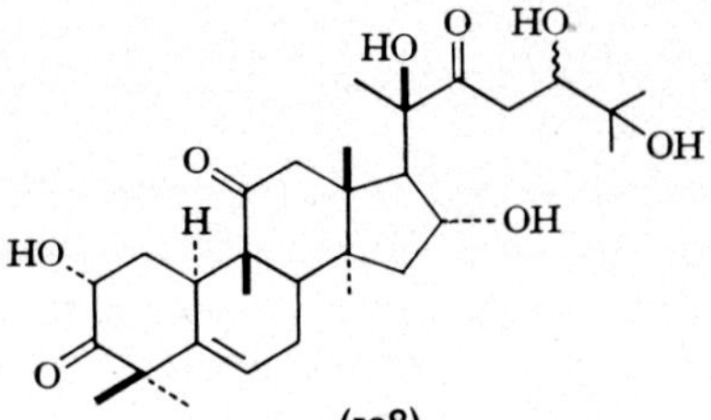

(108) Cucurbitacins G and H

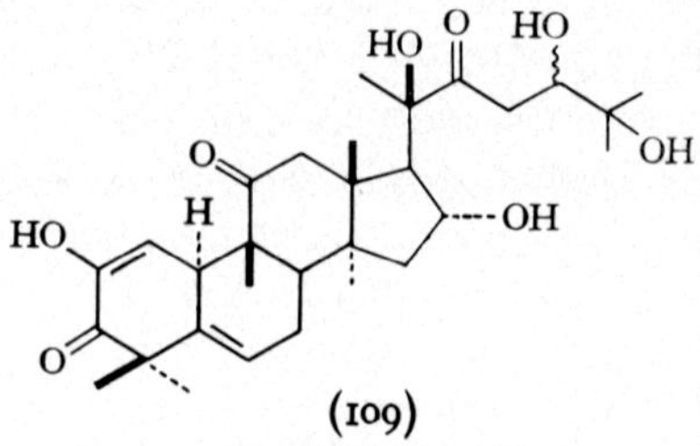

(109) Cucurbitacins J and K

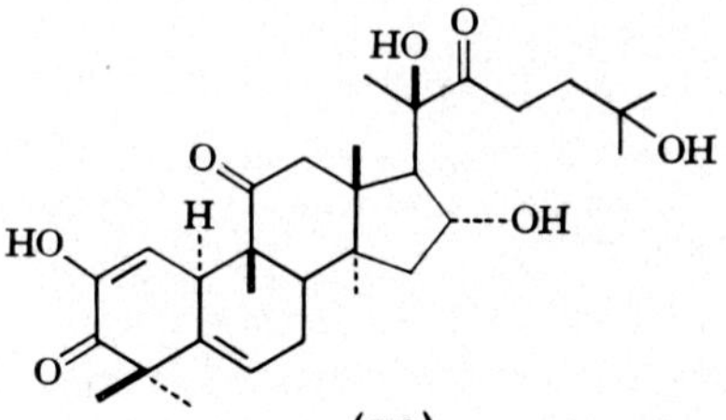

(110) Cucurbitacin L

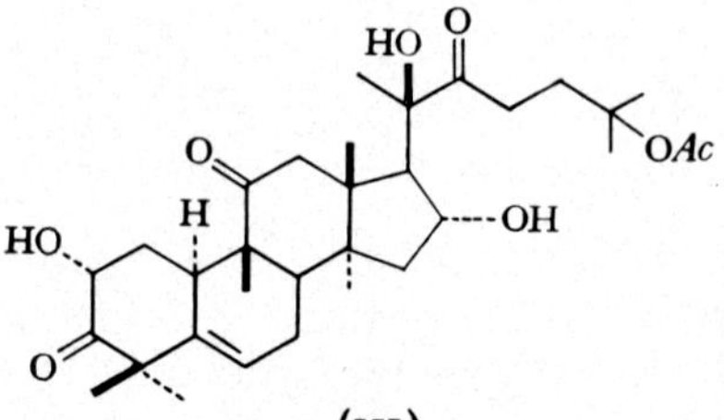

(111) Dihydro-cucurbitacin B

Chart 19. Certain Cucurbitacins

formulae indicate the elements of one additional mole of H_2O compared with cucurbitacin I (7). The assumption that these two compounds are related to (7) in the same way as cucurbitacins G and H are related to cucurbitacin D (6) has been confirmed: acetylation of either J or K afforded cucurbitacin I diacetate (7t).

Treatment of cucurbitacin E (elaterin) (2) with cold alkali induced the reverse reaction, i. e. hydration of the C-23, 24 double bond leading *inter alia* to a mixture of cucurbitacins J and K, together with cucurbitacin I (7), the formation of the latter being due to hydrolysis of the 25-acetate group (*34*). Elateridin (*85*), the material obtained from the cold alkaline treatment of elaterin (2) is probably a mixture of cucurbitacins J and K. The absolute configuration at C-24 in each of these compounds has not yet been determined.

Cucurbitacin L (110) is 23,24-dihydro-cucurbitacin I; its structure has been established by comparison with a compound obtained by hydrogenation of cucurbitacin I with one mole of hydrogen (*34, 90*).

3. Dihydrocucurbitacin B (111)

This compound was isolated from *Marah oreganus* H., in which it occurs together with other cucurbitacins (*61*). The structure (111) has been established by comparison with an authentic sample prepared by the catalytic hydrogenation of cucurbitacin B (5) (*93*).

References, pp. 357—362

4. 22-Deoxocucurbitacin D (112)

This cucurbitacin has been isolated together with 22-deoxoisocucurbitacin D (120) from a hybrid obtained by the crossbreeding of two varieties of *Lagenaria siceraria* (Cucurbitaceae), namely bitter calabash (which contains cucurbitacin B (5) as the main bitter principle), with sweet maranka (which does not contain triterpenoids). It is noteworthy that both 22-deoxocucurbitacins are not bitter (*28*).

Upon treatment with dilute acid, (112) undergoes an interesting rearrangement (see arrows) (Chart 20) involving the elimination of the tertiary 25-hydroxyl and formation of a C-16, C-23 bond which results in a six membered ring ether (93a). Ozonolysis of (112) afforded α-hydroxy-α-methylpropionaldehyde, whereas similar treatment of (93) produced acetone. The absence of the 22-ketone in (112) is evident from the UV spectrum (no absorption for an αβ-unsaturated enone), as well as from the NMR pattern of the 23 and 24 vinylic protons which give rise to a broad signal superimposed on the multiplet due to the 6-H. The transformation of (112) to (93) could be avoided by oxidation of the 16-hydroxyl to the corresponding 16-one (113); thus the involvement of the 16-hydroxy group in the above rearrangement has been demonstrated. Contraction of ring A in (113) according to the scheme outlined for elatericin A (see section IV.1.4) afforded the ring A-norketone (114) in which the tertiary 20-OH was eliminated as well. The spectroscopic data for (114) are in good agreement with the proposed structure.

Compound (115), obtained from (93a) after contraction of ring A, afforded on treatment with acetic anhydride in the presence of p-toluenesulphonic acid, the triene diacetate (116) by opening of the ether ring. Catalytic hydrogenation of (116), followed by alkaline hydrolysis and oxidation of the secondary OH, gave the hydroxy-triketone (117). β-Elimination of the tertiary OH led smoothly to the unsaturated triketone (118) (λ_{max} 255 nm, ε 10800).

The stereochemistry at the asymmetric centers shown in (112) has been deduced by CD measurements on appropriate models in a manner similar to those described in section IV.3: the 10α-H by the CD curve of ring A ketone in (115); the 11-ketone by the CD of the monoketal of (115); the *trans* C/D rings juncture by the CD of the hydroxytriketone (117); the 16α-OH, by molecular rotation differences. Appropriate subtractions were carried out in order to determine the contribution of the relevant carbonyl group (*28*). For conformational studies of ring A, see section IV.4.

5. "β-Elaterin"

The early literature (*17, 93, 101, 102*) refers to a compound named β-elaterin which accompanies in variable proportions the so called α-elaterin (presently referred to as elaterin or cucurbitacin E (2)) on the basis of the available data, the m. p.,

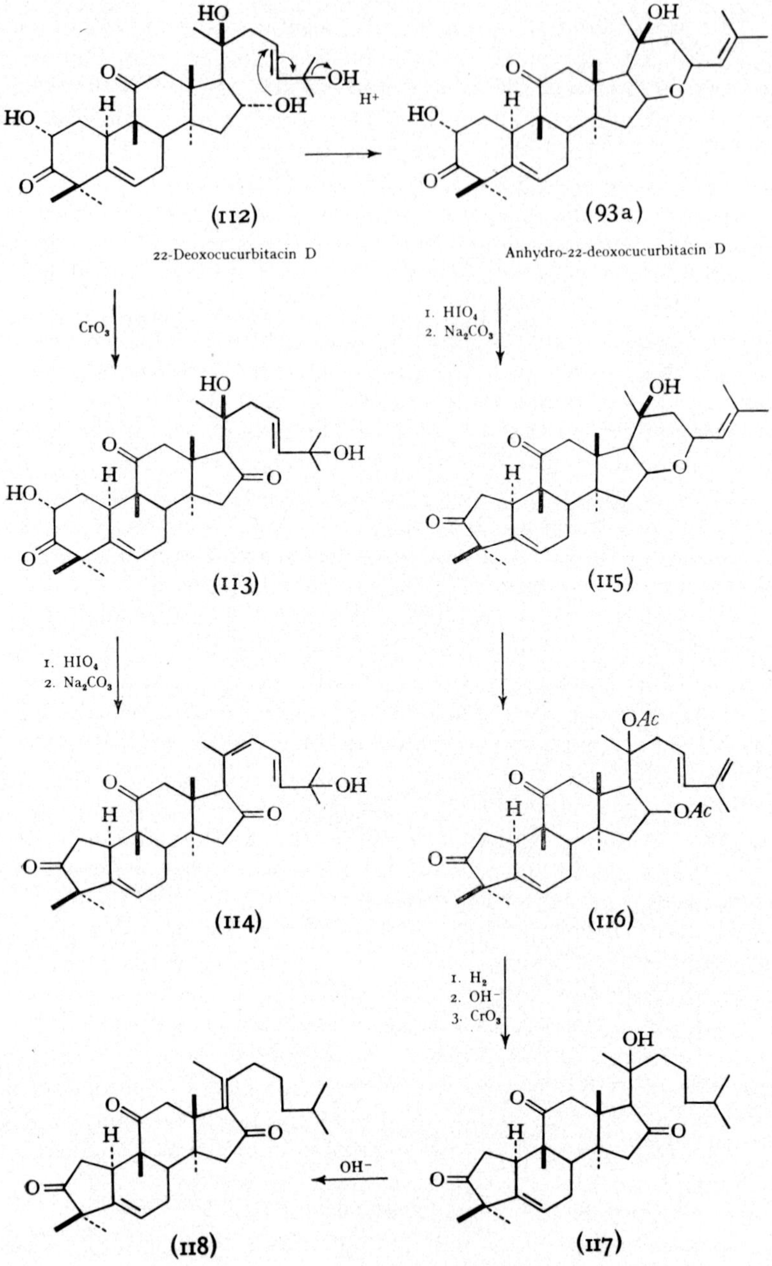

(112) 22-Deoxocucurbitacin D

(93a) Anhydro-22-deoxocucurbitacin D

(113)

(115)

(114)

(116)

(118)

(117)

Chart 20. The Structure of 22-Deoxocucurbitacin D

the positive rotation and the observation that on repeated crystallizations, the mixture yielded consistently additional quantities of α-elaterin (2), it would seem that β-elaterin is actually a mixture of cucurbitacin B (5) and E (2). It is now assumed that during the purification attempts at this earlier period, B underwent autoxidation to the sparingly soluble E which crystallized out of the mixture.

VI. Isocucurbitacin B (119), 22-Deoxoisocucurbitacin D (120) and Tetrahydrocucurbitacin I (121)

The first two compounds are isomeric with cucurbitacin B (5) and 22-deoxocucurbitacin D (112) respectively, the difference being in the substitution pattern of ring A: a 3-hydroxy-2-keto pattern in (119) and (120) against a 2-hydroxy-3-keto structure in (5) and (112). The third compound (121) possesses presumably the same 3-hydroxy-2-keto structure.

1. Isocucurbitacin B (119)

This compound was isolated first from *Luffa echinata* Roxb. along with elaterin (2). At the time it was erroneously assigned a 3-keto-2(axial)-hydroxy structure, and described as "2-epicucurbitacin B" (78). Subsequently, the same compound was isolated from *Marah oreganus* and assigned structure (119) (61) on the basis of reduction experiments carried out with diosphenol-containing systems (see section IV.4, and Cf. ref. 10 in (61)). At a later date, (119) was also isolated from *Luffa operculata* Cogn., the structure being confirmed by the appropriate NMR signal for the 3-H, singlet at δ 3.87 (1). The NMR spectrum points

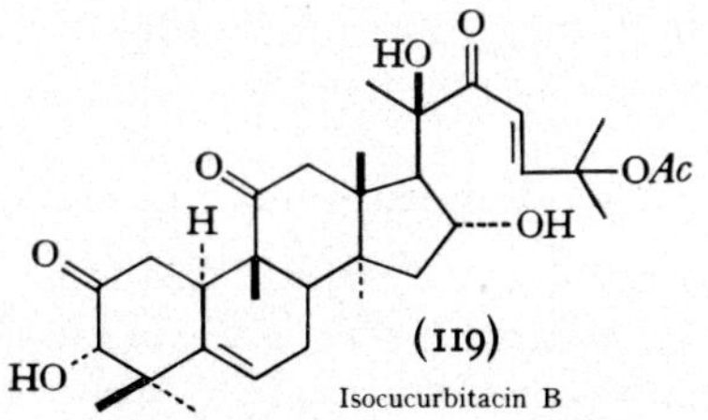

(119)
Isocucurbitacin B

unequivocally to the location of the hydroxyl group at C-3; in addition the proposed ketolic structure is supported by the oxidation of (119) with bismuth oxide to cucurbitacin I (7) (elatericin B). The equatorial 3-OH has been assigned the α-orientation on the basis of a chair conformation for ring A (61). In view of the conformational study discussed earlier (*121*) (see section IV.4) further corroboration of this assignment seems necessary.

2. 22-Deoxoisocucurbitacin D (120)

This is a minor component isolated from a hybrid of *Lagenaria siceraria* (28). Upon treatment with dilute acid, this compound behaved similarly to 22-deoxocucurbitacin D (112) and gave the anhydro-derivative (94a). Acetylation of the latter afforded a monoacetate (94b) which displays a singlet for the 3-H pointing towards a 2-keto-3-hydroxy system (Chart 21).

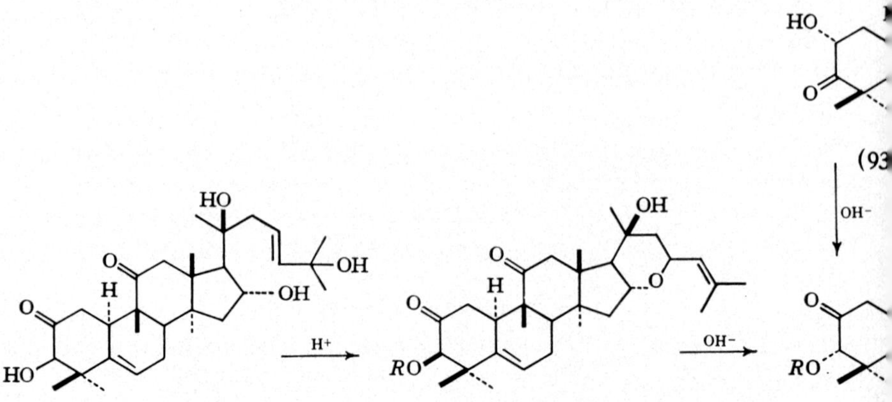

Chart 21. Reactions of 22-Deoxoisocucurbitacin D

Treatment of (94a) and (93a) with deoxygenated methanolic alkali afforded an isomeric ketol characterized as anhydro-22-deoxo-3-epi-isocucurbitacin D (95a). For the configuration of the hydroxyl groups, as well as the conformation of ring A in these compounds, see section IV.4.

3. Tetrahydrocucurbitacin I (121)

The above compound has been isolated (48) from the roots of *Bryonia dioica* Jack. in which it occurs together with cucurbitacins J, K (109) and L (110). It was described as the C-2 epimer (122) of dihydroelatericin A *(dihydro 5)* in analogy with the structure previously assigned to "2-epi-cucurbitacin B" (in fact, isocucurbitacin B (119)). Since (121) has an

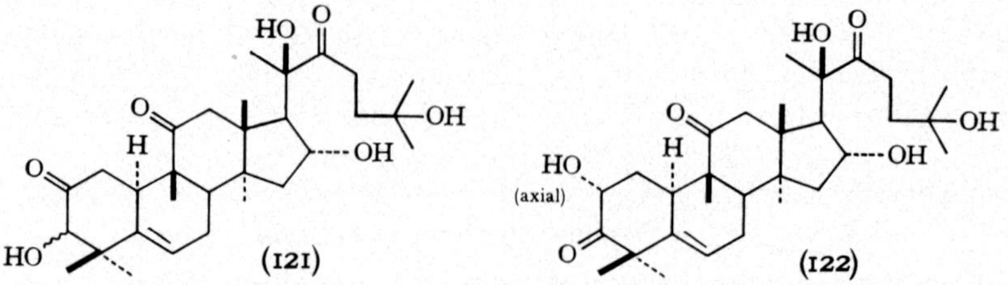

Tetrahydrocucurbitacin I

References, pp. 357—362

optical rotation, $[\alpha]_D + 56.4°$, which is very close to that measured for tetrahydroisoelatericin B (96a), $[\alpha]_D + 59°$, prepared by the catalytic hydrogenation of elatericin B (7) (*67*), it is quite probable that compound (121) has a similar structure and possesses a 3-hydroxy-2-oxo system. Unfortunately, no NMR data are available to confirm this assumption.

VII. Bryodulcosigenin (123), Bryosigenin (124), Bryogenin (125), Gratiogenin (126), and 16-Hydroxygratiogenin (127)

These compounds which exist in nature as glycosides have been grouped together according to their substitution pattern. They all possess an axial hydroxyl group at position 3 and, with the exception of the last compound, lack the C-16 hydroxyl. The eight carbon atom side chain is at a lower oxidation level than in other cucurbitacins.

Enzymatic hydrolysis of the glycoside bryodulcoside isolated from the roots of *Bryonia dioica* Jack. afforded bryodulcosigenin (123) together with a sugar moiety consisting of two moles of glucose and one of rhamnose (*128, 131*). Upon mild acetylation the 3,24-diacetate was formed. The trisubstituted double bond in ring B has been identified by epoxidation and by hydroxylation with osmium tetroxide to the corresponding *cis* glycol. Treatment with chromium trioxide in acetic acid led to cleavage of the C-24 C-25 bond together with concomitant allylic oxidation at C-7 which produced the triketo-carboxylic acid (128) (*127, 129*) (Chart 22).

Bryosigenin (124) has been obtained as a minor companion of (123) following enzymatic hydrolysis of the glycoside mixture (*130*). The structure is based on the analysis of its IR spectrum, the mass spectrum (M⁺ 472, 2 mass units less than compound (123)), and the CD curve which is characteristic for an 11-one in a cucurbitane skeleton. The relationship between (123) and (124) has been established by converting the latter to the former, using sodium borohydride at ice bath temperature for the selective reduction of the 24-ketone. When the hydrolysis of the original glycoside is performed under acidic conditions, bryogenin (125) is obtained along with bryodulcosigenin (123). Since (125) is not formed under enzymatic conditions, it has been suggested (*131*) that bryogenin does not occur in nature but is the product of the dehydration of the secondary-tertiary glycol system present in the side chain of (123). Indeed, a compound obtained by acid treatment of (123) was found identical with bryogenin (125) (*13*).

The elucidation of the structure of bryogenin (125) bore importantly on the constitution of the other cucurbitacins (*14*). The β-axial orientation of the 3-OH is based on the NMR signal of the 3-H (triplet δ 3.47, J = 2.5 Hz). Oxidation affords the 3,11,24-trione (129) which, when submitted to a Wolf-Kishner reduction, affords the monoketone (130). The

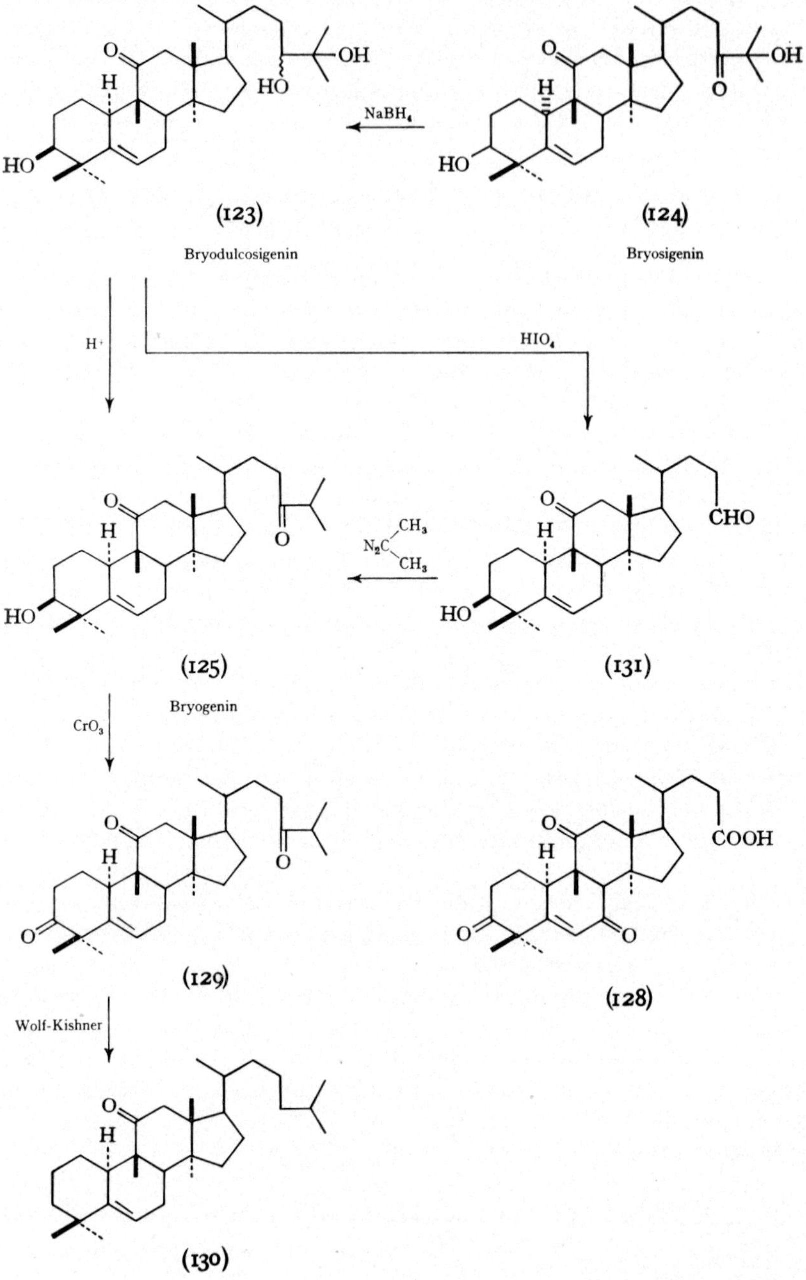

Chart 22. Cucurbitacins from *Bryonia* and Some Reactions

CD curve of this compound (130) is similar to that of the 11-one in a derivative (90) of elatericin A (6) thus supporting the cis B/C juncture. Subtraction of the CD curve of the trione (129) from that of bryogenin (125) gives the contribution of the 3-ketone which has a negative Cotton effect as expected for 10α-stereochemistry (14). The relationship between (123) and (125) has been confirmed (15) by periodic acid cleavage of the glycol in (123) which produced the 24-aldehyde (131). Further reaction with diazo-isopropane afforded bryogenin (125).

Gratiogenin (126) has been obtained (123), together with two moles of glucose, by the acid hydrolysis of gratioside (124), a glycoside isolated from *Gratiola officinalis* L.*. Oxidation of (126) with chromium trioxide in acetic acid produced a five-membered ring lactone as shown in (132), ν_{max} 1761 cm^{-1}, and acetone (Chart 23). The cis juncture of rings B/C in (132) was revealed by CD measurements.

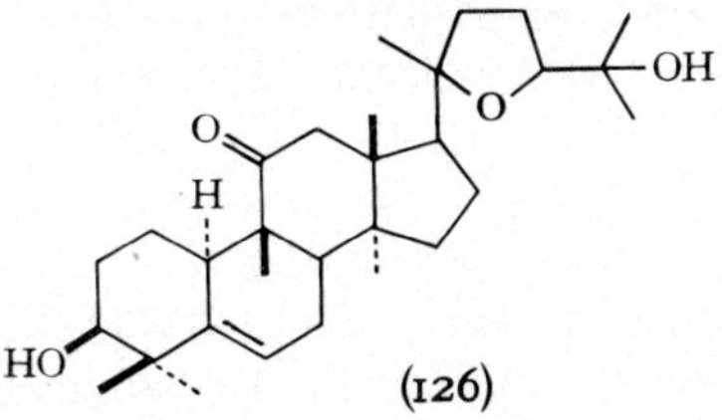

(126) Gratiogenin

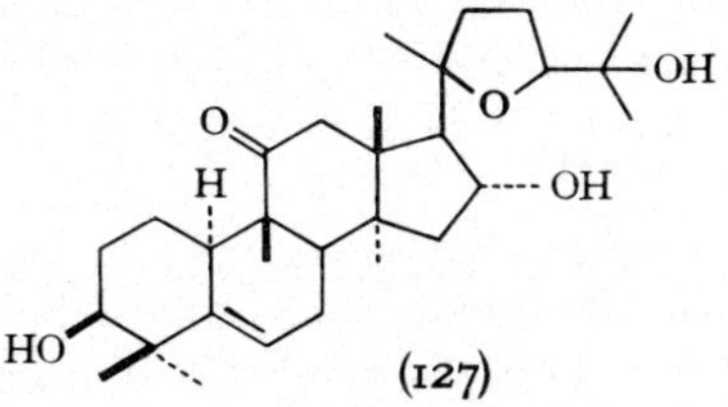

(127) 16-Hydroxygratiogenin

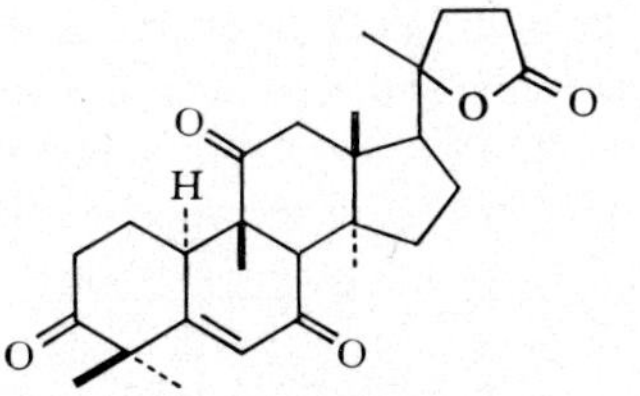

(132)

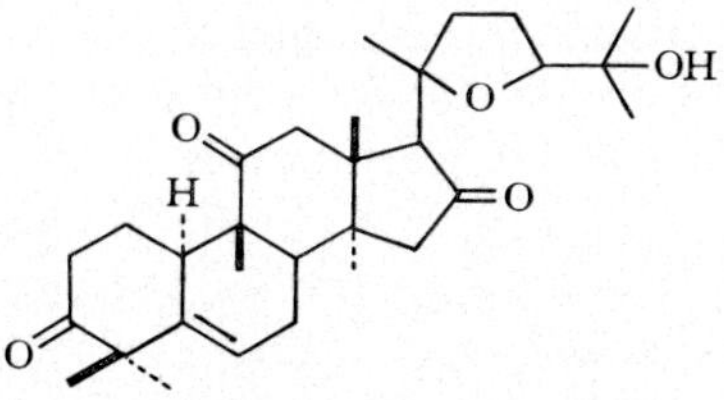

(133)

Chart 23. Gratiogenin and Derivatives

16-Hydroxygratiogenin (127) has been obtained as a minor component during the hydrolysis of the glycosidic mixture obtained from the above plant. Oxidation of this compound with chromium trioxide in acetic acid led to the triketone (133), ν_{max} 1740 cm^{-1}, the position of the five membered ring ketone as well as the *trans* juncture of rings C/D being confirmed by CD measurements. The 16-hydroxyl in the original compound (127) had been assigned the β-orientation (123). This assignment

* Plant family Scrophulariaceae.

which is without precedent in this series has been now shown to be erroneous. Molecular rotation differences between 16-hydroxygratiogenin (**127**), and the corresponding 3,16-diacetate, indicated an α-orientation*, in agreement with the orientation of such a group whenever it is present in cucurbitacins. Fragmentation patterns in the mass spectra of the two compounds (**126**) and (**127**) have been analyzed and conform with the proposed structures.

It is noteworthy that gratiogenin (**126**) possesses the same side chain as ocotillol, a dammarane type triterpene isolated from a desert plant, *Fouquieria splendens* Engelm. (*134*).

VIII. Cucurbitacin F (134a), O (135a), P (136a), and Q (137a)

The distinguishing feature of these cucurbitacins (Chart 24) is the presence of a 2,3-diol system in ring A; they all occur in nature as free aglycones. Cucurbitacin F (**134a**) has been isolated from the leaves of *Cucumis dinteri* Cogn. (*94*), whereas the other three compounds occur in *Brandegea bigelovii* Cogn. (*62*), both of the Cucurbitaceae.

Cucurbitacin F (**134a**) λ_{max} 233 nm (ε 11200) can be reduced by catalytic hydrogenation to the 23,24-dihydro-derivative and consumes two moles of periodic acid. When the substance is first acetylated and then treated with periodic acid, the 20,22-ketol system is cleaved severing the side chain; treatment with alkali to eliminate the 16-hydroxyl and reacetylation yields the hexanor-derivative (**138**). The diaxial orientation of the two acetoxy groups in the latter was deduced by analysis of its NMR spectrum. Furthermore, sodium borohydride reduction of the 23,24-dihydro-derivative of cucurbitacin D (**6**) gave a compound identical with 23,24-dihydrocucurbitacin F. Since the 2-hydroxyl in cucurbitacin D (**5**) is α-oriented and α-ketols are known to give *trans* glycols upon hydride reduction, the stereochemistry of the ring A glycol in (**134**) should be 2α,3β (*62, 94*).

Separation of cucurbitacins O (**135a**) and P (**136a**) was performed by chromatography of the mixture following acetylation. Compound (**135b**) λ_{max} 230 nm (ε 12000 for the αβ-unsaturated carbonyl), displayed in its NMR spectrum the characteristic pattern for the two *trans* vinylic protons of the side chain, doublets at δ 6.68 and 7.18 (J = 15 Hz), whereas in compound (**136b**) these signals were missing, indicating the saturated nature of the side chain. The two compounds differed therefore by two hydrogens, and indeed (**135**) could be converted into (**136**) by catalytic hydrogenation, one mole of hydrogen being absorbed (*62*).

Cucurbitacin B (**5**) was reduced to its tetrahydro derivative which involved reduction of the C-23,C-24 double bond and reduction of the

---

* BIERNROTH and SNATZKE, private communication.

References, pp. 357—362

3-ketone to the corresponding alcohol. Subsequent hydrolysis of the 25-acetate afforded a product identical with (136a), thus interrelating cucurbitacins O and P with the whole series. The only difference between the 23,24-dihydro-derivative of (134a) and compound (136a)

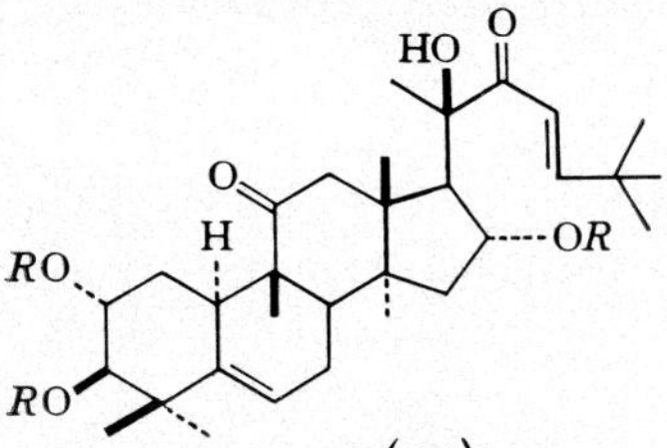

(134) Cucurbitacin F

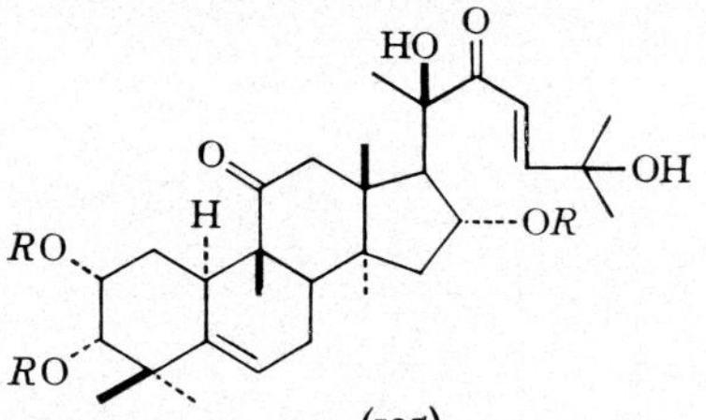

(135) Cucurbitacin O

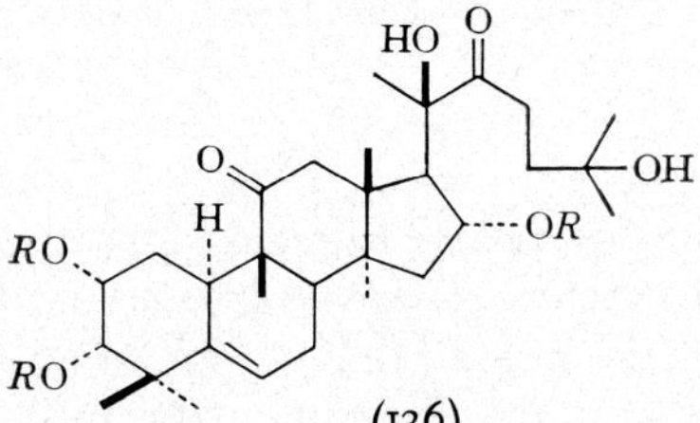

(136) Cucurbitacin P

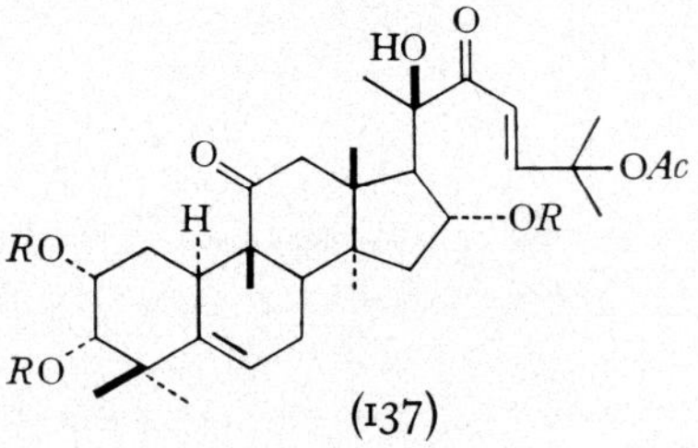

(137) Cucurbitacin Q

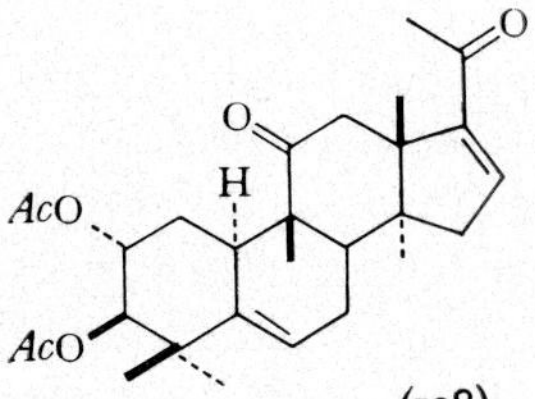

(138)

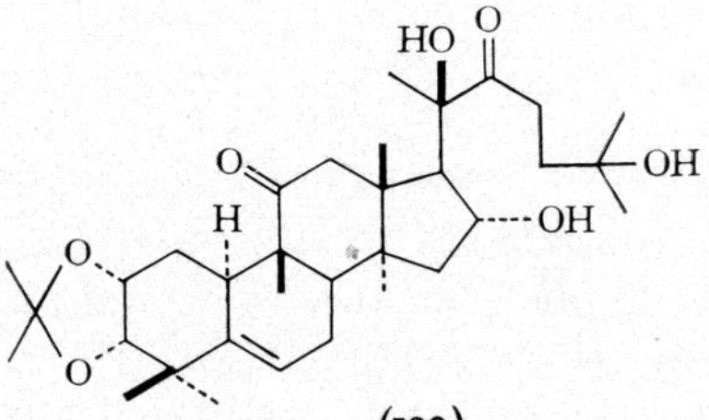

(139)

a) R = H

Chart 24. Cucurbitacins F, O, P, and Q

lies in the orientation of the C-3 hydroxyl group: the ring A diol of (136a) should therefore be *cis* oriented (2α,3α). This orientation was confirmed through the formation of an acetonide (139) (62).

Cucurbitacin Q (137a) is the 25-acetyl derivative of (135a). This structure has been deduced by NMR and mass spectral measurements as well as by catalytic hydrogenation of the C-23,C-24 double bond followed by hydrolysis of the acetate group. This furnished a compound which was identical with (136a) (62).

IX. Cucurbitacins of Unknown Structure

Fabacein has been isolated together with cucurbitacin B (5) from *Echinocystis fabacea* (*112*, *113*). It is a diacetate, $C_{34}H_{48-50}O_9$, and like cucurbitacin B contains an $\alpha\beta$-unsaturated carbonyl group. The two acetates are located at positions 16 and 25, and several reactions are similar to those given by (5). An α-ketol system is present in ring A; however, acetylation of fabacein gives an acetate which differs from that obtained by acetylation of cucurbitacin B (5). No additional information is available.

X. Biogenetic Aspects

The distribution of the cucurbitacins in the Cucurbitaceae plant family, as well as their occurrence in various parts of the plant and the changes taking place in the bitter principle composition in these parts during the development of the plant have been studied (*36*, *39*, *104*, *105*).

It has been suggested that the cucurbitacins are biosynthesized through secondary transformations in lanosterol (*81*) or by rearrangement of a C-9 carbonium ion in a lanostane type structure (*60*). Further oxidative elaborations towards the formation of the differently substituted side chains have been also considered (*96*).

In a recent communication (*135*) dealing with the biosynthetic pathways leading to cucurbitacin B (5), evidence is presented that the latter is produced in the plant through the intermediacy of cycloartenol or parkeol, but not through lanosterol.

Incorporation of $[(4R)\text{-}4\text{-}^3H_1, 2\text{-}^{14}C]$ mevalonic acid into cucurbitacin B in Hubbard squash seedlings afforded labelled cucurbitacin B with a ratio $^3H/^{14}C = 7.55$, as compared to a ratio of 10.8 in the original mevalonic acid. The compound thus obtained was then degraded to the labelled A-norketone (23) ($^3H/^{14}C = 5.91$), and further converted into the isomeric compound (41), the degradation being done in order to confirm that the incorporation of mevalonic acid does not proceed in a random fashion.

The $^3H/^{14}C$ ratio measured for the isotopic cucurbitacin B (5) was consistent (Chart 25) with the intermediate formation of the cation (140) which is subsequently converted into parkeol (141) or cycloartenol (142), but not lanosterol. The migration of the C-19 (10 → 9β) could then be induced in parkeol by an electron deficiency at C-9, or by formation and opening of a 9α,11α-epoxide, the last pathway explaining also the presence of an oxygen at C-11 in all known cucurbitacins.

References, pp. 357—362

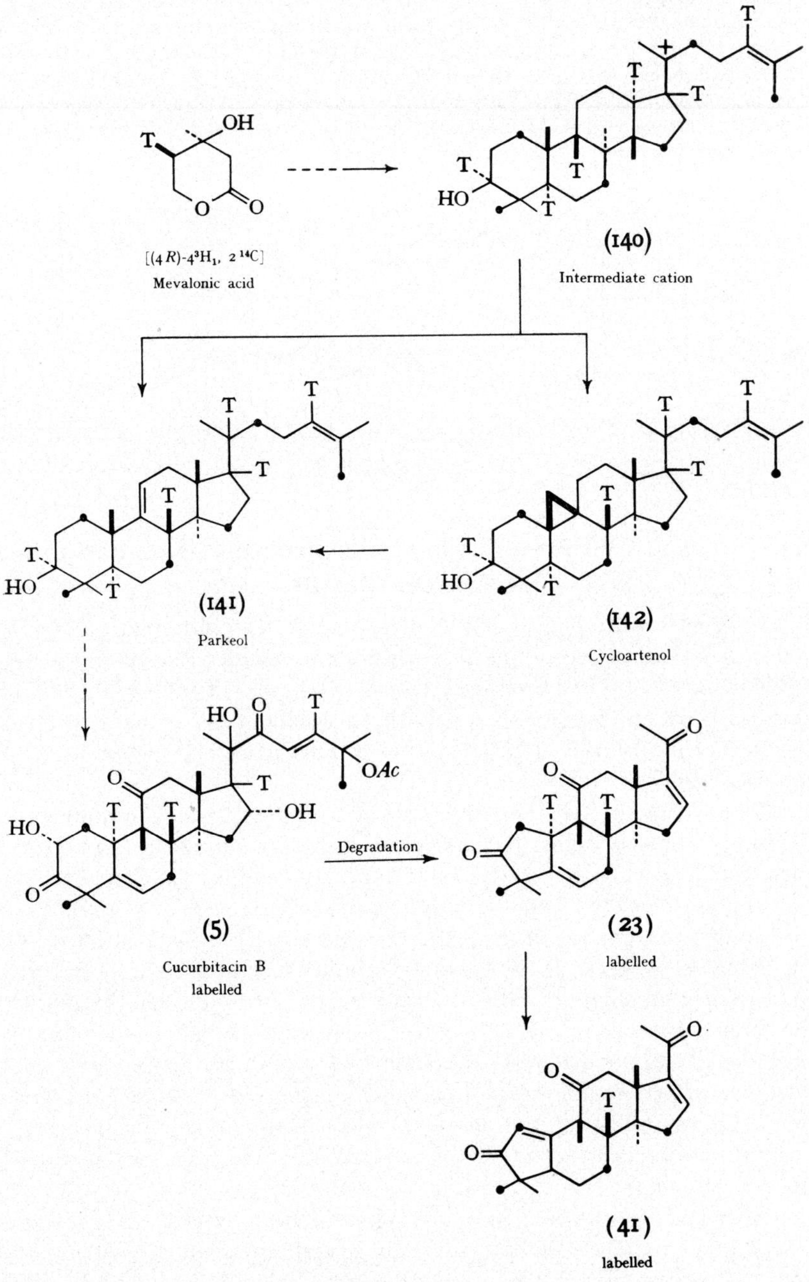

The Cucurbitanes, a Group of Tetracyclic Triterpenes 349

[(4R)-4³H₁, 2¹⁴C]
Mevalonic acid

(140)
Intermediate cation

(141)
Parkeol

(142)
Cycloartenol

(5)
Cucurbitacin B
labelled

Degradation

(23)
labelled

(41)
labelled

Chart 25. Biosynthesis of Cucurbitacin B

The possibility of secondary tranformations in the plant involving the migration of a tertiary methyl induced by the opening of an adjacent epoxide has been illustrated by the transformation of 9,11-epoxy-4,4-dimethylandrost-5-ene-3,17-dione (**104**) into the dienedione (**105**) possessing a 9β-methyl (see section IV.6), also by the conversion of the euphane type compound (**143**) into compound (**144**) belonging to the meliacin series bearing an 8β-methyl group (Cf. D. LAVIE and E. C. LEVY, Tetrahedron Letters, **1968**, 2097).

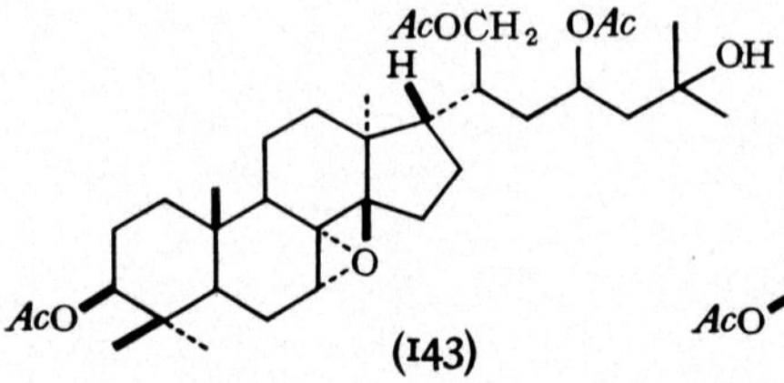

It is therefore our conviction that the natural process leading to the formation of the cucurbitane skeleton involves epoxidation of a parkeol intermediate as proposed in a communication (*135*).

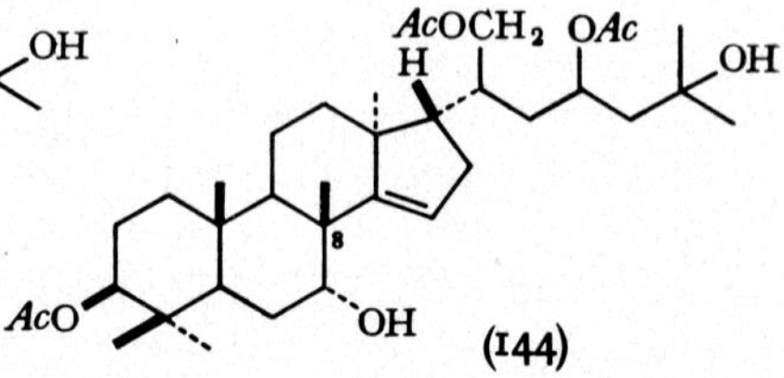

XI. Physical Methods Used in the Structure Elucidation of the Cucurbitacins

As by now must have become quite clear to the reader, spectroscopic methods have been used extensively during the work on the cucurbitacins. NMR spectroscopy has been crucial in the progress of the elucidation of the structures. Reference to NMR data are numerous; in one paper the chemical shifts of the angular methyl groups have been compiled and analyzed (*68*).

IR and UV data, together with physical constants, have been catalogued for several cucurbitacins (*99*) and often the chromophores present have been used as prototypes. ORD and CD curves have been instrumental in the solution of stereochemical problems (*14, 67, 69, 81, 121, 123*).

Mass spectrometric fragmentation patterns have been studied for cucurbitacins A, B, C, D, E, I and bryogenin (*5*); certain fragments have been found to be characteristic of certain parts of the molecule throughout the series. A similar analysis has been performed on cucurbitacins O, P and Q, and the results were compared with B and F; a number of correlations were determined (*62*). An analysis of the mass spectrum of gratiogenin (**126**) and its oxidation product (**132**) has also been published (*123*).

It is noteworthy that no X-ray analysis has been reported for any of the cucurbitacins.

Paper and thin layer chromatography has been extensively used and described in most of the papers dealing with isolation and structural work. However, a number of publications deal more specifically with such techniques used for identification purposes (*31, 47, 48, 50, 58, 104*).

References, pp. 357—362

XII. Biological Properties of the Cucurbitacins

The constituents of the Cucurbitaceae were known for a long time for their purgative action (*10, 11, 17, 40, 103*). The pharmacodynamic action (*2, 3, 42, 54, 55, 56, 100*), and the antineoplastic properties of certain extracts (*10, 41, 89*) have been studied at different times. However, in more recent years experiments were performed on purified substances.

Among the cucurbitacins tested for their cytotoxic and anti-tumor activity in experimental tumors, elatericin A (6), B (7) and elaterin (2) were investigated most thoroughly. They produced growth inhibition of solid tumors *in vivo* as well as characteristic morphological changes of Ehrlich ascites tumor cells *in vitro*. These changes consisting of blistering and thread formation were obtained by incubation of the cells with low concentrations of the compounds (*46*). The 2-methyl ether of elaterin was found to be probably the most effective anti-tumor compound of this group (*46*) and was extensively studied. The methyl group attached to the diosphenol was labelled with ^{14}C and its resorption and metabolism followed in normal and tumor-bearing mice; it was found that excretion from the body was delayed (*43*). The effect of the cucurbitacins on the respiration of tumor cells was also measured (*116*).

Combined treatment with cucurbitacins and X-ray irradiation produced a marked enhancement in activity and better tumor growth inhibition was recorded (*118*). Elatericin A (6) produced morphological changes *in vitro* in human lymphocytes from chronic lymphatic leukemia and lymphosarcoma patients. The lymphocytes taken from normal subjects were less sensitive and required a five fold concentration to produce similar changes. Absorption of the substance by the lymphocytes was studied and demonstrated. This action which is characteristic for chronic lymphatic leukemia cells was proposed as a rapid test for these cells (*117*).

When tested against cell cultures derived from human carcinoma of the nasopharynx (Eagle's KB strain), cucurbitacin B (5), E (2) (61), O (135), P (136) and Q (137) showed cytotoxic activity (*62*). Further *in vivo* evaluation was made on intramuscular Walker carcinoma (rats) and Lewis lung carcinoma (mice). A rather low margin between activity and toxic doses was found (*61*).

The pharmacodynamic activity of elatericin A (6) has been studied on monkeys, cats and dogs and activity detected on arterial blood pressure and respiration. The compound was observed to elicit a strong increase in capillary permeability (*24*).

It has also been observed that the cucurbitacins may behave as insect attractants as shown by experiments carried out with spotted cucumber beetles (*20, 21*); various degress of attractiveness were observed with different compounds of this series. The relation between resistance of the plants to insects and presence of cucurbitacins was considered.

XIII. Tables

Table 1. Occurrence of

Cucurbitaceae	Cucurbitacin A (61)	Cucurbitacin B (5)	Dihydro-B (111)	Iso-B (119)	Cucurbitacin C (69)	Cucurbitacin D (6)	22-Deoxo-D (112)	22-Isodeoxo-D (120)	Cucurbitacin E (2)	Cucurbitacin F (134)	Cucurbitacin G (108)	Cucurbitacin H (108)
1. *Acanthosicyos horrida*		×				×					×	×
2. *Benicasa hispida*									×			
3. *Brandegea bigelovii*												
4. *Bryonia alba*		×				traces			×			
5. *Bryonia dioica*		×				traces			×			
6. *Citrullus colocynthis*		×							×			
7. *Citrullus ecirrhosus*									×			
8. *Citrullus naudinianus*		×				×			×		×	×
9. *Citrullus vulgaris*									×			
10. *Coccinia adoensis*		×				traces						
11. *Coccinia hirtella*		×										
12. *Coccinia quinqueloba*		×										
13. *Coralocarpus sphaerocarpus*		×				×					traces	trac
14. *Cucumis africanus*		×				traces						
15. *Cucumis angolensis (dinteri)*		×				×				×	×	×
16. *Cucumis anguria*		×										
17. *Cucumis dipsaceus*		×				×						
18. *Cucumis heptadactylus*		×				×						
19. *Cucumis hirsutus*		×				×					×	×
20. *Cucumis hookeri*	×	×				×					traces	trac
21. *Cucumis humifructus*		×				traces						
22. *Cucumis kalahariensis*		×				×						
23. *Cucumis leptodermis*	×	traces				traces						
24. *Cucumis longipes (ficifolius)*		×				×					traces	trac
25. *Cucumis melo*		×				×			traces			
26. *Cucumis metuliferus*		traces										
27. *Cucumis myriocarpus*	×	×										
28. *Cucumis prophetarum*		×				×					traces	trac
29. *Cucumis pustulatus*		×				×						
30. *Cucumis sativus* var. *Hanzil*				×								
31. *Cucurbita maxima*		×				×			×			
32. *Cucurbita mixta*		×										
33. *Cucurbita microcarpina*		×							×			
34. *Cucurbita okeechobeensis*						×						
35. *Cucurbita pepo (texana)*									×			
36. *Cucurbita pepo* var. *ovifera*		×				×			×		traces	trac
37. *Cucurbita phoetidissima*		×							×			
38. *Ecballium elaterium*		×				×			×			
39. *Echinocystis fabacea*		×										
40. *Echinocystis lobata*		×							×			

...curbitacins in Nature

Cucurbitacin I (7)	Tetrahydro-I (121)	Cucurbitacin J (109)	Cucurbitacin K (109)	Cucurbitacin L (110)	Cucurbitacin O (135)	Cucurbitacin P (136)	Cucurbitacin Q (137)	Bryodulcosigenin (123)	Bryosigenin (124)	Bryogenin (125)	Gratiogenin (126)	16-Hydroxy-gratiogenin (127)	Fabacein	References
														36, 104
														105
					×	×	×							62
														36, 104
×	×	×	×					×	×	×				15, 36, 48, 50, 104, 131, 133
				×										36, 90, 104, 120
		×	×	×										34, 36, 37, 104
		×	×											36, 104
														36, 104
														36, 104
														36, 104
														36, 104
														36, 104
														36, 104
														36, 37, 94, 104
														105
														36, 104
														36, 104
														36, 37, 53, 104
														29, 36, 104
														36, 104
														36
														29, 36, 104
														36, 104
														105
														36, 104
														29, 36, 104
														36, 104
														36, 104
														36, 104
														105
														36, 104
														105
		traces	traces											36
														36, 104
		traces	traces											36, 104
														36, 104
														17, 36, 85, 86, 104
													×	25, 93
														36, 104

Table 1 (continued)

Cucurbitaceae	(61)	(5)	(111)	(119)	(69)	(6)	(112)	(120)	(2)	(134)	(108)	(108)
41. *Echinocystis wrightii*	×								×			
42. *Gerrardantus* sp.		traces				×					×	×
43. *Kedrostis africana*		×				×			×		×	×
44. *Kedrostis foetidissima* var. *microcarpa*		×				traces			×			
45. *Kedrostis nana*		×				×			×		traces	traces
46. *Kedrostis* sp. (*Toxanthera natalensis*)		×				×					×	×
47. *Lagenaria mascarena*		×				×			traces		traces	traces
48. *Lagenaria siceraria*		×				×					traces	traces
49. *Lagenaria siceraria* (hybrid)							×	×				
50. *Luffa acutangula*		×				traces						
51. *Luffa cylindrica*		×										
52. *Luffa echinata*			×						×			
53. *Luffa operculata*		×	×	×		×						
54. *Marah oreganus*		×	×	×					×			
55. *Melothria punctata*		×				×					traces	traces
56. *Peponium mackenii*		×							×			
57. *Sicyos angulata*		×										
58. *Telfairia pedata*		×				×			×		traces	traces
59. *Trochomeria debilis*		×				×					×	×
60. *Trochomeria sagittata*		×				×					×	×
Cruciferae												
61. *Iberis amara* L.									×			
62. *Iberis coronaria imperialis*									×			
63. *Iberis amara* Hyazinth									×			
64. *Iberis taurica*									×			
65. *Iberis odorata*									×			
66. *Iberis contracta*									×			
67. *Iberis ciliata*									×			
68. *Iberis procumbens*									×			
69. *Iberis pectinata*									×			
70. *Iberis pinnata*									×			
71. *Iberis umbellata*		×				×			traces		traces	traces
72. *Iberis nana*		×				×			traces		traces	traces
73. *Iberis rose*		×				×			traces		traces	traces
74. *Iberis intermedia*		×				traces			traces		traces	traces
Scrophulariaceae												
75. *Gratiola officinalis*									×			

The references are for papers reporting the occurrence and isolation procedures. When sev compounds listed.

The Cucurbitanes, a Group of Tetracyclic Triterpenes

(7)	(121)	(109)	(109)	(110)	(135)	(136)	(137)	(123)	(124)	(125)	(126)	(127)	Fabacein	References
														36, 104
														36, 104
		×	×											*36, 104*
														36, 104
														36, 104
														36, 104
aces														*36, 104*
														36, 104
														28
														9, 36, 104
														104, 105
														78
														1, 57
														61
														36, 104
														36, 104
														36, 104
														36, 104
														36, 104
														36, 104
		traces	traces											*18, 47, 49*
		traces	traces											*49*
		traces	traces											*49*
		traces	traces											*49*
		traces	traces											*49*
		traces	traces											*49*
		traces	traces											*49*
		traces	traces											*49*
			×											*49*
		traces	×											*49*
ces														*47, 49*
ces														*49*
ces														*49*
ces														*49*
											×	×		*50, 123*

rences are given for one plant, this does not necessarily imply that each paper refers to all the

Table 2. Physical Constants of the Cucurbitacins

Name	Other names	M. p. °C (solvent)	$[\alpha]_D$ (solvent)	λ_{max}, nm (ε)	References
Cucurbitacin A (6r)		207–208 (EtOAc)	+97.3 (EtOH)	229 (12200)	29, 38, 59, 60
Cucurbitacin B (5)	Amarin, Fabacein II	184–186 (abs. EtOH)	+87.5 (EtOH)	228 (10500)	25, 60, 84, 91, 93
Dihydrocucurbitacin B (111)		163–164 (acetone-hexane)	+53 (chlf) (61)		
			+57 (chlf) (93)		
Isocucurbitacin B (119)	2-epicucurbitacin B (wrong name)	229–231 (MeOH)	+43 (chlf)	230 (11000)	1, 61, 78
Cucurbitacin C (69)		207–207.5 (EtOAc)	+95.2 (EtOH)	231 (11000)	30, 59, 60
Cucurbitacin D (6)	Elatericin A	151–152 (abs. EtOH)	+52 (EtOH)	230 (10000)	37, 73, 74, 75, 79, 80, 81
22-Deoxocucurbitacin D (112)		non-crystalline	+103 (chlf)		28
22-Deoxoisocucurbitacin D (120)		non-crystalline	+76 (chlf)		28
Cucurbitacin E (2)	Elaterin	232–233 (MeOH)	−59 (chlf)	233 (11700)	79, 80, 81, 84
				267 (8400)	85
Cucurbitacin F (134)		244–245 (chlf)	+38 (EtOH)	232 (11200)	94
Cucurbitacin G (108)		150–152 (aq. MeOH)	+84 (chlf)		53
Cucurbitacin H (108)		non-crystalline	+57 (chlf)		53
Cucurbitacin I (7)	Elatericin B	148–148.5 (aq. MeOH)	−52 (chlf)	234 (11000)	80, 81, 84
				266 (6850)	
Tetrahydrocucurbitacin I (121)	Tetrahydroisoelatericin B	117–126	+56.4 (chlf)		48
Cucurbitacin J (109)		200–202 (EtOH)	−25 (MeOH) (48)	270 (8700)	34, 48
			−36 (chlf) (34)		
Cucurbitacin K (109)		sinters 143°, melts ~193° (aq. MeOH)	−46.8 (EtOH) (48)	270 (8000)	34, 48
			−74 (chlf) (34)		
Cucurbitacin L (110)		sinters 120, melts ~140 (aq. MeOH)	−49 (chlf)		34, 48
Cucurbitacin O (135)		247–248 (EtOAc)		270 (8050)	62
Cucurbitacin P (136)		157–158, resolidifies, then 211–212 (EtOAc)		230 (8250)	62
Cucurbitacin Q (137)		118–135		229 (9000)	62
Bryodulcosigenin (123)	Aglycone m. p. 178	181–182 (Et_2O)	+190 (chlf)		13, 127
Bryosigenin (124)		188–191 (Et_2O)	+144 (chlf)		130
Bryogenin (125)		157			14, 15, 16
Gratiogenin (126)		202–203 (MeOH)	+175 (chlf)		123

References

1. DE ABREU MATOS, F. J., and O. R. GOTTLIEB: Isocucurbitacina B, Constituinte Citotoxico da *Luffa operculata*. An. Acad. Brasileira Ciencias **39**, 245 (1967).
2. ALEXA, E., and P. JITARIU: Actiunea Sucului Natural de *Ecbalium elaterium* Rich. Asupra Catabolismului Glucidic la Soarece. Academia RPR Buletin Stiintific Stiinte Medicale **1**, No. 7, 1 (1949).
3. — — Noi Aspecte ale Actiunii Fiziologice a Sucului din Fructul Copt de *Ecbalium elaterium* la Cobai. Academia RPR Filiala Iasi; Studii Cercetari Stiintifice **2** (No. 1–2), 1 (1951).
4. APSIMON, J. W., and J. M. ROSENFELD: Methyl Migration by Epoxide Cleavage. The Effect of Carbonium Ion Stabilisation by a Neighbouring Double Bond on the Direction of Migration on Cleavage of $9\alpha,11\alpha$-Epoxy-4,4-dimethyl-androst-5-ene-3,17-dione. Chem. Commun. **1970**, 1271.
5. AUDIER, H. E., and B. C. DAS: Mass Spectrometry of Tetracyclic Triterpenes. The Cucurbitacin Group. Tetrahedron Letters **1966**, 2205.
6. AUTERHOFF, H., and R. MÜLLER: Über Inhaltsstoffe der Pharmazeutisch verwendeten Koloquinthen. Dtsch. Apotheker Ztg. **107**, 1353 (1967).
7. BARTON, D. H. R., C. F. GARBERS, D. GIACOPELLO, R. G. HARVEY, J. LESSARD and D. R. TAYLOR: Inter-relationship of Cucurbitacin A with Lanosterol. J. Chem. Soc. (C) (London) **1969**, 1050.
8. BARTON, D. H. R., D. GIACOPELLO, P. MANITTO and D. L. STRUBLE: Conversion of Eburicoic Acid into $4,4,14\alpha$-Trimethylpregn-8-ene-2,7-11,20-tetraone. J. Chem. Soc. (C) (London) **1969**, 1047.
9. BARUA, A. K., S. K. CHAKRABORTI and A. KUMAR RAY: Chemical Examination of the Seeds of *Luffa acutangula* Roxb. J. Indian Chem. Soc. **35**, 480 (1958).
10. BELKIN, M., and D. B. FITZGERALD: Tumor Damaging Capacity of Plant Materials. Plants Used as Cathartics. J. Nat. Cancer Inst. **13**, 139 (1952).
11. BERG, A.: Sur la Formule de l'Elaterine. Bull. soc. chim. France **35** (3), 435 (1906).
12. — Glycoside of *Ecballium elaterium*. Bull. soc. chim. France [IV] **7**, 385 (1910).
13. BIGLINO, G.: Costituenti della Radice della *Bryonia dioica* Jacq. XI. Struttura dell' Aglicone a p. f. $= 178°$ (and references cited therein). Ann. Chim. (Rome) **55** (3), 164 (1965).
14. BIGLINO, G., J. M. LEHN and G. OURISSON: Structure de la Bryogenine. Tetrahedron Letters **1963**, 1651.
15. BIGLINO, G., and G. M. NANO: Sintesi Parziale della Briogenina. Farmaco (Ed. Scient.) **20**, 566 (1965).
16. — — Constituents of the Roots of *Bryonia dioica* Jacq. Farmaco (Ed. Scient.) **22**, 140 (1967).
17. BORSCHE, W., and K. DIACONT: Über Elaterin. Liebigs Ann. Chem. **528**, 39 (1937).
18. BREDENBERG, J. B-SON and R. GMELIN: Über das Vorkommen von Cucurbitacin E und I in *Iberis amara* L. (Cruciferae) und die Identität von „Ibamarin" mit Cucurbitacin I (and references cited therein). Acta Chem. Scand **16**, 1802 (1962).
19. BULL, J. R., and K. B. NORTON: Steroidal Analogues of Unnatural Configuration; $4,4,14\alpha$-Trimethyl-19 (10 $\rightarrow$ 9β) *abeo*-10α-Pregn-5-enes from Cucurbitacins. J. Chem. Soc. (C) (London) **1970**, 1592.
20. CHAMBLISS, O. L., and C. M. JONES: Cucurbitacins: Specific Insect Attractants in Cucurbitaceae. Science **153**, 1392 (1966).
21. — — Chemical and Genetic Basis for Insect Resistance in Cucurbits. Proc. Amer. Soc. Horticultural Science **89**, 394 (1966).

22. CHAUDRY, G. R., and T. G. HALSALL: The Constituents of *Luffa amara* Roxb; the Identity of Amarin with Cucurbitacin B. Chem. and Ind. **1959**, 1119.
23. CHAUDRY, G. R., T. G. HALSALL and E. R. H. JONES: Some Derivatives of 4,4-Dimethylcholestan-3-one. J. Chem. Soc. (London) **1961**, 2725.
24. EDERY, H., G. SCHATZBERG-PORATH and S. GITTER: Pharmacodynamic Activity of Elatericin (Cucurbitacin D). Arch. Intern. Pharmacodyn. **130**, 315 (1961).
25. EISENHUT, W. O., and C. R. NOLLER: Bitter Principles from *Echinocystis fabacea*. J. Organ. Chem. (USA) **23**, 1984 (1958).
26. EL-KHADEM, M., and M. M. A. ABDEL RAHMAN: On the Anticancer Glycoside from *Citrullus colocynthis*. Tetrahedron Letters **1962**, 1137.
27. ENSLIN, P. R.: Observations on the Chemistry of Cucurbitacin A. J. Science Food Agriculture **5**, 410 (1954).
28. ENSLIN, P. R., C. W. HOLZAPFEL, K. B. NORTON and S. REHM: Cucurbitacins from a Hybrid of *Lagenaria siceraria*. J. Chem. Soc. (C) (London) **1967**, 964.
29. ENSLIN, P. R., J. M. HUGO, K. B. NORTON and D. E. A. RIVETT: Cucurbitacin A. J. Chem. Soc. (London) **1960**, 4779.
30. — — — — Cucurbitacin C. J. Chem. Soc. (London), **1960**, 4787.
31. ENSLIN, P. R., T. G. JOUBERT and S. REHM: Paper Chromatography of Bitter Principles and some Applications in Horticultural Research. J. South African Chem. Inst. **7**, 131 (1954).
32. — — — Bitter Principles of the Cucurbitaceae. Elaterase, an Active Enzyme for the Hydrolysis of Bitter Principle Glycosides. J. Science Food Agriculture **7**, 646 (1956).
33. ENSLIN, P. R., and K. B. NORTON: Structure of the Side Chain of the Cucurbitacins. Chem. and Ind. **1959**, 162.
34. — — The Constitutions of Cucurbitacins J, K and L. J. Chem. Soc. (London) **1964**, 529.
35. ENSLIN, P. R., and S. REHM: Die Enzymatische Hydrolyse von Herzgiftglykosiden durch Elaterase. Z. physiol. Chem. (Hoppe Seyler's) **303**, 97 (1956).
36. — — The Distribution and Biogenesis of the Cucurbitacins in Relation to the Taxonomy of the Cucurbitaceae. Proc. Linnean Soc. (London) **1956–57**, 230.
37. ENSLIN, P. R., S. REHM and D. E. A. RIVETT: Bitter Principles of the Cucurbitaceae. The Isolation and Characterization of Six New Crystalline Bitter Principles. J. Science Food Agriculture **8**, 673 (1957).
38. ENSLIN, P. R., and D. E. A. RIVETT: Dehydrogenation of Cucurbitacin A. J. Chem. Soc. (London) **1956**, 3682.
39. — — The Cucurbitacins an Interesting New Group of Natural Products. South African Industrial Chemist **11**, 75 (1957).
40. ERSPAMER, V.: Ricerche sulle Droghe Purgative Nostrane II *Ecballium elaterium*. R. ital. Essenze, Profumi, Piante officinali Olii vegetali, Saponi **28**, 264 (1947).
41. FAUST, R. E., G. E. CWALINA and E. RAMSTAD: The Antineoplastic Action of Chemical Fractions of the Fruit of *Citrullus colocynthis* on Sarcoma-37. J. Amer. Pharm. Assoc. (Sc. Ed) **47**, 1 (1958).
42. FERGUSON, H. C.: The Preliminary Investigation of an Extract of the Root of *Cucurbita foetidissima*. J. Amer. Pharm. Assoc. (Sc. Ed.) **44**, 440 (1955).
43. GALLILY, R., B. SHOHAT, J. KALISH, S. GITTER and D. LAVIE: Further Studies on the Antitumor Effect of Cucurbitacins. Cancer Res. **22**, 1038 (1962).

44. GILBERT, J. N. T., and D. W. MATHIESON: Cucurbitacin E: Some Preliminary Observations. Tetrahedron **4**, 302 (1958).
45. — — The Functional Groupings of Cucurbitacin E (α-Elaterin). J. Pharm. Pharmacol. **10** (suppl.), 252T (1958).
46. GITTER, S., R. GALLILY, B. SHOHAT and D. LAVIE: Studies on the Antitumor Effect of Cucurbitacins. Cancer Res. **21**, 516 (1961).
47. GMELIN, R.: Die Verbreitung von Cucurbitacinen und Cucurbitacinglykosiden in der Gattung *Iberis* (Cruciferae). Arzneimittelforschung **13**, 771 (1963).
48. — Die Bitterstoffe der Wurzeln der roten Zaunrübe, *Bryonia dioica* Jacq. Isolierung der Cucurbitacine L, J, und K, sowie von Tetrahydrocucurbitacin I, einem neuen natürlichen Cucurbitacin. Arzneimittelforschung **14**, 1021 (1964).
49. — Bitterstoffe in der Familie Cruciferae. Planta Medica **1966** (Supplement), 119.
50. — Wirkstoffanalyse von *Gratiola officinalis* L. Vorkommen von Elaterinid und Desacetyl-Elaterinid in der Frischpflanze. Arch. Pharmaz. **300**, 234 (1967).
51. GOTTLIEB, O. R., and D. LAVIE: Chemical Studies on the Stereochemistry of Ring A in the Cucurbitacins. Anais Assoc. Brasil. Quimica **19**, 185 (1960).
52. VON HEMMELMAYR, F.: Über das Elaterin. Ber. dtsch. chem. Ges. **39**, 3652 (1906).
53. HOLZAPFEL, C. W., and P. R. ENSLIN: The Constitutions of Cucurbitacins G and H. J. South African Chem. Inst. **17**, 142 (1964).
54. JITARIU, P., and E. ALEXA: Nouvelles Contributions concernant la Composition Chimique et l'Action Physiologique du Suc Naturel des Fruits d'*Ecballium elaterium*. Academia RPR Filiala Iasi; Studii Cercetari Stiintifice **4**, 389 (1953).
55. JITARIU, P., GH. BADARAU and E. ALEXA: Etude Electrocardiographique de l'Action de l'Extrait Aqueux des Fruits d'*Ecballium elaterium*, sur le Coeur Isolé de Grenouille. Annales Scientifiques Université Jassy (Romania) **30**, 181 (1944).
56. — — — L'Action de l'Extrait Alcoolique du Suc des Fruits d'*Ecballium elaterium* sur la Conductibilité dans le Coeur de quelques Mammifères (Cobaye, Lapin). Annales Scientifiques Université Iassy (Romania) **30**, 185 (1944).
57. KLOSS, P.: Über die Bitterstoffe von *Lufa operculata* Cogn., Arch. Pharmaz. **299**, 351 (1966).
58. KLOSS, P., and H. SCHINDLER: Ein Dünnschicht-chromatographische Methode zur Beurteilung von Cucurbitaceen Tinkturen. Pharmaz. Z. **111**, 772 (1966).
59. DE KOCK, W. T., P. R. ENSLIN, K. B. NORTON, D. H. R. BARTON, B. SKLARZ and A. A. BOTHNER-BY: The Constitutions of the Cucurbitacins. Tetrahedron Letters **1962**, 309.
60. — — — — — — The Constitution of the Cucurbitacins. J. Chem. Soc. (London) **1963**, 3828.
61. KUPCHAN, S. M., A. H. GRAY and M. D. GROVE: The Cytotoxic Principles of *Marah oreganus* H. J. Med. Chem. **10**, 337 (1967).
62. KUPCHAN, S. M., R. M. SMITH, Y. AYNEHCHI and M. MARUYAMA: Cucurbitacins O, P and Q, the Cytotoxic Principles of *Brandegea bigelovii*. J. Organ. Chem. (USA) **35**, 2891 (1970).
63. LAVIE, D.: Observations on the Constituents of *Ecballium elaterium* L. Bull. Research Council Israel **5A**, 106 (1955).
64. — The Oxygen Functions of Elaterin. Chem. and Ind. **1956**, 466.
65. — The Functional Groupings of α-Elaterin. J. Pharm. Pharmacol. **10**, 782 (1958).

66. LAVIE, D., and B. S. BENJAMINOV: Structural Transformations in Rings A and B in the Cucurbitacins. Tetrahedron **20**, 2665 (1964).
67. — — Isomerism in Ring A of the Cucurbitanes. J. Organ. Chem. (USA) **30**, 607 (1965).
68. LAVIE, D., B. S. BENJAMINOV and Y. SHVO: The Nuclear Magnetic Resonance Study of Methyl Group Signals in the Cucurbitanes. Tetrahedron **20**, 2585 (1964).
69. — — — — Elateric Acid. J. Chem. Soc. (London) **1964**, 3543.
70. LAVIE, D., and O. R. GOTTLIEB: Stereochemistry of Ring A in the Cucurbitacins. Chem. and Ind. (London) **1960**, 929.
71. LAVIE, D., and Y. SHVO: A degradation Product of Elatericin A. Proc. Chem. Soc. (London) **1958**, 220.
72. — — The Functions of Elatericin A. J. Amer. Chem. Soc. **81**, 3058 (1959).
73. — — Proposed Structures for Elatericin A and B. Chem. and Ind. (London) **1959**, 429.
74. — — Proposed Structures for Elatericin A and B. J. Amer. Chem. Soc. **82**, 966 (1960).
75. — — The Complete Structures of Elatericin A and Related Cucurbitacins. Chem. and Ind. (London) **1960**, 403.
76. LAVIE, D., Y. SHVO and O. R. GOTTLIEB: The Stereochemistry of the Cucurbitacins. Tetrahedron Letters **1960**, No. 22, 23.
77. — — — Chemical Structure of the Cucurbitacins. Anais. Assoc. Brasil Quimica **21**, 5 (1962).
78. LAVIE, D., Y. SHVO, O. R. GOTTLIEB, R. B. DESAI and M. L. KHORANA: The Occurrence of 2-Epicucurbitacin B in *Luffa echinata*. J. Chem. Soc. (London) **1962**, 3259.
79. LAVIE, D., Y. SHVO, O. R. GOTTLIEB and E. GLOTTER: Constitution of the Cucurbitacins. Tetrahedron Letters **1961**, 615.
80. — — — — The Structures of Elatericin A and Related Cucurbitacins. J. Organ. Chem. (USA) **27**, 4546 (1962).
81. — — — — Stereochemical Problems in the Cucurbitacins. J. Organ. Chem. (USA) **28**, 1790 (1963).
82. LAVIE, D., Y. SHVO and D. WILLNER: Alkaline Degradation of Elaterin and Elatericin A. Chem. and Ind. (London) **1958**, 1361.
83. — — — A Side Chain of Elatericin A and α-Elaterin. J. Amer. Chem. Soc. **81**, 3062 (1959).
84. LAVIE, D., Y. SHVO, D. WILLNER, P. R. ENSLIN, J. M. HUGO and K. B. NORTON: Interrelationships in the Cucurbitacin Series. Chem. and Ind. **1959**, 951.
85. LAVIE, D., and S. SZINAI: α-Elaterin. J. Amer. Chem. Soc. **80**, 707 (1958).
86. LAVIE, D., and D. WILLNER: Elatericin A and B. J. Amer. Chem. Soc. **80**, 710 (1958).
87. — — Structures of α-Elaterin and its Degradation Products. Proc. Chem. Soc. (London) **1959**, 191.
88. — — Proposed Structures for α-Elaterin and its Degradation Products. J. Amer. Chem. Soc. **82**, 1668 (1960).
89. LAVIE, D., D. WILLNER, M. BELKIN and G. HARDY: New Compounds from Plants with Anti-tumor Activity. Acta Unio Internat. contra Cancrum **15** bis, 177 (1959).
90. LAVIE, D., D. WILLNER and Z. MERENLENDER: Constituents of *Citrullus colocynthis* (L) Schrad. Phytochem. **3**, 51 (1964).
91. MELERA, A., M. GUT and C. R. NOLLER: Structure of Cucurbitacin B. Tetrahedron Letters **1960**, No. 14, 13.

92. MELERA, A., and C. R. NOLLER: The Action of Alkali on Alcoholic Solutions of Dihydro-Derivatives of Cucurbitacin B. J. Organ. Chem. (USA) **26**, 1213 (1961).
93. MELERA, A., W. SCHLEGEL and C. R. NOLLER: Structure of the Side Chain of Cucurbitacin B. J. Org. Chem. (USA) **24**, 291 (1959).
94. VAN DER MERWE, K. J., P. R. ENSLIN and K. PACHLER: The Constitution of Cucurbitacin F. J. Chem. Soc. (London) **1963**, 4275.
95. MOORE, C. W.: Note on the Constitution of α-Elaterin. J. Chem. Soc. (London) **97**, 1797 (1910).
96. Moss, G. P.: Some Aspects of Triterpene Bitter Principle biosynthesis. Planta Medica **1966** (Supplement), 86.
97. MÜLLER, R., and H. AUTERHOFF: Über Inhaltsstoffe der pharmazeutisch verwendeten Koloquinthen. Dtsch. Apotheker Ztg. **108**, 1191 (1968).
98. NOLLER, C. R., A. MELERA, M. GUT, J. N. SHOOLERY and L. F. JOHNSON: Concerning Ring A of Cucurbitacin B. Tetrahedron Letters **1960**, No. 15, 15.
99. OURISSON, G., P. CRABBÉ and O. RODIG: The Tetracyclic Triterpenes. Holden Day San Francisco, 1964.
100. PORA, E. A., and P. JITARIU: Action de l'Extrait d'*Ecballium elaterium* Rich. sur le Coeur de Bufo Vulgaris, Retranché de l'Organisme. Annales Scientifiques Université Jassy (Romania) **30**, 1 (1944).
101. POWER, F. B., and C. W. MOORE: Chemical Examination of Elaterium and the Characters of Elaterin. Pharm. J. [IV] **29**, 501 (1909).
102. — — The Constituents of the Fruit of *Ecballium elaterium*. J. Chem. Soc. (London) **95**, 1985 (1910).
103. — — The Constituents of Colocynth. J. Chem. Soc. (London) **97**, 99 (1910).
104. REHM, S., P. R. ENSLIN, A. D. J. MEEUSE and J. H. WESSELS: Bitter Principles of the Cucurbitaceae. The Distribution of Bitter Principles in this Plant Family. J. Science Food Agriculture **8**, 679 (1957).
105. REHM, S., and J. H. WESSELS: Cucurbitacins in Seedlings-Occurrence, Biochemistry and Genetical Aspects. J. Science Food Agriculture **8**, 687 (1957).
106. REICHEL, L., and K. H. EISENLOHR: Zur Kenntnis des α-Elaterins. Liebigs Ann. Chem. **531**, 287 (1937).
107. RIMINGTON, C.: The Toxic Principles of *Cucumis africanus* L. f., *Cucumis myriocarpus* (Naud), and of a New Unnamed *Cucumis* Species. South African J. Science **30**, 505 (1933).
108. RIMINGTON, C., and D. G. STEYN: The Isolation of a Bitter Principle from a South-West African Species of *Cucumis*. South African J. Science **32**, 137 (1935).
109. RIVETT, D. E. A., and P. R. ENSLIN: On the Skeletal Structure of the Cucurbitacins. Proc. Chem. Soc. (London) **1958**, 301.
110. RIVETT, D. E. A., and F. H. HERBSTEIN: Revised Molecular Formulae for the Cucurbitacins. Chem. and Ind. (London) **1957**, 393.
111. SCHLEGEL, W., A. MELERA and C. R. NOLLER: Reduction and Oxidation Products of Cucurbitacin B. J. Organ. Chem. (USA) **26**, 1206 (1961).
112. SCHLEGEL, W., and C. R. NOLLER: The Relation of Fabacein to Cucurbitacin B. Tetrahedron Letters **1959**, No. 13, 16.
113. SCHLEGEL, W., and C. R. NOLLER: The Relation of Fabacein to Cucurbitacin B. J. Organ. Chem. (USA) **26**, 1211 (1961).
114. SCHWARTZ, H. M., S. I. BIEDRON, M. M. VON HOLDT and S. REHM: A Study of some Plant Esterases. Phytochemistry **3**, 189 (1964).
115. SHIMOTSUMA, M.: Inheritance of Several Characters in Watermelons, (and earlier references cited therein). Jap. J. Breeding **13**, No. 4, 31 (1963).

116. SHOHAT, B., S. GITTER and D. LAVIE: Antitumor Activity of Cucurbitacins: Metabolic Aspects. Cancer Chemother. Rept. No. 23, 19 (1962).
117. — — — Action of Elatericin A on Human Leukemic and Normal Lymphocytes. J. Nat. Cancer Inst. 38, 1 (1967).
118. SHOHAT, B., S .GITTER, B. LEVY and D. LAVIE: The Combined Effect of Cucurbitacins and X-ray Treatment of Transplanted Tumors in Mice. Cancer Res. 25, 1828 (1965).
119. SHVO, Y., E. GLOTTER and D. LAVIE: Stereochemical Problems in the Cucurbitacins. Bull. Res. Council Israel 11A, 34 (1962).
120. SIDDIQUI, R. H., I. R. SIDDIQUI and S. MUHAMMAD: Chemical Examination of the Juice of *Citrullus colocynthis*. J. Indian Chem. Soc. 32, 669 (1955).
121. SNATZKE, G., P. R. ENSLIN, C. W. HOLZAPFEL and K. B. NORTON: Stereochemistry of Cucurbitacin Ring A α-Ketols and their Acetates. J. Chem. Soc. (C) (London) 1967, 972.
122. STEYN, D. G.: The Toxicity of Bitter-tasting Cucurbitaceous Vegetables (Vegetable Marrow, Watermelons, etc) for Man. South African Medical J. 1950, 713.
123. TSCHESCHE, R., G. BIERNROTH and G. SNATZKE: Zur Konstitution des Gratiogenins und über weitere Inhaltsstoffe von *Gratiola officinalis* L. (and references cited therein). Liebigs Ann. Chem. 674, 196 (1964).
124. TSCHESCHE, R., and A. HEESCH: Gratiosid, ein Triterpenglykosid aus *Gratiola officinalis* L. (and references cited therein). Ber. dtsch. chem. Ges. 85, 1067 (1952).
125. TUNMANN, P.: Über das Bryoamarid aus der Wurzel von *Bryonia dioica* Jacq. Arzneimittelforschung 14, 1366 (1964).
126. TUNMANN, P., and W. GERNER: Zur chemischen Konstitution des Bryodulcosigenins. Naturwiss. 49, 106 (1962).
127. TUNMANN, P., W. GERNER and G. STAPEL: Konstitution des Bryodulcosigenins. Liebigs Ann. Chem. 694, 162 (1966).
128. TUNMANN, P., and F. K. SCHEHRER: Beitrag zur chemischen Konstitution des Bryodulcosides. Arch. Pharmaz. 292/64, 745 (1959).
129. TUNMANN, P., and G. STAPEL: Zur Konstitution des Bryoducosigenins. Tetrahedron Letters 1964, 2521.
130. — — Bryosigenin, ein neues tetracyclisches Triterpen. Naturwiss. 52, 661 (1965).
131. — — Über das Bryodulcosid. Arch. Pharmaz. 299, 596 (1966).
132. TUNMANN, P., and H. WIENECKE: Über das erste kristalline Bitterstoff-Glykosid der Cucurbitaceen. Isolierung und Eigenschaften des Bryoamarids. Arch. Pharmaz. 293/65, 195 (1960).
133. TUNMANN, P., and G. WOLF: Über die Glykoside der Wurzeln von *Bryonia dioica*. Arch. Pharmaz. 289/61, 459 (1956).
134. WARNHOFF, E. W., and C. M. M. HALLS: Desert Plant Constituents; Ocotillol, an Intermediate in the Oxidation of Hydroxy Isooctenyl Side Chains. Canad. J. Chem. 43, 3311 (1965).
135. ZANDER, J. M., and D. C. WIGFIELD: The Biosynthesis of Cucurbitacin B. Chem. Commun. 1970, 1599.
136. ZELLNER, I., and E. TASCHNER: Studien über die chem. Bestandteile heimischer Arzneipflanzen. Arch. Pharmaz. Ber. dtsch. pharmaz. Ges. 265, 34 (1927).

(Received, February 18, 1971)

Biogenetic-type Synthesis of Terpenoid Systems

By D. GOLDSMITH, Atlanta, Georgia, USA

Contents
	Page
Introduction	363
I. Theory of Polyene Cyclization	364
II. Acid Catalyzed Cyclization	366
III. Oxidative Cyclization	369
IV. Arene and Alkyl Sulfonates, Acetals and Allylic Alcohols	378
V. Cyclopropyl Ketones, Enols, and Tertiary Alcohols	384
VI. Carbonium Ion Catalyzed Cyclization	390
VII. Radical Cyclization	390
References	391

Introduction

Efforts directed toward the synthesis of naturally occurring compounds have, almost from the inception of this kind of work followed two kinds of pathways. The first and for many years the major road travelled was that of total synthesis. Thus the steady accretion of knowlege concerning the structural and stereochemical course of such reactions as base catalyzed condensation, enolate alkylation, and catalytic hydrogenation resulted in the years following the second world war in a series of elegant syntheses of such complex molecules as strychnine, cortisone, chlorophyl, cevine, longifoline, onocerin, germanicol, etc. The alternative approach less well explored until recent years has been the path of biogenetic-type synthesis.

Biogenetic-type synthesis was defined (61) in a previous review in this series as "an organic synthesis designed to follow in at least its major aspects, biosynthetic pathways proved or presumed to be used in the natural construction of the end product". We shall use this term, however, to include work which has also been referred to as the "chemical analog of a biosynthesis" (62); that is, a synthesis which includes reactions patterned after those which occur in enzymatic systems but employing non-biological substrates and often vastly different reaction conditions from biological ones.

364   D. GOLDSMITH.

The impetus for much of the interest in biogenetic-type synthesis in the steroid and terpenoid field in the last 10_15 years has been the continuing elucidation of the biological pathway for the synthesis of cholesterol. The statement *"Omnis ars imitatio est naturae"* (*44*) might well have been made with reference to organic chemistry but for its precedence of some centuries. The remarkable ability of a living system to convert an acyclic, symmetric hydrocarbon to a tetracyclic, dissymmetric unsaturated alcohol containing eight asymmetric centers has fascinated the organic chemist and stimulated him to attempt to emulate the enzymatic process.

I. Theory of Polyene Cyclization

A fundamental process of terpenoid and steroid biosynthesis is olefin alkylation (*10*). This process may in turn be divided into two aspects, the construction of acylic polyenes, and the conversion of these polyenes into cyclic compounds and it is the latter of these processes that we shall be concerned with in this paper. Examples of such cyclization reactions of terpenoid polyenes have been known from the end of the last century (*48*) but a stereoelectronic theory explaining and predicting the course of these cyclizations was first suggested in 1955. This theory is generally referred to as the STORK-ESCHENMOSER (*14, 51*) hypothesis and has two basic postulates. First, the formation of a cyclohexane ring by cationic cyclization of a diene, if it is a concerted process, should occur by an

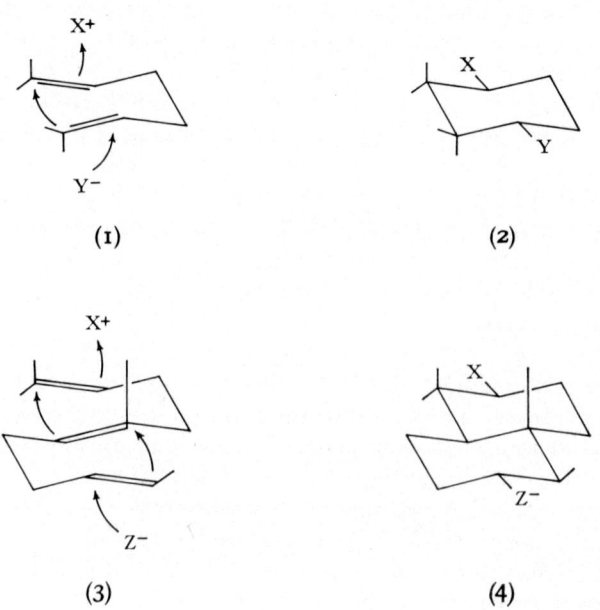

References, pp. 391—394

anti-parallel addition mechanism. Thus, in the conversion of (1) to (2) the entering electrophile X and the nucleophile Y should be found in the product to have a *cis* and (at least initially) diequatorial relationship.

If the nucleophile Y is a π-bond incorporated in an extended trienic chain and if we supply a second nucleophile, Z^-, then the product will be a decalin having a *trans*-ring junction, as in (3) to (4).

Extension of this postulate to a polyene suggests, that cyclization of a system such as (5) would result in a polycyclic product (6) having the familiar *trans-anti-trans-anti-trans*-geometry of many of the naturally occurring terpenoids and steroids.

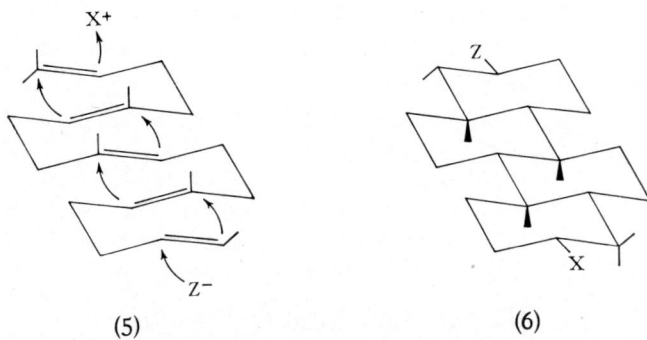

(5)   (6)

A significant part of the biosynthesis of naturally occurring terpenoids can be accounted for on the basis of a concerted cationic cyclization of an appropriate polyene. This postulate, however, cannot account for one of the most important aspects of the biosynthesis of terpenoids, the formation of lanosterol and cholesterol from squalene. Since these molecules are formed biologically (8) by 1,2-rearrangement of methyl groups from an intermediate tetracyclic product the STORK-ESCHENMOSER first postulate would lead to the wrong stereochemistry at, *inter alia*, C-13, 14 of both lanosterol and cholesterol. Thus, the second postulate of this theory is that the principal role of the enzyme in the biological cyclization of polyenes is a conformational one. If the squalene epoxide chain is folded and held as shown in (7) then concerted cyclization will yield an intermediate (8) that upon rearrangement will furnish lanosterol (9).

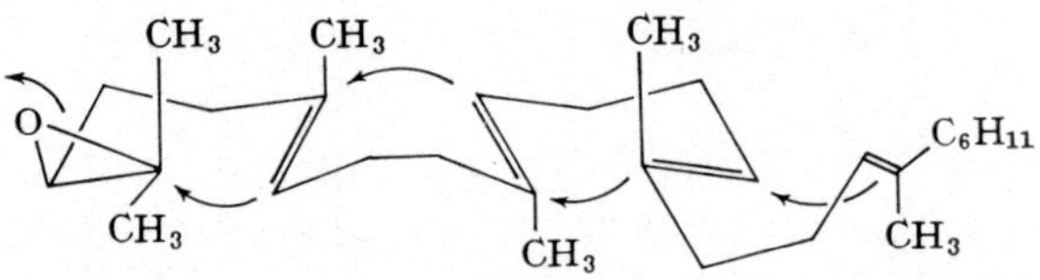

(7)

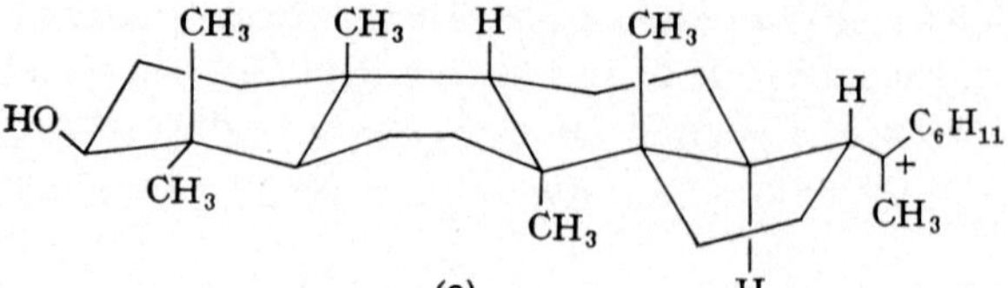

(8)

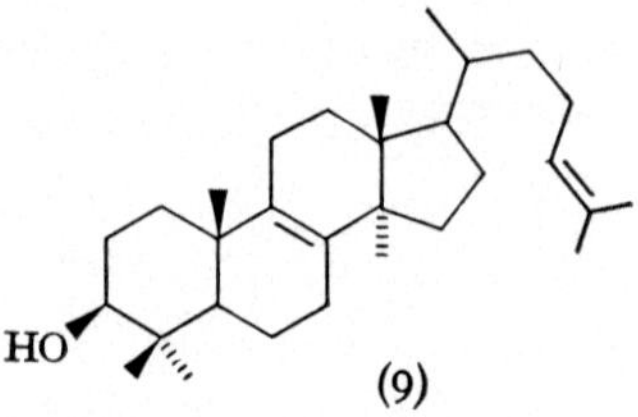

(9)

VAN TAMELEN (*71*) has proposed a partial alternative to this scheme, on the basis that formation of ring C should proceed in a MARKOWNIKOW sense to a five membered ring which then undergoes rearrangement.

II. Acid Catalyzed Cyclization

The initial applications of the STORK-ESCHENMOSER hypothesis to biogenetic-type synthesis were dissapointing. STORK and BURGSTAHLER (*51*) investigated the boron fluoride catalyzed cyclization of *trans-trans*-farnesic acid (10). The bicyclic products of this reaction (11), (12) and (13) all contained a *trans*-ring junction as predicted. The monocyclic acid (14), however, obtained from farnesic acid by cyclization under mild conditions,

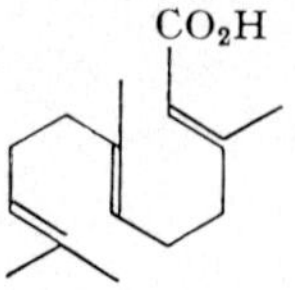

(10)

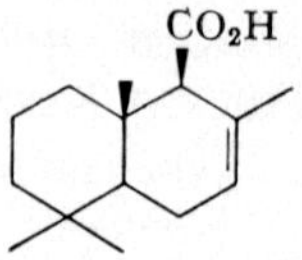

(11)

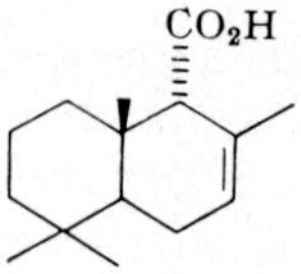

(12)

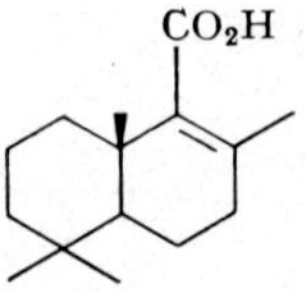

(13)

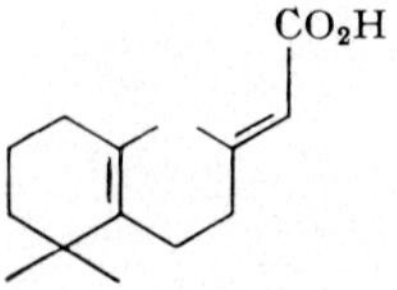

(14)

References, pp. 391—394

also provided these bicyclic products under more vigorous circumstances. This latter result was unexpected since the theory predicts that cyclization of such a monocyclic olefin should produce a *cis*-ring junction. Interestingly this result has been applied to the biogenetic type synthesis of the diterpene lactones α and β-levantenolides (15) and (16) (*39*). Monocylofarnesol (17) was converted to the butenolide (18). Treatment of

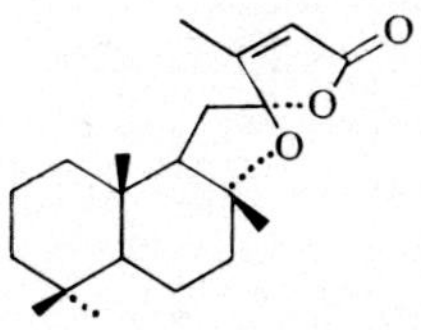

(15)

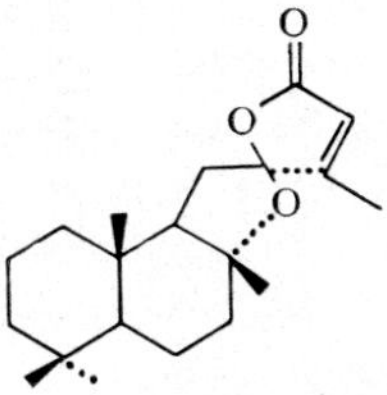

(16)

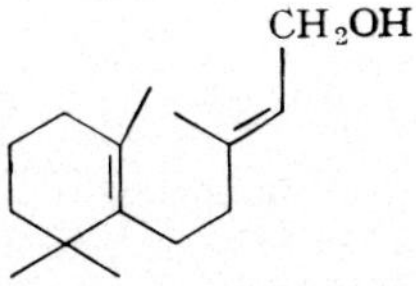

(17)

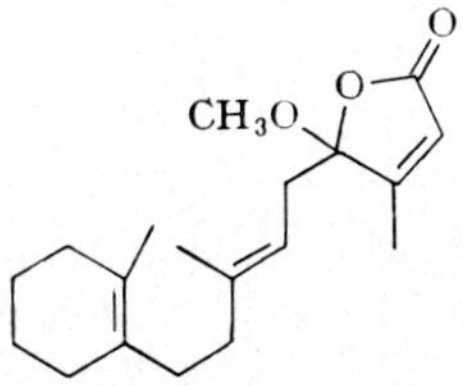

(18)

the latter with stannic chloride in benzene afforded (15) and (16) in 30% and 12% yields respectively. The same products were also produced, though in lower yield, from cyclization of an *acyclic* precurser derived from *trans, trans*-farnesol (*58*). The same type of result as STORK and BURGSTAHLERS was obtained by ESCHENMOSER (*13*). who subjected the *trans-trans*, *cis-trans*, and *trans-cis* isomers of desmethyl farnesic acid (19) as well as the monocyclic compounds (20) and (21) to cyclization conditions using sulfuric and formic acids, and in all cases obtained the *trans*-fused product (22).

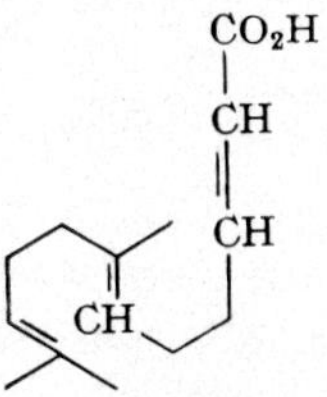

(19)

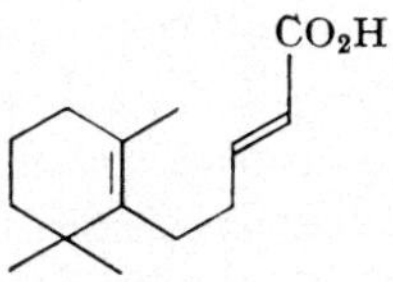

(20)

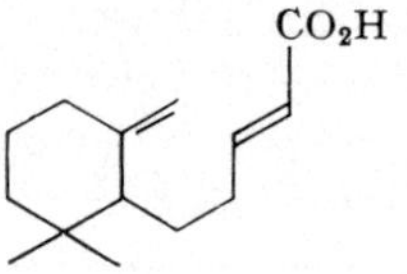

(21)

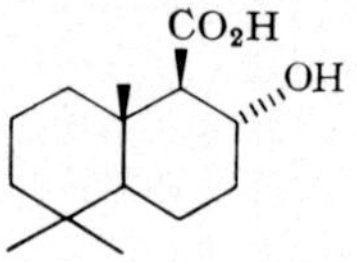

(22)

JOHNSON (*31*) has shown that these results are not caused by interconversion of *cis* and *trans* fused products under the reaction conditions. The acids (23) and (24) were individually subjected to pro-

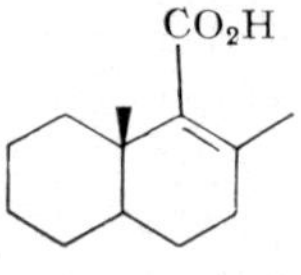

(23)

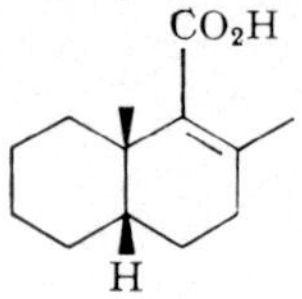

(24)

longed treatment with boron fluoride etherate and recovered unchanged. The explanation for the lack of stereospecificity in acid-catalyzed polyene cyclization must lie in the non-concerted nature of the process. Thus the cyclizations of farnescic acid and desmethylfarnesic acid must involve intermediate cations of sufficient lifetime to dictate the formation of the thermodynamically more stable product, the *trans*-decalin. The formation of an intermediate cation can be accounted for in terms of the poor nucleophilicity of the double bond conjugated with the carboxyl group. Support for this conclusion was adduced by ULERY and RICHARDS (*60*) who showed that the only cyclization product from treatment of the diene (25) with deuteroformic and deuterosulfuric acids was the cyclohexyl formate (26).

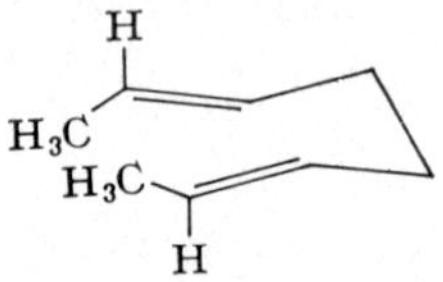

(25)

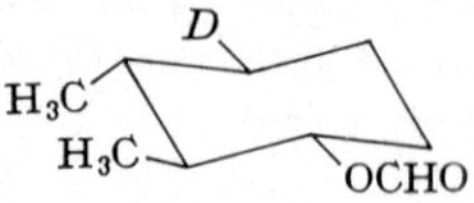

(26)

The stereochemistry of this product is in accord with the STORK-ESCHENMOSER hypothesis. In addition, KUCHEROV and coworkers (*49*) have shown that the *cis* and *trans*-isomers of geranyl acetone (27), lead stereospecifically to the bicyclic ethers (28) and (29) respectively.

References, pp. 391—394

(27) cis and trans   (28)   (29)

JOHNSON has also provided evidence to show that the acid catalyzed cyclization of polyenes does follow the STORK-ESCHENMOSER theory so long as the double bonds are not reduced in nucleophilicity by conjugation with a carboxyl group. Thus while they were unable to reproduce the reported yields on repeating the work of Linstead, JOHNSON and co-workers (*31*) showed that acid-catalyzed cyclization of butenylcyclohexene (30), gave predominantly the *cis-anti*-decalol (31).

(30)   (31)

III. Oxidative Cyclization

Although proton acid-catalyzed cyclization of an acyclic terpenoid polyene may in some cases be the *in vivo* pathway for the synthesis of polycyclic natural products, such a process suffers from a serious drawback as a method for biogenetictype *in vitro* synthesis. A polyenic system may protonate on a variety of olefinic sites and thus be converted to a variegated mixture of cyclized products. Several groups of workers have sought, therefore, to find more specific ways of initiating the cyclization process. These investigations have been aimed at finding not only controlled cyclization methods, but also at finding a biogenetic-type model for the oxidative cyclization pathway common to cholesterol and presumably most triterpene biosynthesis.

The C-3 hydroxyl group of lanosterol, was shown by BLOCH and TCHEN (*59*) to arise *in vivo* from molecular oxygen. It appears to be a reasonable assumption that other terpenoids having an oxygen function in the equivalent position are also produced by oxidative cyclization. The problem of biogenetic-type synthesis of these types of systems was centered for some time, therefore, on finding an *in vitro* equivalent of cationic oxygen. One such equivalent can be found in the reactions of peroxyacids. The conversion of a polyene to an epoxide followed by acid catalyzed opening of the epoxide and synchronous closure of a ring satisfies both the stereochemical and mechanistic requirements of the accepted pathway for biogenesis of lanosterol and, by analogy, a host of other terpenoids. Several examples of the cyclization of unsaturated epoxides were found by BARTON and co-workers. The monoepoxides, (32) and (33), of caryophyllene and isocaryophyllene were both found (*1*) to undergo cyclization and rearrangement upon treatment with aqueous acid to yield (34) and (35) respectively. The naturally occurring sesquiterpenoid epoxide pyre-

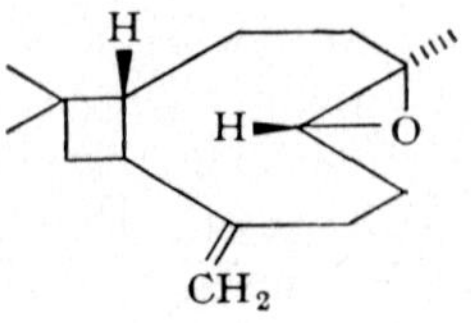

throsin (36) also undergoes cyclization to yield the eudesmane derivative (37) (*2*). Another case of transannular epoxide cyclization is the conversion of (38) to the decalin diol (39) in refluxing water (*12*).

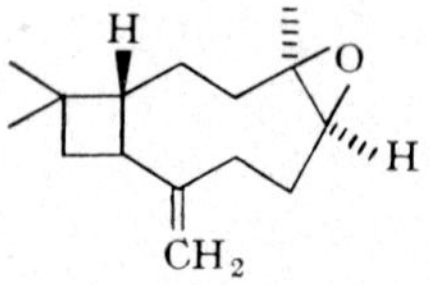

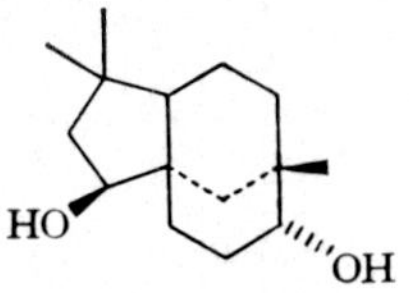

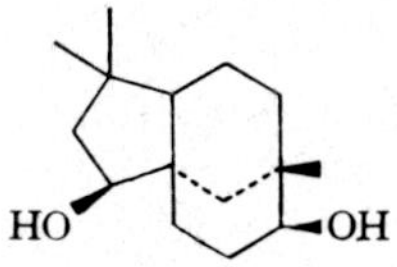

Although these cyclizations were suggestive of the possibility of using an acyclic polyene epoxide for biogenetic-type synthesis they all

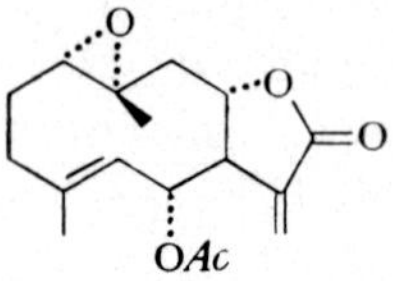

References, pp. 391—394

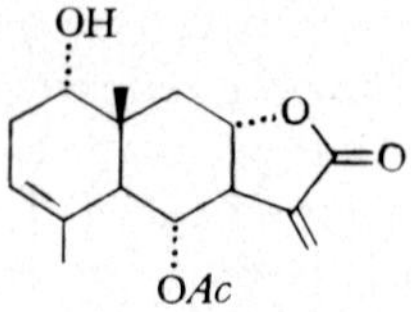

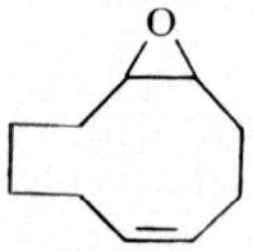

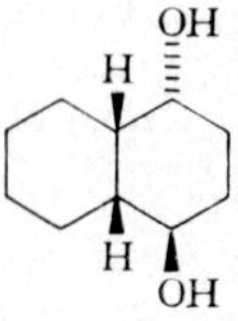

had the advantage of occurring in reasonably rigid ring systems. The first epoxide cyclization which could be directly related to the cyclization of squalene was the boron fluoride or stannic chloride catalyzed reaction of geraniolene mono-epoxide (40) (*17*), Under these conditions (40) afforded three products, the bicyclic ether (41) and the cyclohexenols (42) and (43).

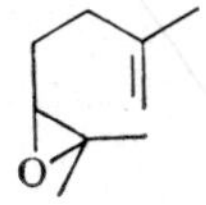

(40)    (41) 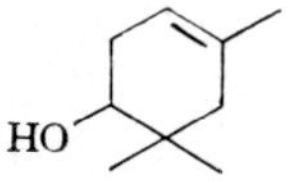   (42) 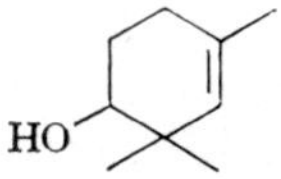   (43)

The ease of cyclization of this acyclic epoxyolefin suggested that the reaction could be extended to biogenetic-type syntheses of a variety of C-3 hydroxylated terpenoids. One problem which had to be overcome, however, was that of introducing an epoxide function at the terminal double bond of a polyene system. Treatment of a polyene such as farnesyl acetate with monoperphthalic acid results in a difficultly separable mixture of terminal and central monoepoxides (*18*). A solution to this problem was found by VAN TAMELEN and coworkers (*64, 71*) who showed that treatment of polyenes with N-bromo-succinimide in an aqueous glyme solution yields primarily a terminal mono-bromohydrin. Thus, squalene is converted by this reagent to the bromohydrin (44) which in turn may be transformed to squalene monoepoxide (45) by treatment with base (*71*).

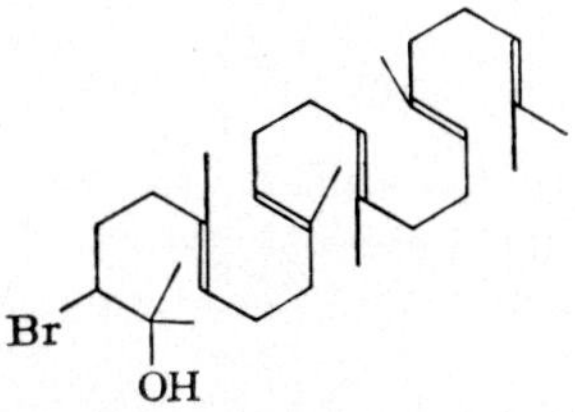

(44) 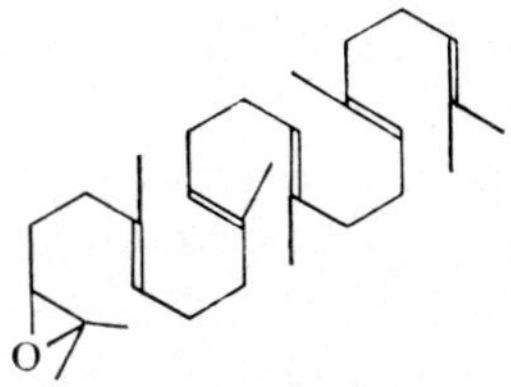   (45)

Application of this oxidation sequence to farnesyl acetate provides the dienic oxide (46), and treatment of (46) with boron fluoride etherate in benzene produces in modest yield the bicyclic diol monoacetates (47) and (48) (*70*). The structures and stereochemistry of these cyclization products were demonstrated by conversion to drimenol and epidrimenol respectively and by comparison with authentic samples of the latter two

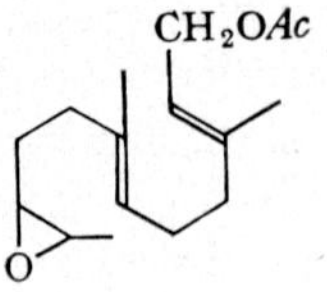

(46) 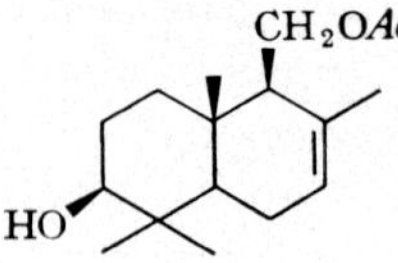   (47) 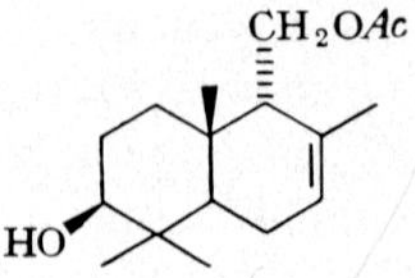   (48)

substances. Minor by-products of this cyclization were the bicyclic ether (49) and the dienol (50). The formation of the latter may be accounted

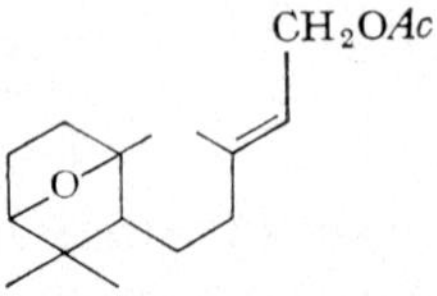

(49) 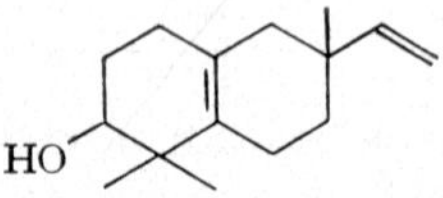   (50)

for by assuming that a monocyclic product (51) undergoes further cyclization via the cation (52), a process reminiscent of the pathway of bioformation of ring-C of many diterpenoids.

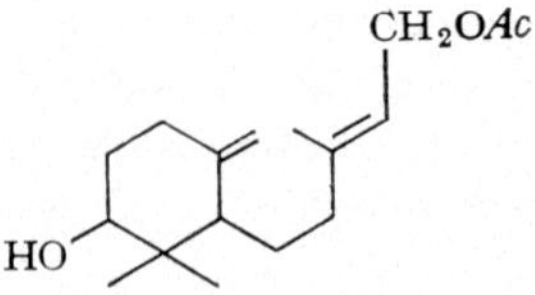

(51) 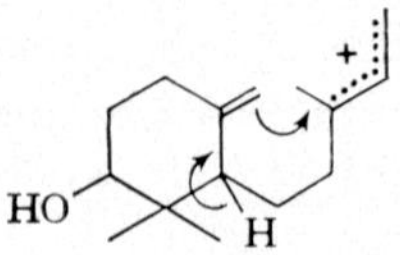   (52)

Several monoterpenoid-like componds have been prepared by employing epoxide cyclization. The unsaturated oxirane (53) obtained from citronellal closes to produce the bicyclic ether (54) as a mixture of stereoisomers (21). Epoxide (55) affords three products, (56), (57), and (58) upon treatment with boron fluoride-etherate (19). Whereas both (56) and (57) arise by opening of the epoxide ring with participation of the

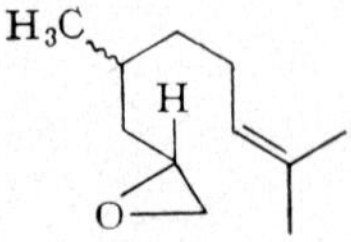

(53) 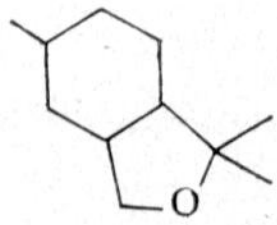   (54) 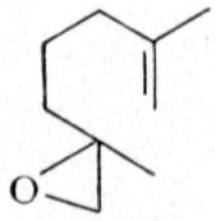   (55) 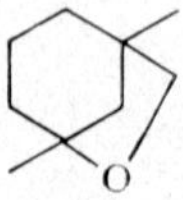   (56)

References, pp. 391—394

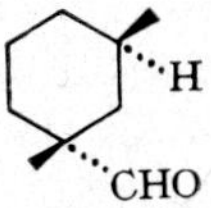

(57) 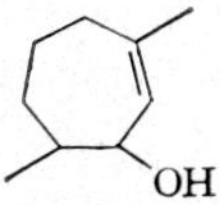 (58) 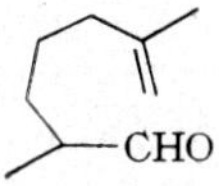 (59)

double bond, the cycloheptenol (58) is the result of the conversion of the starting material to an intermediate aldehyde (59) which then undergoes cyclization. Products related to the bridged bicyclic monoterpenoids are obtained (20) from the acid-catalyzed reactions of (60) derived from α-campholene aldehyde. With phosphoric acid (60) yields the tricyclic ethers (61) and (62) while a Lewis acid catalyst, stannic chloride, converts (60) to both (61) and (62) and to the saturated aldehyde (63) as well.

The first application of the cyclization of an epoxyfarnesyl derivative to a biogenetic-type synthesis of a natural product was the closure of umbelliprenin epoxide (63). Treatment of the *trans, trans* isomer of this epoxide, (64), with boron fluoride yielded, among other products farnesiferol-C, (65). A second member of this class of sesquiterpenes, farnesiferol-A, (66), was obtained as part of the product mixture from cyclization of *trans, cis*-umbelliprenin epoxide (67).

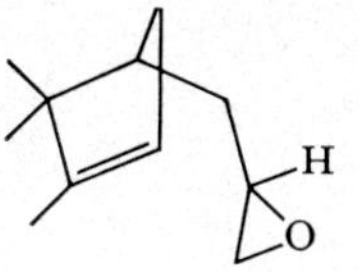

(60) 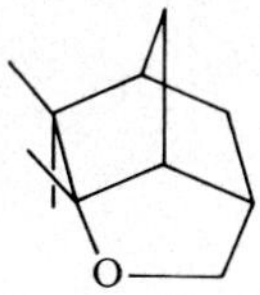 (61) 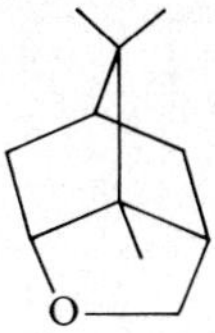 (62) 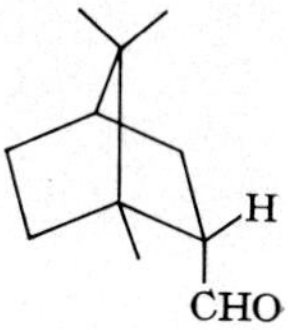 (63)

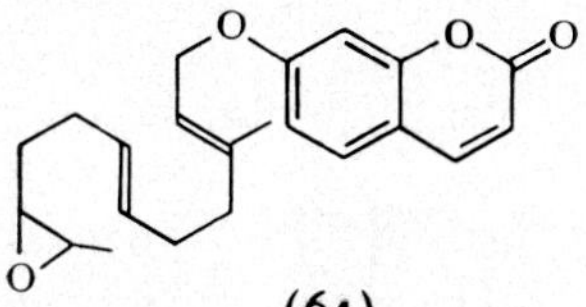

(64) 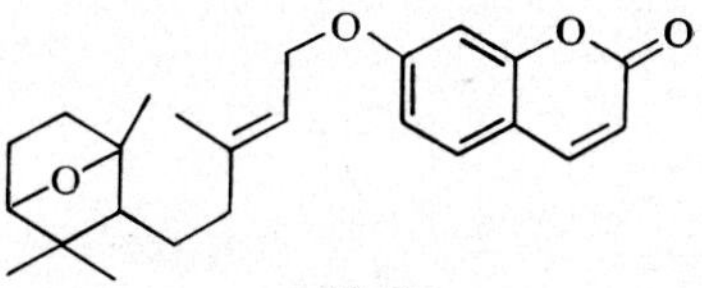 (65)

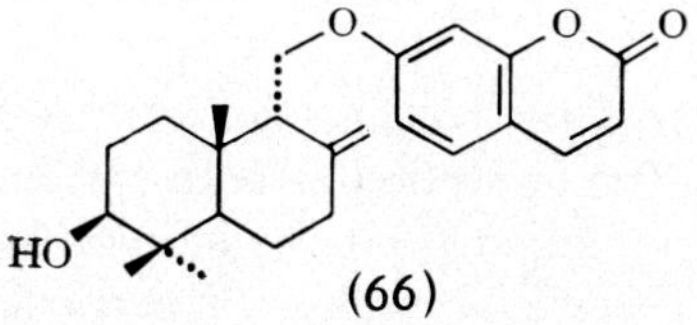

(66) 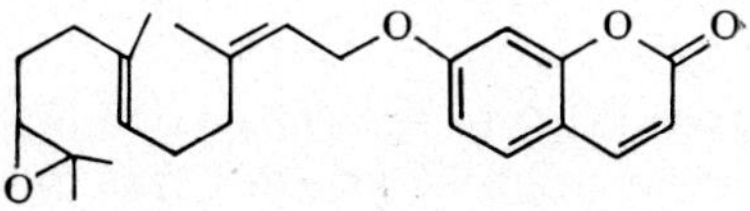 (67)

The cyclization of another farnesyl derivative, methyl *trans, trans*-farnesate 10, 11-epoxide has been employed in the synthesis of the triterpenoids γ-onocerin and hopenone-I (*69*). The conversion of the cyclization product, (68), to β-onocerin diacetate is outlined in Scheme I. At the time this work was done, cyclization of β-onocerin diacetate to the γ-isomer and subsequent conversion to hopenone-I had been carried out by other workers (*3, 16*).

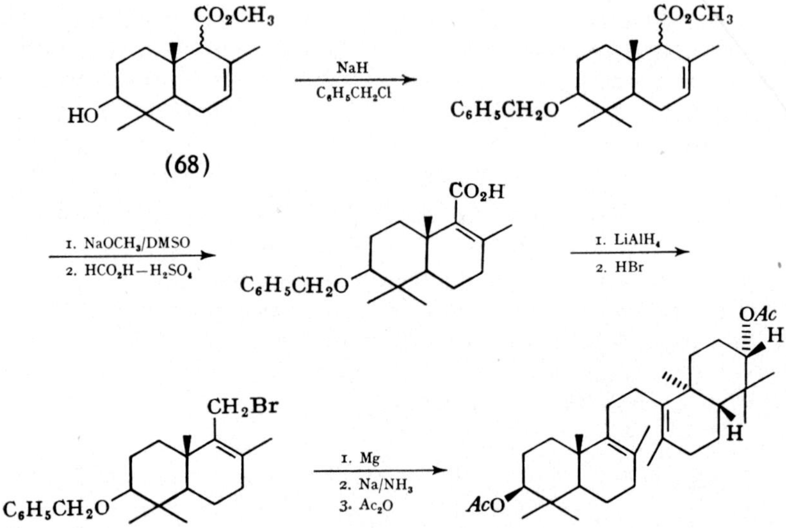

Scheme I

VAN TAMELEN and coworkers (*68, 72*) have extended these epoxide cyclizations to di- and tri-terpene precursors also. For example (*68*), stannic chloride in benzene catalyzed cyclization of geranylgeranyl acetate (69) yielded the tricyclic hydroxy acetate (70). Similar reaction conditions

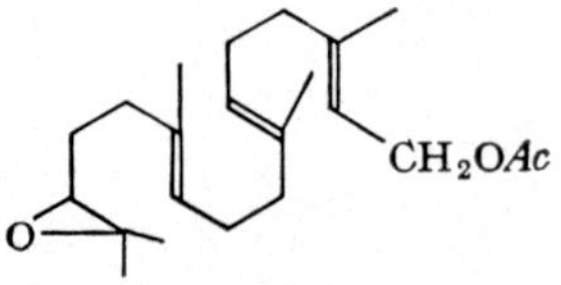

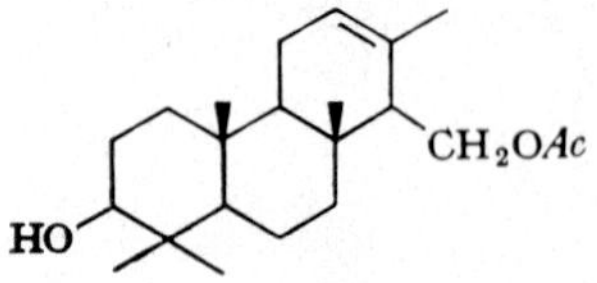

applied (*72*) to squalene oxide (45) produced the bicyclic alcohol (71) an the two tricyclic alcohols (72) and (73). The formation of both (71) and (72) requires methyl migration of the type suggested for the biosynthesis of lanosterol and other triterpenes from squalene oxide. Of particular

References, pp. 391—394

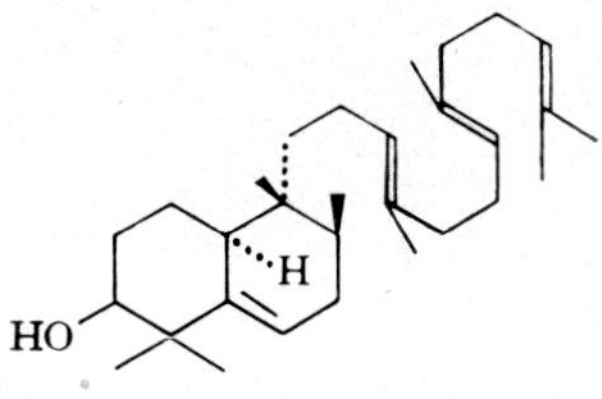

(71)

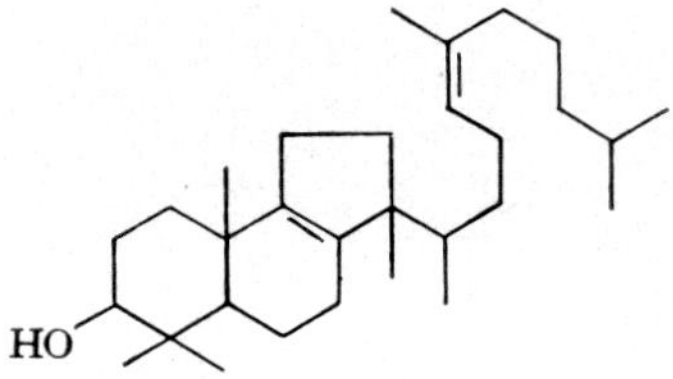

(72)

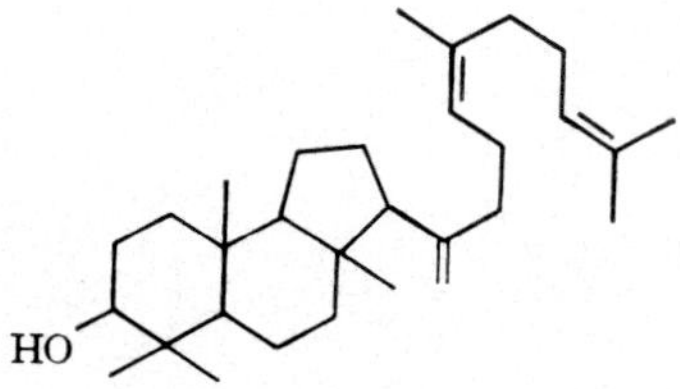

(73)

interest with regard to these non-enzymic cyclization products of squalene oxide is that the naturally occurring triterpenoid malabaricanediol was isolated subsequent to these findings and its structure shown to be (74) (7).

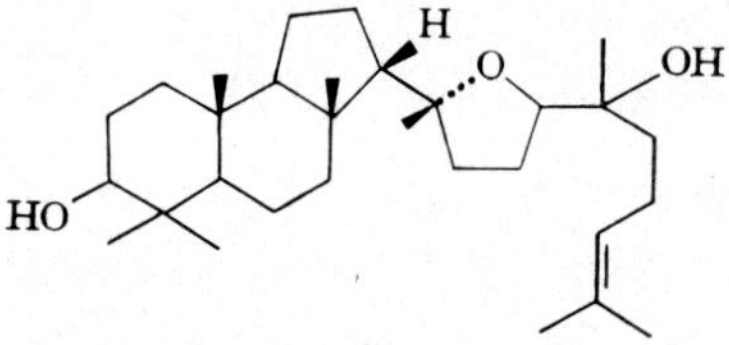

(74)

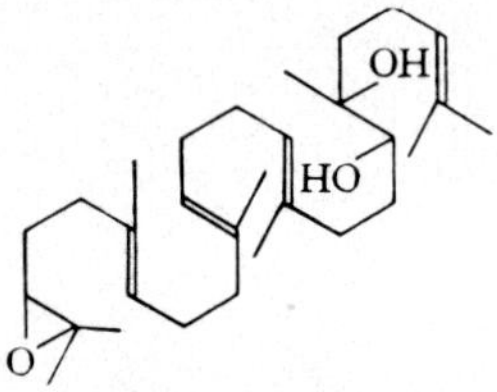

(75)

A biogenetic-type synthesis of this substance from squalene was carried out by Sharpless (45). He converted squalene to the epoxydiol (45) (46) and subjected the latter to treatment with picric acid. This protonic acid in contrast to LEWIS acid catalysts converted (75) to malabaricanediol without further transformations of the latter to unwanted by-products.

Biogenetic-type syntheses of tetracyclic molecules from mono- and bicyclic precursers has been accomplished by VAN TAMELEN (66, 67). The isoeuphenol-like compond (76) was obtained by cyclization of epoxide (77), and the 9 α-H-dihydrolanosterol product (78) was obtained in low yield from the epoxide (79).

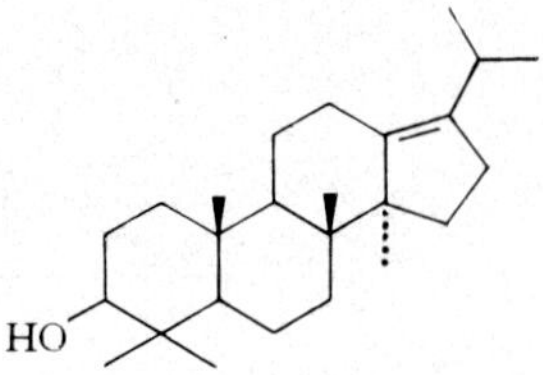

(76) 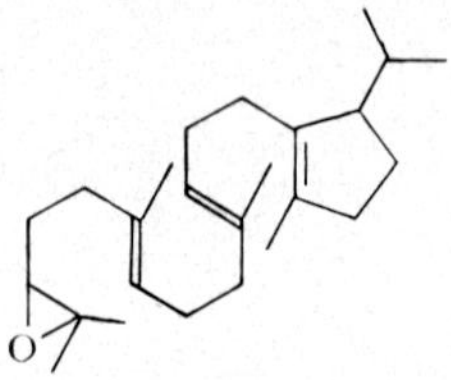   (77)

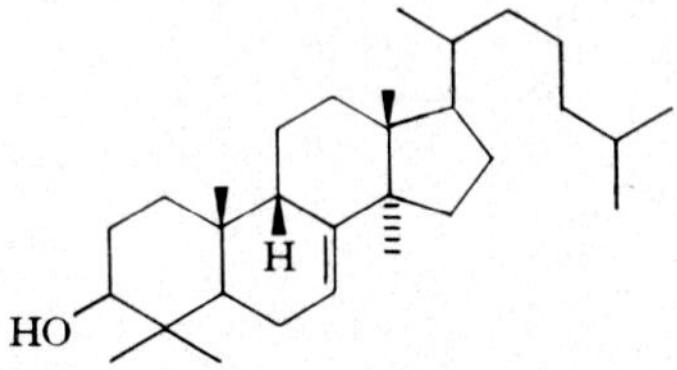

(78) 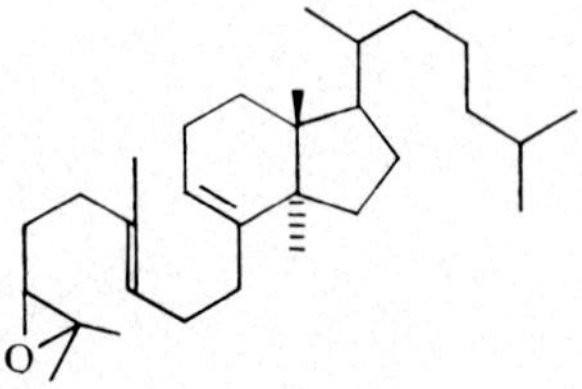   (79)

The cyclization of the latter is of particular interest since it appears to involve a boat-like conformation, (80), of the epoxide substrate in order to account for the stereochemistry of ring-B of the product (16).

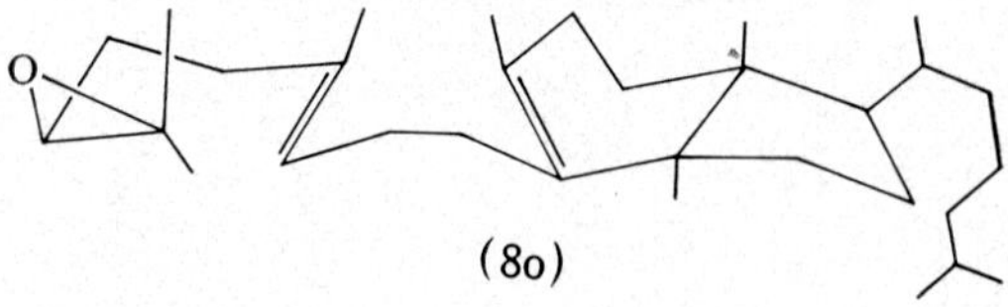

(80)

All of the previous examples of epoxyolefin cyclizations have been carried out with one basic type of substrate, *trans* substituted olefins. In addition, the cyclic products have in every case been found to have *trans*-fused rings. Despite these facts, however, none of these results actually demonstrate whether *in vitro* epoxyolefin cyclization is a stereospecific or stereoselective process. The first such demonstration was supplied by the results of the cyclization reactions of the *cis* and *trans* isomers of (81) (23). Each reaction was shown to occur in a stereospecific manner since *cis*-(81) afforded only the *cis*-tricyclic alcohol (82) and the *trans*-olefin, *trans*-(81), yielded only the *trans* isomer (83). The structure and stereochemistry of (83) was proved by oxidation to the known acid (84) while the structure and stereochemistry assigned to (82) was based on spectroscopic evidence. This type of result has also been found (22) for the conversions of the *trans* and *cis* epoxides (85) and (86) to the *trans*-(81)

cis-(81)

(82)

(83)

(84)

tricyclic alcohols (87) and (88) respectively, and the stereospecific cyclizations of the *cis* and *trans* farnesyl epoxides (89) and (90) to (68) and (91) as well (*65*).

(85)

(86)

(87)

(88)

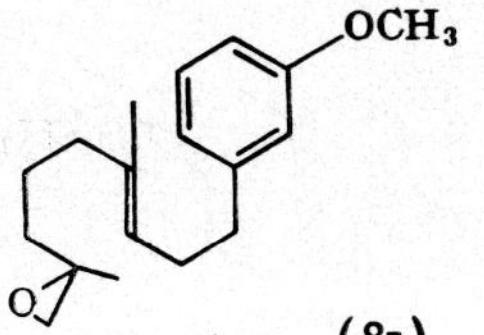

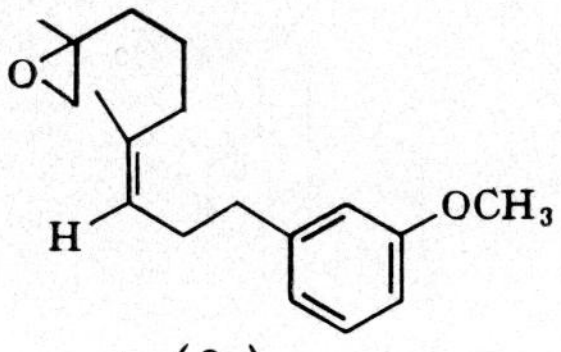

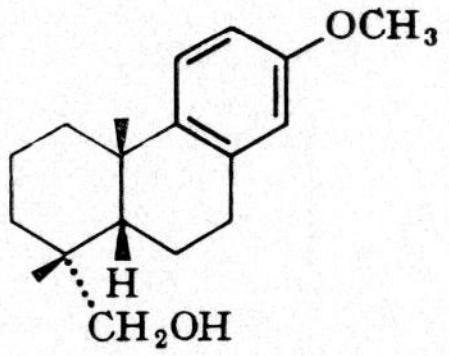

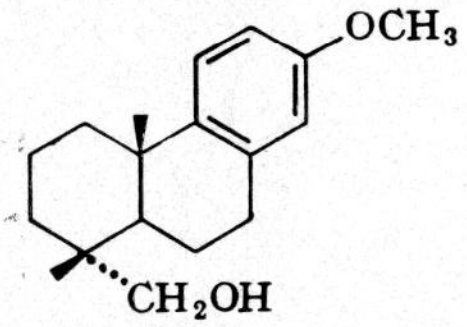

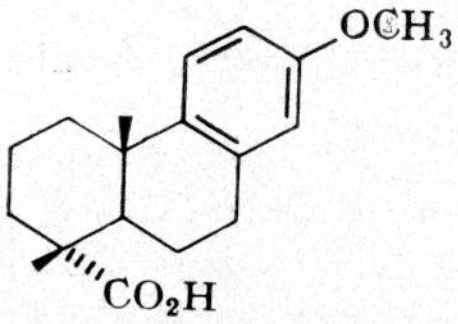

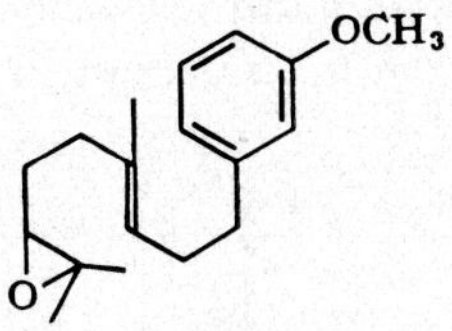

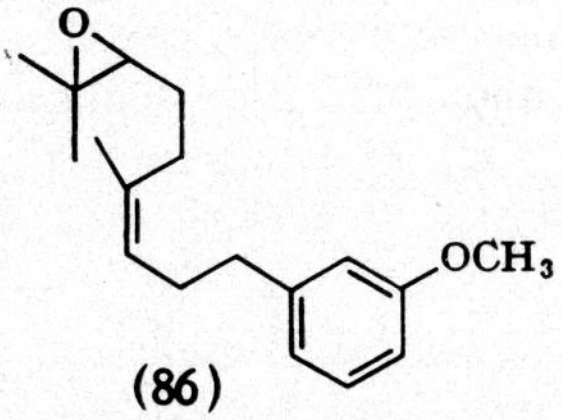

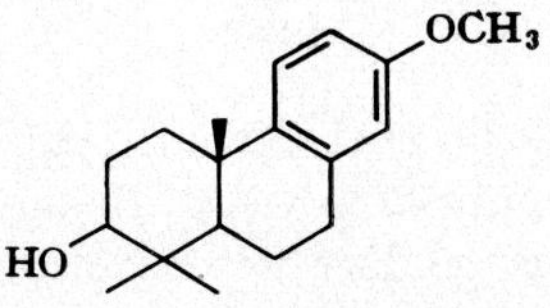

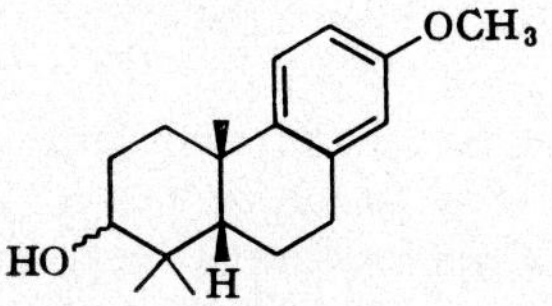

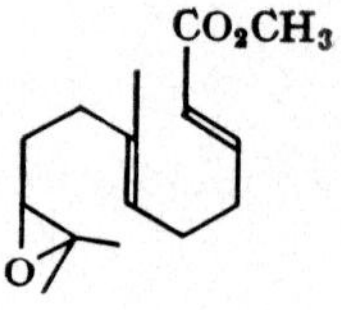

(89) 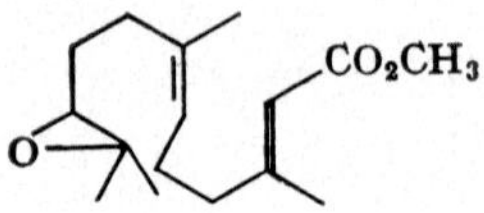   (90)

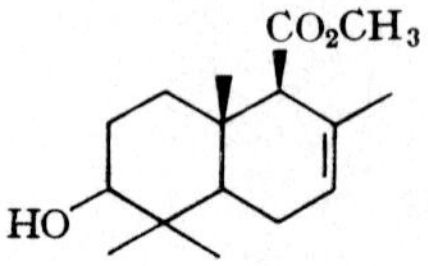

(68) 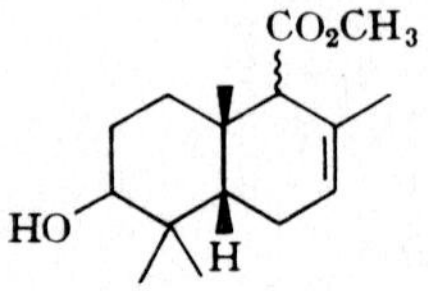   (91)

IV. Arene and Alkyl Sulfonates, Acetals and Allylic Alcohols

The epoxide group of the polyenic compounds described above serves two functions in biogenetic type synthesis. It provides an oxygen function at a characteristic location for a variety of terpenoids and steroids, and it serves as a ready source of the cation needed to initiate cyclization. Other oxygenated functional groups may also serve one or both of these purposes. A number of such systems have been investigated. In particular the reactivity of polyenic alkyl acid arene sulfonates, polyenic acetals and polyenic allylic alcohols have been examined. Much of this work has been reviewed previously (27, 28).

The cyclization of polyenic arene sulfonates occurs in a stereospecific manner but in low yield. For example, the *trans*-p-nitrobenzene sulfonate (92) upon solvolysis in formic acid affords the *trans*-decalol (93) as the

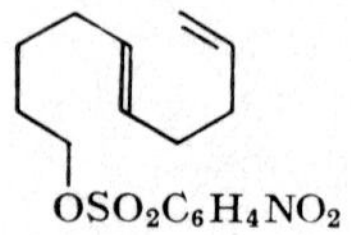

(92) 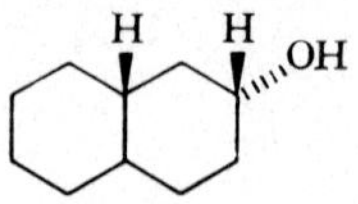   (93)

major bicyclic product (29), but total bicyclic product, all of it possessing *trans* ring fusion geometry, was formed in only 12% yield. In the corresponding *cis* series, (94), all of the bicyclic product was characterized by *cis* ring fusion but was formed in 16% yield only. When the solvolysis-cyclization sequence was applied (33) to a trienic arenesulfonate, (95),

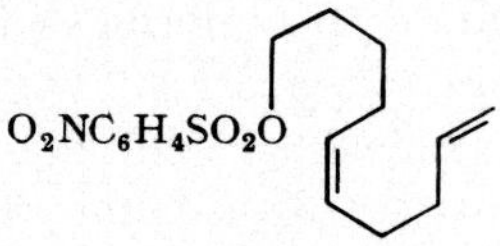

(94)

only 2.8% tricyclic material was obtained. The latter, (96), was, however, the result of stereospecific cyclization.

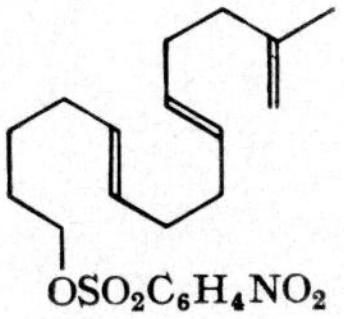

(95) 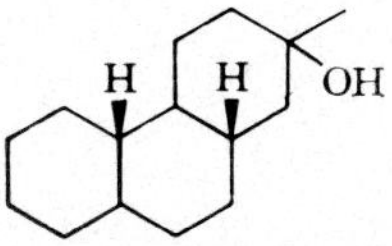   (96)

The cyclization of terpenoid methanesulfonates has been found to be an efficient process. For example, in work directed toward the synthesis of hibaene Herz and co-workers (25) found that mesylate (97) derived from isopimaric acid undergoes cyclization and rearrangement via (98a) to afford (99). Interestingly the diastereoisomeric sulfonate (98b) derived from pimaric acid undergoes the equivalent sequence in what must be a non-concerted manner to yield (100) (26).

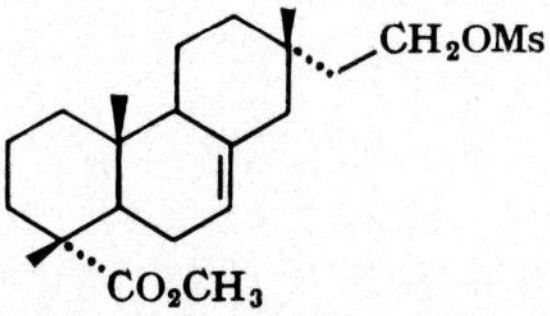

(97)

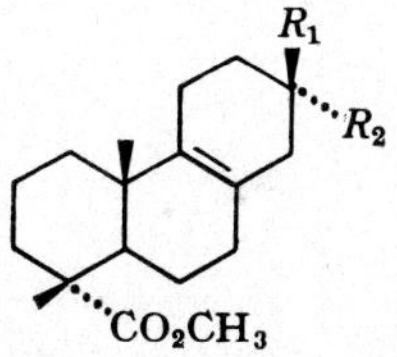

(98a) $R_1 = -CH_3$, $R_2 = -CH_2OMs$
(98b) $R_1 = -CH_2OMs$, $R_2 = -CH_3$

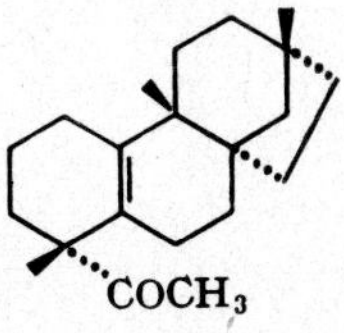

(99) 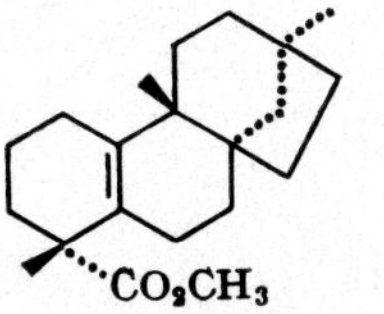   (100)

Acid-catalyzed cyclizations of polyenic actals has proved to be an efficient and stereospecific method for biogenetic-type syntheses of potential steroid and terpenoid substances. For example the trienic acetal (101) (33) upon treatment with stannic chloride in benzene solution underwent cyclization to produce a mixture of products, 87% of which is a monohydric alcohol fraction. The latter was shown to consist principally of the tricyclic alcohols (102) and (103). Removal of the ether

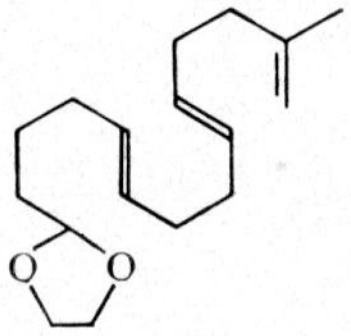

(101) 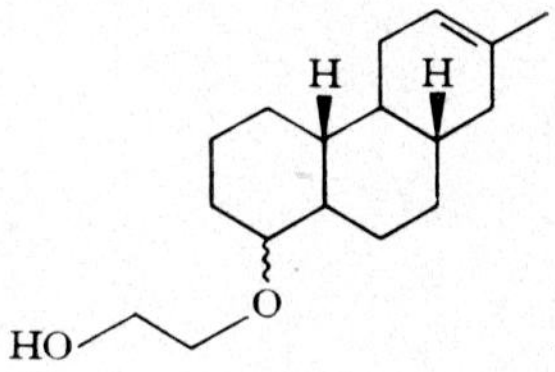 (102) 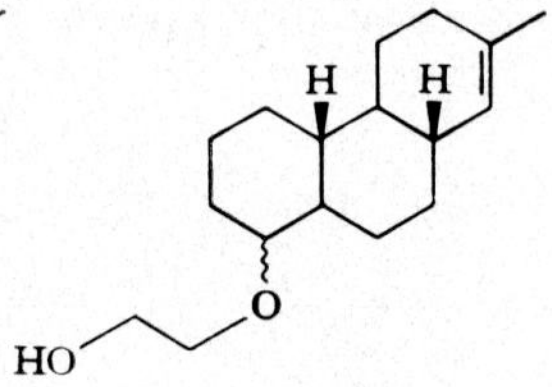 (103)

function by tosylation of the hydroxyl group followed by treatment with zinc and sodium iodide converted the mixture to alcohols. The latter were then oxidized to (104) and (105) with Jones reagent. The stereochemistry of these ketones was demonstrated by comparison with known hydrocarbons following WOLFF-KISHNER reduction.

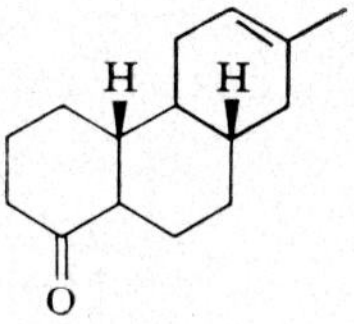

(104) 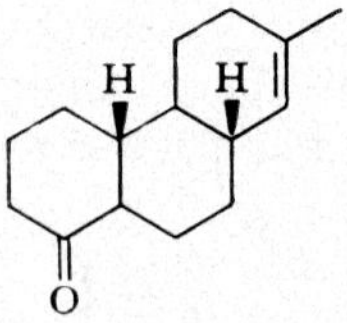 (105)

A more complex example of the cyclization of a polyenic acetal is represented by the conversion of (106) to (107) in which six asymmetric centers are established in a stereospecific manner (37). In contrast to the results of the arene sulfonate cyclizations this latter case produces approximately 30% of tetracyclic product.

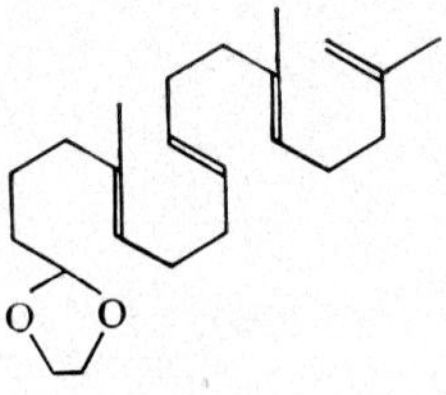

(106) 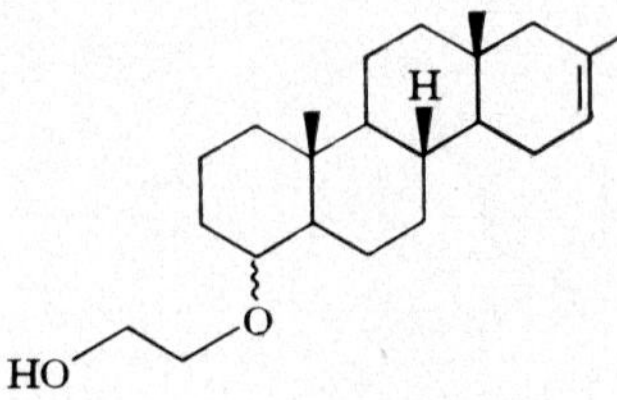 (107)

References, pp. 391—394

The most successful of biogenetic-type synthesis systems developed by JOHNSON has been cyclization of polyenic allylic alcohols. For example, the cyclohexenol (108) is quantitatively transformed by formic acid into

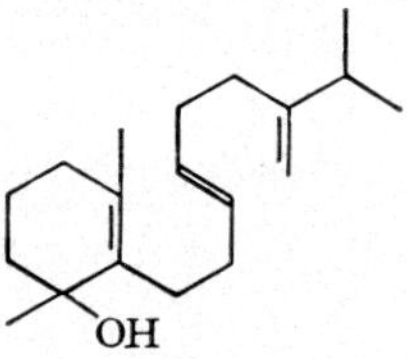

(108)

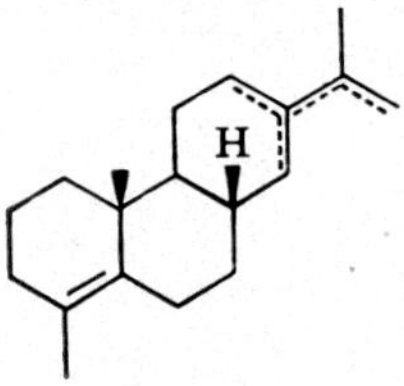

(109)

a mixture of the hydrocarbons (109) and the alcohol (110) (*32*). The stereo- and structural specificity of this cyclization was demonstrated by conversion of the mixture to dl-fichtelite (111) by hydrogenation.

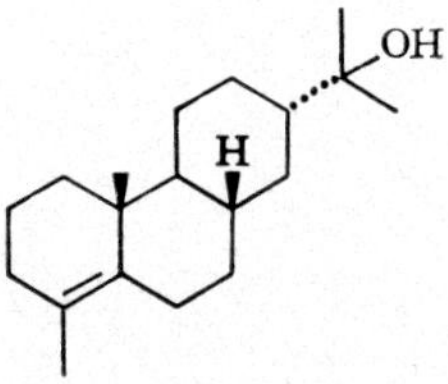

(110)

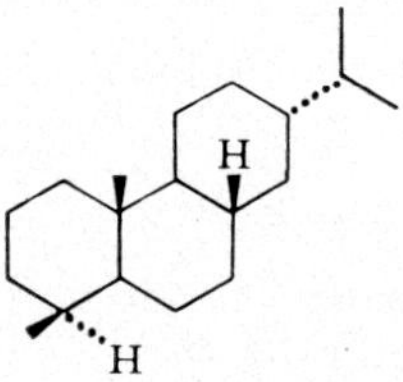

(111)

An even more striking use of the cyclization of an allylic alcohol in synthesis is the conversion of the tetraenol (112) to dl-16,17-dehydroprogesterone (*34, 35*). Treatment of (112) with stannic chloride in nitromethane solution afforded tetracycle (113) in approximately 70% yield. The conversion of the cyclization product to the final steroid was accomplished by osmium tetroxide/lead tetraacetate cleavage of the double bonds of (113) followed by double aldol condensation of the resulting triketoaldehyde. This synthesis of a steroid in contrast to conventional ones established five asymmetric centers in one step; the cyclization. The preparation of (112) requires only two stereoselective reactions, those establishing the *trans* geometry of the acyclic olefinic bonds. Scheme II illustrates the preparation of (112) starting from methallyl vinyl ether. The required *trans* sterochemistry for the two central double bonds of (112) is achieved in one case by $S_N i$ reaction of an allylic acohol with thionyl chloride and in the other by sodium in ammonia reduction of an acetylene, processes found to produce *trans*-trisubstituted olefins.

JOHNSON has also sought an entry into the podocarpene series of diterpenes through use of the tetraenol (114) (*36*). Treatment of this

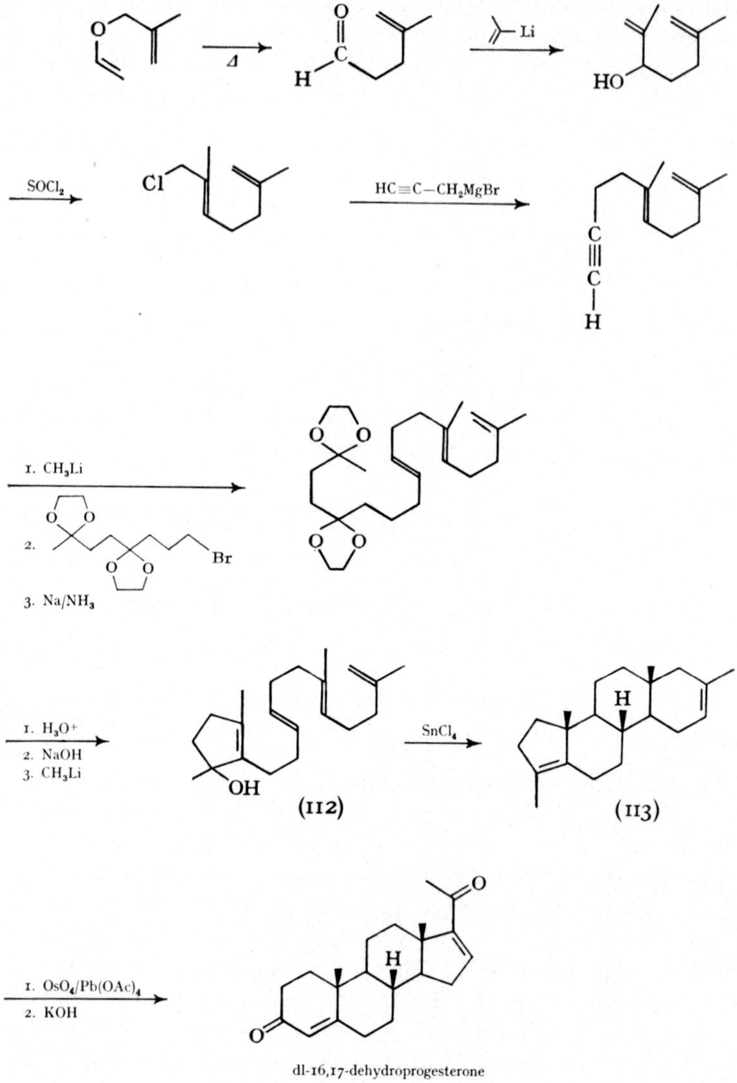

Scheme II

alcohol with trifluoroacetic acid in methylene chloride afforded an over 50% yield of the alcohols (**115a, b**). The structure and stereochemistry of these products was demonstrated by degradation of (**115a**) to (**116**) and (**115b**) to (**117**) and comparison of these to authentic samples obtained from independent syntheses.

References, pp. 391—394

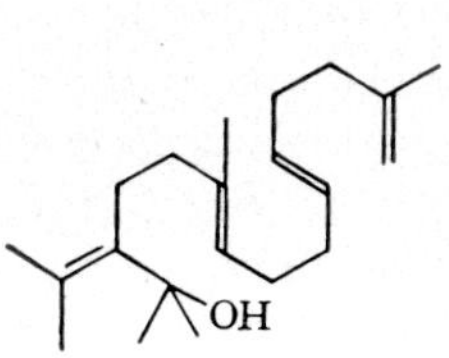

(114) 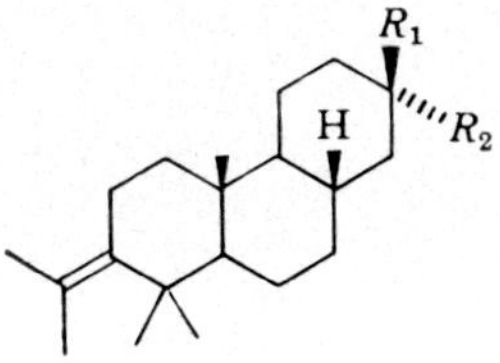 (115)

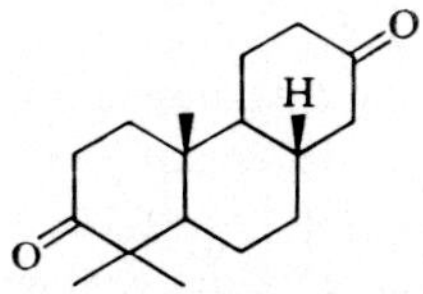

(116) 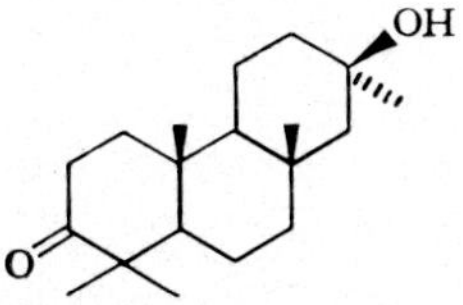 (117)

An allylic alcohol system has been successfully employed in a biogenetic-type step in the synthesis of rosenonolactone (118) (43). Methyl isocupressate (119) on exposure to aqueous acetic-acid-sulfuric acid affords the tricyclic pimaradienes (120) and (121). Further treatment (42) of these with formic acid effects rearrangement to the rosadienes (122) and (123). The reaminder of the synthesis follows non-biogenetic-type steps.

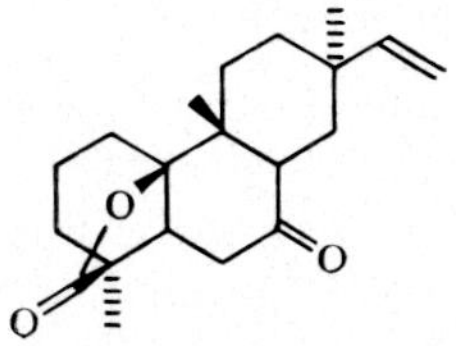

(118) 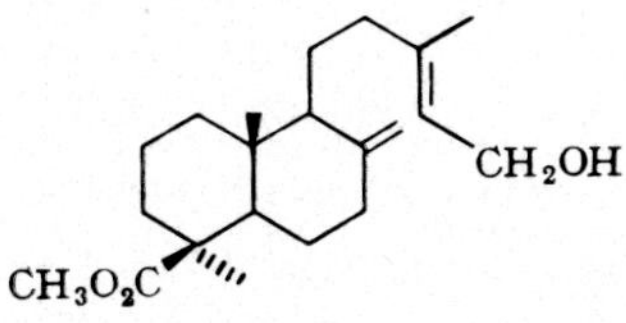 (119)

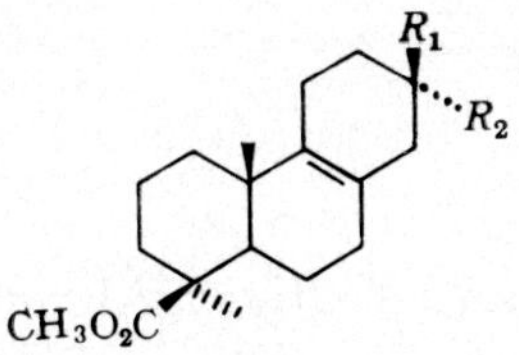

(120) $R_1 = -CH_3$, $R_2 = -CH=CH_2$
(121) $R_1 = -CH=CH_2$, $R_2 = -CH_3$

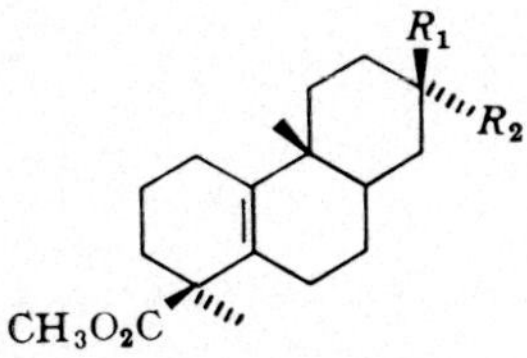

(122) $R_1 = CH_3$, $R_2 = -CH=CH_2$
(123) $R_1 = -CH=CH_2$, $R_2 = -CH_3$

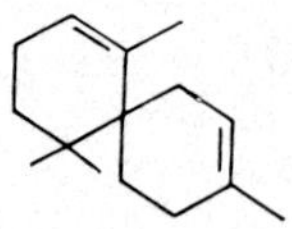

(124)

Several other terpene syntheses are based on cyclization reactions of allylic alcohols. For example, monocyclofarnesol, (17) (*cis* or *trans*) affords α-chamigrene (124) when exposed to iodine in benzene solution (*38*).

In an experiment closely modelled on the actual biosynthetic pathways for both cyclic and acyclic monoterpenoids, WOOD and co-workers (*24*) found that geranyl and neryl diphenyl phosphates (125) and (126) are converted upon standing to a mixture of five terpene hydrocarbons: myrcene (127), *cis*-β-ocimene (128), *trans*-β-ocimene (129), limonene (130) and terpinolene (131). The yield of the cyclic poducts is higher in the case of the neryl compound suggesting that cyclization is an anchimerically assisted process. Cyclization of gernanyl diphenyl phosphate has been suggested to occur only after rearrangement to linaloyl diphenyl phosphate (132).

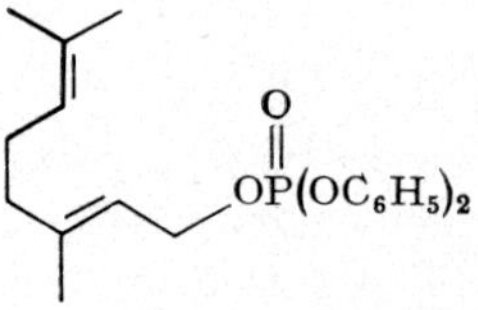

(125) 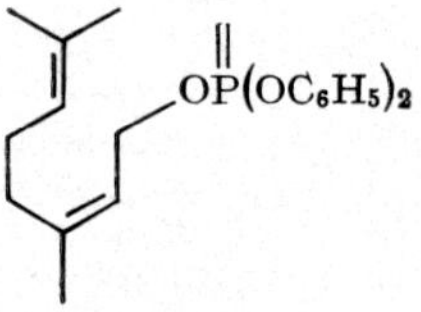   (126)

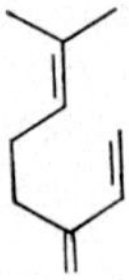

(127) 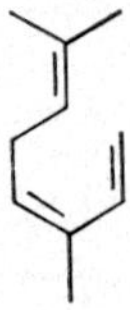   (128) 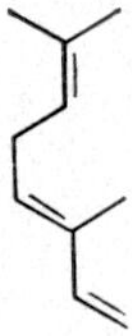   (129) 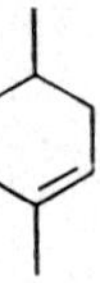   (130) 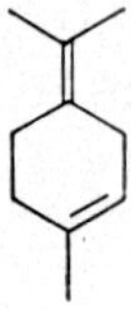   (131)

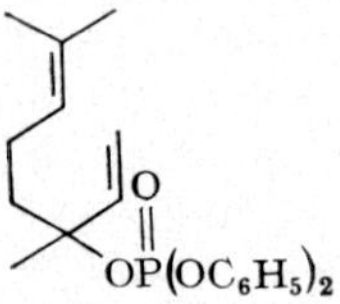

(132)

V. Cyclopropyl Ketones, Enols, and Tertiary Alcohols

An interesting system employing the cyclization reactions of cyclopropyl ketones for the synthesis of terpenoids and steroids has been developed by STORK and co-workers. For example, the unsaturated cyclo-

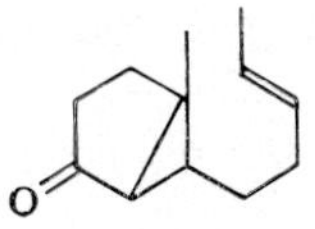

Biogenetic-type Synthesis of Terpenoid Systems

(133) 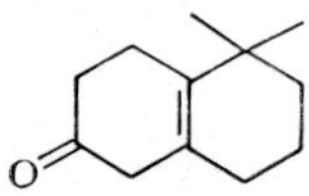 (134) 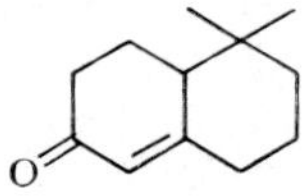 (135) 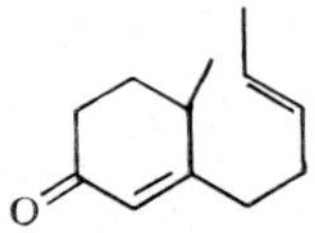 (136)

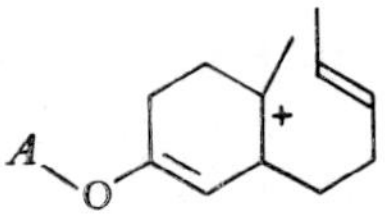

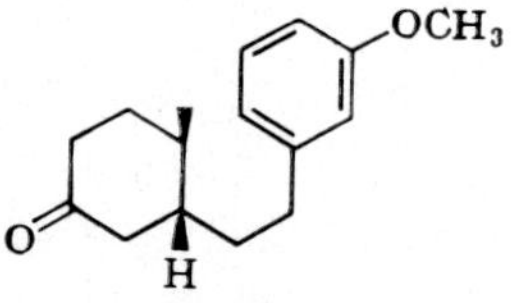

(137) 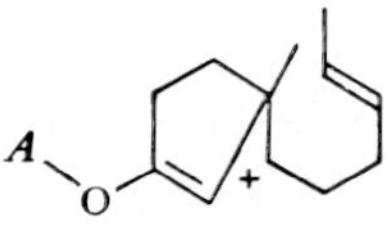 (138)

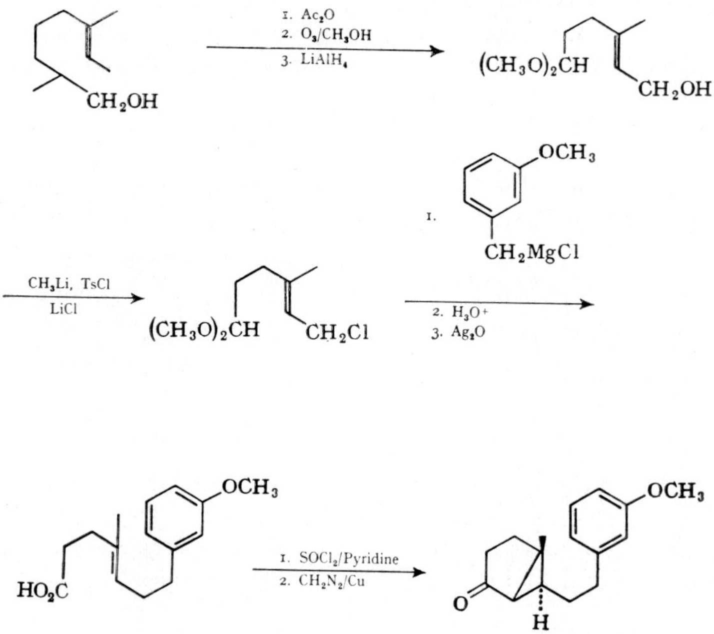

Scheme III

propyl ketone (133) is converted to two bicyclic products, (134) and (135), and the cyclohexenone (136) upon treatment with stannic chloride in benzene (57). Formally these products are all derived from opening of the cyclopropyane ring to a cyclohexyl cation (137) rather than to the alternative cyclopentyl cation (138). The latter might have been expected to be favored because of greater overlap between the orbitals of the bond undergoing cleavage and the carbonyl p-orbitals in the formation of (138) than in the formation of (137) (57).

The formation of rearranged products, (134) and (135), in the cyclization of (133) precludes any conclusions about the stereospecificity of this kind of ring-forming process. To examine this question STORK studied the diastereoisomeric ketones (139) and (140) (53). The ketones were synthesized by internal diazo ketone addition to an olefin (52), a method also used for the preparation of (133). Scheme III illustrates the preparation of (139) from geraniol (54, 56). A similar sequence from nerol affords (140).

The cyclization reactions of both (139) and (140) were shown to be stereospecific. Ketone (139) affords a 5:1 mixture of (141) and (142) both of which have an AB-*trans* fusion, and (140) afford a mixture of the corresponding *cis*-ketones (143) and (144).

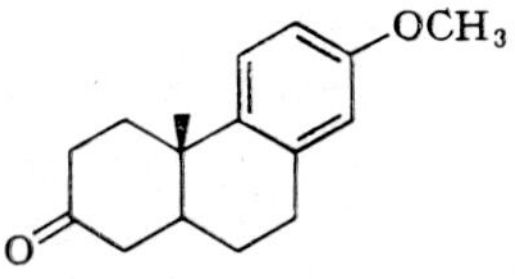

(141)

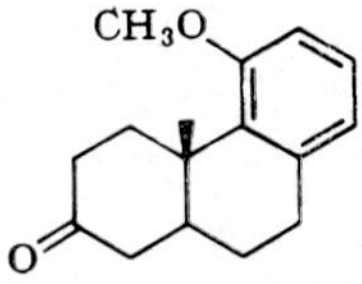

(142)

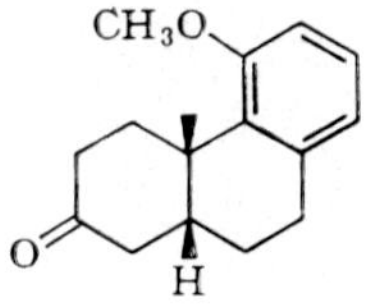

(143)

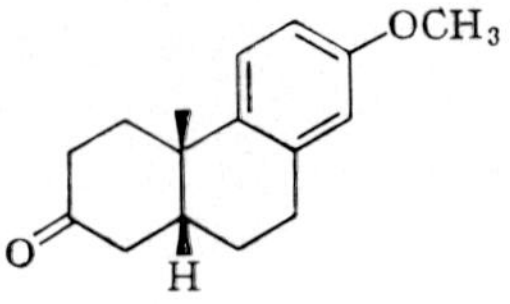

(144)

In acid-catalyzed cyclization reactions termination of the cyclization process often occurs by simple proton loss. Thus the cyclization of an acetal like (106) or an allylic alcohol like (108) leads directly to a polycyclic product with no major intervening rearrangements or side reactions. In a number of cases, however, one or more cation-trapping or rearrangement reactions may occur. For example, as noted earlier the bicyclization of epoxyfarnesyl acetate (46) is interrupted in part by trapping of an intermediate cation to produce (49), and followed in part by rearrangement to (50). Cyclization of epoxide (81) produces not only the tricyclic alcohol (82) but also the bridged ether (145). The cyclization of cyclopropyl ketones also yields products of both rearrangement (eg. (133) to (134)) and cation-trapping. An example of the latter has been found in the acid catalyzed reaction of the *endo* compound (146) (55). Cyclization of (146) with stannic chloride proceeds via the tertiary cation (147) as shown in Scheme IV. Since the cyclization process in this kind of system results in the formation of an *enol* it is not surprising, given the appropriate

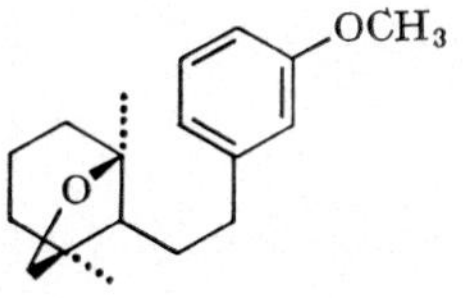

(145) 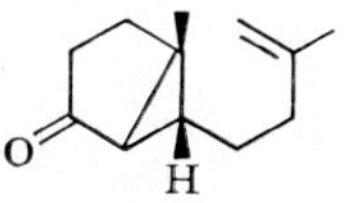   (146)

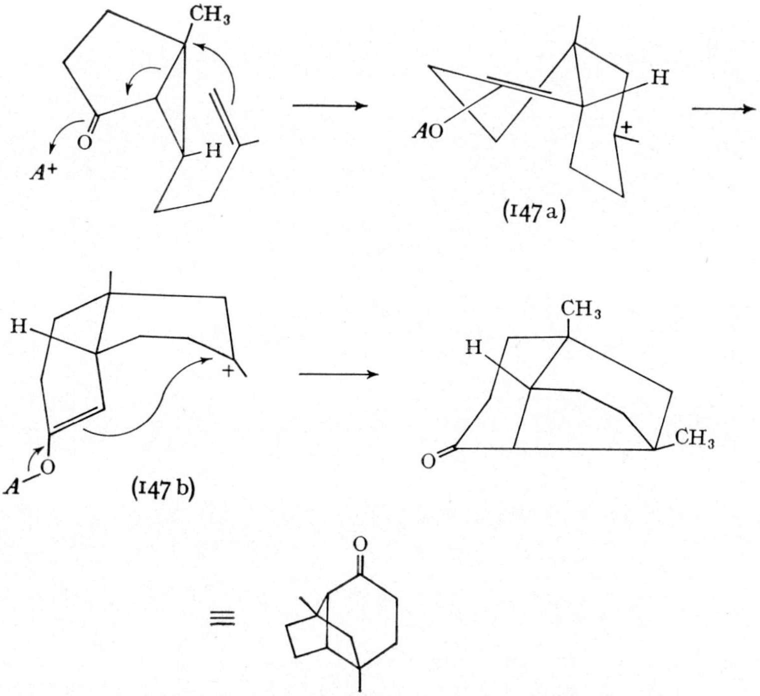

(147 a)

(147 b)

(148)

Scheme IV

stereochemistry, that alkylation of the enol by the carbonium ion should occur producing tricyclic ketone (148).

The alkylation of enols by carbonium ions has been used successfully in several biogenetic-type syntheses. A synthesis of camphor (149) by this method was carried out by MONEY (15). The mixture of enol acetates obtained from dihydrocarvone was separated and isomer (150) was treated with boron fluoride etherate. A 90% yield of camphor is obtained in this cyclization. Interestingly the camphor produced this

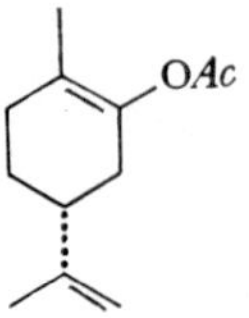

(150) 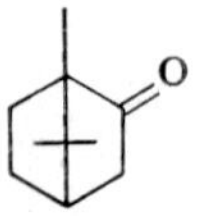 (149) 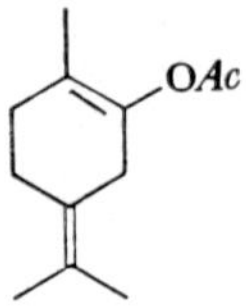 (151)

way is racemic despite the fact that optically active dihydrocarvone was employed as the starting material. One explanation offered for this result is that cyclization may occur only after isomerization of (150) to (151). The latter being symmetric must then yield racemic camphor.

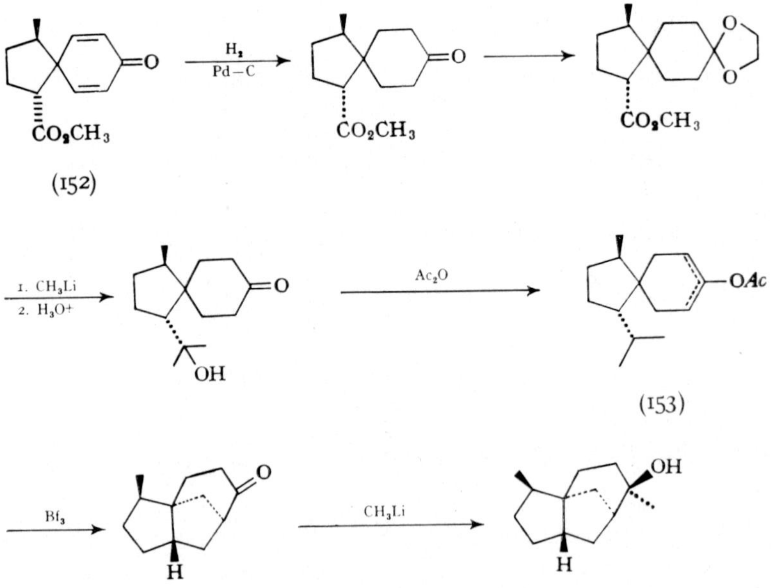

(152)

(153)

(154) (155)

Scheme V

References, pp. 391—394

The alkylation of an enol by a carbonium ion has also been employed in a biogenetic-type synthesis of cedrol by COREY and coworkers (*9*). Dienone (152) was converted to the enol acetates (153) via the steps shown in Scheme V. Treatment of (153) with boron fluoride yielded the tricyclic ketone (154), which upon reaction with methyl lithium afforded cedrol (155).

Several biogenetic-type syntheses of cedrene (156) have also been reported. In one of the COREY syntheses (*9*) diol (157) produces cedrene when exposed to formic acid. Preparation of (157) from (152) was accomplished by catalytic reduction of both double bonds. This was followed by addition of methyl lithium to the ketone and ester functions.

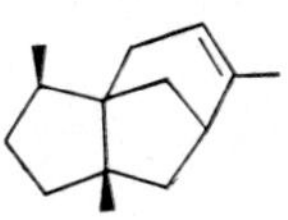

(156)

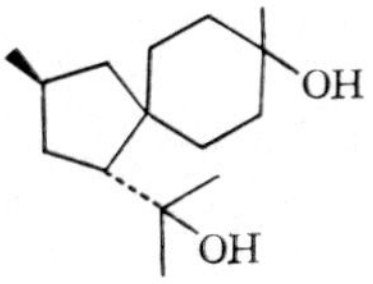
OH
OH

(157)

A similar approach was employed by LAWTON (*11*) who exposed the unsaturated alcohol (158) to formic acid and obtained an approximately 80% yield of cedrene. The preparation of (158) from bromo ester (159) is outlined in Scheme VI.

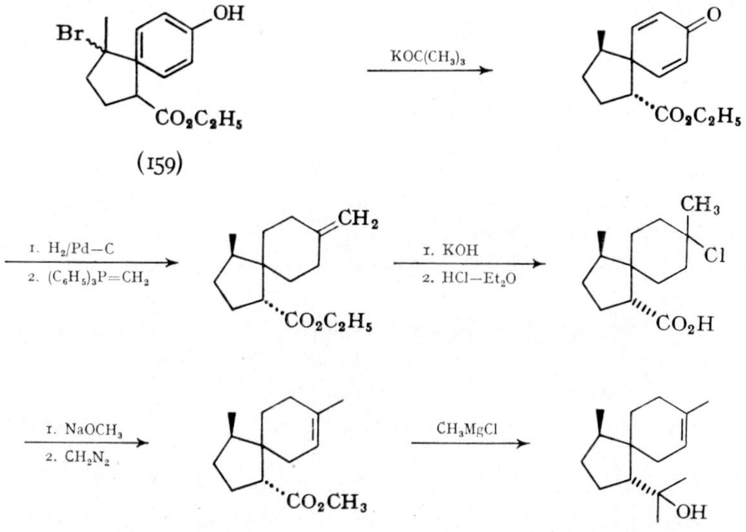
Br
OH
$CO_2C_2H_5$
(159)
$KOC(CH_3)_3$
O
$CO_2C_2H_5$
1. H_2/Pd–C
2. $(C_6H_5)_3P{=}CH_2$
CH_2
$CO_2C_2H_5$
1. KOH
2. $HCl–Et_2O$
CH_3
Cl
CO_2H
1. $NaOCH_3$
2. CH_2N_2
CO_2CH_3
CH_3MgCl
OH
(158)

Scheme VI

VI. Carbonium Ion Catalyzed Cyclization

An interesting variation of the usual methods by which polyene cyclization is initiated has been investigated by KUCHEROV and co-workers (*40*, *41*). While most cyclizations have been catalyzed by protonic acids or the more common Lewis acids, these workers have employed carbonium ion catalysts. Thus for example, treatment of methyl geranate with pivaloyl fluoroborate affords the ketoester (**160**) in 56% yield (*40*). Methoxymethyl fluoroborate also serves to initiate cyclization produces the *trans*-fused ethers (**161**) and (**162**) from ethyl *cis*, *trans*-farnesate (*41*). This cyclization like the proton catalyzed closure of farnesic esters is a two stage process.

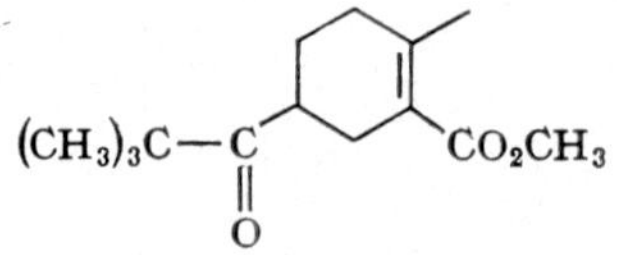

(160) 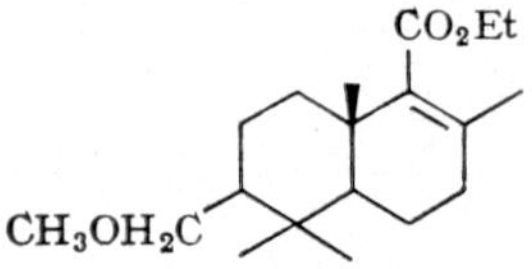   (161)

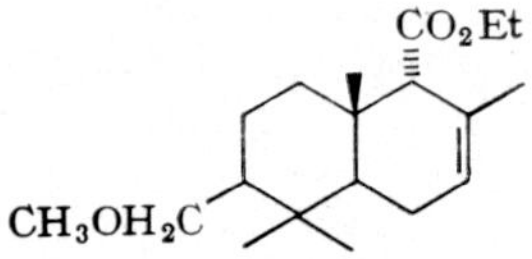

(162)

VII. Radical Cyclization

A contrast to the usual acid catalyzed ionic cyclization of polyenic systems and the assumption (*4*) that this type of process is truly related to actual biosynthetic pathways is found in the work of Breslow. The latter has investigated the cyclization reactions of polyenes in the presence of radicals. For example, treatment of geranyl acetate with benzoyl peroxide in the presence of cuprous chloride and cupric benzoate produces the cyclized product (**163**) in yields of 55–60% (*5*). The process has also been extended to the farnesol series. In particular farnesyl acetate affords 20–30% of (**164**) under similar conditions (*6*).

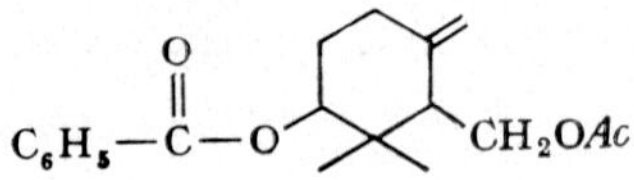

(163) 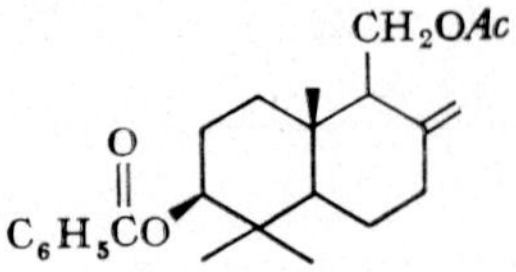   (164)

References

1. AEBI, A., D. H. R. BARTON and A. S. LINDSEY: Sesquiterpenoids. Part. III The Stereochemistry of Caryophyllene. J. Chem. Soc. (London) **1953**, 3124.
2. BARTON, D. H. R., O. C. BROCKMAN and P. DE MAYO: Sesquiterpenoids. Part XII. Further investigations on the Chemistry of Pyrethrosin. J. Chem. Soc. (London) **1960**, 2263.
3. BARTON, D. H. R. and K. H. OVERTON: Triterpenoids. Part XX. The Constitution and Stereochemistry of a Novel Tetracyclic Triterpenoid. J. Chem. Soc. (London) **1957**, 2639.
4. BRESLOW, R., E. BARRETT and E. MOHACSI: Free Radical Additions to Squalene. Tetrahedron Lett. **1962**, 1207.
5. BRESLOW, R., J. T. GROVES and S. S. OLIN: A Novel Terpene Cyclization: Tetrahedron Lett. **1966**, 4717.
6. BRESLOW, R., S. S. OLIN and J. T. GROVES: Oxidative Cyclization of Farnesyl Acetate by a Free Radical Path. Tetrahedron Lett. **1968**, 1837.
7. CHAWLA, A. and S. DEV: A New Class of Triterpenoids from Allanthus Malabarica DC. Derivatives of Malabaracane. Tetrahedron Lett. **1967**, 4837.
8. CLAYTON, R. B.: Biosynthesis of Sterols, Steroids and Terpenoids. Part I. Biogenesis of Cholesterol and the Fundamental Steps in Terpenoid Biosynthesis. Quart. Rev. (London) **19**, 168 (1965).
9. COREY, E. J., N. N. GIROBRA and C. T. MATHEW: Total Synthesis of *dl*-Cedrene and *dl*-Cedrol. J. Amer. Chem. Soc. **91**, 1557 (1969).
10. CORNFORTH, J. W.: Olefin Alkylation in Biosynthesis. Angew. Chem. Internat. Edit. **7**, 903 (1968).
11. CRANDALL, T. G. and R. G. LAWTON: A Biogenetic-Type Synthesis of Cedrene. J. Amer. Chem. Soc. **91**, 2127 (1969).
12. DITTMAN, W. und F. STURZENHOFECKER: Synthese von *racem.* $(1S:45:9R)$ $(1R:4R:9S)$-*cis*-Decalindiol-(1.4). Liebigs Ann. Chem. **688**, 57 (1965).
13. ESCHENMOSER, A., D. FELIX, M. GUT, J. MEIER and P. STADLER: Some Aspects of Acid Catalyzed Cyclizations of Terpenoid Polyenes. Ciba Foundation Symposium on the Biosynthesis of Terpenes and Sterols, p. 217. G. E. W. WOLSTENHOLME and M. O'CONNER, Ed. London: J. and A. Churchill, Ltd. 1959.
14. ESCHENMOSER, A., L. RUZICKA, O. JEGER und D. ARIGONI: Zur Kenntnis der Terpene. 190. Mitt. Eine stereochemische Interpretation der biogenetischen Isoprenregel bei den Triterpen. Helv. Chim. Acta **38**, 1268 (1955).
15. FAIRLIE, J. C., G. L. HODGSON and T. MONEY: Biogenetic-Type Synthesis of $(\pm)$-Camphor. Chem. Comm. **1969**, 1196.
16. FAZAKERLEY, H., T. G. HALSALL and E. R. H. JONES: The Chemistry of Triterpenes and Related Compounds. Part XXXIV. The Structure of Hydroxyhopanone. J. Chem. Soc. (London) **1959**, 1877.
17. GOLDSMITH, D. J.: The Cyclization of Epoxyolefins: The Reaction of Geraniolene Monoepoxide with Boron Fluoride Etherate. J. Amer. Chem. Soc. **84**, 3913 (1962).
18. GOLDSMITH, D. J.: Unpublished observations.
19. GOLDSMITH, D. J. and B. C. CLARK, Jr.: The Cyclization of Epoxyolefins. V. A Model for the Synthesis of the Diterpene Acids. Tetrahedron Lett. **1967**, 1215.
20. GOLDSMITH, D. J., B. C. CLARK, Jr. and R. C. JOINES: The Cyclization of Epoxyolefins. III. The Formation of Bridged Systems: Tetrahedron Lett. **1966**, 1149.
21. GOLDSMITH, D. J., B. C. CLARK, Jr. and R. C. JOINES: The Cyclization of Epoxyolefins. IV. Bicyclic Ethers from a Citronellal Derivative. Tetrahedron Lett. **1967**, 1211.
22. GOLDSMITH, D. J. and J. E. EVANS: Unpublished observations.

23. GOLDSMITH, D. J. and C. F. PHILLIPS: The Structural and Stereochemical Course of *in vitro* Epoxy Olefin Cyclization. Diterpenoid Intermediates. J. Amer. Chem. Soc. **91**, 5862 (1969).
24. HALEY, R. C., J. A. MILLER and H. C. S. WOOD: Phosphate Esters. Part II. The Formation of Monoterpene Hydrocarbons from Geranyl and Neryl Biphenyl Phosphates. J. Chem. Soc. (London) (C) **1969**, 264.
25. HERZ, W., D. MELCHIOR, R. N. MIRRINGTON and P. S. PAUWELS: Resin Acids. II. Cationic Cyclization of Isopimaric Acid Derivatives. Partial Synthesis of Isohibaene. J. Org. Chem. **30**, 1873 (1965).
26. HERZ, W., A. K. PINDER and R. N. MIRRINGTON: Resin Acids. IX. Cationic Cyclization of Pimaric Acid Derivatives. Partial Synthesis of (—)-Hibaene. J. Org. Chem. **31**, 2257 (1966).
27. JOHNSON, W. S.: Nonenzymic Biogenetic-like Olefinic Cyclizations. Accounts Chem. Res. **1**, 1 (1968).
28. JOHNSON, W. S.: Recent Studies on the Synthesis of Homocyclic Systems. Pure and Applied Chem. **7**, 317 (1963).
29. JOHNSON, W. S., D. M. BAILEY, R. OWYANG, R. A. BELL, B. JAQUES and J. K. CRANDALL: Cationic Cyclizations Involving Olefinic Bonds. II. Solvolysis of 5-Hexenyl and *trans*-5,9-Decadienyl p-Nitrobenzenesulfonates. J. Amer. Chem. Soc. **86**, 1959 (1964).
30. JOHNSON, W. S., A. VAN DER GEN and J. J. SWOBODA: New Structural and Stereochemical Aspects of the Cyclization of Olefinic Acetals. J. Amer. chem. Soc. **89**, 170 (1967).
31. JOHNSON, W. S., S. L. GRAY, J. K. CRANDALL and D. M. BAILEY: Cationic Cyclizations Involving Olefinic Bonds. III. On the Mechanism of Formation of *trans*-Fused Rings. J. Amer. Chem. Soc. **86**, 1966 (1964).
32. JOHNSON, W. S., N. P. JENSEN and J. HOOZ: An Efficient Stereospecific Polyolefinic Cyclization. Total Synthesis of d,l-Fichtelite. J. Amer. Chem. Soc. **88**, 3859 (1966).
33. JOHNSON, W. S. and R. N. KINNEL: Stereospecific Tricyclization of a Polyolefinic Acetal. J. Amer. Chem. Soc. **88**, 3861 (1966).
34. JOHNSON, W. S., T. LI, C. A. HARBERT, W. D. BARTLETT: T. H. HERRIN, B. STASKUN and D. H. RICH: Further Developments in the Nonenzymic Biogenetic-like Steroid Synthesis. J. Amer. Chem. Soc. **92**, 4461 (1970).
35. JOHNSON, W. S., M. F. SEMMELHACK, M. U. S. SULTANBAWA and L. A. DOLAK: A New Approach to Steroid Total Synthesis. A Nonenzymic Biogenetic-Like Olefinic Cyclization Involving the Stereospecific Formation of Five Asymmetric Centers. J. Amer. Chem. Soc. **90**, 2994 (1968).
36. JOHNSON, W. S. and T. K. SCHAAF: Entry into the Podocarpene Series through a Biogenetic-like Stereoselective Olefin Cyclization. Chem. Comm. **1969**, 611.
37. JOHNSON, W. S., K. WIEDHAUP, S. F. BRADY and G. L. OLSON: The Nonenzymic, Biogenetic-Like Cyclization of a Tetraenic Acetal. J. Amer. Chem. Soc. **90**, 5277 (1968).
38. KATO, T., S. KANNO and Y. KITAHARA: Cyclization of Polyenes. V. Synthesis of α-Chamigrene by the Cyclization of *cis*- and *trans*-Monocyclofarnesols. Tetrahedron **26**, 4287 (1970).
39. KATO, T., M. TANEMURA, T. SUZUKI and Y. KITAHARA: Biogenetic-type Synthesis of α- and β-Levantenolides. Chem. Comm. **1970**, 28.
40. KRIMER, M. Z., V. A. SMIT, A. V. SEMENOJSKII, V. S. BOGDANOV and V. F. KUCHEROV: Cyclization of Isoprenoid Compounds. Communication 20. Cyclization of the Esters of Geranic Acid under the Influence of Cationic Initiators. Izvestia Akad. Nauk (USSR). Ser. Khim. **1968**, 866 (English Translation).

41. KRIMER, M. Z., V. A. SMIT, A. V. SEMENOVSKII and V. F. KUCHEROV: Cyclization of Isoprenoid Compounds. Communication 21. Cyclization of Farnesylic Esters Under the Action of Cationoid Inutiators. Izvestia Akad. Nauk (USSR) Ser. Khim. **1968**, 1352 (English translation).
42. MCCREADIE, T. and K. H. OVERTON: The Conversion of Labdadienols into Pimara- and Rosadienes. J. Chem. Soc. (C) (London) **1971**, 312.
43. MCCREADIE, T., K. H. OVERTON and A. J. ALLISON: A Synthesis of Rosenonolactone and Deoxyrosenonolactone. J. Chem. Soc. (C) (London) **1971**, 317.
44. SENECA, L'ANNAEUS: Epistolae **65**, 3 (circa 50 AD).
45. SHARPLESS, K. B.: d,l-Malabaricanediol. The First Cyclic Natural Product Derived from Squalene in a Nonenzymic Process. J. Amer. Chem. Soc. **92**, 6999 (1970).
46. SHARPLESS, K. B.: Synthesis of *erythro*-18,19-Dihydroxysqualene-2,3-Oxide and other Internally Oxidized Squalene Derivatives. Chem. Comm. **1970**, 1459.
47. SHARPLESS, K. B. and E. E. VAN TAMELEN: Terpene Terminal Epoxides. Skeletal Rearrangement Accompanying Bicyclization of Squalene 2,3-Oxide. J. Amer. Chem. Soc. **91**, 1848 (1969).
48. SIMONSEN, J. L. and L. N. OWEN: The Terpenes. Vol. I, p. 107 London: Cambridge Univ. Press. 1947.
49. SMIT, V. A., A. V. SEMENOVSKII, B. A. RUDENKO and V. F. KUCHEROV: Cyclization of Isoprenoid Compounds. Communication 8. Mechanism of the Stereospecific Cyclization of Geranylacetone. Izvestiya Akad. Nauk (USSR) Ser. Khim. **1963**, 1782 (English translation).
50. STADLER, P. A., A. ESCHENMOSER, H. SCHINZ and G. STORK: Untersuchungen über den sterischen Verlauf saurekatalysierter Cyclizationen bei terpenoiden Polyenverbindungen. 3. Mitt. Zur Stereochemie der Bicyclofarnesylsäuren. Helv. Chim. Acta **40**, 2191 (1957).
51. STORK, G. and A. W. BURGSTAHLER: The Stereochemistry of polyene Cyclization. J. Amer. Chem. Soc. **77**, 5086 (1955).
52. STORK, G. and J. FICINI: Intramolecular Cyclization of Unsaturated Diazoketones. J. Amer. Chem. Soc. **83**, 4678 (1961).
53. STORK, G. and M. GREGSON: Aryl Participation in Concerted Cyclization of Cyclopropyl Ketones. J. Amer. Chem. Soc. **91**, 2373 (1969).
54. STORK, G., M. GREGSON and P. A. GRIECO: A Convienient Route to *cis* and *trans*-Trisubstituted Olefins from Geraniol and Nerol. Tetrahedron Lett. **1969**, 1391.
55. STORK, G. and P. A. GRIECO: Olefin Participation in the Acid Catalyzed Opening of Acylcyclopropanes. III. Formation of the Bicyclo 2.2.1 heptane System. J. Amer. Chem. Soc. **91**, 2407 (1969).
56. STORK, G., P. A. GRIECO and M. GREGSON: Synthesis of Allylic Halides and 1,5-Dienes from Allylic Alcohols. Tetrahedron Lett. **1969**, 1393.
57. STORK, G. and M. MARX: Six-Membered Rings *via* Olefin Participation in the Opening of Acylcyclopropanones. J. Amer. Chem. Soc. **91**, 2371 (1969).
58. TANEMURA, M., T. SUZUKI, T. KATO and Y. KITAHARA: Synthesis of Levantenolides from Acyclic Progenitor. Tetrahedron Lett. **1970**, 1463.
59. TCHEN, T. T. and K. BLOCH: On the Mechanism of Cyclization of Squalene. J. Amer. Chem. Soc. **78**, 1516 (1956).
60. ULERY, H. E. and J. H. RICHARDS: The Acid-Catalyzed Cyclization of Acyclic Dienes. J. Amer. Chem. Soc. **86**, 3113 (1964).
61. VAN TAMELEN, E. E.: Biogenetic-type Synthesis. Fortschr. Chem. Organ. Naturstoffe **19**, 1961.

62. VAN TAMELEN, E. E.: Bioorganic Chemistry: Sterols and Acyclic Terpene Terminal Epoxides. Accounts Chem. Res. 1, 111 (1968).
63. VAN TAMELEN, E. E. and R. M. COATES: Biogenetic-type Synthesis of (±)-Farnesiferol A and (±)-Farnesiferol C. Chem. Comm. 1966, 413.
64. VAN TAMELEN, E. E. and T. J. CURPHEY: The Selective *in vitro* Oxidation of the Terminal Double Bonds in Squalene. Tetrahedron Lett. 1962, 121.
65. VAN TAMELEN, E. E. and J. P. MCCORMICK: Terpene Terminal Epoxides. Mechanistic Aspects of Conversion to the Bicyclic Level. J. Amer. Chem. Soc. 91, 1847 (1969).
66. VAN TAMELEN, E. E., G. M. MILNE, M. I. SUFFNESS, M. C. RUDLER, R. J. ANDERSON and R. S. ACHINI: Biogenetic-Type Synthesis of the Isoeuphenol System. J. Amer. Chem. Soc. 92, 7202 (1970).
67. VAN TAMELEN, E. E. and J. W. MURPHY: Formation of the Lanosterol System through Biogenetic-type Cyclization. J. Amer. Chem. Soc. 92, 7204 (1970).
68. VAN TAMELEN, E. E. and R. G. NADEAU: Laboratory Cyclization of Geranylgeranyl Acetate Terminal Epoxide. J. Amer. Chem. Soc. 89, 176 (1967).
69. VAN TAMELEN, E. E., M. A. SCHWARTZ, E. J. HESSLER and A STORNI: Biogenetic-type Oxidation-Cyclization in the Total Synthesis of Triterpenoid Systems. Chem. Comm. 1966, 409.
70. VAN TAMELEN, E. E., A. STORNI, E. J. HESSLER and M. SCHWARTZ: The Biogenetically Patterned *in vitro* Oxidation-Cyclization of Farnesyl Acetate. J. Amer. Chem. Soc. 85, 3295 (1963).
71. VAN TAMELEN, E. E., J. D. WILLETT, R. B. CLAYTON and K. E. LORD: Enzymic Conversion of Squalene 2,3-Oxide to Lanosterol and Cholesterol. J. Amer. Chem. Soc. 88, 4752 (1966).
72. VAN TAMELEN, E. E., J. WILLETT, M. SCHWARTZ and R. NADEAU: Nonenzymic Laboratory Cyclization of Squalene 2,3-Oxide. J. Amer. Chem. Soc. 88, 5937 (1966).

(Received, April 21, 1971)

The Biosynthesis of the Diterpenes

By J. R. HANSON, Sussex

Contents

	Page
I. Introduction	395
II. The Biogenesis of the Diterpenes	396
1. The Bicyclic Diterpenes	397
2. The Tricyclic Diterpenes	399
3. The Tetracyclic Diterpenes	401
4. The Macrocyclic Diterpenes	403
III. Biosynthetic Evidence	403
1. The Bi- and Tricyclic Diterpenes	403
2. The Tetracyclic Diterpenes	406
IV. Conclusion	412
References	413

I. Introduction

The diterpenes contain a variety of structural types and functional groups which have provided a rich source of biogenetic speculation. Over the last decade biosynthetic evidence has been accumulating to support and modify some of these speculations. It is our purpose to review some of this evidence in this chapter. First we shall discuss current biogenetic theories and secondly we shall describe the present state of experimental work.

Diterpenoid substances exhibit a range of biological activity, from the bitter principles through antibiotics and tumour inhibitors to the gibberellin plant growth hormones. They represent the major constituents of a number of plant resins, some of which are of commercial importance and which may function in the plant as inhibitors of dehydration and microbial attack. Some transformation products of sclareol and manool find application in the perfumery industry. The chemistry of the diterpenes has been the subject of various reviews (*27, 31, 32, 38, 44*). The application of physical methods, particularly N. M. R. spectroscopy, to the study of natural products has led to a burgeoning of the number of known diterpenes with a consequent interest in the finer details of biogenetic sequences. Indeed this progression from rather general views to the discussion of particular compounds is a characteristic of the development of this area during the last decade. This has been facilitated as a number of fungal systems have become accessible to biosynthetic studies.

II. The Biogenesis of the Diterpenes

Much of the earlier speculation arising from the study of the structural relationship of natural products is summarized in the Biogenetic Isoprene Rule as proposed by RUZICKA (56). Application of this rule to diterpenoid substances led to the rationalization of the major structural types in terms of the stepwise cyclization of a geranylgeraniol pyrophosphate (1) precursor. The major mode of cyclization is initiated by protonation of a double bond and involves the formation of perhydronaphthalene (2) and perhydrophenanthrene (3) derivatives. This mode of cyclization is charac-

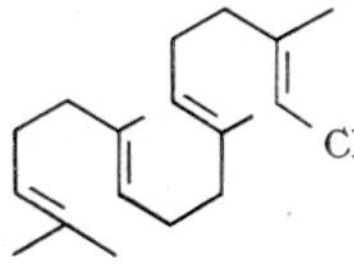

Chart 1

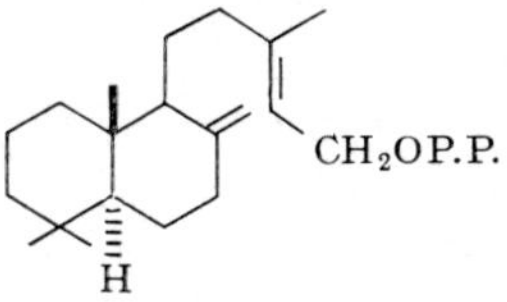

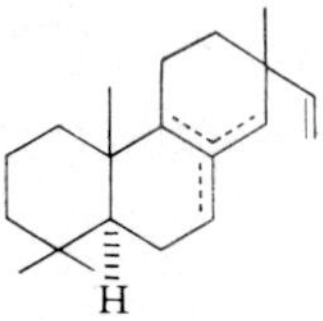

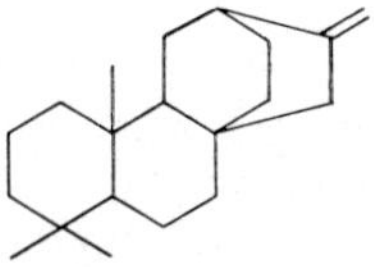

References, pp. 413—416

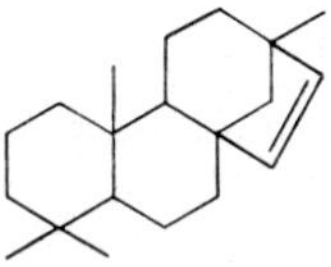

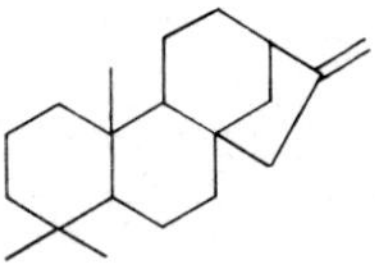

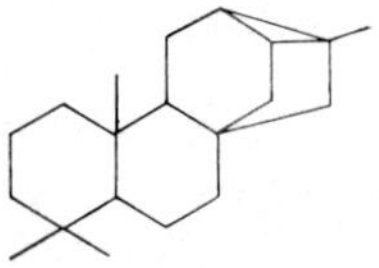

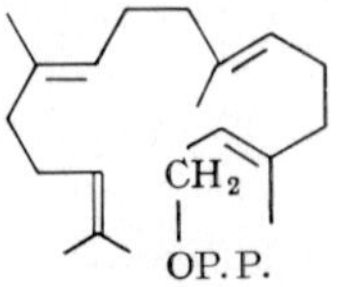

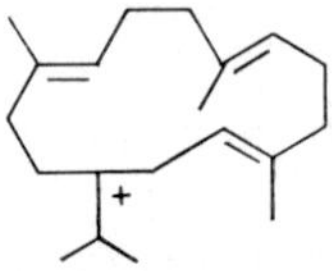

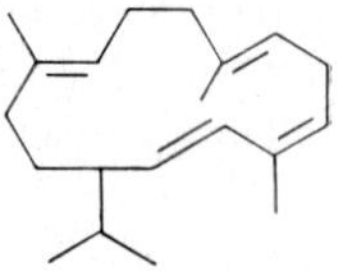

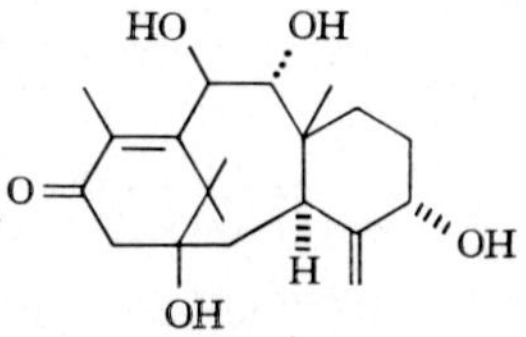

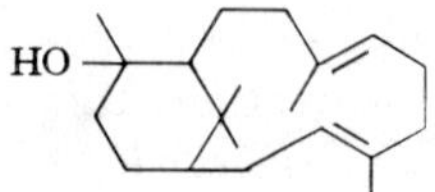

teristic of the higher terpenes and through squalene epoxide, the triterpenes and steroids. There is an alternative mode of cyclization in which the terminal pyrophosphate acts as a leaving group and attacks the distal double bond at the other end of the C_{20} chain. This leads to a series of macrocyclic compounds. This mode of cyclization is characteristic of the lower rather than the higher terpenes and is commonly found amongst the mono- and sesquiterpenes. These overall patterns are summarized in Chart 1.

1. The Bicyclic Diterpenes

A number of discrete phases are involved in the major mode of cyclization. Initially geranylgeraniol pyrophosphate (1) cyclizes to form a bicyclic labdadienol pyrophosphate (2). Immediately one of the first characteristics of the diterpenoid substances becomes apparent; the formation of normal and antipodal ring junctions. The constant absolute stereochemistry which is a feature of the triterpenes and steroids, is not observed in the diterpenes. Examples of both series are quite widespread and there are even reports of their co-occurrence in the same species (e. g. 9). Thus *Agathis australis* has been found (5) to contain both (−)-kaurene (*ent*-kaurene) as well as compounds that possess the normal steroid-like 10β-methyl stereochemistry, such as agathic acid.

The relative stereochemistry of the diterpenoid substances at carbon atoms 5, 8, 9, and 10 is derived (57) from a concerted *trans-anti-trans* addition to the double bonds of the all *trans* acyclic geranylgeraniol pyrophosphate when it is folded in the most stable all-chair conformation. This concept led to the clarification of a number of diterpenoid structures. The cyclization of geranylgeraniol pyrophosphate relies on the relative disposition of the double bonds and does not disturb the prochiral nature of the hydrogen atoms attatched to centres not involved in the cyclization. Consequently information about the stereochemistry of various processes in diterpenoid biosynthesis can be obtained by using information obtained from steroid biosynthesis and in particular using stereospecifically labelled mevalonates.

Subsequent modification of the labdadienol pyrophosphate (2) can lead on the one hand to compounds related to labdanolic acid (14) and on the other hand to compounds of the sclareol (13) series. The rearrangement to compounds of the sclareol series, which parallels the monoterpene geraniol-linalool isomerism, does not always lead to a discrete C(13) stereochemistry. Thus sclareol and 13-*epi*sclareol co-occur (51) in the clary sage, *Salvia sclarea*, and both C(13) epimeric manoyl oxides occur together in the oleorosin of the Norwegian spruce, *Picea abies* (43). Saturation of the labdadiene at C(13) also leads to epimers at this centre as in labdanolic acid and eperuic acid (49). Furan formation between

$C(15)$ and $C(16)$ is a common feature and a sequence of oxygenation can be seen in the sciadin series (59) from *Sciadopitys verticillata*. The isolation (39) of premarrubiin (15) from *Marrubium vulgare* and of a similar compound from *Solidago canadensis* has led to some interesting suggestions regarding the origin of the furan ring of marrubiin (16) and its relationship with the $C(9)$ hydroxyl group. Marrubiin itself is now regarded as an artefact of isolation.

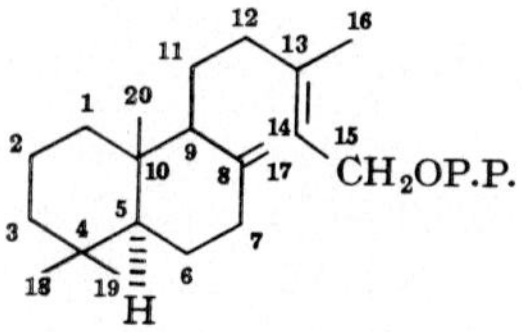

Chart 2

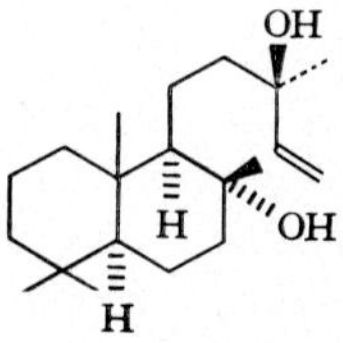

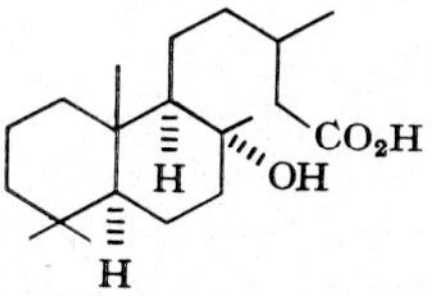

Skeletal variations that are sometimes found involve the migration of a methyl group from $C(10)$ to $C(9)$ and less commonly accompanying this, a second methyl group migration from $C(4)$ to $C(5)$. It is interesting to speculate whether these "friedo" or "backbone" rearrangements represent alternative modes of discharge of the cyclization carbonium ion in which case they may represent the often unconsidered step in biosynthesis of removing the substrate from the enzyme surface. Alternatively they may represent secondary rearrangements. Thus kolavenol (18) (46), from *Hardwickia pinnata*, represents the result of a rearrangement involving $C(8)$, $C(9)$, $C(10)$, $C(5)$, $C(4)$ and the loss of a proton from $C(3)$.

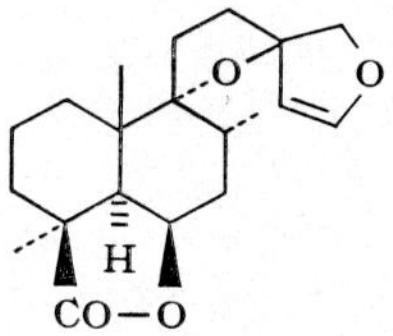

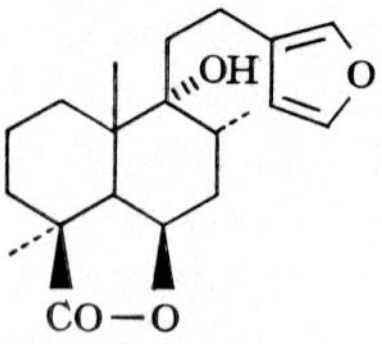

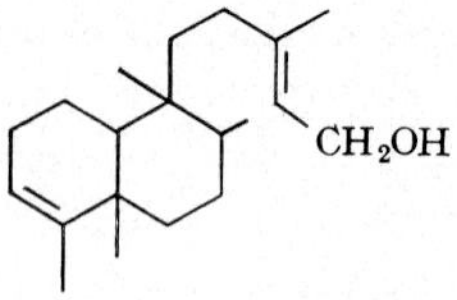

References, pp. 413—416

Oxygenation patterns that are frequently found involve C(15) as an alcohol, acid or part of a furan ring, C(13) as an alcohol or sometimes as part of an ether with C(8) or less commonly C(9), and C(18) or C(19) as an alcohol, acid or lactone. As a consequence of cyclization at the olefin rather than the epoxide oxidation level, the C(3) oxygen function characteristic of the triterpenes and steroids is not a regular feature of this series. Indeed although oxygen functions are found at this centre, they are about as common as oxygenation at other nuclear centres such as C(1), C(2), C(6) and C(7). This variety of oxygenation and modification of a basic carbon skeleton, as exemplified by the bicyclic bitter principles, combines to produce the many facets of diterpenoid chemistry.

2. The Tricyclic Diterpenes

The next phase of diterpenoid biogenesis involves the pyrophosphate acting as a leaving group and the generation of the tricyclic pimaradienes. The existence of epimers at the C(13) vinyl: methyl substitution, which was already apparent at the bicyclic level, is carried through into this series in, for example, pimaric acid (**19**) and sandaracopimaric acid (**20**). Although this would appear to determine the orientation of ring D in the tetracyclic series, a correlation does not yet appear to exist between the co-occurrence of bi-, tri- and tetracyclic diterpenes with a consistent C(13) configuration. The final fate of the nuclear double bond which may lie in the Δ^7, $\Delta^{8(9)}$, or $\Delta^{8(14)}$ positions (e. g. isopimaric acid (**21**)) leads to several series of compounds. Alternatively pimaranes have been isolated (*15*) with a C(8) hydroxyl group. Migration of the C(13) methyl group may lead to the abietadiene (**22**) or cassaine (**23**) diterpenes. Again in the abietic acid series, the fate of the nuclear double bonds leads to a variety of compounds such as levopimaric acid, neoabietic acid and abietic acid, some of which may be artefacts of isolation. Aromatization of the abietadiene series leads to compounds such as ferruginol (**24**) and its congeners whilst the cassaic acid series includes some furans such as vinhaticoic acid (**25**), reminiscent of cyclizations in the labdane series. The co-occurrence of ferruginol (**24**) and totarol (**26**) has led to the interesting suggestion that these compounds are biogenetically related. Indeed the presence af a phenolic hydroxyl at C(13) in totarol suggests that in the aromatic hydroxylation a C(13), C(14) epoxide is formed which opens with a C(13), C(14) isopropyl group shift. This of course may equally account for the formation of sempervirol.

An alternative mode of cyclization involving a "friedo" or "backbone" type of rearrangement may lead to the rosane group of diterpenes and to compounds such as the erythroxydiols (**27**) (*13*) and rimuene (**28**) (*14*). In the formation of the cyclopropane ring exemplified by (**27**), the carbonium ion is discharged with the formation of a cyclopropane ring in a

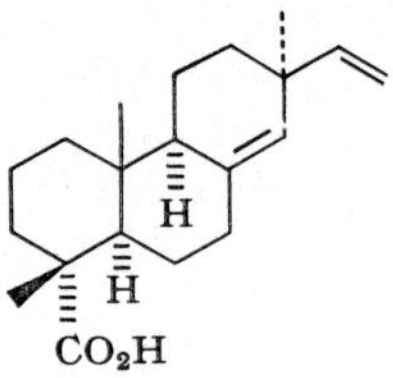

Chart 3

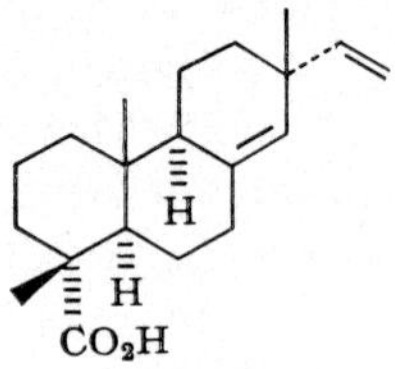

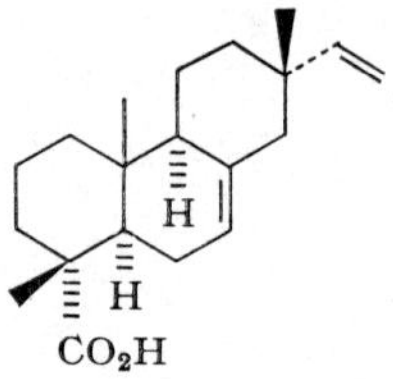

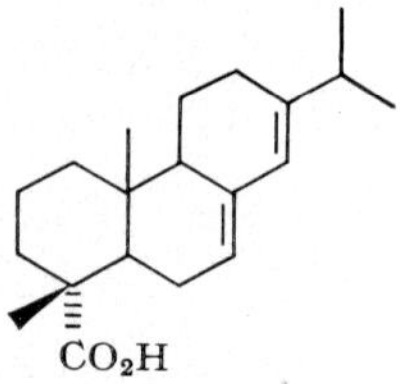

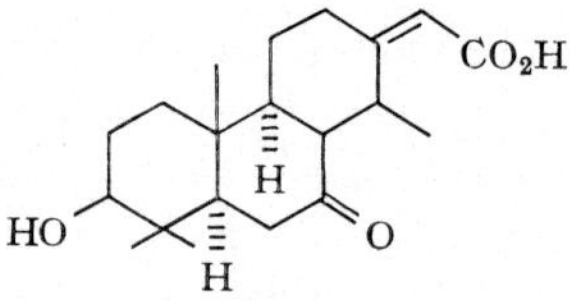

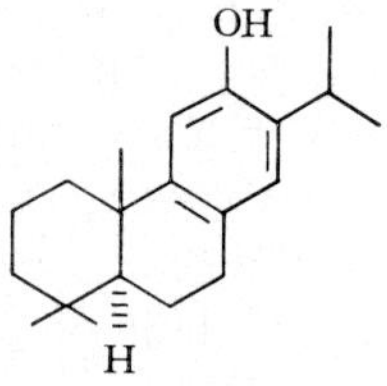

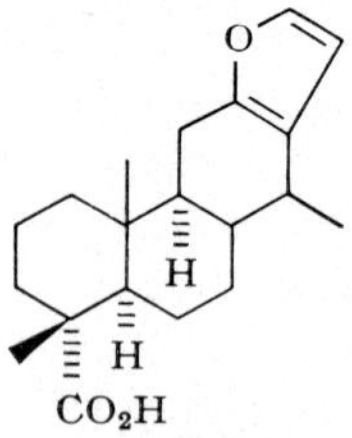

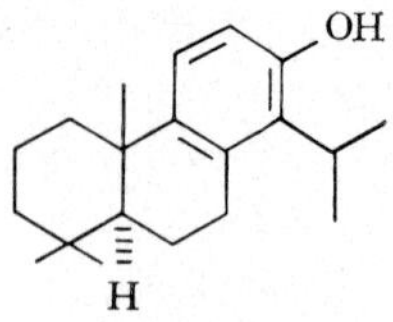

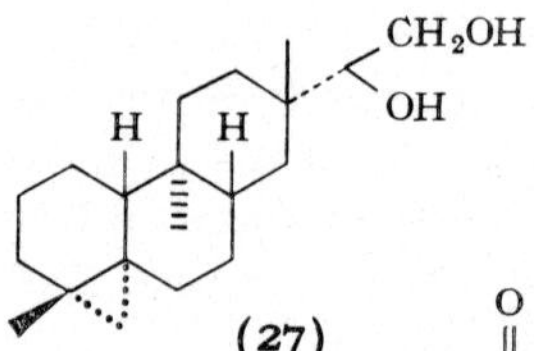

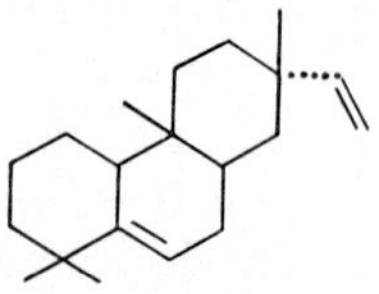

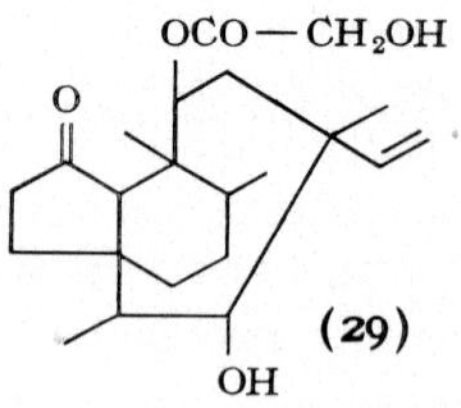

process of which there are an increasing number of terpenoid examples. Oxygenation in the tricyclic series frequently occurs at C(12), C(18) and C(19) although examples are known in which oxygen functions are located at many other centres.

Pleuromutilin (29) represents an unusual tricyclic diterpene (6, 7). Cyclization of geranylgeraniol pyrophosphate with an abnormal boat form for ring B affords a bicyclic precursor with a *cis* relationship of the C(9) and C(10) substituents. The "friedo" rearrangement is accompanied by ring contraction of ring A and the extrusion of the isopropyl group from C(5). This in turn attacks C(13) with the formation of the third ring of pleuromutilin.

3. The Tetracyclic Diterpenes

The tetracyclic diterpenes have provided a fruitful field for speculation as the skeleta are sufficiently closely related to suggest a common biosynthetic pathway. A number of schemes have been proposed for this pathway. These must accomodate various features. Thus, as opposed to the tetracyclic triterpenes and steroids, a large number of these compounds lack a C(3) oxygen function. Secondly the antipodal ring A/B fusion is widespread whilst there invariably exists a *trans* relationship

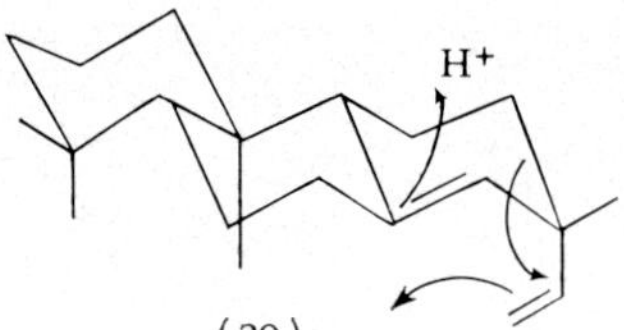

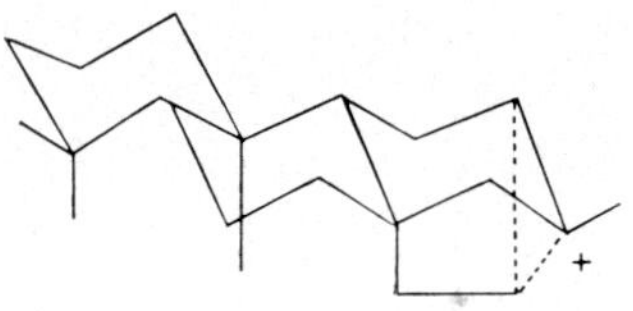

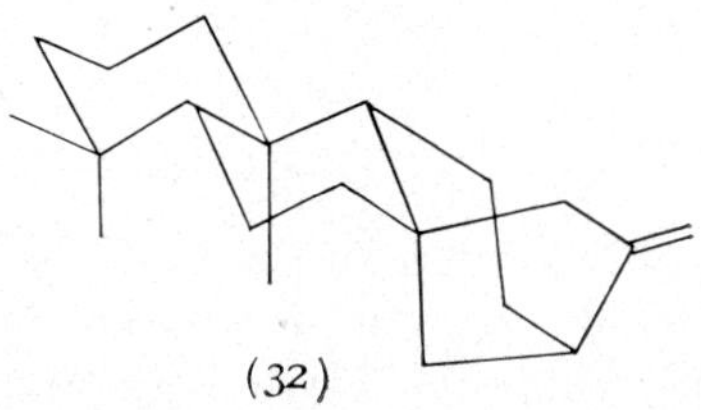

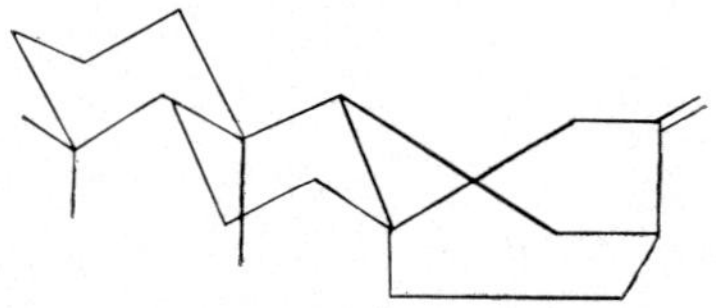

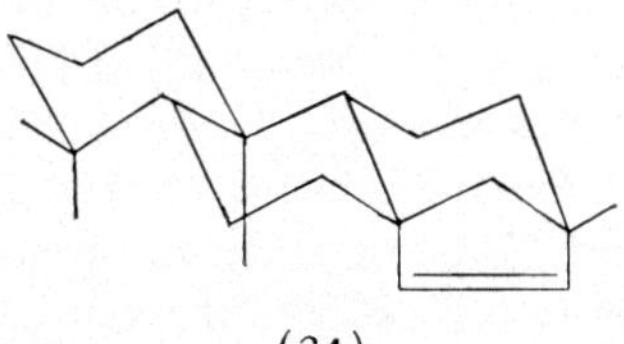

Chart 4

between the angular C(10) substituent and the C(9) hydrogen atom. Thirdly there exist both bicyclo-3,2,2-octane and several bicyclo-3,2,1-octane systems for rings C and D. The scheme must also account for modification of the perhydrophenanthrene backbone to the perhydrofluorene and perhydrobenzazulene skeleta of the gibberellins and the grayanotoxins.

In 1955 WENKERT suggested (*61*) that the tetracyclic diterpenes might arise through cyclization of suitably oriented pimaradienes (30) involving a non-classical carbonium ion (31) which could then collapse in a variety of ways leading to compounds now identified as possessing the phyllocladene and kaurene (32), atisine (33) and beyerene (34) skeleta. Subsequent elaboration (*38, 44, 57, 63*) of this theory have led to the accomodation of compounds possessing the gibbane (35), grayanotoxin (36) and enmein (37) skeleta. The isolation (*41*) of cyclokaurane (trachylobane) (7) diterpenes from *Trachylobium verrucosum* is of considerable interest in this connection. These schemes are illustrated in Chart 5. The gib-

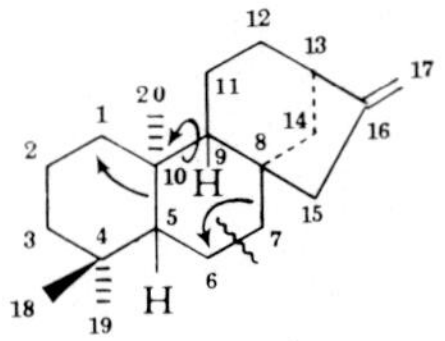

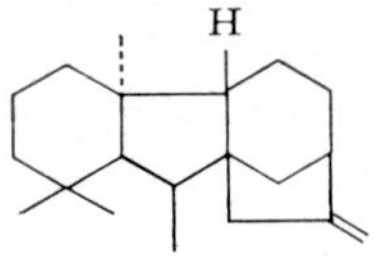

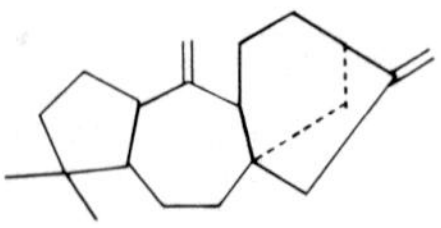

Chart 5

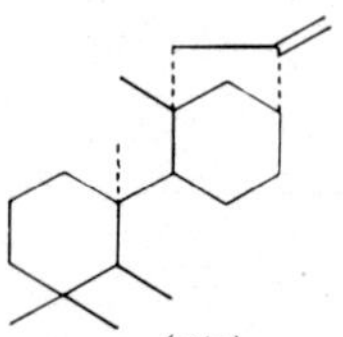

bane skeleton may arise by migration of the C(7)–C(8) bond to C(6) and the extrusion of C(7). On the other hand migration of the C(5)–C(10) bond to C(1) may lead to the grayanotoxins. The furanoid skeleton of cafestol might be formed by a simple WAGNER-MEERWEIN rearrangement at C(18). Cleavage of ring B of (−)-kaurene between C(6) and C(7) followed by rotation about the 9,10 bond leads to the enmein series. The formation of the seven-membered ring B of the lycoctonine alkaloids can be seen in terme of a C(14) equatorial leaving group on an atisine skeleton. Many of these proposals have been the subject of biogenetically patterned transformations in the laboratory. However a discussion of these lies outside the scope of this review.

References, pp. 413—416

4. The Macrocyclic Diterpenes

The formation of macrocyclic diterpenes by cyclization of geranylgeraniol pyrophosphate requires one of the in-chain double bonds to take up a *cis* stereochemistry. Depending upon the particular double bond the macrocycle is either cembrene (thunbergene) (10) or one of the duvatrienes. Verticillol (12) is a bicyclic diterpene which can be derived (26) from this whilst further cyclization leads to the taxane (11) skeleton (*cf.* Chart 1).

III. Biosynthetic Evidence

The origin, mode of linking and stereochemistry of isoprenoid biosynthesis has been the subject of a number of reviews (e. g. *12*). It is not our intention to discuss in detail these general aspects of terpenoid biochemistry but to concentrate on the diterpenoid substances. Mevalonate units have been shown to act as precursors of many diterpenoid substances described in the sequel. An interesting feature of these results contrasts with mono- and sesquiterpenoid biosynthesis. The distal methyl groups of geranylgeraniol pyrophosphate, and consequently the methyl groups of its cyclization products, retain (8) their individuality, *i. e.* only one group possesses the label from 2-^{14}C-mevalonate. The expected pathway through geranyl pyrophosphate, farnesyl pyrophosphate and geranylgeraniol pyrophosphate has been demonstrated (2) for the diterpenoid fungal metabolite, rosenonolactone (40). In addition geranylgeraniol pyrophosphate has been incorporated into pleuromutilin (42) (6), (—)-kaurene (43) and gibberellic acid (45) (*33, 58, 60*).

Most biosynthetic work has been carried out with fungal systems. Work with plant systems has been hampered by low incorporations (*34*) and there are even reports (*40, 47*) of negative incorporations of mevalonate into compounds such as delpheline and sclareol which are clearly diterpenoid. Physiological problems of permeability, transport, time and site of biosynthesis have yet to be overcome. Progress has been made using cell-free preparations from seeds and germinating seedlings in gibberellin biosynthesis. This review of the later stages of diterpenoid biosynthesis will follow the stepwise sequence of cyclizations through bicyclic, tricyclic to tetracyclic diterpenes.

1. The Bi- and Tricyclic Diterpenes

Preliminary results have been reported on the biosynthesis of the bicyclic diterpene, marrubiin (16), using 2-^{14}C-mevalonolactone in white horehound, *Marrubium vulgare*. Although this diterpene is now regarded as an artefact (*39*), the difference between this and premarrubiin is relatively small, involving a facile cleavage of an ether on work-up.

Copalol pyrophosphate (*ent*labdadienol pyrophosphate (2)) which is the primary product of cyclization of geranylgeraniol pyrophosphate in the antipodal series, has been shown to act (*37*) as a precursor of the C(8)–C(13) ether, olearyl oxide.

There has been only a little work on the plant resin acids. The rapid turnover of $^{14}CO_2$ when fed to pine trees and the disappearance of activity from the resin acid fraction (*64*) has led to the conclusion that these are continuously metabolized and are not biologically inert. A number of the phenols exist as dimers. The phenol-coupling of totarol (26) to form the bisditerpene podototarin has been demonstrated (*10*) with an enzyme system found in a number of Podocarpaceae.

In contrast considerable evidence has been obtained by tracer studies on the biosynthesis of the tricyclic diterpenoid fungal metabolite, rosenonolactone (40). This metabolite of *Tricothecium roseum* shows the interesting feature of a methyl group migration. Theory proposes (*8*) that the cyclization of a bicyclic labdane (38) to the tricyclic skeleton is concerted with a C(9)–C(8) hydride shift and a C(10)–C(9) methyl group migration.

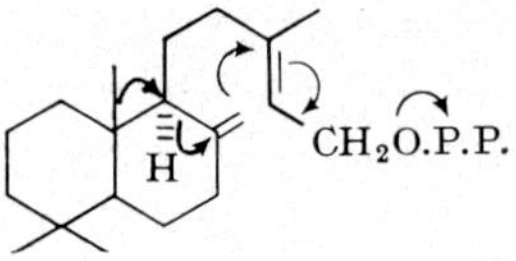

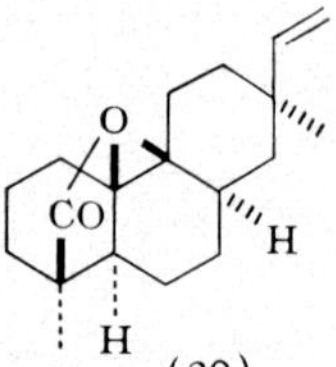

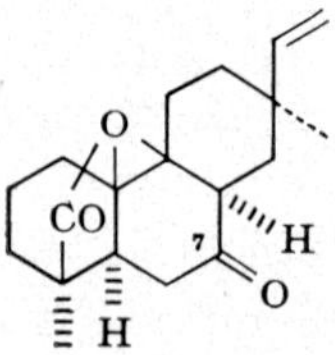

Chart 6

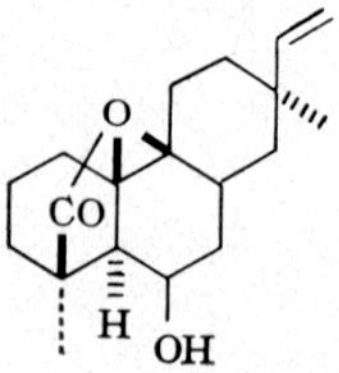

The fate of the C(10) carbonium ion is then open to consideration. The lactone ring is *cis* to the migrating methyl group thus precluding completely concerted lactonization. However the C(1α) and C(5α) protons lie *trans* to the methyl group in the labdane precursor and thus either may be eliminated to form a rosadiene. Subsequent *trans*-lactonization might then lead to the rosane lactones. Aletrnatively a C(10) carbonium ion might possess sufficient stability or be stabilized by a nucleophilic group on the enzyme surface for lactonization to occur in a non-concerted manner.

References, pp. 413—416

In accord with its diterpenoid nature, rosenonolactone incorporated (8) both 1-^{14}C-acetate and 2-^{14}C-mevalonate. The bicyclic labdadienol pyrophosphate (38) was shown (2) to be a specific precursor of rosenonolactone. Furthermore when mevalonate was fed to the mould, activity appeared first in the desoxyrosenonolactone (39) fraction and then later in the more highly oxygenated rosenonolactone (40) and rosololactone (41). Subsequent feeding (3) of labelled desoxyrosenonolactone and 7β-hydroxyrosenonolactone confirmed their intervention in the biosynthesis.

The use of multiply-labelled mevalonates enabled information to be obtained (3) on these cyclization stages. If the postulated hydride shift occurs then a C(9)-hydrogen atom will be retained at C(8). Furthermore if a $\Delta^5(^{10})$-rosadiene is involved in the biosynthesis then a C(5)-hydrogen atom will be lost. Both these correspond to hydrogen atoms that would be expected to arise from the 4(R)-position of mevalonate. 4(R)-4-^{3}H-Mevalonoid hydrogen was indeed located at C(5) and C(8) thus confirming the hydride shift and excluding a $\Delta^5(^{10})$-rosadiene from the biosynthesis. Furthermore both C(1) hydrogens and the C(6) hydrogens of rosenonolactone were mevalonoid (from 2-^{3}H$_2$ and 5-^{3}H$_2$-mevalonate respectively) and thus a $\Delta^1(^{10})$- and a Δ^6 intermediate were eliminated from the biosynthesis. Hence the C(10)-carbonium ion may be stabilized by interaction with a nucleophile on the enzyme surface prior to lactonization. Indeed the lactonization may represent the displacement of the substrate from the enzyme: substrate complex.

Pleuromutilin (42) represents an interesting example (6, 7) of a tricyclic diterpene in which a rearrangement has occurred at the bicyclic stage. The third ring, formed by displacement of the pyrophosphate involves the distal isoprene unit in cyclization. Consequently this diter-

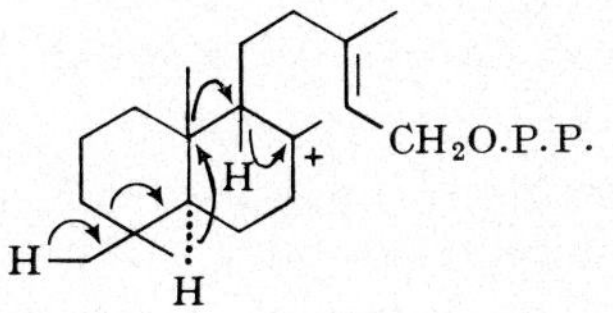

(38)

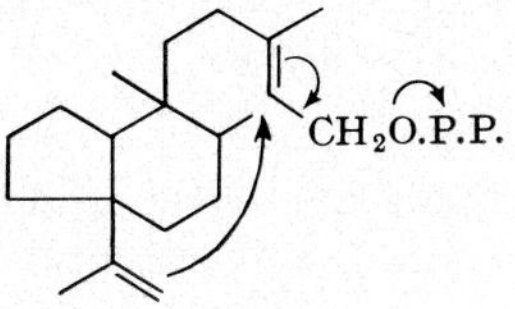

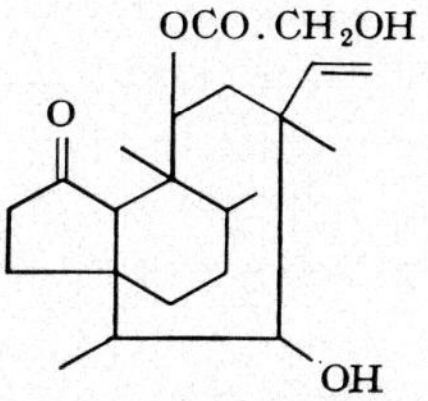

(42)

Chart 7

pene has been the subject of biosynthetic studies which were also of value in structural work (7). As a diterpenoid substance, pleuromutilin was shown to incorporate four molecules of mevalonate and geranylgeraniol pyrophosphate. However in the case of the latter, there was some scrambling of the label by degradation and resynthesis. During the biosynthesis a 4(R)-4-^{3}H-mevalonoid hydrogen was shown to migrate from C(9) to C(8). Since this hydrogen is *cis* to the migrating methyl group, an C(9) enzyme bound intermediate was postulated. Further migration and ring-contraction of ring A must follow this in a second step. An interesting and unexpected feature of the biosynthesis was the final discharge of this carbonium ion by a hydride shift from C(11) to C(4). This was revealed by the mevalonate labelling pattern in which the 5(S)-5-^{3}H-mevalonate hydrogen was shown to migrate.

2. The Tetracyclic Diterpenes

Because of their importance as plant-growth hormones, the biosynthesis of the diterpenoid gibberellins has received considerable attention. The progress of this work has been reviewed on a number of occasions (*16, 32, 38, 62*). The first experimental evidence for the diterpenoid nature of gibberellic acid (44) came (*8*) from a study of the incorporation of 1-^{14}C-acetic acid and 2-^{14}C-mevalonic acid. The quantitative degradative results were in agreement with the incorporation of eight carboxyl units from acetic acid (i. e. 12 acetate units) and four mevalonate units. The degradation from the mevalonate experiment established that the terminal gem-dimethyl group of the geranylgeraniol pyrophosphate was unsymmetrically labelled. The lactone carbonyl carbon atom was derived from the 3-position of the mevalonic acid and the ring A methyl group originated from the 2-position of the mevalonate. The carboxyl carbon atom of the gibberellic acid was also shown to arise by extrusion of the C(7) rather than the C(6) carbon of the cyclic precursor.

Degradation of the acetate labelled material supported the Wenkert mechanism for the formation of the tetracyclic ring system. This scheme requires the migration of the C(12)–C(13) bond to C(15) in a pimarane with the carbon atom that eventually becomes the terminal methylene of the gibberellic acid remaining attatched to an acetatecarboxyl labelled carbon atom. KUHN-ROTH degradation of tetrahydrogibberellic acid showed this to be the case. Alternative proposals involving a methyl migration analogous to the formation of abietic acid from a pimaradiene were therefore ruled out.

There have been proposals (*11, 53*) that senecioic acid ($\beta\beta$-dimethylacrylic acid) was a precursor of gibberellic acid. These were partly based on an increase in yield from the fermentation when senecioic acid was

References, pp. 413—416

added to the medium. The labelling experiments were not supported by degradation and hence the specificity of any incorporations was not clear.

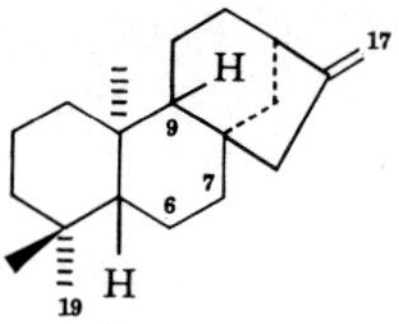

Chart 8

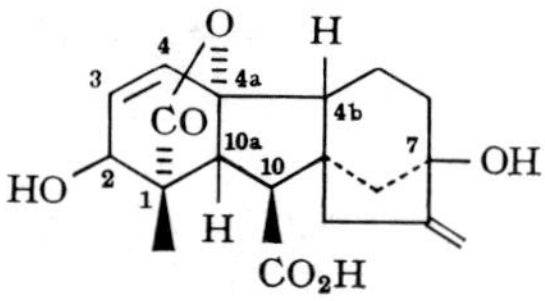

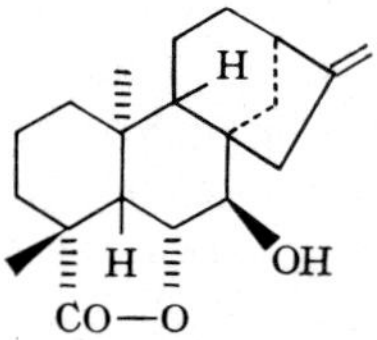

A careful study of the metabolites that co-occur with gibberellic acid (44) revealed (*16, 17*) a number of kauranoid diterpenes (e. g. 45) amongst which was the parent hydrocarbon, (—)-kaurene (43). (—)-Kaurene was labelled (*18*) at C(17) and shown to be specifically incorporated into gibberellic acid to the extent of 5.7%. In an attempt to find evidence for (+)-gibberellic acid, 17-^{14}C-(+)-kaurene was fed (*21*) to *Gibberella fujikuroi*. However autoradiography failed to reveal any transformation products. A number of new gibberellins were isolated both from the fungal studies and from work with higher plants. There are now about 27 known gibberellins (*45*).

Consequently the study of gibberellin biosynthesis may be divided into a number of phases. The first involves the stages leading to the formation of (—)-kaurene, the second involves the hydroxylation of (—)-kaurene, the third the ring-contraction to the gibbane skeleton and the fourth involves the relationship amongst the various gibberellins including the conversion of the C_{20} to the C_{19} series.

Geranylgeraniol, both as the free alcohol and as its pyrophosphate (*33, 48, 60*), has been shown to act as a precursor of gibberellic acid. A soluble enzyme fraction from *Gibberella fujikuroi* capable of converting geranylgeraniol pyrophosphate to (—)-kaurene has been partially purified (*60*). A soluble system that converts both mevalonate and trans geranylgeraniol pyrophosphate into (—)-kaurene has been prepared in a cell-free form from developing seeds of the wild cucumber *(Echinocystis macrocarpa)*. Similar preparations from the seeds of *Pisum sativum* and

Cucurbita pepo have also been reported (*29, 30*). A soluble enzyme preparation from germinating seedlings of *Ricinus communis* converts (*55*) both mevalonate and geranylgeraniol pyrophosphate into (—)-kaurene, (+)-beyerene, (+)-sandaracopimaradiene, trachylobane and casbene. It is interesting to note the absence of pimaradiene which is regarded (*23, 33*) as a possible tricyclic precursor of the kauranoid diterpenes.

The presence of the bicyclic alcohol, copalol pyrophosphate (*ent*-labdadienol pyrophosphate) has been demonstrated (*58*) in cell-free extracts of *Gibberella fujikuroi*. Furthermore it has been shown to act as a precursor of (—)-kaurene in cell-free extracts from *Gibberelηa fujikuroi* and *Echinocystis macrocarpa*. In the system derived from *Ricinus communis*, it also acts as a precursor of the other cyclic diterpene hydrocarbons of the same enantiomeric series. Intact cultures of *Gibberella fujikuroi* convert (*33*) the bicyclic pyrophosphate into (—)-kaurene, the kaurenolides and gibberellic acid.

The putative tricyclic intermediate, (—)-pimaradiene, currently occupies an uncertain position (*23, 33*). By studying the retention of tritium from appropriately labelled mevalonates it was possible to exclude certain bouble bond isomers of (—)-pimaradiene from the biosynthesis. Any isomer which could generate a carbonium ion at C(8) was theoretically a possibility. The C(7) hydrogen atoms arise from the 2-position of mevalonate and the C(9) hydrogen atom arises from the 4-position of mevalonate. When 2-^{3}H$_2$,2-^{14}C-mevalonate was fed to *Gibberella fujikuroi*, all eight tritium atoms corresponding to the four mevalonate units were incorporated into (—)-kaurene thus excluding a $\Delta^7(^8)$-pimaradiene. A similar experiment with 4(R)-4-^{3}H,2-^{14}C-mevalonate showed that mevalonoid hydrogen remained at position 4b in gibberellic acid — the position corresponding to C(9) of a pimaradiene thus excluding a $\Delta^8(^9)$-pimaradiene. $\Delta^8(^{14})$-(—)-Pimaradiene labelled on the vinyl group was specifically incorporated (*33*) into gibberellic acid but in very low yield (0.024%). Another report (*23*) of this incorporation was withdrawn when it was found that the (—)-pimaradiene used in the study was contaminated with labelled (—)-kaurene. The most satisfactory explanation of the low incorporation is that the tricyclic stage is enzyme bound, possibly as an undischarged but solvated C(8) carbonium ion. The formation of the tetracyclic hydrocarbon involving attack of the vinyl group at C(8) then includes the displacement of the molecule from the enzyme surface.

(—)-Kaurene has been shown to act as the parent hydrocarbon of gibberellic acid, the kaurenolides and steviol 40 (*34*). In all three cases the next stage in the biosynthesis involves the oxidation of the 19-methyl group to the level of a carboxylic acid possibly by a mixed-function oxidase system. It is interesting to note that in gibberellin and in steviol biosynthesis there is an enzyme system that is capable of inserting a bridgehead hydroxyl group after the formation of ring D. 17-^{14}C-(—)-Kaur-16-en-19-ol was specifically incorporated (*19*) into gibberellic acid in 4.9% yield whilst it was also a precursor of gibberellin A_{13}. This evidence was used to define part of the stereochemistry of gibberellin A_{13}. 17-^{14}C-(—)-Kaur-16-en-19-al and (—)-kaur-16-en-19-oic acid (46) both act as precursors of gibberellic acid (44), the kaurenolides (48) and (51), fujenal (52), and steviol (49) (*28, 34, 37*). These compounds have been identified as products of the mevalonate metabolism in the endosperm homogenate of *Echinocystis macrocarpa*. A mixed-function oxidase reponsible for this conversion has been partially characterized (*24, 62*). In this connection it is worth noting that (—)-kaurene, (—)-kaur-16-en-19-ol,(—)-kaur-16-en-19-oic acid and steviol show gibberellin-like activity in the dwarf mutants, d-5 and an-1, of *Zea mays* (*42*).

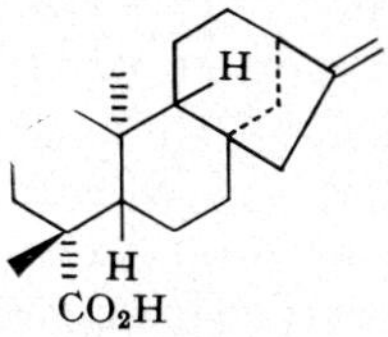

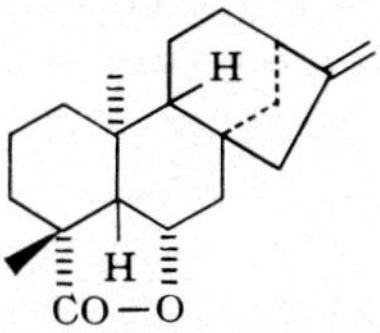

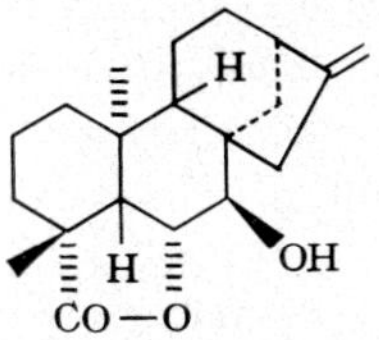

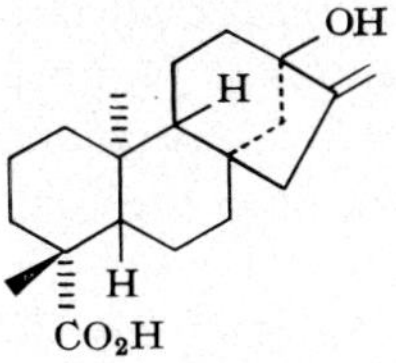

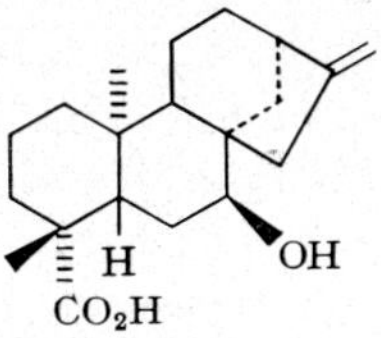

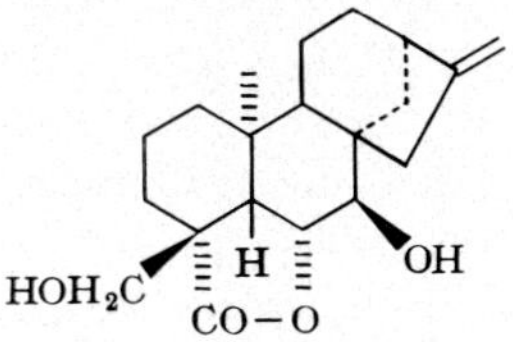

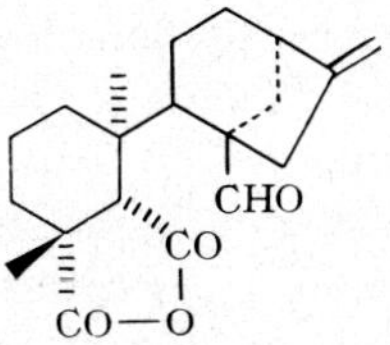

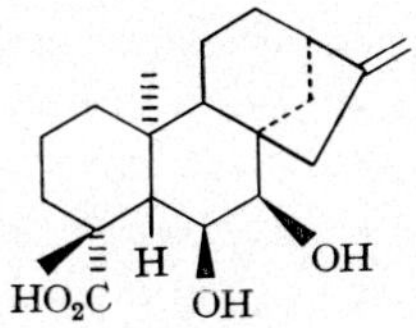

Chart 9

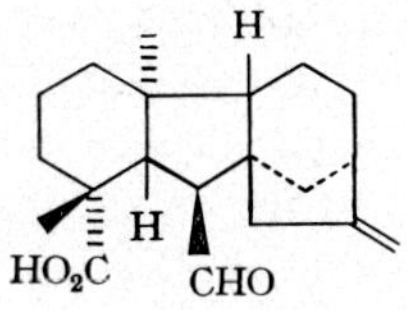

The hydroxylation of ring B involves a number of stages. Apart from the gibberellins, *Gibberella fujikuroi* elaborates a further group of diterpenoid metabolites, the kaurenolides, which possess a diaxial $6\alpha,7\beta$-diol on ring B. Their presence raises a number of points, firstly the order of oxidation of ring B, secondly the oxidation level of the ring contraction and thirdly the stereochemistry of these stages. 7β-Hydroxy-(—)-kaur-16-en-19-oic acid (50) has been isolated by radiochemical dilution of the metabolites of *Gibberella fujikuroi* (35) and *Echinocystis macrocarpa* (62). Labelled 7β-hydroxy-(—)-kaur-16-en-19-oic acid was specifically incorporated into gibberellic acid and the kaurenolides by *Gibberella fujikuroi*. However kaurenolide (42) was not formed by *Gibberella fujikuroi*. One-half of the radio-activity from 1-^{3}H$_2$,2-^{14}C-geranyl pyrophosphate was specifically retained in gibberellic acid at the 10 position which corresponds to C(6) of (—)-kaurene. Consequently ring contraction takes place without enolization into the gibbane ring. Compounds containing a 6α-hydroxyl group and 7α or 7β-hydroxyl groups were not incorporated into gibberellic acid. However 7β-hydroxykaurenolide (48) was converted very efficiently into 7,18-dihydroxykaurenolide (51). The stereochemistry of these hydroxylations was studied using stereospecifically labelled mevalonates. Stereochemical information using 2(R),2(S),4(R) and 5(R)-^{3}H-mevalonoid hydrogen based on results in the steroid series can be related to this series. (36). Thus 7β-hydroxykaurenolide retains a 2(R)-mevalonoid hydrogen at position 7 and a 5(R)-mevalonoid hydrogen at position 6 but loses a 2(S)-mevalonoid hydrogen from position 7. Consequently hydroxylation must have proceeded with retention of configuration. On the other hand, although gibberellic acid retains a hydrogen atom from position C(6) of the kaurene skeleton, it does not retain a 5(R)-^{3}H-mevalonoid hydrogen in ring B. Thus during the ringcontraction an equatorial hydrogen atom is lost. The leaving group may be either a 6β-alcohol (perhaps as its pyrophosphate) or ring contraction may be initiated by the abstraction of hydrogen. Recently 6β,7β-dihydroxy-(—)-kaur-16-en-19-oic acid (53) has been synthesized (22, 37) but both groups have shown that it is not incorporated into gibberellic acid. Ring contraction is accompanied by the migration of the 6β-hydrogen atom to C(7) and the extrusion of this atom as a primary alcohol.

The product of ring contraction is a gibbane alcohol. This and the aldehydic acid (54) have been shown to be formed from 7β-hydroxy-(—)-kaur-16-en-19-oic acid (50) and to be converted into gibberellic acid (44) by *Gibberella fujikuroi*. The corresponding acid, gibberellin A$_{12}$ (55), has been isolated from *Gibberella fujikuroi* and shown (19) to be incorporated into gibberellic acid. Suprisingly the corresponding diol is also incorporated into gibberellic acid. Gibberellin A$_{14}$ (56) which possess a ring A hydroxyl group is also converted into gibberellic acid.

Many of the C$_{20}$ gibberellins have the angular methyl group oxidized to the level of an alcohol, aldehyde or carboxylic acid. Information has been obtained on the loss of this angular group in the formation of the

C_{19} gibberellins by the use of multiply-labelled mevalonates. Gibberellic acid retains mevalonoid hydrogen at C(4b) and at C(10a). Furthermore the C_{20} gibberellin A_{13} (52) and the ring A saturated C_{19} gibberellin A_4 (59) both contain six tritium atoms from 2-^{3}H$_2$,2-^{14}C-mevalonate. In particular two tritium atoms must remain on ring A at C(4) during the decarboxylation. Hence it is likely that the angular group is lost through a Baeyer-Villiger type of oxidation rather than through the decarboxylation of a $\beta\gamma$-unsaturated acid. This retention of hydrogen contrasts with steroid biosynthesis in which skeletal hydrogen is lost during the removal

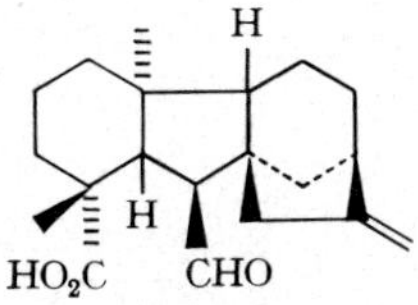

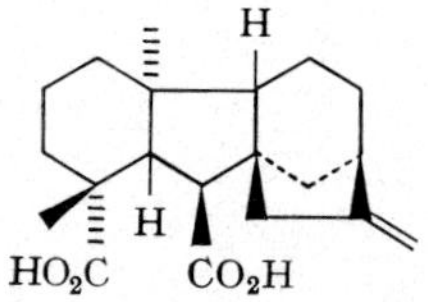

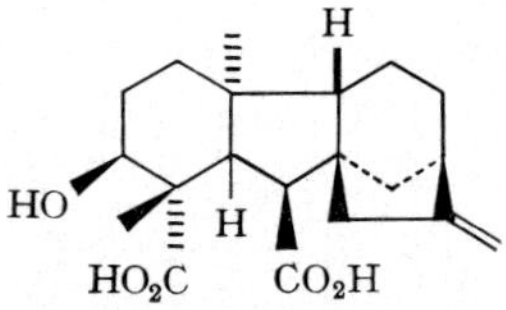

Chart 10

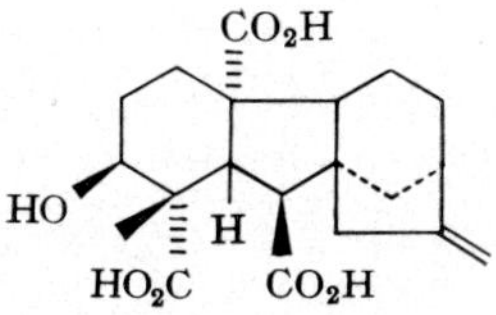

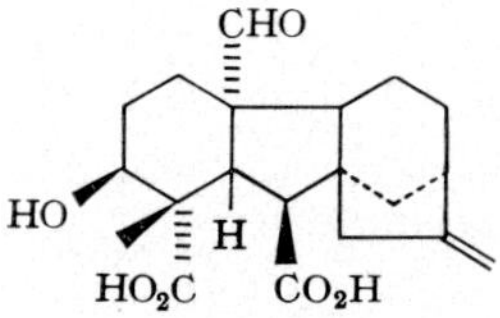

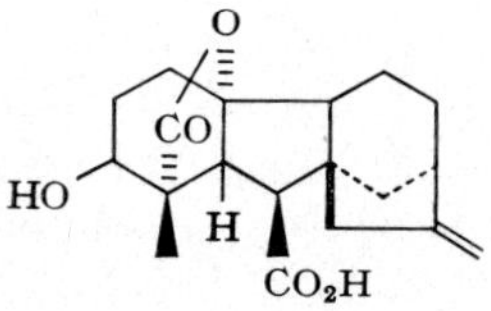

of the various methyl groups. Labelling experiments suggest that gibberellin A_{13} itself does not appear to be an intermediate in the biosynthesis–possibly gibberellin A_{24} (58) in which the angular substituent is at the aldehyde oxidation level, fills this role (52).

The interrelationship between the various C_{19}-gibberellins has been studied. Thus gibberellin A_9 (60) does not appear to act as a precursor of gibberellic acid but is converted to gibberellin A_{10} (61) (19). On the other hand in a time course study (28) with labelled (−)-kaur-16-en-19-oic acid (46), radioactivity was first found in the gibberellin A_4/A_7 (62) and (65) fraction and subsequently in the gibberellin A_1/A_3 (63) and (64) fraction. The study of the incorporation of 2(R),2(S) and 5(R)-mevalonoid hydrogen into gibberellic acid has led to the conclusion that the dehydrogenation step to form the Δ^3-double bond involves the overall *cis* elimination of hydrogen from the "α" face of a saturated gibberellin. However there remain many aspects of gibberellin biosynthesis that are incomplete. In particular there is evidence based on the structural relationships of naturally occurring gibberellins which suggests that the biosynthetic pathways in the higher plants differ in detail from those in *Gibberella fujikuroi*.

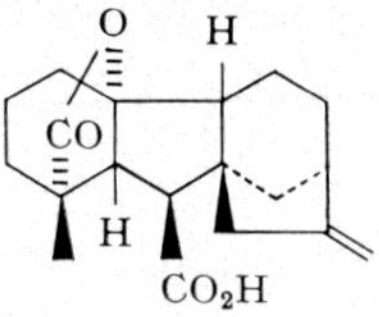

(60) 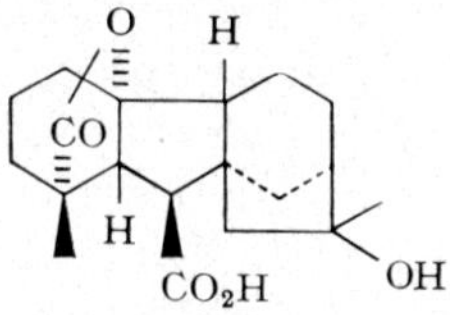 (61) 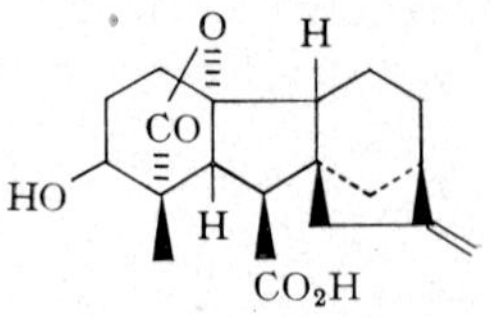 (62)

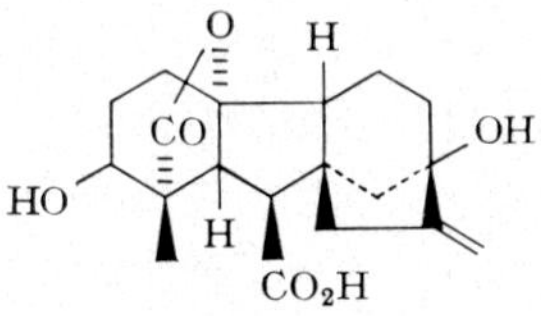

(63) 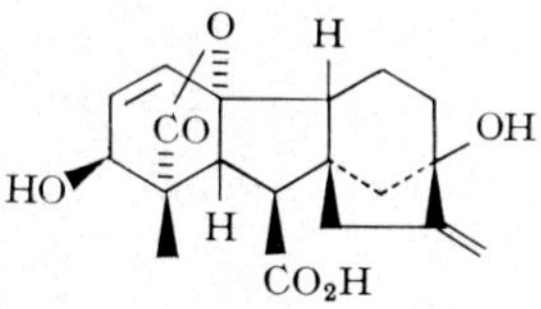 (64) 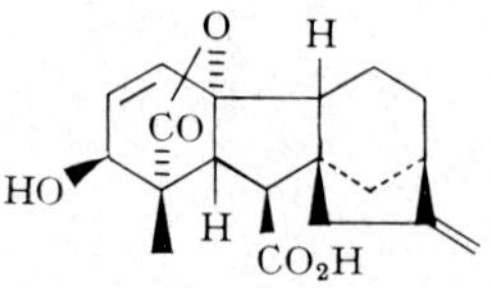 (65)

Chart 11

A number of plant growth retardants such as (2-chloroethyl)-trimethylammonium chloride (cycocel) and 2′-isopropyl-4′-(trimethylammonium chloride)-5′-methylphenyl piperidine-1-carboxylate (Amo 1618) reduce stem elongation and thus exhibit an anti-gibberellin activity. This effect can be reduced by the application of gibberellins. Addition of these substances to growing cultures of *Gibberella fujikuroi* at the beginning of the gibberellic acid production phase completely suppresses the biosynthesis of gibberellic acid and other diterpenoid metabolites (20, 54). It is not clear what stages in the biosynthesis are inhibited. In the *Echinocystis macrocarpa* system there is some evidence that Amo-1618 inhibits the cyclization of geranylgeraniol pyrophosphate (25). On the other hand 2-diethylaminoethyl-2,2-diphenylvalerate (SKF 525) appears (54) to inhibit later stages such as the conversion of (—)-kaur-16-en-19-al to (—)-kaur-16-en-19-oic acid.

IV. Conclusion

In this review we have tried to set out the current state of biosynthetic evidence for the diterpenes and to show the way in which the use of multiply-labelled mevalonates has complemented studies with larger molecules. The interrelationship of many of the more highly oxygenated diterpenes remains to be established. Hitherto this has been hindered by the absence of satisfactory techniques of handling biosynthesis in higher plants. The use of germinating seeds may permit the solution of some of these problems. It is now becoming clear which functional groups mark essential features of the biosynthesis and what functionality is merely secondary to the overall biogenetic pattern. How these substances may be fitted into schemes of biogenetic evolution is a fascinating area for speculation. It also suggests structures of natural products that may be biogenetic missing links. This area also reveals a number of relatively unusual biosynthetic processes such as ring-contraction whose enzymology and co-enzymes represent unknown problems. Finally the control of diterpene biosynthesis, particularly of the gibberellins, is a problem which has economic as well as academic interest.

References

1. ABBONDANZA, A., R. BADIELLO, and A. BRECCIA: On the Biosynthesis of the Terpene, Marrubiin from 1,4-^{14}C-Succinic Acid and 2-^{14}C Mevalonolactone. Tetrahedron Letters **1965**, 4337.
2. ACHILLADELIS, B., and J. R. HANSON: The Biosynthesis of the Metabolites of Tricothecium roseum. Phytochemistry **7**, 589 (1968).
3. — — The Biosynthesis of Rosenonolactone. J. Chem. Soc. (London) C **1969**, 2010.
4. ANDERSON, J. D., and T. C. MOORE: The Biosynthesis of (—)-Kaurene in Cell-free Extracts of Immature Pea Seeds. Plant Physiol. **42**, 1527 (1967).
5. APLIN, R. T., R. C. CAMBIE, and P. S. RUTLEDGE: The Taxonomic Distribution of some Diterpene Hydrocarbons. Phytochemistry **2**, 205 (1963).
6. ARIGONI, D.: Some Studies in the Biosynthesis of Terpenes and Related Compounds. Pure and Applied Chemistry **17**, 331 (1968).
7. BIRCH, A. J., C. W. HOLZAPFEL, and R. W. RICKARDS: The Structure and some Aspects of the Biosynthesis of Pleuromutilin. Tetrahedron, Suppl. **8**, 359 (1966).
8. BIRCH, A. J., R. W. RICKARDS, H. SMITH, A. HARRIS, and W. B. WHALLEY: Rosenonolactone and Gibberellic Acid. Tetrahedron **7**, 241 (1959).
   BIRCH, A. J., R. W. RICKARDS, H. SMITH, J. WINTER, and W. B. TURNER: The Allogibberic Acid-Gibberic Acid Rearrangement. Chem. and Ind. **1960**, 401.
9. BEVAN, C. W. L., D. E. U. EKONG, and J. I. OKOGUN: The Diterpenes of *Oxystigma oxyphyllum*. J. Chem. Soc. (London) C **1968**, 1067.
10. BOCKS, S. M., and R. C. CAMBIE: The Enzymic Coupling of Totarol. Proc. Chem. Soc. (London) **1963**, 143.
11. BUNSOW, R.: Senecioic Acid in the Biosynthesis of Gibberellins in Hugher Plants. Naturwissenschaften **48**, 411 (1961).
12. CLAYTON, R. B.: The Biosynthesis of Sterols, Steroids and Terpenoids. Quart. Rev. Chem. Soc. (London) **19**, 168, 201, (1965).
13. CONNOLLY, J. D., R. MCCRINDLE, R. D. H. MURRAY, A. J. RENFREW, K. H. OVERTON, and A. MELERA: Erythroxydiols X, Y, and Z, Two Novel Skeletal Types of Diterpenoid. J. Chem. Soc. (London) C **1966**, 268.
14. CONNOLLY, J. D., R. MCCRINDLE, R. D. H. MURRAY, and K. H. OVERTON: The Constitution and Stereochemistry of Rimuene. J. Chem. Soc. (London) C **1966**, 273.

15. CORBETT, R. E., and R. A. J. SMITH: Diterpenes from the Volatile Oil of *Dacrydium Colensoi*. J. Chem. Soc. (London) C **1967**, 300.
16. CROSS, B. E.: The Biosynthesis of the Gibberellins. Progr. in Phytochemistry 1, 195 (1968).
17. CROSS, B. E., R. H. B. GALT, J. R. HANSON, and W. KLYNE: Some New Metabolites of *Gibberella fujikuroi* and the Stereochemistry of (—)-Kaurene. Tetrahedron Letters **1962**, 145.
18. CROSS, B. E., R. H. B. GALT, and J. R. HANSON: (—)-Kaurene as a Precursor of Gibberellic Acid. J. Chem. Soc. (London) **1964**, 295.
19. CROSS, B. E., R. H. B. GALT, and K. NORTON: The Biosynthesis of the Gibberellins II. Tetrahedron **24**, 231 (1968).
20. CROSS, B. E., and P. L. MYERS: The Effect of Plant Growth Retardants on the Biosynthesis of Diterpenes by *Gibberella fujikuroi*. Phytochemistry **8**, 79 (1969).
21. CROSS, B. E., K. NORTON, and J. C. STEWART: An Attempt to Find Evidence for the Existance of (+)-Gibberellic Acid. Phytochemistry **7**, 83 (1968).
22. CROSS, B. E., K. NORTON, and J. C. STEWART: Biosynthesis of the Gibberellins–III. J. Chem. Soc. (London) C **1968**, 1054.
22a. CROSS, B. E., J. C. STEWART, and J. L. STODDART: $6\beta,7\beta$-Dihydroxy-(—)-kaurenoic acid: Its Biological Activity and Possible Role in the Biosynthesis of Gibberellic Acid. Phytochemistry **9**, 1065 (1970).
23. CROSS, B. E., and J. C. STEWART: (—)-Pimaradiene–A New Precursor of the Gibberellins. Tetrahedron Letters **1968**, 5195, 6321.
24. DENNIS, D. T., and C. A. WEST: Conversion of (—)-Kaurene to (—)-Kaur-16-en-19-oic Acid in the Endosperm of *Echinocystis macrocarpa*. J. Biol. Chem. **242**, 3293 (1967).
25. DENNIS, D. T., C. D. UPPER, and C. A. WEST: An Enzymic Site of Inhibition of Gibberellin Biosynthesis by Amo 1618 and other Plant Growth Retardants. Plant Physiol. **40**, 948 (1965).
26. ERDTMANN, H., T. NORIN, M. SUMIMOTO, and A. MORRISON: Verticillol, A Novel Type of Conifer Diterpene. Tetrahedron Letters **1964**, 3879.
27. FUJITA, E.: The Chemistry of the Diterpenoids. Bull. Inst. Chem. Research (Kyoto) **43**, 278 (1964); **44**, 239 (1965); **45**, 229 (1966).
28. GEISSMAN, T. A., A. J. VERBISCAR, B. O. PHINNEY, and G. CRAGG: Studies on the Biosynthesis of Gibberellins from (—)-Kaurenoic acid in Cultures of *Gibberella fujikuroi*. Phytochemistry **5**, 933 (1966).
29. GRAEBE, J. E.: Biosynthesis of Kaurene, Squalene and Phytoene from 2-^{14}C Mevalonate in a Cell-free System from Pea fruits. Phytochemistry **7**, 2003 (1968).
30. GRAEBE, J. E.: Enzymic Preparation of ^{14}C-Kaurene using *Curcubita pepo*. Planta **85**, 171 (1969).
31. GRIGOREVA, N. Y., and V. F. KUCHEROV: The Gibberellins. Russian Chem. Rev. **35**, 850 (1966).
32. GROVE, J. F.: The Gibberellins. Quart. Rev. Chem. Soc. (London) **15**, 56 (1961).
33. HANSON, J. R., and A. F. WHITE: The Biosynthesis of the Kaurenolides and Gibberellic Acid. J. Chem. Soc. (London) C **1969**, 981.
34. HANSON, J. R., and A. F. WHITE: Biosynthesis of Steviol. Phytochemistry **7**, 595 (1968).

35. HANSON, J. R., and A. F. WHITE: The Oxidative Modification of the Kauranoid Ring B during Gibberellin Biosynthesis. Chem. Comm. 1969, 410.
36. HANSON, J. R., and A. F. WHITE. The Stereochemistry of some Stages in Gibberellin Biosynthesis. Chem. Comm. 1969, 1071.
37. HANSON, J. R., J. HAWKER, and A. F. WHITE: unpublished results.
38. HANSON, J. R.: The Tetracyclic Diterpenes. Oxford: Pergamon Press. 1968.
39. HENDERSON, M. S., and McCRINDLE: Pre-marrubiin. A Diterpenoid from *Marrubium vulgare.* J. Chem. Soc. (London) C 1969, 2014.
40. HERBERT, E. J., and G. W. KIRBY: Mevalonic Acid and Delpheline Biosynthesis. Tetrahedron Letters 1963, 1505.
41. HUGEL, G., L. LODS, J. M. MELLOR, D. W. THEOBALD, and G. OURISSON: Diterpenes of Trachylobium. Bull. Soc. Chim. (France) 1965, 2882.
42. KATSUMI, M., B. O. PHINNEY, P. R. JEFFERIES, and C. A. HENRICK: The Growth Response of the d-5 and an-1 Mutants of Maize to some Kaurene Derivatives. Science 144, 849 (1964).
43. KIMLAND, B., and T. NORIN: The Oleorosin of the Norwegian Spruce, *Picea abies.* Acta Chem. Scand. 21, 825 (1967).
44. McCRINDLE, R., and K. H. OVERTON: The Diterpenoids, Triterpenoids and Sesterterpenoids. Chemistry of Carbon Compounds (Ed. E. H. RODD) vol. 2b. Chap. 14, London: Elsevier. 1969.
45. MACMILLAN, J., and N. TAKAHASHI: Names of the Gibberellins. Nature 217, 170 (1969).
46. MISRA, R., R. C. PANDEY, and S. DEV: Chemistry of the Oleorosin of *Hardwickia pinnata.* Tetrahedron Letters 1964, 3751.
47. NICHOLAS, D. J.: Biosynthesis of Sclareol,β-Sitosterol and Oleanolic Acid from 2-^{14}C-Mevalonic Acid. J. Biol. Chem. 237, 1481 (1962).
48. OSTER, M. O., and C. A. WEST: Biosynthesis of trans-Geranylgeranyl Pyrophosphate in the Endosperm of *Echinocystis macrocarpa.* Arch. Biochem. Biophys. 127, 112 (1968).
49. OVERTON, K. H., and A. J. RENFREW: The Configuration at C-13 in Labdanolic and Eperuic Acids. J. Chem. Soc. (London) C 1967, 931.
50. PELLETIER, S. W.: The Chemistry of the C_{20} Diterpene Alkaloids. Quart. Rev. Chem. Soc. (London) 21, 525 (1967).
51. POPA, D. P., and G. V. LAZURVESKI: Sclareol and 13-*epi*-Sclareol from Clary Sage. Zh. Obshch. Khim. 33, 303 (p. 296 in translation) (1963).
52. PRYCE, R. J., J. MACMILLAN, and A. McCORMICK: The Identification of Bamboo Gibberellin in *Phaseolus Multiflorus* by Combined Gas Chromatography-Mass Spectrometry. Tetrahedron Letters 1967, 5009.
53. REDEMAN, C. T., and L. J. MEULE: Senecioic Acid as a Precursor of Gibberellic Acid in *Fusarium monoliforme.* Naturwissenschaften 46, 382 (1959).
54. REID, W. W.: Effect of SKF 7997 and SKF 525 on Diterpene and Sterol Biosynthesis in *Gibberella fujikuroi* from 2-^{14}C Mevalonate. Biochem. J. 113, 37p (1969).
55. ROBINSON, D. R., and C. A. WEST: Biosynthesis of Cyclic Diterpenes in Extracts from Seedlings of *Ricinus communis.* Biochemistry 9, 70, 80 (1969).
56. RUZICKA, L.: History of the Isoprene Rule. Proc. Chem. Soc. (London) 1959, 341.

57. Scott, A. I., F. McCapra, F. Comer, S. A. Sutherland, D. W. Young, G. W. Sim, and F. Ferguson: The Structure and Stereochemistry of some Polycyclic Diterpenoids. Tetrahedron **20**, 1339 (1964).
58. Schechter, I., and C. A. West: The Biosynthesis of Cyclic Diterpenes from Geranylgeranyl pyrophosphate. J. Biol. Chem. **244**, 3200 (1969).
59. Sumimoto, M., Y. Tahaka, and K. Matsufiji: The Heartwood Constituents of *Sciadoptys verticillata*. Tetrahedron **20**, 1427 (1964).
60. Upper, C. A., and C. A. West: The Enzymic Cyclization of Geranylgeranyl pyrophosphate to (−)-Kaurene. J. Biol. Chem. **242**, 3285 (1967).
61. Wenkert, E.: Structural and Biogenetic Relationships in the Diterpene Series. Chem. and Ind. **1955**, 282.
62. West, C. A., M. Oster, D. Robinson, F. Lew and P. Murphy: Biochemistry and Physiology of Plant Growth Substances. Ed. F. Wightman and G. Setterfield, p. 313. Ottawa: Runge Press. 1968.
63. Whalley, W. B.: The Stereochemistry and Biosynthesis of Certain Complex Diterpenes. Tetrahedron **18**, 43 (1962).
64. Zhukov, L.: Quoted by H. J. Nicholas in P. Bernfeld, Biogenesis of Natural Compounds, p. 675. Oxford: Pergamon Press. 1963.

(Received, November 16, 1970)

Chemistry of Natural Products Derived from Marine Sources

By E. PREMUZIC, Colton, California

With 1 Figure

Contents

	Page
I. Introduction	417
II. Steroids	418
III. Sapogenins of Marine Origin	425
IV. Bile Alcohols and Bile Acids	431
V. Terpenes and Related Hydrocarbons	435
VI. Halogen-Containing Compounds	446
VII. Non-Proteinoid Nitrogen-Containing Substances	451
VIII. Quinonoid and Related Pigments	460
IX. Carbohydrates	467
X. Related Topics	469
Addendum	469
References	472

Acknowledgments. I wish to thank the National Science Foundation Sea Grant (GH-34) for financial support of this work. Thanks are also due to Dr. A. I. SCOTT, Sterling Laboratories, Yale University and Dr. J. H. GREEN, Mrs. B. B. PREMUZIC, Mr. C. KELLY, Mrs. M. NORMAN and Mrs. L. M. FARMER, formerly all of New England Institute, Ridgefield, Connecticut.

I. Introduction

The chemistry of substances derived from plants, fungi and bacteria has in past several decades received a great deal of attention. Indeed, many natural products and their synthetic analogs play an important part in several major industries today. However, a brief glance at the literature dealing with marine sources reveals a gross lack of chemical activity. We may describe investigations in marine science as zoological, biological, pharmacological, clinical and chemical—with activity decreasing in that order.

Several outstanding works available today describe marine sources, even crude extracts of which show real and potential pharmacological activity; in general, however, the chemical information provided by the reports, is not arranged systematically. Nevertheless one fact is clear: marine species will yield many unique and useful compounds. It is not unreasonable to think that marine species as a source of unique compounds may eventually compete with their terrestrial counterparts. The above mentioned key references are Halstead's and Courvilles's Poisonous and Venomous Marine Animals of the World (*112*), NIGRELLI and coworkers' Substances of Potential Biomedical Importance from Marine Organisms (*216*), Der Marderosian's Marine Pharmaceuticals (*182*), Scheuer's The Chemistry of Toxins Isolated from Some Marine Organisms (*253*, *255*) and Baslow's Marine Pharmacology (*38a*). While there are many other references, it is the opinion of this reviewer that the quoted references are the most comprehensive and up-to-date sources of information related to the field under discussion.

One may speculate as to the reasons why there is still a relative lack of chemical investigation in this area. It may be due to difficulties in procurement of starting materials, isolation and chemical characterization. However, this seems unlikely. After all, plants, for example, have presented and still present difficulties. There are many marine biological stations which already possess and are certainly capable of furnishing large collections of marine species.

It is hoped that this review will bring up to date the knowledge of the chemistry in this field and hopefully, promote interest among chemists to focus their attention to this, as yet, wide open field.

II. Steroids

The classical reviews by BERGMANN (*42*) dealing with steroids derived from marine invertebrates and the review by MILLER (*196*) dealing with steroid distribution in algae, indicate that in addition to cholesterol, a number of closely related steroids are distributed in a variety of marine species. The major differences are in the *degree* of unsaturation and the number of positional isomers in ring B and the side chain. The occurrence of C-24 substituents (e. g. methyl, ethyl, methylene) as well as the C-20 methyl configuration (i. e. α or β) (*240*) suggest a biogenetic relationship within a given phylum. Selected steroids of marine origin are listed in Table 1. Caution must be, however, exercised in any attempts at establishing biogenetic relationships: careful examination and in some cases reexamination of suggested structures by modern physical methods has to be carried out.

This is clearly demonstrated by IDLER and coworkers (*144*) in their re-examination of sterols in red algae (Rhodophyceae). The non-saponifiable fraction of the hot acetone extract of the algae yielded a mixture of steroids. Cholesterol [see Fig. 1], 22-dehydrocholesterol, 24-dehydrocholesterol, 24-methylenecholesterol, β-sitosterol, stigmasterol and fucosterol were isolated. Characterization of these compounds was

References, pp. 472—488

based on their retention times in gas liquid chromatography, NMR, IR and mass spectroscopic data. It is evident from the examples in Fig. 1 that problems associated with isolation and identification of these steroids are very complex. For example, isomerisation during hydrolysis under

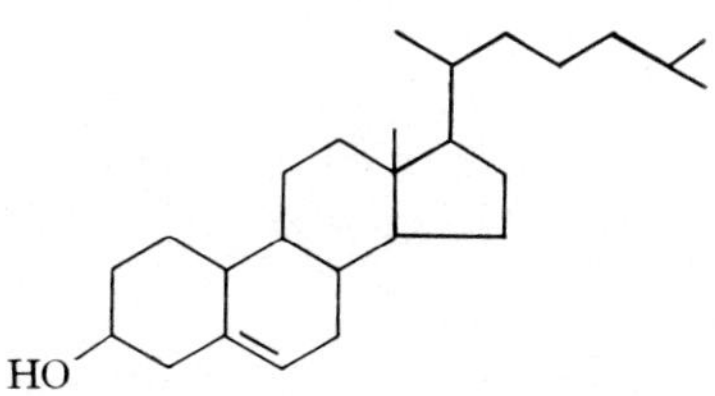

Cholesterol

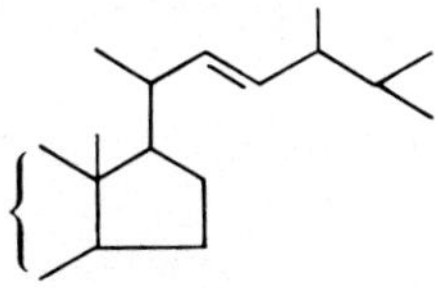

Brassicasterol
(24α-Methylcholesta-5,22-diene-3β-ol)

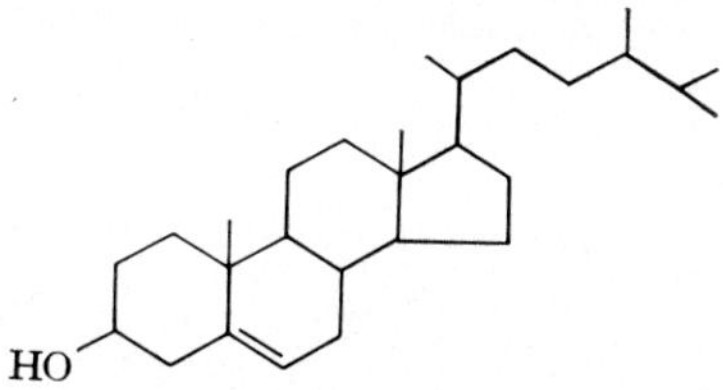

β-sitosterol (Δ⁵-stigmastene-3β-ol)

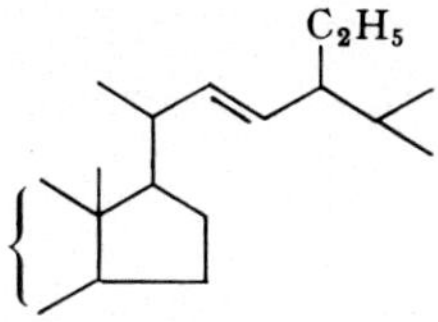

Stigmasterol

Fig. 1

alkaline conditions may lead to artifacts. Systematic work of TSUDA and co-workers (*146*) on green and brown algae, GUPTA and SCHEUERS work on echinoderm (*106*) and coelenterate (*107*) sterols and MEUNIER et al work on Rhodophyta (*107a*) has brought the research in this field much nearer to tabulation and classification of marine steroids.

GUPTA (*108*) has recently reported the presence of coprostanol (**2**) in whale oil.

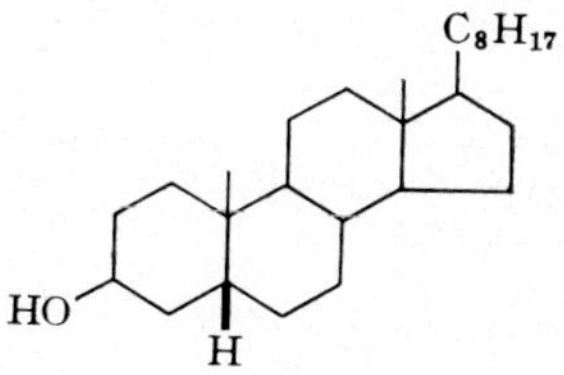

(**2**) Coprostanol

BERGMANN (*42*) has isolated from the gorgonian *Plexaura flexuosa* a very unusual sterol whose structure has only been recently elucidated (*110*, *111*). Gorgosterol (**3**) contains a cyclopropane ring at C-22 and C-23.

Table 1

Type	Structure	Characteristic Constants	Source
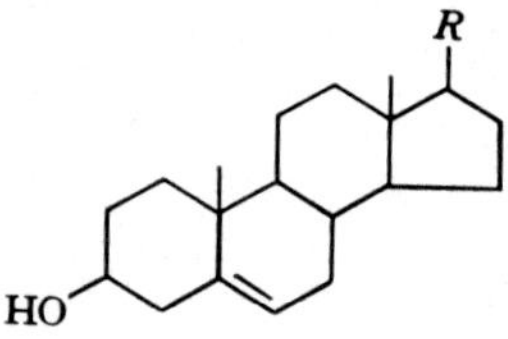 (steroid nucleus with R, HO-)	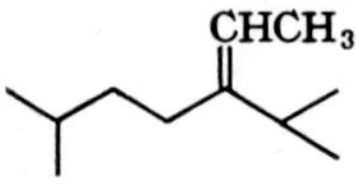 R = CHCH$_3$ group; C-20 Methyl β $\Delta^{5,24(28)}$ diene: Fucosterol	m. p. 124° C $(\alpha)_D = -38°$	Algae
	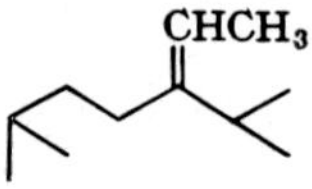 CHCH$_3$ group; C-20 Methyl α $\Delta^{5,24(28)}$ diene: Sargasterol	m. p. 130° C $(\alpha)_D = -47,5°$	Algae
	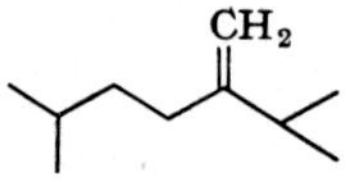 CH$_2$ group; 24-Methylene cholesterol	m. p. 142° C $(\alpha)_D = -35°$	Sea anemone Sponges bivalves
	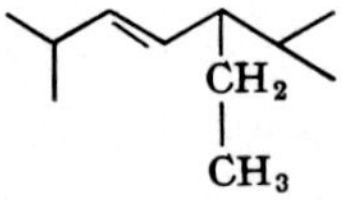 CH$_2$–CH$_3$ group; C-24 ethyl α Poriferasterol	m. p. 156° C $(\alpha)_D = -49°$	Sponges
	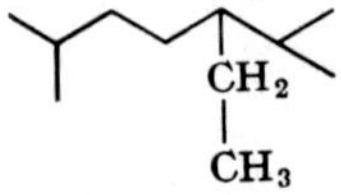 CH$_2$–CH$_3$ group; C-24 Ethyl α γ-Sitosterol	m. p. 138° C $(\alpha)_D = -42°$	Marine Invertebrates
	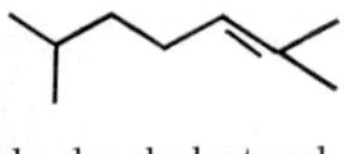 24-Dehydrocholesterol (Desmosterol)	m. p. 121° C $(\alpha)_D = -41°$	Barnacle
	22-Dehydrocholesterol	m. p. 135° C $(\alpha)_D = -57°$	

References, pp. 472—488

Table 1 (continued)

Structure		Characteristic Constants	Source
R = (C-24 Me β Campesterol)	C-24 Me β Campesterol	m. p. 158° C (α)D = − 33°	Sponges, Mollusks
(C-24 Me α Brassicasterol)	C-24 Me α Brassicasterol	m. p. 148° C (α)D = − 64°	Sponges Mollusks, Sponges Coelenterates Mollusks
$\Delta^{5,7,22(23)}$ Ergosterol		m. p. 166° C (α)D = − 132°	Marine animals, Mollusks, Echinodermata, Liver oils, Annelids Echinoderms Sponges
24-ethyl-cholestan 3 β-ol Poriferastanol		m. p. 143° C (α)D = + 25	Sponges
$\Delta^{7,22(23)}$ diene, 24 ethyl Chondrillasterol		m. p. 169° C (α)D = − 2°	Sponges
Zymosterol, $\Delta^{8,24}$-cholestadiene-3β-ol		m. p. 111° C (α)D = + 49°	

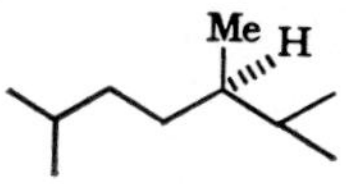

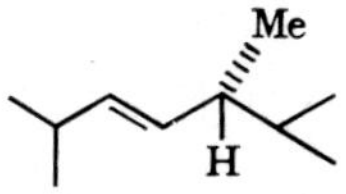

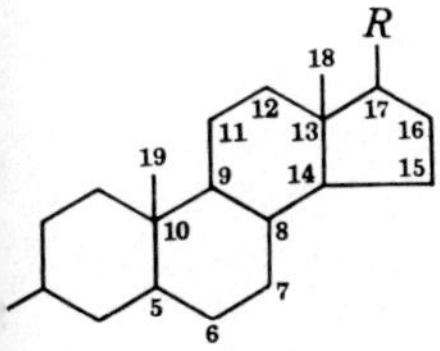

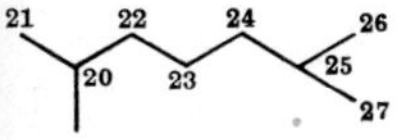

Structural analysis and x-ray data are consistent with (3) being (22 R, 23 R, 24 R)-22,23-methylene-23,24-dimethylcholest-5-en-3β-ol. Gorgonians have also yielded another sterol containing a cyclopropane ring in

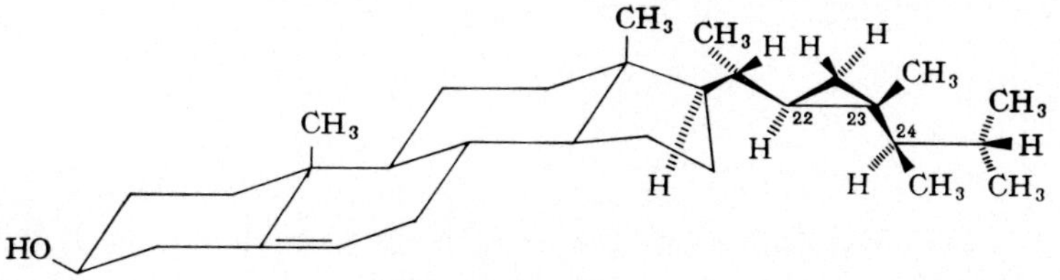

(3) m. p. 186.5°–188° C
[α]D = − 45°

the side chain, but missing the C-23 methyl (*260a*). The biosynthesis and the biological activity of these steroids is unknown; however, the presence of an alternative biosynthetic sequence involving a cyclopropane intermediate in the alkylation of the side chain of sterols is suggested.

BROTZU (*53*) has isolated from the sea near a Sardinian sewage outfall an antibiotic producing organism belonging to the *Cephalosporium* species. ABRAHAM and his collaborators (*2*) subsequently isolated from cultures of cephalosporium species, cephalosporin P_1 (*4*), together with other antibiotics which will be discussed later, (*4*) inhibited the growth *in vitro* of staphylococci, cornyebacteria and Cl. tetani.

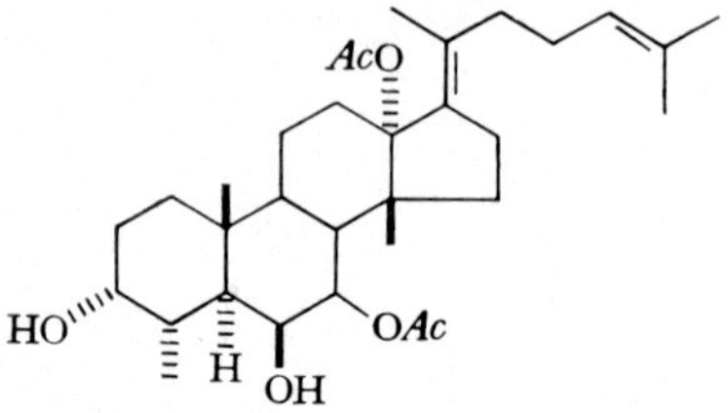

(4)

Within the past few years, the study of moulting hormones has become a very active field (*93, 99, 159, 269*). The chemistry and biochemistry of moulting hormones has been reviewed by BERKOFF (*46*) and that of arthropod moulting hormones by HIKINO and HIKINO (*131*).

Work of HORN and co-workers (*98, 114, 134, 268*) led to the isolation and characterization of crustacean moulting hormones: crustecdysone (*5*) and deoxycrustecdysone (*6*) were both isolated from crayfish *Jasus*

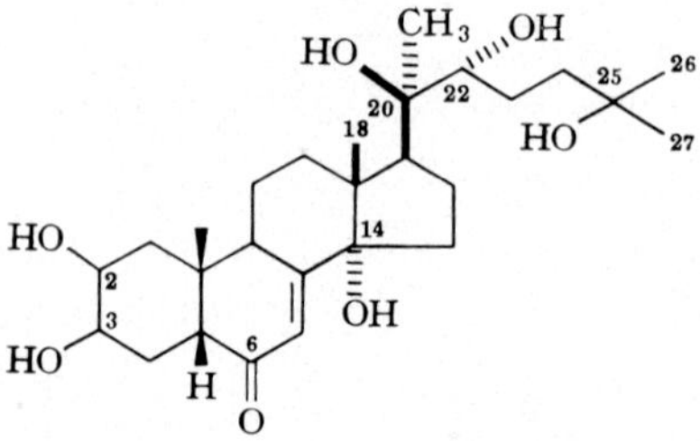

(5) Crustecdysone m. p. 241–242.5° C 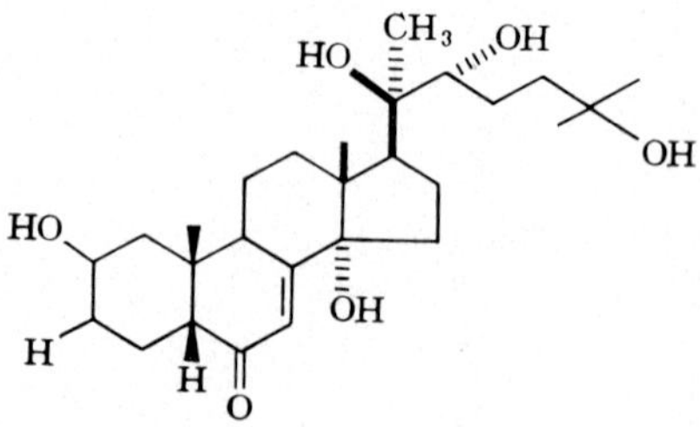   (6) Deoxycrustecdysone m. p. 232–235° C

lalandei. The crustacean moulting hormones are highly active in the Calliphora test, normally used for insect moulting hormones (*91, 157, 158*). Crustecdysone (*5*) is structurally closely related to the known moulting hormone ecdysone (*7*); however, (*5*) has an additional hydroxyl group

References, pp. 472—488

at C-20. The structure of 20-hydroxyecdysone (5) was further proved by synthesis (*142, 165, 268*).

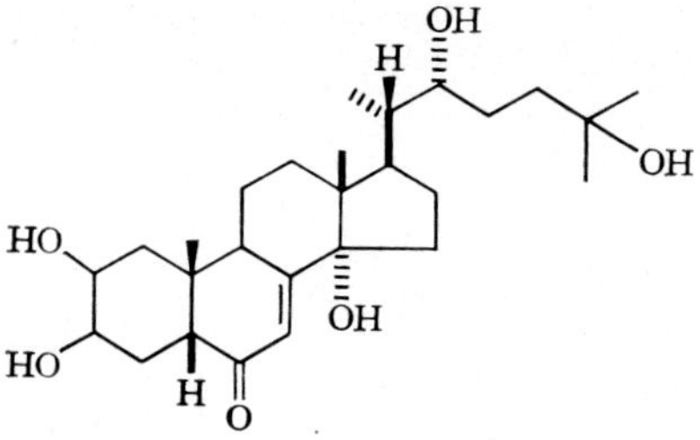

(7) Ecdysone

The 20 R : 22 R configuration is consistent with the C-20 biochemical hydroxylation of cholesterol in warm blooded animals.

A convenient synthesis of crustecdysone has been reported (*165*). In this synthesis progesterone was converted to 20β-hydroxy-5α-pregnan-3,6-dione (8). The dione (8) on nitration at C-20, followed by bromination,

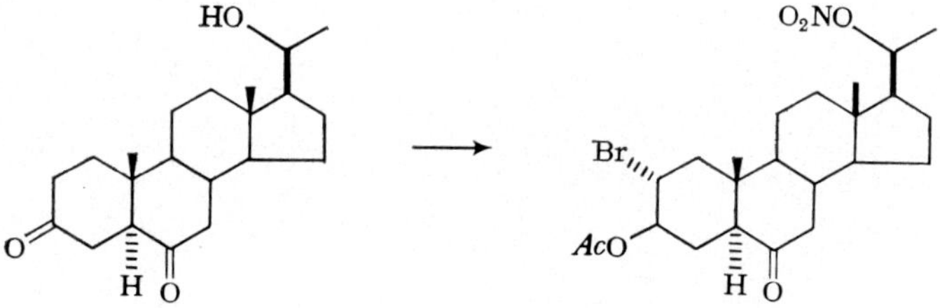

(8)   (9)

selective reduction with LiAl(t-BuO)$_3$H and acetylation yielded the 5α-bromo-3β-acetoxy C-20 nitro analog (9). Treatment of (9) with silver acetate in acetic acid furnished the 2,3-diacetate. Reductive removal of nitrato group, introduction of the Δ^7 double bond, followed by bromination gave the 7α-bromo ketone (10), which on oxidation and removal of hydrogen bromide afforded the 6,20 dione (11). Selective degradation of

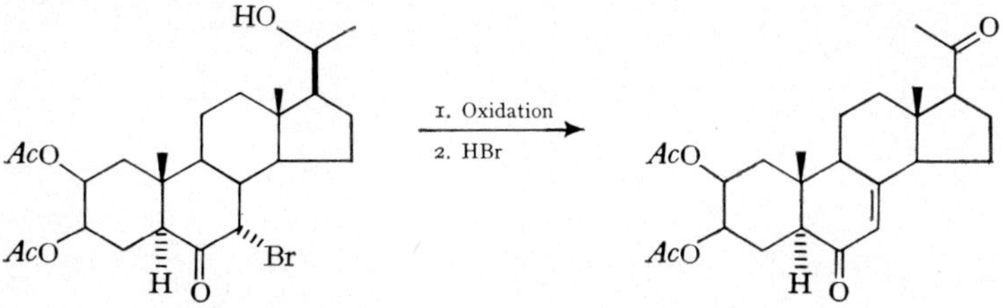

(10)   1. Oxidation   2. HBr   (11)

the dione with 4-(tetrahydropyranyloxy)-4-methyl-1-pentinyl magnesium bromide gave a mixture of C-20 isomers. The predominant isomer has the 20 R configuration (12) and was separated by chromatography.

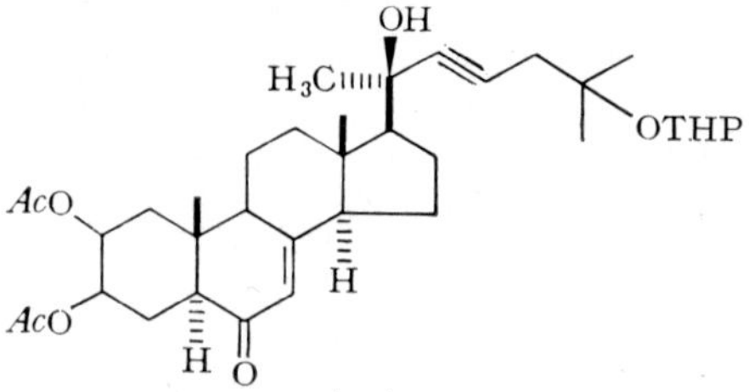

(12)

Removal of tetrahydropyranyl ether (THP) afforded the corresponding dione which on treatment with $Hg(OAc)_2$-BF_3 etherate in methanol yielded specifically the C-22 ketone (13). Allylic oxidation of (13) with

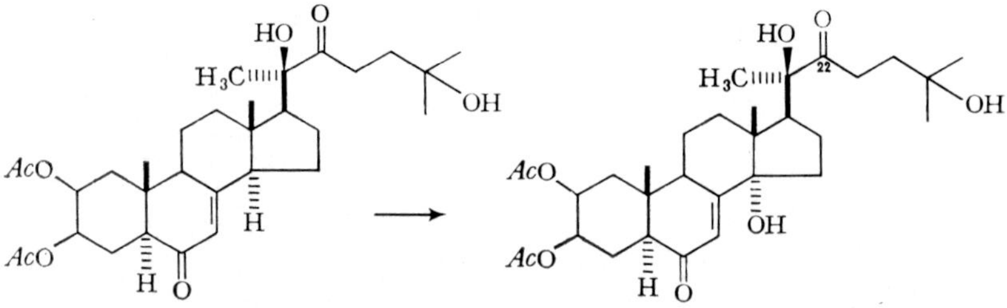

(13) → (14)

SeO_2 yielded the triol (14). Alkaline hydrolysis and simultaneous inversion of configuration at C-5 led to the 5β-pentol. Selective reduction of the C-22 carbonyl with LiAl(t-BuO)$_3$H afforded a mixture of C-22

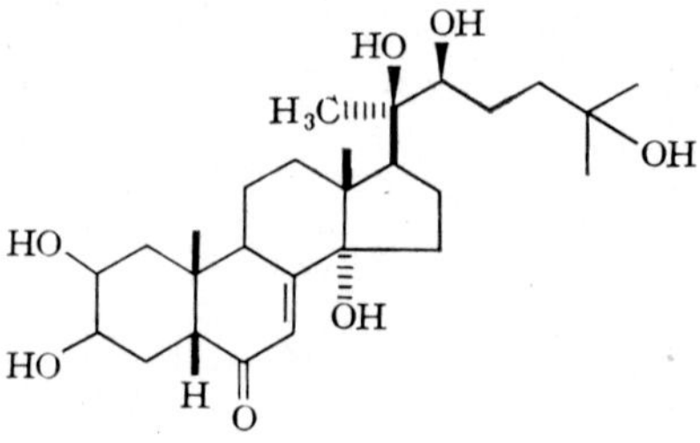

(15) Isocrustecdysone
m. p. 259–260° C $\varepsilon_{242} = 11700$

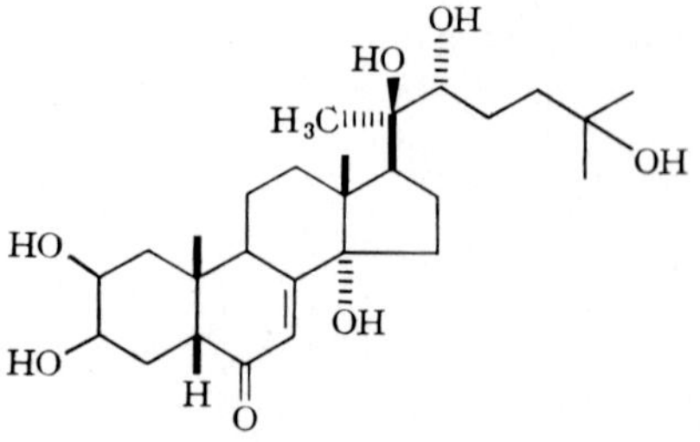

(16) m. p. 240–242° C
λEtOH max 240 mμ (ε 12,670), nmr
(pyridine-d5) 1.07 (19-H), 1.20 (18-H),
1.36 (26 H, 27 H), 1.56 (21 H) and
6.17 (7-H)[28, 29]

References, pp. 472—488

isomeric alcohols (15) and (16) which were separated by thin layer chromatography. Mixed melting point, UV, IR and NMR data showed that (16) was identical with the naturally occurring crustecdysone. Careful NMR analysis of the isomers is required, because the major difference between (15) and (16) is the stronger de-shielding of the C-21 methyl signal (1.66 ppm) in the spectrum of isocrustecdysone compared to the C-21 methyl signal of crustecdysone (1.55 ppm). An alternative synthesis of the crustacean moulting hormone has been recently reported by HÜPPI and SIDDALL (*143*).

III. Sapogenins of Marine Origin

In the extracts of Cuvierian glands of sea cucumbers (Echinodermata, Holothurioidea), several toxic principles have been found. Their pharmacological and physiological action has been investigated by a number of workers (*61, 62, 92, 215, 242, 246, 324*). These substances are glycosides, the principal monosaccharide components being D-glucose, D-xylose, quinovose and 3-O-methylglucose (e. g. Holothurin A isolated from *Actinopyga agassizi* (*61*)). The toxic principles also fall into the category of saponins and can be divided into two sub-groups, i. e. those that can be precipitated by cholesterol and those that cannot. A further sub-division into two groups is based on the nature of the aglycone. The first group on hydrolysis yields steroidal sapogenins and the second triterpenoid sapogenins (*189*). Presence of steroidal saponins is not limited to sea cucumbers. Their presence in starfish and other echinoderms has been reported (*326, 327*).

Holothurin isolated from *Actinopyga agassizi* (*61*) is a mixture of several glycosides containing the already mentioned monosaccharides *(vide infra)*, sulfuric acid, and several steroidal aglycones for which CHANLEY, SOBOTKA and co-workers (*63, 64*) have suggested the name of Holothurinogenins. The two most abundant components, 22,25-oxidoholothurinogenin (17) and its 17-desoxy derivative (18) have been characterized:

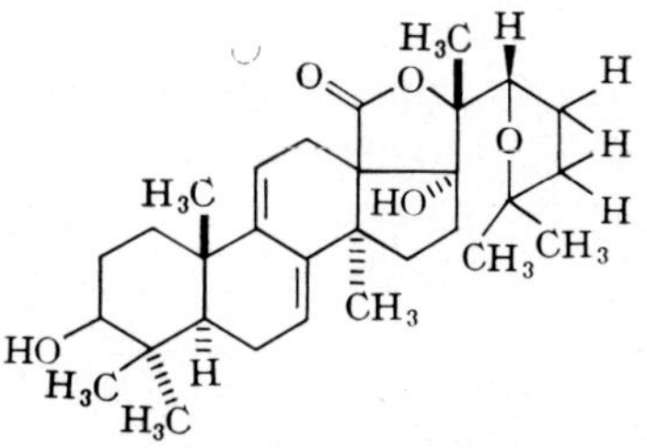

(17) m. p. 315.2–315.8° C
$[\alpha]_D^{25} = -21.2°$ (CHCl$_3$)

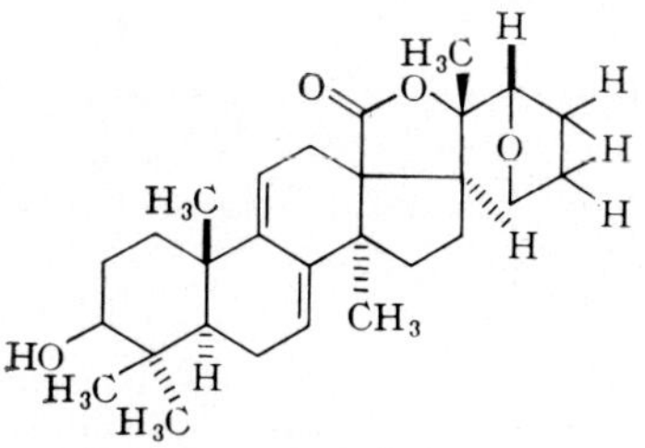

(18) m. p. 285.8–286.4° C
$[\alpha]_D^{25} = -9.3°$

3β, 17α, 20ξ-trihydroxy-5α-lanosta $7:8$, $9:11$-diene-18-carboxylic acid lactone ($18 \rightarrow 20$).

The structure and stereochemistry of (**17**) and (**18**) had been thoroughly elucidated primarly by means of UV, IR, NMR and ORD studies (*63, 64*). The bulk of the remaining aglycones associated with the extract have the basic skeleton of (**17**), but differ in the structure of the side chain. Results presented by CHANLEY and co-workers suggest the following possible structures (**19**).

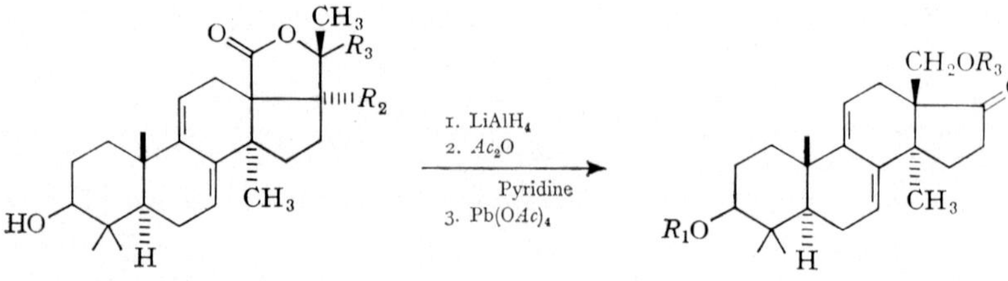

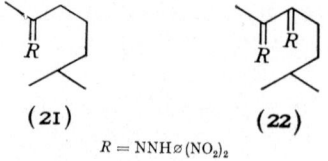

(**19**) $R_2 =$ H or OH

$R_3 =$ OH and/or and/or

(**20**) R_1, $R_3 = Ac$

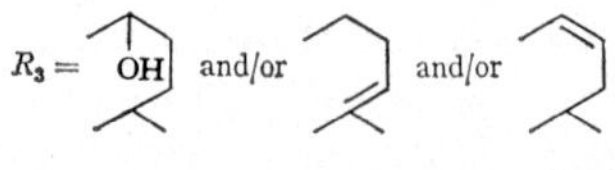

The C-22 ketone (**20**) is identical with the ketone derived from (**17**) under identical experimental conditions which confirms the common basic skeleton. In addition to (**20**) the reaction mixture also contained ketonic fragments which yielded two known 2,4-dinitrophenylhydrozones, viz. 6-methylheptan-2-one (**21**) and 6-methyl-heptan-2,3-dione (**22**).

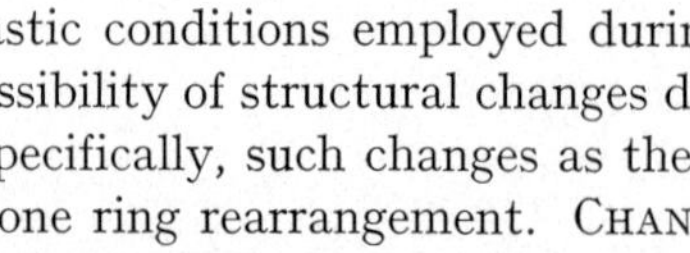

(**21**)   (**22**)

$R = \text{NNH}\emptyset(\text{NO}_2)_2$

In view of the drastic conditions employed during the hydrolysis of the sapogenins, the possibility of structural changes during this procedure cannot be excluded, specifically, such changes as the introduction of the diene system and lactone ring rearrangement. CHANLEY and co-workers (*63, 65*) have, indeed, shown that changes do occur. Mild hydrolysis of holothurin A ($CH_3OH-HCl$) removes the sugar residues and gives a mixture of mono- and dimethoxylated neo-holothurinogenins. These do not contain a conjugated diene system and are, therefore, consistent with a native holothurin in which the diene system is also absent. The mixture of neo-holothurinogenins consists of 12β-methoxy-7,8-dihydro-22,25 oxidoholothurinogenin (**24**), the 17-desoxy analog (**25**), 12β-methoxy-

References, pp. 472—488

(23)

	R_1	R_2	R_3
(24)	O*Ac*	OH	
(25)	O*Ac*	H	

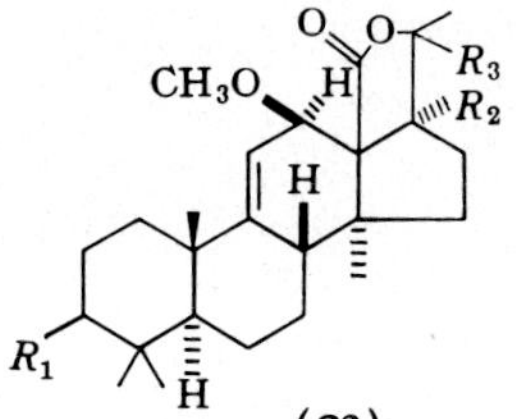

7,8-dihydro-24,25-dehydro, (26), 12β, 25-dimethoxy-7, 8-dihydro (27) and 12β-methoxy-22-acetoxy-7, 8-dihydroholothurinogenin (28). Strong

	R_1	R_2	R_3	
(26)	O*Ac*	OH		R', R'' (see text)
(27)	O*Ac*	OH		
(28)	O*Ac*	OH		

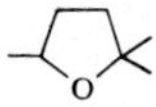

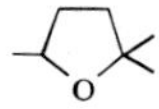

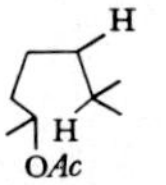

acid treatment of these derivatives yielded the known holothurinogenins (e. g. (17), (18), and griseogenin (29), thus furnishing further evidence that the neo-holothurinogenins correspond in structure to the aglycons of native holothurin A. In addition, native holothurin A has only one OMe group: that associated with 3-methoxyglucose in the sugar residue (*63, 65*). This means that during the mild hydrolysis, the C-12 hydroxyl is methoxylated.

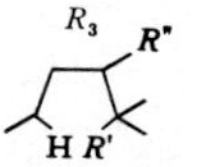

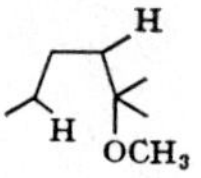

The stereochemistry at C-12 was deduced in the following manner: enzymatic hydrolysis (glusulase) of desulfated holothurin A afforded, in addition to the known holothurinogenins (24) and (28) obtained by MeOH/HCl hydrolysis, 12α-methoxy-17-desoxy-7,8-dihydro-22,25-oxido-holothurinogenin-3-acetate (30), 12α-acetoxy-7,8-dihydro-24,25-dehydro-holothurinogenin-3-acetate (31), and the 3β-xyloside of 12β-methoxy-7,8-dihydro-24,25-dehydroholothurinogenin (32). Isolation of (31) was taken as further evidence for the 12α-OH (axial) configuration in the aglycone moiety of the native holothurin. The fast rate of methoxylation of the OH at C-12 is also consistent with this configuration. In view of this evidence, it is quite reasonable to assume that formation of the 12β isomer (32) occurs during the hydrolysis as the result of acid catalysis.

Investigation of an extract of Cuvier glands of the sea cucumber *Halodeima Grisea* L. led also to the isolation of 22,25-oxidoholothurino-genin (17) (*297*). In addition, the extracts of the body wall of *Halodeima*

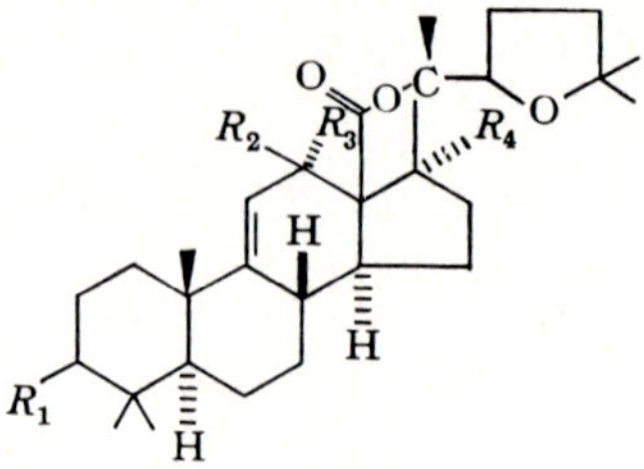

(30) $R_1 = OAc$, $R_2 = H$, $R_3 = OCH_3$, $R_4 = H$

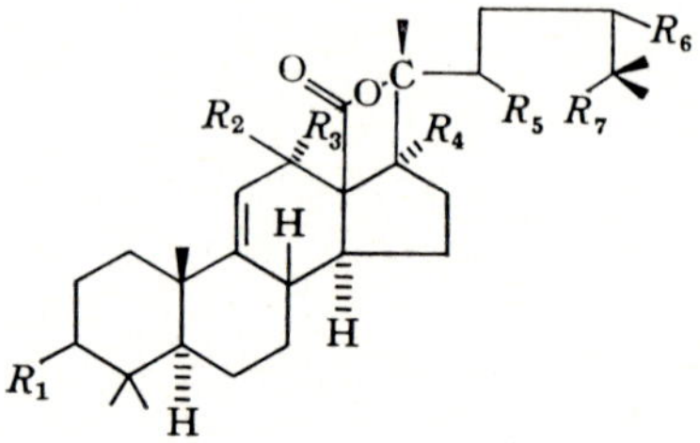

(31) $R_1 = OAc$, $R_2 = H$, $R_3 = OAc$, $R_4 = OH$, $R_5 = H$, R_6, $R_7 ====$
(32) $R_1 = \beta$-xylose, $R_2 = OCH_3$, $R_3 = H$, $R_4 = OH$, $R_5 = H$ R_6, $R_7 ====$

Grisea and *Halothuria Vagabunda* yielded a triterpenoid sapogenin griseogenin (29). The fundamental difference between griseogenin and

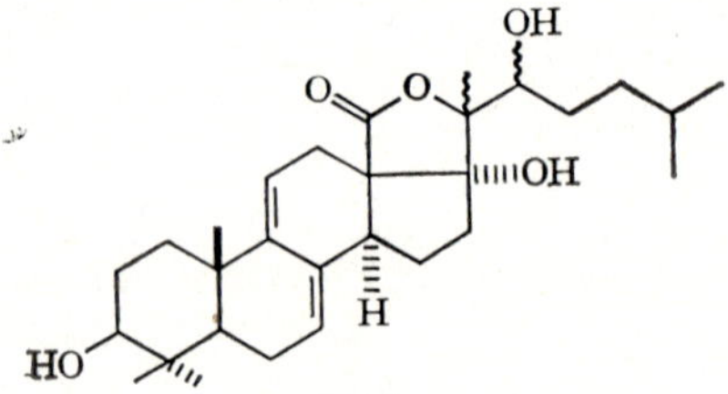

(29) m. p. 285–287°

22,25-oxidoholothurinogenin is the absence of oxygenation at C-25. The presence of the common lanostane skeleton in (29) is strongly supported by NMR evidence e. g. there is no proton signal due to an alcoholic lactone terminal, indicating that the lactone is attached to a tertiary carbon. There is only one position for such a lactone bridge in the lanostane skeleton, i. e. that shown in structures (29) and (17).

DJERASSI and co-workers (244) have reported the isolation and structure determination of three holothurinogenins (33), (34) and (35) from the sea cucumber *Bohadschia Koellikeri*.

References, pp. 472—488

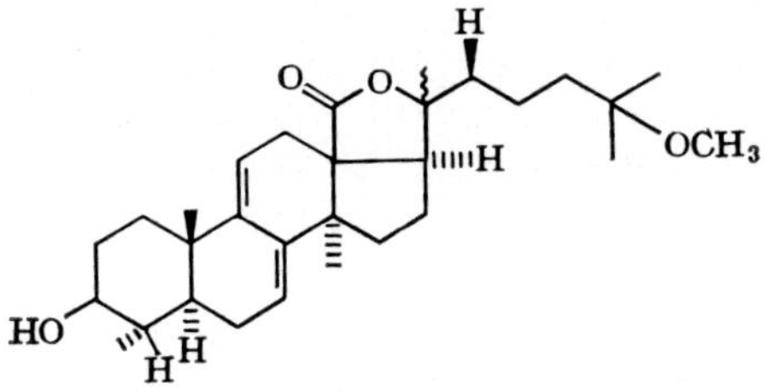

(33) Ternaygenin m. p. 239–242° $[\alpha]_D = +2°$

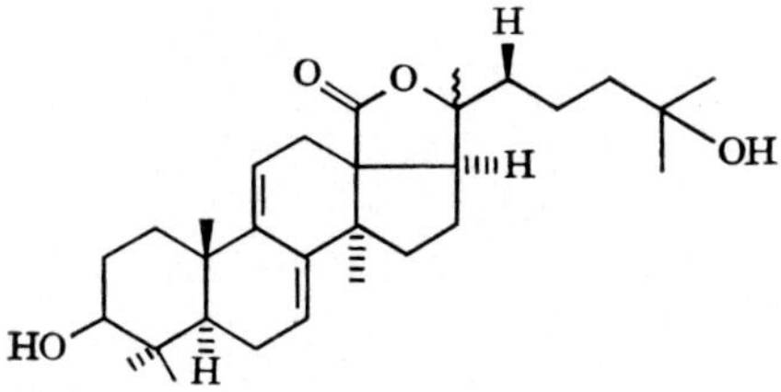

(34) Koellikerigenin m. p. 213–214° $[\alpha]_D = -8°$

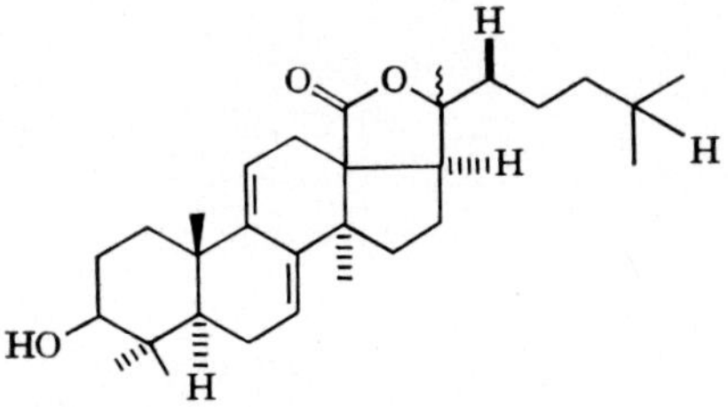

(35) Seychellogenin m. p. 234–238° $[\alpha]_D = -7°$

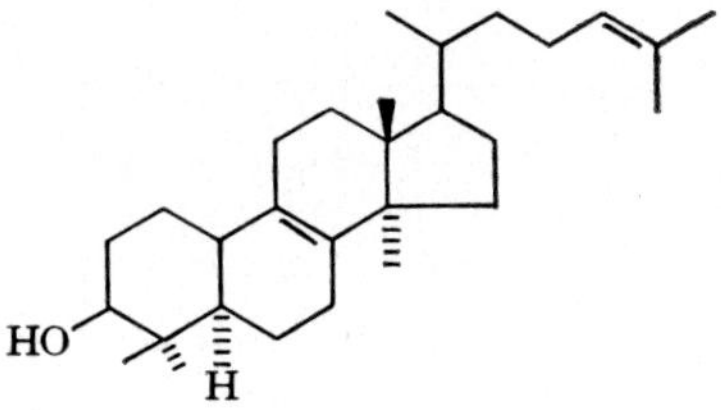

(36) Lanosterol

All three compounds possess the basic lanosterol skeleton; the chemical relationship between seychellogenin (35) and lanosterol (36) was established unambiguously by degradation (244) which led to the preparation from (35) and (36) of the same triol (37):

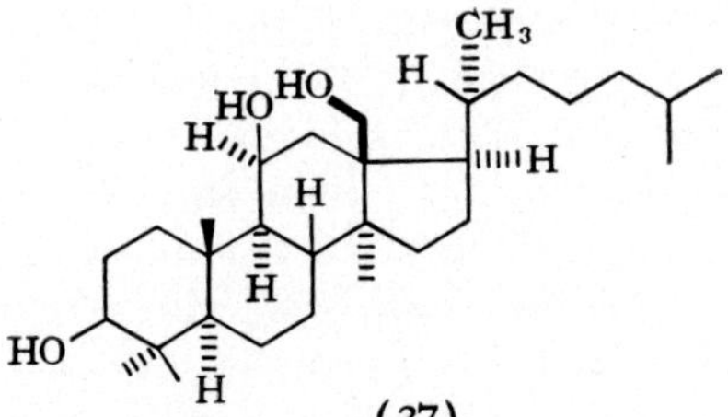

(37)

Bohadschia koellikeri also yielded (*296*) a small amount of a triterpenoid sapogenin, which was named praslinogenin (38). By analogy with a reaction exhibited by coellikerigenin (34), methylation of a tertiary hydroxyl may have occurred during acid hydrolysis in methanol and consequently (38) may be an artifact produced from a C-25-OH precursor. The stereochemistry at C-20 has yet to be elucidated.

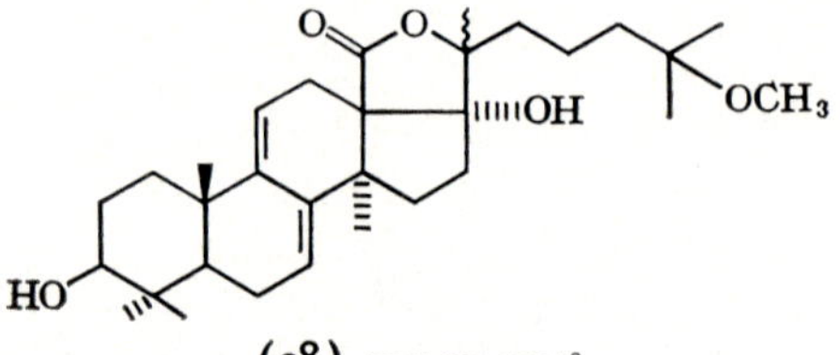

(38) m. p. 290–291.5°

In the methanolic extracts of the Far-Eastern Sea cucumber, the *trepang (Stychopus Japonicus Selenka)* (*84*) two glycosides, stichoposide A and C have been identified. The aglycones, stichopogenin A_2 (39) and A_4 (40) are tetracyclic triterpenes, the proposed structures being consistent with those of other sapogenins in this series. The major difference is the

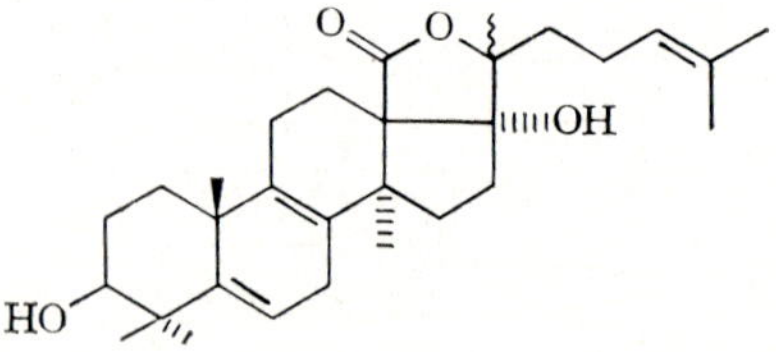

(39) m. p. 238–240°
monoacetate m. p. 216–219°
$[\alpha]_D = -36.3°$ (CHCl$_3$)

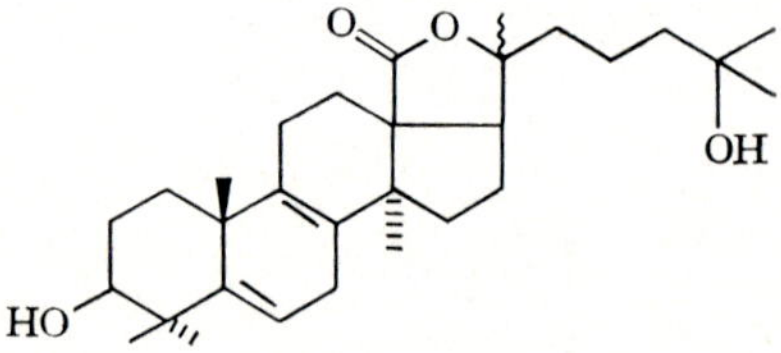

(40) m. p. 238–240°
monoacetate
m. p. 221–223°

absence of the heteroanular diene system and its replacement by unsaturation at $\Delta^{5,6}$ and $\Delta^{8,9}$ and, in stichopogenin A_2, the presence of a double bond at C-24 and absence of a C-25 tertiary hydroxyl.

A preliminary note by HABERMEHL and VOLKWEIN (*109*) indicates that there are geographic differences in holothurin content and suggests an overall biogenetic relationship of these compounds with lanosterol. These authors have investigated the following Mediterranean echinodermata: *Holothuria tubulosa, H. polii, H. forskali, H. sanctori, Stichopus regalis,* and *Cucumaria planci.* Of these only *H. forskali* and *H. sanctori* possess Cuvierian glands. *H. tubulosa* contained 22,25-oxidoholothurinogenin (17) and the 17-desoxy analog (18). However in contrast to *Actinopyga agassizi (vide infra)* the glycosidic portion consisted only of an aldopentose, and the hydrolysate of the saponin extracts did not contain

References, pp. 472—488

any free sulfuric acid. Chromatography of the hydrolysates of *H. polii* and *H. forksali* revealed substances whose R_f values corresponded to those of (17) and (18). However, hydrolysates of *H. sanctori, Stichopus regalis* and *Cucumaria planci* had no such components. All the species investigated contained cholesterol and lanosterol, providing additional evidence for biogenetic relationships within this class of compounds.

Systematic investigation of the holothurins in the skin of *Holothuria polii (Delle Chiaje)*, has revealed (*109a*) that in addition to the known holothurinogenins (17) and (18) three novel holothurinogenins (40a), (40b) and (40c) were present. All possess the heteroannular diene grouping,

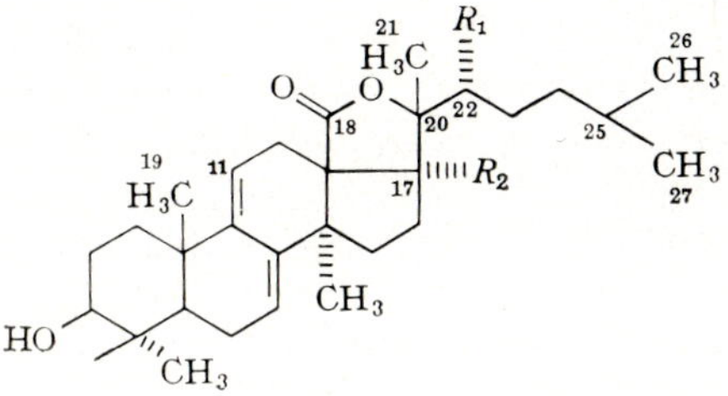

(40a) $R_1 = H, R_2 = OH$

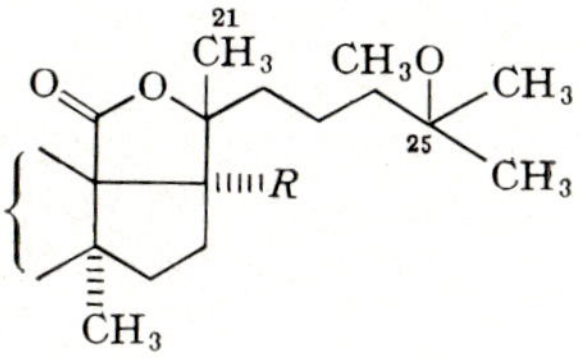

(40b) $R = OH$
(40c) $R = H$

with (40a) closely resembling (29). HABERMEHL and VOLKWEIN (*109a*) have concluded from their NMR analysis of the 17α-OH and C-21 methyl region of (17) and the C-20 methoxyholothurinogenin (40b), that since the C-21 methyl signals are identical, the C-21 methyl group must be above the plane of the molecule. This means that it is *trans*- with respect to the 17α-hydroxy group, and consequently the geometry at C-20 is *R*.

IV. Bile Alcohols and Bile Acids

HASLEWOODS reviews (*121, 122, 124*) of the biochemistry and chemistry of bile salts and HOSHITAS and KAZUNOS review (*123*) dealing with the chemistry and metabolism of bile alcohols and bile acids covers the literature in this field until 1968. The work cited illustrates the utility of bile alcohols and bile acids for comparative studies and indeed as possible handles for chemical taxonomy. Although the present review is primarily concerned with natural products of marine origin, the occurrence of bile alcohols and bile acids in a variety of species warrants a brief mention of general methods used for the isolation of these substances.

Crude ethanolic extracts of bile *(vide infra)* are frequently complex mixtures of bile salts requiring careful work up and involved chromatography. The development by Scandinavian chemists of paper chromatography (*274, 275*), partition chromatography (*44, 276*), reversed phase partition chromatography (*45*) and column chromatography (*218*) of free and conjugated bile acids and the contribution of Japanese workers (*137, 138, 139, 163*) to further development of thin layer, paper and column chromatography facilitated isolation of pure compounds for structural studies.

Historically speaking, the earliest bile alcohol from a marine source was scymnol (**41**) whose isolation was reported by HAMMARSTEN (*113*) in 1898 from the bile of the shark *Scymnus borealis*. Scymnol has since been reported to be present in other elasmobranch fish, e. g. ray fish *(Dasyatis akajei)* (*83*), blue skate *(Raia batis)* (*76*), grey dogfish *(Squalus acanthias)* (*76*); in all some ten other species of sharks and rays, and is now considered a characteristic component of the bile of elasmobranch fish (*43, 52, 76, 79*). Although scymnol has been known for a considerable period, its correct structure has only been recently established. For a long time, it was thought that scymnol contained an epoxide ring in the side chain as in (**42**). Early structural work (*34, 43, 312*) established the

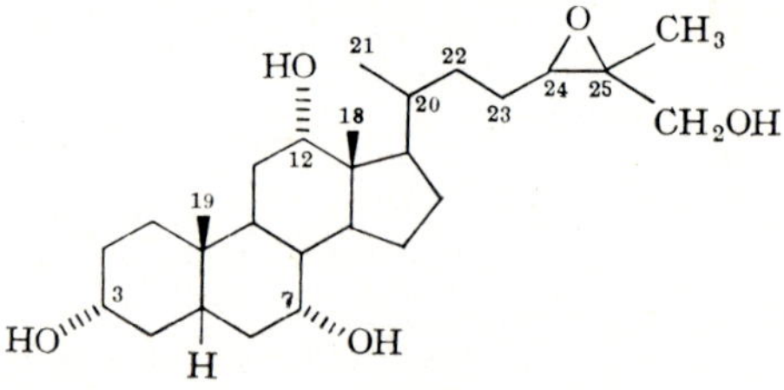

(42)

presence of one primary and three secondary hydroxyl groups. The oxide ring could be opened by hydrogen chloride to give a chlorohydrin and was subsequently regenerated by the action of alkali. Acetylation of scymnol gave a tetraacetate which on oxidation, followed by hydrolysis yielded cholic acid (**43**). FIESER and FIESER (*88*), however, pointed out that the ethylene oxide grouping of (**42**) would not withstand the drastic method adopted for the oxidation of the tetraacetate and that a four-membered oxide ring as in (**44**) would be more consistent with the experimental data.

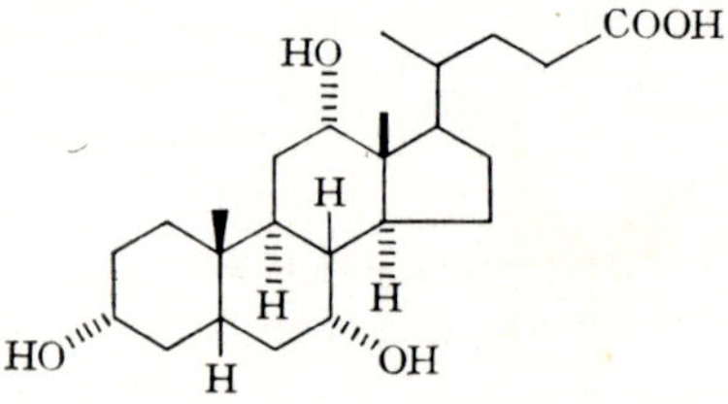

(43) 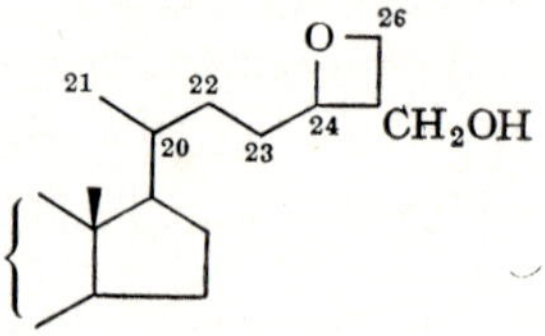   (44)

References, pp. 472—488

Thorough study by CROSS (79) involving both degradative work and NMR analysis established firmly that the material earlier thought to be scymnol, was really that of an anhydro derivative (45) and is an artifact

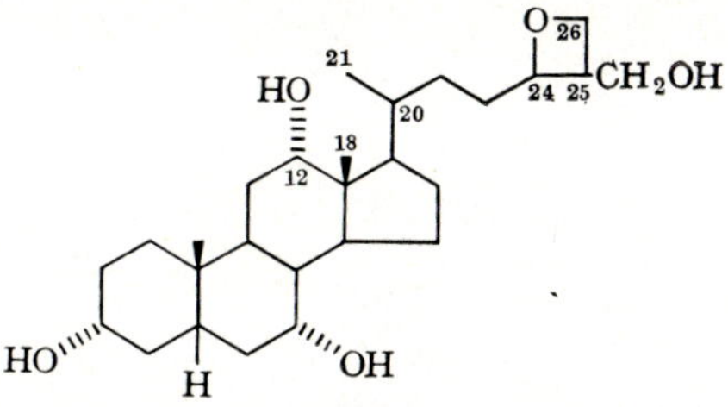

(45)

formed (52) during the alkaline hydrolysis, by elimination of SO_4^{2-} ion between an —OH and an —O·SO_3^- group. This means that the natural bile salt is the sulfate of the hexahydric alcohol scymnol (41) (3α, 7α, 12α, 24ξ–26, 27 hexahydroxycoprostane).

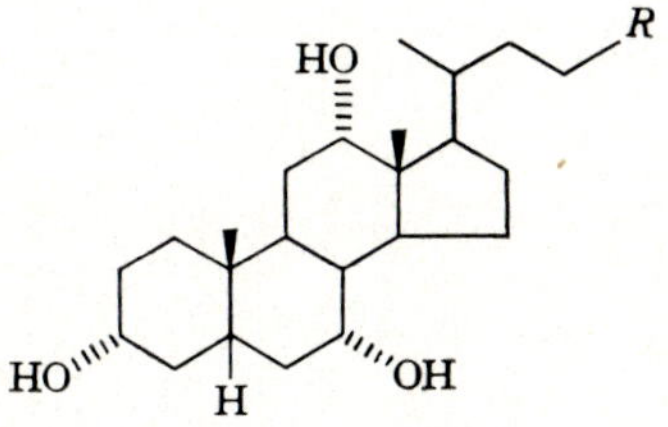

(41) natural bile salt

$R = -\text{CH(OH)CH}\underset{24}{\underset{|}{}}\begin{smallmatrix}\text{CH}_2\text{OH}\\\text{CH}_2\text{OH}\end{smallmatrix}$

Scymnol
m. p. of the dihydrate 190° C
$[\alpha]_D = +34°$ (EtOH)

$R = -\text{CH(OH)CH}\underset{24}{\underset{|}{}}\underset{25}{}\begin{smallmatrix}\text{CH}_2\text{OH}\\\text{CH}_2\text{O.SO}_3^\ominus\end{smallmatrix}$

Cold chromic acid oxidation of the sulfate obtained from shark bile gave the known dehydrocholic acid (46) substantiating further the correctness of structure (41, 52, 89). Partial synthesis of scymnol from cholic acid has also been reported by HASLEWOOD (52).

The presence of β-phocaecholic acid (3α: 7α: 23 trihydroxycholanic acid (47)) in the bile of leopard seal and Californian sea lion has been reported (123).

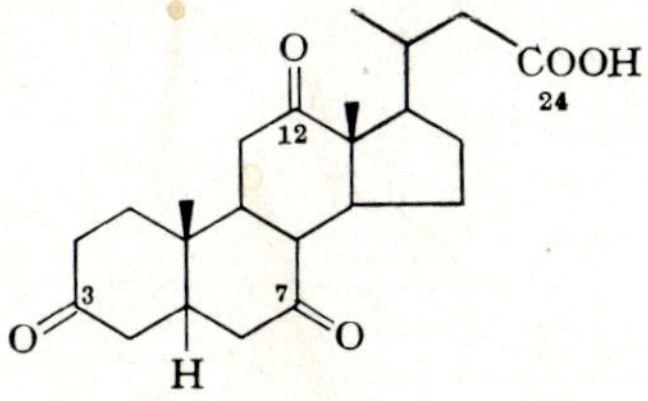

(46)
m. p. 237° C $[\alpha]_D = +26°$ (EtOH)

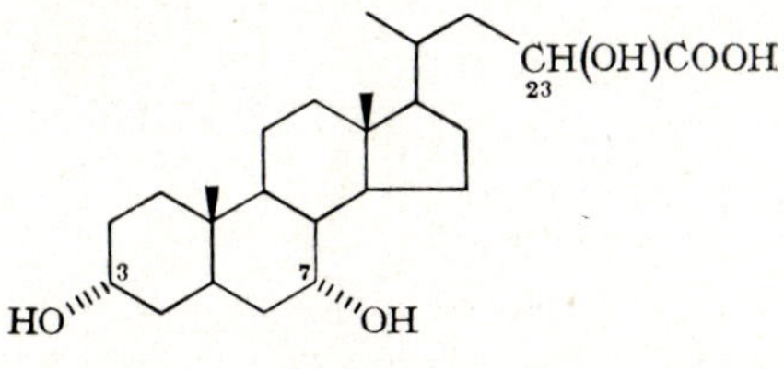

(47)
m. p. 222–224° C $[\alpha]_D = +10.8$ (EtOH)

From alkaline hydrolysates (e. g. 2N NaOH, sealed metal bomb at 158° C for 18 hours) of ethanolic extracts obtained from the bile of hagfish *Eptatretus stoutii* (Pacific) and the Atlantic *Myxine glutinosa*, two prin-

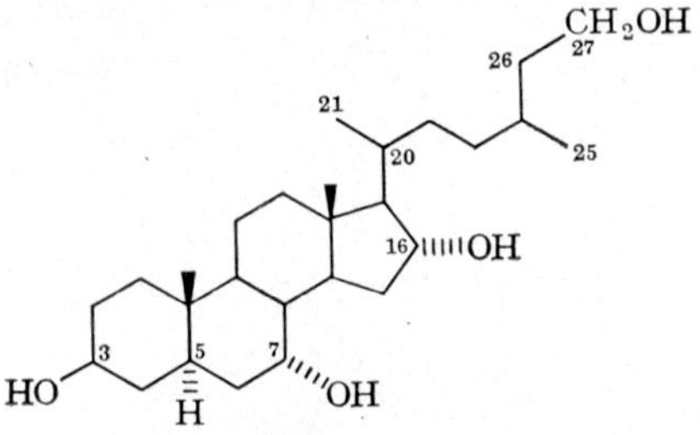

(48)

m. p. 204–206° C $[\alpha]_D = -15°$ (EtOH)

cipal bile alcohols have been isolated: myxinol (48) and 16-Deoxymyxinol (49) (*31, 80, 123, 125, 126*). IR, NMR, mass spectral and ORD data

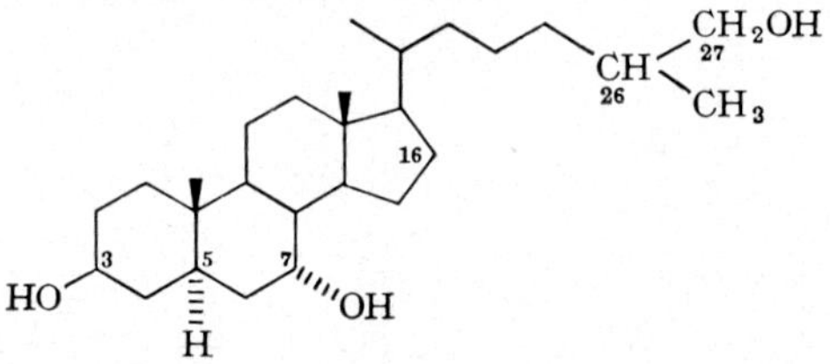

(49)

m. p. 218–219° C $[\alpha]_D = +13°$ (EtOH)

indicate (*31*) that the structure of myxinol is consistent with that of (48) (3β, 7α, 16α, 26(27)-tetrahydroxy-5α-cholestane), the native bile salt being the 3,26(27)-disulphate ester of (48). Further work up and chro-

Table 2

Source	Solvent	m. p./b. p.
Plexaura crassa	pentane or ether	b. p. 71° C/2.3 mm Hg
Eunicea mammosa	pentane or ether	m. p. 150–152° C
Xiphigorgia anceps (*Pterogorgia anceps*)	pentane	m. p. 91.5–92° C
Briareum asbestinum	pentane	m. p. 120–127° C
Pseudopterogorgia americana	pentane	
Leptogorgia setacea	pentane	m. p. 192–196° C

References, pp. 472—488

matography on celite (32) of a gum obtained together with myxinol disulfate from the ethanolic extract yielded 16-deoxymyxinol. Mass spectroscopic and NMR studies have shown (295) that the structure of 16-deoxymyxinol is consistent with that of (49) (3β, 7α, 26(27)-trihydroxy-5α-cholestane).

V. Terpenes and Related Hydrocarbons

A pentacyclic triterpenoid alcohol tetrahymanol has been isolated from the nonsaponifiable lipid fraction of the ciliated protozoan *Tetrahymnena pyriformis* (180). The structure deduced from chemical, NMR and mass spectral data is (50).

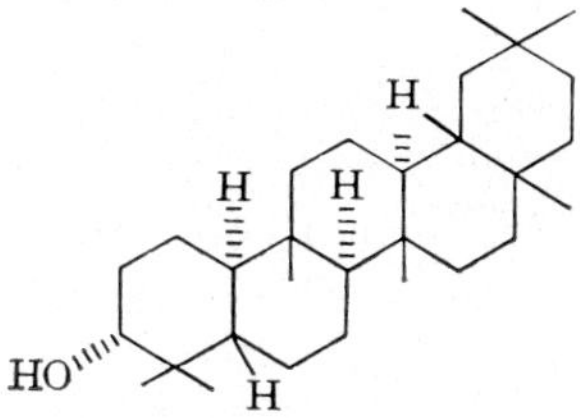

(50) *Tetrahymanol*
m. p. 312.5–314.5° C

Triterpenoids of this type were previously known only in the plant kingdom.

Systematic chemical investigation of coelenterates (gorgonians) has yielded a number of extracts possessing antibiotic activity (54, 73). The substances isolated from these extracts were terpenoids and are listed in Table 2. (All the substances listed in Table 2 were obtained from extracts of gorgonian cortex.)

The structure of β-gorgonene was established by X-ray analysis of its silver nitrate adduct (m. p. 132.5–133.5) (140, 303). β-Gorgonene is a

Suggested Name	$[\alpha]_D$	Ref.
cadinene	+ 24.5° (c = 6.0 CHCl$_3$)	73, 74
eunicin	− 95° (c = 2.0 EtOH)	73, 74
ancepsenolide	+ 13.2 (c = 2.8 CHCl$_3$)	73, 74, 259, 260
	− 135° (c = .516 CHCl$_3$)	73, 74
β-gorgonene	+ 13.9° (neat)	73, 74, 140, 303
		73, 74

bicyclic sesquiterpene (51), is non-isoprenoid in the sense that it does not follow the head-to-tail isoprene rule and is the predominant constituent of the pentane extract of *Pseudopterogorgia americana*. The minor components are 9-aristolene (52), 1(10)-aristolene (53), and (+)-γ-maaliene (54), their structures being deduced by comparison with mass and NMR spectra of authentic sesquiterpenes of this type.

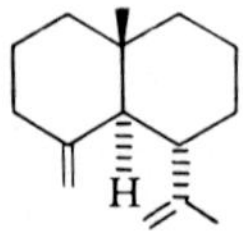

(51) (+)-β-gorgonene 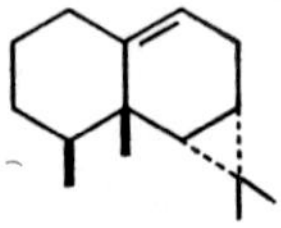 (52) 9-aristolene 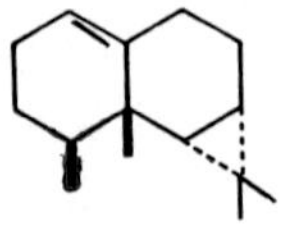 (53) 1(10)-aristolene 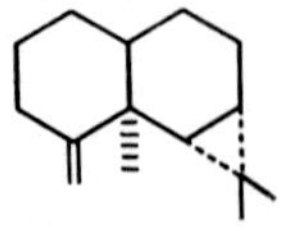 (54) (+)-γ-maaliene

A possible pathway for the biogenesis of gorgonene has been suggested by WEINHEIMER and co-workers (*303*), and involves an intramolecular hydride migration in the farnesyl derived cyclic ion (55). Rearrangement of the intermediate species (56), and loss of a proton yields a monocyclic intermediate (57). Cyclization of (57) would then yield β-gorgonene (51).

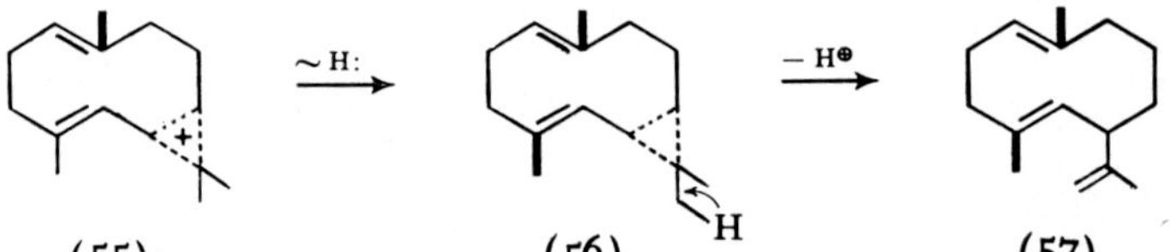

(55)   (56)   (57)

Cadinene (58) has been isolated from the gorgonian *Plexaura crassa*.

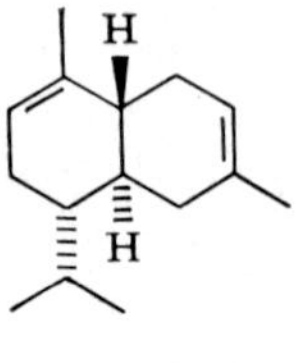

(58) Cadinene

Ancepsenolide (59) and hydroxyancepsenolide (60) which were isolated from *Pterogorgia anceps* (*259, 260*) are *bisbutenolides*, in which the two lactone rings are joined by a twelve-carbon hydrocarbon chain.

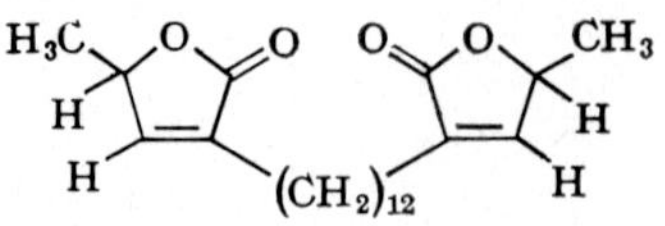

(59) m. p. 91.5–92.0° 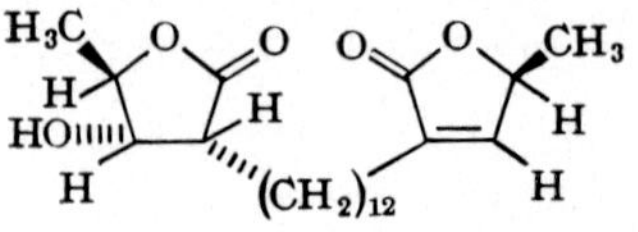   (60) m. p. 122.5–123.7°

References, pp. 472—488

Single unsaturated γ-lactone rings appear in many natural products and indeed a number of antibiotics are derivatives of these lactones, for example, penicillic acid (61). It is not unreasonable to assume that gorgonians may become a source of antibiotics in the future.

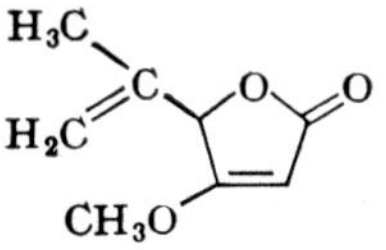

(61)

The petroleum ether extract of the gorgonian *Eunicella stricta* afforded a novel diterpenoid eunicellin (*164*) whose structure, based on the X-ray analysis of its di-bromide, is (62).

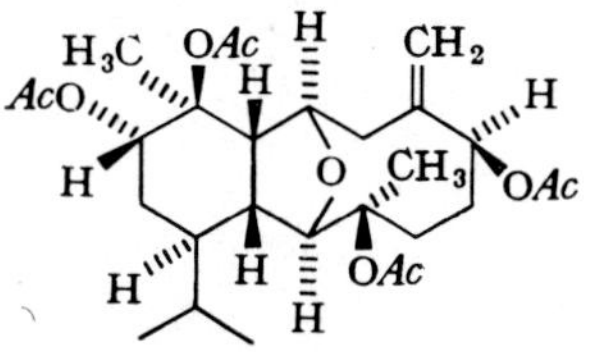

(62) Eunicellin m. p. 186–188° [α]$_D$ = − 36°

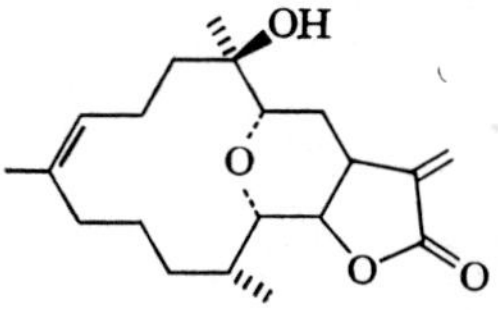

(63) Eunicin m. p. 155° [α]$_D$ = − 89°

The analysis of the dibromide indicates that the six-membered ring is in the chair form and is *cis*-fused to the 10-membered ring. Eunicin (63) (*141, 305*) an antibacterial diterpene isolated from the gorgonian *Eunicea mammosa Lamouroux* incorporates a lactone ring as well as an ether linkage.

Other terpenoid hydrocarbons reported to be present in gorgonians are listed in Table 3. Furoventaline (see Table 3) presents two interesting possibilities: it is either formed during isolation, i. e. during the steam distillation by a combination of thermal and solvolytic elimination from a parent precursor, or it may be biosynthesized by dimethylallylation at the γ-position of a piperitone-like *structure*. Subsequent steps would be furan formation and aromatization. The validity of the proposed structure (64) has recently been established by synthesis (*304*) and a biosynthetic pattern analogous to the piperitone-pulegone-menthofuran sequence has been proposed by WEINHEIMER and WASHECHEK (*304*).

In view of the known thermal lability of the naturally occurring germacrenes which rearrange (*145*) at elevated temperatures to β-elemene carefully-controlled extraction and work-up of the hexane extract of air dried *E. mammosa* was required to yield pure (−) germacrene-A (*305*).

Table 3 (304)

Source	$[\alpha]_D$	Name	Structure
Eunicea mammosa (Bimini)	+ 15.1°	(+)β-elemene	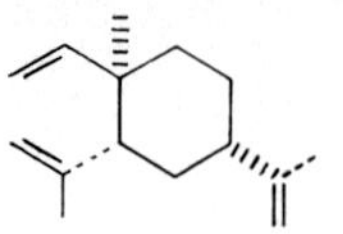
Eunicea mammosa (Bimini)	− 3.2°	(−)germacrene-A	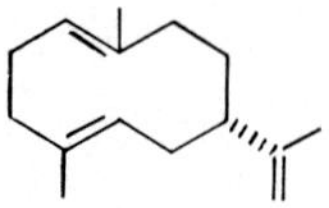
Pseudoplexaura porosa *Eunicea palmeri* *Plexaurella dichotoma*	+ 67.1°	(+) α-maurolene	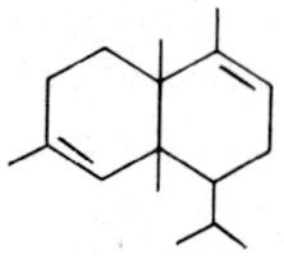
Pseudoplexaura porosa *Eunicea palmeri*	+ 9.2–10.6°	(+) β-ylangene	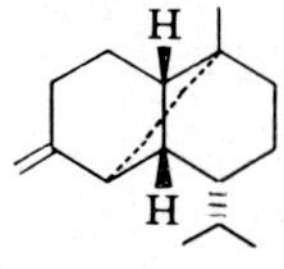
Pseudoplexaura porosa	+ 23.6°	(+) α-cubebene	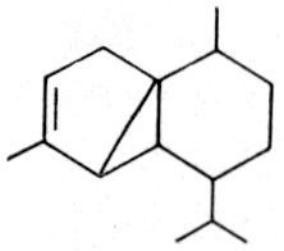
Pseudoplexaura porosa	+ 55.4°	(+) calamene	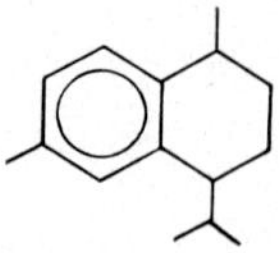
Plexaurella dichotoma	+ 83.1°	(+) β-bisabolene	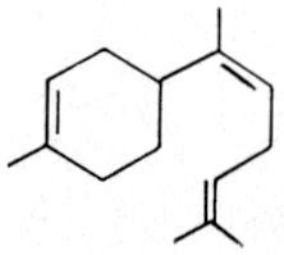
Gorgonia ventalina		furoventalene	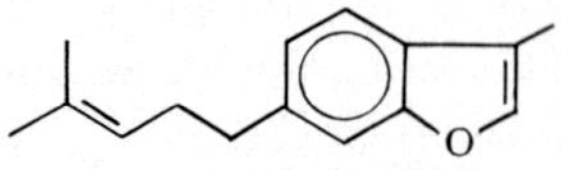
Pseudoplexaura porosa wagenaari flagellosa	+ 70.4°	crassin acetate	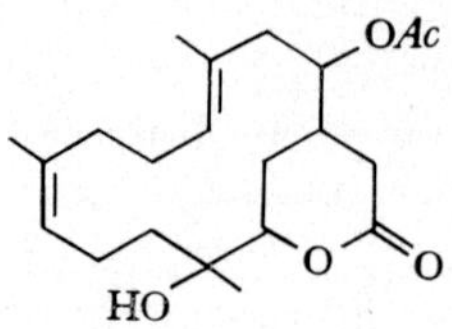

References, pp. 472—488

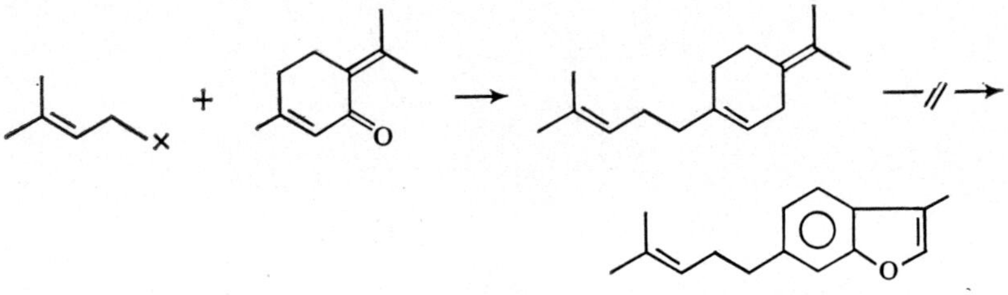

(64)

WEINHEIMER and collaborators (306) have elucidated the absolute configuration of (−) germacrene-A as (65).

The presence of (−)β-selenene (66) in the extract of *E. mammosa* has also been reported (306).

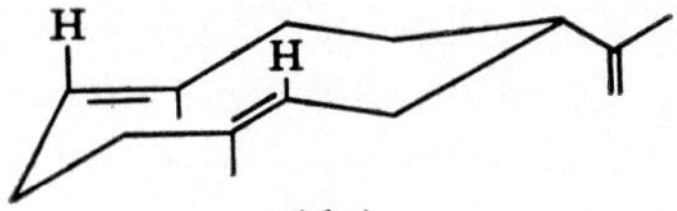

(65) 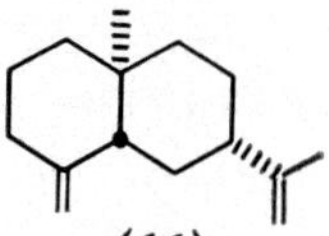   (66)

The remarkable variety of natural products formed by gorgonians has been further exemplified by the recent isolation from the gorgonian *Plexaura homomalla* (Esper) of two prostaglandin derivatives in high yield (0.2% and 1.3% resp. from the air dried cortex) (307). These compounds, 15-epi-PGA$_2$ (67) and its diester (68), are identical with mammalian prostaglandin, PGA$_2$ (69), except in the configuration

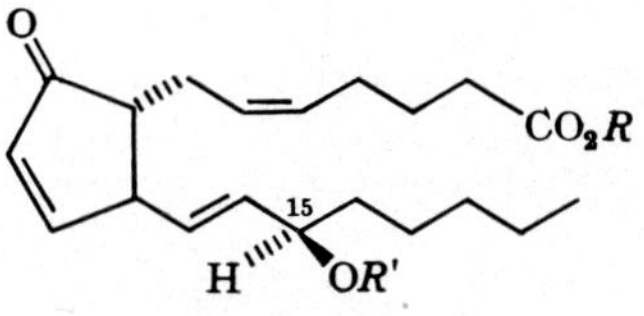

(67) R, R' = H
(68) R = Me, R' = Ac

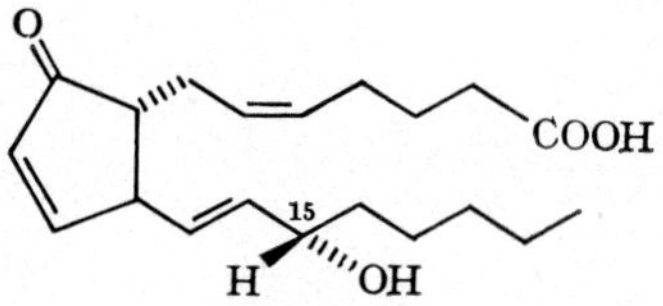

(69)

at C-15, which was established as (R) by degradative and physical methods. Thus ozonolysis of (68) afforded, together with monomethyl glutarate, (+)-α-acetoxy-heptanoic acid which on hydrolysis yielded (−)-α-hydroxy heptanoic acid whose absolute configuration is known (219).

Methanol extracts of air-dried sea-weed, *Dictyopteris divaricata* (154) have yielded a number of sesquiterpenes: (−) γ-cadinene (70), (−)-copaene (71), (−) δ-cadinol (72), cadalene (73), and (+) β-elemene (see Table 3).

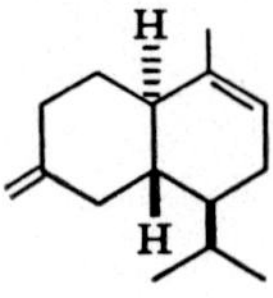

(70) $[\alpha]_D = -19.6°$ 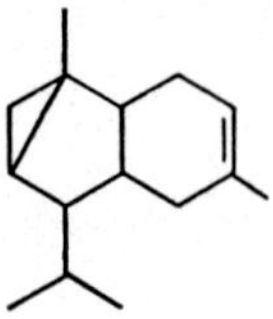   (71) $[\alpha]_D = -26.1°$

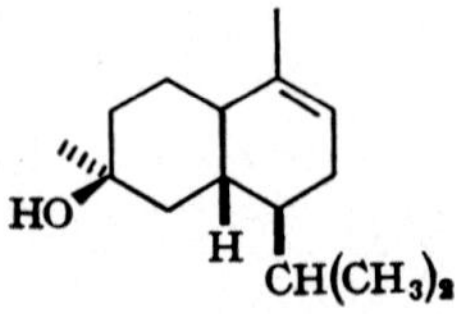

(72) $[\alpha]_D = -108°$ 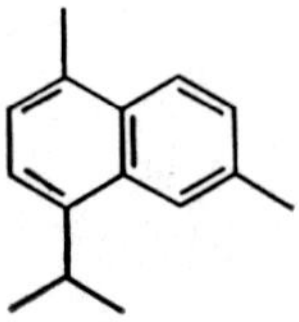   (73)

The essential oil of genus *Dictyopteris* has also yielded (*231a*) two related hydrocarbons, an optically inactive, *trans,cis,cis*-undeca-1,3,5,8-tetraene (74a), and dictyopterine B, an optically active hydrocarbon,

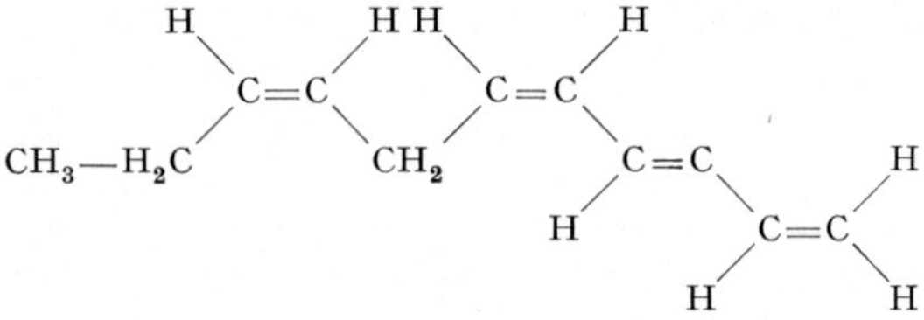

(74a)

whose structure corresponds to *trans*-1-(*trans,cis*-hexa-1',3'-dienyl)-2-vinyl cyclopropane (74b). The structure has also been confirmed by synthesis (*29a*).

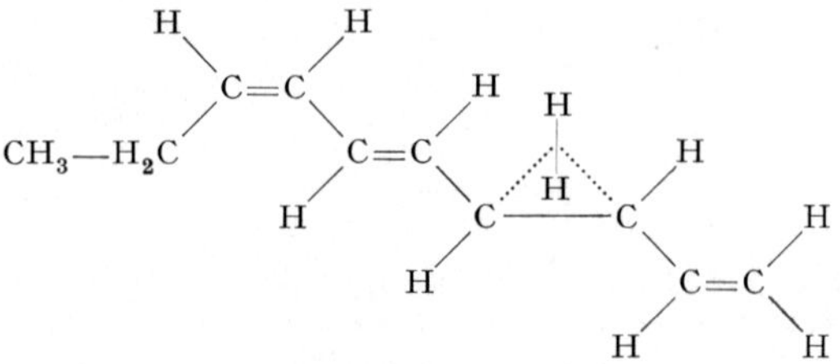

(74b) $[\alpha]_D = -43°$

Recently, HALSALL and HILLS (*111a*) have isolated from the non-saponifiable fraction of brown alga *Fucus vesiculosus*, two polyolefins, *cis,cis,cis,cis,cis*-heneicosa-1,6,9,12,15,18-hexaene (74c) and *cis,cis,cis,cis*-heneicosa-1,6,9,12,15-pentaene (74d).

References, pp. 472—488

$MeCH_2 \cdot CH=CH \cdot [CH_2 \cdot CH=CH]_4 \cdot [CH_2]_3 \cdot CH=CH_2$

(74 c)

$Me \cdot [CH_2]_3 \cdot [CH_2 \cdot CH=CH]_4 [CH_2]_3 CH=CH_2$

(74 d)

Volatile constituents obtained from dried algae have been reviewed by KATAYAMA (*160*). A Honolulu group has recently isolated a rather unusual hydrocarbon from the essential oil of algae of the genus *Dictyopteris* (*202*). The structure of this substance, dictyopterene A (74), *trans*-1-(*trans*-1-hexenyl)-2-vinylcyclopropane has been confirmed by synthesis (*202, 223*).

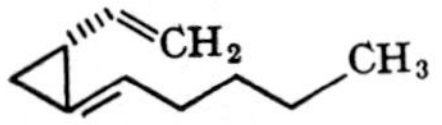

(74)

The neutral extract of *L. Nipponica Yamada* has also yielded (*151*) a sesquiterpene hydrocarbon, laurene (75). Laurene is very unstable to acidic medium which converts it readily to isolaurene (76). The absolute

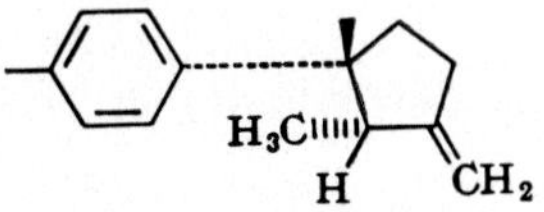

(75) b. p. 131–133° (21 mm)
[α]$_D$ = + 48.7°

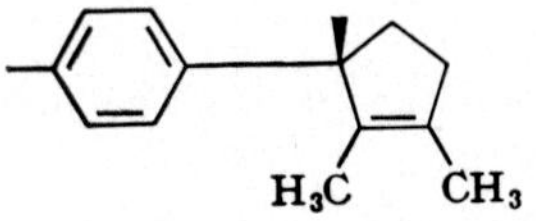

(76) b. p. 140–142° (21 mm)
(x-C$_7$H$_7$) [α]$_D$ = + 108.7°

configuration of laurene has been established by correlation with (+) cuparene in the following manner.

Oxidation of (75) with osmium tetroxide (ether-pyridine) followed by treatment with periodic acid yielded a ketone (77). Ketone (77) on treatment with ethyl formate ans sodium methoxide in benzene afforded (79). When the latter was refluxed with n-butyl-mercaptan and p-toluene-sulfonic acid (benzene), thioether (80) was obtained.

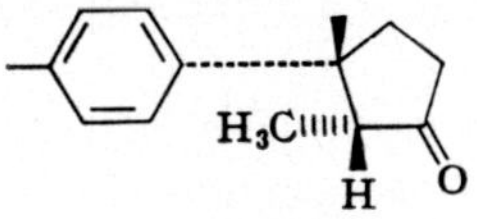

(77)

[α]$_D$ = + 70° (CHCl$_3$), + ve Cotton effect
Semicarbazone m. p. 227–229°

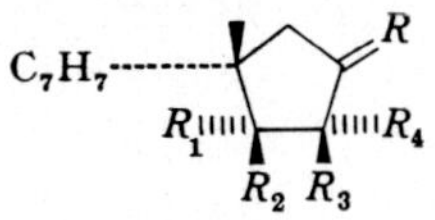

(78)

(79) $R_1 = CH_3$, $R_2 = H$, R_3, $R_4 = O$, $R = CHOH$

(80) $R_1 = CH_3$, $R_2 = H$, R_3, $R_4 = O$, $R = CHSC_4H_9$

Methylation of (80), followed by alkaline treatment in diethylene glycol afforded a ketone identical with α-cuparenone (81), whose stereochemistry is known (*151*).

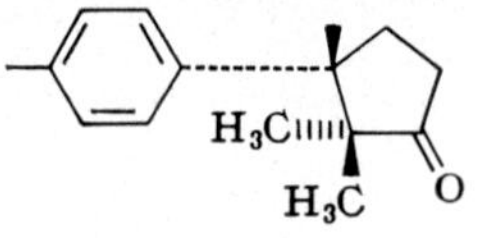

(81) m. p. 52—53° [α]$_D$ = + 170°

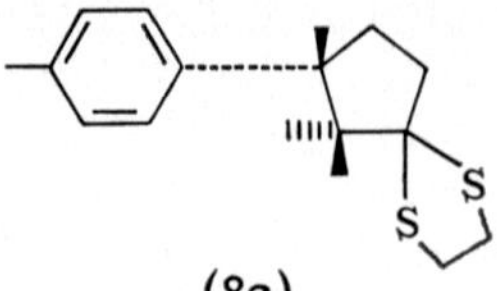

(82)

Treatment of the thioketal (82), with Raney nickel in ethanol afforded a product identical with the known naturally-occurring (+)-cuparene (*105*). Attempts to convert (77) to laurene (75) led only to formation of isolaurene (76) and epilaurene (83).

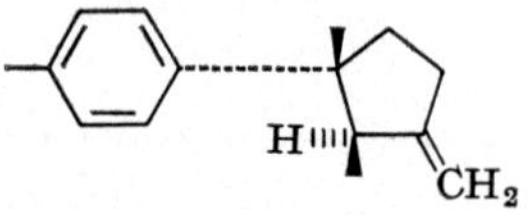

(83) [α]$_D$ = — 3.1°

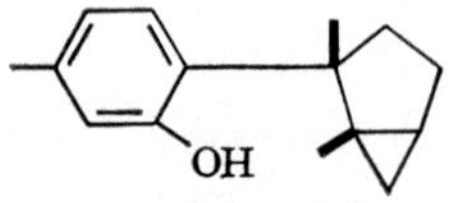

(84)

Other *Laurencia* species (e. g. *L. Okamurai* Yamada) (*153, 153a*) have also yielded aplysin (103), aplysinol (104), debromoaplysin (118), and debromolaurinterol (84).

Isoprenoid quinones have been reviewed extensively (*207, 232, 291*). Ubiquinones (85) are widely distributed in marine species just as they are in terrestrial species. Their structure is well known and has been con-

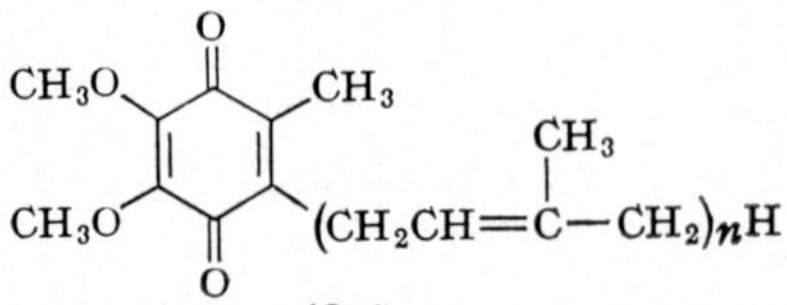

(85)

where n = 6—10

firmed by synthesis (*208*). Examples of the distribution of ubiquinones in marine species are given in Table 4.

There are several well-documented reviews (*39, 68, 95, 97, 102, 116, 117, 129, 130, 155, 156, 178, 208, 226, 251, 294, 302*) dealing with distribution, chemistry, and biochemistry of tocopherols, e. g. α-tocopherol (86); vitamin D, e. g. D$_3$ (87), vitamin A, e. g. retinol (88) and carotenoids, e. g. β-carotene (89). Such compounds are also widely distributed in marine species. For example,

References, pp. 472—488

Table 4

Source	Ubiquinone*
Codfish (liver, heart)	n = 10
Sea Lamprey (heart)	n = 10
Eel (heart)	n = 10
Char (heart)	n = 10
Walleyed pike (muscle)	n = 9
Shark (liver, heart, muscle)	n = 6, 9, 10
Annelida	n = 10
Echinodermata	n = 10
Crustacea	n = 10
Protozoa: *Tetrahymena pyriformia*	n = 8
Crithidia fasciculata, Stigomonas	n = 9
Fasciculata and *Astasia klebsii*	n = 9

* For detailed description, see (232).

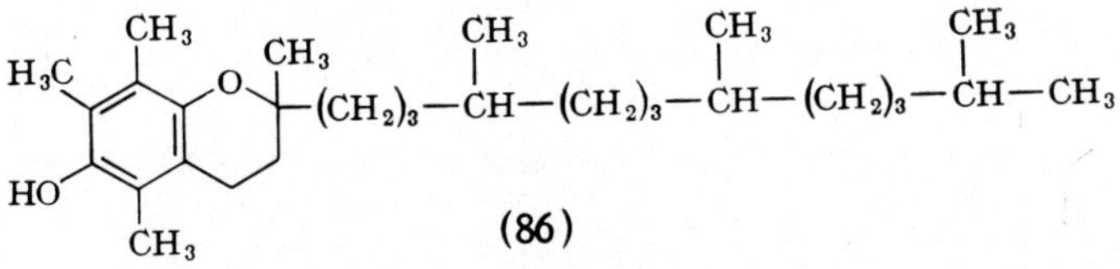

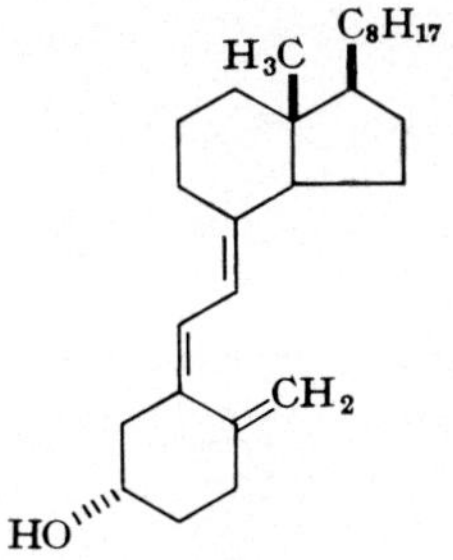

β-carotene (89), lutein (90), and astaxanthin (91) are found in many marine species but with the devolpment of more sophisticated methods of isolation and characterization, other unique cartenoids are being isolated from marine sources (117, 318–323). Thus from the red sponge *Reniera japonica*, three carotenoids containing aromatic rings have been isolated, isorenieratene (92), renieratene (93), and renierapurpurin (94).

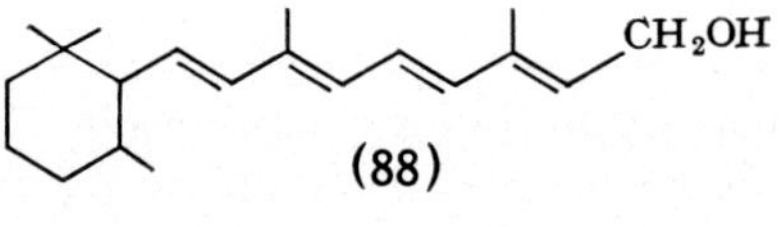

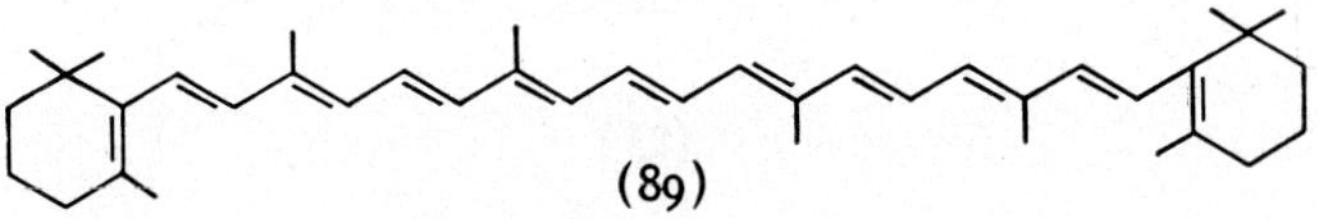

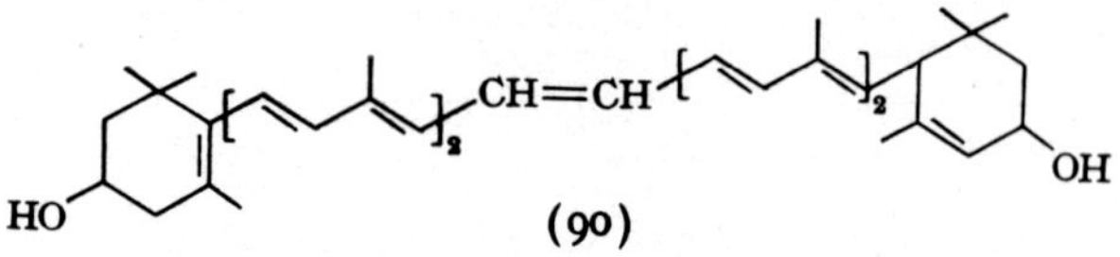

(90)

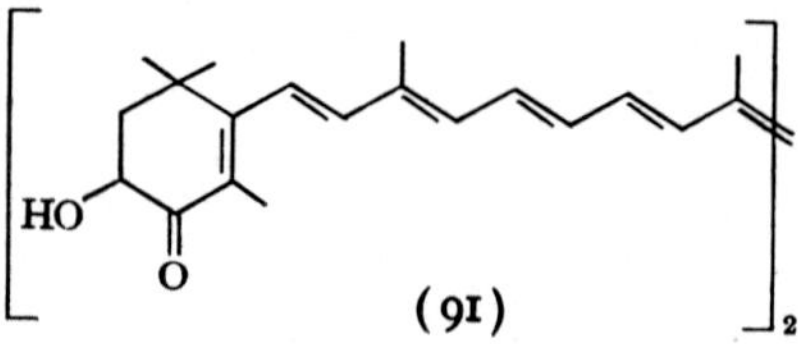

(91)

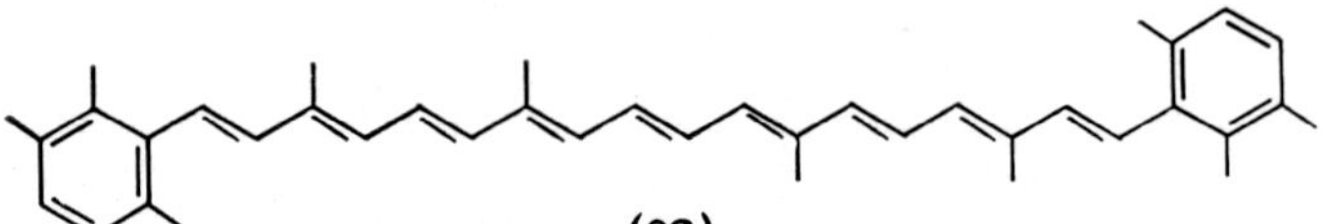

(92)

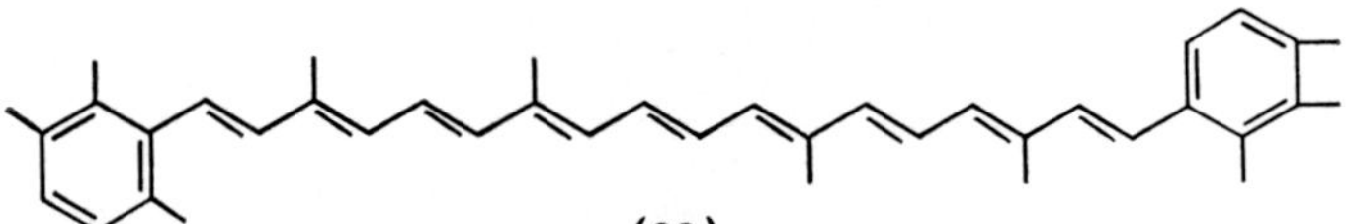

(93)

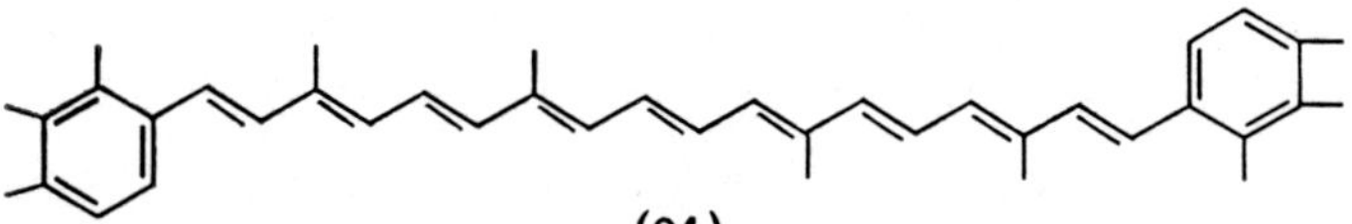

(94)

Thomas and Goodwin (*289*) have recently reviewed the distribution of carotenoids in Xanthophyta algae. The marine dinoflagellate, *Glenodinium foliaceum* Stein, contains fucoxanthin (95) as the major xanthophyll (*181*).

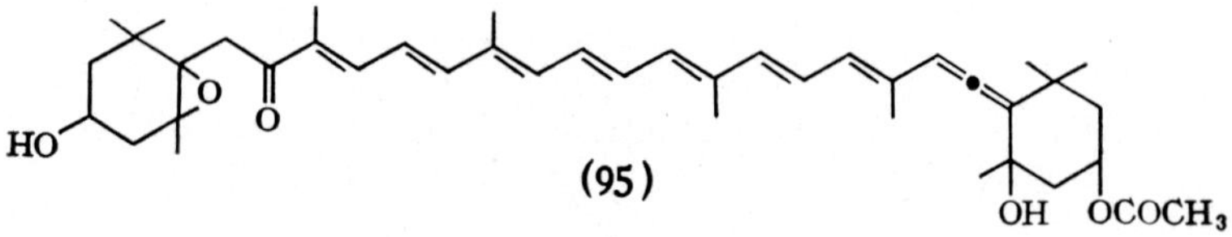

(95)

Siphonous green algae contain not only known plant carotenoids but also (*166*, *167*) two new substances, siphonaxanthin (96), and siphonein, for which structure (97) has been proposed.

Kleinig (*168*) has recently published a chemotaxonomical study covering fifty species of nine orders of siphonean algae, and Healey (*128*) has similarly investigated four species of blue-green algae. The diatoms,

References, pp. 472—488

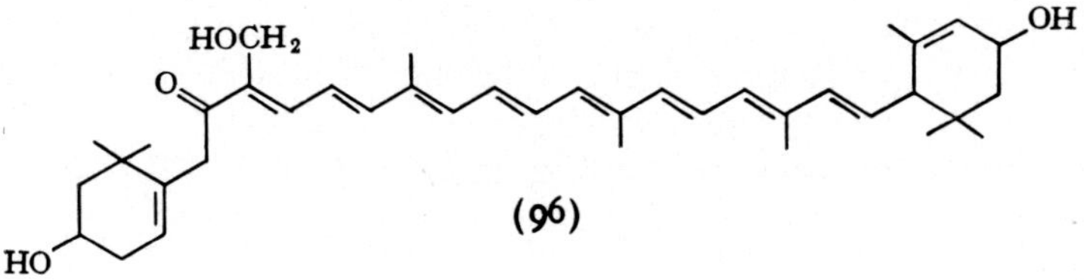

(96)

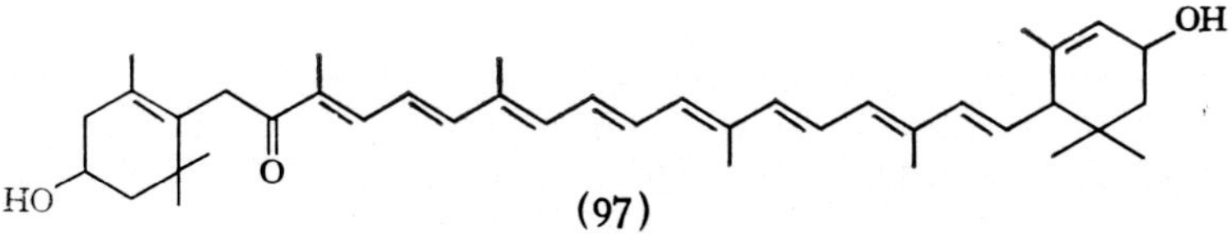

(97)

Nitzschia closterium F. *minutissima* and *Euglena gracilis*, strain Z., have yielded (27) a novel xanthophyll, diadinoxanthin (5',6'-epoxy-3, 3'-dihydroxy-7,8-dehydro-β-carotene) (98).

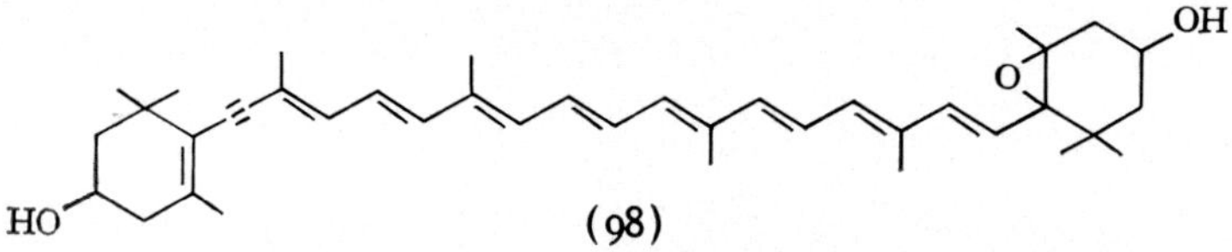

(98)

Green algae, *Scenedesmus obliquus* and *Chlorella vulgaris*, have yielded a new xanthophyll which appears to be characteristic for this species (28). This compound, loroxanthin (99), although similar to lutein (90), differs in the substitution pattern in the rings and the side chain.

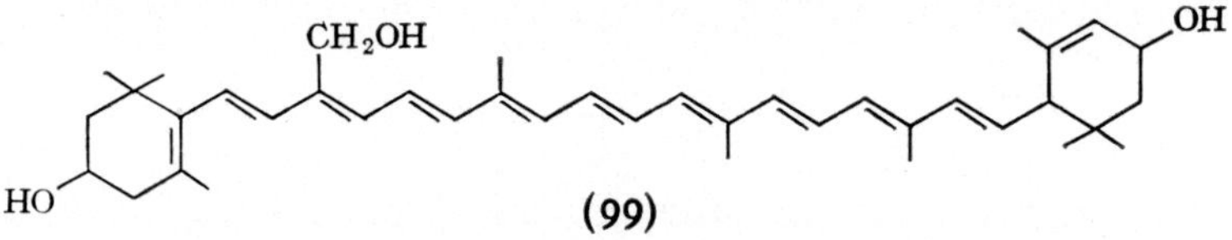

(99)

In conclusion, it is to be noted that cartenoids of marine species frequently occur complexed with proteins. The nature of these carotenoproteins has recently been reviewed by CHEESMAN and collaborators (70).

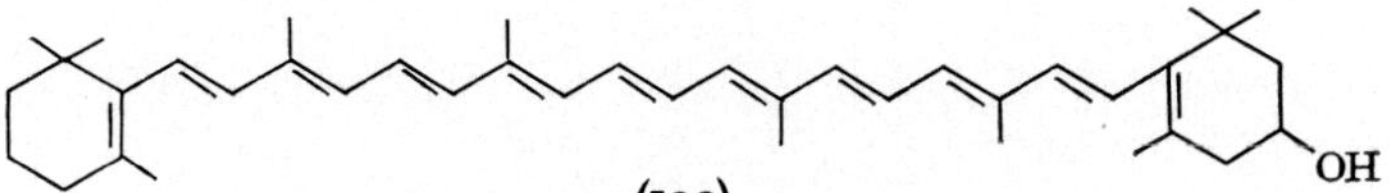

(100)

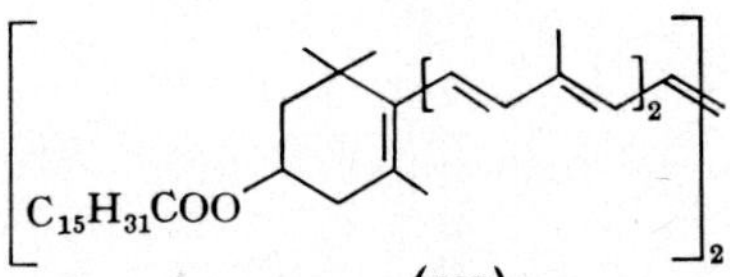

$$\left[C_{15}H_{31}COO \cdots \right]_2$$

(101)

Fox and Crozier (*96*) in a study of carotenoids occurring in some lower fishes, have found that colored carotenoids are absent from the scales and liver of the coelacanth, *Latimeria chalumnae*. They were also not present in the skin, immature eggs and the liver of the Pacific hagfish *Eptatretus stoutii*. On the other hand, zeaxanthin (100) either free or as an ester, e. g. as the palmitate (*101*), was present either in the skin or the liver as the only carotenoid so far identified in the thornback ray *Platyrhinoides triseriata*, and the horned shark *Heterodontus francisci*. The skin of the pacific mako shark, *Isurus glaucus*, like the hagfish, yielded no carotenoids.

VI. Halogen-Containing Compounds

Free and bound halides occurring naturally have been reviewed by Roche, Fontain and Leloup (*243*). The literature contains a number of reports on the antibiotic activity of ether extracts of seaweed against several species of gram positive and gram negative pathogens (*71, 161, 191*). Likewise, alcohol, acetone and chloroform extracts obtained from a variety of species of algae (*55, 212, 213, 225, 229, 249, 270, 313*) have shown antibiotic activity. Although in most cases no chemistry was done, occasionally simple halogen-containing substances such as bromophenols were shown to be responsible for this activity. The red algae *Odonthalia dentata (L) Lyngbye* and *Rhodemela confervoides (Hudson)* (*77*), yielded 3,5-dibromo-4-hydroxybenzyl alcohol (102).

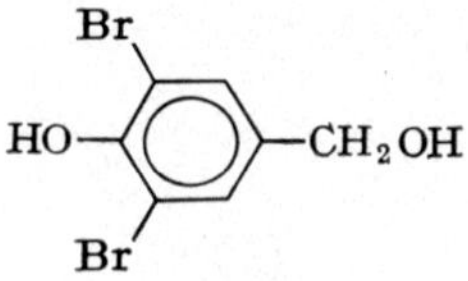

The unsaponifiable fraction of the ether extract of dried sea hare *(Aplysia kurodai)* (*325*) afforded two novel naturally-occurring bromo compounds, aplysin (103) and aplysinol (104). Degradation and spectroscopic data led Yamamura and Hirata (*325*) to structures (103) and (104).

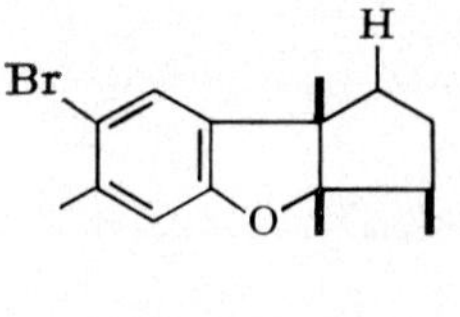

(103) m. p. 85–86°
$[\alpha]_D = -85.4°$

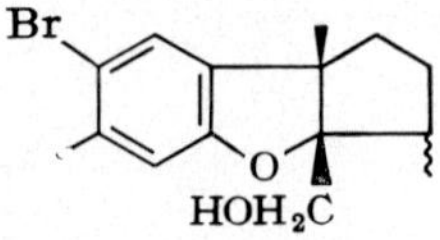

(104) m. p. 158–160°
$[\alpha]_D = -55.6°$

The biogenesis of these sesquiterpenes is considered to be similar to that of cuparene (105) and trichothecin (106), and consequently structure (104) for aplysinol was preferred (325) to the alternative possibility (107).

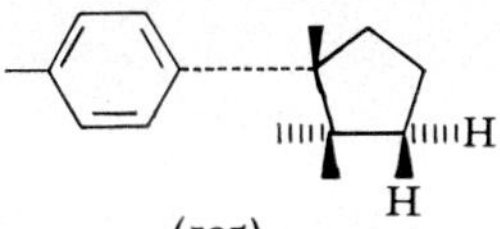

(105)

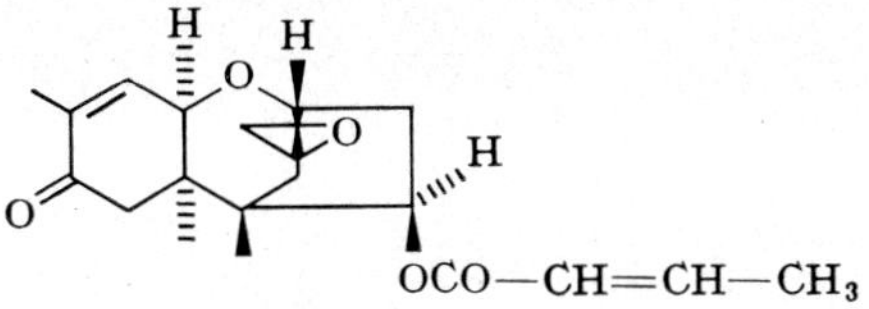

(106)

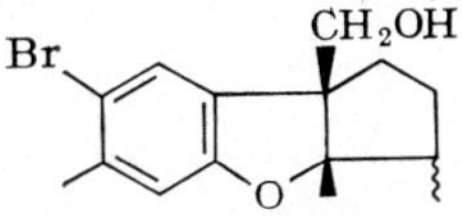

(107)

Aplysia Kurodai afforded also aplysin-20 (108) a bicyclic diterpene with an axial C-8 hydroxyl group and an equatorial C-3 bromine (108).

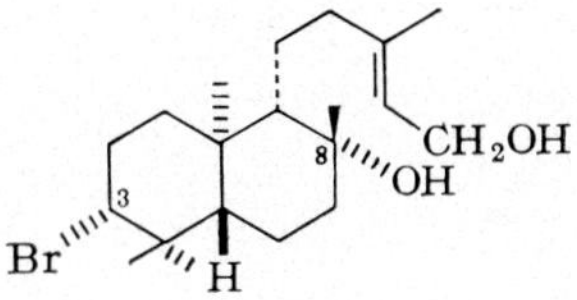

(108)

m. p. 146–147° $[\alpha]_D = -78.1°$

The structure of racemic aplysin and debromoaplysin (118) has been confirmed by synthesis (315, 316), the route being outlined below.

Over the past several years reports have appeared (for reviews see 182, 216) which describe the occurrence of a number of antibiotic-producing marine organisms. Systematic study by BURCKHOLDER and coworkers (56, 57) has led to the isolation and characterization of a number of marine sources possessing antibiotic activity. Ether and methanol extracts of sponges (217, 264) possess broad spectrum antibiotic properties. SHARMA and BURCKHOLDER (265) have isolated, characterized and synthesized a bromine-containing broad spectrum antibiotic 2,6-dibromo-4-acetamido-4-hydroxycyclohexadienone (119) from the marine sponge *Verongia cauliformis*. Several closely related substances were also present in the extracts (264, 266). A methanolic extract of the marine sponge *Verongia fistularis* (265a) yielded a closely related compound

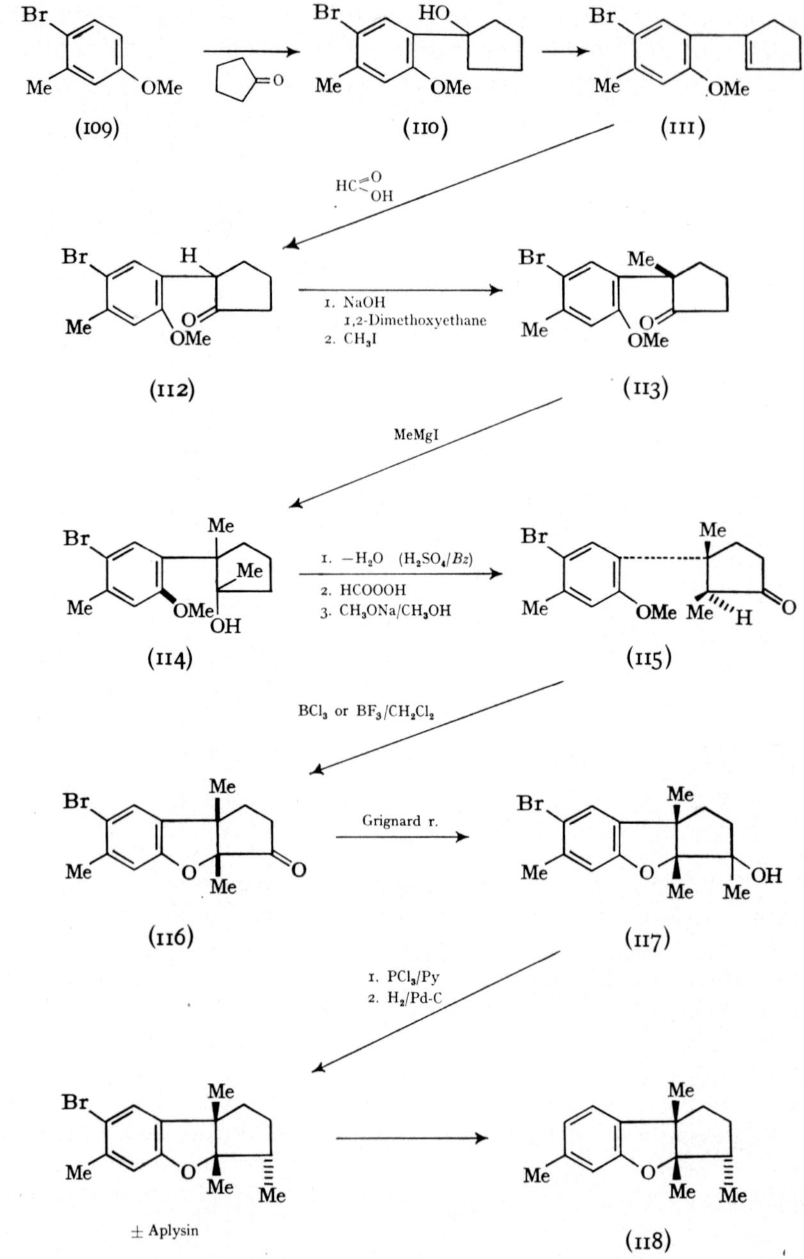

(109) → (110) → (111)

(112) → (113)

Reagents: cyclopentanone; HC(=O)OH; 1. NaOH, 1,2-Dimethoxyethane; 2. CH₃I

(114) → (115)

MeMgI; 1. −H₂O (H₂SO₄/Bz); 2. HCOOOH; 3. CH₃ONa/CH₃OH

(116) → (117)

BCl₃ or BF₃/CH₂Cl₂; Grignard r.

± Aplysin → (118)

1. PCl₃/Py
2. H₂/Pd-C

(119a). The structure of this broad spectrum antibiotic has been established by spectroscopic data and synthesis (265a). MINALE and coworkers (86a) have isolated from the sponge *Aplysina aerophoba* a closely related bromine-containing antibacterial substance aeroplysinin-I (119b). This antibacterial substance has also been isolated from the sponge of *Ianthella ardis*. Its structure was established by X-ray analysis as 3,5-dibromo-1(ax), 6(ax)-dihydroxy-4-methoxycyclohexa-2,4-diene-1-acetonitrile (76a). These authors have also found (86b) in the acetone extracts of sponges *Aplysina aerophoba* and *Verongia thiona* a tetrabromo-spirocyclohexadienylisoxazole, aerothionin. Spectroscopic data for aerothionin are consistent with structure (119c).

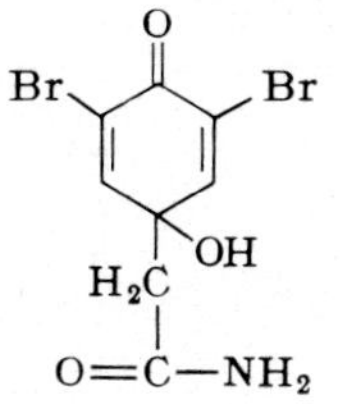

(119) m.p. 193–195°

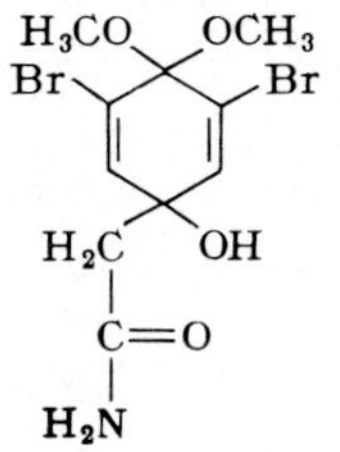

(119a) m.p. 191°

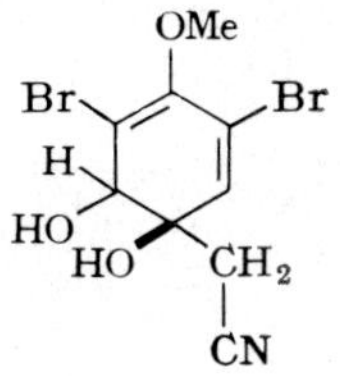

(119b) m.p. 120°

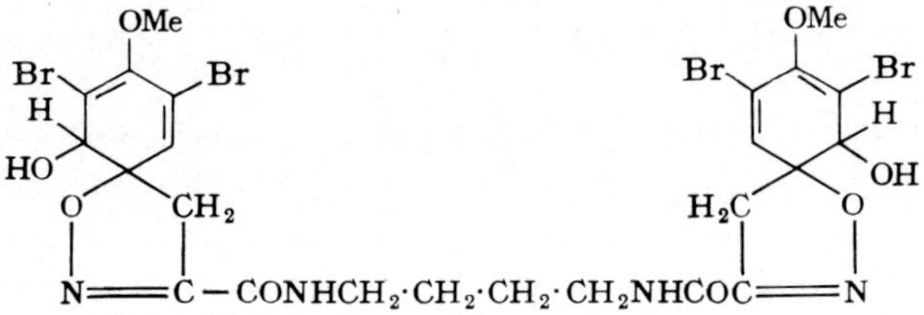

(119c) m.p. 134–137°

A bromine-containing antibiotic was also isolated from a marine bacterium, *Pseudomonas bromoutilis* (56) and was characterized spectroscopically, by synthesis (115), as well as by X-ray analysis (176). The antibiotic, 2-(3,5-dibromophenyl)-3,4,5-tribromopyrrole (120), showed

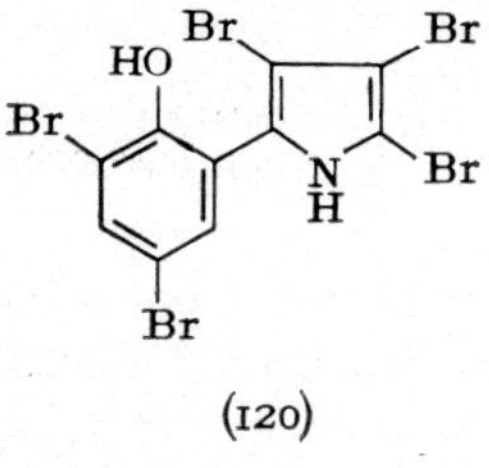

(120)

substantial activity *in vitro* against gram-positive bacteria and *Mycobacterium tuberculosis* H 37 R, while being inactive against gram-negative organisms.

In mice, an intravenous dose of 25 mg/kg was non-toxic, however, a 50 mg/kg proved to be toxic. Subcutaneously, 250 mg/kg dose protected mice against infection with *Staphylococcus aureus UC-76*.

Seaweed, *Laurencia nipponica* Yamada, proved to be a rich source of natural products. Thus, the neutral fraction from methanolic extracts of dried seaweed yielded laureatin (121) which was shown (*147*) to be a bicyclic bromine-containing compound possessing unsaturation in the side chain only and having the (S) configuration at C-9 and C-10, thus differing from a second closely related material, laurencin (122), isolated from the seaweed *L. glandulifera* Kützing (*148, 149*).

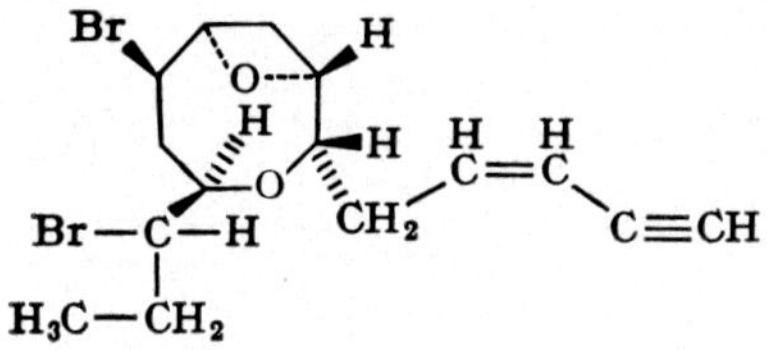

(121) m. p. 82–83°
$[\alpha]_D = +96°$ (CCl$_4$)

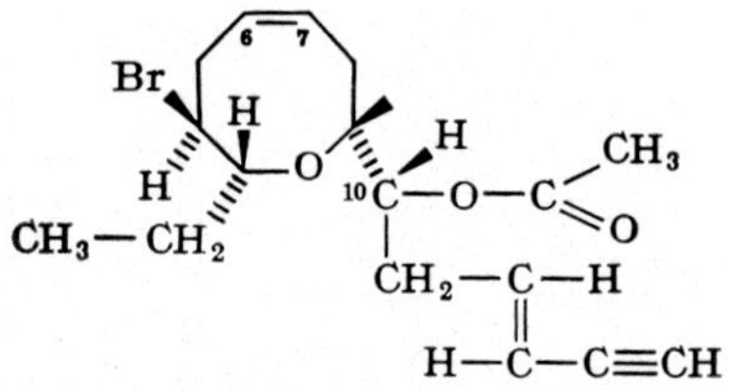

(122) m. p. 73–74°
$[\alpha]_D = +70.2°$ (CHCl$_3$)

(122) possesses an eight-membered ring with unsaturation at C-6, contains one bromine atom only, and has the (R) configuration at C-9 and C-10. The same source also yielded an isomeric bromo compound isolaureatin (123), possessing the (S) configuration at C-9 and C-10 (*150*). The complete stereochemistry of (121), (122), and (123) has recently been elucidated (*150a*).

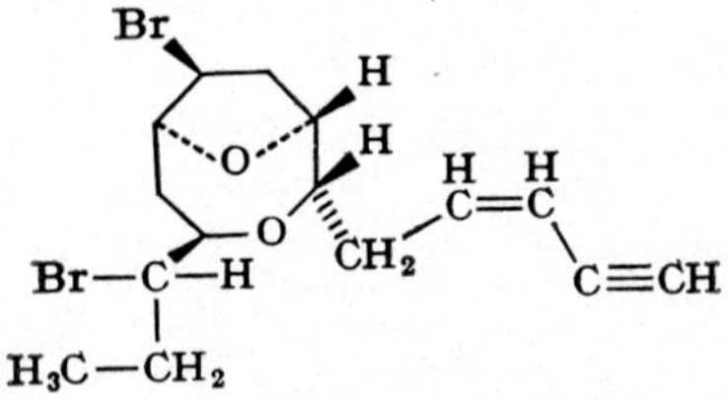

(123) m. p. 83–84°
$[\alpha]_D = +40°$ (CCl$_4$)

The number of compounds that *Laurencia nipponica* Yamada offers has been increased, by isolation from the neutral methanolic extract of the dried seaweed, of two bromine-containing sesquiterpenoid hydro-

References, pp. 472—488

carbons laurenisol and laurinterol (*152, 153a*). Chemical and spectroscopic data led to (**124**) and (**124a**) for laurenisol and laurinterol, respectively.

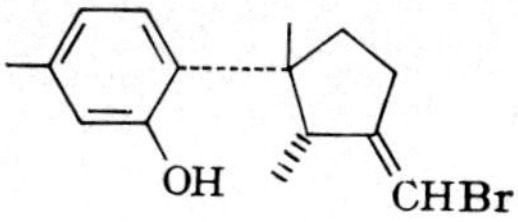

(**124**) $[\alpha]_D = +85.9°$

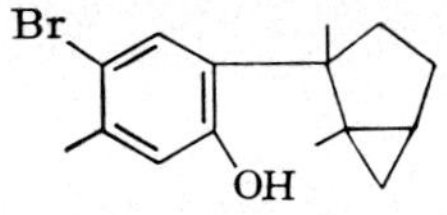

(**124a**) $[\alpha]_D = +13.3°$

VII. Non-Proteinoid Nitrogen-Containing Substances

The literature dealing with the distribution and biological effect of toxic species of marine origin is very large indeed, but very little is known about the chemistry of the toxic principles involved. In general, many of the bound nitrogen-containing substances so far isolated from marine species have been found to be toxic. Comprehensive reviews dealing with toxicology, isolation, pharmacology and chemistry have already been mentioned in the Introduction but our current knowledge is so limited that no systematic tabulation or correlation of toxic principles can be done on a chemical basis. However, certain general features are evident. Extracts of toxic principles obtained from sponges, hydroids, jellyfishes, sea anemones, corals and a variety of fish contain 5-hydroxytryptamine (serotonin (**125**), homarine (**126**) (*9, 11, 26*), quaternary ammonium compounds and proteins of variable molecular weight (*112, 186, 187, 308, 309, 310*).

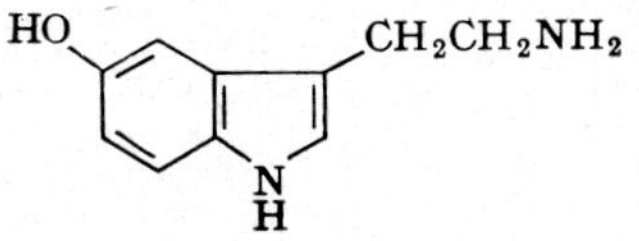

(**125**)

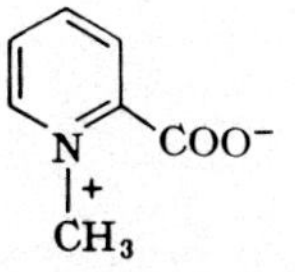

(**126**)

It has been known for a long time that puffer fish (particularly *tora fugu*) contain a powerful toxic principle. The first report on the chemical and physical properties of this toxic principle was published in 1889 (*112*). It took a great deal of work by many workers, before the structure of tetrodotoxin (**127**) was elucidated in 1964 (*293, 314*). The free base (**127**) is a zwitterion which exists in two tautomeric forms in an equilibrium which is very sensitive to pH changes. The chemistry of tetrodotoxin has recently been reviewed by SCHEUER in an article (*255*) which should be consulted for details.

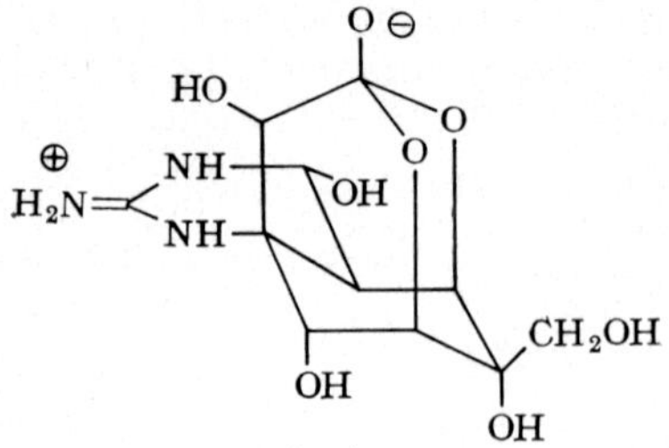

(127)

It is interesting to note that the neurotoxin of the California newt, *Taricha torosa*, is identical with tetrodotoxin (*58, 209*).

Saxitoxin is a potent toxic substance reponsible for paralytic shellfish poisoning. This substance is also associated with "red tides", which occur during the blooming of the dinoflagellate *Gonyaulax catenella*. Isolation, characterization and structure determination of saxitoxin has been carried out by SCHANTZ and co-workers (*112, 193, 194*) and by RAPOPORT (*233, 261*). The proposed structure for saxitoxin is (128). A recent report (*185*)

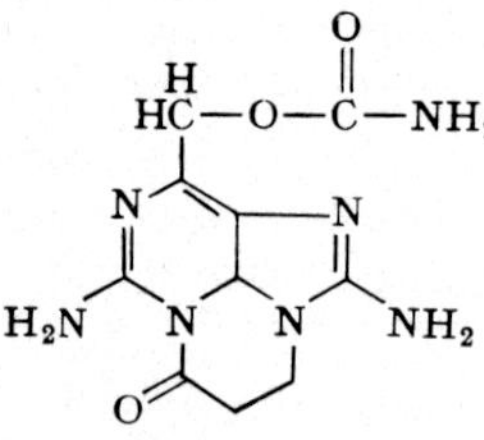

(128) Hydrochloride
$[\alpha]_D = +130°$

on the toxic principle of another red tide organism, *Gymnodinium breve*, suggests that two other chemically distinct toxins may also be present. Aqueous acid extracts of the marine annelide, *Lumbriconereis heteropoda* Marenz, contain a toxic principle called nereistoxin whose structure has been elucidated by HASHIMOTO and OKAICHI (*119, 229*) 4-N,N-dimethyl-amino-1,2-dithiolane (129).

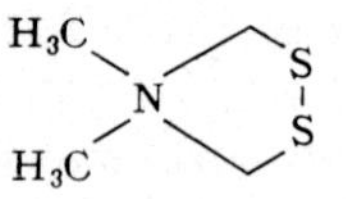

(129)
free base b. p. 212–213°
oxalate, m. p. (decomp.) 168–170°

References, pp. 472—488

KONISHI (*170*) has reported a synthesis of nereistoxin in good yield which begins with lithium aluminium hydride reduction of the oxime (130) in ether to the amine (132). ESCHWEILER-CLARKE N,N-dimethylation of (132) yielded the dimethylamine (131). N-formylation of (132) followed by reduction with lithium aluminium hydride afforded the methyl amine (133). Debenzylation of the dimethylamine (131) in refluxing butanol in the presence of metallic sodium, followed by air oxidation in the presence of ferric chloride afforded nereistoxin (129) which is a strong insectiside.

$$
\begin{array}{c}
\text{NOH} \\
\parallel \\
\text{C} \\
\diagup \quad \diagdown \\
\text{H}_2\text{C} \quad \text{CH}_2 \\
| \quad | \\
\text{Ph·CH}_2\text{·S} \quad \text{S·CH}_2\text{·Ph}
\end{array}
\quad
\begin{array}{c}
\text{R}_1 \quad \text{R}_2 \\
\diagdown \diagup \\
\text{N} \\
| \\
\text{CH} \\
\diagup \quad \diagdown \\
\text{H}_2\text{C} \quad \text{CH}_2 \\
| \quad | \\
\text{Ph·CH}_2\text{·S} \quad \text{S·CH}_2\text{·Ph}
\end{array}
\longrightarrow (129)
$$

(130)

(131) $R_1 = R_2 = CH_3$
(132) $R_1 = R_2 = H$
(133) $R_1 = CH_3, R_2 = H$

Ciguatera is a pathological condition caused by eating toxic fish in certain geographical areas. It is widespread in tropical waters (*36*); it is thought that the fish becomes toxic as a consequence of its diet. Ciguatera-causing poisons have been found in certain oysters *(Crassostria virginica)* and clams *(Venus mercenaria campechiensis)* (*195*), in the red snapper (*Lutjans bohar*, Forskål), in the liver of the shark *Carcharhinus menisorrah*, the moray eel, *Gymnothorax javanicus* and in blue-green algae, *Schizothrix calciocola (Agardh) Gamont* (*37*). The major contribution to our knowledge of the ciguatera toxin is that of the University of Hawaii group. It is known that there is species differentiation: there are sources which yield toxins possessing anticholinesterase activity and there are others that do not possess this activity (*37*).

SCHEUER and co-workers (*254*) have recently reported on the chemical nature of ciguatoxin isolated from the moray eel, *Gymnothorax javanicus*. The raw flesh was extracted with acetone and the acetone extract was subjected to fractionation and extensive chromatography to yield TLC homogeneous material. The empirical formula of this material was $C_{35}H_{65}NO_8$, consistent with the presence of one nitrogen. The substance was saturated (end absorption in the UV), possesses a significant NMR signal at 1.25 ppm in the NMR spectrum, and had IR bands at 2924, 2849, 1460 and 1379 cm^{-1}. A positive Dragendorff test and the negative ninhydrin reaction was consistent with the presence of a quaternary nitrogen function. An IR band at 3390 cm^{-1} indicated the presence of a hydroxyl group; weak ultraviolet absorption at 270 nm, and a positive dinitrophenylhydrazine and ferric hydroxamate test suggested the presence of a ketone.

Hydrolysis of ciguatoxin with 2N HCl-methanol produced two fractions: non-hydroxylic long-chain fatty acids soluble in petroleum ether and a chloroform-soluble fraction, which gave a positive test for quaternary nitrogen and a positive dinitrophenylhydrazine test for carbonyl function. The presence of glycerol in the hydrolysate was also established. On the basis of the available data, SCHEUER et al. (254) suggested that ciguatoxin is a lipid containing a quaternary nitrogen, hydroxyl functions and a cyclopentanone moiety. The pharmacology of ciguatoxins has been discussed by KOSAKI and ANDERSON (171).

Boxfish, *Ostracion lentiginosus*, excretes a toxic substance pahutoxin (134) which has been recently isolated, characterized as (134) and synthesized by SCHEUER and BOYLAN (49). Rapid extraction of the aqueous solution of the original secretion with n-butanol followed by chromatography on silicic acid and Dowex 1-X4 yielded pure pahutoxin. The synthesis of pahutoxin has been accomplished as follows. Oxidation of

$$CH_3-(CH_2)_{12}-\overset{\overset{H}{|}}{\underset{\underset{OCOCH_3}{|}}{C}}-CH_2-CO_2(CH_2)_2-\overset{+}{N}(CH_3)_3Cl^-$$

(134)

m. p. 74°—75° $[\alpha]_D = +3.05°$ (c = 2.3, CH_3OH)

tetradecanol (135) with lead tetraacetate gave the aldehyde (136).

A Reformatsky reaction (with Zn/bromoacetate) on (136) yielded ester (137) which on hydrolysis yielded intermediate (138), which was acetylated to (*139*). The latter on conversion to the acid chloride and esterification with choline yielded (±) pahutoxin (134).

It is appropriate to point out at this stage that investigation of toxic principles may well lead to discovery of valuable drugs. For example it has been mentioned that some clams contain ciguatera toxin. The common edible clam, *Mercenaria mercenaria* and *M. campechiensis*, has yielded extracts which inhibited growth of sarcoma 180 and Krebs-2 ascites tumors in mice (256, 257). The structure of the active principle, called mercenene has not as yet been elucidated. However, judging by the data available, it seems very probable that it is a glycopeptide (258).

Ethanolic extracts of the hypobranchial body of mollusks, particularly *Murex trunculus*, have yielded a rather unusual choline analogue, murexine (140). This substance, β-[imidazolyl-(4)]-acrylcholine, has been synthetized by PASINI, ERSPAMER and co-workers (85, 227).

References, pp. 472—488

(135) $\quad CH_3-(CH_2)_{12}CH_2OH$

$\quad\quad\quad\quad\downarrow Pb(OAc)_4$

(136) $\quad CH_3-(CH_2)_{12}-CHO$

$\quad\quad\quad\quad\downarrow BrCH_2CO_2C_2H_5/Zn$

(137) $\quad CH_3-(CH_2)_{12}-\underset{H}{\overset{OH}{C}}-CH_2-COOC_2H_5$

$\quad\quad\quad\quad\downarrow KOH$

(138) $\quad CH_3-(CH_2)_{12}-\underset{H}{\overset{OH}{C}}-CH_2-COOH$

$\quad\quad\quad\quad\downarrow (CH_3CO)_2O$

(139) $\quad CH_3-(CH_2)_{12}-\underset{H}{\overset{OCOCH_3}{C}}-CH_2-COOH$

$\quad\quad\quad\quad\downarrow \begin{array}{l}1.\ PCl_3 \\ 2.\ [HOCH_2CH_2\overset{+}{N}(CH_3)_3]OH^-\end{array}$

(134) $\longleftarrow$

$$\text{imidazolyl}-CH=CH-COOCH_2-CH_2-\overset{+}{N}-(CH_3)_3\ OH^-$$

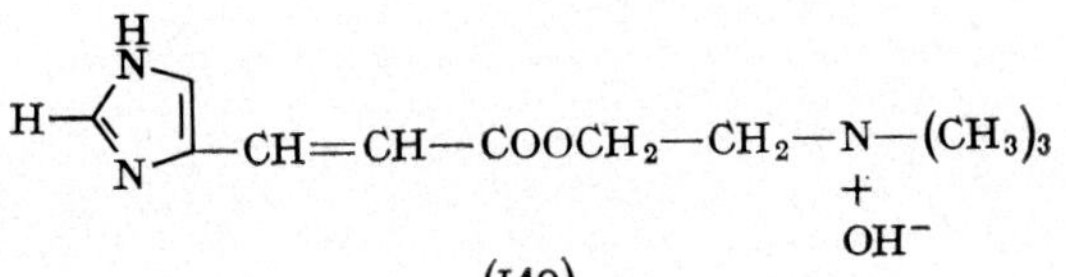

(140)

It may be said that marine sources so far have yielded a number of unusual nitrogen-containing compounds. A guanidine type compound, asterubin (141), has been isolated from sea star, *Asterias rubens L.* (North Sea) and *Asterias glacialis L.* (Mediterranean) (12). Asterubin has been synthetized by ACKERMANN and MÜLLER as follows (13, 14). Combination of cystamine (142) and dimethylcyanamide at room temperature yielded tetramethyldiguanylcystamine (143) which on oxidation with hydrogen peroxide in presence of ferric salts furnished asterubin (141).

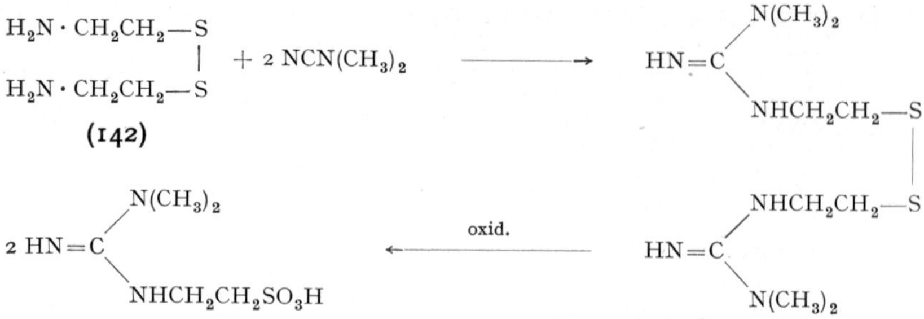

H$_2$N·CH$_2$CH$_2$—S
$\quad\quad\quad\quad$ | $\quad$ + 2 NCN(CH$_3$)$_2$ $\quad\longrightarrow\quad$ HN=C$\diagup^{\text{N(CH}_3)_2}_{\diagdown\text{NHCH}_2\text{CH}_2\text{—S}}$
H$_2$N·CH$_2$CH$_2$—S
(142)

$\quad\quad\quad\quad\quad\quad\quad\quad\quad\quad\quad\quad\quad\quad\quad\quad\quad$ |

2 HN=C$\diagup^{\text{N(CH}_3)_2}_{\diagdown\text{NHCH}_2\text{CH}_2\text{SO}_3\text{H}}$ $\quad\xleftarrow{\text{oxid.}}\quad$ HN=C$\diagup^{\text{NHCH}_2\text{CH}_2\text{—S}}_{\diagdown\text{N(CH}_3)_2}$

(141) $\quad\quad\quad\quad\quad\quad\quad\quad\quad\quad\quad\quad\quad\quad\quad\quad\quad\quad\quad\quad\quad\quad$ **(143)**

Extracts of the mussel *Pecten megallanicus* and the octopus *Eledone moschata* (*15*) have yielded a novel iminodicarboxylic acid, octopin (**144**) which is related to arginine. ACKERMANN and collaborators (*16, 17, 18,*

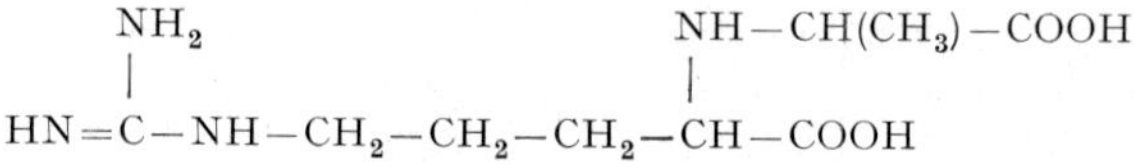

$$\text{HN=C}-\underset{\underset{\text{NH}_2}{|}}{\text{NH}}-\text{CH}_2-\text{CH}_2-\text{CH}_2-\underset{\underset{\text{NH—CH(CH}_3)-\text{COOH}}{|}}{\text{CH}}-\text{COOH}$$

(144)

23, 24, 25) have isolated spinacine (**145**) from the liver of the shark *Acanthias vulgaris* and the crab *Crango vulgaris*. This substance is 4 : 5 : 6 : 7-tetrahydroimidazo[5,4-c]pyridine-5-carboxylic acid and is closely related to spinaceamine (**146**), 4 : 5 : 6 : 7-tetrahydroimidazo[5,4-c]pyridine which was isolated by ERSPAMER and co-workers (*86*) from acetone extracts of the skin of South American amphibian *Leptodactylus pentadactylus labyrinthicus*. The extract also contained the 6-methyl analog.

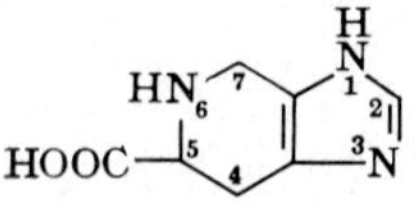

(145) m. p. 265°
$[\alpha]_D = -174.6°$

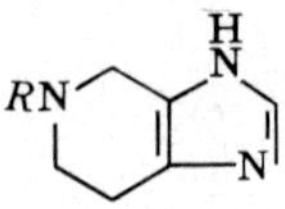

(146) m. p. 277–279° R = H
m. p. 272–274° R = CH$_3$

Anthozoe, *Anemonia sulcata*, and the mussel, *Arca noae* yielded homarine (**126**), trigonelline (**147**) and anemonine (**148**) whose structure was confirmed by synthesis (*8, 18*).

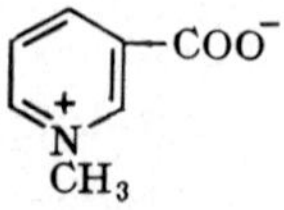

(147) 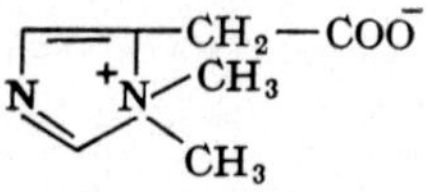 $\quad\quad\quad\quad\quad\quad\quad\quad$ **(148)**

References, pp. 472—488

The seasquirt *Cionia intestinalis* (*10*) afforded a diamine, spermin (**149**). From the seaanemone, *Anemonia sulcata*, and the sponge, *Geodia*

$$H_2N(CH_2)_3NH(CH_2)_4NH(CH_2)_3 \cdot NH$$
(**149**)

gigas, the new purine bases zooanemonine (**150**) and herbipoline (**151**) have been isolated (*6, 19, 20, 21*). Similarly, two closely related purine

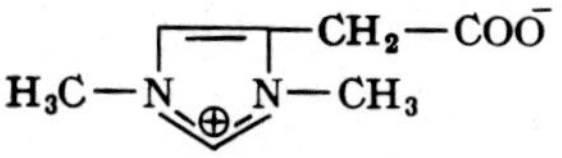

(**150**)

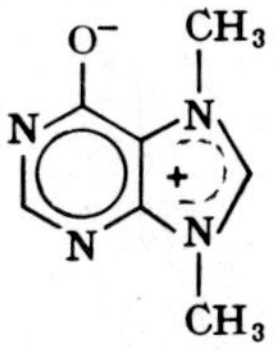

(**151**)

bases ergothionenine (**152**) and herzynine (**153**) have been isolated from the crab *Limulus polyphemus L.* (*22*).

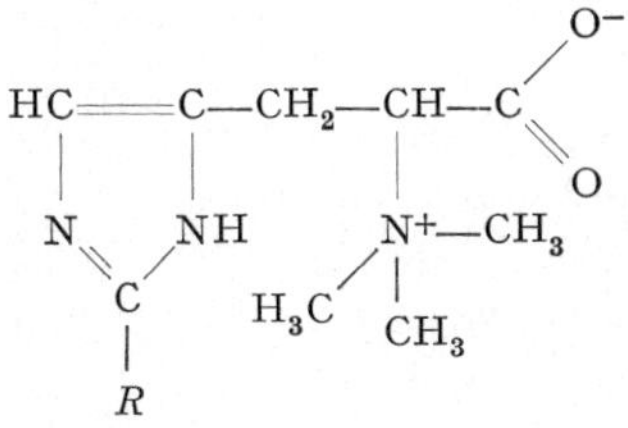

(**152**) R = SH
(**153**) R = H

A marine antibiotic-producing organism belonging to *Cephalosporium species*, has already been mentioned (see section 2, p. 422, structure [4]). ABRAHAM and his collaborators (*2*) subsequently isolated from cultures of *Cephalosporium* in addition to cephalosporin P_1 (*4*), penicillin N (**154**) and cephalosporin C (**155**). In general, (*4*) is not very active against

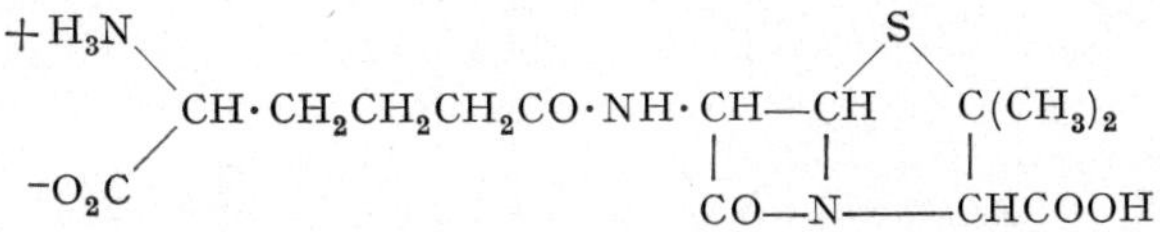

(**154**) D-(4-amino-4-carboxybutyl)-penicillin

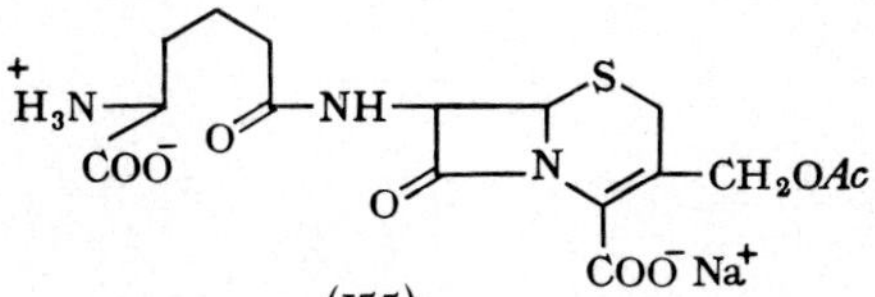

(**155**)

gram-negative bacteria. On the other hand, penicillin N is more effective against gram-positive bacilli than common penicillins and is highly effective against *Salmonella* infections. Cephalosporin C (**155**) possess greater stability to acid and penicillinase activity than other β-lactam antibiotics. Mild hydrolysis of (**155**) yields 7-aminocephalosporanic acid (**156**). The latter product and its N-acetylated derivative possess greatly enhanced antimicrobial activity.

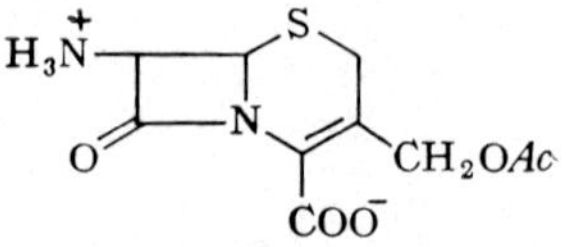

(**156**)

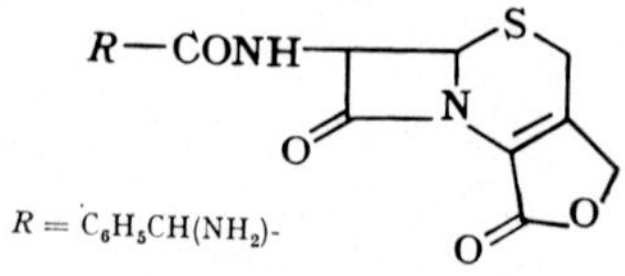

(**157**) $R = C_6H_5CH(NH_2)$-

A group at Eli Lilly has developed (*206*) a synthesis of (**156**) which eventually led to the synthesis of lactone derivatives of cephalosporanic acid of the type (**157**). These compounds (e. g., cephalotin® and cephaloridine®) are broad spectrum antibiotics. In addition to being resistant to penicillinase activity, cephalosporanic acids are also nontoxic compared with the corresponding penicillins. The chemistry and the properties of cephalosporins have been thoroughly reviewed by ABRAHAM (*2, 3*) and by MORIN and JACKSON (*205 a*).

Polypeptide-type toxins have been isolated from sea urchins (e. g., *Tripneustes gratilla* [*Linnaeus*]) (*29*) bony marine fish and carp *(Cyprinus carpio)* (*4*), ray *(Raia clavata)* (*5*), Elasmobranchs (*231*), stonefish and scorpionfish venoms (*35, 81, 252*) and soapfish *(Rypticus Saponaceus)* (*183*). Some of these polypeptides also possess hormonal activity within the species.

Red algae, *Digenea simplex* Ag. and *Chondria armata* (Kützing) Okamura, have yielded two compounds possessing anthelmintic properties: α-Kainic acid (**158**) and domoic acid (**159**). Efforts of Japanese workers (*197, 204, 205, 210, 282, 283, 298, 299*) have led to the elucidation of structure of α-kainic acid, as L-arabo-2-carboxy-3-carboxymethyl-4-isopropenylpyrrolidine (**158**) and its synthesis (*298*).

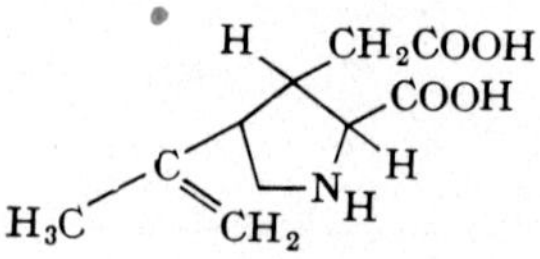

(**158**) m. p. (decomp.) 250°
$[\alpha]_D = -15°$ (c = 1, H_2O)

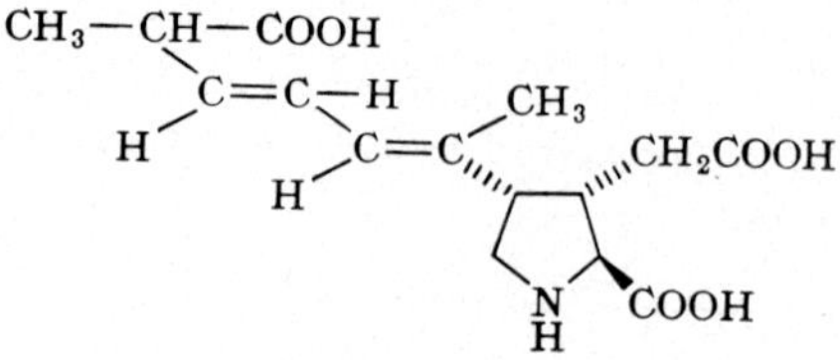

(**159**) m. p. 217° (decomp.)
$[\alpha]_D^{12} = -109.6°$ (H_2O)

References, pp. 472—488

TAKEMOTO and co-workers (*284*) assigned to domoic acid the structure of L-arabo-2-carboxy-4-(1-methyl-5-carboxy-*trans* : *trans* : S-*trans*-1,3-hexadienyl)-3-pyrrolidineacetic acid (**158**). The same group has also isolated from the marine algae, *Laminaria angustata* a hypotensive agent laminine and its oxalate (**160**) (*285, 286, 287*). The anthelmintic agents (**158**) and (**159**) are now commercially available pharmaceuticals.

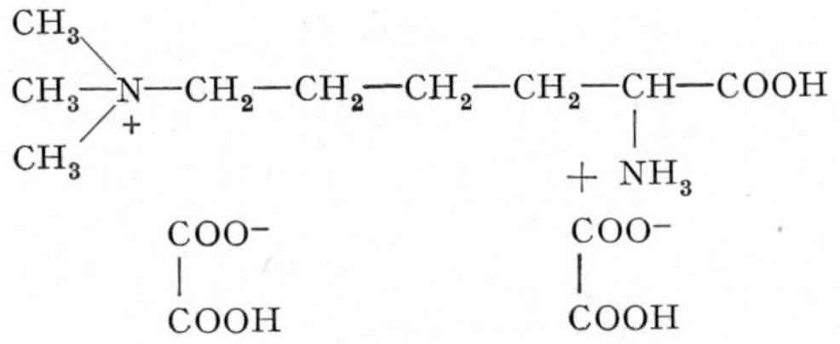

$$CH_3\diagdown$$
$$CH_3-\underset{CH_3}{\overset{+}{N}}-CH_2-CH_2-CH_2-CH_2-CH-COOH$$
$$+ NH_3$$
$$\underset{COOH}{COO^-} \qquad \underset{COOH}{COO^-}$$

(**160**)

m. p. 122—124° $[\alpha]_D^{18} = +10.8°$ (H_2O)
Trimethyl-(5-amino-5-carboxy-pentyl)-ammonium dioxalate

The chemistry and distribution of pteridines (*94*) and porphyrins (*59, 241*) has been discussed elsewhere and will not be dealt with here. These families of compounds possess valuable characteristics from a biological and chemotaxonomic point of view, but systematic investigation of pteridines and pyrrole pigments, for example, in marine sources is greatly lacking. The usefulness of some of these compounds is evident from some recently published data (*222*). Thus, the bile pigment aplysioviolin (**161**) excreted by *Aplysia limacina*, a seahare (*72, 245*), is closely

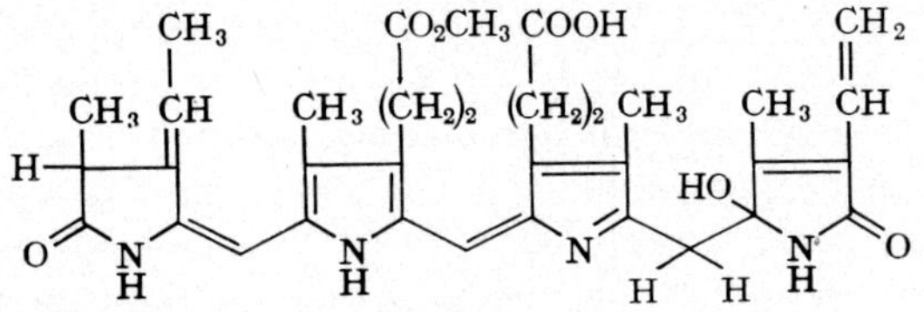

(**161**)

related to the prosthetic groups of algal biliproteins (*67, 220*), e. g., phycoerythrobilin (**162**). CHAPMAN and FOX (*66*) have shown that aplysioviolin is derived from phycoerythrin of red algae. Feeding experi-

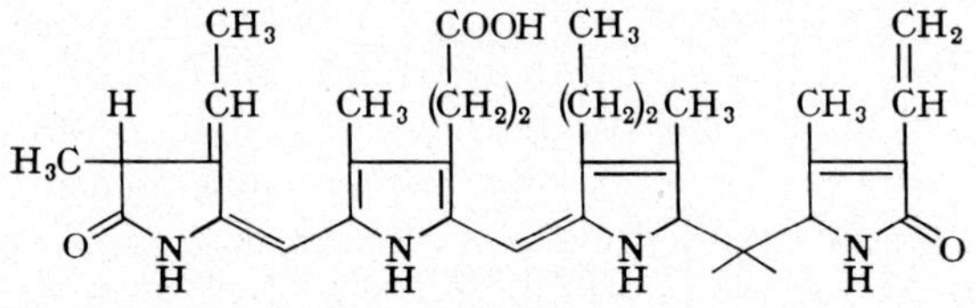

(**162**)

ments with the seahare *Aplysia californica* revealed that when it is fed on a diet of brown algae it becomes de-inked. The process is reversed on addition of red algae to the diet. Such relationships are bound to play a vital part in environmental studies.

The structure of the pigment phycocyanobilin (**163**) of blue-green algae has been elucidated (*69, 75, 78, 221*). These results suggest areas

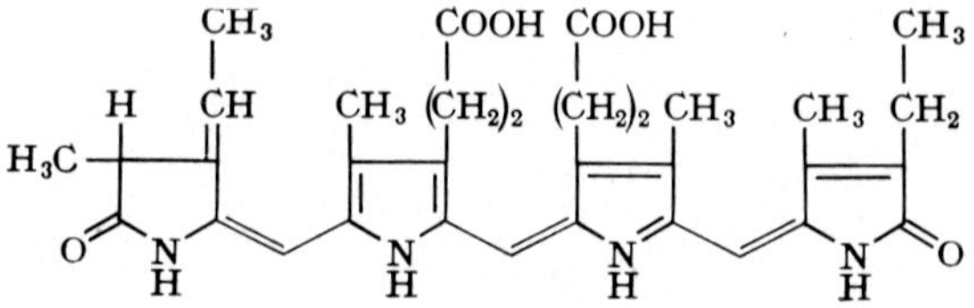

$$\text{structure (163)}$$

(163)

for future study, and indicate that the need for more information on the occurrence of such pigments before we can embark upon a biological and chemotaxonomic correlation of the nitrogen-containing pigments in marine species.

VIII. Quinonoid and Related Pigments

Terpenoid quinones have been discussed in Section V (p. 435). The occurrence and distribution of other quinones in nature has recently been reviewed by THOMSON (*290*). There are also several reviews (*103, 175, 328*) dealing with the distribution and nature of quinonoid pigments isolated from needles and shells of sea urchins. Problems of isolation prior to the introduction of modern methods are evident (*288*) throughout the literature (for example, see (*118, 172, 173, 174, 211, 288, 300, 301*)).

The possible significance of quinones to chemotaxonomic studies has been realized for a long time, the first publication in this field, by MACMUNN (*174*), appearing in 1885. More recently GOODWIN and SRISUKH (*104*) have reported on the pigments of sea urchins, *Echinus esculentus* L. and *Paracentrotus lividus* Lamarck and SMITH and THOMSON (*278*) on the pigments of *Hemicentrotus pulcherrimus* (Ag). Systematic work of several groups, notably those of THOMSON (*30, 290*) and SCHEUER (*60, 199, 200, 201*) who developed methods of isolation and studied model compounds, in order to apply nuclear magnetic resonance and mass spectrometry (*40*) to structure elucidation has clearly shown that quinonoid pigments play an important rote in chemical evolution and can be used to shed light on evolutionary relationships. Results of studies on marine species at the time of this writing show clearly that the predominant pigments are structurally related to naphthazarin (**164**) and juglone (**165**).

References, pp. 472—488

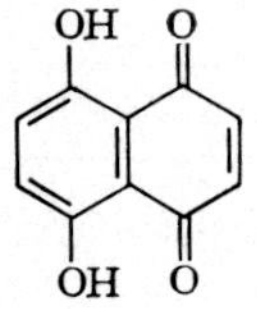

Chemistry of Natural Products Derived from Marine Sources 461

(164) 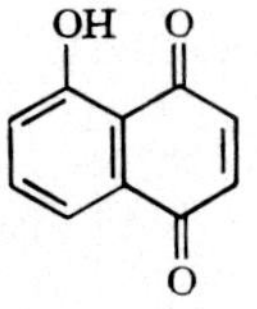 (165)

The known structures of these pigments, which have been isolated mainly from spines and tests of *Echinoidea*, are shown in Table 5. A comparison of quinonoid pigments reviewed by THOMSON (*290*) with the more

Table 5

Pigment	Structure	Source	Ref.
Echinochrome A (167)	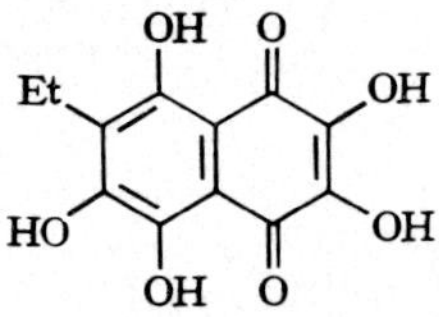	Sea urchins *Echinothrix diadema* L.	(*199*)
Spinochrome A (168)	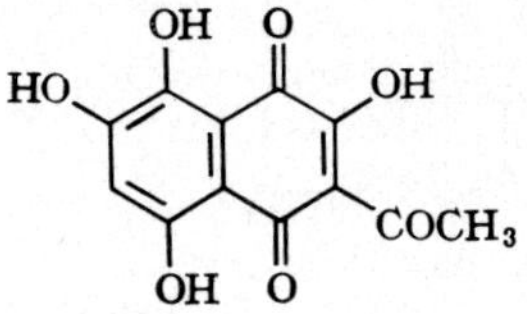	*Echinothrix calamaris* Pall.	(*199*)
Spinochrome B (169)	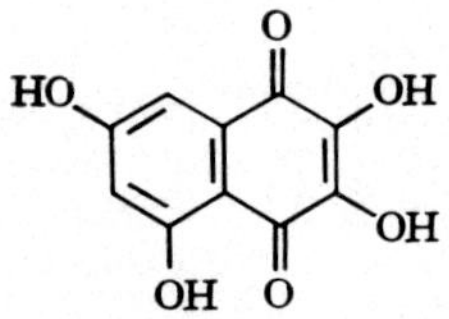	*Echinothrix calamaris* Pall.	(*199*)
Spinochrome C (170)	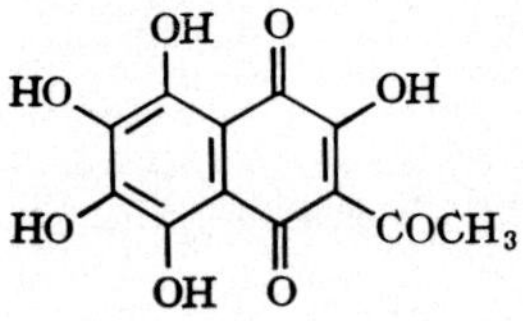	*Echinothrix calamaris* Pall.	(*199*)
Spinochrome D (171)	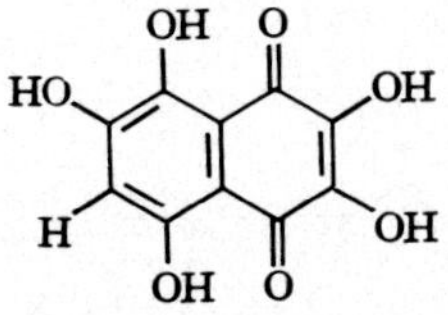	*Echinothrix calamaris* Pall.	(*199*)

Table 5 (continued)

Pigment	Structure	Source	Ref.
Spinochrome E (**172**)	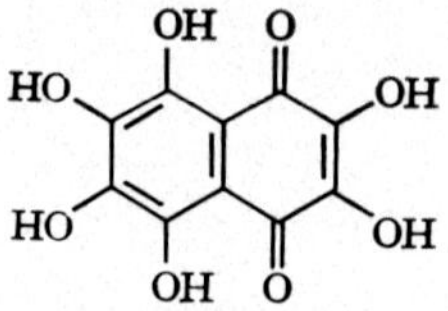	Echinothrix calamaris Pall.	(199)
Namakochrome	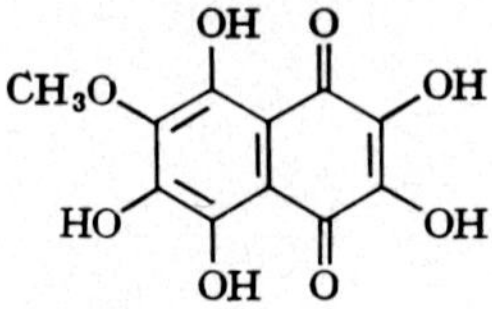	Holothurian, Polycheria rufesceus Brandt	(272)
2,6-dihydroxy-3,7-dimethoxy naphthazarin	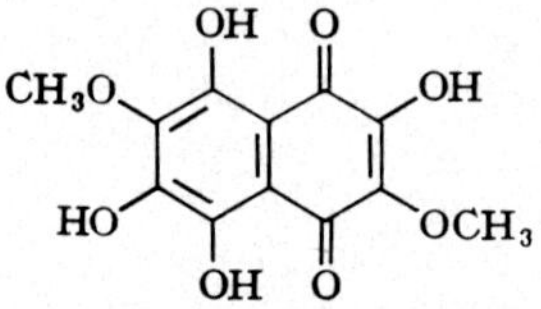	Spines of the sea-star Acanthaster planci L.	(272)
2,7-dihydroxy-3,6-dimethoxy naphthazarin	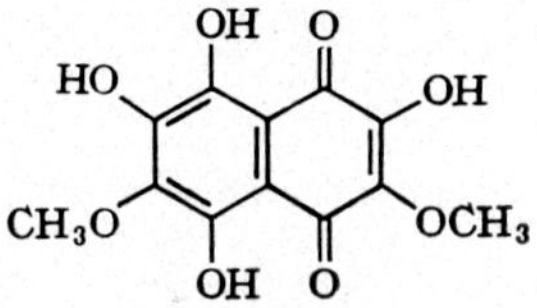	Acanthaster planci L.	(272)
2-hydroxy-3-ethyl-naphthazarin	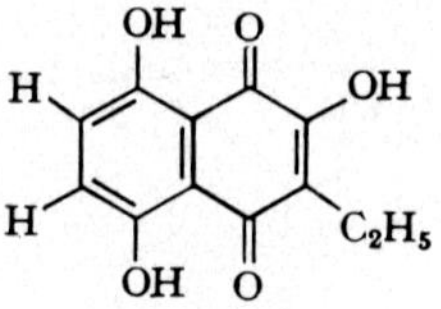	Spines of ophiuroids, O. erinaceus Müller and Troschel, O. insularia Lyman Echinoids, Echinothrix	(272, 199)
2-hydroxy-3-acetyl-naphthazarin	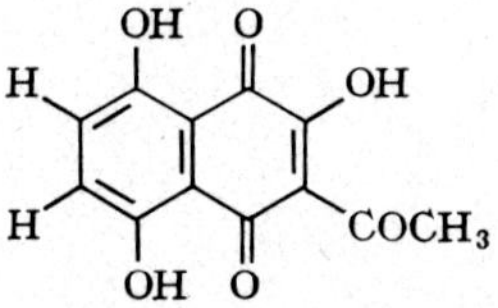	O. erinaceus Müller and Troschel O. insularia Lyman Echinoids, Echinothrix	(272, 199)
6-Ethyl-2-hydroxy-naphthazarin	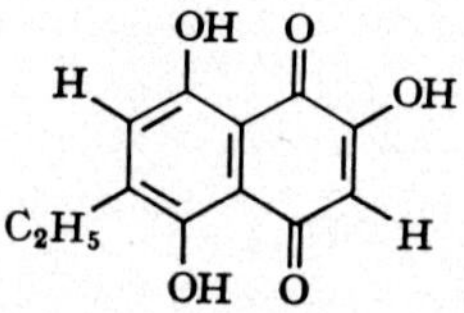	idem	(272, 199)

References, pp. 472—488

Table 5 (continued)

Pigment	Structure	Source	Ref.
Spinochrome A	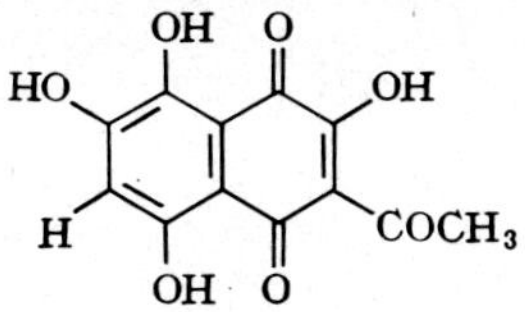	idem	(272, 199)
2-hydroxy-3-acetyl-7-methoxy-naphthazarin	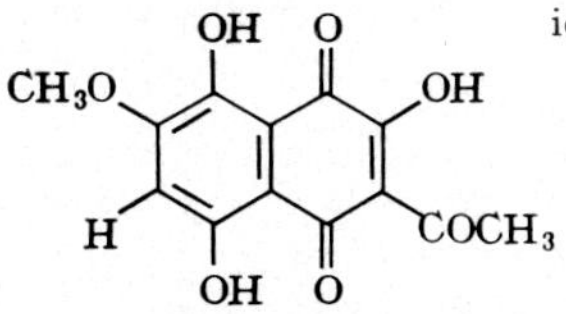	idem	(272, 199)
2,7-dihydroxy-3-ethyl-naphthazarin	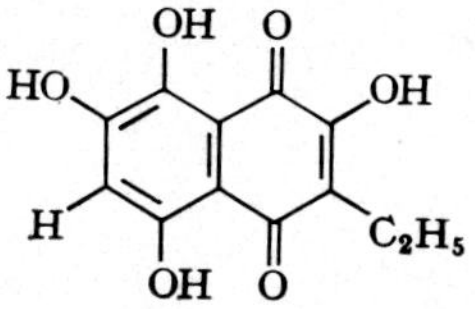	O. erinaceus	(272)
2,6,7-trihydroxy-3-ethyljuglone	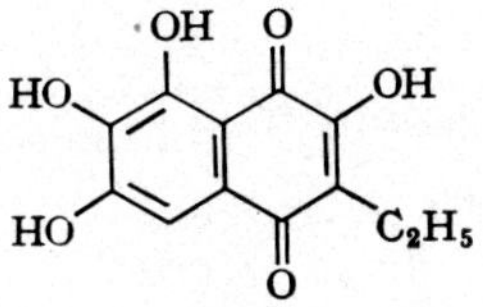	O. erinaceus O. insularia	(272)

2,7-dihydroxy-naphthazarin	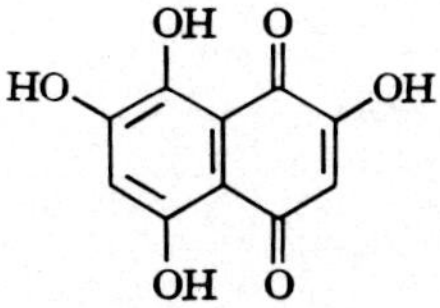	Crinoid, genus
		  Antedon
		Spines echinoids
		Echinometra oblonga
		  Blainville
		Tripneustes gratilla
		  Linn.
			(272, 199)

Rhodo-comatulin	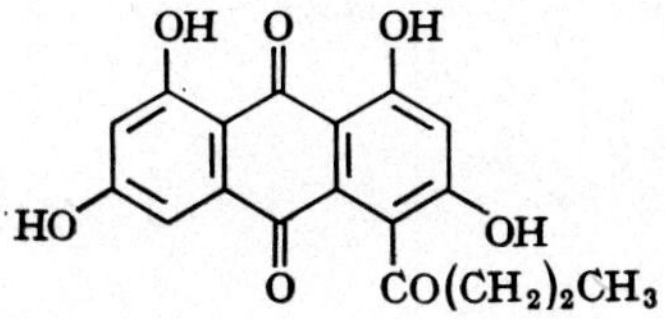	Crinoid, Comatula pectinata Linnaeus	(281, 177)
2-hydroxy-6-ethyl-juglone	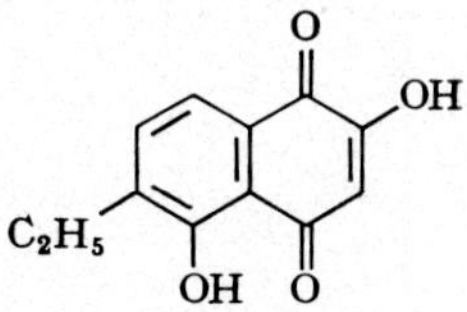	Echinoids, Echinothrix diadema L. E. calamaris Pall.	(199)

Table 5 (continued)

Pigment	Structure	Source	Ref.
2-hydroxy-6-ethyl-naphthazarin	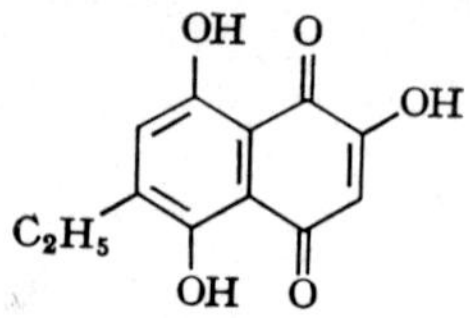	idem.	(199)
Naphthopurpurin	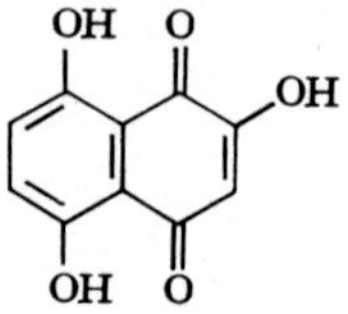	idem.	(199)
2,5-dihydroxy-3-ethyl-benzoquinone	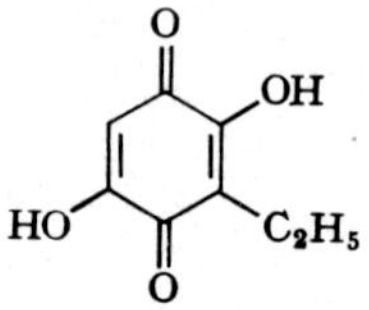	idem.	(199)
2,7-dihydroxy-6-acetyljuglone	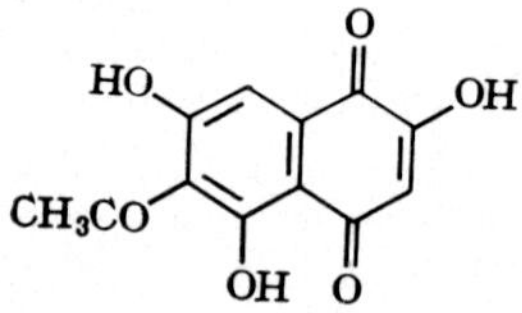	Echinoids, *Echinothrix diadema* L. *E. calamaris* Pall.	(199)
2,7-dihydroxy-3-ethyl-naphthazarin	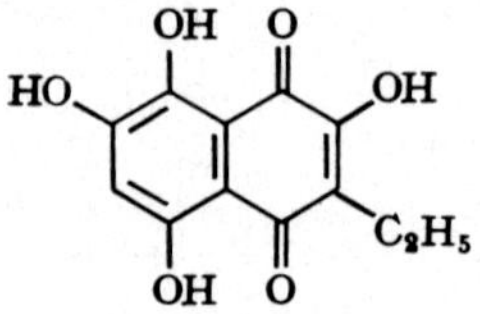	idem.	(199)
2-hydroxy-6-acetyl-naphthazarin	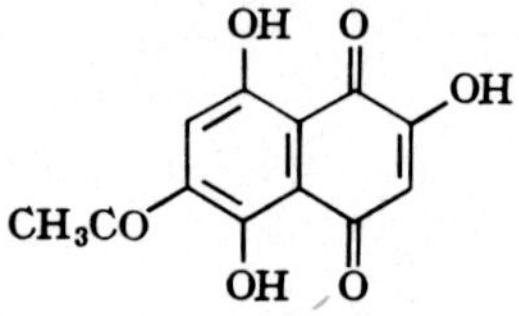	idem.	(199)
2,3,7-trihydroxy-6-acetyljuglone	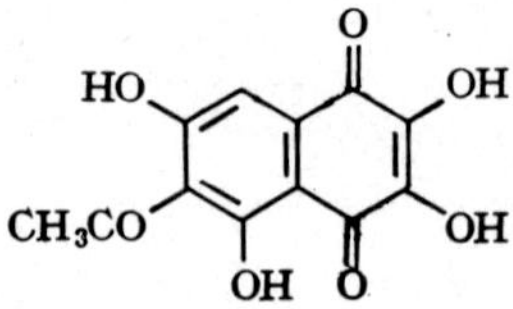	idem.	(199)

References, pp. 472—488

recently characterized structures reveals that marine species, in addition to closely related napththazarine (**164**) and juglone (**165**) analogs, can also elaborate single ring structures, e. g., 2,5-dihydroxy-3-ethylbenzoquinone (**173**) (*199*) and structures incorporating an anthraquinone moiety, such as analogs of rhodocomatulin (**174**) (*281*).

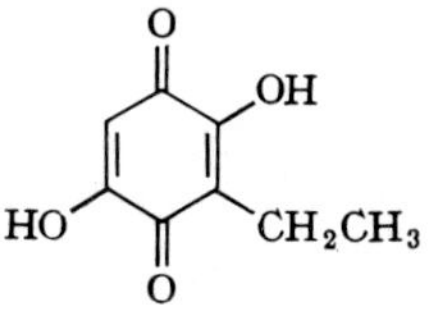

(173) 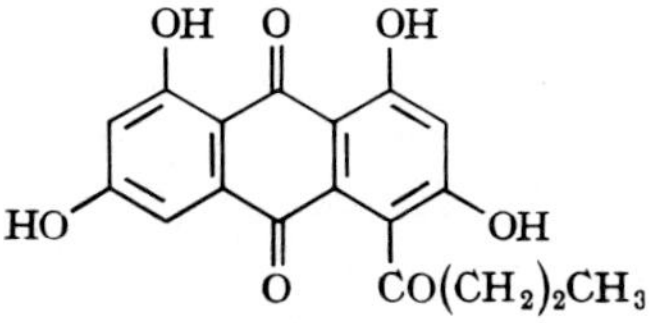   (174)

WALLENFELS and GAUHE (*300*) reported the first synthesis of echinochrome A (**167**) (Table 5). Later on extensive synthetic and structural studies of polyhydroxynaphthoquinones by MOORE, SCHEUER and collaborators (*198*) and THOMSON and his group (*30*) led to the synthesis of spinochromes. The Hawaii group (*273*) developed syntheses of spinochromes A, C, D and E in good yields from essentially common starting materials.

Thus, condensation of chloromaleic anhydride and 1,2-dihydroxy-3,4-dimethoxybenzene yields as the major product 6-chloro-2,3-dimethoxynaphthazarin (**175**), which on treatment with sodium methoxide followed by concentrated hydrobromic acid yields spinachrome D (**171**). Similar condensation with dichloromaleic anhydride leads to 2,3-dimethoxy-6,7-dichloronaphthazarin (**176**).

Sodium methoxide treatment of (**176**) in methanol, followed by chromatography, afforded 2,3,6-trimethoxy-7-hydroxynaphthazarin (**177**).

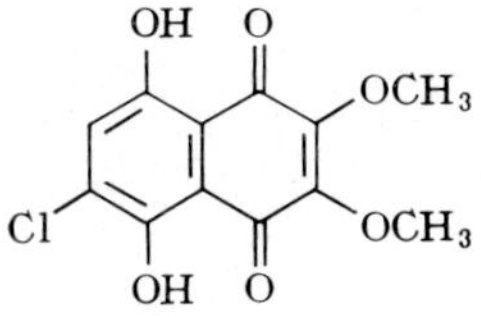

(175) 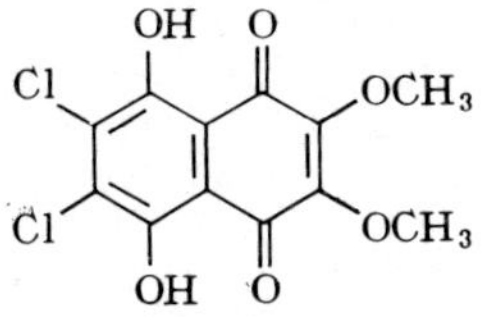   (176) 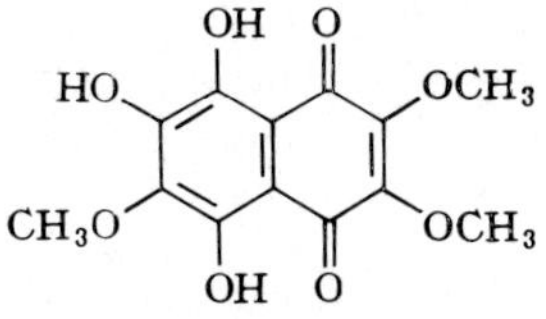   (177)

Treatment of the latter with diazomethane followed by hydrobromic acid yielded spinochrome E (**172**).

Analogously, use of acetylated naphthazarin (**178**) and polyacetoxynaphthalenes (**179**) and (**180**) led to the synthesis of spinochrome A (**168**) and spinochrome C (**170**) (see Table 5).

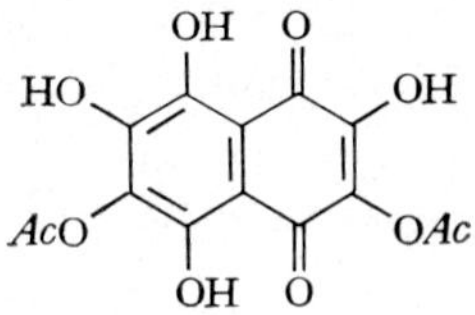

(178) 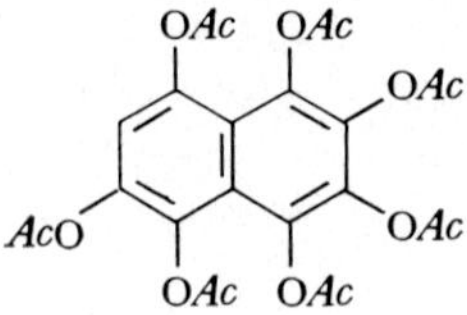   (179) 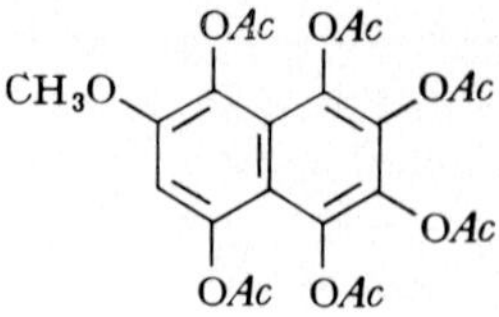   (180)

The existence of hydroxyanthraquinone pigments (174) in the present-day crinoid *Comatula pectinata* L. (*211*) and their presence in fossil crinoids, such as the Jurassic crinoid (*Apiocrinus* sp.) (*48a*) suggests a common origin. The pigments found in fossilized crinoids have been characterized as fringelite D (181), E (182), F (183) and H (184).

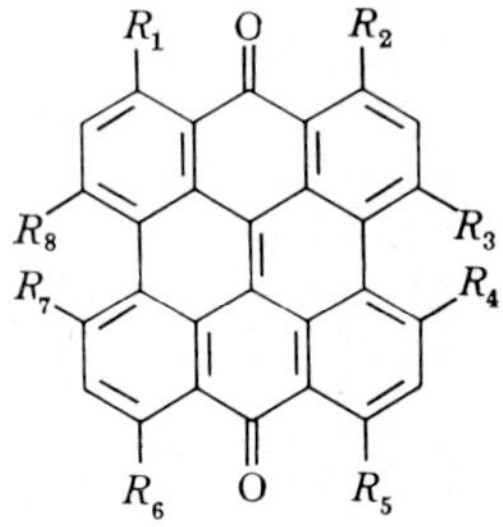

	R_1	R_2	R_3	R_4	R_5	R_6	R_7	R_8
(181)	OH	OH	OH	OH	OH	OH	OH	OH
(182)	OH	OH	OH	OH	OH	OH	H	OH
(183)	OH	OH	OH	OH	OH	OH	H	H
(184)	OH	OH	H	H	OH	OH	H	H

In their discussion of echinoderm pigments SCHEUER et al. (*272*) pointed out that echinoderms synthesize predominantly polyhydroxy-naphthoquinones possessing free hydroxy groups and frequently *two-carbon side chains*. This contrasts with the naphthoquinone pigments found in higher plants and fungi, which often contain a one-carbon side chain.

The recent isolation of a pyranonaphthazarin pigment from the spines of the sea urchin *Echinothrix diadema* L. (*203*) provides an exception to this generalization. The structure of this pigment has been elucidated by SCHEUER and collaborators (*203*) as 2-methyl-8-hydroxy-2 H-pyrano

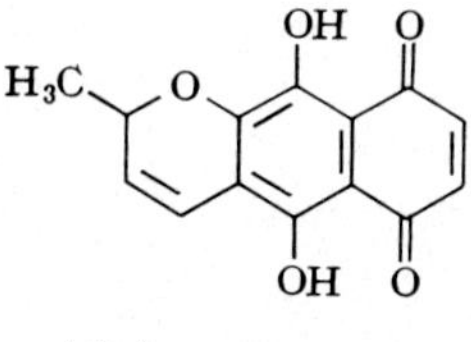

(185) m. p. 165–172° (decomp.)

[3,2-g] naphthazarin (185), the first pigment in this species which possesses a one-carbon side chain and a four carbon unit attached to the naphthoquinone system.

References, pp. 472—488

It should also be borne in mind that other parts of the body (for example, other than spines and the shell) may well contain different and unique pigments. Systematic investigation of distribution, chemistry and biochemistry of echinoderm pigments, both living and fossilized, may well lead to an understanding of the origin of these species (*38, 214, 250, 272*).

The first flavonoid in a marine species has been found by MARKHAM and PORTER (*184*) in the green algae (chlorophyta), *Nitella hookeri* (F. *Characeae*). Preliminary evidence, led these authors to suggest that the flavonoids present are of the vicenin (**186**) and lucenin (**187**) type. These findings sould be interpreted as providing further evidence for a biochemical relationship between chloropythes and the higher plants.

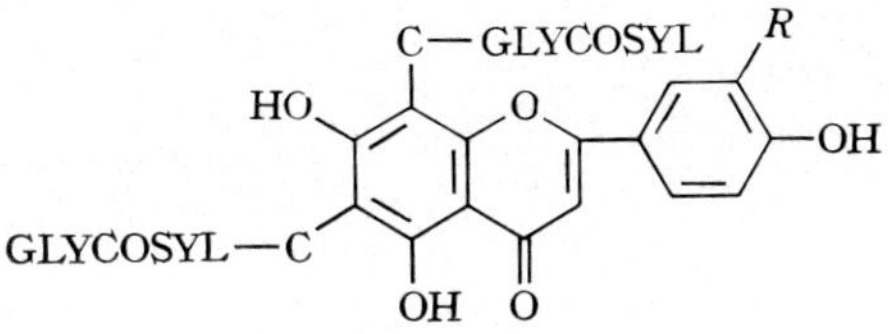

(**186**) $R = H$
(**187**) $R = OH$

IX. Carbohydrates

BELL (*41*) and BERNFELD (*47*) in their discussions of mono- and polysaccharides have also reviewed carbohydrates from marine sources. Free carbohydrates present in sea water have been discussed by Duursma (*83*). More recently, polysaccharides of marine algae have been reviewed by PEAT and TURVEY (*228*) and by PERCIVAL and McDOWELL (*230*). A great deal of the research in this field was concentrated on the investigation of the carbohydrate moiety of *Chlorophyceae* (green algae), *Rhodophycae* (red algae) and *Phacophycae* (brown algae), all of which contain commercially useful substances such as alginic acid, agar and carragenan. Although names of these substances have become household words, complete structures of these substances have not as yet been elucidated.

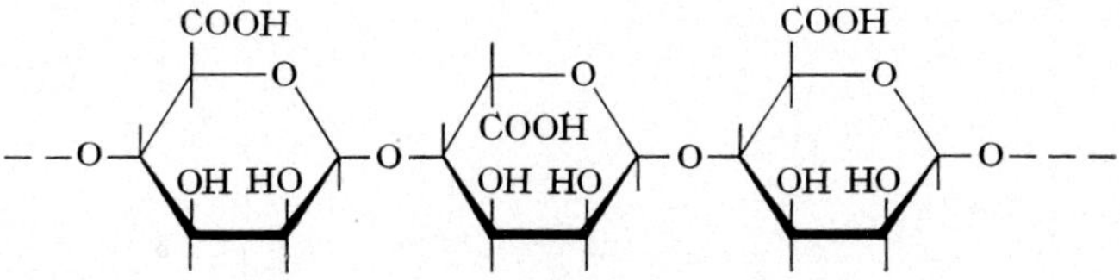

The major polysaccharide in the brown algae which produce algin is a polymer of D-mannuronic acid and L-guluronic acids (**188**) connected by β-(1 → 4) glycosidic linkages. Red seaweed polysaccharides on the other hand are composed of galactose and substituted galactose units connected by glycosidic linkages having an alternating α-(1 → 3), β-(1 → 4) configuration. In Table 6 some of the marine polysaccharides and their predominant sugar units are listed.

Table 6

Polysaccharide	Predominant Sugar Unit
λ-carragenan	D-galactose-2-sulfate D-galactose-2,6-disulfate
χ-carragenan	D-galactose-4-sulfate 3,6-anhydro-D-galactose
ι-carragenan	D-galactose-4-sulfate 3,6-anhydro-D-galactose-2-sulfate
Agarose	D-galactose 3,6-anhydro-L-galactose
Porphyran	3,6-anhydro-L-galactose D-galactose 6-O-methyl-D-galactose L-galactose-6-sulfate 3,6-anhydro-L-galactose
Fucellaran	D-galactose, D-galactose-sulfate, 3,6-anhydro-D-galactose

A — B — A — B — A — B — A — B — A — B.

A =

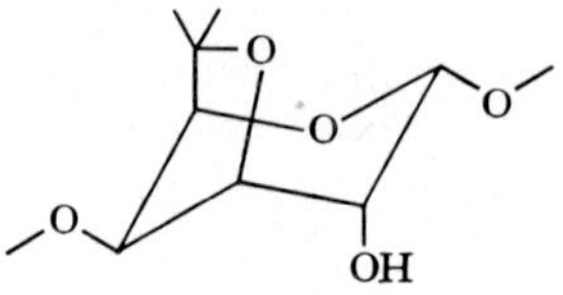

3,6-Anhydro-D-galactose -or its 2-sulfate

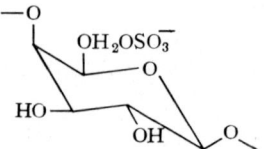

*D-galactose-6-sulfate -or its 2-sulfate

B =

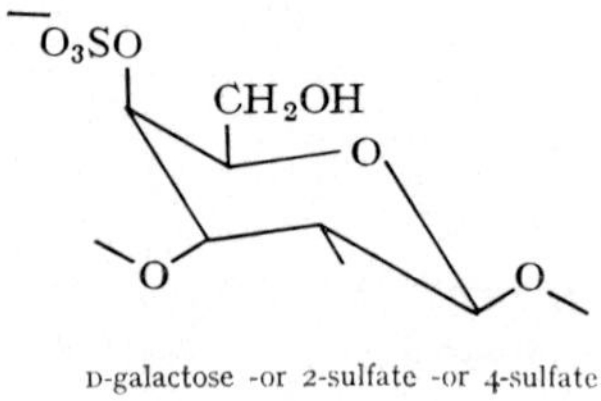

D-galactose -or 2-sulfate -or 4-sulfate

(189)

References, pp. 472—488

Properties of alginates have been recently discussed by McDowell (*192*) and Rees (*234, 235*). Structural studies have been reviewed elsewhere (*280, 329*). On the basis of presently available data, the following structures of type (**189**) have been proposed for the carragenans.

Brown algae (e. g. the *Laminaria* group) elaborate a polysaccharide, laminaran which contains D-glucose as the predominant sugar unit. Rees (*235, 236, 237, 238*) has reviewed our present knowledge on the glycosyl-aminoglycans which occur in connective tissue and on the agar-carragenan types of seaweed polysaccharides, while Kochetkov and Bockhov (*169*) have carried out recently a successful synthesis of a β-(1 $\rightarrow$ 3)-linked D-glucan closely related to the G-chain of insoluble laminaran.

X. Related Topics

Toxic principles occurring in amphibians such as toads and frogs and in sea snakes have been discussed elsewhere (*90, 182*), and will not be dealt with here. However a few related topics will be mentioned briefly.

Ethanolic extracts of hagfish *(Myxine glutinosa)* islets and those of the digestive tissue of the purple starfish *(Pisaster ochraceous)* have afforded high yields of insulin (*311*). It is to be noted that the hagfish yielded an extract which possessed 11 units/gram of tissue of insulin-like activity, i. e. twice the yield normally obtained from bovine pancreas. The amino acid sequence in insulin isolated from cod has been elucidated by Reid and Grant (*239*).

Commercial products from marine sources have been thoroughly reviewed elsewhere, as for example, fish oils (*132, 292*). The occurrence and distribution of fatty acids has been discussed by Shorland (*267*). Products obtained from algae, e. g., algin and alginates, have been reviewed by Steiner and McNeely (*279*) and McDowell (*192*). The role of algae in medicine has been reviewed by Schwimmer and Schwimmer (*262*) and excellent sources of information on seaweed and algae are to be found in the Proceedings of International Seaweed Symposia (*50, 51, 82, 329*). Standard references are also to be found in "Fish and Food" (*48*).

Addendum

Unrau and co-workers (*330*) have recently reported a systhesis of a C_{26} sterol (**190**) isolated from the scallop, *Placopecten magellanicus* (Gmelin) and thus established the stereochemistry at C-20.

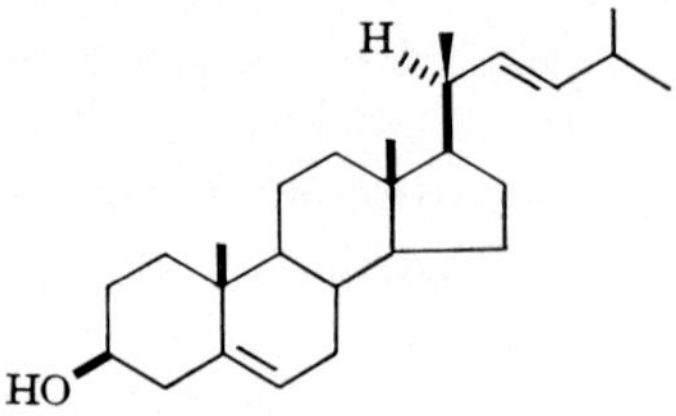

(190)

m. p. 143—144° C [α]$_D$ —65° (C, 2.7)

Hawaiian algae, *Dictyopteris plagiogramma* and *D. australis* have afforded two sulfur containing hydrocarbons, bis-(3-oxoundecyl) trisulfide (191) and bis-(oxoundecyl) tertrasulfide (192) (*331*).

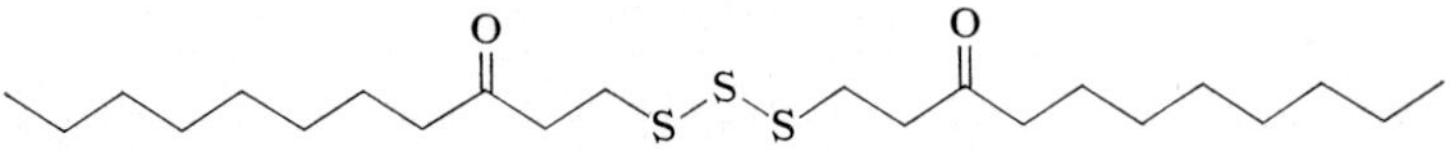

(191)

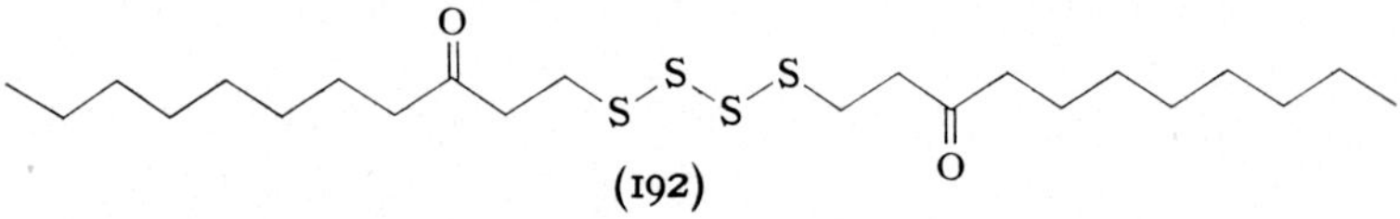

(192)

The marine alga *Taonia atomaria,* has yielded (*332*) a unique phenolic compound, taondiol (193). GONZÁLES and co-workers (*332*) suggested that (193) is the first naturally occurring cyclic side chain derivative of tocopherol.

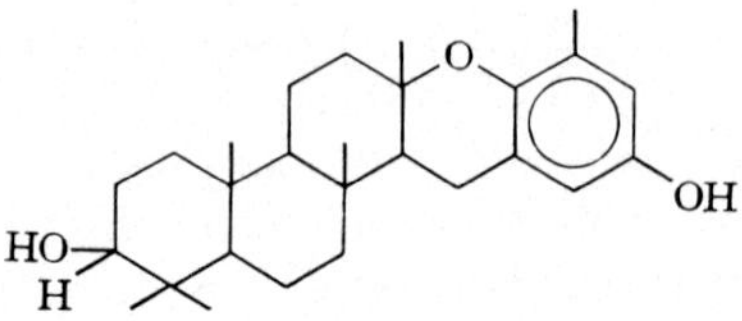

(193)

m. p. 283—284° C [α]$_D$ = —76° (C, 0,3, CHCl$_3$)

In a recent report, HABERMEHL and VOLKWEIN (*333*), have substantiated their earlier finding (*109*) that holothurinogenins are present in mediterranean holothurians.

SIMS and collaborators (*334*), have isolated from *Laurencia pacifica* a novel chlorine and bromine containing compound pacifenol (194).

References, pp. 472—488

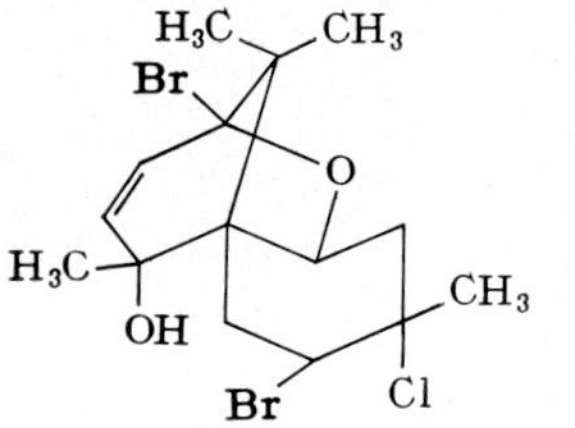

(194)
m. p. 149—150.5° C

This alga also yielded laurinterol (124a).

The sponge *Angelas Orides* has yielded the novel bromopyrrole derivatives (195) and (196), and an aminoimidazole (197) (*335*).

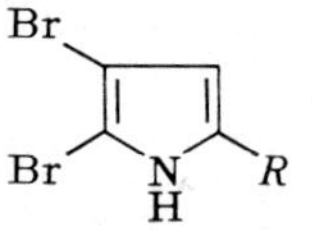

R = —COOH
R = —CN
R = —CO, NH$_2$

(195)

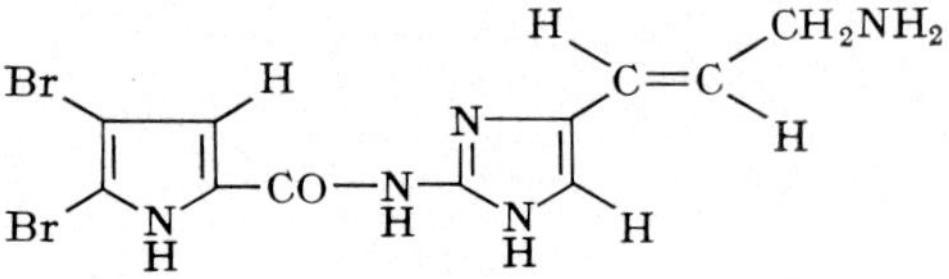

(196)

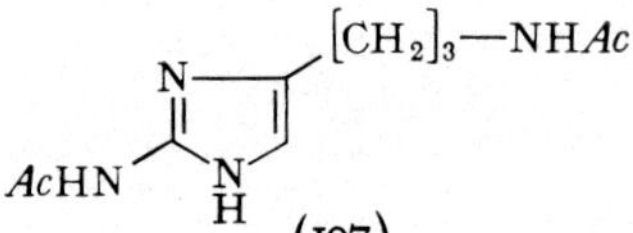

(197)

RAPOPORT and coworkers (*336*) reported the degradation of saxitoxin by treatment with dilute sodium hydroxide and 0.8% hydrogen peroxide, followed by catalytic decomposition of peroxide and chromatography to a crystalline product (198) which retains nine of the ten carbon atoms and six of the seven nitrogen atoms of saxitoxin. Subsequently (*337*) evidence was reported which permitted reformulation of saxitoxin as (199).

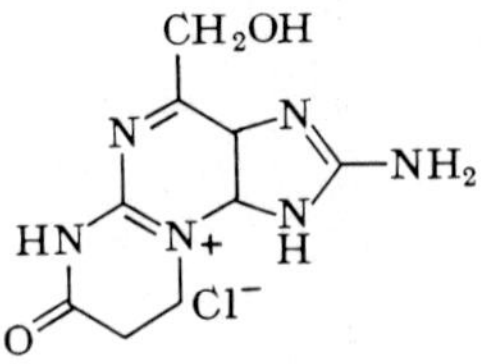

(198)

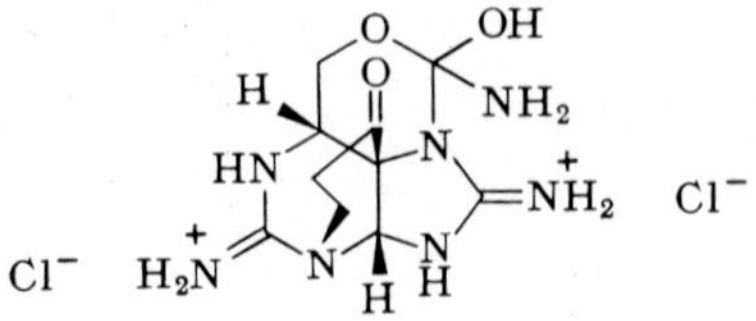

(199)

Methanolic extracts of the Mediterranean sponge *Spongia nitens* (*338*) have yielded two unusual furanoterpenes nitenin (200) and dihydro-nitenin (201). These C-21 furanoterpenes may have been formed biosynthetically by degradation of higher terpenoids built from isoprene units linked head to tail.

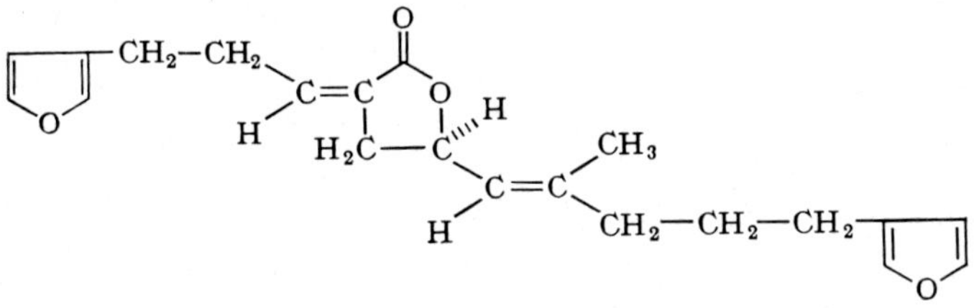

(200)

$[\alpha]_D$ —45.4°, λ_{max} 221 (ε 14 000) nm

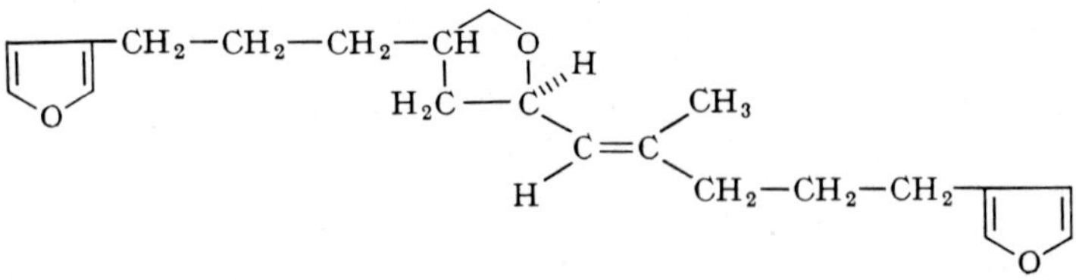

(201)

$[\alpha]_D$ —25.2°, λ_{max} 222 (ε 5230) nm

References

1. AASEN, A. J. and S. L.-JENSEN: Carotenoids of Flexibacteria, IV. Carotenoids of Two Further Pigment Types. Acta Chem. Scand. **20**, 2322 (1966).
2. ABRAHAM, E. P.: The Cephalosporins. Pharmac. Rev. **14**, 473 (1962).
3. ABRAHAM, E. P.: The Cephalosporin C Group. Quart. Rev. **21**, 231 (1967).
4. ARCHER, R., J. CHAUVET, M. T. CHAUVET, D. CREPY: Characterisation des Hormones Neurophypophysaires, d'un Poisson Osseux d'eau douce, La Capre (*Cyprinos Carpio*). Comparaison avec les Hormones des Poissons Osseux Marins. Comp. Biochem. Physiol. **14**, 245 (1965).
5. — Phylogenie des Peptides Neurophypophysaires. Biochem. Biophys. Acta. **107**, 393 (1965).

6. ACKERMANN, D. and P. H. LIST: Spongopurin, eine in der Natur neue Purinbase. Naturwiss. **48**, 74 (1961).
7. ACKERMANN, D. and S. SKRAUP: Endgültige Konstitutionsermittlung und Synthese des Spinacins. Ibid, **284**, 129 (1949)
8. ACKERMANN, D.: Über das Vorkommen von Homarin, Trigonellin und einer neuen Base Anemonin in der Anthozoe *Anemonia Sulcata*. Ibid. **295**, 1 (1953).
9. ACKERMANN, D. and P. H. LIST: Über das Vorkommen von Trimethylaminoxyd, Homarin, Trigonellin und einer Base $C_4H_9O_2N$ in der Krabbe *Crangon vulgaris*. Ibid. **306**, 260 (1957).
10. ACKERMANN, D. and R. JANKA: Erstmalige Beobachtung von Spermin bei Avertebraten, *Cionia intestinalis*. Ibid, **296**, 279 (1954).
11. ACKERMANN, D.: Über das Vorkommen von Homarin, Taurocyamin, Cholin, Lysin und anderen Aminosäuren sowie Bernsteinsäure in dem Meereswurm *Arenicola marina*. Z. Physiol. Chem. **302**, 80 (1955).
12. ACKERMANN, D.: Asterubin, eine Schwefelhaltige Guanidinverbindung in der belebten Natur. Ibid. **232**, 206 (1935).
13. ACKERMANN, D.: Synthese des Asterubins. Ibid, **249**, 208 (1935).
14. ACKERMANN, D. and E. MÜLLER: Zweite Synthese des Asterubins. Ibid. **235**, 233 (1935).
15. ACKERMANN, D. and M. MOHR: Über das Vorkommen von Octopin, Agmantin und Arginin in der Octopodenart *Eledone moschata*. Ibid. **250**, 243 (1937).
16. ACKERMANN, D. and E. MULLER: Spinacin, ein Bestandteil der Selachierleber. Ibid. **268**, 277 (1941).
17. ACKERMANN, D.: Nachweis des Imidazolkernes in Spinacin. Ibid. **276**, 268 (1942).
18. ACKERMANN, D. and R. JANKA: Konstitution und Synthese des Anemonins. Ibid. **294**, 93 (1953).
19. ACKERMANN, D. and P. H. LIST: Konstitutionsermittlung des Herbipolins, einer neuen tierischen Purinbase. Ibid. **309**, 286 (1957).
20. ACKERMANN, D. and P. H. LIST: Zur Konstitution des Zooanemonins und des Herbipolins. Ibid. **318**. 281 (1960).
21. ACKERMANN, D. and H. G. MENSSEN: N-haltige Inhaltsstoffe des Pferdeschwammes *Hippospongia equina*, Ibid. **322**, 198 (1960).
22. ACKERMANN, D. and P. H. LIST: Über das Vorkommen von Ergothionein und Herzynin in *Limulus polyphemus* L. Naturwiss. **45**, 131 (1958).
23. ACKERMANN, D.: Über das Vorkommen von Spinacin in der Krabbe *Crango Vulgaris*, Z. Physiol. Chem. **328**, 275 (1962).
24. ACKERMANN, D. and G. HOPPE-SEYLERS: Vergleich zweier biologischer, von Histidin und Histamin ableitbarer, isomerer, Ringsysteme (Spinacin und Zapotidin). Ibid. **336**, 283 (1964).
25. ACKERMANN, D. and M. MOHR: Über stickstoffhaltige Bestandteile der Leber des Haifisches *(Acanthias vulgaris)*. Z. Biol. **98**, 37 (1937).
26. ACKERMANN, D. and R. JANKA: Über das Vorkommen von Homarin, Glykokollbetain, Cholin, Arginin, Mytilit und d,l-Milchsäure in der Meeresschnecke *Patella sp.* Z. Physiol. Chem. **298**, 5 (1954).
27. AITZETMÜLLER, K., W. A. SVEC, J. J. KATZ and M. H. STRAIN: Structure and Chemical Identity of Diadinoxanthin and the Principal Xanthophyll of Euglena. Chem. Comm., **1968**, 32.
28. AITZETMÜLLER, K., H. H. STRAIN, W. A. SVEC, M. GRANDOLFO and J. J. KATZ: Loroxanthin, A Unique Xanthophyll from *Scenedesmus Obliquus* and *Chlorella Vulgaris*, Phytochem. **8**, 1761 (1969).

29. ALEXANDER, C. B., G. A. FEIGEH and J. T. TOMITA: Isolation and Characterization of Sea Urchin Toxin. Toxicon **3**, 9 (1965).
29a. ALI, A., D. SARANTAKIS and B. WEINSTEIN: Synthesis of the Natural Product (−)-Dictyopterene B. Chem. Comm. 940 (1971).
30. ANDERSON, K. A., J. SMITH and R. H. THOMSON: Naturally Occurring Quinones. Part VI. Spinochrome D. J. Chem. Soc. **1965**, 2141 and references therein.
31. ANDERSON, I. G., G. A. D. HASLEWOOD, A. D. CROSS and L. TÖKES: New Evidence for the Structure of Myxinol, Biochem. J. **104**, 1061, (1967).
32. ANDERSON, I. G. and G. A. D. HASLEWOOD: Comparative Studies of Bile Salts, 16-Deoxymyxinol, Biochem. J. **112**, 763 (1969).
33. ASHIKARI, H.: Über die Galle des „Akajei"-Fisches *(Dasyatis Akajei)* und die Konstitution des Scymnols. Biochem. (Japan) **29**, 319 (1939).
34. ASHIKARI, H.: Über die Galle des „Akajei"-Fisches (Dasyatis Akajei) und die Konstitution des Scymnols. Biochem. (Japan) **29**, 319 (1939).
35. AUSTIN, L., R. G. GILLIS and G. YOUATT: Stonefish Venom, Some Biochemical and Chemical Observations, Austr. J. Exptl. Biol. Med. Sci. **43**, 79 (1965).
36. BANNER, A. H., P. HELFRICH, P. J. SCHEUER and T. YOSHIDA: Research on Ciguatera in the Tropical Pacific, Proc. Gulf Caribbean Fish Institute, 16th Ann. Session, 1962.
37. BANNER, A. H.: Marine Toxins from the Pacific, I. Advances in the Investigation of Fish Toxins, Proc. of 1st Int. Symp. on Animal Toxins, Atlantic City, p. 157, 1966, Ed. by F. E. RUSSELL and P. R. SAUNDERS, Pergamon Press.
38. BARRACLOUGH-FELL, H.: The Evolution of the Echinoderms, Smithsonian Annual Report, 457–490 (1962).
38a. BASLOW, MORRIS H.: Marine Pharmacology, The Williams & Wilkins Co., Baltimore, 1969.
39. BAXTER, J. G., VITAMIN A, Ch. 6, p. 169 in Comprehensive Biochemistry, Ed. by M. FLORKIN and E. H. STOTZ, Vol. 9, Elsevier, 1963.
40. BECHER, D., C. DJERASSI, R. E. MOORE, H. SINGH and P. J. SCHEUER: Mass Spectrometry in Structural and Stereochemical Problems. CXI. The Mass Spectrometric Fragmentation of Substituted Naphthoquinones and its Aplication to Structural Elucidation of Echinoderm Pigments, J. Org. Chem. **31**, 3650 (1966).
41. BELL, D. J.: Natural Monosaccharides and Oligosaccharides: Their Structure and Occurrence, Ch. 7, p. 287 in Comparative Biochemistry, ed. by M. FLORKIN and H. S. MASON, Vol. III, 1962, Academic Press, New York, London.
42. BERGMANN, W.: Comparative Biochemical Studies on the Lipids of Marine Invertebrates, with Special Refernce to the Sterols, J. Marine Res. **8** (2), 137, 1949. BERGMANN, W.: Sterols: Their Structure and Distribution, Ch. 2, p. 103 in Comparative Biochemistry, ed. by M. FLORKIN and H. S. MASON, Vol. III, 1962, Academic Press, New York.
43. BERGMANN, W. and W. T. PACE: Scymnol. J. Am. Chem. Soc. **65**, 477 (1943).
44. BERGSTROM, S. and A. NORMAN: Synthesis of Conjugated Bile Acids, Acta Chem. Scand **7**, 1126 (1953).
45. BERGSTROM, S. and J. SJÖVALL: Separation of Bile Acids with Reversed Phase Partition Chromatography. Acta Chem. Scand. **5**, 1267 (1951).
46. BERKOFF, C. E.: The Chemistry and Biochemistry of Insect Hormones. Quart. Rev. **33**, 372 (1969).
47. BERNFELD, P.: Polysaccharides, Ch. 8, p. 355 in Comparative Biochemistry Ed. M. FLORKIN and H. S. MASON, Vol. III. New York: Academic Press. 1962.

48. BORGSTROM, G., ed.: Fish and Food. New York: Academic Press. 1961 (and references therein).
48a. BLUMER, M.: Organic Pigments: Their Long-Term Fate. Science 149, 722 (1965).
49. BOYLAN, D. B. and P. J. SCHEUER: Pahutoxin, A Fish Poison. Science 155, 52 (1967).
50. Proceedings of the First International Seaweed Symposium, Ed. W. A. P. BLACK and E. T. DEWAR, Institute of Seaweed Research, Inveresk, Scotland, 1953.
51. Proceedings of the Second International Seaweed Symposium, Ed. T. BRAARUD and N. A. SORENSEN. New York: Pergamon Press. 1956.
52. BRIDGWATER, R. J., T. BRIGGS and G. A. D. HASELWOOD: Comparative Studies of "Bile Salts" Part 14 Isolation from Shark Bile and Partial Synthesis of Scymnol. Biochem. J. 82, 285 (1962).
53. BROTZU, G.: Richerche su di un nuovo antibiotico. Lav. Ist. Ig. Cagliari, 1948.
54. BURKHOLDER, P. R. and L. M. BURKHOLDER: Antimicrobial Activity of Horny Corals, Science 127, 1174 (1958).
55. BURKHOLDER, P. R., L. M. BURKHOLDER and L. R. ALMODOVAR: Antibiotic Activity of Some Marine Algae of Puerto Rico. Botanica Marina II, Fasc. I/2.
56. BURKHOLDER, P. R., R. M. PFISTER and F. H. LEITZ: Production of a Pyrrole Antibiotic by a Marine Bacterium. Applied Microbiology 14, 649 (1966).
57. BURKHOLDER, P. R.: Antimicrobial Substances from the Sea, in Transactions of the Drugs from the Sea Symposium, Ed. H. W. YOUNGKEN, Jr., Rhode Island, 1967, and BURKHOLDER, P. R., Ibid. p. 255, 1969.
58. BUCHWALD, H. D., L. DURHAM, H. G. FISCHER, R. HARADA, H. S. MOSHER, C. Y. KAO, F. A. FUHRMAN: Identity of Tarichatoxin and Tetradotoxin, Science 143, 474 (1964).
59. CALVIN, M.: Chemical Evolution, Oxford University Press, 1969.
60. CHANG, C. W. J., R. E. MOORE and P. J. SCHEUER: Spinochromes A (M) and C (F). Tetrahedron Lett. 1964, 3557.
61. CHANLEY, J. D., R. LEDEEN, J. WAX, R. F. NIGRELLI and H. SOBOTKA: Holothurin I, The Isolation, Properties and Sugar Components of Holothurin A. J. Am. Chem. Soc. 81, 5180 (1959).
62. CHANLEY, J. D., J. PERLSTEIN, R. F. NIGRELLI, H. SOBOTKA: Further Studies on the Structure of Holothurin. Ann. N. Y. Acad. Sci. 90, Art. 3, 902 (1960).
63. CHANLEY, J. D. and C. ROSSI: The Holothurinogenins II, Methoxylated Neo-Holothrurinogenins. Tetrahedron 25, 1897 (1969).
64. CHANLEY, J. D., T. MEZZETTI and H. SOBOTKA: The Holothurinogenins. Tetrahedron 22, 1857 (1966).
65. CHANLEY, J. D. and C. ROSSI: The Neo-Holothurinogenins III. Tetrahedron 25, 1911 (1969).
66. CHAPMAN, D. J. and D. L. FOX: Bile Pigment Metabolism in the Sea-Hare Aplysia. J. Exp. Mar. Biol. Ecol. 4, 71 (1969).
67. CHAPMAN, D. J., W. Y. COLE and H. W. SIEGELMAN: The Structure of Phycoerythrobilin. J. Am. Chem. Soc. 89, 5976 (1967).
68. CHAPMAN, D. J.: Three New Carotenoids Isolated from Algae. Phytochem. 5, 1331 (1966).
69. CHAPMAN, D. J., W. J. COLE and H. W. SIEGELMAN: Cleavage of Phycocyanobilin from C-Phycocyanin. Biochim. Biophys. Acta 153, 692 (1968).
70. CHEESMAN, D. F., W. L. LEE and P. F. ZAGAESKY: Carotenoproteins, in Invertebrates. Biol. Rev. 42, 132 (1967).

71. CHESTERS, C. G. C. and J. A. STOTT: The Production of Antibiotic Substances by Seaweeds, Proc. 2nd International Seaweed Symposium, 1956, pp. 49–54.
72. CHRISTOMANOS, A.: Nature of the Pigment of *Aplysia depilans*. Nature **175**, 310 (1955).
73. CIERESZKO, L. S., D. H. SIFFORD, A. J. WEINHEIMER: Chemistry of Coolenterates I. Ann. N. Y. Acad. Sci. **90**, Art. 3, Nov. (1960).
74. CIERESZKO, L. S., D. H. ATTAWAY and M. A. WOLF: Hydrocarbons in Coral Reef Organisms. 8th Annual Report on Research, The Petroleum Research Fund, 1963.
75. COLE, W. J., D. J. CHAPMAN, H. W. SIEGELMAN: The Structure of Phycocyanobilin. J. Am. Chem. Soc. **89**, 3643 (1967).
76. COOK, J. W.: Bile Acids of Elasmobranch Fish. Nature **147**, 388 (1941).
76a. COSULICH, D. B. and F. M. LOVELL: An X-Ray Determination of the Structure of an Antibacterial Compound from the Sponge, *Ianthella ardis*. Chem. Comm. 397 (1971).
77. CRAIGIE, J. S. and D. E. GRUENIG: Bromophenols from Red Algae. Science **157**, 1058 (1967).
78. CRESPI, M. L., L. J. BOUCHER, G. D. NORMAN, J. J. KATZ and R. C. DOUGHERTY: Structure of Phycocyanobilin. J. Am. Chem. Soc. **89**, 3642 (1967).
79. CROSS, A. D.: Scymnol Sulphate and Anhydroscymnol. J. Chem. Soc. 2817 1961.
80. CROSS, A. D.: Nuclear Magnetic Resonance and Mass-Spectral Study of Myxinol Tetra-acetate. Biochem. J. **100**, 238 (1966).
81. DEAKINS, D. E. and P. R. SAUNDERS: Purification of the Lethal Fraction of the Venom of the Stonefish *Synanceja Horrida* (L.). Toxicon **4**, 257 (1967).
82. Proceedings of the Fourth International Seaweed Symposium, Ed., Ad. Davy DE VIRVILLE and J. FELDMANN, New York: Pergamon Press and MacMillan Co. 1964.
83. DUURSMA, E. K.: The Dissolved Organic Constituents of Sea Water, Ch. 11, p. 433 in Chemical Oceanography, Ed. J. P. RILEY and G. SKIRROW, Vol. 1. New York: Academic Press. 1965.
84. ELYAKOV, G. B., T. A. KUZNETSOVA, A. K. DZIZENKO and YU. N. ELKIN: A Chemical Investigation of the Trepang. Tetrahedron Lett. **1969**, 1151.
85. ERSPAMER, V. and O. BENATI: Identification of Murexine as β[Imidazo(4)]-Acryl Choline. Science **117**, 161 (1953).
86. ERSPAMER, V., T. VITALI, M. ROSEGHINI and J. M. CEI: Occurrence of New Imidazolealkylamines (Spinaceamine and 6-Methyl-Spinaceamine), in Skin Extracts of *Leptodactylus pentadactylus labyrinthicus*. Experientia **19**, 346 (1963).
86a. FATTORUSSO, E., L. MINALE and G. SODANO: Aeroplysinin-I, a New Bromo-Compound from *Aplysina aerophoba*. Chem. Comm. **1970**, 751.
86b. — Aerothionin, a Tetrabromo Compound from *Aplysina aerophoba* and *Verongia thiona*. Chem. Comm. **1970**, 752.
87. FAUX, A., D. H. S. HORN and E. J. MIDDLETON: Molting Hormones of a Crab During Ecdysis. Chem. Comm. **1969**, 175.
88. FIESER, L. F. and M. FIESER: Natural Products Related to Phenanthrene, 3rd Ed., pp. 112–113. New York: Reinhold. 1949.
89. FIESER, L. F. and M. FIESER: Steroids, pp. 55–56. New York: Reinhold. 1967.
90. FIESER, L. F. and M. FIESER: Steroids, p. 787. New York: Reinhold. 1967.
91. FRAENKE, G.: A Hormone Causing Pupation in the Blowfly *Calliphora erythrocephala*. Royal Soc. London, Proc. Series B **118** (1935).

92. FRIESS, S. L., F. G. STANDAERT, E. R. WHITCOMB, R. F. NIGRELLI, J. D. CHANLEY, H. SOBOTKA: Some Pharmacological Properties of Holothurin A, A Glycosidic Mixture From the Sea Cucumber, Ann. N. Y. Acad. Sci., **90**, Art. 3, 893 (1960) (and references therein).
93. FURLENMEIER, A., A. FÜRST, A. LANGEMANN and G. WALDVOGEL: Die Synthese des Ecdysons. Tetrahedron Lett. **1966**, 1387.
94. FORREST, H. S.: Pteridins: Structure and Metabolism, Ch. 13, p. 615 in Comparative Biochemistry, Ed. M. FLORKIN and H. S. MASON, Vol. IV. New York: Academic Press. 1962.
95. Fox, D. L.: in "The Physiology of Fishes" (M. E. BROWN, Ed.) Vol. 2, p. 375. New York: Academic Press. 1957.
96. Fox, D. L. and G. F. CROZIER: Absence or Singular Specificity of Carotenoids in Some Lower Fishes. Science **150**, 771 (1965).
97. Fox, D. L. and R. A. LEWIN: A Preliminary Study of the Carotenoids of Some Flexibacteria. Can. J. Microbiol. **9**, 753 (1963).
98. GALBRAITH, M. N., D. H. S. HORN, E. J. MIDDLETON and R. J. HACKNEY: Structure of Deoxycrustecdysone. Chem. Comm. **1968**, 83.
99. GALBRAITH, M. N., D. H. S. HORN and E. J. MIDDLETON, R. J. HACKNEY: The Structure of Podecdysone B. Chem. Comm. **1969**, 402 (and references therein).
100. GALBRAITH, M. N., H. S. HORN, E. J. MIDDLETON and R. J. HACKNEY: Molting Hormones of Insects and Crustaceans: The Synthesis of 2-Deoxy-3-epicrustecdysone. Aust. J. Chem. **22**, 1059 (1969).
101. GALBRAITH, M. N. and D. H. S. HORN: Insect Molting Hormones, Crustecdysone from *Podocorpus Elatus*. Aust. J. Chem. **22**, 1045 (1969).
102. GIANOTTI, C., B. C. DAS and E. LEDERER: Sur la constitution chimique du Kitol, dimère de la Vitamine A. Bull. Soc. Chim. **9**, 3299 (1966).
103. GLASER, R. and E. LEDERER: Échinochrome et spinochrome; dérivés méthylés; distribution; pigments associés. Compt. Rend. **208**, 1939 (1939).
104. GOODWIN, T. W. and S. SRISUKH: A Study of the Pigments of the Seaurchins, *Echinus esculentus · L.* and *Paracentrotus lividus Lamarck*, Biochem. J. **47**, 69 (1950) (and references therein).
105. GOODWIN, T. W.: Carotenoids: Structure, Distribution, and Function, Ch. 14 in Comparative Biochemistry, Ed. M. FLORKIN and H. S. MASON, Vol. IV., p. 643. New York: Academic Press. 1962.
106. GUPTA, K. C. and P. J. SCHEUER: Echinoderm Sterols. Tetrahedron **24**, 5831 (1968).
107. GUPTA, L. C. and P. J. SCHEUER: Zoanthid Sterols. Steroids **13**, 343 (1969).
107a. MEUNIER, H., S. ZELENSKI and L. WORTHEN: Comparison of the Sterol Content of Certain Rhodophyta, Proceedings of the Conference on Food-Drugs from the Sea, p. 319. Rhode Island, 1969.
108. GUPTA SEN, A. K.: Presence of Coprostanol in Whale Oil. Z. Physiol. Chem. **348**, 1688 (1967).
109. HABERMEHL, G., and G. VOLKWEIN: Über Gifte der Mittelmeerischen Holothurien. Naturwiss. **55**, (2), 83 (1968).
109a. HABERMEHL, G., and G. VOLKWEIN: Die Aglyka der Toxine von *Holothuria polii*. Ann. Chem. **731**, 53 (1970).
110. HALE, R. L., J. LECLERQ, B. TURCH, C. DJERASSI, R. A. GROSS, Jr., A. J. WEINHEIMER, K. GUPTA and P. J. SCHEUER: Demonstration of a Biogenetically Unprecedented Side Chain in the Marine Sterol, Gorgosterol. J. Am. Chem. Soc. **92**, 2179 (1970).

111. HALE, R. L., N. C. LING and C. DJERASSI: The Structure and Absolute Configuration of the Marine Sterol Gorgosterol. J. Am. Chem. Soc. **92**, 5281 (1970).
111a. HALSALL, T. G. and I. R. HILLS: Isolation of Heneicosa-1,6,9,12,15,18-hexaene and -1,6,9,12,15-pentaene from the Alga *Fucus vesiculosus*, Chem. Comm. 448 (1971).
112. HALSTEAD, B. W., D. A. COURVILLE: Poisonous and Venomous Marine Animals of the World, Vol. I–III. Washington, D. C.: U. S. Government Printing Office. 1965.
113. HAMMARSTEN, O.: Über eine neue Gruppe gepaarter Gallensäuren. Z. Physiol. Chem. **24**, 322 (1898).
114. HAMPSHIRE, F. and D. H. S. HORN: Structure of Crustecdysone, A Crustacean Molting Hormone, Chem. Comm. **1966**, 37.
115. HANESSIAN, S., I. S. KALTENBRONN: Synthesis of a Bromine-Rich Marine Antibiotic. J. Am. Chem. Soc. **88**, 4509 (1966).
116. HARRIS, R. S.: Vitamins E., Ch. 7, p. 187 in Comprehensive Biochemistry, Ed. M. FLORKIN, E. H. STOTZ, Vol. IX. Elsevier, 1963.
117. L.-JENSEN, S., A. JENSEN: Recent Progress in Carotenoid Chemistry, Vol. 8, Part 2, p. 133 in Progress in Chemistry of Fats and Lipids. Pergamon Press, 1965.
118. HARTMANN, M., O. SCHARTAU, R. KUHN and U. WALLENFELS: Über die Sexualstoffe der Seeigel. Naturwiss. **27**, 433 (1939).
119. HASHIMOTO and T. OKAICHI: Some Chemical Properties of Nereistoxin. Ann. N. Y. Acad. Sci. **90**, 667 (1960).
120. HASLEWOOD, G. A. D.: Bile Salts of Fish. Biochem. Soc. Symp. **6**, 83 (1951).
121. HASLEWOOD, G. A. D.: Recent Developments in Our Knowledge of Bile Salts. Physiol. Rev. **35**, 178 (1955) (and references therein).
122. HASLEWOOD, G. A. D.: Bile Salts. London: Methuen. 1967.
123. HASLEWOOD, G. A. D.: Comparative Studies of Bile Salts, Part 13. Biochem. J. **78**, 352 (1961).
124. HASLEWOOD, G. A. D.: Bile Salts: Structure, Distribution, and Possible Biological Significance as a Species Character, Comparative Biochemistry, Vol. III, Ch. 2, p. 205. New York-London: Academic Press. 1962.
125. HASLEWOOD, G. A. D.: A Bile Alcohol Sulphate from the Hagfish, *Eptatretus Stoutii*. Biochem. J. **78**, 30 (1961).
126. HASLEWOOD, G. A. D.: Comparative Studies of Bile Salts, Myxinol Disulfate. Biochem. J. **100**, 233 (1966).
127. HASLEWOOD, G. A. D.: Bile Salts, Ch. 4, p. 205 in Comparative Biochemistry, Ed: M. FLORKIN, H. S. MASON. New York: Academic Press. 1962.
128. HEALEY, F. P.: The Carotenoids of Four Blue-Green Algae. J. Phycol. **41**, 126 (1968).
129. HERTZBERG, S. and S. L.-JENSEN: The Carotenoids of Blue-Green Algae-I., The Carotenoids of *Oscillatoria rubescene* and an *Arthrospria sp*. Ann. Chem. **696**, 1187 (1966).
130. HERTZBERG, S. and S. L.-JENSEN: The Carotenoids of Blue-Green Algae-II. The Carotenoids of *Aphanizomenon flosaquea*. Phytochem. **5**, 557 (1966).
131. HIKINO, H. and Y. HIKINO: Athropod Molting Hormones. Progress in the Chemistry of Organic Natural Products. **29**, 256 (1970).
132. HILDITCH, T. P.: The Chemical Constitution of Natural Fats, 3rd New York: Wiley. 1956.
133. HOCKS, P. and R. WIECHERT: 20-Hydroxy-Ecdyson, isoliert aus Insekten. Tetrahedron Lett. **1966**, 2989.
134. HORN, D. H. S., E. J. MIDDLETON and J. A. WUNDERLICH: Identity of the Molting Hormones of Insects and Crustaceans. Chem. Comm. **1966**, 339.

135. HORN, D. H. S., S. FABBRI, F. HAMPSHIRE and M. E. LOWE: Isolation of Crustecdysone (20 R-Hydroxyecdysone) from a Crayfish. Biochem. J. **109**, 399 (1968).
136. HOSHITA, T. and T. KAZUNO: Chemistry and Metabolism of Bile Alcohols and Higher Bile Acids, Adv. in Lipid Res. **6**, 207 (1968). New York & London: Academic Press. 1968.
137. HOSHITA, T., T. SASAKI, Y. TANAKA, S. BETSUKI and T. KAZUNO: Stero-bile Acids and Bile Sterols, LXXIV. Biochem. (Japan), **57**, 751 (1965).
138. HOSHITO, T., S. HIROFUJI, T. NAKAGAWA and T. KAZUNO: Stero-bile Acids and Bile Alcohols XCVI. Ibid. **62**, 62 (1967).
139. HOSHITA, T., K. AMIMOTO, T. NAKAGAWA and T. KAZUNO: Stero-bile Acids and Bile Alcohols, XCIV. Ibid. **61**, 750 (1967).
140. HOSSAIN, M. B. and D. VAN DER HELM: The Crystal Structure of β-Gorgonene Silver Nitrate. J. Am. Chem. Soc. **90**, 6607 (1968).
141. HOSSAIN, M. B., A. F. NICHOLAS and D. VAN DER HELM: The Molecular Structure of Eunicin Iodoacetate. Chem. Comm. **1968**, 385.
142. HÜPPI, G. and J. B. SIDDALL: Synthetic Studies on Insect Hormones, Part VI, The Synthesis of Ponasterone A and its Stereochemical Identity with Crustecdysone. Tetrahedron Lett. **1968**, 1113.
143. HÜPPI, G. and I. B. SIDDALL: Synthetic Studies on Insect Hormones V. J. Am. Chem. Soc. **89**, 6790 (1967).
144. IDLER, D. R., A. SAITO and P. WEISMAN: Sterols in Red Algae (Rhodophyceae). Steroids **11**, 465 (1968).
145. IGUCHI, M., A. NISHIYAMA, S. YAMAMURA and Y. HIRATA: Conversion of Elemene-type Sesquiterpenes into Cadinene-type Compounds and Formation of Ten-membered Germacrone type Intermediates. Tetrahedron Lett. **1969**, 4295.
146. IKEKAWA, N., N. MORISAKI, K. TSUDA and T. YASHID): Sterol Compositions in Some Green Algae and Brown Algae. Steroids **12**, 41 (1968) (and references therein).
147. IRIE, T., M. IZAWA and E. KUROSAWA: Laureatin, A Constituent from *Laurencia Nipponica* Yamada, Tetrahedron **1968**, 2091.
148. IRIE, T., M. SUZUKI and T. MASAMUNE: Laurencin, A Constituent from *Laurencia Species*, Ibid. **1965**, 1091.
149. IRIE, T., M. SUZUKI and T. MASAMUNE: Laurencin, A Constituent of *Laurencia Glandulifera* Kützing, Tetrahedron **24**, 4193 (1968).
150. IRIE, T., M. IZAWA and E. KUROSAWA: Isolaureatin, A Constituent from *Laurencia Nipponica* Yamada, Tetrahedron Lett. **1968**, 2735.
150a. IRIE, T., M. IZAWA and D. KUROSAWA: Laureatin and Isolaureatin, Constituents of *Laurencia Nipponica* Yamada. Tetrahedron **26**, 851 (1970). See also FERGUSON, G., Structure and Absolute Configuration of Laurencin. J. C. S. (B), 559 (1969).
151. IRIE, T., T. SUZUKI, Y. YASUNARI, E. KUROSAWA and T. MASAMUNE: Laurene, A Sesquiterpene Hydrocarbon from *Laurencia* Species. Ibid. **25**, 459 (1969).
152. IRIE, T., A. FUKUZAWA, M. IZAWA and E. KUROSAWA: Laurenisol, A New Sesquiterpenoid Containing Bromine from *Laurencia Nipponica* Yamada, Tetrahedron Lett. **1969**, 1343.
153. IRIE, T., M. SUZUKI and Y. HAYAKAWA: Isolation of Aplysin, Debromoaplysin, and Aplysinol from *Laurencia Okamurai* Yamada. Bull. Chem. Soc. Japan **42**, 843 (1969).
153a. IRIE, T., M. SUZUKI, E. KUROSAWA and T. MASAMUNE: Laurinterol and Debromolaurinterol, Constituents of *Laurencia Intermedia*. Tetrahedron Lett. **1966**, 1837.

154. IRIE, T., K. YAMAMOTO and T. MASAMUNE: Sesquiterpenes from *Dictyopteris Divaricata* I. Bull. Chem. Soc. Japan **37**, 1053 (1964).
155. ISLER, O. and P. ZELLER: Total Synthesis of Carotenoids, Vitamins and Hormones **15**, 33. New York: Academic Press. 1957.
156. JACQUOT, R., Organic Constituents of Fish and Other Aquatic Animal Foods, Ch. 6, 1453, in G. BORGSTROM, Ed., Fish as Food. New York: Academic Press. 1961.
157. KARLSON, P.: Chemische Untersuchungen über die Metamorphosehormone der Insecten, Annales des Sciences Naturelles, Paris, Zoologie et biologie animale **18**, 125 (1956).
158. KARLSON, P.: Chemie und Wirkungsweise der Insektenhormone, Max Planck Inst. Biochem. Proc. (Munich) 4th Intern. Congress of Biochem. (Vienna) **12**, 1959.
159. KARLSON, P., H. HOFFMEISTER, H. HUMMEL, P. HOCKS, U. KERB, G. SPITELLER and R. WIECHERT: Reaktionen des Ecdysonmoleküls. Chem. Ber. **98**, 2394 (1965).
160. KATAYAMA, T.: Physiology and Biochemistry of Algae, p. 467, R. A. Lewin, Ed. New York: Academic Press. 1962.
161. KATAYAMA, T.: Chemical Studies on Volatile Constituents of Seaweeds, Antibacterial Action of *Enteromorpha sp*. Bull. of the Japanese Society of Scientific Fisheries **22**, 248 (1956).
162. KATAYAMA, T.: Chemical Studies on Volatile Constituents of Seaweeds, X, Antibacterial Action of *Enteromorpha sp*. Ibid. **22**, 248 (1956).
163. KAZUNO, T. and T. HOSITA: Stero-Bile Acids and Bile Alcohols, LVII, Chromatography of Bile Alcohols. Steroids **3**, 55 (1964).
164. KENNARD, O., D. G. WATSON, L. R. DI SANSEVERINO, B. TURSCH, R. BOSMANS and C. DJERASSI: Chemical Studies of Marine Invertebrates IV. Tetrahedron Lett. **1968**, 2879.
165. KERB, U., R. WIECHERT, A. FURLENMEIER and A. FÜRST: Über eine Synthese des Crustecdysones. Ibid. **1968**, 4277.
166. KLEINIG, H. and K. EGGER: Zur Struktur von Siphonaxanthin und Siphonein, den Hauptcarotinoiden Siphonaler Grünalgen. Phytochemistry **6**, 1681 (1967).
167. KLEINIG, H., H. NITSCHE and K. EGGER: The Structure of Siphonaxanthin. Tetrahedron Lett. **1969**, 4139.
168. KLEINIG, H.: Carotenoids of Siphonous Green Algae: A Chemotaxonomical Study. J. Phycol. **5**, 281–284 (1969).
169. KOCHETKOV, N. K. and A. F. BOCHKOV: The Synthesis of a $(1 \to 3)$ Glucan Related to Laminaran. Carbohydrate Res. **9**, 61 (1969).
170. KONISHI, K.: Studies on Organic Insecticides, Synthesis of Nereistoxin and Related Compounds, III. Agr. Biol. Chem. **32**, 1199 (1968).
171. KOSAKI, T. I. and H. H. ANDERSON: Marine Toxins from the Pacific, IV, Pharmacology of Ciguatoxins. Toxicon **6** 55–58 (1968).
172. KUHN, R. and K. WALLENFELS: Echinochrome als prosthetische Gruppen hochmolekularer Symplexe in den Eiern von *Arbacia Pustulosa*. Chem. Ber. **73**, 458 (1940).
173. KUHN, R. and K. WALLENFELS: Über die chemische Natur des Stoffes, den die Eier des Seeigels *(Arbacia Pustulosa)* absondern, um die Spermatozoen anzulocken. Chem. Ber. **72**, 1407 (1939).
174. KUHN, R. and K. WALLENFELS: Über den Stachelfarbstoff von Arbacia. Chem. Ber. **74**, 1594 (1941).
175. LEDERER, E.: Sur les Pigments Naphtoquinonique des Epines et du Test des Oursins *Paracentrotus Lividus* et *Arbacia Pustulosa*. Biochim. Biophys. Acta **9**, 92 (1952).

176. LOVELL, F. M., The Structure of a Bromine-Rich Marine Antibiotic. J. Am. Chem. Soc. **88**, 4510 (1966).
177. LOW, T. F., R. J. PARK, M. D. SUTHERLAND and I. VESSEY: Pigments of Marine Animals, III, The Synthesis of Some Substituted Polyhydroxyanthraquinones. Aust. J. Chem. **18**, 182 (1965) (and references therein).
178. MACKINNEY, G.: Carotenoids and Vitamin A, Ch. 13, 221 in D. M. GREENBERG (Ed.) Metabolic Pathways 3rd Ed. **2**. New York and London: Academic Press. 1968.
179. MACMUNN C. A.: Quart. J. Micr. Sci. **25**, 469 (1885).
180. MALLORY, F. B., J. T. GORDON, R. L. CONNER: The Isolation of a Pentacyclic Triterpenoid Alcohol from a Protozoan. J. Am. Chem. Soc. **85**, 1362 (1963).
181. MANDELLI, E. F.: Carotenoid Pigments of the *Dinoflagellate Glenodinium Foliaceum* Stein. J. Phycol. **4**, 347–348 (1968).
182. MARDEROSIAN, A. D.: Marine Pharmaceuticals. J. Pharmaceutical Sciences **58**, 1 (1969).
183. MARETZKI, A. and J. DEL CASTILLO: A Toxin Secreted by the Soapfish *Rypticus Saponaceus*. Toxicon **4**, 245 (1967).
184. MARKHAM, K. R. and L. J. PORTER: Flavonoids in the Green Algae (Chlorophyta). Phytochem. **8**, 1777 (1969).
185. MARTIN, D. F. and A. B. CHATTERJEE: Isolation and Characterization of a Toxin from the Florida Red Tide Organism. Nature **221**, 59 (1969).
186. MATHIAS, A. P., D. M. ROSS and M. SCHACHTER: Distribution of Histamine, 5-Hydroxytryptamine, Tetramethylammonium, and Other Substances in Coelenterates Possessing Nematocysts, Proceedings of the Physiological Society. J. Physiol. **142**, 561 (1958).
187. MATHIAS, A. P., D. M. ROSS, M. SCHACHTER: Identification and Distribution of 5-Hydroxytryptamine in a Sea Anemone. Nature **180**, 658 (1957).
188. MATSUDA, H. and Y. TOMII: The Structure of Aplysin-20. Chem. Comm. **1967**, 898.
189. MATSUNO, T. and T. YAMANOUCHI: A New Triterpenoid Sapogenin of Animal Origin (Sea Cucumber). Nature **191**, 75 (1961).
190. MATSUNO, T. and I. IBA: Studies on the Saponins of the Sea Cucumber, Yakugaku Zasshi. **86**, 637–638 (1966).
191. MAUTNER, H. G. , G. M. GARDNER and R. PRATT: Antibiotic Activity of Seaweed Extracts II. *Rhodomela Larix*. J. A. Pharm. Assoc. **42**, 294 (1953).
192. McDowell, R. H.: Properties of Alginates, Alginate Industries Ltd., Publication London, 1968.
193. McFARREN, E. F., E. J. SCHANTZ, J. E. CAMPBELL and H. K. LEWIS: Chemical Determination of Paralytic Shellfish Poison in Clams. J. of the A. O. A. C. **41**, 168 (1958).
194. — Ibid. **42**, 399 (1959).
195. McFARREN, E. F., H. TANABE, F. J. SILVA, W. B. WILSON, J. E. CAMPBELL and K. H. LEWIS: The Occurence of a Ciguatera-like Poison in Oysters, Clams, and *Gymnodinium breve* Cultures. Toxicon **3**, 111 (1965).
196. MILLER, J. D. A.: Physiology and Biochemistry of Algae, p. 357, Ed. R. A. LEWIN. New York: Academic Press. 1962 (and references therein).
197. MIYASAKI, M.: Studies on the Components of Digenea Simplex Ag. IV. Studies on the Structure of Kainic Acid. Yakugaki Zasshi **75**, 692 (1955).
198. MOORE, R. E., H. SINGH, C. W. J. CHANG and P. J. SCHEUER: Polyhydroxynaphthoquinones. Preparation and Hydrolysis of Methoxyl Derivatives. Tetrahedron **23**, 3271 (1967) (and references therein).

199. MOORE, R. E., H. SINGH and P. J. SCHEUER: Isolation of Eleven New Spirochromes from Echinoids of the Genus *Echinotrix*. J. Org. Chem. **31**, 3645 (1966).
200. MOORE, R. E. and P. J. SCHEUER: Nuclear Magnetic Resonance Spectra of Substituted Naphthoquinones. Influence of Substituents on Tautomerism, Anisotropy and Stereochemistry in the Naphthazarin System. J. Org .Chem. **31**, 3272 (1966).
201. MOORE, R. E., H. SINGH, C. W. J. CHANG and P. J. SCHEUER: Sodium Borohydride Reduction of Spinochrome A. Romoval of Phenolic Hydroxyls in the Naphthazarin System. J. Org. Chem. **31**, 3638 (1966).
202. MOORE, R. E., J. A. PETTUS, Jr., M. S. DOTY: Dictyopterine A. Tetrahedron Lett. **1968**, 4787. K. C. DAS and B. WEINSTEIN: The Synthesis of ($\pm$) Dictyopterene A. Tetrahedron Lett. **1969**, 3459.
203. MOORE, R. E., H. SINGH, P. J. SCHEUER: A Pyranonaphthazarin Pigment from the Sea Urchin *(Echinothrix Diadema)*. Tetrahedron Lett. **1968**, 4581.
204. MORIMOTO, H. and R. NAKAMORI: Studies on the Active Components of Digenea Simplex Ag. XXXVI. Isomerization and Isomers of Kainic Acid. Yakugaki Zasshi **76**, 294 (1956) (and references therein).
205. MORIMOTO, H. and R. NAKAMORI: Stereochemical Structures of Kainic Acid and its Isomers. Proc. Jap. Acad. Sci. **32**, 41 (1956).
205a. MORIN, R. B. and B. G. JACKSON: Chemistry of Cephalosporin Antibiotics Progress in the Chemistry of Organic natural Products **29**, 343 (1970).
206. MORIN, R. B., B. G. JACKSON, E. H. FLYNN and R. W. ROESKE: Chemistry of Cephalosporin Antibiotics, I. 7-aminocephalosporanic Acid from Cephalosporin C. J. Am. Chem. Soc. **84**, 3400 (1962).
207. MORTON, R. A.: Quinones, Ch. 9, in Comprehensive Biochemistry, Ed. M. FLORKIN and E. H. STOTZ, p. 110–210. Elsevier, 1963.
208. MORTON, R. A.: Biochemistry of Quinones. New York: Academic Press. 1965.
209. MOSHER, H. S., F. A. FUHRMAN, H. D. BUCHWALD, H. G. FISCHER: Tarichatoxin- tetrodotoxin, A Potent Neurotoxin. Science **144**, 1100 (1964).
210. MURAKAWI, S., T. TAKEMOTO and Z. SHIMIZU: Studies on the Effective Principles of *Digenea Simplex Ag. I*. Yakugaki Zasshi **73**, 1026 (1953).
211. MUSAJO, L. and M. MINCHILLI: Lo Spinocromo P. Gazz. Chim. Ital. **70**, 287 (1940).
212. NADAL, N. G. M., L. V. RODRIGUEZ and C. CASILLAS: Isolation and Characterization of Sarganin Complex, Antimicrobial Agents and Chemotherapy, p. 131, 1964.
213. NADAL, N. G. M., L. V. RODRIGUEZ and C. CASILLAS: Sarganin and Chronalgin, New Antibiotic Substances from Marine Algae from Puerto Rico, Antimicrobial Agents and Chemotherapy, p. 68, 1963.
214. NICHOLS, D.: Echinoderms. London: Hutchinson University Library. 1966.
215. NIGRELLI, R. F., J. D. CHANLEY, S. K. KOHN, H. SOBOTKA: The Chemical Nature of Holothurin, A Toxic Principle from the Sea-Cucumber. Zoologica **40**, 47 (1955).
216. NIGRELLI, R. F., M. F. STEMPIEN, Jr., G. D. RUGGIERI, V. R. LIGUORI and J. T. CECIL: Substances of Potential Biomedical Importance from Marine Organisms. Federation Proc. **26**, 1197 (1967).
217. NIGRELLI, R. F., S. JAKOWSKA and Y. CALVENTI: Ectyonin, An Antimicrobial Agent from the Sponge *Microciona prolifera Verrill*. Zoologica (N. Y. Zool. Soc.) **44**, 173 (1959).
218. NORMAN, A.: Separation of Conjugated Acids by Partition Chromatography Bile Acids and Steroids. 6. Acta Chem. Scand. **7**, 1413 (1953).

219. NUGTEREN, D. H., D. A. VAN DORP, S. BERGSTRÖM, M. HAMBERG and B. SAMUELSSON: Absolute Configuration of the *Prostaglandins*. Nature **212**, 38 (1966).
220. O'CARRA, P. and C. O'HEOCHA: Structure of Phycoerythrobilin and Phycocyanobilin. Nature **215**, 1477 (1967).
221. O'HEOCHA, C.: Spectral Properties of the Phycobilins, I. Phycocyanobilin. Biochem. **2**, 375 (1963).
222. O'HEOCHA, C.: Comparative Biochemical Studies of the Phycobilins, Arch. Biochem. Biophys. **73**, 207 (1958).
223. OHLOFF, G. and W. PICKENHAGEN: Synthese von ($\pm$)-Dictyopteren A. Helv. Chim. Acta **52**, 880 (1969).
224. OKAICHI, T. and Y. HASHIMOTO: The Structure of Nereistoxin, Agr. and Biol. Chem. (Tokyo) **26**, 224 (1962).
225. OLESEN, P. E., A. MARETZKI, L. A. ALMODOVAR: An Investigation of Antimicrobial Substances from Marine Algae, Botanica Marina **6**, 224 (1964).
226. OLSON, J. A.: The Biosynthesis and Metabolism of Carotenoids and Retinol (Vitamin A), J. Lipid Res. **5**, 281 (1964).
227. PASINI, C., A. VERCELLONE and V. ERSPAMER: Synthesis of Murexine, Ann. Chem. **578**, 6 (1952).
228. PEAT, S. and J. R. TURVEY: Polysaccharides of Marine Algae, p. 1 in Progress in the Chemistry of Organic Natural Products, L. ZECHMEISTER, Vol. 23. Wien-New York: Springer. 1965.
229. PÉRGUY, M.: Phenolic Compounds in Rhodomelacae. Compt. Rend. IV Congress Intern. des Algues Marines, **1964**, 366.
230. PERCIVAL, E. and R. H. MCDOWELL: Chemistry and Enzymology of Marine Algae Polysaccharides. London: Academic Press. 1967.
231. PERKS, A. M.: Pharmacological and Chromatographic Studies of the Neurohypophysical Activities of the Pituitary of Further Elasmobranch Species. Gen. and Comp. Endocrinology **6**, 428 (1966).
231 a. PETTUS, J. A. JUN and R. E. MOORE: Isolation and Structure Determination of an Undeca-1,3,5,8-tetraene and Dictyopterene B from Algae of the genus *Dictyopteris*. Chem. Comm. 1093 (1970).
232. RAMASARMA, T.: Lipid Quinones. Adv. Lipid Res. **6**, 107–180, New York: Academic Press. 1968.
233. RAPOPORT, H., M. S. BROWN, R. OESTERLIN and W. SCHUETT: Saxitoxin, Abstracts 147th Meeting Am. Chem. Soc., 3N, 1964.
234. REES, D. A. and G. P. MUELLER: Current Structural Views of Red Seaweed Polysaccharide Transactions of the Drugs from the Sea Symposium University of Rhode Island, 27–29, August 1967, p. 241 (and references therein).
235. REES, D. A. and E. CONWAY: The Structure and Biosynthesis of Porphyran: A Comparison of Some Samples. Biochem. J. **84**, 411 (1962).
236. REES, D. A.: Carbohydrate Sulfates. Ann. Reviews Chem. Soc. **62**, 469 (1965) (and references therein).
237. REES, D. A. and R. J. SKERETT: Conformational Analysis of Cellobiose, Cellulose and Xylan. Carbohydrate Res. **7**, 334 (1968).
238. REES, D. A.: Conformational Analysis of Polysaccharides. Part II. Alternating Copolymers of the Agar-Carrageenan-Chondroitin type by Model Building in the Computer with Calculation of Helical Parameters. J. Chem. Soc. B, **1969**, 217.
239. REES, D. A.: Structure, Conformation and Mechanism in the Formation of Polysaccharide Gels and Networks. Adv. Carbohydrate Chem. **24**, 267 (1969).

240. REID, K. B. M., P. T. GRANT and A. YOUNGSON: The Sequence of Amino Acids in Insulin Isolated from Islet Tussue of the Cod *(Gadus callarias)*. Biochem. J. 110, 289 (1968) (and references therein).
241. REVISED Tentative Rules for Nomenclature of Steroids. Biochem. Biophys. Acta 164, 453 (1968).
242. RIMINGTON, C. and G. Y. KENNEDY: Porphyrins: Structure, Distribution, and Metabolism, ch. 12, p. 557 in *Comparative Biochemistry*, M. FLORKIN and M. S. MASON, Vol. IV. 1962. New York: Academic Press. 1962.
243. RIO, G. J., M. F. STEMPIEN, Jr., R. F. NIGRELLI and G. D. RUGGIERI: Echinoderm Toxins I, Some Biochemical and Physiological Properties of Toxins from Several Species of Asteroidea. Toxicon 3, 147 (1965).
243a. ROCHE, J., M. FONTAINE and J. LELOUP: Halides, Ch. 6, p. 493 in *Comparative Biochemistry*, Ed., M. FLORKIN and M. S. MASON, Vol. V, Academic Press. New. York: 1963.
244. ROLLER, P., C. DJERASSI, R. CLOETEUS and B. TURSCH: Terpenoids, LXIV, Chemical Studies of Marine Invertebrates V, The Isolation of Three New Holothurinogenins and Their Chemical Correlation with Lanosterol. J. Am. Chem. Soc. 91, 4918 (1969).
245. RÜDIGER, W.: Aplysioviolin, ein neuartiger Gallenfarbstoff. Naturwiss. 53, 613 (1966).
246. RUGGIERI, G. D.: Echinoderm Toxins II, Animalizing Action in Sea Urchin Development. Toxicon 3, 157 (1965).
247. RUSSELL, F. E.: Marine Toxins and Venomous and Poisonous Marine Animals. Adv. Mar. Biol. 3, 255 (1965).
248. RUSSELL, F. E.: Comparative Pharmacology of Some Animal Toxins. Fed. Proc. 26, 1206 (1967).
249. SAITO, KANAME and MUNEO SAMESHIMA: Studies on the Antibiotic Action of Algae Extracts, III. J. Agr. Chem. Soc. Japan 29, 427 (1955).
250. SALAQUE, A., A. BARBIER and E. LEDERER: Sur La Biosynthese de L'Echinochrome A par L'Oursin *Arbacia pustulosa*. Bull. Soc. Chem. Biol. 49, 277 (1967).
251. SANDERS, G. M., J. POT and E. HAVINGA: Some Recent Results in the Chemistry and Stereochemistry of Vitamin D and its Isomers, p. 131, Vol. 27, Progress in the Chemistry of Organic Natural Products, Wien-New York: Springer. 1969.
252. SAUNDERS, P. R.: Pharmacological and Chemical Studies of the Venom of the Stonefish *(Genus Synanceja)* and other Scorpion fishes. Ann. N. Y. Acad. Sci. 90, Art. 3, 798 (1960).
253. SCHEUER, P. J.: The Chemistry of Toxins Isolated from Some Marine Organisms, Vol. 22, p. 265. Progress in the Chemistry of Organic Natural Products. Wien-New York: Springer. 1964.
254. SCHEUER, P. L., W. TAKAHASHI, J. TSUTSUMI, B. T. YOSHIDA: Ciguatoxin: Isolation and Chemical Nature. Science 155, 1267 (1967).
255. SCHEUER, P. J., The Chemistry of Some Toxins Isolated from Marine Organisms, Vol. 27, p. 322. Progress in the Chemistry of Organic Natural Products. Wien-New York: Springer. 1969.
256. SCHMEER, M. ROSARII: Growth-Inhibiting Agents from *Mercenaria* Extracts: Chemical and Biological Properties. Science 144, 413 (1964).
257. SCHMEER, M. ROSARII: Mercenene, Growth Inhibiting Agent. Ann. N.Y. Acad. Sci. 136, 213 (1966).
258. SCHMEER, M. ROSARII, DEREK HORTON and AKIO TANIMURA: Mercenene, A Tumor Inhibitor from *Mercenaria*, Purification and Characterization Studies. Life Sciences 5, 1169 (1966).

259. SCHMITZ, F. J., E. K. LORANCE and L. S. CIERESZKO: Chemistry of Coelenterates, XII. Hydroxyancepsenolide, A Dilactone from Octocoral, *Pterogorgia anceps*. J. Org. Chem. **34**, 1989 (1969).
260. SCHMITZ, F. J., K. W. KRAUS, L. S. CIERESZKO, D. H. SIFFORD and A. J. WEINHEIMER: Ancepsenolide: A Novel Bisbutenolide of Marine Origin, Tetrahedron Lett. **1966**, 97.
260a. SCHMITZ, F. J. and T. PATTABHIRAMAN: New Marine Sterol Possessing a Side Chain Cyclopropyl Group: 23-Demethyl gorgosterol. J. Am. Chem. Soc. **92**, 6073 (1970).
261. SCHUETT, W. and H. RAPOPORT: Saxitoxin, The Paralytic Shellfish Poison. J. Am. Chem. Soc. **84**, 2266 (1962).
262. SCHWIMMER, D. and M. SCHWIMMER: Algae in Medicine, in Algae and Man. Ed. D. F. JACKSON, p. 368. Plenum Press. 1964.
263. SESTAK, Z. and M. BASLEROVA: in Studies on Microalgae and Photosynthetic Bacteria, p. 423, Pub. by the Japanese Society of Plant Physiology, U. of Tokyo Press, Tokyo, Japan, 1963.
264. SHARMA, G. M. and P. R. BURKHOLDER: Studies on Antimicrobial Substances of Sponges I. J. Antibiotics, Series A **20**, 200 (1967).
265. SHARMA, G. M. and P. R. BURKHOLDER: Studies on the Antimicrobial Substances of Sponges II, Structure and Synthesis of a Bromine Containing Antibacterial Compound from a Marine Sponge. Tetrahedron Lett. **1967**, 4147.
265a. SHARMA, G. M., B. VIG and P. R. BURKHOLDER: Studies on the Antimicrobial Substances of Sponges. IV. Structure of a Bromine-Containing Compound from a Marine Sponge. J. Org. Chem. **35**, 2823 (1970).
266. SHARMA, G. M., B. VIG and P. R. BURKHOLDER: Antimicrobial Substances of Marine Sponges IV. Proceedings of the Conference on Food-Drugs from the Sea, p. 307. New York. 1969.
267. SHORLAND, F. B.: The Comparative Aspects of Fatty Acid Occurrence and Distribution, Ch. 1, p. 1 in Comparative Biochemistry, Ed., M. FLORKIN and H. S. MASON, Vol. 3. New York: Academic Press. 1962.
268. SIDDALL, J. B., D. H. S. HORN and E. J. MIDDLETON: Synthetic Studies on Fused Hormones, The Synthesis of a Possible Metabolite of Crustedysone, Chem. Comm. **1967**, 899.
269. SIDDALL, J. B., A. D. CROSS, J. H. FRIED: Synthetic Studies on Insect Hormones II. J. Am. Chem. Soc. **88**, 862 (1966) (and ref. therein).
270. SIEBURTH, J. McN.: Antibacterial Substances Produced by Marine Algae, Developments in Industrial Microbiol. **5**, 124 (1964) (and ref. therein).
271. SIMIDU, W.: Non-Protein Nitrogenous Compounds, Ch. 11, p. 353 in Fish as Food, G. BORGSTRÖM Ed. New York: Academic Press. 1961.
272. SINGH, H., R. E. MOORE and P. J. SCHEUER: The Distribution of Quinone Pigments in Echinoderms. Experientia **23**, 624 (1967) and references therein.
273. SINGH, I., R. E. MOORE, C. W. J. CHANG, R. T. OGATA and P. J. SCHEUER: Spinochrome Synthesis. Tetrahedron **24**, 2969 (1968).
274. SJÖVALL, JAN: Separation of Conjugated and Free Bile Acids by Paper Chromatography. Acta Chem. Scand. **8**, 339 (1954).
275. SJÖVALL, JAN: Separation of Bile Acids by Paper Chromatography, Acta Chem. Scand. **6**, 1552 (1952).
276. SJÖVALL, JAN: On the Separation of Bile Acids by Partition Chromatography. Acta Physiol. Scand. **29**, 232 (1953).
277. SMITH, J. and R. H. THOMSON: Spinochrome E. Tetrahedron Lett. **1960**, 10.
278. SMITH, J. and R. H. THOMSON: Naturally Occurring Quinones. J. Chem. Soc. **1961**, 1008, and references therein.

279. STEINER, A. B. and W. H. MCNEELY: Algin in Review. Adv. in Chem. **7**, 68 (1954).
280. The Structure of Red Seaweed Polysaccharides, Technical Bulletin 3, 1967, Marine Colloids, Inc., Springfield, N. J.
281. SUTHERLAND, M. D. and J. W. WELLS: Anthraquinone Pigments from the *Crinoid Comatula* Pectinata, Chem. and Industry **1959**, 231.
282. TAKEMOTO, T. and K. DAIGO: Über die Inhaltsstoffe von *Chondria armata* und ihre pharmakologische Wirkung. Arch. Pharmazie **293**, 627 (1960).
283. TAKEMOTO, T., T. NAKAJIMA and K. DAIGO: Die fliegentötenden Bestandteile von *Chondria armata*. Jap. J. of Pharm. & Chem. **34**, 21 (1963).
284. TAKEMOTO, T., K. DAIGO, Y. KONDO and K. KONDO: Studies on the Constituents of *Chondria Armata* VIII, On the Structure of Domoic Acid. Yakugaki Zasshi **86**, 874 (1966) (and references therein).
285. TAKEMOTO, T., K. DAIGO and N. TAKAGI: Studies on the Hypotensive Constituents of Marine Algae I. A New Basic Amino Acid, "Laminine" and other Basic Constituents Isolated from *Laminaria Augustata*. Ibid. **84**, 1176 (1964).
286. TAKEMOTO, T., K. DAIGO and N. TAKAGI: Studies on the Hypotensive Constituents of Marine Algae II. Yakugaku Zasshi **84**, 1180 (1964).
287. TAKEMOTO, T., K. DAIGO and N. TAKAGI: Ibid. **85**, 37 (1965).
288. TAYLOR: Crystalline Echinochrome and Spinochrome: Their Failure to Stimulate the Respiration of Eggs and of Sperm of Strongylocentrotus. Proc. Nat. Acad. Sci. **25**, 523 (1939).
289. THOMAS, D. M. and T. W. GOODWIN: Nature and Distrubution of Carotenoids in the Xanthophyta (Heterovontae). J. Phycol. **1**, 118 (1965).
290. THOMSON, R. H.: Quinones: Structure and Distribution, Ch. 12, pp. 686–696, in Comparative Biochemistry, Ed. M. FLORKIN and H. S. MASON, Vol. III., New York: Academic Press. 1962.
291. THOMSON: Quinones: Structure and Distribution, Ch. 12, pp. 631–725, in Comparative Biochemistry, Ed. M. FLORKIN and H. S. MASON, Vol. III. New York: Academic Press. 1962.
292. TRESSLER, D. K. and J. McW. LEMON: Marine Products of Commerce, 2nd Ed. New York: Reinhold. 1951.
293. TSUDA, K., S. IKUMA, M. KAWAMURA, R. TACHIKAWA, K. SAKAI, C. TAMURA and O. AWAKASU: Tetrodotoxin VII, On the Structures of Tetrodotoxin and its Derivatives. Chem. Pharm. Bull. (Tokyo) **12**, 1357 (1964) (and references therein).
294. TSUKUDA, N. and K. AMANO: Studies on the Discoloration of Red Fishes. I. Content of Carotenoid Pigments in Eighteen Species of Red Fishes. Bull. Jap. Soc. Sci. Fisheries **32**, 334 (1966).
295. TÖKES, L.: N. M. R. and Mass Spectral Study of 16-Deoxymyxinol, Biochem. J. **112**, 765 (1969).
296. TURSCH, B., R. CLOETENS and C. DJERASSI: Chemical Studies of Marine Invertebrates, VI. Terpenoids LXV. Praslinogenin, A New Holothurinogenin from the Indian Ocean Sea Cucumber *Bohadschia koellikeri*. Tetrahedron Lett. **1970**, 467.
297. TURSCH, B., I. S. DESOUZA GUIMARAES, B. GILBERT, R. T. APLIN, A. M. DUFFIELD and C. DJERASSI: Chemical Studies of Marine Invertebrates II, Griseogenin. Tetrahedron **23**, 761 (1967).
298. UENO, Y., K. TANAKA, J. UEYANAGI, H. NAWA, Y. SANNO, M. HONJO, R., NAKAMORI, T. SUGAWA, M. UCHIBAYASHI, K. OSUGI and S. TATSUOKA: Studies on the Active Components of *Digenea Simplex Ag.*, and Related Compounds V. Synthesis of α-Kainic Acid. Proc. Jap. Acad. Sci. **33**, 53 (1957).
299. — Yakugaki Zasshi **77**, 618 (1957).

300. WALLENFELS, K. and A. GAUHE: Synthese von Echinochrom A. Chem. Ber. **76**, 325 (1943).
301. WALLENFELS, K.: Der Farbstoff der roten Blutzellen des Seeigels *Arbacia pustulosa* Ber. **76**, 323 (1943).
302. WEEDON, B. C. L.: Spectroscopic Methods for Elucidating the Structures of Carotenoids, Vol. 27, p. 81. Progress in the Chemistry of Organic Natural Products. Wien-New York: Springer. 1969.
303. WEINHEIMER, A. J., P. H. WASHECHECK, D. VAN DER HELM and M. BILAYET HOSSAIN: The Sesquiterpene Hydrocarbons of the Gorgonian *Pseudopterogorgia Americana*. Chem. Comm. **1968**, 1070.
304. WEINHEIMER, A. J., F. J. SCHMITZ and L. S. CIERESZKO: Transactions of the Drugs from the Sea Symposium, University of Rhode Island, 27–29 August, 1967, p. 135. WEINHEIMER, A. J. and P. H. WASHENCHECK, The Structure of the Marine Benzofuran, Furoventalene, A Non-Farnesyl Sesquiterpene. Tetrahedron Lett. **1969**, 3315.
305. WEINHEIMER, A. J., R. E. MIDDLEBROOK, J. O. BLEDSOE, Jr., W. E. MARSICO and T. K. B. KARNS: Eunicin, An Oxa-bridged Cembranolide of Marine Origin. Chem. Comm. **1968**, 384.
306. WEINHEIMER, A. J., W. W. YOUNGBLOOD, P. H. WASHECHECK, T. K. B. KARNS and L. S. CIERESZKO: Isolation of the Elusive (−)-germacrene-A from the Gorgonian, *Eunicea Mammosa*, Chemistry of Coelenterates XVIII. Tetrahedron Lett. **1970**, 497.
307. WEINHEIMER, A. J. and R. L. SPRAGGINS: The Occurrence of Two New Prostaglandins, Derivatives (15-epi-PGA$_2$ and its Acetate, Methyl Ester) in the Gorgonian *Plexaura Homomalia*, Chemistry of Coelenterates XV. Tetrahedron Lett. **1969**, 5185. See also: WEINHEIMER, A. J. and R. L. SPRAGGINS: Two New Prostaglandins Isolated from The Gorgonian Plexaura Homomalia. Proceeding of the Conference on Food-Drugs from the Sea, p. 311, Oklahoma, 1969.
308. WELSH, J. H., Composition and Mode of Action of Some Invertebrate Venoms. Ann. Rev. of Pharmacol. **4**, 293 (1964).
309. WELSH, J. H.: 5-Hydroxytryptamine in Coelenterates. Nature **186**, 811 (1960).
310. WELSH, J. H.: Compounds of Pharmacological Insterest in Coelenterates in the Biology of Hydra and of Some Other Coelenterates, H. LENHOFF and W. F. LOOMIS. U. of Miami Press, 1961.
311. WILSON, S. and S. FALKMER: Starfish Insulin. Can. J. Biochem. **43**, 1615 (1965).
312. WINDAUS, A., W. BERGMANN and G. KÖNIG: Über einige Versuche mit Scymnol. Z. Physiol. Chem. **189**, 148 (1930).
313. WOLTERS, B.: Antibiotic and Toxic Agents from Algae and Mosses. Planta Medica **12**, 85 (1964).
314. WOODWARD, R. B.: The Structure of Tetrodotoxin, Pure Appl. Chem. **9**, 49 (1964).
315. YAMADA, K., H. YAZAWA, M. TODA, Y. HIRATA: The Synthesis of ($\pm$) Aplysin and ($\pm$) Debromoaplysin. Chem. Comm. **1968**, 1432.
316. YAMADA, K., H. YAZAWA, D. UEMURA, M. TODA and Y. HIRATA: Total Synthesis of ($\pm$) Aplysin and ($\pm$) Debromoaplysin. Tetrahedron Lett. **1969**, 3509.
317. YAMAGUCHI, M.: Carotenoids of the Sponge *Reiniera japonica*. Bull. Chem. Soc. Jap. **30**, 111 (1957).
318. YAMAGUCHI, M.: Chemical Constitution of Renieratene. Bull. Chem. Soc. Jap. **30**, 979 (1957).
319. YAMAGUCHI, M.: Chemical Constitution of Isorenieratene. Bull. Chem. Soc. Jap. **31**, 51 (1958).

320. YAMAGUCHI, M., Y. IWAKARI and S. TOKINCHI: Pigments of Pearl-oyster Shells, Mem. Fac. Sci. Kyushu Univ. Ser. C. Chem. **3**, 161 (1960).
321. — Renieratene, A New Carotenoid Containing Benzene Rings, Isolated from a Sea Sponge. Bull. Chem. Soc. Jap. **31**, 739 (1958).
322. — Pigments of Marine Animals VIII. Total Synthesis of Isorenieratene. Bull. Chem. Soc. Jap. **32**, 1171 (1959).
323. — Total Synthesis of Renieratene and Reniera Purpurin. Bull. Chem. Soc. Jap. **33**, 1560 (1960).
324. YAMANOUCHI, T.: On the Poisonous Substance Contained in Holothurians, Publ. Seto Mar. Biol. Lab. **IV** (2–3) 25 (1955).
325. YAMAMURA, S. and Y. HIRATA: Structures of Aplysin and Aplysinol, Naturally Occurring Bromo-compounds. Tetrahedron **19**, 1485 (1963).
326. YASUMOTO, T. and Y. HASHIMOTO: Properties and Sugar Components of Asterosaponin A Isolated from Starfish. Agr. Biol. Chem. (Japan), **29**, 804 (1965).
327. YASUMOTO, T., M. TANAKA and Y. HASHIMOTO: Distribution of Saponin in Echinoderms. Bull. Jap. Soc. Sci. Fisheries **32**, 673 (1966).
328. YOSHIDA, M.: Naphthoquinone Pigments in *Psammechinus milliaris* (Gmelin). J. Mar. Biol. Assoc. U. K. **38**, 455 (1959).
329. Proceedings of the Fifth International Seaweed Symposium, Ed. E. G. YOUNG and J. L. MCLACHLAN. New York: Pergamon Press. 1965.
330. FRYBERG, M., A. C. OEHLSCHLAGER and A. M. UNRAU: Synthesis of a Novel C_{26} Marine Sterol. Chem. Comm. **1971**, 1194.
331. MOORE, R. E.: Bis-(3-oxoundecyl) Polysulphides in *Dictyopteris*. Chem. Comm. **1971**, 1168.
332. GONZÁLEZ, A. G., J. DARIAS and J. D. MARTIN: Taondiol. A New Component from *Taonia Atomaria*. Tetrahedron Lett. **1971**, 2729.
333. HABERMEHL, G. and GERT VOLKWEIN: Aglycons of the Toxins from the Cuvierian organs of *Holothuria Forskåli* and a New Nomenclature for the Aglycons from *Holothurioideae*. Toxicon **9**, 319–326, 1971.
334. SIMS, J. J., W. FENICAL, R. M. WING and P. RADLICK: Marine Natural Products. I. Pacifenol, a Rare sesquiterpene Containing Bromine and Chlorine from the Red Alga, *Laurencia pacifica*. J. Am. Chem. Soc. **93**, 3774 (1971).
335. FORENZA, S., L. MINALE, R. RICCIO and E. FATTORUSSO: New Bromo-pyrrole Derivatives from the Sponge *Angelas oroides*. Chem. Comm. **1971**, 1129.
336. WONG, J. L., M. S. BROWN, K. MATSUMOTO, R. OESTERLIN and H. RAPOPORT: Degradation of Saxitoxin to a Pyrimido [2, 1-b] purine. J. Am. Chem. Soc. **93**, 4633 (1971).
337. WONG, J. L., R. OESTERLIN and H. RAPOPORT: The Structure of Saxitoxin. J. Am. Chem. Soc. **93**, 7344 (1971).
338. FATTORUSSO, E., L. MINALE, G. SODANO and E. TRIVELLONE: Isolation and Structure of Nitenin and Dihydronitenin, New Furanoterpenes from *Spongia Nitens*. Tetrahedron **27**, 3909 (1971).

(Received, December 21, 1970)

Namenverzeichnis. Author Index

Kursiv gedruckte Seitenzahlen beziehen sich auf Literaturverzeichnisse

Page numbers printed in *italics* refer to References

AASEN, A. J. *472*.
ABASHIAN, D. V. 57.
ABBONDANZA, A. *413*.
ABDEL RAHMAN, M. M. A. *358*.
ABERHART, D. J. 237, 276, *288*.
ABLONDI, F. *42*.
ABOLINŠ, L. *128*, *137*.
ABRAHAM, E. P. 422, 457, 458, *472*.
ABRAHAM, R. J. *42*.
ACHILLADELIS, B. *413*.
ACHINI, R. S. *394*.
ACTITELLI, M. A. 187, 188, *202*, *206*.
ACKERMANN, D. 455, 456, *473*.
ADAM, G. 49, *58*.
ADITYACHAUDHURY, N. *51*.
ADOLPHEN, G. *42*, *57*.
AEBI, A. *391*.
AGARWAL, S. C. 271, *288*.
AGETA, H. 297, *305*.
AGHORAMURTHY, K. 237, *288*.
AHMADJIAN, V. 285, 287, *288*, 297.
AITZETMÜLLER, K. *473*.
ÅKERMARK, B. 220, 222, 223, 271, *288*.
ALEXA, E. *357*, *359*.
ALEXANDER, C. B. *474*.
ALI, A. *474*.
ALI, A. A. E. R. *46*.
ALLISON, A. J. *393*.
ALMODOVAR, L. A. *483*.
ALMODOVAR, L. R. *475*.
ALSOP, P. A. 297.
ALSTON, R. R. *52*.
AMANO, K. *486*.
AMESZ, J. *131*.
AMIMOTO, K. *479*.
ANDERSON, H. A. 234, *288*.
ANDERSON, H. H. 454, *480*.
ANDERSON, I. G. *474*.

ANDERSON, J. D. *413*.
ANDERSON, K. A. *474*.
ANDERSON, R. J. *394*.
ANDO, T. *294*.
ANET, F. A. L. *202*.
APLIN, R. T. 413, *486*.
APPEL, H. H. 36, *42*, *43*, *44*, *46*, *57*.
APSIMON, J. W. *357*.
ARAI, A. *304*.
ARATA, Y. 32, 33, *42*.
ARCHER, R. *472*.
ARDENNE, M. v. *295*.
ARIGONI, D. *391*, *413*.
ARK, P. A. 287, *289*.
ARKLEY, V. *289*.
ARNDT, R. R. *42*, *54*, *57*.
ARNOLD, W. *128*.
AROJAN, A. A. *51*.
ARSHAD, M. 235, 238, *289*.
ARTHINGTON, W. *49*.
ARTHUR, H. R. 20, *42*.
ASAHINA, Y. 209, 210, 228, 231, 285, *289*.
ASHIKARI, H. *474*.
ASMINO, K. *205*.
ASLANOV, X. A. 32, *42*, *53*, *55*.
ASPEN, A. J. *43*.
ATAL, C. K. *43*, *57*.
ATKINSON, J. E. *289*.
ATTAWAY, D. H. *476*.
AUDA, H. 36, *43*.
AUDIER, H. E. *357*.
AUSTIN, L. *474*.
AUTERHOFF, H. *357*, *361*.
AWAKASU, O. *486*.
AYER, W. A. *43*.
AYNEHCHI, Y. *359*.

BACH, S. R. 223, 305.
BACHELOR, F. W. 248, 289.
BADA, J. L. 128.
BADARAU, GH. 359.
BADER, G. 134.
BADIELLO, R. 413.
BAECKSTRÖM, P. 228, 289.
BAILEY, D. M. 392.
BAILLIE, A. C. 293.
BAKER, B. R. 17, 43.
BAKER, P. M. 228, 289.
BALENOVIC, K. 10, 43.
BALL, K. 293.
BALTHIS, V. M. 293.
BANDONI, R. J. 287, 289.
BANGA, S. S. 43.
BANNER, A. H. 474.
BARBALIĆ, L. 289.
BARBER, M. 295.
BARBIER, A. 484.
BARBIER, M. 135, 137, 138.
BARCHET, R. 33, 43.
BARCROFT, J. 134.
BARONDES, S. H. 202, 203.
BARRETT, E. 391.
BARRACLOUGH-FELL, H. 474.
BARTLETT, W. D. 392.
BARTON, D. H. R. 244, 262, 278, 289, 334, 357, 359, 370, 391.
BARTZ, Q. R. 184, 203, 208.
BARUA, A. K. 357.
BASKOY, L. 203.
BASLEROVA, M. 485.
BASLOW, M. H. 474.
BATTERSBY, A. R. 43.
BAUER, H. 136.
BAUERSCHMIDT, E. 47.
BAUMGARTNER, H. 131.
BAXTER, J. G. 474.
BAXTER, R. M. 45.
BEACH, N. A. 134, 139.
BEARD, C. 202.
BECHER, D. 295, 474.
BEECHAM, A. F. 43.
BEECKEN, H. 289.
BERNSTEIN, H. J. 42.
BELKIN, M. 357, 360.
BELL, D. J. 467, 474.
BELL, R. A. 392.
BENATI, O. 477.
BENDZ, G. 289, 290.
BENES, J. 206.

BENJAMINOV, B. S. 360.
BENNETT, L. L. 202.
BENNHOLD, H. 128.
BENZ, G. 55.
BERG, A. 357.
BERG, P. 49.
BERGMANN, W. 418, 419, 474, 487.
BERGSTROM, S. 474, 483.
BESSHO, K. 47.
BERGY, M. E. 202.
BERKOFF, C. E. 422, 474.
BERNARD, T. 217, 290.
BERNASCONI, R. 28, 43.
BERNFELD, P. 51, 416, 467, 474.
BERTRAND, J. A. 293.
BERNS, D. S. 117, 129, 133.
BERNSTEIN, H. J. 205.
BERTH, P. 56.
BERTILSSON, L. 243, 290.
BESSMO, K. 47.
BETSUKI, S. 479.
BEVAN, C. W. L. 43, 413.
BEVING, H. F. G. 290.
BEYERMAN, H. C. 10, 12, 14, 44.
BHACCA, N. S. 202.
BHAKUNDI, D. S. 44.
BHUTANI, S. B. 290.
BIANCHI, E. 44.
BIEDRON, S. I. 361.
BIERNROTH, G. 362.
BIGLINO, G. 357.
BILAYET HOSSAIN, M. 487.
BILLING, B. H. 91, 129, 130, 133.
BIRCH, A. J. 413.
BLACK, W. A. P. 475.
BLAHA, K. 51.
BLANCO, A. N. 54.
BLANK, H. 287, 290.
BLAUER, G. 93, 129.
BLEDSOE, J. O. JR. 487.
BLINKS, L. R. 116, 132.
BLOCH, K. 369, 393.
BLOOMER, J. L. 275, 290.
BLOOMFIELD, V. A. 129.
BLUM, M. S. 52.
BLUMER, M. 475.
BLUNDEN, G. 44.
BOBBITT, J. N. 50, 55, 57.
BOCHKOV, A. F. 469, 480.
BOCKS, S. M. 413.
BOGDANOV, V. S. 392.
BOHLMANN, F. 44.

BOHMAN, G. *289, 290.*
BOHONOS, N. *146, 208.*
BOIT, H.-G. *44.*
BOLL, P. M. *290.*
BONNER, B. A. *138.*
BONNETT, R. *129.*
BORÉN, H. B. *290.*
BORGSTROM, G. *475, 480.*
BORSCHE, W. *357.*
BORTHWICK, H. A. *124, 129, 132.*
BOSMANS, R. *480.*
BOSTOGANASHVILL, V. S. *44.*
BOTHNER-BY, A. A. *359.*
BOTTINI, A. T. *289.*
BOUCHER, L. J. *130, 476.*
BOULANGER, P. *44.*
BOYLAN, D. B. *454, 475.*
BRAARUD, T. *475.*
BRACHET, J. *290.*
BRADY, S. F. *392.*
BRANDÄNGE, S. *3, 44.*
BRAUN, F. *55, 56.*
BRECCIA, A. *413.*
BREDENBERG, J. B. *357.*
BRESLOW, R. *390, 391.*
BREWER, J. D. *241, 290.*
BRIDGWATER, R. J. *475.*
BRIGGS, M. H. *290.*
BRIGGS, T. *475.*
BRIGGS, W. R. *129.*
BRIGHT, A. *46.*
BROCKMAN, J. A. *50.*
BROCKMAN, O. C. *391.*
BROCKMAN, R. W. *202.*
BROCKMANN, H. JR. *129.*
BRODERSEN, R. *129.*
BROOKS, R. C. *123, 129.*
BROTHERTON, J. *290.*
BROTZU, G. *422, 475.*
BROWN, C. J. *255, 290.*
BROWN, R. T. *44.*
BROWN, S. *483.*
BRUCKENSTEIN, S. *134.*
BRUNNER, E. *131.*
BRUNNER, R. *290.*
BRUTKO, L. I. *44.*
BRUUN, T. *229, 235, 251, 268, 289, 291.*
BRYAN, R. F. *332.*
BÜCHEL, K. H. *10, 44.*
BUCHWALD, H. D. *475, 482.*
BUDZIKIEWICZ, H. *44, 132, 137.*
BULL, J. R. *357.*

BULLOCK, E. *228, 289.*
BULLOCK, W. K. *203.*
BUNDSCHUH, W. *56.*
BUNSOW, R. *413.*
BURGSTAHLER, A. W. *366, 367, 393.*
BURKHARDT, K. *55.*
BURKHOLDER, L. M. *475.*
BURKHOLDER, P. R. *287, 291, 447, 475, 485.*
BUTLER, W. L. *129.*
BUTRUILLE, D. *47.*
BY, A. W. *50.*

CAGLIOTI, L. *135.*
CAHN, R. S. *129.*
CAIN, B. F. *291.*
CAIOLA, S. M. *179, 205.*
CALLAHAN, E. W. JR. *129.*
CALVENTI, Y. *482.*
CALVIN, M. *475.*
CAMBIE, R. C. *245, 291, 413.*
CAMPBELL, J. E. *481.*
CAMPBELL, M. *139.*
CANO, L. *203.*
CARDINAL, R. *134.*
CARLSON, A. A. *169, 189, 191, 204, 207.*
CARR, N. G. *130.*
CARROLL, D. M. *135.*
CASILLAS, C. *482.*
CASINOVI, C. G. *44.*
CASTLE, H. *285, 291.*
CATION, D. *202.*
CECIL, J. T. *482.*
CEI, J. M. *476.*
CERU, J. *203.*
CHAKRABORTI, S. K. *357.*
CHALLEN, S. B. *44, 46.*
CHAMBLISS, O. L. *357.*
CHAN, D. *155, 208.*
CHAN, R. P. K. *20, 42, 52.*
CHAN, T. H. *49.*
CHANDA, N. B. *291.*
CHANDRASENAN, K. *291.*
CHANG, C. W. J. *475, 481, 482, 485.*
CHANLEY, J. D. *425, 426, 475, 477, 481.*
CHAPMAN, D. J. *66, 120, 130, 138, 459, 475, 476.*
CHAPUT, M. *12, 52.*
CHATTERJEE, A. *45.*
CHATTERJEE, A. B. *481.*
CHAUBAL, M. G. *45.*
CHAUDRY, G. R. *358.*
CHAUVET, J. *472.*

CHAUVET, M. T. *472*.
CHAWLA, A. *391*.
CHEESMAN, D. F. *445, 475*.
CHESTERS, C. G. C. *476*.
CHIANG, H. C. *250, 303*.
CHIBBER, S. *290*.
CHOU, T. Q. *45*.
CHRISTMAN, D. R. *57*.
CHRISTOMANOS, A. *476*.
CHURCHILL, B. W. *202*.
CIERESZKO, L. S. *476, 485, 487*.
CLARK, B. C. JR. *391*.
CLARK, D. E. *290*.
CLARK-LEWIS, J. W. *5, 45*.
CLAYTON, D. A. *227, 297*.
CLAYTON, R. B. *391, 394, 413*.
CLOETENS, R. *484, 486*.
CLUGSTON, D. M. *45*.
COATES, R. M. *394*.
COHEN, H. D. *202, 203*.
COHEN, M. *320*.
COKE, J. L. *23, 45, 54*.
COLE, P. G. *91, 129, 130*.
COLE, W. J. *66, 120, 130, 138*.
COLE, W. Y. *475, 476*.
COLLERAN, E. *135*.
COLLINS, S. *134*.
COMER, F. *50, 416*.
COMPERNOLLE, F. *131*.
CONNER, R. L. *481*.
CONNOLLY, J. D. *159, 202, 413*.
CONTI, S. F. *132*.
CONWAY, E. *483*.
COOK, J. W. *476*.
COOKE, G. A. *45, 47*.
COOKE, J. R. *130*.
CORBELLA, A. *288*.
CORBETT, R. E. *266, 267, 291, 414*.
CORDELL, G. A. *45*.
COREY, E. J. *157, 202, 389, 391*.
CORNFORTH, J. W. *391*.
CORRAL, R. A. *53*.
CORRELL, D. L. *130, 137*.
COSULICH, D. B. *476*.
COURVILLE, D. A. *478*.
COWGER, M. L. *134*.
CRABBÉ, P. *361*.
CRAGG, G. *414*.
CRAIG, I. W. *130*.
CRAIG, J. C. *52*.
CRAIGIE, J. S. *476*.
CRANDALL, J. K. *392*.

CRANDALL, T. G. *391*.
CREASER, E. H. *291*.
CREPY, D. *472*.
CRESPI, H. L. *119, 129, 130, 132, 476*.
CROMWELL, B. T. *11, 45, 54*.
CROOKS, P. A. *45*.
CROSS, A. D. *433, 474, 476, 485*.
CROSS, B. E. *414*.
CROSS, D. R. *130*.
CROZIER, G. F. *446, 477*.
CUDLIN, J. *58, 202, 206, 208*.
CULBERSON, CH. F. *210, 247, 248, 250, 273, 285, 286, 291, 292*.
CULBERSON, W. L. *286, 292*.
CULLEN, W. P. *174, 180, 206*.
CUMMING, S. D. *267, 291*.
CURPHEY, T. J. *394*.
CWALINA, G. E. *358*.

DAIGO, K. *486*.
DAINIS, J. *45*.
DARIAS, J. *488*.
DAS, B. C. *54, 357, 477*.
DAS, K. C. *482*.
DAS GUPTA, A. K. *174, 206*.
DAVIDSON, T. A. *259, 292*.
DAVIES, A. P. *45*.
DAVIS, D. *245, 292*.
DAVIS, E. *134, 136, 139*.
DAWSON, R. F. *57*.
DE ABREU MATOS, F. J. *357*.
DEAKINS, D. E. *476*.
DEAN, F. M. *244, 289, 292*.
DEAN, P. D. G. *305*.
DECKER, M. *136*.
DEFLORIN, A. M. *289*.
DE KLONIA, H. *59*.
DE KOCK, W. T. *359*.
DEL CASTILLO, J. *481*.
DELLE MONACHE, F. *44*.
DE MAYO, P. *289, 391*.
DENNIS, D. T. *414*.
DESAI, R. B. *360*.
DESLONGCHAMPS, P. *45, 58*.
DE SOUZA GUIMARAES, I. S. *486*.
DEV, S. *391, 415*.
DE VIRVILLE, C. *476*.
DE WAAL, H. L. *54, 57*.
DEVLIN, J. P. *289*.
DEWAR, E. T. *475*.
DHAR, M. L. *44*.
DHAR, M. M. *44*.
DIACONT, K. *357*.

DICKINSON, E. M. 45.
DIETRICH, S. M. C. 45, 50.
DIETZSCH, K. 45.
DION, H. W. 184, 203, 208.
DI SANSEVERINO, L. R. 480.
DITTMAN, W. 391.
DJERASSI, C. 44, 45, 46, 53, 132, 134, 151, 202, 203, 295, 428, 474, 477, 478, 480, 484, 487.
DOBENECK, H. V. 131.
DOLAK, L. A. 392.
DOLEŽILOVA, L. 50, 58, 204, 206, 208.
DOMINGUEZ, J. 28, 46.
DÖPKE, W. 46.
DORFMAN, L. 54, 55.
DOTY, M. S. 482.
DOUGHERTY, R. C. 130, 476.
DOWNS, R. J. 134.
DRAKE, B. 292.
DREIDING, A. S. 39, 52, 59.
DREW, E. A. 292, 293.
DRILLIEN, G. 46.
DRING, M. J. 131.
DUBOST, M. 179, 206.
DUFFIELD, A. M. 486.
DUMMER, G. 55, 56.
DU PLESSIS, L. M. 42.
DUQUETTE, L. G. 177, 203.
DURHAM, L. 475.
DUTTA, C. P. 45.
DUURSMA, E. K. 476.
DUYSENS, L. M. N. 131.
DYER, J. R. 258, 293.
DZIZENKO, A. K. 476.

EBIHARA, K. 306.
EBIZUKA, Y. 293.
EBLE, T. E. 202.
EDER, W. R. 290.
EDERY, H. 358.
EDWARDS, A. G. 17, 49.
EDWARDS, J. L. 130.
EDWARDS, M. R. 129.
EDWARDS, O. E. 289.
EENSHUISTRA, J. 43.
EGAWA, Y. 174, 187, 188, 191, 193, 194, 195, 203, 205, 208.
EGGER, K. 480.
EGGERS, S. H. 297.
EHL, K. 136.
EHRENSVÄRD, U. 273, 299.
EILERS, N. J. 207.

EISENBRAUN, E. J. 43, 46, 203.
EISENHUT, W. O. 358.
EISENLOHR, K. H. 361.
EISNER, U. 46.
EKONG, D. E. U. 413.
ELIX, J. A. 241, 245, 290, 292, 302.
EL-KHADEM, M. 358.
ELKIN, YU N. 476.
ELLIS, R. C. 293.
EL-OLEMY, M. M. 16, 46.
ELYAKOV, G. B. 476.
EMERSON, R. L. 146, 208.
ENNIS, H. L. 147, 203.
ENSLIN, P. R. 358, 359, 360, 361, 362.
ERDTMAN, H. 288.
ERDTMANN, H. 414.
ERLICH, J. 204.
ERSPAMER, V. 358, 454, 456, 476, 483.
ESCHENMOSER, A. 366, 367, 368, 369, 391, 393.
ESCHRICH, W. 293.
EVANS, A. W. 287, 291.
EVANS, C. A. 292.
EVANS, D. A. 46.
EVANS, E. 292.
EVANS, J. E. 391.
EVANS, J. S. 203.
EVELEENS, W. 43.

FABBRI, S. 479.
FAGER, J. 298.
FAIRBAIRN, J. W. 10, 46.
FAIRLIE, J. C. 391.
FALES, H. M. 49, 52, 279, 300.
FALKMER, S. 487.
FARMER, L. M. 417.
FARNSWORTH, N. R. 15, 46.
FATTORUSSO, E. 476.
FAUGERAS, G. 46.
FAUST, R. E. 358.
FAUX, A. 476.
FAZAKERLEY, H. 391.
FEHLHABER, H. W. 58, 296.
FEIGE, B. 274, 275, 293, 303.
FEIGEH, G. A. 474.
FEIST, W. 48.
FELDMANN, J. 476.
FELDOTT, G. 297.
FELIX, D. 391.
FENICAL, W. 488.
FERGUSON, F. 416.
FERGUSON, G. 225, 293.
FERGUSON, H. C. 358.

Fett, H. *56.*
Fevery, J. *131.*
Ficini, J. *393.*
Field, J. B. *203.*
Fieser, L. F. *476.*
Fieser, M. *476.*
Finney, K. F. *49.*
Firer, E. M. 124, *138.*
Fischer, H. 62, 67, *131.*
Fischer, H. G. *475, 482.*
Fisher, M. W. *204.*
Fitzgerald, D. B. *357.*
Fitzgerald, J. S. *46.*
Flasch, H. *294.*
Fleming, M. *293.*
Flodgaard, H. *129.*
Florkin, M. *474, 477, 478, 482, 483, 485, 486.*
Flynn, E. H. *482.*
Fog, J. *131.*
Fodor, G. *45, 47.*
Folkers, K. *50.*
Follmann, G. *286, 287, 293, 295, 296.*
Foltz, C. M. *5, 59.*
Fontaine, M. *446, 484.*
Ford, J. H. *150, 153, 157, 203, 205.*
Forist, A. A. 146, *203.*
Fork, D. C. *129.*
Forrest, H. S. *477.*
Forrest, T. P. *33, 43, 294.*
Forsen, S. *293.*
Fowden, L. *5, 8, 47, 57.*
Fox, C. H. *257, 273, 285, 287, 293, 294.*
Fox, D. L. *130, 131, 446, 475, 477.*
Fox, J. A. *202.*
Fraenke, G. *476.*
Francis, T. *292.*
Franck, B. 12, 14, 15, 19, 20, 47, 233, *294.*
French, C. S. *131.*
French, J. C. *203.*
Fridrichson, 36, *47.*
Fried, J. H. *485.*
Friedrich, H. 223, *305.*
Friess, S. L. *477.*
Fritsch, G. *46.*
Fritz, G. 67, *138.*
Fritzsche, W. *49.*
Frohardt, R. P. 182, *203.*
Frohofer, H. *52.*
Fröwis, W. *137.*
Frybeg, M. *488.*

Fu, F. Y. *45.*
Fuhrman, F. A. *475, 482.*
Fuji, K. *47.*
Fujikawa, F. *294.*
Fujimori, E. *131, 133, 135.*
Fujisawa, M. *294.*
Fujita, E. 24, 47, *414.*
Fujita, Y. *131.*
Fujiwara, T. 299, *305.*
Fukuoka, F. 299, 300, *303.*
Fukushima, S. *305.*
Fukuzawa, A. *479.*
Furlenmeier, A. *477, 480.*
Fürst, A. *477, 480.*
Furumai, T. *206.*
Furuya, T. 211, 294, *303.*
Fuzikawa, F. 231, *289.*

Galbraith, M. M. *204.*
Galbraith, M. N. *477.*
Gallily, R. 358, *359.*
Galt, R. H. B. *414.*
Gams, E. *56.*
Gantt, E. *131, 132.*
Garay, A. S. *47.*
Garbarino, J. A. *44.*
Garbers, C. F. *357.*
Gardner, G. M. *481.*
Garegg, P. J. *290.*
Garrett, E. R. 146, 168, 189, *203.*
Gauhe, A. 465, *487.*
Geers, J. M. *133.*
Geissmann, T. A. *414.*
Geller, A. *203.*
Gellert, E. *47.*
Gerner, W. *362.*
Giacapello, D. *357.*
Gianotti, C. *477.*
Gilbert, B. *486.*
Gilbert, J. N. T. *359.*
Gilbertson, T. J. *51.*
Gillis, R. G. *474.*
Gilman, R. E. 12, *47.*
Giovannozzi-Sermanni, G. *44.*
Girotra, N. N. *391.*
Gitter, S. 358, 359, *362.*
Giudici de Nicola, M. *300.*
Gizycki, U. v. *289.*
Glaser, R. *477.*
Glass, M. A. W. *134.*
Glombitza, K.-W. *294.*
Glotter, E. 307, 360, *362.*

GMELIN, R. 5, 47, 58, 357, 359.
GOAS, G. 290.
GOLDBERG, S. I. 47.
GOLDSMITH, D. J. 363, 391, 392.
GONZALES, A. G. 47, 470, 488.
GOODWIN, T. W. 132, 135, 136, 138, 444, 460, 477, 486.
GORDON, J. T. 481.
GORDON, M. 298.
GORDON, S. 42.
GOTTLIEB, D. 53, 57, 58.
GOTTLIEB, O. R. 357, 359, 360.
GOTTSCHALK, E.-H. 289.
GOUTAREL, R. 23, 54.
GOVINDACHARI, T. R. 23, 47.
GRADY, J. E. 207.
GRAEBE, J. E. 414.
GRANDOLFO, M. 473.
GRANDOLINI, G. 44.
GRANT, P. T. 469, 484.
GRAY, A. H. 359.
GRAY, C. H. 64, 130, 132, 133, 134, 136, 138.
GRAY, S. L. 392.
GREAM, G. E. 294.
GREEN, J. H. 417.
GREENBERG, D. M. 481.
GREGSON, M. 393.
GREUELL, H. M. 133.
GREWE, R. 8, 48.
GRIECO, P. A. 393.
GRIFFIN, C. E. 298.
GRIFFITH, G. D. 48.
GRIFFITH, T. 48.
GRIGOREVA, N. Y. 414.
GRIPENBERG, J. 269, 294.
GROBBELAAR, N. 48.
GRÖGER, D. 48.
GROLLMANN, A. P. 147, 203.
GROS, E. G. 51.
GROSS, D. 1, 48.
GROSS, R. A. JR. 477.
GROVE, J. F. 414.
GROVE, M. D. 359.
GROVER, P. K. 237, 294.
GROVES, J. T. 391.
GRUENIG, D. E. 476.
GSCHWEND, H. W. 48.
GUPTA, L. C. 419, 477.
GUPTA, R. N. 7, 13, 16, 30, 48, 50.
GUPTA SEN, A. K. 477.
GUROWITZ, W. D. 180, 188, 204.

GUT, M. 360, 361, 391.
GUTTMAN, J. L. 332.

HAARSTAD, V. 184, 208.
HABERLAND, H. W. 67, 131.
HABERMEHL, G. 56, 430, 470, 477, 488.
HABGOOD, T. E. 43.
HACKNEY, R. J. 477.
HAGLID, F. 52.
HALBACH, H. 131.
HALE, M. E. JR. 286, 294.
HALE, R. L. 477, 478.
HALEWOOD, P. 292.
HALEY, R. C. 392.
HALLS, C. M. M. 362.
HALSALL, T. G. 358, 391, 440, 478.
HALSTEAD, B. W. 478.
HAMBERG, M. 483.
HAMILTON, J. M. 203.
HAMMARSTEN, O. 432, 478.
HAMMERSTEDT, R. H. 54.
HAMMOUDA, Y. 48.
HAMNER, C. L. 203.
HAMPSHIRE, F. 478, 479.
HANESSIAN, S. 478.
HANSON, J. R. 395, 413, 414, 415.
HARADA, R. 475.
HARBERT, C. A. 392.
HARDY, G. 360.
HARMATZ, D. 129.
HARPER, S. H. 294, 297.
HARRIS, A. 413.
HARRIS, G. 48.
HARRIS, R. L. N. 132.
HARRIS, R. S. 478.
HART, N. K. 15, 31, 46, 48.
HARTING, J. 298.
HARTMANN, K. M. 132.
HARTMANN, M. 478.
HARVEY, R. G. 357.
HASHIMOTO, K. 133, 139.
HASHIMOTO, Y. 452, 478, 483, 488.
HASLEWOOD, G. A. D. 431, 433, 474, 475, 478.
HASLINGER, E. 53.
HASSALL, C. H. 45, 49.
HASSE, K. 49.
HATA, T. 207.
HATANAKA, S. 49.
HATANO, K. 59.
HATTORI, A. 131, 132.
HAUPT, W. 132.

HAUSCHILD, G. *294*.
HAUSMANN, W. K. *52*.
HAVINGA, E. *484*.
HAWKER, J. *415*.
HAWKINSON, V. *139*.
HAXO, F. T. 116, *132*.
HAYAKAWA, Y. *479*.
HEESCH, A. *362*.
HEALEY, F. P. 444, *478*.
HEGARTY, M. P. *49*.
HEIRWEGH, K. P. M. *131*.
HELFRICH, P. *474*.
HEMMELMAYR, F. VON *359*.
HENDERSON, L. M. *54*.
HENDERSON, M. S. *415*.
HENDRICKS, S. B. 124, *129*, *132*, *138*.
HENDRICKSON, J. B. 256, *294*.
HENNIS, H. E. 177, *203*.
HENRICK, C. A. *415*.
HENRICKSSON, E. 285, *294*.
HERBERT, E. J. *415*.
HERBSTEIN, F. H. *361*.
HERR, R. R. 174, 175, 177, *202*, *204*.
HERRIN, T. H. *392*.
HERTZBERG, S. *478*.
HERZ, W. 59, 379, *392*.
HESS, D. 287, *295*.
HESS, R. *131*.
HESSE, M. *295*.
HESSE, O. *295*.
HESSLER, E. J. *394*.
HIGHET, P. F. 22, *49*.
HIGHET, R. J. 22, *49*, *53*, 180, *204*.
HIKIJI, T. *206*.
HIKINO, H. 422, *478*.
HIKINO, Y. 422, *478*.
HILDITCH, T. P. *478*.
HILLEGAS, A. B. *204*.
HILL, R. K. 10, 17, 18, *49*.
HILLS, I. R. 440, *478*.
HINKEL, H. 55, *56*.
HIRAI, K. *294*.
HIRATA, Y. 446, *479*, *487*, *488*.
HIRAYAMA, T. *294*.
HIROFUJI, S. *479*.
HIROSE, H. *132*.
HIRST, E. L. *291*.
HOCKS, P. 478, *480*.
HODGSON, G. L. *391*.
HOFFMAN, W. F. *290*.
HOFFMEISTER, H. *480*.
HÖHN, M. *56*.

HÖHNE, E. 49, *58*.
HOLDEN, J. T. *57*.
HOLDT, M. M. VON *361*.
HOLLINSHEAD, W. H. *139*.
HOLLIS, D. P. *291*.
HOLZAPFEL, C. W. 358, 359, 362, *413*.
HONDA, Y. *304*.
HONJO, M. *486*.
HOOZ, J. *392*.
HOPPE-SEYLER, G. *473*.
HÖRHAMMER, L. *49*.
HORN, D. H. S. 422, 476, 477, 478, 479, *485*.
HORNING, E. C. 51, *58*.
HORTON, D. *484*.
HOSENEY, R. C. *49*.
HÖSS, H. G. 11, *55*.
HOSHITA, T. 431, 479, *480*.
HOSSAIN, M. B. *479*.
HOUVENAGHEL-CREVECOEUR, N. *137*.
HUANG, W. N. *49*.
HUGEL, G. *415*.
HUGO, J. M. 358, *360*.
HULME, A. C. *49*.
HUMMEL, H. *480*.
HUNECK, S. 209, 211, 212, 213, 217, 235, 237, 239, 273, 285, 286, *288*, *293*, 294, 295, 296, 297, 300, *304*.
HÜPPI, G. 425, *479*.
HUTTERER, C. 111, *133*.
HUTZLER, A. *55*.
HYLIN, J. W. *49*.

IBA, I. *481*.
IDA, Y. *303*.
IDLER, D. R. 418, *479*.
IGARASHI, S. 186, *204*.
IGUCHI, M. *479*.
IITAKA, Y. *297*.
IIZUKA, H. 294, *303*.
IKEKAWA, N. 211, 297, *479*.
IKUMA, S. *486*.
INATOMI, H. *53*.
INAYAMA, S. 57, *59*.
INGOLD, C. *129*.
INUKAI, F. 49, *53*.
IRIE, T. 479, *480*.
ISKANDAROV, S. *49*.
ISLER, O. *480*.
ISOBE, K. *305*.
ISSELBACHER, K. J. 91, *132*.
IVANOV, I. C. 32, *52*.

Iwakari, Y. *488*.
Iwata, K. *297, 305*.
Izawa, M. *479*.
Izumi, S. *306*.

Jackson, A. H. *132, 133*.
Jackson, B. G. *458, 482*.
Jackson, W. G. *202*.
Jacksonhill, D. *301*.
Jacobsen, J. G. *133*.
Jacquot, R. *480*.
Jain, V. K. *161, 162, 170, 190, 206*.
Jakowska, S. *482*.
Jakubowski, Z. L. *203*.
Janka, R. *473*.
Jannke, P. J. *302*.
Jansen, H. *133*.
Jaques, B. *44, 392*.
Jarvik, M. E. *203*.
Jayalakshmi, V. *237, 297*.
Jeffcoat, A. R. *59*.
Jefferies, P. R. *415*.
Jeger, O. *391*.
Jellum, E. *131*.
Jenner, E. L. *134*.
Jennings, B. R. *129*.
Jensen, A. *478*.
Jensen, L. *478*.
Jensen, N. P. *392*.
Jensen, S.-L. *472, 478*.
Jindra, A. *57*.
Jirsa, M. *133*.
Jitariu, P. *357, 359, 361*.
Johansson, N.-G. *223, 288*.
Johns, S. R. *43, 46, 48*.
Johnson, A. W. *132*.
Johnson, F. *140, 156, 157, 162, 168, 169, 171, 177, 180, 187, 188, 189, 203, 204, 207*.
Johnson, L. F. *46, 361*.
Johnson, R. B. *297*.
Johnson, W. S. *368, 369, 381, 392*.
Johnston, R. L. *203*.
Joines, R. C. *391*.
Jones, C. M. *357*.
Jones, E. R. H. *358, 391*.
Jones, G. *45, 49*.
Jones, P. M. *132*.
Jones, R. F. *133*.
Jones, R. G. *204*.
Jones, S. *134*.
Joseph, J. B. *43*.

Joshi, B. S. *9, 49, 232, 297*.
Joubert, T. G. *358*.
Joule, J. A. *46, 49*.
Juneau, K. N. *18, 51*.
Juneja, H. R. *43*.
Junge, H. *133*.

Kaczmarek, F. *49*.
Kalish, J. *358*.
Kaltenbronn, S. *478*.
Kamalitdinov, D. *49*.
Kamat, V. N. *49*.
Kamiya, K. *300*.
Kamp, D. M. *131*.
Kanao, M. *206*.
Kandatsu, M. *53*.
Kaneko, T. *50*.
Kaneko, Y. *58, 303*.
Kanno, S. *392*.
Kao, C. Y. *475*.
Kao, O. *133*.
Kao, Y. S. *45*.
Kari, S. *5, 58*.
Karlson, P. *480*.
Karns, T. K. B. *487*.
Kasche, V. *130, 134*.
Katayama, T. *441, 480*.
Kato, T. *392, 393*.
Katritzky, A. R. *136*.
Katsumi, M. *415*.
Katz, J. J. *119, 129, 130, 132, 473, 476*.
Kauffmann, T. *56*.
Kawamata, T. *59*.
Kawamura, M. *486*.
Kawasaki, I. *50*.
Kay, I. T. *132, 133, 134*.
Kays, W. R. *43*.
Kazuno, T. *431, 479, 480*.
Keglević, D. *300*.
Kelleher, W. J. *46*.
Kelling, K. L. *56*.
Kelly, C. *417*.
Kemp, R. J. *207*.
Kennard, O. *480*.
Kennedy, G. Y. *484*.
Kennedy, R. V. Jr. *174, 206*.
Kenner, G. W. *132, 133*.
Keogh, M. F. *13, 17, 20, 50, 53*.
Kerb, U. *480*.
Kersten, S. *48*.
Khanna, K. L. *50, 57*.
Kharatyan, S. *50, 197, 204*.

Khorana, M. L. *360*.
Khuong-Huu, Q. *54*.
Killilea, S. D. *133, 135*.
Kimland, B. *415*.
King, F. E. 5, *50*.
King, T. E. 93, *129, 134*.
King, T. J. *50*.
Kinnel, R. N. *392*.
Kinraide, W. T. B. 287, *297*.
Kirby, G. W. *415*.
Kirchmeier, O. *50*.
Kiryukhin, V. K. *53*.
Kisaki, T. 25, *50*.
Kitagawa, I. *306*.
Kitahara, Y. *392, 393*.
Klein, E. *294*.
Klein, W. H. *130*.
Kleinig, H. 444, *480*.
Klinga, K. *133*.
Klöden, D. *58*.
Klomparens, W. *203*.
Klose, W. 73, 76, *137*.
Kloss, P. *359*.
Kloss, R. A. 227, *297*.
Klyne, W. *132, 414*.
Knöfel, D. *56*.
Knüsel, F. *53*.
Kobayashi, N. 297, 300, *301*.
Koch, U. 271, *301*.
Kochetkov, N. K. 469, *480*.
Kochiyama, Y. *133, 139*.
Koepfli, J. B. 17, *50*.
Kohan, S. *139*.
Kohberger, D. L. *204*.
Kohn, S. K. *482*.
Kojitani, K. *294*.
Koleoso, O. A. *50*.
Kometani, K. *58*.
Komiya, T. 254, 285, *297*.
Kompis, I. *57*.
Komzak, A. *55*.
Kondo, K. *486*.
Kondo, Y. 50, *486*.
König, G. *487*.
Konishi, K. 453, *480*.
Koo, S. H. *50*.
Koppernock, F. *56*.
Kornfeld, E. C. 147, 164, 170, 182, *204*.
Korte, F. 10, *44*.
Kortekangas, A. E. 297, *305*.

Kosaki, T. I. 454, *480*.
Köst, H.-P. *133, 137*.
Kotera, K. *205*.
Kovacs, P. *57*.
Kovar, J. *51*.
Kowitz, F. *58*.
Koyama, M. *294*.
Kramer, V. *137*.
Krämer, H. *289*.
Kraus, G.-J. 5, *50*.
Kraus, K. W. *485*.
Krauth, C. A. *207*.
Kreibich, K. *56*.
Kress, R. *56*.
Krimer, M. Z. *392, 393*.
Kristinsson, H. D. 210, 250, *292*.
Kroes, H. H. *133*.
Krog, H. 287, *297*.
Krogh Hansen, J. *129*.
Krueger, W. C. *134*.
Krueger, W. R. *130*.
Krüger, G. *56*.
Kubsch, F. 285, *291*.
Kucherov, V. F. 368, 390, 392, 393, *414*.
Kuehl, F. A. Jr. *50*.
Kuenzle, C. C. *133*.
Kuhn, R. 478, *480*.
Kulczycka, A. *132*.
Kullik, R. K. *205*.
Kumano, S. *132*.
Kumar, Ray A. *357*.
Kump, W. G. 4, *53*.
Kupchan, S. M. 50, 332, *359*.
Kuroda, Y. *58*.
Kurokawa, S. 254, *297*.
Kurosawa, D. *479*.
Kurosawa, E. *479*.
Kurze, J. *136*.
Kutney, J. P. 45, *46*.
Kuwada, Y. *300*.
Kuwata, S. *50*.
Kuznetsova, T. A. *476*.

Lääke, Oy *297*.
Lack, R. E. *207*.
La Londe, R. T. 33, *59*.
Lambert, B. F. 54, *55*.
Lamberton, J. A. 43, 46, *48*.
Langemann, A. *477*.
Lardy, H. A. *297*.
Large, C. M. *202*.

LATHE, G. H. 91, *129*, *130*.
LAVIE, D. 307, *358*, *359*, *360*, *362*.
LAVIGNE, R. *52*.
LAWES, B. C. 154, 155, 156, 187, 188, *204*, *205*.
LAWTON, R. G. 389, *391*.
LAZUREVSKI, G. V. *415*.
LEACH, B. E. *203*, *205*.
LEARY, J. D. *50*, *57*.
LECLERQ, J. *477*.
LEDEEN, R. *475*.
LEDERER, E. 111, *133*, *138*, *477*, *480*, *484*.
LEDVINA, M. *133*.
LEE, J. J. *129*, *133*.
LEE, W. L. *475*.
LEETE, E. 4, 11, 18, 24, 26, 27, 37, 50, *51*.
LEGGE, J. W. *134*.
LEHFELD, J. *56*.
LEHN, J.-M. *296*, *297*, *357*.
LEITZ, F. H. *475*.
LELOUP, J. *446*, *484*.
LEMAY, L. *52*.
LEMBERG, R. 62, 64, *133*, *134*.
LEMEN, J. *48*.
LEMIEUX, R. A. 165, *205*.
LEMIN, A. J. 150, 153, 157, *205*.
LEMMON, G. A. *132*.
LEMON, J. McW. *486*.
LEONTEV, V. B. *53*.
LERCH, U. *136*.
LESSARD, J. *357*.
LESTER, R. *134*.
LETCHER, R. M. 213, 215, 294, *297*.
LEVITA, B. *135*.
LEVY, B. *362*.
LEW, F. *416*.
LEWIN, R. A. *477*, *481*.
LEWIS, D. H. *301*.
LEWIS, H. K. *481*.
LEWIS, J. R. *289*.
LI, T. *392*.
LICHTAROWICZ-KULSZYCKA, A. *132*.
LIEBISCH, H. W. 13, 39, *51*.
LIGHTNER, D. A. *134*, *136*, *139*.
LIGOURI, V. R. *482*.
LINDBERG, B. 297, *298*.
LINDSEY, A. S. *391*.
LINDSTEDT, G. *51*.
LINDSTEDT, S. *51*.
LING, N. C. *478*.
LINSCHEID, P. *296*.

LINSCHITZ, H. *130*, *134*.
LIPKIN, A. H. *47*.
LIPSCOMB, W. N. 10, *59*.
LIST, P. H. *473*.
LITVINOV, M. A. *302*.
LIWSCHITZ, Y. *295*.
LLOYD, H. A. 29, *51*, *53*.
LOCKSLEY, H. D. *298*.
LODER, J. W. *51*.
LODS, L. *415*.
LOEFER, J. B. *205*.
LORANCE, E. K. *484*.
LORD, K. E. *394*.
LÖSEL, W. 212, *304*.
LOUNASMAA, M. 270, *298*.
LOVELL, F. M. *476*, *481*.
LOW, T. F. *481*.
LOWE, M. E. *479*.
LOWRY, P. T. *134*, *139*.
LOWY, P. H. *51*, *57*.
LUCAS, R. A. *55*.
LUCHETTI, M. A. *51*.
LUKES, R. *51*.
LUMMIS, W. L. *207*.
LÜNING, B. 3, *44*.
LYNEN, F. *298*,
LYTHGOE, D. 23, *52*.

MAASS, W. S. G. 271, 276, 280, 281, 282, 294, *298*.
MAASSEN, D. *289*.
MAAT, L. 43, *44*.
MABRY, T. J. 39, 52, *59*.
MACCOLL, R. *129*.
MACCONNELL, J. G. 22, *52*.
MACHEK, G. *290*.
MACHOLAN, L. 18, *52*.
MACKEY, I. R. *293*.
MACKENZIE, S. 243, *298*.
MACKINNEY, G. *481*.
MACLEAN, D. B. *45*.
MACMILLAN, J. *415*.
MACMUNN, C. A. 460, *481*.
MADONO, K. *132*.
MAISACK, H. *49*.
MALLORY, F. B. *481*.
MALONEY, T. E. *206*.
MANAKTALA, S. K. 228, *298*.
MANDELLI, E. F. *481*.
MANGONI, L. *135*.
MANITTO, P. *357*.
MANN, P. J. G. *52*.

Manners, D. J. *291, 293.*
Manske, R. H. F. *30, 44, 45, 50, 54, 56, 59.*
Marčenko, E. *300.*
Marderosian, A. D. *481.*
Marekov, N. *51.*
Maretzki, A. *481, 483.*
Marini-Bettolo, G. B. *35, 44, 52.*
Marion, L. *4, 12, 15, 47, 52, 54, 59.*
Markham, K. R. *467, 481.*
Markó, M. *299.*
Marshak, A. *287, 298.*
Marsiko, W. E. *487.*
Martin, D. F. *481.*
Martin, J. D. *488.*
Martin, J. H. *52.*
Martin, R. O. *12, 45, 50.*
Maruyama, M. *359.*
Marver, H. S. *138.*
Marx, M. *393.*
Masamune, T. *479, 480.*
Mason, H. S. *474, 477, 478, 484, 485, 488.*
Massagetov, P. S. *44.*
Mathew, C. T. *391.*
Mathey, A. *296.*
Mathias, A. P. *481.*
Mathieson, A. M. *36, 47.*
Mathieson, D. W. *359.*
Matney, T. S. *205.*
Matschiner, B. *51.*
Matsuami, N. *294.*
Matsuda, A. *306.*
Matsuda, H. *481.*
Matsudaira, H. *299.*
Matsufiji, K. *416.*
Matsui, M. *59, 297.*
Matsumae, A. *207.*
Matsumoto, K. *488.*
Matsuno, T. *478.*
Matsuo, M. *305.*
Matsutani, S. *50.*
Matsuura, F. *133, 139.*
Matte, H. O. *139.*
Matthies, H.-G. *289.*
Mautner, H. G. *481.*
Mauzerall, D. *134.*
McCapra, F. *416.*
McCarthy, E. A. *91, 132.*
McConnel, W. B. *53.*
McCormick, A. *415.*
McCormick, J. P. *394.*
McCreadie, T. *393.*
McCrindle, R. *159, 202, 413, 415.*

McDonagh, A. F. *129.*
McDowell, R. H. *467, 469, 481, 483.*
McEvoy, F. J. *43.*
McFarren, E. F. *481.*
McLachlan, J. L. *488.*
McLure, T. T. *205.*
McNeely, W. H. *469, 485.*
McPherson, J. *297.*
Mead, J. F. *50.*
Meeuse, A. D. J. *361.*
Meier, E. *137.*
Meier, H. *297.*
Meier, J. *391.*
Meinwald, J. *187, 206.*
Meister, A. *43.*
Melchior, D. *392.*
Melera, A. *360, 361, 413.*
Mellor, J. M. *415.*
Mengel, G. D. *203.*
Menssen, H. G. *473.*
Merenlender, Z. *360.*
Merkel, W. *56.*
Merlis, V. M. *28, 52, 54.*
Meule, J. L. *415.*
Meunier, H. *419, 477.*
Meyruey, M. H. *46.*
Mezzetti, T. *475.*
Michel, K.-H. *51, 55.*
Michl, M. *52.*
Middlebrook, R. E. *487.*
Middleton, E. J. *476, 477, 478, 485.*
Mieras, G. A. *134.*
Miller, C. O. *134.*
Miller, E. V. *287, 298.*
Miller, H. E. *39, 52.*
Miller, J. A. *392.*
Miller, J. D. *418, 481.*
Miller, L. L. *6, 55.*
Millott, N. *131.*
Milne, G. M. *394.*
Minale, L. *52, 449, 476.*
Minchilli, M. *482.*
Mireles, A. *203.*
Mirrington, R. N. *392.*
Misiorny, A. *297.*
Mislow, K. *134.*
Misra, R. *415.*
Mittal, P. P. *272, 298, 299.*
Mitusaki, S. *58.*
Miura, Y. *299.*
Miyasaki, M. *481.*
Miyashita, S. *304, 305.*

MIZUSAKI, S. 50.
MOFFAT, J. 50.
MOHACSI, E. 391.
MOHR, H. 134.
MOHR, M. 473.
MOISSEJEWA, E. N. 302.
MOLE, T. 205.
MÖLLER, H. 138.
MOLLOV, N. M. 32, 52.
MONEY, T. 388, 391.
MONGKOLSUK, S. 292.
MONSEUR, X. 54.
MONTGOMERY, J. A. 207.
MOORE, C. W. 361.
MOORE, R. E. 465, 474, 475, 481, 482, 483, 485.
MOORE, T. C. 413.
MORELL, D. B. 89, 90, 134.
MORGAN, K. 300.
MORI, N. 49, 53.
MORIKUBO, Y. 206.
MORIMOTO, H. 482.
MORIMOTO, K. 306.
MORIN, R. B. 458, 482.
MORISAKI, N. 479.
MORITA, Y. 306.
MORRIS, C. J. 58.
MORRISON, A. 414.
MORRISON, R. I. 52.
MORTIMER, P. I. 12, 45, 52.
MORTON, J. 42.
MORTON, R. A. 482.
MOSBACH, K. 273, 276, 279, 285, 294, 299.
MOSCOWITZ, A. 130, 134, 136, 139.
MOSER, C. 57.
MOSHER, H. S. 475, 482.
MOSS, G. P. 361.
MOTHES, K. 48, 51, 52, 55, 56.
MOTL, O. 304.
MUELLER, G. P. 483.
MUHAMMAD, S. 362.
MUKHAMEDZHANOV, S. Z. 42, 53, 55.
MÜLLER, B. 42.
MÜLLER, E. 56, 455, 473.
MÜLLER, H. 55.
MÜLLER, R. 357, 361.
MULLER, Y. M. F. 14, 43, 44.
MÜLLER-ENOCH, D. 129.
MUMFORD, F. E. 134, 137.
MURAKAMI, T. 53, 231, 299.
MURAKAWI, S. 482.

MURPHY, J. W. 394.
MURPHY, P. 416.
MURRAY, I. G. 298.
MURRAY, R. D. H. 413.
MURTY, T. K. 299.
MUSAJO, L. 482.
MUSSO, H. 228, 298.
MUSTAFA, M. G. 134.
MYERS, P. L. 414.

NADAL, N. G. M. 482.
NADEAU, R. G. 394.
NADOR, K. 299.
NAEGELI, P. 53.
NAEGELI, R. 53.
NAGAI, M. 294.
NAGASAWA, M. 53.
NAKAGAVA, M. 287, 293.
NAKAGAWA, T. 479.
NAKAI, T. 294.
NAKAJIMA, O. 134.
NAKAJIMA, T. 486.
NAKAMORI, R. 482, 486.
NAKAMURA, S. 47.
NAKAMURA, T. 294.
NAKAMURA, Y. 299.
NAKANISHI, M. 303.
NAKANISHI, T. 299, 306.
NAKANO, T. 37, 53.
NAKARAGAN, K. 47.
NANO, G. M. 357.
NATORI, S. 297.
NAWA, H. 486.
NEELAKANTAN, S. 231, 248, 251, 258, 291, 297, 298, 299.
NEISH, A. C. 281, 298.
NEMEC, P. 57.
NEPARSTEK, A. N. 129.
NEUFELD, G. J. 134.
NEUMANN, H. 48.
NG, C. Y. 54.
NICHOL, A. W. 89, 90, 134.
NICHOLAS, A. F. 479.
NICHOLAS, D. J. 415.
NICHOLAS, H. J. 416.
NICHOLS, D. 482.
NICHOLSON, D. C. 130, 132, 133, 134, 136.
NICOLAUS, R. A. 52, 132, 135.
NIEMANN, G. 131.
NIGAM, S. N. 53.
NIGRELLI, R. F. 418, 475, 477, 482.

NILSSON, M. *293*.
NISHIKAWA, Y. *299, 300, 303, 305*.
NISMIMAKI, T. *294*.
NISHIO, S. *294*.
NISHIYAMA, A. *479*.
NITSCHE, H. *480*.
NOLLER, C. R. *358, 360, 361*.
NOMURA, T. *138*.
NORIN, T. *52, 414, 415*.
NORMAN, A. *474, 482*.
NORMAN, G. D. *130, 476*.
NORMAN, M. *417*.
NORMATOV, M. *34, 53*.
NORRIS, K. H. *129*.
NORRMAN, B. *219, 300*.
NORTON, K. *414*.
NORTON, K. B. *357, 358, 359, 360, 362*.
NOTARI, R. E. *146, 168, 179, 189, 203, 205*.
NÜESCH, J. *53*.
NUGTEREN, D. H. *483*.
NUNO, M. *251, 300*.
NURIDDINOV, R. N. *53*.

O'CARRA, P. *120, 133, 135, 137, 483*.
O'CONNER, M. *391*.
O'DONOVAN, D. G. *13, 17, 20, 50, 53*.
OEHLSCHLAGER, A. C. *488*.
OESTERLIN, R. *483, 488*.
OGAN, A. U. *43*.
OGATA, R. T. *485*.
OGIHARA, Y. *297, 300, 301, 303*.
OHASHI, T. *42*.
O'HEOCHA, C. *61, 120, 130, 135, 137, 483*.
OHLOFF, G. *483*.
OHNSORGE, U. *294*.
OKAICHI, T. *452, 478, 483*.
OKUDA, T. *150, 153, 154, 156, 161, 164, 170, 188, 193, 203, 205, 206, 208*.
OKURA, T. *59*.
OKOGUN, J. I. *413*.
OLESEN, P. E. *483*.
OLIN, S. S. *391*.
OLLIS, W. D. *238, 289, 290*.
OLSON, G. L. *392*.
OLSON, J. A. *483*.
OLSON, J. O. *51*.
ONODERA, R. *53*.
OPPENHEIMER, J. R. *128*.
ORAZI, O. O. *53*.
ORECHOFF, A. *28, 53*.
ORR, J. C. *289*.
ORTH, H. *131*.

OSATO, T. *206*.
OSHIMA, K. *304*.
OSHIMA, S. *174, 203*.
OSIECKI, J. *203*.
OSONO, T. *206*.
OSTER, M. O. *415, 416*.
OSTEUX, R. *44*.
OSUGI, K. *486*.
OTROSHCHENKO, O. S. *59*.
OURISSON, G. *357, 361, 415*.
OVERTON, K. H. *42, 288, 296, 391, 393, 413, 415*.
OWEN, L. N. *393*.
OWYANG, R. *392*.
OZAWA, M. *53*.

PACE, W. T. *474*.
PACHL, H. R. *203*.
PACHLER, K. *361*.
PADMANSANI, R. *299*.
PAECH, K. *295, 302*.
PAILER, M. *4, 53*.
PALFI, G. *53*.
PALMER, C. *206*.
PANDEY, R. C. *415*.
PARIS, R. *46*.
PARISH, D. H. *53*.
PARK, R. J. *481*.
PARKE, T. V. *204*.
PARMENTIER, G. *58*.
PASINI, C. *454, 483*.
PASSANA-VUILLAUME, M. *135*.
PATON, A. C. *189, 204*.
PATTABHIRAMAN, T. *485*.
PAUL, R. *179, 180, 206*.
PAULMANN, L. *135*.
PAUWELS, P. S. *302*.
PEARSON, L. C. *285, 294*.
PEAT, S. *300, 467, 483*.
PECCI, J. *131, 135*.
PELLETIER, S. W. *415*.
PENROSE, W. R. *58*.
PENTILLA, A. *279, 300*.
PERCIVAL, E. *467, 483*.
PÉRGUY, M. *483*.
PERKS, A. M. *483*.
PERLIN, A. S. *300*.
PERLMAN, D. *287, 300*.
PERLSTEIN, J. *475*.
PERSSON, B. *253, 300*.
PETERS, R. *287, 293*.
PETRYKA, Z. J. *132, 134, 136, 139*.

PETTIT, G. R. *174*, *206*.
PETTUS, J. A. *482*, *483*.
PFISTER, R. M. *475*.
PHILLIPS, C. F. *392*.
PHILLIPS, D. D. *187*, *206*.
PHILLIPS, D. M. *53*.
PHINNEY, B. O. *414*, *415*.
PIATAK, D. M. *174*, *206*.
PIATELLI, M. *52*, *300*.
PICKENHAGEN, W. *483*.
PIERSON, W. G. *54*, *55*.
PIKE, D. G. *300*.
PINDER, A. R. *392*.
PINHEY, J. T. *36*, *53*.
PLAT, M. *48*.
PLIENINGER, H. *131*, *136*.
POELT, J. *239*, *300*.
POKORNY, M. *274*, *300*.
POLAN, C. E. *54*.
POLLARD, J. K. *48*.
POLLOCK, J. R. A. *48*.
POPA, D. P. *415*.
PORA, E. A. *361*.
PORTER, L. J. *467*, *481*.
POT, J. *484*.
POWER, F. B. *361*.
PRASAD, K. B. *54*.
PRATT, R. *481*.
PREMUZIC, B. B. *417*.
PREMUZIC, E. *417*.
PRELOG, V. *129*, 180, *204*.
PRIDHAM, J. B. *45*, *47*.
PROSKURNINA, N. F. *28*, *52*, *53*, *54*.
PRUD'HOMME, J. *179*, *206*.
PRYCE, R. J. *415*.
PUEYO, G. *300*.
PUZA, M. *50*, *58*, *202*, *204*, *206*, *208*.

QUINKERT, G. *289*.

RADLICK, P. *488*.
RAJAGOPALAN, T. R. *300*.
RALL, G. J. H. *23*, *54*, *57*.
RAMAKRISHNAN, S. *300*, *301*.
RAMANANTHAN, S. *297*.
RAMASARMA, T. *483*.
RAMAUT, J. L. *300*.
RAMSAY, M. V. J. *256*, *294*.
RAMSTAD, E. *358*.
RANGASWAMI, S. *299*.
RAO, K. V. *174*, *175*, *176*, 180, *206*.
RAPOPORT, H. *54*, *452*, *471*, *483*, *485*, *488*.

RASHID, M. H. *58*.
RATHJEN, C. *289*.
RATLE, G. *54*.
RAUSCH, R. *56*.
READ, G. *301*.
REDCLIFFE, A. H. *46*.
REDEMAN, C. T. *415*.
REES, D. A. *469*, *483*.
REHM, S. *358*, *361*.
REICHEL, L. *361*.
REID, K. B. M. *469*, *484*.
REID, W. W. *415*.
REINBOTHE, H. *5*, *50*.
REINHOLD, L. *295*.
REININGER, W. *233*, *304*.
REMIZIYE, S. H. *54*.
RENFREW, A. J. *413*, *415*.
RESTIVO, R. J. *332*.
RIBAS, M. *54*.
RIBAS-MARQUES, I. *28*, *29*, *46*, *54*.
RICE, W. Y. *23*, *45*, *54*.
RICH, D. H. *392*.
RICHARDS, J. H. *368*, *393*.
RICHARDSON, D. H. S. *301*.
RICKARDS, R. W. *202*, *208*, *413*.
RIEDL, W. *301*.
RIGGS, A. F. *134*.
RIGGS, N. V. *294*.
RILEY, J. P. *476*.
RIMINGTON, C. *361*, *481*.
RIO, G. J. *484*.
RIPPERGER, H. *54*.
RITCHIE, E. *53*, *54*.
RIVETT, D. E. A. *358*, *361*.
ROBERTS, J. C. *301*.
ROBERTS, L. B. *130*.
ROBERTS, M. F. *11*, *12*, *45*, *54*.
ROBERTSON, A. *289*, *292*.
ROBERTSON, A. L. *54*.
ROBINSON, B. *45*.
ROBINSON, B. P. *208*.
ROBINSON, D. R. *415*, *416*.
ROBINSON, R. *10*, *18*, *55*.
ROBISON, M. M. *30*, *54*, *55*.
ROBUSTELLI, F. *203*.
ROCHE, J. *446*, *477*.
RODIG, O. *361*.
RODRIGUEZ, D. D. *47*.
RODRIGUEZ, L. V. *482*.
ROESKE, R. W. *482*.
ROESLER, H. *52*.
ROKOHL, R. *55*.

ROLLER, P. *484*.
RONDON, Y. *287*, *301*.
ROSEGHINI, M. *476*.
ROSENFELD, J. M. *357*.
ROSS, D. M. *481*.
ROSS, F. *278*, *302*.
ROSS, M. E. *138*.
ROSSI, C. *475*.
ROTHER, A. 50, 55, 57.
ROTHSTEIN, M. 6, 55.
ROUX, D. G. 55.
ROVIJEN, A. v. 133.
RUDENKO, B. A. *393*.
RÜDIGER, H. *136*.
RÜDIGER, R. *136*.
RÜDIGER, W. 60, 73, 76, *136*, *137*, *128*, *484*.
RUDLER, M. C. *394*.
RUGGIERI, G. D. *482*, *485*.
RUHLAND, W. *302*.
RUNGE, F. *271*, *301*.
RUNECKLES, V. C. *52*.
RUPPERT, J. *136*.
RUSSELL, F. E. *474*, *484*.
RUTLEDGE, P. S. *413*.
RUZICKA, L. *391*, *396*, *415*.
RYAN, J. J. *300*.
RYDER, A. *203*.
RYHAGE, R. *291*.

SAAYMAN, H. M. 55.
SACH, G. S. *133*.
SADYKOV, A. S. *42*, *53*, 55, *59*.
SAITO, A. *479*.
SAITO, K. *484*.
SAKAI, K. *486*.
SAKSENA, A. K. *49*.
SALAQUE, A. *484*.
SAMEK, Z. *304*.
SAMESHIMA, M. *484*.
SAMUELSSON, B. *483*.
SANDBERG, F. 3, 14, 25, *52*, 55.
SANDERS, G. M. *484*.
SANKARA SUBRAMANIAN, S. 217, *287*, *299*, *300*, *301*.
SANKAWA, U. *293*, *297*, *301*, *303*, *304*.
SANNO, Y. *486*.
SANTESSON, J. 210, 211, 213, 215, 224, 225, 227, 234, 235, 237, 238, 239, 253, 260, *289*, *290*, *296*, *300*, *301*, *302*.
SARANTAKIS, D. *474*.
SARGENT, M. V. *229*, *302*.

SARMA, K. G. *288*, *302*.
SASAKI, T. *479*.
SATO, N. 53.
SATO, Y. 44.
SAUNDERS, P. R. *474*, *476*, *484*.
SAVAGE, G. M. 207.
SAVICZ, V. P. *302*.
SBOROO, V. E. *139*.
SCHAAF, T. K. *392*.
SCHACHTER, D. *137*.
SCHACHTER, M. *481*.
SCHADE, A. 220, *302*.
SCHAEFERS, T. *298*.
SCHAEFFER, H. J. 161, 162, 170, 190, 206, 207.
SCHANTZ, E. J. *452*, *481*.
SCHARTAU, O. *478*.
SCHATZBERG-PORATH, G. *358*.
SCHAUB, R. E. *43*.
SCHECHTER, I. *416*.
SCHEHRER, F. K. *362*.
SCHENK, W. 5, 8, 55.
SCHENKENBERGER, E. 56.
SCHEUER, P. J. *419*, *451*, *453*, *454*, *460*, *465*, *466*, *474*, *475*, *477*, *481*, *482*, *484*, *485*.
SCHIEDT, U. 11, 55.
SCHINDLER, H. *359*.
SCHINZ, H. *393*.
SCHLEGEL, W. *361*.
SCHLUNEGGER, E. 25, 55, 57.
SCHMEER, M. R. *484*.
SCHMID, R. *129*, *137*, *138*, *295*.
SCHMITZ, F. J. *485*, *487*.
SCHNEIDER, M. J. *129*.
SCHNEIDER, W. G. 205.
SCHNYDER, D. 57.
SCHNYDER, F. 57.
SCHÖPF, C. 14, 16, 19, 20, 28, 29, 55, 56, *278*, *302*.
SCHRAM, B. L. *137*.
SCHREIBER, K. *49*, *54*, 56, *58*, *296*.
SCHUETT, W. *483*, *485*.
SCHUMANN, D. 44.
SCHÜTTE, H. R. 5, 7, *48*, *51*, *52*, 55, 57.
SCHWARTING, A. E. 12, 16, *46*, 50, 57.
SCHWARTZ, H. M. *361*.
SCHWARTZ, M. A. *394*.
SCHWARTZ, S. 94, *137*.
SCHWEET, R. S. 7, 57.
SCHWIMMER, D. *469*, *485*.
SCHWIMMER, M. *469*, *485*.

SCOTT, A. I. 259, 292, 300, 416, 417.
SCOTT, J. A. 476.
SEELIG, G. 7, 56.
SEITZ, W. 220, 303.
SEKITA, T. 57.
SEMENOVSKII, A. V. 392, 393.
SEMMELHACK, M. F. 392.
SENECA, L'ANNAEUS 393.
SENEVIRATNE, A. S. 57.
SEO, S. 301, 303.
SESHADRI,T. R. 231, 237, 250, 272, 273, 288, 290, 291, 294, 297, 298, 299, 300, 302.
SESTAK, Z. 485.
SETTERFIELD, G. 416.
SHAH, G. D. 294.
SHAH, R. C. 294.
SHAMMA, M. 45, 46.
SHARMA, G. M. 447, 485.
SHARMA, R. K. 302.
SHARPLESS, K. B. 393.
SHAW, P. D. 53, 57, 58.
SHAW, S. C. 54.
SHEALY, Y. F. 207.
SHIBATA, S. 210, 211, 219, 228, 245, 259, 276, 285, 289, 293, 294, 297, 299, 300, 301, 302, 303, 304, 305, 306.
SHIMIZU, H. 303.
SHIMIZU, Z. 482.
SHIMOTSUMA, M. 361.
SHOHAT, B. 358, 359, 362.
SHOJI, J. 303.
SHOOLERY, J. N. 46, 361.
SHOPPEE, C. W. 177, 207.
SHORLAND, F. B. 469, 485.
SHROPSHIRE, W. JR. 130.
SHVO, Y. 360, 362.
SICHER, J. 10, 23, 57, 58, 202.
SIDDALL, I. B. 425, 479, 485.
SIDDIQUI, I. R. 362.
SIDDIQUI, R. H. 362.
SIDISUNTHORN, P. 289.
SIEBURTH, J. McN. 485.
SIEDEL, W. 62, 137, 138.
SIEGEL, M. R. 57, 141, 204.
SIEGELMAN, H. W. 66, 119, 120, 124, 129, 130, 134, 138, 475, 476.
SIFFORD, D. A. 476, 485.
SIGEL, C. W. 332.
SILVA, F. J. 481.
SILVANDER, B.-G. 298.
SIM, G. W. 416.
SIMS, J. J. 470, 488.
SIMIDU, W. 485.
SIMON, H. 52.
SIMONIS, W. 293, 303.
SIMONSEN, J. L. 393.
SINGH, H. 474, 481, 482, 485.
SINGH, I. 485.
ŠIPOŠ, F. 202.
SISLER, H. D. 57, 141, 207.
SJÖVALL, J. 474, 485.
SKERETT, R. J. 483.
SKEWES, G. 134.
SKIRROW, G. 476.
SKLARZ, B. 359.
SKRAUP, S. 473.
SMALBERGER, T. M. 54, 57.
SMIDT, J. 43.
SMIT, V. A. 392, 393.
SMITH, C. F. 207.
SMITH, C. G. 207.
SMITH, D. C. 274, 292, 293, 301, 303.
SMITH, D. O'N. 302.
SMITH, G. F. 33, 44, 45, 46.
SMITH, G. N. 45, 46.
SMITH, H. 413.
SMITH, H. H. 57.
SMITH, J. 460, 474, 485.
SMITH, K. M. 133.
SMITH, R. A. J. 291, 414.
SMITH, R. M. 359.
SMITH, U. 130.
SMITH, W. G. 54.
SMITHIES, W. R. 52.
SMOGROVICOVA, H. 57.
SMOLIKOVÁ, J. 304.
SNATZKE, G. 58, 296, 362.
SNEEN, R. A. 157, 202.
SOBOTKA, H. 425, 475, 477, 482.
SODANO, G. 476.
SÖDERBERG, U. 287, 303.
SOKOLSKI, W. T. 207.
SOLBERG, Y. J. 217, 223, 253, 303, 304.
SOLT, M. L. 57.
SONDHEIMER, E. 203.
SONN, A. 227, 304.
SÖRENSEN, N. A. 291, 475.
ŠORM, F. 46.
SOVIAR, K. 237, 304.
SPENCER, C. F. 50.
SPENSER, I. D. 7, 13, 16, 30, 48, 50, 58.
SPITELLER, G. 23, 57, 480.
SPITELLER-FRIEDMANN, M. 23, 57.
SPIZEK, J. 50, 206.

SPIŽEK, T. *204*.
SPRAGGINS, R. L. *487*.
SRISUKH, S. *460*, *477*.
STADLER, P. *391*, *393*.
STAF, B. *55*.
STAHL, E. *138*.
STANDAERT, F. G. *477*.
STAPEL, G. *362*.
STAPLEFORD, K. S. J. *44*, *46*.
STARKOVSKY, N. A. 169, 180, 188, 189, 191, *204*, *207*.
STASKUN, B. *392*.
STEERS, E. *130*.
STEGLICH, W. 213, 233, 237, *304*.
STEINEGGER, E. 25, 27, 28, *43*, *49*, *55*, *57*.
STEINER, A. B. *469*, *486*.
STEINER, M. *294*.
STEINERT, M. *304*.
STEINFELDER, K. *295*.
STEINSTRÄSSER, R. *136*.
STEMPIEN, M. F. JR. *482*, *484*.
STENSIÖ, K.-E. *304*.
STEPANOVA, N. L. *3*, *59*.
STERN, A. *131*.
STERNHELL, S. *207*.
STEWARD, F. C. *48*, *59*.
STEWART, J. C. *414*.
STEYN, D. G. *361*, *362*.
STIMAC, N. *10*, *43*.
STODDART, J. L. *414*.
STOLL, M. S. *138*.
STORK, G. 366, 368, 369, 384, 386, *243*, *304*, *393*.
STORNI, A. *394*.
STOTT, J. A. *476*.
STOTZ, E. H. *474*, *478*, *482*.
STRAIN, H. H. *473*.
STRAIN, M. H. *473*.
STREETER, M. P. *57*.
STROMBERG, V. L. *58*.
STRUBLE, D. R. *357*.
STRUCK, R. F. 196, *207*.
STRUHAL, H. 287, *290*.
STURZENHOFECKER, F. *391*.
SUFFNESS, M. I. *394*.
SUGAWA, T. *486*.
SUGAWARA, R. 184, 186, *207*.
SULTANBAWA, U. S. *392*.
SUMI, A. *47*.
SUMIMOTO, M. *414*, *416*.
SUNDHOLM, G. 237, 238, *302*.

SUNDSTRÖM, G. 228, *289*.
SUTHERLAND, M. D. *481*, *486*.
SUTHERLAND, S. A. *416*.
SUWAL, P. K. *46*.
SUYAMA, Y. *49*, *53*.
SUZUKI, M. 170, 172, 188, 193, 196, *203*, 205, 206, 207, *208*, *479*.
SUZUKI, S. *300*.
SUZUKI, T. *392*, *393*, *479*.
SVEC, W. A. *473*.
SWAIN, T. *45*, *47*.
SWAMY, M. N. *301*.
SWOBODA, J. J. *392*.
SZEGI, J. *299*.
SZINAI, S. *360*.
SZKOLNIK, M. *203*.

TACHIKAWA, R. *486*.
TAGUCHI, H. 245, 279, *303*, *304*.
TAKAKA, Y. *416*.
TAKAGAWA, M. *294*.
TAKAGI, N. *486*.
TAKAHASHI, H. 206, *208*.
TAKAHASHI, K. 267, *303*, *304*, *305*.
TAKAHASHI, N. *415*.
TAKAHASHI, R. *303*, *305*.
TAKAHASHI, W. *484*.
TAKANI, M. *305*.
TAKASHITA, M. *208*.
TAKEDA, T. 299, *303*, *305*.
TAKEMOTO, T. 459, *482*, *486*.
TALAFANT, E. *138*.
TALLENT, W. H. *18*, *58*.
TAMAKI, E. *50*, *58*.
TAMURA, C. *486*.
TANABE, H. *481*.
TANAKA, K. *47*, *486*.
TANAKA, M. 300, *303*, *488*.
TANAKA, O. *303*, *305*.
TANAKA, Y. *303*, *479*.
TANEMURA, M. *392*, *393*.
TANIMURA, A. *484*.
TAPIA, A. *136*.
TARTTER, A. *139*.
TASCHNER, E. *362*.
TATSUOKA, S. *486*.
TAYLOR, A. O. *138*.
TAYLOR, D. A. *44*.
TAYLOR, D. R. *357*.
TAYLOR, W. C. *53*, *54*.
TAYLORSON, R. B. *129*.
TCHELITCHEFF, S. 177, 180, *206*.

TCHEN, T. T. 369, *393*.
TENENBAUM, J. *130*.
TENHUNEN, R. *138*.
TERASHIMA, T. 18, *58*.
THEAL, S. 146, *203*.
THEOBALD, D. W. *415*.
THOMAS, D. M. 444, *486*.
THOMPSON, J. F. *58*, *59*.
THOMPSON, J. P. *289*.
THOMPSON, P. E. *204*.
THOMSON, R. H. *288*, *305*, 460, 461, 465, 474, *485*, *486*.
TICHY, M. 10, 23, 57, *58*, *202*.
TIWARI, H. P. *58*.
TIXIER, R. *138*.
TJERNBERG-NELSON, M. 55.
TODA, M. *487*.
TÖKES, L. *471*, *486*.
TOKINCHI, S. *488*.
TOKUTAKE, N. *303*.
TOMASELLI, R. *285*, *305*.
TOMII, Y. *481*.
TOMISAKI, K. *294*.
TOMITA, H. *58*.
TOMITA, J. T. *474*.
TOMITA, K. *299*, *305*.
TOMKO, J. 49, *58*.
TOOTE, V. K. *129*.
TOWE, K. M. *130*.
TOWERS, G. N. H. 47, *287*, *289*, *298*.
TOYOTA, T. *294*.
TRACEY, M. V. *295*, *302*.
TREIBS, A. 67, *138*.
TRESSLER, D. K. *486*.
TROTET, G. *296*.
TROXLER, R. F. *134*.
TRŠKA, P. *296*.
TSCHESCHE, R. 19, 20, 30, 31, *58*, *362*.
TSUCHIYA, Y. *138*.
TSUDA, E. *306*.
TSUDA, K. 419, *479*, *486*.
TSUDA, Y. *264*, *305*.
TSUJI, T. *131*.
TSUJI, Y. *294*.
TSUKUDA, N. *486*.
TSUTSUMI, J. *484*.
TÜMMLER, R. *295*, *297*.
TUMUR, B. 55.
TUNMANN, P. *362*.
TURNER, B. C. *138*.
TURNER, W. B. *413*.

TURSCH, B. *137*, *477*, *480*, *483*, *486*.
TURVEY, J. R. *300*, 467, *483*.

UCHIBAYASHI, M. *486*.
UEDA, Y. *304*, *305*.
UEMURA, D. *487*.
UENO, Y. *486*.
UEYANAGI, J. *486*.
ULERY, H. E. 368, *393*.
ULLRICH, H. *296*.
UMEZAWA, S. 174, 194, 195, *203*, *206*, *208*.
UNGER, R. 14, 56.
UNRAU, A. M. *469*, *488*.
UPPER, C. A. *416*.
URASAKI, M. *294*.
UTKIN, L. M. 44.

VÄHÄTALO, M. L. *58*.
VALENTA, Z. 45, *58*.
VAN DER GAN, A. *392*.
VANDERHAEGHE, H. *58*.
VAN DER HELM, T. *479*, *478*.
VAN DER MERWE, K. J. *361*.
VAN DORP, D. A. *484*.
VANEK, Z. 50, *58*, *202*, *204*, *206*, *208*.
VAN LEEUWEN, M. 43.
VAN TAMELEN, E. E. 184, *208*, *223*, *305*, *366*, *371*, *374*, *375*, *393*, *394*.
VAN VEEN, A. 43, 44.
VARTIA, K. C. *287*, *305*.
VASSOVA, A. *58*.
VAUGHN, J. R. *208*.
VAZQUEZ, D. *58*.
VEAL, P. L. *290*.
VEČERECK, B. *133*.
VEGA, J. 28, 46, 54.
VELAIRE, C. D. *208*.
VENKATARAMAN, K. *297*.
VENKATASUBRAMANIAN, G. B. 250, *302*.
VERBISCAR, A. J. *414*.
VERCELLONE, A. *483*.
VERNENGO, M. J. 23, 52.
VESSEY, I. *481*.
VIDAL, A. 29, 54.
VIEL, C. 46.
VIG, B. *485*.
VILLAGRÁN, V. *293*.
VINING, L. C. *301*.
VIRTANEN, A. I. 5, *58*.
VIRTANEN, O. E. *297*, *305*.
VISSER, B. J. *59*.
VISWANATHAN, N. 47.

VITALI, T. *476.*
VOLKWEIN, G. *430, 431, 470, 477, 488.*
VONDRÁČEK, M. *58, 202, 206, 208.*
VON PHILLIPSBORN, W. *44.*
VUILLAUME, M. *137, 138.*

WACHTMEISTER, C. A. *243, 258, 288, 289, 290, 293, 297, 298, 304, 305.*
WADA, S. *186, 204.*
WAGNER, H. *49, 223, 305.*
WALBE, R. *56.*
WALDVOGEL, G. *477.*
WALKER, G. C. *45.*
WALL, R. A. *134.*
WALLENFELS, K. *465, 480, 487.*
WALLENFELS, U. *478.*
WALLER, G. R. *43.*
WALTER, W. G. *57.*
WARD, V. L. *202.*
WARNHOFF, E. W. *362.*
WARNOCK, W. D. C. *42.*
WARWICK, A. J. *50.*
WASHECHECK, P. H. *437, 487.*
WATANABE, K. *59.*
WATANABE, M. *294.*
WATSON, C. J. *94, 133, 134, 136, 137, 139.*
WATSON, D. G. *480.*
WAX, J. *475.*
WEBER, P. *57.*
WEBER, W. A. *296.*
WEBSTER, D. E. *54.*
WEEDON, B. C. L. *487.*
WEIMER, M. *133, 134, 136, 139.*
WEINHEIMER, A. J. *436, 437, 439, 476, 477, 485, 487.*
WEINSTEIN, B. *474, 482.*
WEISMAN, P. *479.*
WEYGAND, F. *52.*
WELCH, J. J. *208.*
WELLS, H. D. *208.*
WELLS, J. W. *288, 486.*
WELSH, H. J. *487.*
WENKERT, E. *59, 402, 406, 416.*
WESSELS, J. H. *361.*
WEST, C. A. *414, 415, 416.*
WHALLEY, W. B. *292, 293, 413, 416.*
WHEELER, R. E. *289.*
WHELAN, W. J. *300.*
WHIFFEN, A. J. *146, 205, 208.*
WHITCOMB, E. R. *477.*
WHITE, A. F. *414, 415.*

WHITEHEAD, A. *204.*
WHITEHOUSE, M. W. *305.*
WICKBERG, B. *298.*
WICKBERG, G. *298.*
WICKY, K. *57.*
WIECHERT, R. *478, 480.*
WIEDHAUP, K. *392.*
WIEHLER, G. *4, 59.*
WIELAND, H. *139.*
WIENECKE, H. *362.*
WIETERS, E. *56.*
WIGFIELD, D. C. *362.*
WIGHTMAN, F. *416.*
WIGHTMAN, R. H. *58.*
WILCOX, M. E. *52.*
WILDMAN, W. C. *49, 53.*
WILKINS, M. B. *134, 138.*
WILKINSON, S. *52.*
WILLETT, J. D. *394.*
WILLIAMS, D. H. *159, 202.*
WILLIAMS, J. H. *42, 43.*
WILLIAMSON, W. R. N. *189, 208.*
WILLIG, A. *139.*
WILLNER, D. *360.*
WILLSTAEDT, H. *139.*
WINDAUS, A, *487.*
WILSON, E. M. *49, 59.*
WILSON, J. L. *211, 305.*
WILSON, J. M. *132.*
WILSON, J. S. *45, 58.*
WILSON, S. *487.*
WILSON, W. B. *481.*
WINTER, J. *413.*
WISSE, J. H. *59.*
WITH, T. K. *139.*
WITKOP, B. *5, 59.*
WOHLPART, A. *52.*
WOLF, G. *362.*
WOLF, M. A. *476.*
WOLINSKY, J. *155, 208.*
WOLSTENHOLME, G. E. W. *391.*
WOLTERS, B. *487.*
WONG, C. F. *33, 59.*
WONG, J. L. *488.*
WOO, P. W. K. *184, 208.*
WOOD, H. C. S. *384, 392.*
WOODWARD, R. B. *487.*
WORTHEN, L. *477.*
WROBEL, J. T. *50, 59.*
WUNDERLICH, J. A. *478.*
WÜST, W. *56.*
WYLER, H. *39, 52, 59.*

YAMADA, K. *487*.
YAMADA, Y. *59*.
YAMAGUCHI, K. *133, 139*.
YAMAGUCHI, M. *487, 488*.
YAMAKI, M. *306*.
YAMAMOTO, K. *480*.
YAMAMURA, S. *446, 479, 488*.
YAMANOUCHI, T. *481, 488*.
YAMAUCHI, H. *299, 305, 306*.
YAMAZAKI, S. *206*.
YANAGITA, M. *38, 59*.
YANAI, H. S. *10, 59*.
YANG, D.-M. *303*.
YANG, N. P. *49*.
YANO, K. *206*.
YASHIDO, T. *479*.
YASSI, J. *54*.
YASUDA, S. *42, 294*.
YASUMOTO, T. *488*.
YASUNARI, Y. *479*.
YAZAWA, H. *487*.
YEN, C. C. *174, 206*.
YONEMITSU, M. *42*.
YOSHIDA, B. T. *484*.
YOSHIDA, M. *488*.
YOSHIDA, T. *474*.

YOSHIKAWA, T. *294*.
YOSIOKA, I. *235, 254, 261, 262, 264, 306*.
YOUATT, G. *474*.
YOUNG, D. W. *416*.
YOUNG, E. G. *488*.
YOUNG, H. *266, 291*.
YOUNG, V. K. *131*.
YOUNGBLOOD, W. W. *487*.
YOUNGKEN, H. W. *475*.
YOUNGSON, A. *484*.
YUNUSOV, S. Y. *34, 49, 53*.
YUNUSOV, T. K. *3, 59*.
YURASHEWSKI, N. K. *3, 59*.

ZACHARIUS, R. M. *59*.
ZAGAESKY, P. F. *475*.
ZAHORSZKY, U. I. *289*.
ZANDER, J. M. *362*.
ZECHMEISTER, L. *483*.
ZELENSKI, S. *477*.
ZELLER, P. *480*.
ZELLNER, I. *362*.
ZHUKOV, L. *416*.
ZOPF, W. *224, 225, 229, 237, 248, 253, 306*.
ZWEISTRA, A. *43, 44*.

Sachverzeichnis. Subject Index

Von · By

A. Siegel, Wien

Aal *(Anguilla japonica)* 105, 106.
Abietadiene diterpenes 399.
Abietic acid series 399.
—, formation 406.
Acacia-Arten 4, 5, 8.
Acacia homalophylla 8.
Acanthaster planci 462.
Acanthias vulgaris 456.
Acanthosicyos horrida 352.
Acaranosäure 224.
—, Synthese 225.
Acarospora chlorophana 224.
Acetaldehyde 182, 184.
Acetals in terpenoid biosynthesis 378.
Acetanhydrid 91, 164, 231, 339, 374, 385, 455, 462.
Acetat 36, 42.
1-^{14}C-Acetat 11, 18, 275, 276, 405.
2-^{14}C-Acetat 276.
Acetat-Malonat-Stoffwechsel 40.
Acetic acid 28, 87, 88, 99, 103, 170, 235, 238, 260, 319, 321, 325, 345, 383, 423.
[1-^{14}C] Acetic acid 201, 277, 278, 408.
Acetic anhydride s. Acetanhydrid.
Acetoacetat 12.
Acetoacetyl-CoA 13.
Acetoin 312, 313.
Aceton 71, 88, 186, 194, 195, 202, 260, 323, 339, 345, 418, 453, 456.
Acetondicarbonsäure 12, 17.
Acetone s. Aceton.
Acetophenone 195.
2-Acetoxycyclohexanone 196.
12-α-Acetoxy-7,8-dihydro-24,25-dehydroholothurinogenin-3-acetate 427.
6α-Acetoxy-7α,22-dihydroxyhopan 267.
(+)-α-Acetoxyheptanoic acid 439.
7β-Acetoxy-22-hydroxyhopan 266.
15α-Acetoxy-22-hydroxyhopan 267.

28-Acetoxy-22-hydroxyhopan-23-säure 267.
Acetyl coenzyme A 200, 281, 284.
Acetyldihydro-pbrombenzoyl-portentol 215, 225.
Acetylendicarbonsäure dimethylester 242.
Acetylated naphthazarin 465.
N-Acetylhystrin 27.
6α-Acetylleucotylin 264.
16β-Acetylleucotylsäure 264.
N_δ-Acetyl-lysin 16.
Acetylmesitolmethyläther 225.
2-Acetyl-5-nitrofuran 195.
Acetylportentol 225.
—, Biosynthese 276.
O-Acetyl-streptimidone 184.
N-Acetyl-Δ^2-tetrahydro-anabasin → Ammodendrin 27, 28.
Achillea atrata 4.
Achillea moschata 4.
Acroscyphus sphaerophoroides 231, 259.
Actias atremis 110.
Actias selene 110.
Actidione → Cycloheximide 140, 142, 146.
Actinidia polygama 36.
Actinidin 36.
Actinomycetes ETH 7796 144, 180.
Actinopyga agassizi 425, 430.
Actiphenol C-73, 144, 145, 180f.
—, 2,4-dinitrophenylhydrazone 181.
—, IR-spectrum 181.
—, isolation 180.
—, methylation 181.
—, monoacetate 181.
—, physicochemical data 181.
—, structure 180.
—, synthesis 180, 186, 193.
—, treatment with alkali 181.

Sachverzeichnis. Subject Index

Actiphenol C-73, UV-spectrum 181.
Actogenine 220 f.
Acylglucuronide 91.
Acyloin rearrangement 313, 324.
Adenocarcinoma 755 196.
Adenocarpin 28, 29.
Adenocarpus-Arten 28, 29.
Adenocarpus viscosus 28.
Adipedatol 266.
Adsorbosil I 87.
Aeroplysinin-I 449.
Aerothionin 449.
Agar 467.
Agarose 468.
Agathis australis 397.
Aglycones, isolated from *Cucurbitaceae* 309.
Alangiaceae 37.
Alangin A 37.
Alangium lamarckii 37.
Alanin 86, 121, 123.
Albizzia-Arten 4, 5.
Albizzia lophantha 5.
Aldit 218.
Aldol condensation 381.
Aldopentose 430.
Aldotripiperidein 14, 28.
Alectoria nigricans 253.
Alectoria virens 257.
Alectorialsäure 253.
—, methylester 254.
Algae 458, 459, 460, 467.
—, antibiotic activity of extracts 446.
—, Chromoproteide 61.
—, steroid distribution 418, 419, 420.
—, terpenoids 440 f.
Algen s. Algae.
Algin 467.
Alginates 469.
Alginic acid 467.
Alkaloide.
    Anabasin-Alkaloide 24.
    Areca-Alkaloide 6.
    Azima-Alkaloide 23 f.
    Caria-Alkaloide 23 f.
    Cassia-Alkaloide 227.
    Chinolizidin-Typ 2, 24, 25, 27, 28, 31, 32.
    Conium-Alkaloide 2, 9f, 14, 16, 42.
    Dipiperidyl-Alkaloide 27.
    Haloxylon-Alkaloide 14.
    Hydroxypiperidin-Alkaloide 10.

Alkaloide.
    Indol-Alkaloide 33.
    Lobelia-Alkaloide 14, 16, 31.
    Lupinen-Alkaloide 16.
    Lythraceen-Alkaloide 24.
    Monoterpenoide Piperidinalkaloide 34, 42.
    Nicotiana-Alkaloide 26.
    Nuphar-Alkaloide 32 f.
    Ormosia-Alkaloide 31.
    Ornithin-Alkaloide 17.
    Piper-Alkaloide 2, 8, 9.
    Piperidin-Alkaloide 17 f, 34, 36.
    Piperidon-Alkaloide 37.
    Prosopis-Alkaloide 10, 22, 23.
    Punica-Alkaloide 12, 14, 16.
    Pyridin-Alkaloide 2.
    Secodin-Alkaloide 33.
    Secophenanthrochinolizidin-Alkaloide 15.
    Sedum-Alkaloide 13, 16, 31.
    Sesquiterpen-Alkaloide 38.
    Tetrahydroanabasin-Alkaloide 24.
    Tropan-Typ 2.
    Urticaceen-Alkaloide 15.
    Withania-Alkaloide 13, 16 f.
Alkyl sulfonates in terpenoid biosynthesis 378.
Allo-Phycocyanin 121, 126.
(—)-Allosedamin 14, 15, 19.
2-O-Allyl-D-arabinit 218.
Allylbromid 218.
Allylic alcohols in terpenoid biosynthesis 378.
Allylic bromination 319.
Allylic oxidation 343.
—, with SeO_2 424.
erythro-2-Allyl-3-propyl-bernsteinsäure 223.
Alumina, acid-deactivated 150.
Aluminium amalgam reduction 324.
Aluminiumchlorid 237.
Aluminium isopropoxide 166.
Aluminiumoxid, desaktiviertes 86.
Amaranthin 39.
Amarin → Cucurbitacin B 356.
Amebicides 181.
Ameisensäure 18, 104, 247, 248, 253, 374, 378, 381, 383, 389, 448.
^{14}C-Ameisensäure 278.
Amine in Flechten 217, 284.
Aminoacetessigsäure 18.

α-Aminoadipinsäure, ^{15}N-markiert 7.
α-Aminoadipinsäure-δ-semialdehyd 13, 16.
D-(4-Amino-4-carboxybutyl)penicillin 457.
7-Aminocephalosporanic acid 458.
—, N-acetyl derivative 458.
Aminoimidazole 471.
ε-Amino-α-ketocapronsäure 6, 7, 8, 13, 16.
δ-Aminolävulinsäure 110.
D-β-Amino-β-phenylpropionsäure 217.
4-Aminopipecolinsäure 5, 8.
2-Aminopyridine 174.
Aminosäuren 284.
—, in Flechten 216f, 286.
Aminovaleraldehyd 18, 27.
(+)-Ammodendrin → Sphärodendrin 27, 28.
Ammodendron conollyi 28.
Ammonia 18, 87, 149, 174, 180, 181, 182, 374, 381.
—, liquid 329.
Ammoniak s. Ammonia.
Ammoniumsulfat 111, 112.
Amo 1618 412.
Amphibians, toxic principles 469.
Amyl acetate 145.
Amylase 273.
α-Amyrin 268.
Anabaena sp. 117.
Anabasamin 25.
Anabasin 13, 14, 25, 26, 27, 42.
—, Biosynthese 26.
Anabasin-Alkaloide 24.
Anabasis aphylla 25, 26, 32.
Anabasis salsa 3.
Anacystis nidulans 117, 118.
Anaferin 13, 16, 17.
Anahygrin 16.
Anämie, hämolytische 99.
Anaptychia fusca 217.
Anaptychia obscurata 229, 235, 260.
(−)-Anatabin 25.
—, Biosynthese 27.
Anatallin 25.
—, Biosynthese 27.
Ancepsenolide 435, 436.
Anchimerically assisted process 384.
Anemonia sulcata 456, 457.
Anemonine 456.
Angelas Orides 471.
Anguilla japonica → Aal 105, 106.
Anhydro-22-deoxocurcubitacin D 332, 340.

Anhydro-22-deoxo-3-epi-isocucurbitacin D 332, 342.
Anhydro-22-deoxoisocucurbitacin D 332, 342.
—, monoacetate 342.
3,6-Anhydro-D-galactose 468.
—, 2-sulfate 468.
3,6-Anhydro-L-galactose 468.
Anhydroisocycloheximide 169, 188, 189.
—, catalytic reduction 169, 170.
—, formation 167.
—, reduction 169.
Anhydroleucotylin 264.
Anhydroleucotylsäure-methylester 264.
Anhydronupharamin 32, 33.
Anilin 91.
Anissäure 271.
Annelida 421, 443.
Antedon sp. 463.
Anthelmintic properties 458, 459.
Antherea pernyi 110.
Anthrachinone, chlorhaltige 229.
Anthrachinone, hydroxylierte 212.
Anthrachinone in Flechten 229f, 234.
Anthranilsäure 18.
—, diazotierte 67.
Anthranilsäureester, diazotierter 91.
Anthraquinone moiety 465.
Anthrone in Flechten 229, 284.
Antibiotic activity.
—, of algae extracts 446.
—, of coelenterate extracts 435.
—, of seaweed extracts 446.
Antibiotic-producing marine organisms 457.
Antibiotics 139f, 395, 422, 437, 458.
—, from marine sources 447.
—, structure-acticity relationships 141, 147.
Antibiotika mit Piperidinring 40.
Antibiotische Wirkung von Flechtenstoffen 287.
Anticholinesterase activity 453.
Antifungal activity of glutarimide antibiotics 145, 146, 174.
—, mechanism 147.
Antigibberellin activity 412.
Antimalariawirkung 17.
Antimicrobial activity 458.
Antineoplastic compounds 174.
Antineoplastic properties 351.

Antitumor activity s. Antitumorwirksamkeit.
Antitumorwirksamkeit 174, 220, 287, 351.
Anziasäure 245.
—, dimethyläther 245.
—, 2-methyläther 259.
Aphyllinsäuremethylester 31, 32.
Apiocrinus sp. 466.
Aplysia californica 460.
Aplysia-Farbstoffe 108, 111 f.
Aplysia kurodai (sea hare) 446, 447.
Aplysia limacina (Kiemenschnecke) 111, 112, 459.
Aplysin 442, 446.
Aplysin 20 447.
Aplysina aerophoba 449.
Aplysinol 442, 446, 447.
Aplysioverdin 111, 112, 114.
Aplysioviolin 70, 459.
—, Analysendaten 112.
—, Chromatabbau 70.
—, Chromsäureabbau 113.
—, Cotton-Effekt 113.
—, Dünnschichtchromatographie 87.
—, Elektronenspektrum 113.
—, Massenspektrum 113.
—, Optische Aktivität 77, 113.
—, ORD-Spektrum 79, 113, 114.
—, Strukturermittlung 113.
Aplysioviolin-dimethylester 120.
Aplysiourobilin 114.
Apocynaceae 33, 34.
Apothecien 234.
—, Farbstoff 228.
Arabinose 218.
L-Arabo-2-carboxy-3-carboxymethyl-4-isopropenylpyrrolidine → α-kainic acid 458.
L-Arabo-2-carboxy-4-(1-methyl-5-carboxy-*trans* : *trans* : S-*trans*-1,3-hexadienal)-3-pyrrolidine-acetic acid → domoic acid 458, 459.
D-Arabit 274, 275, 217.
Arca noae 456.
Areca-Alkaloide 6.
Areca catechu 6.
Arecaidin 6.
Arecolin 6.
Arene sulfonates in terpenoid-biosynthesis 378.
Arginine 456.

9-Aristolene 436.
1(10)-Aristolene 436.
Armeria maritima 5.
Arthonia impolita 225, 247.
Arthoniasäure 247.
Arthotelin 238.
Arthothelium pacificum 238.
Arzneimittel aus Flechten 287.
Ascorbinsäure 273.
—, Photooxydation 116.
Asparagin 274.
Asparaginase 273.
Asparaginsäure 121, 123, 274.
Aspartocin 40.
Aspergillus nidulans 7.
Astasia klebsii 443.
Astaxanthin 443.
Asterias glacialis 455.
Asterias rubens 455.
Asterubin 455.
Astrocasia phyllanthoides 29.
Astrocasin 29.
Astrophyllin → Dihydroisoorensin 29.
Äthanol 86, 87, 88, 161.
Äthylacetat s. Essigsäureäthylester.
2-Äthyl-3-[2-(3-äthylpiperidin)äthyl]-indol 33.
Äthylendichlorid 88.
Äthylmethylketon 88.
(2 R : 8 S)-(+)-8-Äthyl-norlobelol 14.
(+)-2 S-Äthyltetradecansäure 222.
Atisine 402.
Atranol 254.
Atranorin 245, 251, 264, 286.
—, Biosynthese 278.
—, Halogenierung 252.
—, radioaktives 278.
Attacidae 110.
Avena sativa 125.
Averythrin-6-monomethyl-äther 234.
Azacyclopentadienon-Ring 94.
Azcarpin 24.
Azima tetracantha 24.
Azimin 24.
Aziminsäure 24.
Azopigmente aus Bilirubinglucuronaten 91, 92.

„Backbone" rearrangements 398, 399.
Baeomyces roseus 285.
Baeyer-Villiger type of oxidation 411.
Baikiaea plurijuga 5.

Baikiain 5, 8.
Bakterienproteine, Blockierung der Synthese 287.
Bangiales 115.
Barbatolsäuremethylester 254.
Barnacle 420.
Beckmann rearrangement 322.
Beetles, specific attraction 308.
Belone belone 106.
Benicasa hispida 352.
Benzaldehydes, p-substituted 196.
Benzene s. Benzol.
Benzidin 71.
—, bis-diazotiertes 211.
Benzilic acid rearrangement 314, 323, 324.
Benzoesäure 271, 280.
—, -phenylester 273.
Benzol 87, 88, 384.
Benzoylameisensäure 271, 280.
—, -methylester 271.
Benzoylessigsäure 20.
Benzoyl peroxide 390.
Benzylamine 149, 172, 173, 182, 186.
Benzyl chloride 374.
Benzylcyanid 271.
Benzyldimethyl-{2-[2-(p-1,1,3,3-tetramethylbutylphenoxy-)-äthoxy-]-äthyl}-ammoniumusneat → Usno 287.
Benzylester 284.
—, Derivate in Flechten 253f.
Benzyl-3,4-O-isopropyliden-β-arabinopyranosid 218.
Benzyl-2-O-allyl-β-D-arabinopyranosid 218.
1,4-^{14}C-Bernsteinsäure 275.
Betacyane 38.
Betalaine 38f.
Betalaminsäure 39.
Betanidin 39.
Betanin 38, 39.
Betaxanthine 38.
Beyerene 402, 408.
Bichromat-Schwefelsäure 242.
Bicyclo-3,2,1-octane systems 402.
Bicyclo-3,2,2-octane systems 402.
Biladiene s. Biladiene-(a, b) und Biladiene-(a, c).
—, R$_F$-Werte 88.
—, III α Struktur 102.
—, IX α Struktur 102.
—, XIII α Struktur 102.

Biladiene-(a, b) 62, 65, 66.
—, Asymmetriezentrum 76.
—, Dünnschichtchromatographie 87, 88.
—, Elektronenspektrum 73, 74.
—, Entstehung durch Oxydation von Bilanen 99.
—, Massenspektrum 82.
—, NMR-Spektrum 86.
Biladiene-(a, c) 62, 64, 65.
—, Bildung von Bilatrienen 104.
—, Dünnschichtchromatographie 87.
—, Elektronenspektren 74.
—, Kupplung mit Diazoniumsalzen 67.
—, Massenspektren 79, 80.
Biladiene, unsymmetrisch substituierte, Bildung von Azofarbstoffen 67.
Biladienone 64, 65, 126.
—, Dünnschichtchromatographie 87.
—, R$_F$-Werte 87, 88.
Bilan → Tetrapyrran 62, 65.
—, asymmetrische C-Atome 76.
—, Elektronenspektrum 73.
Bilane 76, 88, 94.
—, asymmetrische C-Atome 76.
—, aus Bilirubin 94.
—, Oxydationsprodukte 99.
—, R$_F$-Werte 88.
—, säurekatalysierte Spaltung 68.
Bilatriene 64, 65, 66, 104f, 126.
—, bei Invertebraten 108f.
—, bei Lepidopteren 108f.
—, bei Vertebraten 104f.
—, Dünnschichtchromatographie 87, 88, 104.
—, Elektronenspektren 73, 74.
—, Entstehung durch Oxydation von Bilanen 99.
—, Entstehungsmechanismus aus Biladienen-(a, c) 104.
—, Massenspektren 79.
—, NMR-Spektren 86, 104.
—, Optische Aktivität 76, 77.
—, R$_F$-Werte 88.
—, Strukturermittlung 104.
Bile acids 431f.
Bile alcohols 431f, 434.
Bile, ethanolic extracts 431.
Bilen-(b)-dion(a, c) 64.
Bilene 62, 65, 66.
—, asymmetrische C-Atome 76.
—, aus Bilirubin 94.
—, Dünnschichtchromatographie 87, 88.

Bilene, Elektronenspektren 73, 74.
—, Entstehung durch Oxydation von Bilanen 99.
—, FeCl$_3$-Reaktion 102.
—, Konfiguration 78.
—, Massenspektren 82, 84.
—, ORD-Spektren 77.
—, R$_F$-Werte 87, 88.
Bilenon 64.
Bile pigments 61, 459 s. Gallenfarbstoffe.
Bile salts 431 f.
Bilidien 62.
Bilien 62.
Biline 61.
—, Nomenklatur 62.
Bilinoide 62.
Biliproteide 60f, 116, 459.
s. Gallenfarbstoffe.
—, Chromsäureabbau 121.
—, Fluoreszenz 114.
—, in Algen 114.
—, in Pflanzenpigmenten 111.
—, photosynthetisch aktive 116.
Biliproteins s. Biliproteide.
Bilipurpurin 87.
Bilirubin → 1′,8′-Dioxo-1,3,6,7-tetramethyl-2,8-divinylbiladien-(a, c)-dipropionsäure-(4,5) 61, 64, 89f.
—, Bestimmung durch Diazoreaktion 66.
—, Dehydrierung zu Biliverdin 104.
—, Dünnschichtchromatographie 87.
—, Elektronenspektren 89, 93.
—, IR-Spektrum 89.
—, katalytische Hydrierung 96.
—, NMR-Spektrum 89, 90.
—, Optische Aktivität 93.
—, ORD-Spektrum 93.
—, Totalsynthese 89.
—, Umwandlungsprodukte 94f.
—, Vorkommen im Gallensaft von Wirbeltieren 104.
Bilirubin-IXα 64.
Bilirubindiglucuronid 91.
Bilirubindimethylester 89, 90.
Bilirubindimethylester-dimethyläther 89.
Bilirubin-glucuronide 94.
Bilirubin-Konjugate 67, 90f.
Bilirubin-monoglucuronid 91.
Bilirubin-Proteide 91 f.
Bilirubin-Serumalbumin-Komplex, Optische Drehung 78.

Bilirubinsulfat 91.
Bilitrien 62.
Biliverdin 104 f.
—, Abbau u. Aufbau bei Invertebraten 108.
—, Bildung durch Dehydrierung aus Bilirubin 104.
—, Dünnschichtchromatographie 87.
—, im Korallenpigment 108.
—, Vorkommen bei Vertebraten 104.
Biliverdin IXα 104, 106.
Biliverdin-Proteide 104 f.
—, bei Fischen 105.
—, bei Vertebraten 104, 106.
Binan 287.
Biogenetic isoprene rule 396.
Biologische Wirkung d. Flechtenstoffe 287.
Biotin 273.
(+)β-Bisabolene 438.
Bisanthrachinone in Flechten 259f, 284.
Bisanthronyle in Flechten 259, 284.
Bisbutenolides 436.
Bismuth oxide 311, 314, 341.
Bis-(3-oxoundecyl) tetrasulfide 470.
Bis-(3-oxoundecyl) trisulfide 470.
Bisxanthone in Flechten 261, 284.
Bitter calabash 339.
Bitter glycosides 337.
Bitter principles 395.
—, bicyclic 399.
Blaualgen → Cyanophyten 114, 115.
Blaue Koralle (Heliopora coerulea) 108.
Bleitetraacetat s. Lead tetraacetate.
Blue pigment 118.
Blue skate (Raia batis) 432.
Blutzuckerregulation 287.
Boehmeria cyclindrica 15.
Boehmeria platyphylla 15.
Bohadschia Koellikeri 428, 430.
Boron trichloride 448.
Boron trifluoride 366, 371, 373, 389, 448.
—, etherate 368, 371, 372, 388, 424.
Bortribromid 241.
Boxfish (Ostracion lentiginosus) 454.
Brandegea bigelovii 346, 352.
Brassicasterol 419, 421.
Briareum asbestinum 434.
Brom 273.
Bromacetaldehyd 241.
16β-p-Brombenzoyl-6-ketoleucotylin, Röntgenstrukturanalyse 215, 264.
6α-p-Brombenzoylzeorin, Röntgenstrukturanalyse 215, 262.

Sachverzeichnis. Subject Index

Bromcyan, Spaltung von Peptiden 117.
Bromkresolgrün 211.
4-Bromnephroarctin, Röntgenstrukturanalyse 251.
Bromacetic acid, ethyl ester 455.
Bromopyrrole derivative 471.
N-Bromosuccinimide 180, 231, 233, 242, 319, 371.
Bromwasserstoff 104.
Brucin für Antipodenspaltung 97, 98, 210, 244.
Brushite 125.
Bryodulcoside 343.
Bryodulcosigenin 343.
—, -3,24-diacetate 343.
—, occurrence in nature 353, 355.
—, physical constants 356.
Bryogenin 343, 345, 350.
—, occurrence in nature 353, 355.
—, physical constants 356.
Bryonia alba 352.
Bryonia dioica 337, 342, 343, 352.
Bryosigenin 343.
—, occurrence in nature 353, 355.
—, physical constants 356.
Buellia-Arten 238.
Butanol 145.
n-Butanol 454.
2-Buten-1,4-diol 228, 249.
Butenylcyclohexene 369.
t-Butyl alcohol 319, 336.
t-Butylhypochlorit 71.
n-Butyl-mercaptan 441.
2-t-Butyl-4-methylcyclohexanone 158, 159.
4-t-Butyl-2-methylcyclohexanone 158, 159.
C-73 → Actiphenol 180.
Cadalene 439.
Cadaverin 11, 13, 26, 27, 30, 31.
Cadinene 435, 436, 439.
(−)δ-Cadinol 439.
Cafestol 402.
Calamene 438.
Calcium, reduction with 329.
Calciumoxalat in Flechten 220.
Calciumphosphat, für Chromatographie 125.
Californian sea lion 433.
California newt *(Taricha torosa)* 452.
Calliphora test 422.
Caloplaca arenaria 229.

Caloplaca-Arten 231, 234.
Caloplaca bryochrysion 231.
Caloplaca cinnamomea 231.
Caloplaca leucoraea 231.
Caloplaca percrocata 229.
Caloplaca tetraspora 231.
Calothrix 274.
Calothrix membranacea 117.
Calycin 271, 272, 281, 285.
—, Biosynthese 281.
Campanula medium 37.
Campedin 37.
Campesterol 421.
α-Campholene aldehyde 373.
Camphor, synthesis 388.
Candelariella vitellina 285.
Capillary permeability 351.
Carausius morosus 108.
Carbohydrates in marine organisms 467f.
Carbon dioxide s. Kohlendioxid.
Carbonium ion catalyzed cyclization 390.
3-Carboxymethylglutarimide 187.
Carcharhinus menisorrah 453.
Cardiotonische Aktivität 287.
Caria-Alkaloide 23, 24.
Cariacaceae 23.
Caria papaya 23.
Carnavallin 10, 23.
Carotene 268, 442, 443.
Carotenoids 284, 442, 443, 444, 445, 446.
Carotenoproteins 445.
Carotin s. Carotene.
Carotinoide in Flechten 268.
Carpain 23, 24.
—, Biosynthese 24.
Carpaminsäure 24.
Carragenans 467, 468, 469.
Caryophyllene, monoepoxide 370.
Cassaic acid series 399.
Cassaine diterpenes 399.
Cassia-Alkaloide 22.
Cassia carnaval 23.
Cassia excelsa 22.
Cassin 10, 22, 23.
Catopsilia florella 110.
Cedrene 389.
Cedrol 389.
Celite 435.
Cellulase 273.
Cembrene → Thunbergene 403.
Centrospermen, anthocyanfreie 38.
Cephaloridine 458.

Cephalosporanic acid 458.
—, lactone derivatives 458.
Cephalotin 458.
Cephalosporins 422, 457, 458.
Cephalosporinum sp. 422, 457.
Ceramiales 115.
Cerin 268.
Cetraria islandica 223, 274, 275, 279.
Cetraria nivalis 268.
Cevine 363.
Chaetomium cochliodes 276.
α-Chamigrene 384.
Ch'an San (chinesische Droge) 17.
Char 443.
Charaxes dilitus 110.
Charaxes eupale 110.
Charaxes Khaldeni 110.
Charaxes subornatus 110.
Charcoal, adsorption on 145.
Charcoal-diatomaceous earth, chromatography on 145.
Chavicin → *cis,cis*-Piperin 8.
Cheilinus undulatus 106.
Chenopodiaceae 3, 14, 25.
Cherry leaf spot 146.
4-Chinazolon 17, 18.
Chinolinsäure 4.
Chinolizidine 24.
Chnolizidin-Typ von Alkaloiden 2, 24.
Chinone 104, 210, 284.
Chiodecton sanguineum 237.
Chiodectonsäure 237.
Chloranil 269.
Chloratranorin 278.
8-Chlor-5,7-dihydroxy-2,6-dimethylchromon → Sordidon → Rupicolon 235, 236, 285.
Chlorella vulgaris 445.
2-Chloremodin 234.
7-Chloremodin 229.
—, -1,6-dimethyläther 229.
—, -1-methyläther 229.
7-Chlor-4-hydroxyemodin 229.
7-Chlor-5-hydroxyemodin → Papulosin 229.
Chlorine, Abbau mit Chromsäure 70.
2-Chlornorlichexanthon 238.
Chloroacetyl chloride 191.
6-Chloro-2,3-dimethoxy-naphthazarin 465.
(2-Chloroethyl)-trimethylammonium chloride → Cycocel 412.

Chloroform 77, 87, 88, 103, 118, 120, 145.
Chloromaleic anhydride 465.
p-Chlormercuribenzoat 117.
Chlorophyceae 467.
Chlorophyll 61, 116, 363.
Chloroplastendrehung, lichtabhängige 124.
5-Chlorsellinsäure 227.
5-Chlorparietin 229.
$\Delta^{8, 24}$-Cholestadiene-3-ol → Zymosterol 421.
Cholesterol 418, 419, 423, 431.
—, biological pathway for the synthesis 364, 365, 369.
—, C-20 biochemical hydroxylation 423.
Cholic acid 432, 433.
Choline 454, 455.
—, analogs 454.
Cholinschwefelsäure 220.
Cholinsulfat 284.
Chondria armata 458.
Chondrillasterol 421.
Chromatographie.
  Azofarbstoffe 67.
  Biliproteide 61, 66.
  Ciguatoxin 453.
  Cucurbitacins 346.
  Cyclohexanone derivatives 175.
  Endocrocin 234.
  Gallenfarbstoffe 61, 68, 86f.
  Glutarimide antibiotics 145.
  Holothurins 431.
  Mesobiliviolin 99.
  Myxinol 435.
  Naphthazarin derivatives 465.
  Oxydationsprodukte von Bilanen 65.
  Pahutoxin 454.
  Proteinabtrennung 112.
  Protomycin 184.
  Phytochrom 125, 126.
  Steroid derivatives 424.
  Streptimidone 182.
Chromic acid, oxidation with 149, 151, 157, 165, 169, 170, 171, 179, 191, 215, 433.
Chromium trioxide 260, 267, 320, 322, 323, 326, 343, 345.
Chromone 210, 284.
—, Alkalischmelze 215.
—, in Flechten 235f.
Chromophore, verdrillte 93.

Chromoproteide 111, 112.
—, Abbau mit Chromsäure 72.
—, Abbau mit Kaliumpermanganat 70.
Chromous chloride reduction 334.
Chromsäure-Abbau von Chromoproteiden 72.
Chromsäure-Abbau von Pyrrolfarbstoffen 70.
Chromsäure-Oxydation s. Chromic acid, oxidation with
Chromtrioxid s. Chromium trioxide.
Chroococales 115.
Chrysopa carnea 108.
Chrysophanol 231.
Ciguatera-causing poisons 453.
Ciguatoxin 453.
N-*cis*-Cinnamoyl-3(S)-[2'(R)-piperidyl]-piperidin → Astrophyllin → Dihydroisoorensin 29.
N-*trans*-Cinnamoyl-β-tetrahydroanabasin → Adenocarpin 28, 29.
Cionia intestinalis 457.
Circulardichroismus 117, s. Cotton-Effekt.
Citrat 41.
Citreorosein 231.
—, -triacetat 231.
Citronellal 372.
Citrullus colocynthis 337, 352.
Citrullus ecirrhosus 337, 352.
Citrullus naudinianus 352.
Citrullus vulgaris 352.
Cladonia alpestris 278.
— *chlorophaea* 285, 286.
— *conistea* 286.
— *convoluta* 220, 275.
— *cristatella* 285.
— *cryptochlorophaea* 250, 286.
— *deformis* 220, 268.
— *endiviaefolia* 274.
— *grayi* 257, 286.
— *merochlorophaea* 250, 286.
— *mitis* 245, 278.
— *perlomera* 250, 286.
— *pleurota* 245.
— sp. 228.
— *subconistea* 286.
— *submitis* 245.
— *sylvatica* 245.
Clary sage *(Salvia sclarea)* 397.
Clemmensen reduction 180.

Clitumnus extradentatus 108.
Clostridien-Arten 94.
Clostridium tetani, growth inhibition 422.
^{14}C-markierte Verbindungen.
Acetat 11, 18.
1-^{14}C-Acetat 40, 275, 276, 405.
2-^{14}C-Acetat 276, 280.
1-^{14}C-Acetic acid s. Essigsäure.
Ameisensäure 278.
δ-Aminolävulinsäure 110.
1,4-^{14}C-Bernsteinsäure 200, 275.
1,5-^{14}C-Cadaverin 13, 27, 28.
^{14}CO$_2$ s. Kohlendioxid.
Cucurbitacin B 348.
[1-^{14}C]-3,5-Dimethoxyphthalsäureanhydrid 233.
Elaterin-2-methyl-ether 351.
10-^{14}C-Endocrocin 233.
Essigsäure 201, 277, 278.
Formaldehyd 104.
1-^{3}H$_2$-2-^{14}C-Geranyl pyrophosphate 410.
Glucose 274, 275, 280.
Glutamic acid 197.
Glycerin 275.
Glycin 110.
Gyrophorsäure 277.
7β-Hydroxy-(−)-kaur-16-en-19-oic acid 410.
6-^{14}C-δ-Hydroxylysin 8.
17-^{14}C-(+)-Kaurene 407.
17-^{14}C-(−)-Kaur-16-en-19-al 409.
17-^{14}C-(−)-Kaur-16-en-19-oic acid 409, 411.
17-^{14}C-(−)-Kaur-16-en-19-ol 409.
Kohlendioxid 12, 40, 277, 404.
Lecanorsäure 278.
Lysin 7, 8, 13, 15, 16, 17, 20, 26, 30, 197.
1-^{14}C-Malonat 276.
1,3-^{14}C-Malonat 40.
1-^{14}C-Malonsäure 277.
2-^{14}C-Malonsäure 277, 278, 406.
1^{14}C-Malonsäurediäthylester 277.
Mannit 274.
10-^{14}C-2-Methylemodin 233.
Methyl-^{14}C-methionine 197, 276.
Methylphloracetophenon 278.
Mevalonate 276, 397, 403, 405, 408, 410, 411.
Mevalonic acid s. Mevalonsäure.
2-^{14}C-Mevalonolactone 403.

¹⁴C-markierte Verbindungen.
  Mevalonsäure 36, 197, 348, 406.
  1-¹⁴C-Natriumacetat 197, 200.
  2-¹⁴C-Natriumacetat 278.
  Natriumformiat 278.
  Natriumhydrogencarbonat 274.
  Octansäure 11.
  Oxooctansäure 11.
  Phenylalanin 20, 30, 279, 280, 281.
  Phenylmilchsäure 281.
  CO¹⁴CH₃-Phloracetophenon 278.
  6-¹⁴C-$\it\Delta$-Piperidein 26.
  1-¹⁴C-Propionat 197, 276.
  2-¹⁴C-Propionat 276.
  1-¹⁴C-Propionic acid 200.
  Protoporphyrin IX 110.
  Pulvinsäurelacton 280.
  Sodium acetate s. Natriumacetat.
  Sodium carbonate 201.
  Sodium[1,3-¹⁴C]-malonate 200.
  1,4-¹⁴C-Succinic acid s. Bernsteinsäure.
  Vulpinsäure 279.
  2-¹⁴C-Zimtsäure 280.
Coccinia sp. 352.
Cocciferae 228.
Coccomyxa 274.
Codfish 443.
Coelacanth *(Latimeria chalumnae)* 446.
Coelenterates (Gorgonians) 435, 437, 439.
Coelenterate sterols 419, 421.
Coelidium fourcadei 28.
Colensoinsäure 257.
Collema tenax 285.
Cololabis saira 106.
Comatula pectinata 463, 466.
Compositae 4.
Confluentinsäure 247.
(+)-Conhydrin 9, 10.
—, stereochemische Aufklärung 10.
DL-Coniin 9, 11, 12.
—, Biosynthese 11.
γ-Conicein 9, 10, 11, 12.
Conium-Alkaloide 1, 9f.
—, Biosynthese 10.
Conium maculatum 9, 10, 12.
Constictinsäure 254.
(−)-Copaene 439.
Copalol pyrophosphate → *(ent)*Labdadienol pyrophosphate 404, 408.
Coprostanol 419.
Coralocarpus sphaerocarpus 352.

Cora pavonia 275.
Cornyebacteria, growth inhibition 422.
Corrine, Abbau mit Chromsäure 70.
—, Numerierung 62.
Corsevin 37.
Cortisone 363.
Cotton-Effekt 78, 79, 113, 151, 154, 196, 224, 259, 331, 339, 343, 345, 350, 441.
—, axial halo-ketone dispersion rule 154.
—, Glutarimide antibiotics 142, 143, 154.
—, Hudson-Klyne-Regel 224.
Cottus scorpius 106.
Counter-current distribution.
  Protomycin 184.
  Streptovitacins 175.
Crango vulgaris 456.
Crassin acetate 438.
Crassostria virginica 453.
Crassulaceae 12.
Crayfish *(Jasus lalandei)* 422.
Crenilabrus pavo 106.
Crinoids 463, 466.
Crithidia fasciculata 443.
Cruciferae 309, 337, 354.
Crustaceae 443.
Crustacean moulting-hormones 422.
Crustecdysone 422, 423.
Cryptocarya pleurosperma 15.
Cryptochlorophaesäure 250, 251, 286.
Cryptomonadales 115.
Cryptomonad-Phycocyanin 115.
Cryptomonad-Phycoerythrin 115.
—, enzymatische Hydrolyse 123.
—, von *Rhodomas* 123.
Cryptonemiales 115.
Cryptophyta 111, 114, 115.
Cryptopleurin 15.
(+)-α-Cubebene 438.
Cucumaria planci 430, 431.
Cucumis dinteri 346.
Cucumis hirsutus 337.
Cucumis sp. 352.
Cucurbitaceae 308, 339, 346, 348, 351, 352.
Cucurbitacins 309f.
—, biogenetic aspects 348.
—, biological properties 351.
—, catalytic reduction of the diosphenol system 334.
—, CD-data 350.
—, chemistry 311f.
—, cytotoxic activity 351.
—, degradation to a monoketone 329.

Cucurbitacins, distribution in plants 348.
—, effect on the respiration of tumor cells 351.
—, formation of a „pseudo-glycol" 329.
—, 16-hydroxy-group 322.
—, insect attractiveness 351.
—, interrelationship between A, B and C 329.
—, IR-spectra 350.
—, mass spectra 350.
—, 19-methyl group 316.
—, NMR-spectra 310, 350.
—, occurrence in nature 352 f.
—, ORD-spectra 350.
—, paper chromatography 350.
—, physical constants 356.
—, physical methods in the structure elucidation 350 f.
—, ring A substituents 314.
—, ring B double bond 317.
—, ring C carbonyl 316.
—, side chain 312 f.
—, skeleton 312.
—, stereochemistry 329.
—, stereochemistry of ring A ketols 332.
—, structure determination 311 f, 350 f.
—, thin layer chromatography 350.
—, UV-spectra 350.
—, X-ray measurements 311.
Cucurbitacin A.
—, conversion to a derivative of eburicoic acid 316.
—, mild acid hydrolysis 325.
—, occurrence in nature 352, 354.
—, ozonolysis 314.
—, physical constants 356.
—, reactions 326.
—, selenium dehydrogenation 309, 312.
—, structure determination 325 f.
—, transformation into a tetraketone 334.
Cucurbitacin A acetate.
—, reaction with lead tetraacetate 322.
—, transformation to a hydroxytriketone 329.
Cucurbitacin A triacetate, conversion into a dienedione 334.
Cucurbitacin B 311, 339, 341.
—, biosynthetic pattern 348, 349.
—, catalytic dehydrogenation 314.
—, cleavage with periodic acid 316.
—, cytotoxic activity 351.
—, hydrogenation 338.

Cucurbitacin B, occurrence in nature 352, 354.
—, oxidation 312.
—, oxidative degradation 319.
—, physical constants 356.
—, reactions of ring B 318, 320.
—, reduction 314, 346.
—, tetrahydroderivative 346.
Cucurbitacin B acetate.
—, cleavage of the C_{20}-C_{22} bond 323.
—, oxidation 320, 321, 322.
—, reactions of rings A and B 321.
—, reduction 329.
Cucurbitacin C.
—, cleavage of the side chain 328.
—, conversion to a aldehydotriketone 329.
—, occurrence in nature 352, 354.
—, ozonolysis 314.
—, physical constants 356.
—, reactions 327.
—, structure determination 325 f.
Cucurbitacin C acetate, oxidation with CrO_3 328.
Cucurbitacin D 311, 329, 337, 338.
—, 23,24-dihydro derivative 346.
—, occurrence in nature 352, 354.
—, physical constants 356.
Cucurbitacin D diacetate 337.
Cucurbitacin E → Elaterin 311, 329, 341.
—, cytotoxic activity 351.
—, occurrence in nature 352, 354.
—, ozonolysis 314.
—, physical constants 356.
—, treatment with cold alkali 338.
Cucurbitacin E acetate, reaction with lead tetraacetate 322.
Cucurbitacin F 346, 347.
—, 23,24-dihydro derivative 346.
—, occurrence in nature 352, 354.
—, physical constants 356.
Cucurbitacin G 337, 338.
—, occurrence in nature 352, 354.
—, physical constants 356.
Cucurbitacin H 337, 338.
—, occurrence in nature 352, 354.
—, physical constants 356.
Cucurbitacin I 311, 329, 337, 338, 341.
—, occurrence in nature 353, 355.
—, physical constants 356.
Cucurbitacin I diacetate 338.
Cucurbitacin J 313, 337, 338, 342.
—, occurrence in nature 353, 355.

Cucurbitacin J, physical constants 356.
Cucurbitacin K 313, 337, 338, 342.
—, occurrence in nature 353, 355.
—, physical constants 356.
Cucurbitacin L → 23,24-Dihydrocucurbitacin I 337, 338.
—, occurrence in nature 353, 355.
—, physical constants 356.
Cucurbitacin O 346, 347.
—, 25-acetyl derivative → Cucurbitacin Q 347.
—, cytotoxic activity 351.
—, occurrence in nature 352, 354.
—, physical constants 356.
Cucurbitacin P 346, 347.
—, cytotoxic activity 351.
—, occurrence in nature 353, 355.
—, physical constants 356.
Cucurbitacin Q 346, 347.
—, cytotoxic activity 351.
—, occurrence in nature 353, 355.
—, physical constants 356.
Cucurbitanes 307f.
—, carbon skeleton 309f.
—, interrelationship with the lanostane series 334.
—, nomenclature 310.
—, synthesis of the skeleton 336.
5α-Cucurbitane 310.
32-nor-Cucurbitane-skeleton, syntheses 336.
Cucurbita pepo 408.
Cucurbita spp. 352.
10α-Cucurbit-5-ene 310.
Cucurbitone A 325.
Cucurbitone B 320, 321.
Cucurbitone C 328.
(+)-Cuparene 441, 442, 447.
α-Cuparenone 442.
Cupric benzoate 390.
Cuprous chloride 390.
Cuskhygrin 13, 16.
Cuvierian glands 425, 427, 430.
Cyanidium caldarium 115, 117.
Cyanoacetic acid 187.
Cyanocobalamin, Chromatabbau 70.
Cyanophyta 111, 114, 115.
Cyclizations.
—, acid catalyzed 366f, 380.
—, boron fluoride catalyzed 366, 371.
—, carbonium ion catalyzed 390.
—, cyclopropyl ketones 384, 387.

Cyclizations.
—, oxidative 369f.
—, polyenic acetals 380.
—, polyenic allylic alcohols 381.
—, polyenic arene sulfonates 378.
—, radical 390.
—, stannic chloride catalyzed 371, 387.
—, steroids 397.
—, terpenoids 396f, 399.
Cycloaliphatische Verbindungen in Flechten 225.
Cycloartenol 348, 349.
Cyclodopa 39.
Cyclohexane 72, 175.
Cyclohexane ring, cleavage of 172.
Cyclohexanone 194, 196.
—, formation by cationic cyclization of a diene 364.
Cycloheximide 140f.
—, absolute configuration 151.
—, analogs of 193f.
—, acetate 147, 165.
—, —, catalytic hydrogenation 164.
—, —, nitrogen mustard derivative 174.
—, —, NMR-spectrum 160.
—, —, pyrolysis 169.
—, —, reduction 164.
—, anhydro product 168.
—, —, catalytic reduction 169, 170.
—, —, reduction 169.
—, antifungal activity 146, 174.
—, antitumor activity 146, 175.
—, biological activity 146.
—, biosynthesis 197, 200.
—, catalytic reduction 147.
—, chemical evidence of the structure 149.
—, commercial production 145.
—, chloroacetate 191.
—, —, NMR-spectrum 160.
—, configuration of the α-hydroxyl group 164.
—, degradative studies 147.
—, dehydration by P_2O_5 161.
—, dehydration products 168.
—, dideoxy derivative 161.
—, electrometric titration 147.
—, epimerization 155.
—, esters 174.
—, estimation 145, 146.
—, homologs of 193f.
—, hydrogenation 161, 164.

Cycloheximide, inhibition of protein synthesis 147, 175.
—, IR-spectrum 147, 157.
—, isomerization 155, 156, 191.
—, mass spectrum 180.
—, mono-p-nitrobezoate 147.
—, monoxime 147.
—, myristate 174.
—, NMR-spectra 260.
—, ORD-spectrum 154.
—, orientation of the α-hydroxyl-group 167.
—, oxime 174.
—, palmitate 174.
—, reduction products 170f.
—, —, NMR-spectra 165.
—, reduction with aluminium isopropoxide 166.
—, reduction with diphenyltin dihydride 165.
—, resorcinol color reaction 189.
—, semicarbazone 147, 174.
—, stearate 174.
—, stereochemistry 152f, 161.
—, stereoselective synthesis 189.
—, structure 146f, 152f, 161.
—, structure-activity relationships 147.
—, synthesis 186f.
—, thiosemicarbazone 174.
—, tosylate 161, 168.
—, —, NMR-spectrum 160.
—, treatment with thionyl chloride 161.
—, UV-absorption 147.
ψ-Cycloheximide 141, 172.
—, treatment with ammonia 174.
gem-Cycloheximide 189.
Cyclokaurane → trachylobane 402.
Cyclopentanone moiety 454.
Cyclopropane ring 419, 421.
—, formation 399.
Cyclopropyl ketones, cyclization reactions 384, 387.
Cyclotetrapeptide 217.
Cycocel 412.
Cyprinus carpio 458.
Cystamine 455, 456.
Cystein 121, 123.
Cytochrom c 121.
Cytotoxic activity 351.

Dammaran 267.
Dammaran type triterpene 346.

Darmbakterien 95.
Dasyatis akajei (Ray fish) 432.
Deacetoxycurcurbitacin B 314.
DEAE-Sephadex 125.
Debenzylation 453.
Debromoaplysin 442, 447.
Debromolaurinterol 442.
Debye-Scherrer-Diagramm 98.
Decalin diol 370.
cis-anti-Decalol 369.
trans-Decalol 378.
Decarboxylation of β,γ-unsaturated acids 411.
Decarboxyportentol 225.
Decarboxyportenton 225.
Decarboxyschizopeltsäure 241.
Dehydration inhibitors in plants 395.
22-Dehydrocholesterol 418, 420.
24-Dehydrocholesterol → Desmosterol 418, 420.
Dehydrocholic acid 433.
Dehydrocycloheximide 149, 151, 174, 180.
—, hydrogenation 190.
—, NMR-spectrum 160.
—, reaction with hydrazine 174.
—, synthesis 190.
Dehydroisocycloheximide 150.
—, hydrogenation 191.
—, NMR-spectrum 160.
—, reduction 191.
—, synthesis 191.
dl-16,17-Dehydroprogesterone 381, 382.
Δ^7-Dehydroskytanthin 35.
Dehydrosolenopsin B u. C 22.
Delpheline 403.
22-Deoxocucurbitacin D 339, 341, 342.
—, occurrence in nature 352, 354.
—, physical constants 356.
—, reactions 342.
—, structure 340.
22-Deoxoisocucurbitacin D 339, 341f.
—, occurrence in nature 352, 354.
—, physical constants 356.
Deoxycrustecdysone 422.
Deoxyisocycloheximide 169, 170.
16-Deoxymyxinol 434, 435.
Depside 210, 211, 284.
—, in Flechten 245f, 284.
—, Katabolismus 273.
—, Massenspektren 213, 214.
—, O-Methylderivate 286.

Depside, Nachweis von Esterbrücken 215.
—, NMR-Spektren 212.
—, Synthese 148, 149.
Depsid-hydrolysierende Esterase 273.
Depsidone 210, 211.
—, in Flechten 254, 284.
—, Massenspektren 214.
—, NMR-Spektrum 212.
—, O-Methylester 286.
—, Spaltung von Ätherbrücken 215.
—, Spaltung von Esterbrücken 215.
Depsone 210, 284.
—, in Flechten 258f, 284.
Dermatiscum thunbergii 220.
4-O-Desmethylbarbatinsäure 251.
Desmethyldecarboxyschizopeltsäure 241.
Desmethyl farnesic acid 367, 368.
4-O-Desmethylmerochlorophaesäure 251.
4-O-Desmethyl-2′-O-methyl-microphyllinsäure 247.
Desmosterol → 24-Dehydrocholesterol 418, 420.
6-Desoxy-16β-acetylleucotylin 264.
6-Desoxyleucotylin 264.
1-Desoxy-1-nitro-D-glycero-D-taloheptit 218.
Desoxyrosenonolactone 405.
Deuterierte Trifluoressigsäure 84, 85.
Deuterochloroform 84, 85, 159, 160.
Deutero-dimethylsulfoxid 85.
Deuteroformic acid 368.
Deuteropyridin 84, 85.
Deuterosulfuric acid 368.
Deutin 94.
Diacetylanhydroleucotylin 262.
Diacetylisousninsäure 245.
Diacetylleucotylin 262.
6α,16β-Diacetylleucotylin 264.
Diacetylparietin 231.
Diacetylpyxinol 267.
O,O-Diacetylusninsäure 244, 245.
Diadinoxanthin 445.
Dialyse 126.
o-Dianisidin 211.
Diäthylamin 223.
cis-8,10-Diäthylnorlobelidion 20.
Diatoms 444.
2-Diazopropane 345.
Diazomethan 149, 241, 243, 248, 272, 385, 389, 465.
Diazoniumsalze 104.

Diazoreaktion für Bilirubinbestimmung 66.
Dibenzochinon 269.
Dibenzofurane 210, 211, 284.
—, Alkalischmelze 215.
—, in Flechten 240.
Dibenzoylperoxid 242.
Di-O-benzylorsellinsäure 228.
3β,12β-Di-p-brombenzoyl-pyxinol, Röntgenstrukturanalyse 215, 267.
(+)-Dibromdehydrotetrahydrorugulosin, Röntgenstrukturanalyse 215, 259.
Dibromdihydroorsellinsäure-äthylester 227.
2,6-Dibromo-4-acetamido-4-hydroxycyclohexadienone 447.
3,5-Dibromo-1(ax),6(ax)-dihydroxy-4-methoxycyclohexa-2,4-diene-1-acetonitrile 449.
3,5-Dibromo-4-hydroxybenzyl alcohol 446.
2-(3,5-Dibromophenyl)-3,4,5-tribromopyrrole 449.
Dibromorsellinsäure 227.
—, -äthylester 227.
3,4-Dichlor-4-O-benzyl-orsellinsäure 257.
5,7-Dichlor-1,3-dihydroxy-6-methoxy-8-methylxanthon 238.
5,7-Dichloremodin 229.
3,5-Dichloreverninsäure 238.
2,4-Dichlororcin 257.
—, -3-methyläther 255.
3,5-Dichlorlecanorsäure-methylester 245.
—, Hydrolyse und Synthese 246.
2,5-Dichlorlichexanthon 238.
2,7-Dichlorlichexanthon 238.
Dichlormethan 88.
2,4-Dichlornorlichexanthon 238.
2,7-Dichlornorlichexanthon 238.
Dichloromaleic anhydride 465.
3,5-Dichlororsellinsäure 245.
—, -dimethyläther 255.
4,5-Dichlorparietin 229.
Dichroa febrifuga 17.
Dickdarmbakterien 94.
Dictyopterine A 441.
Dictyopterine B 440.
Dictyopteris divaricata (sea weed) 439, 440.
Dictyopteris sp. 441, 470.
N,N′-Dicyclohexylcarbodiimid 249.
Didymsäure 285.

Diels-Alder-Reaktion 220.
Diethylamine 188.
2-Diethylaminoethyl-2,2-diphenylvalerate → SKF 525 412.
Diethylene glykol 442.
Diffractasäure, Biosynthese 279.
m-Digallussäure 273.
Digenea simplex 458.
Dihydroactidione 147.
—, diacetate 147.
7,8-Dihydrobiladiene, Massenspektrum 82.
1,2-Dihydrobilene 78, 79.
—, hydrochlorid, Optische Drehung 79.
Dihydrocarvone 388.
Dihydrocucurbitacin B 338, 314.
—, occurrence in nature 352, 354.
—, physical constants 356.
23,24-Dihydrocucurbitacin I → Cucurbitacin L 337, 338, 353, 355, 356.
5,6-Dihydrocucurbitone A 325.
Dihydrocycloheximide.
—, acetate 164.
—, acid lactone 201.
—, borate ester, optical activity 164.
—, diacetate 164.
—, formation of an acetonide 166.
—, methylation 149.
—, oxidation 191.
—, stereochemistry 164.
—, synthesis 190.
Dihydrodeacetoxycucurbitacin B 314.
3,4-Dihydro-3,7-dimethoxy-9-methyldibenzofuran-1,2-dicarbonsäuredimethylester 242.
23,24-Dihydroelaterin 316.
Dihydroelatericin A 334, 342.
—, diacetate 334.
Dihydrohämatinsäure-imid-methylester 72.
Dihydroisoorensin → Astrophyllin 29.
9α-H-Dihydrolanosterol 375.
Dihydronitenin 472.
Dihydroorsellinsäure-äthylester 227.
(+)-Dihydroprotolichesterinsäure 223.
allo-Dihydroprotolichesterinsäure 224.
Dihydrosecamin 33.
16,17-Dihydrosecodin 33.
16,17-Dihydrosecodin-17-ol 33.
Dihydrostreptimidone 182, 183.
(—)-Dihydrousninsäure 244, 245.
2,7-Dihydroxy-6-acetyljuglone 464.

1,8-Dihydroxyanthrachinone, NMR-Spektren 213.
5,8-Dihydroxyanthrachinone in Flechten 234.
1,7-Dihydroxydibenzofuran 240.
5,6-Dihydroxydihydroindol-2-carbonsäure 39.
1,2-Dihydroxy-3,4-dimethoxybenzene 465.
2,6-Dihydroxy-3,7-dimethoxynaphthazarin 462.
2,7-Dihydroxy-3,6-dimethoxynaphthazarin 462.
2,5-Dihydroxy-3-ethylbenzoquinone 464, 465.
2,7-Dihydroxy-3-ethylnaphthazarin 463, 464.
11α,22-Dihydroxyhopan 267.
15α-22-Dihydroxyhopan 266.
16β-22-Dihydroxyhopan-23-säure 264.
3,5-Dihydroxy-n-heptylbenzol 235.
6α,16β-Dihydroxyisohopan 262.
6β,7β-Dihydroxy-(—)-kaur-16-en-19-oic acid 410.
7,18-Dihydroxykaurenolide 410.
2,4-Dihydroxy-6-methyl-isophthalsäure 227.
2,7-Dihydroxynaphthazarin 463.
3,4-Dihydroxyphenylalanin (Dopa) 39.
3,5-Dihydroxyphenylessigsäure 235.
2,5-Dijodhydrochinondimethyläther 270.
α-Diketones, enolized → diosphenols 314.
1,4-Diketones, base-catalyzed autoxidation to enediones 319.
Dimedon 68.
threo-3,4-Dimethoxycarbonyl-6-oxoheptansäure 220.
erythro-3,4-Dimethoxycarbonyl-pentansäure 223.
α,β-Dimethoxycarbonyl-γ-tridecyl-γ-butyrolacton 223.
2,3-Dimethoxy-6,7-dichloronaphthazarin 465.
12β,25-Dimethoxy-7,8-dihydroholothurinogenin 427.
1,2-Dimethoxyethane 448.
3,5-Dimethoxy-2-jodtoluol 241.
3,5-Dimethoxy-4-jodtoluol 241.
3,7-Dimethoxy-9-methyldibenzofuran-1,2-dicarbonsäure-dimethylester 242.
3,4-Dimethoxyphenol 269.

[1-¹⁴C]-3,5-Dimethoxyphthalsäure-anhydrid 233.
Dimethyl acetonedicarboxylate 187.
β,β-Dimethylacrylic acid → Senecioic acid 406.
Dimethylamin 217.
4-N,N-Dimethylamino-1,2-dithiolane → Nereistoxin 452, 453.
4,4-Dimethylandrostane derivatives 336.
Dimethyläther-cryptochlorophaesäure-methylester 250.
Dimethyläther-olivetolcarbonsäure 247, 250.
Dimethyläther-pannarsäure-methylester 241.
2,4-Dimethyläther-6-n-pentyl-pyrogallolcarbonsäure-methylester 250.
Dimethyläther-perlatolinsäure 247.
3,4-Dimethyl-5-benzyl-Δ³-pyrrolon-(2), Elektronenspektrum 73.
4,4-Dimethylcholestane derivatives 310.
Dimethylcyanamide 455, 456.
1,3-Dimethylcyclohexanol 175.
2,4-Dimethylcyclohexanon 40, 148, 149, 151, 154, 156, 157, 191, 194, 200.
—, Cotton effect 154.
—, 1-menthhydrazone 148.
cis-2,4-Dimethylcyclohexanone 188, 189, 190, 195, 196.
trans-2,4-Dimethylcyclohexanone 189.
—, enamine 190.
2,4-Dimethyl-2-cyclohexenone 175.
Dimethyl formamide 161.
3,5-Dimethyl-3,5-heptadiene-2-one 186.
3,5-Dimethyl-2-heptanone 182, 186.
Dimethyl glutaconate 187.
3-[2-{3(S),5(S)-Dimethyl-2-oxocyclohexyl}-2(R)-hydroxyethyl] glutarimide → cycloheximide 141.
1,4-Dimethylnaphthalene 312.
4,4-Dimethyl-Δ(2)-norcholest-5-en-3-one 329.
2,3-Dimethylphenol 233.
2,6-Dimethylpiperidin 3.
3,4-Dimethylpyrrolon, Elektronenspektrum 73.
—, NMR-spektrum 85.
Dimethyl sulfate 181, 235, 243.
Dimethylsulfon 220, 284.
Dimethylsulfoxid 71, 231,
Dimroths-Reagens 240.
Dinitrophenylhydrazin 71, 454.

Diosphenols 311, 314.
—, NMR-spectra 309.
—, reaction with $FeCl_3$ 314.
Diosphenol system in cucurbitacins, catalytic reduction 334.
—, reduction experiments 341.
Dioxan 88, 114, 161.
6,16,-Dioxoisohopan 262.
2,6-Dioxopiperidin → Glutarimid 2, 40.
1',8'-Dioxo-1,3,6,7-tetramethyl-2,8-divinyl-biladien-(a,c)-dipropionsäure-(4,5) → Bilirubin 61, 64, 89f.
2,2'-Di-n-pentyl-4,6,4',6'-tetrahydroxydiphenyl 259.
(—)-cis-8,10-Diphenyl-lobelidiol 20.
(2 S : 6 R : 8 S)-(—)-8,10-Diphenyl-lobelionol → (—)-Lobelin 19, 20.
Diphenyltin dihydride 165, 169, 171.
(+)-α,β-Dipiperidyl 29.
Dipiperidyl-Alkaloide 27.
Diploicin 255.
—, Synthese 255, 256, 257.
Diploschistessäure 249.
Dipyrromethane, Bildung von Azofarbstoffen 67.
—, Massenspektren 80.
Dipyrromethen, „klassischer Typ" 74.
—, „Neotyp" 74.
Dirina repanda 225.
Disaccharid-Konjugate 91.
Diterpene 211, 395f.
—, bicyclic 397f, 403.
—, —, mode of cyclization 397.
—, biological activity 395.
—, bioformation of ring-C 372.
—, cooccurrence of bi, tri- and tetracyclic 399.
—, in Flechten 261.
—, macrocyclic 403.
—, tetracyclic 401f, 406.
—, tricyclic 399f, 403.
Divaricatinsäure 247, 250.
—, -methyläther 250.
Dodecylsulfat 117, 126.
Domoic acid 458, 459.
Dopa 39.
Dopachinon 39.
Double bond, cleavage 381.
Dowex 1-X4 454.
Dragendorff test 453.
Drimenol 371.
Duboisia myoporoides 12, 25.

Dünnschichtchromatographie.
  Abbauprodukte von Pyrrolfarbstoffen 71.
  Anthrachinone 234.
  Azopigmente 91.
  Bilatriene 108.
  Bile acids 432.
  Chinone 210.
  Chromone 210.
  Crustecdysone derivatives 425.
  Cucurbitacins 350.
  Depside 210.
  Depsidone 210.
  Depsone 210.
  Dibenzofurane 210.
  Flechtensäure 210.
  Flechtenstoffe 210f.
  Flechtenxanthone 239.
  Gallenfarbstoffe 86, 87, 88.
  Imide 71, 72.
  Phycobiline 119.
  Pyrroldialdehyde 71, 72.
  Racemattrennung 210.
  Usninsäure 210.
  Xanthone 210.
Duvatrienes 403.

E-73 174.
—, acid hydrolysis 176.
—, antitumor activity 174.
—, catalytic reduction 179.
—, estimation 145.
—, inhibition of protein synthesis 175.
—, -monoacetate 176.
—, -mono-p-benzoate 176.
—, -monosemicarbazone 176.
—, -monoxime 176.
—, NMR-spectrum 160, 177.
—, oxidation with chromic acid 179.
—, spectral properties 175.
—, structure determination 175, 177.
Eagle's KB strain 351.
Eagles KKB cells 196.
Eburicoic acid 316, 334.
Ecballic acid 312, 314, 323.
—, treatment with alkali 323.
Ecballium elaterium 352.
Ecdysone 422.
Echinochrome A 461.
—, synthesis 465.
Echinocystis fabacea 348, 352.
Echinocystis lobata 352.

Echinocystis macrocarpa (wild cucumber) 407, 408, 409, 410, 412.
Echinocystis wrightii 354.
Echinodermata 443.
Echinoderm pigments 466, 467.
Echinoderms, sapogenins 425.
—, sterols 419, 421.
Echinoidea 461, 462, 463, 464.
Echinometra oblonga 463.
Echinothrix calamaris 461, 462.
Echinothrix diadema 461, 466.
Echinus esculentus 460.
Eel s. Aal 443.
Ehrlich ascites tumor cells, morphological changes 351.
Eisen(III)-chlorid 99, 101, 102, 104, 157, 181, 215, 285, 314, 453.
—, Oxydation von Bilanen 65.
Elasmobranch fish 432.
Elasmobranchs 458.
Elastin 94.
Elaterase 308, 337.
Elateric acid 323.
—, aluminium amalgam reduction 324.
—, formation 323, 324.
—, IR-spectrum 324.
—, methyl ester 324.
Elatericin A → Cucurbitacin D 311, 337, 356.
—, acetate, periodic acid treatment 322.
—, antitumor activity 351.
—, cleavage of the C-20 C-22 bond 323.
—, cytotoxic activity 351.
—, effect on human lymphocytes 351.
—, ozonolysis 314.
—, pharmacodynamic activity 351.
—, selenium dehydrogenation 312.
—, treatment with periodic acid 312, 316.
Elatericin B → Cucurbitacin I 311, 341, 356.
—, antitumor activity 351.
—, cytotoxic activity 351.
—, -diacetate, molecular rotation 332.
—, hydrogenation 334, 343.
—, molecular rotation 332.
—, treatment with hot alkali 314.
Elateridin 313, 338.
Elaterin → Cucurbitacin E 309, 310, 311, 338, 339, 341, 356.
—, antitumor acetivity 351.
—, cytotoxic activity 351.
—, -2-methyl ether 351.

Elaterin, selenium dehydrogenation 312.
—, treatment with alkali 312, 314, 323.
α-Elaterin 339, 341.
β-Elaterin 339, 341.
Elbs-Oxydation 234.
Electrometric titration 147, 148.
Eledone moschata 456.
Elektronenanlagerungs-Massenspektrometrie 213.
Elektronenmikroskopie, Phytochrom 126.
Elektronenspektren.
   Aplysioviolin 113.
   Bilirubin 89, 93.
   Gallenfarbstoffe 73, 127.
   Mesobilirhodin (a) 103.
   Mesobilirhodin (b) 103.
   Mesobiliviolin (a) 103.
   Mesobiliviolin (b) 103.
   Phycobiline 120.
   Phycobiliproteide 115, 116, 117.
   Phycocyanine 116, 117.
   Phycoerythrin 116, 117.
   Phytochrom 126.
   Porphyrine 73.
   Urobilin (a) 103.
   Urobilin (b) 103.
   Zinkkomplexe von Gallenfarbstoffen 75.
Elektronenstoß-Massenspektrometrie 213.
Elektrophorese 220.
β-Elemene 437, 438, 439.
Emodin 231, 234, 235.
Emodinaldehyd 231.
Emodinsäure 231.
Emu 106.
Enamines of 2-substituted cyclohexanones 189.
Endocrocin 232, 233, 234.
Enediones, formation by autoxidation of 1,4-diketones 319.
Enmein 402.
Enol, alkylation by a carbonium ion 389.
Enols, cyclization reactions 384.
Enolization 157.
ent-Labdadienol pyrophosphate 404, 408.
Entothein 261, 264.
Enzyme in Flechten 273.
Enzyminduktion 124.
Enzymrepression 124.
Epanorin 273, 284.
—, UV-Spektrum 212.

Eperuic acid 397.
2-Epicucurbitacin B 341, 342, 356.
α-Epi derivatives of cycloheximide 145.
Epidrimenol 371.
α-Epicycloheximide 141.
—, NMR-spectrum 160.
α-Epiisocycloheximide 154, 189.
—, NMR-spectrum 160.
—, ORD-spectrum 192.
—, stereochemistry 191, 192.
—, synthesis 191.
Epilaurene 442.
22,26-Epiminocholestan 37.
3-Epinuphamin 32, 33.
13-Episclareol 397.
Epoxide cyclization, transannular 370.
Epoxides by conversion of polyenes 369.
Epoxides, unsaturated, cyclization 370.
5′,6′-Epoxy-3,3′-dihydroxy-7,8-dehydro-β-carotene → Diadinoxanthin 445.
9,11-Epoxy-4,4-dimethylandrost-5-ene-3,17-dione 350.
Epoxyfarnesyl derivative, cyclization 373.
9α,11α-Epoxylanostan-3β-ol 337.
Epoxyolefin cyclizations, stereochemistry 376.
Eptatretus stoutii (Pacific hagfish) 434, 446.
Erbsenextrakte 27.
Ergochrom AA 261.
Ergochrom AB 261.
Ergosterin 268, 421.
Ergosterol s. Ergosterin.
Ergothionenine 457.
Erythrin 249.
Erythrit 274, 275.
meso-Erythrit 217.
Erythroglaucin 231.
Erythroxydiols 399.
Eschweiler-Clarke-N,N-dimethylation 453.
Essigester 71.
Essigsäure s. Acetic acid.
Essigsäure-^{14}C s. ^{14}C-Acetic acid.
Essigsäure-äthylester 72, 87, 88, 145, 175, 191, 271.
Ethanol s. Äthanol.
Ether function, removal 380.
Ethyl acetate s. Essigsäure-äthylester.
24-Ethylcholestan-3β-ol 421.
Ethyl cycloheximide carbonate 169.
Ethyl *cis,trans*-farnesate 390.

Ethyl formate 441.
Ethyl glutarimide 149.
6-Ethyl-2-hydroxynaphthazarin 462.
(−)-2-Ethyl-1-propanol 151.
Eudesmane derivative 370.
Eugenitin 237, 285.
Eugenitol 235, 285.
Euglena gracilis 445.
Eunicea mammosa 434, 437, 438, 439.
Eunicea palmeri 438.
Eunicella stricta 437.
Eunicellin 437.
—, dibromide 437.
Eunicin 435, 437.
Euphorbiaceae 29, 37.
Evernia mesomorpha 278.
Everninaldehyd 241.
Everninsäure 248.
Evernsäure 273.

Fabacein 348.
—, occurrence in nature 353, 355.
—, physical constants 356.
Fallacinal 231, 234.
Fallacinol → Teloschistin 231, 234.
trans,trans-Farnesate 10,11-epoxide 374.
Farnesic acid 366, 368.
Farnesic esters 390.
Farnesiferol-A 373.
Farnesiferol-C 373.
Farnesol 367.
Farnesyl acetate 371, 390.
Farnesyl epoxides 377.
—, Synthese 231, 232.
Farnesylpyrophosphat 281, 403.
Febrifugin 10, 17, 18.
—, Biosynthese 18.
Fermicidin 144, 186.
Ferric chloride s. Eisen(III)-chlorid.
Ferric hydroxamate 454.
Ferrimycin A_1 40.
Ferruginol 399.
Fettsäuren 284.
—, aus Flechten 220f.
Feuerameise *(Solenopsis saevissima)* 22.
dl-Fichtelite 381.
Fische, Vorkommen von Biliverdin 106.
Flavonoids in marine species 467.
Flavoobscurin A 260.
Flavoobscurin B_1 260.
Flavoobscurin B_2 260.

Flavoobscurine, oxydative Spaltung 261.
Fluoreszenz 116, 117.
—, der Zinkkomplexe von Bilenen, Biladienen und Bilatrienen 66.
—, von Biliproteiden 114, 115.
Flechten.
—, Arzneimittel aus Fl. 287.
—, biologische Wirkung 287.
—, Chemotaxonomie 210, 285.
—, Kohlenhydratstoffwechsel 274.
Flechteninhaltsstoffe.
—, Anthrachinone 229.
—, Anthrone 229.
—, antibiotische Wirkung 287.
—, Benzylester-Derivate 253.
—, biologische Wirkung 287.
—, Biosynthese 210, 273f.
—, Bisanthrachinone 259f.
—, Bisanthronyle 259f.
—, Bisxanthone 261f.
—, Calciumoxalat 220.
—, Carotinoide 268.
—, chemische Methoden zur Strukturermittlung 215.
—, chemische Nachweismethoden 215.
—, Chromone 235.
—, cycloaliphatische Verbindungen 225.
—, Depside 245f.
—, Depsidone 254.
—, Depsone 258.
—, Dibenzofurane 240.
—, Diterpene 261.
—, Dünnschichtchromatographie 210f.
—, Einteilung 216.
—, Fettsäuren 220f.
—, Gaschromatographie 211.
—, IR-Spektroskopie 211.
—, γ-Lactonsäurederivate 223.
—, Massenspektroskopie 213.
—, Mevalonat-Derivate 261f.
—, Monosaccharide 218.
—, Nachweismethoden 210f.
—, Naphthochinone 228.
—, NMR-Spektren 212, 213.
—, Oligosaccharide 218.
—, Orcinderivate 227.
—, Papierchromatographie 211.
—, Phenylalanin-Derivate 269.
—, Polyole 217.
—, Polysaccharide 218.
—, Primärstoffwechsel, Produkte 216f.
—, Proteine 217.

Flechteninhaltsstoffe, Pulvinsäure-
    Derivate 271.
—, Röntgenstrukturanalyse 215.
—, schwefelhaltige organische Verbindungen 220.
—, Steroide 268.
—, Strukturaufklärung 210, 216f.
—, Triterpene 262.
—, UV-Spektren 211, 212.
—, Xanthone 237.
—, Zuckeralkohol-Glykoside 218.
Flechten-Massenspektrometrie 215, 234.
Flechtensäuren, aliphatische 210.
Flechtensäuren, Einfluß auf Flechtenmycobionten 287.
Folinsäure 273.
Folsäure 273.
Formaldehyde 68, 182, 184, 223, 325, 326.
—, markiert 104.
Formazan 314.
Formic acid s. Ameisensäure.
N-Formylation 453.
Formyl group transfer 328.
3-Formyl-4-hydroxy-2 H-pyran 12.
Fouquieria splendens 346.
Fragilin 229, 234.
—, -dimethyläther 229.
—, Synthese 229, 230.
Friedelan-3β-ol 268.
Friedelin 268.
Friedel-Carafts-Kondensation 233.
„Friedo" rearrangements 398, 399, 401.
Fries rearrangement 193.
Fringelites 466.
D-Fructose 218.
Fucellaran 468.
Fucosterol 418, 420.
Fucus vesiculosus 440.
Fucoxanthin 444.
Fujenal 409.
Fumarprotocetrarsäure 286.
Fumarsäuredimethylester 220.
Fungi, inhibition of the growth 140.
Fungisterin 268.
Furan ring, formation 398.
Furanoterpenes 472.
Furoventaline 437.

Gaillardia pulchella 38.
2-O-β-D-Galaktofuranosyl-D-arabinit →
    Umbilicin 218.
3-O-β-D-Galaktofuranosyl-D-mannit →
    Peltigerosid 218.

D-Galactose 218, 468.
—, -2-sulfate 468.
—, -4-sulfate 468.
—, -2,6-disulfate 468.
L-Galactose-6-sulfate 468.
Gallenfarbstoffe 60f.
—, Abbau mit Chromsäure 70, 88.
—, Abbau mit Permanganat 69.
—, Abbaureaktionen 68f.
—, Absorptionsmaxima 127.
—, Biogenese 82.
—, chemische Untersuchungsmethoden 64f.
—, Elektronenspektren 73f.
—, Farbreaktionen 64f.
—, Massenspektren 79f.
—, mit Äthylidengruppe 111f.
—, NMR-Spektren 84f.
—, Nomenklatur 62f.
—, physikalische Untersuchungsmethoden 73f.
—, Strukturaufklärung 68.
Gallenflüssigkeit 91, 104.
Gartenkresse *(Lepidium sativum)* 287.
Gaschromatographie.
    Depside 211.
    Flechtenstoffe 211.
Gas-liquid chromatography, steroids 419.
Gelfiltration, Phytochrom 125, 126.
Genista-Arten 27.
Genista hystrix 27, 28.
Genista lusitanica 28.
Genista transcaucasica 16.
Gentiobiose 219.
Gentiotetraose 219.
Gentiotriose 219.
Geodia gigas 457.
Geraniol-linalool isomerism 397.
Geraniol monoepoxide 371.
Geranyl acetate 390.
Geranyl acetone, cis- und trans-isomers 368.
Geranyl diphenyl phosphate 384.
Geranylgeranyl acetate 374.
Geranylgeraniol 284, 407.
Geranylgeraniol pyrophosphate 396, 397, 401, 403, 404, 406, 407, 408, 412.
Geranyl pyrophosphate 403, 410.
(—)-Germacrene-A 437, 438, 439.
Germacrenes 437.
Germanicol 363.
Gerrardantus sp. 354.

Gibbane 402.
Gibbane skeleton, formation by ring contraction 407.
Gibberella fujikuroi 407, 408, 409, 410, 412.
Gibberellic acid 403, 411.
—, biosynthesis 406, 408, 409, 410, 412.
Gibberellin A_1 411.
— A_3 411.
— A_4 411.
— A_7 411.
— A_9 411.
— A_{10} 411.
— A_{12} 410.
— A_{13} 409, 411.
— A_{14} 410.
— A_{24} 411.
Gibberellins 395, 402.
—, biosynthesis 403, 406.
Gibbssche Reaktion 241.
Gigartinales 115.
Girgensohnia diptera 3.
Girgensohnia oppositiflora 3.
Glaucobilin → Mesobiliverdin 104.
Glenodinium foliaceum 444.
β-(1 → 3)-linked D-Glucan 469.
3-O-β-D-Glucopyranosyl-D-mannit 218, 274.
4-O-β-D-Glucopyranosyl-D-mannose 218.
Glucose 145, 218, 219, 274, 343, 345, 425, 469.
^{14}C-Glucose 274, 275.
Glucose-U-^{14}C 280.
Glucose-Konjugate 91.
L-Glucuronic acid 467.
β-Glucuronidase 91.
Glucuronsäure 91.
Glusulase 427.
Glutamat-glyoxylat-Transaminase 273.
Glutamat-hydroxypyruvat-Transaminase 273.
Glutamat-oxalacetat-Transaminase 273.
Glutamic acid s. Glutaminsäure.
Glutamin 274.
Glutaminsäure 41, 123, 197, 274.
Glutarimid 2, 40, 148.
Glutarimide antibiotics 140f.
—, antifungal activity 145.
—, biosyntheses 186f.
—, chemical assay methods 145f.
—, chemistry 141f.
—, Cottoneffect 142, 143.
—, determination 145.

Glutarimide antibiotics, isolation 145.
—, nomenclature 141f.
—, optical activity 142, 143, 144.
—, ORD-spectra 142, 143, 144.
—, synthesis 186f.
Glutarimide ring, fission 172, 174.
(1 R; 5 R; 10 R)-1-(Glutarimidomethyl)-chinolizidin → Lamprolobin 31.
3-Glutarimidyl-acetaldehyde 193, 194.
3-Glutarimidylacetyl chloride 190.
Glutathion 117.
(R)-Glyceraldehyde 152.
Glycerin 217, 454.
^{14}C-Glycerin 275.
Glycerol s. Glycerin.
D-Glycero-D-taloheptose 218.
Glycin 110, 121, 123.
β-Glycosidase 308.
Glycosides 425.
Glycosyl-aminoglycans 469.
Glycopeptide 454.
Gmelin-Reaktion 64, 65.
Golden Hubbard Squash, fresh juice 311.
Gonyaulax catenella 452.
β-Gorgonene 435, 436.
—, biosynthesis 436.
—, silver nitrate adduct 435.
Gogonian cortex 345.
Gorgonians (Coelenterates) 421, 435, 437, 439.
Gorgonia ventalina 438.
Gorgosterol 419.
Gramineen 4.
Graphium leonidas 110.
Graphium policenes 110.
Graphium sarpedon 110.
Gratiogenin 343, 345, 346, 350.
—, occurrence in nature 353, 355.
—, physical constants 356.
Gratiogenin derivatives 332.
Gratiola officinalis 345, 354.
Gratioside 345.
Grayanoldicarbonsäure 257.
Grayanolsäure 257.
Grayanotoxins 402.
Grayansäure 257, 286.
—, Hydrolyse 256, 257.
Grey dogfish *(Squalus acanthias)* 432.
Grisan 257.
Griseogenin 427, 428.
Guanidin 117, 174.
Guanidine type compound 455.

Guvacin 6.
Guvacolin 6.
Gymnodinium breve 452.
Gymnothorax javanicus 453.
Gyrophora esculenta 219.
Gyrophorsäure 273.
—, Biosynthese 277.

Haemathamnol 253.
Haemathamnolsäure 253.
Haematomma coccineum 240.
Haematomma porphyrium 240.
Haematommsäure-4-methyläther 253.
Hagfish *(Eptatretus stoutii)* 434.
Halbstercobilin 95, 98.
—, Chromsäureabbau 99, 101.
—, Dehydrierung 99.
—, FeCl$_3$-Reaktion 102.
—, Massenspektrum 99.
Halbstercobilinogen 95, 99.
Halbwertszeiten beim Chromatabbau von Pyrrolpigmenten 70.
Halodeima grisea 427.
Halogen-containing compounds from marine sources 446f.
Halosalin 14.
Haloxin 14.
Haloxylon-Basen 14.
Haloxylon salicornicum 3, 14, 25.
Hämatinsäureimid 99, 101, 103, 106, 113.
—, -methylester 72.
Hardwickia pinnata 398.
Harnstoff 126.
Heliopora coerulea 108.
Helioporobilin 108.
Hemicentrotus pulcherrimus 460.
cis,cis,cis,cis,cis-Heneicosa-1,6,9,12,15,18-hexaene 440.
cis,cis,cis,cis-Hencicosa-1,6,9,12,15-pentaene 440.
(−)-1-(2 R :3 R)-Heptadecantricarbonsäure → Norrangiformsäure 222.
n-Heptan 88.
Herbipoline 457.
Herzynine 457.
Heterodontus francisci 446.
Heuschrecken, Vorkommen von Bilin 108.
trans-Hexadec-2-ensäure-methylester 223.
—, Epoxid 223.
trans-1-*(trans,cis*-Hexa-1′,3′-dienyl)-2-vinylcyclopropane → Dictyopterine B 440.

Hexahydroanabasin 29.
3α,7α,12α,24ξ−26,27-Hexahydroxycoprostane → Scymnol 433.
1,2,5,6,7,8-Hexahydroxy-3-methylanthrachinon 234.
Hexane 437.
trans-1-*(trans*-1-Hexenyl)-2-vinylcyclopropane → Dictyopterin A 441.
Hibaene 379.
(+)-Himandravin 36.
Himantandraceae 36.
Himbacin 36.
(+)-Himbelin 36.
Himgravin 37.
Holothorioideae (sea cucumbers) 425.
Holothuria sp. 430, 431, 470.
Holothuria vagabunda 428.
Holothurin A 425.
—, enzymatic hydrolysis 427.
—, mild hydrolysis 426.
Holothurins 431.
—, biogenetic realtionship with lanosterol 430.
—, geographic differences 430.
Holothurinogenins 425, 427, 431, 470.
Homarine 451, 456.
Homodiploschistessäure-methylester 250.
Homoorsellinsäure 273.
Homosekikasäure 250, 286.
Homostachydrin 4.
—, Biosyntheseversuche 4.
—, Konfigurations-Ermittlung 4.
Hopan 262, 264.
—, Derivate 283.
Hopen-1 264, 265.
Hopenone-1 374.
Hormonal activity of polypeptide type toxins 458.
Horned shark *(Heterodontus francisci)* 446.
Huang-Minlon-Reduktion 267.
Hubbard squash seedlings 348.
Hudson-Klyne-Regel 224.
Humulus lupulus 4.
Hund 106.
Hyalococcus 274.
Hydrangea umbellata 17.
Hydrazine 174.
Hydrazinhydrat 229.
Hydrogenation 164, 381.
Hydrogen bonds 153, 164.
Hydrogen bromide 374.

34*

Hydrogen peroxide, oxidation 455, 456.
Hydroids 451.
2-Hydroxy-3-acetyl-7-methoxynaphtha-
zarin 463.
2-Hydroxy-3-acetylnaphthazarin 462.
2-Hydroxy-6-acetylnaphthazarin 464.
Hydroxyancepsenolide 436.
Hydroxyanthraquinone pigments 466.
4-Hydroxy-2,4-dimethylcyclohexanone
176.
20-Hydroxyecdysone 423.
3-(β-Hydroxyethyl)glutarimide 182.
2-Hydroxy-6-ethyljuglone 463.
2-Hydroxy-3-ethylnaphthazarin 462.
2-Hydroxy-6-ethylnaphthazarin 464.
Hydroxyfettsäuren aus Flechten 223.
16-Hydroxygratiogenin 343, 345.
—, -3,16-diacetate 346.
—, occurrence in nature 353, 355.
—, physical constants 356.
(—)-α-Hydroxyheptanoic acid 439.
22-Hydroxyhopan 264, 267.
Hydroxyhopanon 265.
2-Hydroxyisobutyraldehyde 313, 314.
(—)-16α-Hydroxykauran 261, 281.
7β-Hydroxy-(—)-kaur-16-en-19-oic acid
410.
7β-Hydroxykaurenolide 410.
α-Hydroxyketones, formation of a formazan 314.
—, triphenyltetrazolium chloride test 337.
β-Hydroxyketones 194.
Hydroxylamine 91, 146.
Hydroxylapatit 125.
Hydroxylysin 8, 27.
δ-Hydroxylysin-(6-^{14}C) 8.
2-Hydroxymethylene-4,6-dimethyl-cyclo-
hexanone 188.
trans-4-Hydroxy-4-methyl-pent-2-enoic
acid 312.
—, acetate 312.
α-Hydroxy-α-methylpropionaldehyde
339.
α-Hydroxy-γ-methylthiobuttersäure 274.
3β-Hydroxy-norhopan-22-on 265.
2-Hydroxy-6-(3-oxo-n-pentyl)-anissäure
248.
3-Hydroxypicolinsäure 40.
4-Hydroxypipecolinsäure 5, 8.
5-Hydroxypipecolinsäure 5, 8.
Hydroxypiperidin-Alkaloide 10.
20β-Hydroxy-5α-pregnan-3,6-dione 423.

Hydroxypyrrole 89.
7β-Hydroxyrosenonolactone 405.
4-Hydroxyskytanthin 35.
7-Hydroxyskytanthin 35.
4a-Hydroxyskytanthin 35.
7a-Hydroxyskytanthin 35.
5-Hydroxytryptamine 451.
1-Hydroxyxanthone 215.
—, Tüpfeltest 240.
Hyperbilirubinämie 93.
Hypogymnia physodes 223, 275.
Hypoprotocetrarsäure 257, 258.
Hystrin 27.

Ianthella ardis 449.
Iberis amara 337.
Iberis sp. 354.
Ikterus 94.
Imbricarsäure 286.
β-[Imidazolyl-(4)]-acrylcholine 454.
Imide, Dünnschichtchromatographie 72.
Iminodicarboxylic acid 456.
Inactone 144, 179 f.
—, alkaline degradation 180.
—, catalytic hydrogenation 179.
—, dihydroderivatives 179.
—, —, alkaline degradation 180.
—, —, antifungal activity 179.
—, —, oxidation 180.
—, IR-spectrum 179.
—, isolation 179.
—, mass spectrum 180.
—, reduction 180.
—, structure 179.
Indicaxanthin 38, 39.
Indolalkaloide, piperidinhaltige 33.
—, Biosynthese 34.
Infrarot-Spektren s. IR-Spektren.
myo-Inosit 217.
Insect attractants 351.
Insecticide 453.
Insekten, Vorkommen von Biliproteiden
108.
Insektoverdin 108.
Insulin 469.
Insulin-like activity 469.
Invertase 273.
Iodine 384.
IR-Spektren.
   Actiphenol 181.
   Bilirubin 89.
   Bryosigenin 343.

Sachverzeichnis. Subject Index 533

IR-Spektren.
  Ciguatoxin 453.
  Crustecdysone 425.
  Cucurbitacins 350.
  Cycloheximide 147.
  Elateric acid 324.
  αEthyl-glutarimide 149.
  βEthyl-glutarimide 149.
  Flechtenstoffe 211.
  Flechtenxanthone 239.
  Holothurinogenins 426.
  trans-4-Hydroxy-4-methyl-pent-2-enoic acid 312.
  Inactone 179.
  Isoelateric acid 324.
  2-Methyl-3-dodecylbernsteinsäure 222.
  Myxinol 434.
  Naramycine-B 151.
  Neocycloheximide 157.
  PC-3 220.
  Phenolcarbonsäuren 211.
  Protomycin 184.
  Reaction products from cucurbitacins 319, 332, 333.
  Roccellsäure 222.
  Steroids 419.
  Streptimidone 182.
  Triketone from elaterin 317.
  Usninsäure 242.
  Xanthone 211.
Isoammodendrin 28.
Isocaproic acid 313, 314.
—, -methyl ester 314.
Isocaprolactone 312, 313.
Isocaryophyllene, epoxide 370.
Isocrustecdysone 424, 425.
Isocucurbitacin B 341, 342.
—, occurrence in nature 352, 354.
—, physical constants 356.
Isocycloheximide 142, 154, 157.
—, acetate, NMR-spectrum 160.
—, activity on *Saccharomyces* 150.
—, anhydro product 167.
—, catalytic reduction 166.
—, chromic acid oxidation 150.
—, dehydration 167.
—, isomerization 156.
—, NMR-spectra 160.
—, ORD-spectra 154.
—, orientation of the α-hydroxyl-group 168.

Isocycloheximide, reduction products, NMR-spectra 165.
—, reduction with aluminium isopropoxide 166.
—, relation to cycloheximide 153.
—, structure 146f, 150, 161.
—, synthesis 186, 187f, 189, 191.
—, thermal degradation 156.
—, toxicity 150.
Isodihydrousninsäure 244, 245.
Isoelateric acid 323.
—, formation 323, 324.
—, IR-spectrum 324.
—, -methyl ester 324.
Isoeuphenol-like compound 375.
Isofebrifugin 17.
Isohopan 262, 264.
Isolaureatin 450.
Isolaurene 441, 442.
Isoleucotylsäuremethylester 264.
Isolichenin 218.
Isomesobiliviolin → Mesobilirhodin 99.
Isonorlobaridon 254.
Isoorensin 28, 29.
Isopelletierin 12, 16, 17, 42.
—, Biosynthese 13.
Isopimaric acid 379, 399.
Isopinastrinsäure 271.
—, ozonolytischer Abbau 272.
Isopren 33, 220.
Isoprenoid biosynthesis 403.
Isoprenoid quinones 442.
Isopropanol 87.
2′-Isopropyl-4′-(trimethylammonium chloride)-5′-methylphenylpiperidine-1-carboxylate → Amo 1618 412.
Isorenieratene 443.
Isotoma longiflora 20.
Isotopen-Hybride 86.
Isotripiperidein 28, 29.
Isousninsäure 242f, 245, 284.
—, Abbau und Hydrierung 246.
Isozeorinin 211.
Isozeorininon 211.
Isurus glaucus (Pacific mako shark) 446.

Jaffe-Schlesinger-Reaktion von Bilenen, Biladienen und Bilatrienen 66.
Jasus lalandei 422.
Jelly fishes 451.
5-Jod-2-hydroxyhydrochinon-trimethyläther 270.

Jodacetylvicanicin, Röntgenstrukturanalyse 215, 258.
5-Jodorsellinsäure 227.
Jodwasserstoffsäure 254.
—, reduktiver Abbau von Gallenfarbstoffen 68.
Jones reagent 380.
Juglone 460.
Juglone analogs 465.
Julocrotin 37.
Julocroton montevidensis 37.

α-Kainic acid 458.
Kaktusfeige *(Opuntia ficus-indica)* 38.
Kaliumhexacyanoferrat(III) 244.
Kaliumpermanganat 240, 241.
Kaliumperoxydisulfat 235.
Katalase 273.
Katsuwonus pelamis 106.
Kauranoid diterpenes 407, 408.
Kaurene 402, 407, 408.
(—)-Kaurene (*ent*-Kaurene) 397, 403, 409.
17-^{14}C-(+)-Kaurene 407.
(—)-Kaur-16-en-19-al 409, 412.
17-^{14}C-(—)Kaur-16-en-19-al 409.
(—)-Kaur-16-en-19-oic acid 409, 412.
17-^{14}C-(—)-Kaur-16-en-19-oic acid 409.
(—)-Kaur-16-en-19-ol 409.
17-^{14}C-(—)-Kaur-16-en-19-ol 409.
Kaurenolides 408, 409, 410.
Kedrostis sp. 354.
Ketal formation 166, 167.
Ketols, oxidation to diosphenols 311.
α-Keto-γ-methylthiobuttersäure 274.
16-Keto-13β,14α-steroids, negative Cotton effect 331.
Kieselgel D 5 88.
Kieselgel G 72, 87, 88.
—, weinsaures 234.
Kodak Chromagramfolie K 301 R 2 210.
Koellikerigenin 429, 430.
Kohlendioxid 274.
^{14}C-Kohlendioxid 277, 404.
^{14}C-Kohlendioxid-Kurzzeitversuche 12.
Kohlenhydratstoffwechsel bei Flechten 274.
Kohlweißling *(Pieris brassicae)* 110.
Kolavenol 398.
Korolkowia severzowii 38.
Krebs-2-ascites tumor, growth inhibition 454.
Kuhn-Roth degradation 406.

Labdadiene 397.
Labdadienol pyrophosphate 397, 405.
Labdane, bicyclic 404.
Labdane series 399.
Labdanolic acid 397.
Labriden (Lippfische) 106.
Lackmus-Farbstoff 228.
β-Lactam antibiotics 458.
γ-Lactone rings, unsaturated 437.
trans-Lactonization 404.
γ-Lactonsäuren 284.
—, Derivate in Flechten 223.
Lagenaria mascarena 354.
Lagenaria siceraria 339, 341, 354.
Laminaran 469.
Laminaria angustata 459.
Laminaria sp. 469.
Laminine 459.
—, oxalate 459.
Lamprolobin 31, 32.
Lamprolobium fructicosum 31.
19(10 → 9β)-*abeo*-5α-Lanostane 310.
Lanostane series, interrelationship with the cucurbitane series 334.
Lanostane skeleton 428.
Lanostane-type tetracyclic triterpene 309.
19(10 → 9β)-*abeo*-10α-Lanost-5-ene 310.
Lanosterol 348, 365, 369, 370, 374, 429, 431.
Lasallia papulosa 219, 229.
Lasallia pustulata 219, 273, 277.
Latimeria chalumnae (Coelacanth) 446.
Lauraceae 15.
Laureatin 450.
Laurencia glandulifera 450.
Laurencia nipponica Yamada 441, 450.
Laurencia sp. 442, 470.
Laurencin 450.
Laurene 441, 442.
Laurenisol 451.
Laurera purpurina 234.
Laurinsäure 223.
Laurinterol 451, 470.
Lead tetraacetate 265, 321, 322, 329, 381, 382, 426, 454.
Lecanora-Arten 228, 238.
Lecanora carpinea 235.
Lecanora muralis 264.
Lecanora reuteri 238.
Lecanora rupicola 235, 285.
Lecanora saligna 245.
Lecanora sarcopis 245.

Lecanora straminea 238.
Lecanora vinetorum 239.
Lecanorsäure 210, 237.
—, Halogenierung 251, 252.
—, -methylester 245.
—, UV-Spektrum 212.
^{14}C-Lecanorsäure 278.
Lecidea-Arten 238.
Lecidea carpathica 239.
Lecidea coarctata 285.
Lecidea leucophaea 248.
Lecidea lilienstroemii 248.
Lecidea lithophila 247.
Lecidea piperis 234.
Lecidea plana 247, 285.
Lecidea silacea 240.
Leguminosae 4, 22 25, 28, 31.
Leontice alberti 25.
Leontice lenotopetalum 32.
Leontiformin 31, 32.
Leopard seal 433.
Lepidium sativum (Gartenkresse) 287.
Lepidopteren (Schmetterlinge).
—, Phylogenie 110.
—, Vorkommen von Bilatrienen 108.
—, Vorkommen von Gallenfarbstoffen 110.
Leprantha impolita 225.
Lepranthin 225.
Lepraria-Arten 272.
Lepraria latebrarum 237.
Lepraria membranacea 241.
Leprarin 237.
Leprarinsäure 272.
—, -methyläther 272.
Leprarsäure 237.
—, Abbaureaktionen 236.
Leptodactylus pentadactylus labyrinthicus 456.
Leptogorgia setacea 434.
Letharia vulpina 279.
Leucaena glauca 4, 5.
Leucin 86, 121.
Leucotylin 262, 264.
—, Derivate 264.
Leucotylsäure 264, 265.
—, -methylester 264.
Leukemia 1210 system 174, 196.
α-Levantenolide 367.
β-Levantenolide 367.
Levopimaric acid 399.
Lewis lung carcinoma (mice) 351.

Lichenase 273.
Lichenin 219.
(—)-Lichesterinsäure, Konfiguration 224.
Lichexanthon 237.
Liliaceae 237.
Limonene 384.
Limulus polyphemus 457.
Linaloyl diphenyl phosphate 384.
Liparia parva 28.
Liparia sphaericarpa 28.
Lipase 273.
Lippfische (Labriden) 106.
Lithium chloride 385.
Lithium aluminium hydride 171, 242, 265, 267, 314, 316, 374, 385, 426, 453.
Lithium aluminium tri-t-butoxyhydride s. Lithium tri(t-butoxy)-aluminium hydride.
Lithium tri(t-butoxy) aluminium hydride 164, 170, 171, 191.
Lobaria amplissima 250.
Lobaria-Arten 248.
Lobaria dissecta 248.
Lobaria laetevirens 217, 273.
Lobaria pulmonaria 217, 273, 274.
Lobaria scrobiculata 217, 250.
Lobaria verrucosa 250.
Lobariol 254.
Lobelia-Alkaloide 14, 19f.
—, Biosynthese 20.
Lobelia-Arten 14, 20, 30.
Lobelia cardinalis 30.
Lobelia inflata 14, 19, 20.
Lobelia syphilitica 20, 30.
Lobelidiole 19, 20.
Lobelidione 19, 20.
Lobelin 19, 20, 42.
—, Biosynthese 20, 21.
Lobelionole 19.
Lobinalin 29f, 42.
—, Biosynthese 30.
—, Konfigurationsbestimmung 30.
Locusta migratoria 108.
Longifoline 363.
Loroxanthin 445.
Loxodin → Norlobarsäuremethylester 254.
Lucenin type of flavonoids 467.
Luffa acutangula 354.
Luffa cylindrica 354.
Luffa echinata 341, 354.
Luffa operculata 341, 354.

Lumbriconereis heteropoda 452.
Lupanin 25, 28.
Lupeol 268.
Lupinenalkaloide 16.
Lupinenextrakte 27.
Lutein 443, 445.
2,6-Lutidin 87.
Lutjans bohar 453.
Luzerne 4.
Lycoctonine alkaloids 402.
Lymphatic leukemia 351.
Lymphocytes, human, morphological changes 351.
Lysiloma bahamense 5.
Lysin 18, 20, 24, 26, 27, 30, 31, 32, 38, 41, 42, 197.
—, Einbau in Pipecolinsäure 6f.
—, oxydative Desaminierung 6.
—, Precursor für Piperidinbasen 4.
—, Stoffwechsel 3, 5, 17, 26.
—, Vorstufe von Coniin 10.
Lysine, s. Lysin.
Lythraceen-Alkaloide 24.
Lythramin 24.
Lythranidin 24.
Lythranin 24.
Lythrum anceps 24.

(+)-γ-Maaliene 436.
Mackinlaya macrosciadea 25.
Mackinlaya sublata 25.
Magnesium 374.
Magnesiummethylcarbonat 227.
Malabaricanediol 375.
Maleinimide 70, 71, 72.
Malonamyl coenzyme A 200.
1-^{14}C-Malonat 276.
1-^{14}C-Malonsäure 277.
2-^{14}C-Malonsäure 277.
1-^{14}C-Malonsäurediäthylester 277.
Malonsäuredimethylester 223.
Malonyl coenzyme-A 200, 284.
Mangandioxid, aktives 257, 259.
Manganese dioxide, oxidation 336.
D-Mannit 217, 274, 275.
—, -glykosid 275.
^{14}C-Mannit 274.
D-Mannose 218.
D-Mannuronic acid 467.
Manool 395.
Manoyl oxides 397.
Mantis religiosa 108.

Marah oreganus 314, 338, 341, 354.
Marine invertebrates, steroids derived from 418, 420, 421.
Markownikow rule 366.
Marrubiin 398.
—, biosynthesis 403.
Marrubium vulgare (White horehound) 398, 403.
Massenspektrometrie.
 Anthrachinone 215.
 Aplysioviolin 113.
 Aristolene 436.
 Arthoniasäure 247.
 Bilene(b) 102.
 Bryosigenin 343.
 Carpain 23.
 Cucurbitacin Q 347.
 Cucurbitacins 350.
 Cycloheximide 180.
 16-Deoxymyxinol 435.
 Depside 213, 214.
 Depsidone 214.
 4-O-Desmethylbarbatinsäure 251.
 Flechtenstoffe 213, 214, 215.
 Gallenfarbstoffe 79f.
 Gratiogenin 346.
 Halbstercobilin 99.
 16-Hydroxygratiogenin 346.
 Inactone 180.
 Lobinalin 30.
 Maaliene 436.
 Mesobilirhodin 103.
 Mesobiliviolin 101, 103.
 Myxinol 434.
 Phycobiline 119.
 Phycobiliverdin 120.
 Pulvinsäurederivate 215.
 Quinoid pigments 460.
 Rhodocladonsäure 228.
 Streptimidone 180.
 Steroids 419.
 Tetrahymanol 435.
 Urobilin 94.
 Xanthone 215.
Medicago sativa 4.
Meerrettichperoxidase 279.
Mercaptoäthanol 117.
Mercenaria campechiensis 454.
Mercenaria mercenaria 454.
Mercenene 454.
Merochlorophaesäure 250, 286.
Melothria punctata 354.

1-Menthydrazide 148.
Mesitol 225.
Mesobiladiene.
  Dünnschichtchromatographie 88.
  NMR-Spetrum 85.
Mesobiladien-(a,b)-IX α 70.
—, Chromatabbau 70.
—, -dimethylester, Massenspektrum 83.
—, Elektronenspektren 74.
—, pK_a-Wert 75.
Mesobilan 65.
—, Elektronenspektren 73.
Mesobilan-IX α 65.
—, Dünnschichtchromatographie der Oxydationsprodukte 88.
Mesobilatrien-IX α 118.
—, Chromatabbau 70.
—, Elektronenspektren 74.
—, pK_a-Wert 75.
Mesobilatriene 104.
—, Dünnschichtchromatographie 88.
—, NMR-Spektrum 85, 104.
Mesobilen-(b)-IX → Urobilin 64.
—, Chromatabbau 70.
—, Dünnschichtchromatographie 87.
—, Elektronenspektrum 74.
—, pK_a-Wert 75.
Mesobilen-(b)-IX α-dimethylester, Massenspektrum 83.
Mesobilipurpurin, Dünnschichtchromatographie 87.
Mesobilirhodin 76, 99, 102, 120.
—, Chromsäureabbau 103.
—, Dünnschichtchromatographie 87.
—, Elektronenspektrum 103.
—, Massenspektrum 103.
—, Struktur 103.
Mesobilirubin 91.
—, Dehydrierung zu Mesobiliverdin 104.
—, Dünnschichtchromatographie 87.
Mesobilirubin-IX α, Resorcinspaltung 104.
Mesobiliverdin 104 f, 120.
—, Bildung durch Dehydrierung aus Mesobilirubin 104.
—, Dünnschichtchromatographie 87.
Mesobiliviolin 76, 82, 99, 101, 113, 119.
—, Absorptionsmaxima 127.
—, Dünnschichtchromatographie 87.
Mesobiliviolin-IX α-dimethylester, Massenspektrum 83.
Mesoporphyrin IX 64.

Methallyl vinyl ether 381.
Methanetriacetic acid 149, 180, 187, 200.
Methanol 78, 86, 87, 99, 103, 113, 118, 119, 120, 121, 126, 151, 174, 237, 385, 424, 454.
Methionin 13, 16, 40, 86.
—, Metabolismus in Flechten 274.
12β-Methoxy-22-acetoxy-7,8-dihydroholothurinogenin 427.
p-Methoxybenzoylameisensäure 271.
—, -methylester 271.
o-Methoxybenzylcyanid 271.
12 α-Methoxy-17-desoxy-7,8-dihydro-22,25-oxidoholothurigenin-3-acetate 427.
12 β-Methoxy-7,8-dihydro-24,25-dehydroholothurinogenin 427.
—, 3β-xyloside 427.
12β-Methoxy-7,8-dihydro-22,25-oxidoholothurinogenin 426.
—, 17-desoxy analog 426.
C-20-Methoxyholothurinogenin 431.
6-Methoxy-4-methylbenzofuran-2-aldehyd 241.
Methoxymethylentriphenyl-phosphoran 242.
Methoxymethyl fluoroborate 390.
6-Methoxy-4-methyl-2-(β-methoxyvinyl)-benzofuran 242.
2-(p-Methoxyphenyl)-5-phenyl-3,4-dioxoadiponitril 272.
3-(p-Methoxyphenyl)-2-(1′-piperidyl)-n-propanol → Alangin A 37.
Methylacetat 87, 88.
Methylacrolein 313, 314.
(+)-3-Methyladipic acid 151.
Methylamine 149.
N-Methylanabasin 25.
N-Methylanatabin 25.
N-Methylaniline 195, 196.
o-Methyläther-olivetolcarbonsäure 247.
p-Methyläther-olivetonid 247.
Methyl-äthyliden-succinimid 72, 113, 114, 127, 128.
Methyl-äthyl-maleinimid 72, 99, 101, 103, 127.
3-Methyl-4-äthyl-pyrroldialdehyd-(2,5) 72.
Methyl-äthyl-succinimid 72, 96, 98, 99, 101, 103.
24 α-Methylcholest-5,22-diene-3β-ol 419.
N-Methylconiin 9, 11.

4-O-Methylcryptochlorophaesäure 250, 286.
Methyl cyanoacetate 187.
2-Methylcyclohexanone 194, 196.
4-Methylcyclohexanone 194, 196.
4-Methyl-4-cyclohexen-1,2-dicarbonsäure 222.
—, dimethylester 220.
N-Methylcycloheximide acetate, NMR-spectrum 160.
Methylcyclopentan 35.
trans-N-Methyl-dekahydrochinolin 29.
3-O-Methyl-2,5-dichlornorlichexanthon 238.
6-O-Methyl-2,4-dichlornorlichexanthon 239.
(+)-2 S-Methyl-3 R-dodecylbernsteinsäure → (+)-Roccellsäure 222.
threo-2-Methyl-3-dodecylbernsteinsäure 222.
Methyl ecballate 323.
Methyl elaterate 324.
[10-^{14}C]-2-Methylemodin 233.
Methylene chloride 145, 382.
24-Methylenecholesterol 418, 420.
(22 R,23 R,24 R)-22,23-Methylene-23,24-dimethylcholest-5-en-3β-ol → Gorgosterol 421.
Methyl ethyl ketone 182, 184, 194.
1-O-Methylfragilin 234.
Methyl geranate 390.
3-O-Methylglucose 425, 427.
β-Methyl-glutaconsäure 237.
β-Methyl-glutarsäure 237.
Methyl group migration 404, 406.
4-O-Methylgyrophorsäure 248.
—, Hydrolyse und Methylierung 249.
6-Methylheptan-2,3-dione 426.
6-Methylheptan-2-one 426.
(+)-2-4-Methyl-2-hexanone 151.
2(S)-Methyl-3(S)-hydroxy-6(R)-(7′-carboxyheptyl)-piperidin → Carpaminsäure 24.
2(S)-Methyl3(S)-hydroxy-6(R)-(5′-carboxypentyl)-piperidin → Aziminsäure 24.
2(R)-Methyl-3(R)-hydroxy-6(S)-(11-oxododecyl)-piperidin → Cassin 22.
2-Methyl-8-hydroxy-2 H-pyrano[3,2-g]naphthazarin 466.
N-Methyl-2-hydroxypyrrolidin 17.
Methyl iodide s. Methyljodid.

Methyl isocupressate 383.
N-Methylisopelletierin 12.
—, Biosynthese 13.
Methyl isopropyl ketone 194.
Methyljodid 243, 448.
Methyl lithium 382, 385, 389.
$N_δ$-Methyl-lysin 16.
Methyl magnesium chloride 389.
[Methyl-^{14}C]-methionine 197, 276.
Methyl-(2-methoxycarbonyl-äthyl)-maleinimid 72.
3-Methyl-4-(2-methoxycarbonyl-äthyl)-pyrroldialdehyd-(2,5) 72.
2-Methyl-3-(2-methoxycarbonyl-äthyl)-succinimid 72.
(+)-8-Methylnorlobelol → (+)-Sedridin 14.
(+)-4-Methyl-6-oxoheptanoic acid 151.
(+)-2 S-Methylpentadecansäure 222.
trans-8-Methyl-10-phenyl-4,5-dehydrolobelidiol → Sedinin 20.
Methylphloracetophenon 244, 245, 278, 279.
—, -CO-^{14}CH$_3$ 278.
4-O-Methylphysodsäure 257.
L-(−)-N-Methylpipecolinsäure 25.
N-Methylpiperidin 3.
—, -N-oxid 3.
Methylpiperidine.
—, Biosynthese 4.
—, natürliches Vorkommen 3.
(−)-cis-2-Methyl-6-(2-propenyl)-piperidin 18.
Methylpropionat 88.
erythro-2-Methyl-3-propyl-bernsteinsäure 223.
Methylpropylmaleinimid 223.
3-Methyl-pyrroldialdehyd-(2,5) 72.
4-O-Methylsalazinsäure 254.
6-Methylsalicylsäure 273.
Methylsuccinimid 72.
Methyl-2,3,4,6-tetra-O-methyl-D-glucopyranosid 220.
3-O-Methyl-2,5,7-trichlornorlichexanthon 238.
Methyl-2,4,6-tri-O-methyl-D-glucopyranosid 220.
7-O-Methylusninsäure 243.
8-Methylusninsäure 243.
Methylvinylmaleinimid 72, 106, 110, 113, 128.

3-Methyl-4-vinyl-pyrroldialdehyd-(2,5) 72.
Mevalonat 41.
^{14}C-Mevalonat 276.
Mevalonat-Derivate in Flechten 261.
2-^{3}H$_2$-Mevalonate 405.
5-^{3}H$_2$-Mevalonate 405, 406.
Mevalonate metabolism 409.
Mevalonates, stereospecifically labelled 397.
Mevalonate units as precursors in biosynthesis of diterpenoids 403, 406, 408.
(2-^{14}C)-Mevalonic acid 197.
[(4 R)-4-^{3}H,2-^{14}C]-Mevalonic acid 348.
Mevalonsäure 36, 42, 281, 284.
Michael addition 187, 336.
Mimosaceae 23.
Mimosa pudica 4.
Miriquidisäure 248.
Mitochondrien-Schwellung 93.
Mixed function oxidase 408, 409.
Molecular distillation 184.
Molekulargewichts-Bestimmungsmethoden 117.
Mollusks.
—, hypobronchial body 454.
—, steroids 421.
Monocyclofarnesol 367.
Monomethylamin 217.
Monomethyl glutarate 439.
Monoperphthalic acid 371.
Monosaccharide in Flechten 218.
Monoterpenes 397.
—, biosynthesis 403.
Monoterpenoide Piperidinalkaloide 34f.
Monoterpenoids, bridged bicyclic 373.
Montagnetol 227, 249.
—, Synthese 227, 228.
Morpho catenarius 110.
Morpholine 190, 195, 196.
Morus alba 4, 5.
Mougeotia 124.
Moulting hormones 422.
Möve 106.
Murexine 454.
Murex trunculus 454.
Mutterkornfarbstoff 261.
(+) α-Muurolene 248.
Mycobacterium tuberculosis H 37 R 450.
Mycobionten 210, 218.
—, isolierte Verbindungen 285.

Mycoblastus sanguinarius 234.
Mykamycin 40.
Mylothris sp. 110.
Myoxocephalus scorpioides 106.
Myrcene 384.
Myriogramme spectabilis 116.
Myristinsäure 220.
Myrmecia 274.
Myxine glutinosa 434, 469.
Myxinol 434, 435.
—, -disulfate 434, 435.

NAD 2.
Namakochrome 462.
Naphthazarin 460.
—, analogs 465.
Naphthochinone in Flechten 228.
Naphthopurpurin 464.
Naphthoquinone pigments 466.
Naramycin-A → cycloheximide 142, 146.
Naramycin-B 142.
—, -acetate, NMR-spectrum 160.
—, alkaline degradation 151.
—, anhydro product 167.
—, dehydration 167.
—, IR-spectrum 151.
—, isomerization 156.
—, NMR-spectrum 160.
—, ORD-spectrum 154.
—, orientation of the α-hydroxylgroup 168.
—, structure 146f, 150f, 153, 161, 164.
—, synthesis 186, 187, 191.
—, thermal degradation 156.
Nasopharynx, human carcinoma of 351.
Natriumacetat-2-^{14}C 278.
Natriumamalgam 94.
—, Reduktion von Gallenfarbstoffen 65, 74.
Natriumborhydrid 218, 237.
Natriumdithionit 229.
Natriumdodecylsulfat-Polyacrylamidgel-Elektrophorese 117.
Natriumformiat-^{14}C 278.
Natriumhydrogencarbonat-^{14}C 274.
Natriumhypochlorit 215.
Natural products derived from marine sources 417f.
—, antibiotics 447.
—, bile acids 431f.
—, bile alcohols 431f.
—, carbohydrates 467.

Natural products derived from marine sources, halogen-containing compounds 446.
—, non-protenoid nitrogen-containing substances 451.
—, quinoid and related compounds 460f.
—, sapogenins 425f.
—, steroids 418f.
Nemalionales 115.
Neoabietic acid 399.
Neocycloheximide 157, 188, 189.
—, -acetate, NMR-spectrum 160.
—, chromic acid oxidation 157.
—, IR-spectrum 157.
—, structure 161.
Neodihydroprotolichesterinsäure 224.
—, -methylester 223.
Neo-holothurinogenins 427.
—, methoxylated 426.
Neopterobilin 110.
Neopulchellidin 38.
Nepheronia thalassina 110.
Nephroarctin 251.
Nephroma arcticum 251.
Nephroma laevigatum 229, 231.
Nephromopsinsäure 224.
Nereistoxin 452, 453.
Neryl diphenyl phosphate 384.
Neugeborenen-Serum, Bilirubingehalt 93.
—, β-Lipoproteid-Fraktion 94.
Neurospora crassa, Lysin-Mangelmutante 7.
Nicotellin 26, 27.
Nicotiana-Alkaloide 26.
Nicotiana-Arten 25, 27.
Nicotiana glauca 26.
Nicotiana rustica 26.
Nicotiana tabacum 3, 4.
Nicotin 13, 27.
Nicotinsäure 4, 6, 27, 273.
—, N-Methylierung 4.
Nielsen condensation 188, 196.
Nigrifactin 18.
Niromycin-A 143, 186.
Niromycin-B 143, 186.
Nitella hookeri 467.
Nitenin 472.
Nitraria schoberi 34.
Nitrarin 33, 34.
p-Nitroacetophenone 195.
Nitroketones, aromatic 195.
Nitromethane 381.

Nitzschia closterium F. *minutissima* 445.
^{15}N-markierte Verbindungen.
α-Aminoadipinsäure 7.
Lysin 6.
(2-^{14}C,ε-^{15}N)-Lysin 26.
NMR-Doppelresonanzexperimente 248.
NMR-Spektren.
O-Acetylstreptimidone 184.
Aristolene 436.
Arthoniasäure 247.
Bilatriene 104.
Bilirubin 89, 90.
Bryogenin 343.
2-t-Butyl-4-methylcyclohexanone 158, 159.
4-t-Butyl-2-methylcyclohexanone 158, 159.
Ciguatoxin 453.
Crustecdysone 425.
Cucurbitacin F derivative 346.
Cucurbitacin O and P derivatives 346.
Cucurbitacin Q 347.
Cucurbitacins 310, 350.
Cucurbitone A 325.
Cycloheximide 157, 158, 159, 160.
Cycloheximide derivatives 160.
Cycloheximide isomers 160.
22-Deoxocucurbitacin D 339.
Deoxyisocycloheximide 169.
16-Deoxymyxinol 435.
Depside 212.
Depsidone 212.
4-O-Desmethylbarbatinsäure 251.
1,8-Dihydroxyanthrachinone 213.
Diosphenol system 309.
Diterpenoid substances 395.
E 73 160, 177.
α-Epiisocycloheximide 191.
Flechtenstoffe 212.
Flechtenxanthone 239.
Gallenfarbstoffe 84f.
Griseogenin 428.
Himgravin 37.
Holothurinogenins 426, 431.
Hydrogenation products of cucurbitacins 334.
Isocrustecdysone 425.
Isocucurbitacin B 341.
Leprarsäure 237.
Maaliene 436.
Mesobilatriene 104.
2-Methylcyclohexanone 159.

NMR-Spektren.
4-Methylcyclohexanone 159.
Methyl elaterate and isoelaterate 324.
Myxinol 434.
Naramycin-B 160.
Phenarctin 251.
Phycobiline 119.
Phycobiliverdin 120.
Protobilatriene 104.
Quinoid pigments 460.
Reaction products from cucurbitacins 319, 321, 322, 325, 332, 333.
Reduction product of inactone 180.
Reduction products of cycloheximide, isocycloheximide and naramycine-B 165.
Scymnol 433.
Steroids 419.
Streptovitacin A 160, 177.
Tetrahymanol 435.
Usninsäure 242.
Xanthone 213.
Non-protenoid nitrogen-containing substances from marine sources 451.
(+)-Norallosedamin 14, 19.
Nordiploicin 257.
—, Synthese 255.
Norlichexanthon 237, 238, 240.
—, -trimethyläther 237.
Norlobaridon 254.
—, -monomethyläther 254.
Norlobarsäuremethylester → Loxodin 254.
N-Normethylskytanthin 35.
(±)-Norrangiformsäure, Synthese 220, 221, 222.
(−)-Norrangiformsäure → (−)-1-(2 R : 3 R)-Heptadecantricarbonsäure 222.
—, -monomethylester → Rangiformsäure 222.
Norsolorinsäure 234.
Norwegian spruce *(Picea abies)* 397.
Nostocales 115.
Novochlorophaesäure 250.
Nuclear magnetic resonance spectra s. NMR-Spektren.
Nucleinsäurestoffwechsel 287.
Nuphamin 32, 33.
Nuphar-Alkaloide 32.
(−)-Nupharamin 32.
Nupharidin 32.

Nuphar japonicum 32, 33.
Nuphar luteum 33.
Nuphar variegatum 33.
Nuphenin 33.
Nymphaceae 32.
Nymphalidae 110.

cis-β-Ocimene 384.
trans-β-Ocimene 384.
Ocotillol 346.
2,4,5,2′,5′,2″,4″,5″-Octamethoxy-p-terphenyl 270.
2-Octanone 194.
Octansäure 11.
Octopin 456.
Octopus 456.
Odonthalia dentata 446.
Oedipoda coerulescens 108.
Olearyl oxide 404.
Olefin alkylation 364.
Oleorosin 397.
Oligosaccharide in Flechten 218.
Olivetol 228.
—, -monomethyläther 247.
Olivetol-carbonsäure 248.
—, -methylester 247.
Onocerin 363.
γ-Onocerin 374.
β-Onocerin-diacetate 374.
Ophiuroids 462, 463.
Opisthobranchien 111.
Opsopyrrolcarbonsäure 89.
Optical rotation disperson spectra s. ORD-spectra.
Optische Aktivität.
Aplysioviolin 113.
Bilirubin 93.
Cucurbitacins 356.
Dihydrocycloheximide borate 164.
Gallenfarbstoffe 76.
16-Hydroxy-gratiogenin 346.
Terpenoids 438.
Tetrahydrocucurbitacin I 343.
Tetrahydroisoelatericin B 343.
Opuntia ficus-indica 38.
Orchidaceen 3.
Orcin 227.
Orcin-Derivate in Flechten 227, 284.
β-Orcincarbonsäure 251, 278.
—, -methylester 227.
ORD-spectra.
Aplysioviolin 113.

ORD-spectra.
  Bilene-(b) 77.
  Bilirubin 93.
  Cucurbitacin derivatives 329, 331.
  Cucurbitacins 350.
  Cyclohexanone derivatives 196.
  Cycloheximide 154, 177.
  Deoxyisocycloheximide 169.
  α-Epiisocycloheximide 154, 192.
  Holothurinogenins 426.
  Isocycloheximide 154, 177.
  Methyl elaterate and isoelaterate 324.
  Myxinol 434.
  Naramycin-B 154.
  PC-3 220.
  Reduction product of elateric acid 324.
  Streptovitacin-A 177.
Orensin 29.
Ormojin 31.
Ormosajin 31.
Ormosanin 31.
Ormosia-Alkaloide 31.
Ormosia-Arten 31.
Ormosia jamaicensis 31.
Ormosia panamensis 31.
Orsellinsäure 237, 238, 248.
—, -äthylester 227.
—, Biosynthese 277.
—, enzymatische Decarboxylierung 273.
—, methylester 245, 248.
—, Synthese 227.
—, tritiummarkierte 278.
p-Orsellinsäure 227.
Orsellindecarboxylase 273.
Osmiumtetroxid 228, 319, 343, 381, 382, 441.
Ostracion lentiginosus 454.
Ostreogrycine 40.
Oxalsäure 280.
—, -diäthylester 271.
Oxalylchlorid 271.
Oxidase 273.
22,25-Oxidoholothurinogenin 425, 427, 430.
—, 17-desoxy derivative 425, 430.
2,3-Oxidosqualen 281.
6-Oxo-isohopan 262.
threo-2-Oxo-4,5-nonadecandicarbonsäure 220.
5-Oxooctanal 11.
5-Oxo-octansäure 11.

7-Oxo-phenoxazon-2-Chromophore 228.
4-Oxopipecolinsäure 8, 40.
Oxygenation patterns in terpenoid chemistry 399.
Oxyskyrin 260.
Ozonolysis 220, 223, 244, 245, 271, 314, 339.

Pacifenol 470.
Pacific hagfish *(Eptatretus stoutii)* 434, 446.
Pacific mako shark *(Isurus glaucus)* 446.
Pahutoxin 454.
Paludossäure 251.
Pankreatin 123.
Pannarol 241.
Pannarsäure 241.
—, -dimethyläther 241.
Panthothensäure 273.
Paper chromatography.
  Bilatriene 86.
  Bile acids 431.
  Cucurbitacins 350.
  Flechtenstoffe 210, 211.
  Gallenfarbstoffe 87.
  Pyrroldicarbonsäuren 69.
Papierchromatographie s. Paper chromatography.
Papilio graphium antheus 110.
Papilio graphium tynderaeus 110.
Papilionaceae 31.
Papilionidae 110.
Papilio phorcas 110.
Papulosin 229, 234.
Paracentrotus lividus 460.
Parietaria officinalis 10.
Parietin 231, 234, 285.
—, Chlorierung 229.
—, UV-Spektrum 212.
Parietinsäure 231, 234.
Parkeol 348, 349.
Parmelia-Arten 254, 264.
Parmelia aurulenta 261.
Parmelia caperata 220, 278, 279.
Parmelia centrifuga 223.
Parmelia conspersa 254.
Parmelia entotheiochroa 261, 264.
Parmelia leucotyliza 264.
Parmelia livida 257.
Parmelia persidians 261.
Parmelia subaurulenta 261.
Parmelia tinctorum 278.

Partition chromatography.
　Bile acids 431.
　Streptovitacins 145, 175.
Pathogenic bacteria, activity against 140.
PC-3 220.
—, Methyläther 220.
Pecten megallanicus 456.
Peepuloidin 8.
Pelletierin 12.
Peltigera aphthosa 248, 267.
Peltigera-Arten 248, 274.
Peltigera polydactyla 274.
Peltigerin → Tenuiorin 248.
Peltigerosid 218.
Penicillic acid 437.
Penicillinase activity 458.
Penicillin N 457, 458.
Penicillins 457, 458.
Penicillium baarnense 276.
Pentane 434, 436.
6-n-Pentylpyrogallolcarbonsäure 250.
Peponium mackenii 354.
Peptide, Spaltung durch Bromcyan 117.
Perchloric acid 325.
Perhydrobenzazulene skeleton 402.
Perhydrofluorene skeleton 402.
Perhydronaphthalene derivatives 396.
Perhydrophenanthren derivatives 396, 402.
Periodate oxidation 175, 220.
Periodic acid 312, 316, 321, 322, 328, 329, 340, 345, 346, 441.
Perjodatoxydation s. Periodate oxidation.
Perjodsäure s. Periodic acid.
Perlatolinsäure 257.
Perlatolsäure 286.
Permanganat, Abbau von Gallenfarbstoffen 69.
Peroxidase 93, 273.
Peroxy acids 369.
Pertusaria amara 258.
Pertusaria-Arten 238.
Pertusaria lutescens 239.
Pertusaria rhodesiaca 253.
Pertusaria wulfenii 239.
Petrosimonia monandra 3.
Phacophycae 467.
Phaeophorbid a, Chromatabbau 70.
Pharmacodynamic action 351.
Phaseolus angularis 4.
Phaseolus aureus 4.

Phaseolus vulgaris 4, 7.
Phenacetylchlorid 271.
Phenarctin 251.
Phenolase 273.
Phenolcarbonsäuren 211, 227, 273.
Phenylalanin 16, 24, 30, 36, 280, 284.
Phenylalanin-1-^{14}C 279, 280.
Phenylalanin-2-^{14}C 20, 280.
Phenylalanin-2,3-^{14}C 20.
Phenylalaninderivate in Flechten 269.
Phenylalanin, ringmarkiert 280.
Phenylbrenztraubensäure 281.
o-Phenylendiamin 273.
p-Phenylendiamin 211, 215.
(2 S : 8 R)-(—)-8-Phenyllobelol → (—)-Allosedamin 14, 15, 19.
(—)-8-Phenyllobelol II → (—)-Sedamin 15.
Phenylmilchsäure-^{14}C 281.
(2 S : 8 R)-(+)-8-Phenylnorleobelol I → (+)-Norallosedemin 14, 19.
Phlebsäure A 267.
—, Korrelation mit Adipedatol 266.
Phlebsäure B 267.
—, Korrelation mit 22-Hydroxyhopan 267.
—, -methylester 267.
Phloracetophenon-CO-^{14}CH$_3$ 278.
Phloroglucin 237, 238.
Phloroglucinyl-orsellinat 238.
β-Phocaecholic acid 433.
Phormidium luridum 117.
Phosphoric acid 373.
Phosphoroxychlorid 238, 262.
Phosphorus pentoxide 149, 151, 161.
Phosphorus tribromide 161.
Phosphorylierung, oxydative 93.
—, Wirkung der Usninsäure 287.
Photomorphogenese 124.
Photomorphosen, Photoreversibilität 124.
Photosynthese, Wirkungsspektrum 116.
Phycobilin-630 120.
Phycobilin-655 120.
Phycobiline 118 f.
—, Chromsäureabbau 120.
—, Dünnschichtchromatographie 88, 119.
—, Elektronenspektren 120.
—, kovalente Bindungen an das Protein 123.
—, Massenspektren 119.
—, Nebenvalenzbindungen an das Protein 123.

Phycobiline, NMR-Spektren 84, 119.
—, Strukturermittlung 84, 120.
Phycobiliproteide 114f.
—, alkalische Hydrolyse 118.
—, Chromophore 118.
—, Dissoziation 117.
—, Molgewichte 117.
—, Spektraldaten 115.
—, Verbreitung 115.
Phycobilisom 116.
Phycobiliverdin 118, 119.
—, Absorptionsmaxima 127.
—, Cotton-Effekt 79.
—, -dimethylester 120.
—, Dünnschichtchromatographie 87.
—, Massenspektren 120.
—, NMR-Spektrum 84, 120.
—, Optische Aktivität 77, 79.
Phycobiliviolin 119, 120.
—, Absorptionsmaxima 127.
—, -dimethylester, NMR-Spektrum 84.
Phycobiont 210.
Phycocyanin 114, 115, 121.
—, Absorptionsbanden 116, 117.
—, deuteriertes 86.
—, enzymatische Hydrolyse 121.
—, Fluoreszenzemission 116.
—, Molgewicht 117.
—, NMR-Spektren 86.
—, Reaktion mit conz. HCl 120.
R-Phycocyanin, Dissoziationsprodukte 118.
Phycocyanobilin 118, 119, 121, 460.
—, Absorptionsmaxima 127.
—, Protonenaustausch 86.
—, Struktur 120.
Phycoerythrin 111, 114, 115, 120, 121, 459.
—, Struktur 123.
R-Phycoerythrin.
—, Absorptionsspektrum 116, 117.
—, Dissoziationsprodukte 118.
—, Molgewicht 118.
Phycoerythrobilin 111, 118, 119, 121, 123, 459.
—, Absorptionsmaxima 127.
—, Dünnschichtchromatographie 87.
—, Struktur 120.
—, Vorkommen in Phycoerythrinen 120.
Phycourobilin, Vorkommen in Phycoerythrinen 121.
Phyllocladene 402.

Phylloerythrin 61.
Phytochrom 111, 116, 124f.
—, Chromatographie 125.
—, Chromophor 126.
—, Chromsäureabbau 127.
—, Elektronenmikroskopie 126.
—, Gelfiltration 125, 126.
—, Isolierung aus Haferkeimlingen 124.
—, Isolierung aus Roggenkeimlingen 125.
—, Molgewicht 126.
—, Struktur 128.
—, Wirkung bei der Photomorphogenese 124.
Phytochromobilin 111, 126.
—, Chromatabbau 127.
—, Chromatographie 126.
—, Dünnschichtchromatographie 87.
Picea abies 397.
Picric acid 375.
Picroroccellin 269, 284.
Picrolicheninsäure 258.
—, Abbau 258, 259.
—, Synthese 259.
—, UV-Spektren 212.
Pieridae 110.
Pieris brassicae 110.
Pimaradienes 383, 399, 402, 408.
enantio-Pimaradien 281.
Pimaranes 399.
Pimarane, migration of the $C(12)-C(13)$-bond 406.
Pimaric acid 379, 399.
Pinastrinsäure 271, 287.
—, ozonolytischer Abbau 271, 272.
—, Synthese 272.
Pinidin 18, 19, 42.
—, Biosynthese 18.
Pinus-Arten 3.
Pinus jeffreyi 18.
Pinus sabiniana 3, 18.
Pinus torreyana 18.
(+)-α-Pipecolin 3.
—, Bildung aus Acetat 4.
—, S-Konfiguration 3.
Pipecolinsäure 2, 40, 42.
—, Biosynthese 6, 16.
—, N-Methylbetain 4.
—, pflanzliche Derivate 5.
—, tritiiert 13.
Piper-Alkaloide 2, 3, 8.
Piper-Arten 8.
Piperettin 8.

Δ^1-Piperidein 2, 3, 10, 11, 12, 16, 17, 20, 26, 27, 31.
Δ^2-Piperidein 10.
Piperideincarbonsäuren 4, 5, 6, 26.
Piperidide 8.
Piperidine 8, 14, 195, 196.
—, Biosynthese 4.
—, natürliches Vorkommen 3.
Piperidin-carbonsäuren 4.
Piperidinverbindungen, natürliche 1 f.
—, aliphatisch disubstituierte Piperidine 18 f.
—, aliphatisch monosubstituierte Piperidinbasen 8 f.
—, Antibiotika 40 f.
—, einfache Piperidinderivate 3 f.
—, Entstehungsprinzip 2.
—, heterocyclisch substituierte Piperidine 24 f.
—, monoterpenoide Piperidinalkaloide 34 f.
—, α-substituierte Piperidine 9.
—, N-substituierte Piperidine 8.
Piperidon-Alkaloide 37.
Piperin 8.
cis,cis-Piperin → Chavicin 8.
trans,trans-Piperin 8.
Piperinsäure 8.
Piperitone-pulegone-menthofuran sequence 437.
Piper longum 9.
Piperlonguminin 8.
Piperolein 9.
Piplartin → Piperlongumin 9.
Piptanthin 31.
Piptanthus nanus 31.
Pisaster ochraceous 469.
Pisum sativum 407.
Pivaloyl fluoroborate 390.
Placopecten magellanicus 469.
Planasäure 247.
—, -methylester, Massenspektrum 214.
Plant growth hormones 395, 406.
Plant growth retardants 412.
Plant resin acids 404.
Plant resins 395.
Platyrhinoides triseriata 446.
Plectonema calothricoides 117.
Pleopsidsäure 224.
Pleuromutilin 401, 403.
—, biosynthesis 405.
Pleurospermin 15.
—, O-Methylderivat 15.

Plexaura crassa 434, 436.
Plexaura flexuosa 419.
Plexaura homomalla 439.
Plexaurella dichotoma 438.
Podocarpaceae 404.
Podocarpene series of diterpenes 381.
Podototarin 404.
Polyacetoxynaphthalenes 465.
Polyamid ITLC 87.
Polycheria rufesceus 462.
Polyene arene sulfonates, cyclization 378.
Polyene cyclization theory 364 f.
Polyenes, conversion to epoxides 369.
Polyenes, treatment with N-bromosuccinimide 371.
Polyenic acetals, cyclization 380.
Polyenic allylic alcohols, cyclization 381.
Poly-β-D-glucopyranose → Lichenin 219.
Poly-β-D-glucosan aus Getreidesamen 219.
Polyhydroxynaphthoquinones 465, 466.
Polyketid aus Flechten 225.
Polyketide 276, 278.
Polyketidkette 42.
Polyole in Flechten 217.
Polypeptid-Antibiotikum 40.
Polypeptide-type toxins from sea urchins 458.
Polyporsäure 269, 280, 281, 282, 284.
—, Antitumorwirkung 287.
—, Chinoniminderivat 281.
Polysaccharide 284.
—, in Flechten 218.
Polysaccharides of marine algae 467.
Polysaccarides of red seaweed 478, 469.
Poriferasterol 420, 421.
Porphobilinogen 110.
Porphyran 468.
Porphyridales 115.
Porphyridium aerugenium 117.
Porphyridium cruentum 116, 117.
Porphyrilsäure 241.
—, -dimethyläther 240.
Porphyrine 62, 65, 459.
—, Elektronenspektren 73.
—, Numerierung 62.
—, Optische Aktivität 76.
—, Strukturuntersuchung 68.
Porphyrinogene 104.
Portentol 225, 284.
—, Abbau 226.
Potassium-t-butoxide 319, 336.

Potassium dihydrogen phosphate 145.
Potassium permanganate 149.
Praslinogenin 430.
Premarrubiin 398, 403.
Prephensäure 284.
Priestleya elliptica 25.
Primärstoffwechsel, Coenzyme 2.
Progesterone 423.
Prolin 39, 217.
Pronase 123.
2-(2'-Propenyl)-Δ^1-piperidein 12.
Propinyl magnesium bromide 382.
Propionaldehyd-2,2-diacetic acid 149.
1-^{14}C-Propionat 276.
2-^{14}C-Propionat 276.
[1-^{14}C]Propionic acid 200.
(2 R : 8 R)-(−)-8-Propylnorlobelol I → (−)-Halosalin 14.
Prosopin 23.
Prosopinin 23.
Prosopis africana 23.
Prosopis-Alkaloide 10, 22, 23.
Prostaglandin derivatives 439.
Protease 273.
Proteine 284.
—, in Flechten 217.
Protein synthesis, inhibition of 175, 181.
—, —, by cycloheximides 147.
Protobiladien-(a,c)-IX α → Bilirubin IX α 64.
—, Bildung von Azofarbstoffen 67.
—, Massenspektrum 82.
—, NMR-Spektrum 85.
Protobilatrien IX α 104, 108.
—, Chromatabbau 70.
—, -dimethylester, Massenspektrum 80.
—, in Insekten 108.
—, in Korallenpigment 108.
Protobilatrien IX γ 110.
Protobilatriene 104.
—, NMR-Spektren 85, 104.
—, Strukturen 104.
Protoblastenia-Arten 231.
Protocetrarsäure, Biosynthese 279.
Protolichesterinsäure 223, 224, 286.
—, Biosynthese 275, 276, 279.
—, ^{14}C-markierte 275.
—, Hydrierung 223.
—, Synthese 223.
allo-Protolichesterinsäure 224.
Protomycin 144, 145, 181 f.
—, alkali degradation 186.

Protomycin, antifungal activity 181.
—, biosynthesis 201.
—, chromatography 184.
—, counter-current distribution 184.
—, estimation 145.
—, hydrogenation 184.
—, IR-spectrum 184.
—, isolation 184.
—, molecular distillation 184.
—, structure 186.
—, tetrahydroderivative 184.
—, UV-spectrum 184.
Protonen-Austauschversuche in Phycobilin 86.
Protoporphyrin IX 64, 110.
—, Chromatabbau 70.
Protozoa 443.
Pseudocarpain 23.
Pseudoconhydrin 9, 10.
—, stereochemische Aufklärung 10.
Pseudoconiin 11.
Pseudocyphellaria billardierii 266.
Pseudocyphellaria crocata 271, 280, 281.
Pseudocyphellaria intricata 266.
Pseudo-glycol, formation 329.
Pseudomonas bromoutilis 449.
Pseudoplexaura porosa 438.
Pseudopterogorgia americana 434, 436.
Pseudotropin 13, 16.
Psilocaulon absimile 3.
Psoromsäure 286.
Pteridines 459.
Pterobilin 108, 110.
—, Biosynthese 109, 110.
—, Chromatabbau 110.
Pterogorgia anceps 434, 436.
Puffer fish 451.
Pulchellidin 38.
Pulvinsäure 271, 273, 285.
—, Aminosäurekonjugate 281.
Pulvinsäureamid 271, 281.
Pulvinsäure-Derivate 210, 211, 284.
—, Biosynthese 282.
—, in Flechten 271, 284.
—, Massenspektren 215.
Pulvinsäurelacton 271, 285.
—, Abbau 280.
—, Biosynthese 280, 281.
—, cardiotonische Aktivität 287.
—, radioaktiv markiert 280.
Punica-Alkaloide 12 f, 14.
Punicaceae 12.

Punica granatum 12, 13.
Purgative action of Cucurbitaceae 351.
Purine bases 457.
Purple pigment 118.
Pustulan 219.
Pyranonaphthazarin pigment 466.
Pyrethrosin 370.
Pyridin-2,6-dicarbonsäure 39.
Pyridine 159, 160, 164, 262, 271, 385.
—, hydrochloride 150.
Pyridinring, Biosynthese 2.
Pyridinverbindungen, natürliche, Entstehungsprinzip 2.
Pyridoxalphosphat 2, 281.
Pyrrolderivate, Kupplung mit Diazoniumsalzen 67.
Pyrroldialdehyde-(2,5) 71.
—, Dünnschichtchromatographie 72.
Pyrroldicarbonsäuren-(2,5) 69, 70.
Pyrrole, alkylsubstituierte, Elektronenspektren 73.
Pyrrole pigments 459.
Pyrrolidine 195, 196.
Pyrrolone 89.
—, Elektronenspektrum 73.
Pyrus malus 4.
Pyruvaldehyde 184.
Pyxiferin 269, 284.
Pyxine coccifera 269.
Pyxine endochrysina 265, 267.
Pyxinsäure 265.
—, -methylester 265.

Quecksilber(II)-acetat 424.
Quinonoid and related pigments in marine organisms 460.
Quinovose 425.

Radiochemical dilution 410.
Ray fish *(Dasyatis akajei)* 432.
Raia batis 432.
Raia clavata 458.
Ramalina-Arten 245, 258, 261.
Ramalinaceae 286.
Ramalina crassa 285.
Ramalina fastigiata 275.
Ramalina maciformis 275.
Ramalina paludosa 251.
Ramalina siliquosa 286.
Ramalina stenospora 247.
Ramalina subdecipiens 251.
Ramalina yasudae 285.

Raney nickel reduction 321, 322, 442.
Raney nickel desulfurization 329.
Rangiformsäure 222, 286.
Red algae (Rhodophyceae) 418.
Reduction, selective with $LiAl(t-BuO)_3H$ 423, 424.
Red tide 452.
Reniera japonica 443.
Renierapurpurin 443.
Renieratene 443.
Resorcin, reduktiver Abbau von Gallenfarbstoffen 68.
Reformatsky reaction 454, 455.
Reinkella parishii 241.
Resorcinol 146.
—, color reaction of cycloheximide 189.
Resorcinspaltung 104.
β-Resorcylsäure 273.
Retama raetam 28.
Retama sphaerocarpa 28.
Retinol 442.
Retro-aldol cleavage 313.
Retro-aldol reaction 172.
—, thermal 154.
Reverse aldol reaction 325.
Reversed phase partition chromatography 431.
R_F-Werte, Gallenfarbstoffe 87.
Rhapis excelsa 5.
Rhazia orientalis 33.
Rhazia stricta 33.
Rhizocarpsäure 273, 284.
Rhodemela confervoides 446.
Rhodocladonsäure 228, 237, 285.
Rhodocomatulin 463, 465.
Rhodophyceae 111, 114, 115, 418, 467.
Rhodymemiales 115.
Ribit 217, 274, 275.
Riboflavin 273.
Ribonuclease 273.
Ricinus communis 408.
Rimuene 399.
Rinderserum-Albumin 93.
Ring contraction, enzymology 413.
Roccella canariensis 217.
Roccella fuciformis 225, 237, 269, 276.
Roccella galapagoensis 225.
Roccella maderensis 225.
Roccella portentosa 225.
Roccellaria mollis 223.
(+)-Roccellarsäure 223, 224.
—, -ester, Cotton-Effekt 224.

Roccella sp. 220, 228.
Roccella vicentina 217.
Roccellina condensata 225.
Roccellina luteola 241.
(+)-Roccellsäure 222, 285.
—, Konfigurationsbestimmung 221.
—, Synthese 222, 223.
Rodent-repellant 146.
Roggenkeimlinge 125.
—, Isolierung von Phytochrom 125.
—, Koleoptilen 125.
Röntgenstrukturanalyse.
    Antibiotic from *Pseudomonas* 449.
    4-Brom-nephroarctin 251.
    Cucurbitacins 311.
    Eunicellin dibromide 437.
    Flechtenstoffe 215, 225.
    β-Gorgonene 435.
    Jodacetylvicanicin 258.
    Steroids 421.
Rosadiene 383, 404, 405.
Rosane lactones 404.
Rosenmund reduction 188.
Rosenonolactone 383, 403, 404.
—, tracer studies on the biosynthesis 404, 405.
Rosololactone 405.
Rotalgen s. *Rhodophyceae*.
Rote Koralle 108.
Rote Rübe 38.
Rugulosin 259.
Rupicolon 235, 236.
Rypticus saponaceus 458.

Saccharomyces cerevisiae 145, 194.
Saccharomyces pastorianus 145, 150.
Saccharomyces sake 145, 151.
Saccharose 218, 274.
Salazinsäure 251, 285.
—, UV-Spektrum 212.
Salicylsäuren 215.
Salix fragilis 5.
Salmonella infections 458.
Salvadoraceae 24.
Salvia sclarea 397.
Sandaracopimaradiene 408.
Sandaracopimaric acid 399.
Santiaguin 29.
Sapogenins of marine origin 425f.
Sarcoma 180 196.
—, growth inhibition 219, 454.

Sarkomwachstum, Inhibitorwirkung 219, 454.
Sargasterol 420.
Säulenchromatographie 86, 431.
Saxifragaceae 17.
Saxitoxin 452, 471.
Scenedesmus obliquus 445.
Schizopelte californica 241.
Schizopeltsäure 241.
—, UV-Spektrum 212.
Schizothrix calciocola 453.
Schwefelhaltige organische Verbindungen in Flechten 220.
Sciadin 398.
Sciadopitys verticillata 398.
Sclareol 395, 397, 403.
Scrobiculin 250.
Scrophulariaceae 309, 345, 354.
Scymnol → $3\alpha,7\alpha,12\alpha,24\xi$-26,27-Hexahydroxycoprostane 432, 433.
Scymnus borealis 432.
Scytonema 274.
Sea anemones 451.
—, steroids 420.
Sea cucumbers (Echinodermata Holothurioidea) 425, 427, 428.
Sea lamprey 443.
Sea urchins 461.
—, polypeptide toxins 458.
Sea weed *(Dictyopteris divaricata)* 439, 440.
—, antibiotic activity of extracts 446.
Secale cereale 125.
Secalonsäure A 261.
Secalonsäure C 261.
Secamine 33, 34.
—, Biosynthese 34.
Secodinalkaloide 33.
Secodine 33, 34.
—, Biosynthese 34.
Secophenanthrochinolizidin-Alkaloid 15.
Sedamin 13, 19, 20.
—, Biosynthese 13.
(—)-Sedamin → (—)-8-Phenyl-lobelol II 15.
Sedimentation für Molekulargewichtsbestimmung 117.
Sedinin 20.
(+)-Sedridin → (+)-8-Methylnorlobelol 13, 14.
Sedum acre 4, 7, 14, 15, 16.
Sedum-Alkaloide 13, 19f.

Sedum-Arten 12, 19.
Sedum sarmentosum 13.
Sekikasäure 250, 286.
—, methylester 251.
(−)-β-Selenene 439.
Selenium dehydrogenation 309, 312.
Sempervirol 399.
Senecioic acid → β,β-Dimethylacrylic acid 406.
Sephadex G-25 112.
Serin 121, 123.
Serotonin 451.
Serumalbumin 91, 93.
Sesquiterpenalkaloide 38.
Sesquiterpenes, biosynthesis 403.
—, mode of cyclization 397.
Seychellogenin 429.
Shark 433, 443.
Shellfish 452.
Sicyos angulata 354.
Sideromycine 40.
Silica gel, chromatography on 145.
Silicic acid, chromatography on 454.
Silver acetate 423.
Silver oxide 385.
Sinapis alba 124.
Siphonaxanthin 444.
Siphonein 444.
Siphonous green algae 444.
Siphula ceratites 217, 235.
Siphulin 235.
D-Siphulit → 1-Desoxy-D-glycero-D-taloheptit 217.
β-Sitosterin s. Sitosterol.
β-Sitosterol 268, 418.
γ-Sitosterol 420.
SKF 525 → 2-Diethylaminoethyl-2,2-diphenylvalerate 412.
Skyrin 259, 260.
Skyrinol 260.
Skytanthine 34, 42.
—, Biosynthese 35.
—, Konfigurationsermittlung 34.
β-Skytanthin-N-oxid 34.
Skytanthus acutus 34, 36.
Soapfish (Rypticus saponaceus) 458.
Sodium 381.
Sodium borohydride reductions 164, 182, 334, 335, 343, 346.
Sodium iodide 380.
Sodium[1,3-^{14}C]malonate 200.
Sodium methoxide 374, 441, 465.

Sodium periodate 314.
Solanaceae 12, 16.
Solanum-Arten 37.
Solanum tomatillo 37.
Solasodin 37.
Solenopsine A, B u. C 22.
Solenopsis saevissima 22.
Solidago canadensis 398.
Solorina crocea 234, 275.
Solorinsäure 234.
Sorbit 274.
Sordidon 235, 236.
—, UV-Spektrum 212.
Soybean meal 145.
Spartein 25, 28.
Spektrometrische pH-Titration 75, 113.
Spermin 457.
Sphaerocarpin 28.
Sphaerophorus fragilis 229.
Sphaerophorus globosus 229.
Sphaerophorus sp. 245.
Sphodomantis 108.
Spinaceamine 456.
Spinacine 456.
Spinochromes 461, 462, 463.
—, synthesis 465.
Spirosolan 37.
Sponges 447, 451.
—, steroids 420, 421.
Spongia nitens 472.
Squalene 281, 284, 365, 375.
—, cyclization 371.
—, epoxide 365, 371, 374, 375, 397.
Squalus acanthias 432.
Stannic chloride 367, 371, 373, 374, 380, 381, 387.
Staphylococci, growth inhibition 422.
Staphylococcus aureus UC-76 450.
Staphylomycin S 40.
Starfish 425.
Stecklingsbewurzelung, Hemmung 287.
Stenosporsäure 247.
—, -methylester 247.
Stercobilin 94, 95, 96.
—, Chromatabbau 96, 98.
—, Dünnschichtchromatographie 87.
—, -hydrochlorid 77.
—, Konfiguration 78, 98.
—, Oxydationsbeständigkeit 102.
—, Totalsynthese 98.
Stercobilinogen 95.
Stereocaulon-Arten 257.

Stereocaulon colensoi 257.
Stereocaulon ramulosum 245.
Sterine in Flechten, Biosynthese 281, 284.
Steroide 211.
—, in Flechten 268.
—, stickstoffhaltige 37.
Steroids from marine sources 418f.
Steroids, mode of cyclization 397.
17β-20-one-Steroids, ORD-spectrum 331.
Sterols 418.
Stichopogenin A_2 and A_4 430.
Stichoposide A and C 430.
Stichopus regalis 430, 431.
Sticta billardierii 266.
Stictaceen 217.
Sticta colensoi 269.
Sticta coronata 269.
Sticta fuliginosa 217.
Sticta limbata 217.
Sticta mougeotiana 267.
Sticta sylvatica 217, 273.
Stictinsäure, Massenspektrum 214.
Δ^5-Stigmastene-3β-ol → β-Sitosterol 268, 418, 419.
Stigmasterol 418, 419.
Stigomonas fasciculata 443.
Stonefish 458.
Stork-Eschenmoser hypothesis 364, 366, 368, 369.
Strelitzia reginae 5.
Strepsilinmethyläther 241.
—, Synthese 241, 242.
Streptimidone 40, 41, 144, 145, 180, 181f.
—, activity against yeast strains 181.
—, alkaline degradation 182, 186.
—, biosynthesis 200, 201.
—, catalytic reduction 182.
—, chromatography 182.
—, herbicide activity 182.
—, inhibition of protein synthesis 181.
—, isolation 182.
—, monoacetate 182.
—, monoxime 182.
—, ozonolysis 182, 184.
—, stereochemistry 184.
—, structure 184.
—, UV-spectrum 182.
B-Streptogramine 40.
Streptomyces albulus 143, 175, 180.
Streptomyces albus 142, 186.

Streptomyces griseolus 144, 186.
Streptomyces griseus 40, 140, 142, 143, 144, 145, 146, 174, 179.
Streptomyces naraensis 142, 146, 151.
Streptomyces noursei 40, 142, 197.
Streptomyces reticuli var. protomycicus 184.
Streptomyces rimosus 41, 144, 182.
Streptomyces Stamm FFD-101 18.
Streptomyces violaceus 40.
Streptovitacin-A 142, 174.
—, acid catalyzed dehydration 175, 179.
—, alkaline degradation 175.
—, antitumor activity 175.
—, catalytic reduction 178.
—, inhibition of protein synthesis 175.
—, monoacetyl derivative → E 73 174.
—, NMR-absorption 160, 177.
—, ORD-curve 177.
Streptovitacin-B 143.
—, acid catalyzed dehydration 175.
—, alkaline degradation 175.
Streptovitacin-C_1 143, 175.
Streptovitacin-C_2 143.
—, acid catalyzed dehydration 175.
—, alkaline degradation 175.
—, periodate oxidation 175.
—, structure 175.
Streptovitacin-D 143, 175.
Streptovitacin-E 143, 175.
Streptovitacins 174f, 186.
—, antifungal properties 174.
—, antitumor properties 174.
—, biosynthesis 202.
—, estimation 145.
—, partition chromatography 175.
—, separation 145, 175.
—, stereochemistry 177.
—, structure determination 175.
Strophantus scandens 4, 5, 8.
Structure-activity relationships of antibiotics 141, 147.
Strychnine 363.
Stychopus japonicus selenka → Trepang 430.
[1,1-^{14}C]Succinic acid 200.
Succinimide 70, 72.
Sulfuric acid 383.
Sulfurylchlorid 227, 235, 238, 245.
Sweet maranka 339.
Synechococcus lividus 117.
Syphilobin A 30, 31.

Syphilobin F 30, 31.
—, Biosynthese 31.

Tabernamontana cumminsii 33.
D-Tagatose 218.
Talkum für Säulenchromatographie 86.
Tannase 273.
Taondiol 470.
Taonia atomaria 470.
Taraxeren 268.
Taricha torosa 452.
Taxane skeleton 403.
Tecomanin 35.
Tecoma stans 35.
Tecostanin 35.
Telfairia pedata 354.
Teloschistes-Arten 231.
Teloschistes flavicans 258.
Tenuiorin 267.
Ternaygenin 429.
Terpene in Flechten, Biosynthese 281, 283, 284.
Terpenes from marine sources 435.
Terpenoid methanesulfonates, cyclization 379.
Terpenoid polyenes, cyclization reactions 364.
Terpenoids, C-3 hydroxylated 371.
Terpenoid systems, biogenetic-type synthesis 363 f.
Terphenylchinon 269.
Terpinolene 384.
Tetraacetylthelephorsäure 270.
1,3,4,5-Tetra-O-benzyl-D-arabinit 218.
Tetrabromo-spiro-cyclohexadienyl-isoxazole 449.
1,2,6,7-Tetracarboxy-3,9-dimethoxydibenzofuran 241.
Tetrachlorkohlenstoff 72, 85, 87, 88, 237.
Tetracyclic ring system, formation 406.
Tetracyclic triterpenes 307 f.
— —, degradation 312.
— —, negative Cotton effect 331.
Tetradecanol 454, 455.
Tetrahydroanabasin 27, 28.
Tetrahydroanabasin-Alkaloide 24.
1,2,7,8-Tetrahydrobilane 76.
Tetrahydrobilene-(b), Konfiguration 78.
Tetrahydrocucurbitacin I 341, 342 f.
—, occurrence in nature 353, 355.
—, optical rotation 343.
—, physical constants 356.
Tetrahydroelatericin B 334.

Tetrahydrofuran 171, 235, 237, 238, 244, 245.
Tetrahydrogibberellic acid 406.
4 : 5 : 6 : 7-Tetrahydroimidazo-[5,4-c]-pyridine 456.
—, -5-carboxylic acid 456.
—, —, 6-methyl analog 456.
Tetrahydroisoelatericin B → Tetrahydrocucurbitacin I 343, 356.
Tetrahydropyranyl ether (THP), removal 424.
4-(Tetrahydropyranyloxy)-4-methyl-1-pentinyl magnesium bromide 424.
Tetrahydrosecamin 33.
16,17,15,20-Tetrahydrosecodin 33.
—, -17-ol 33.
Tetrahydrostreptimidone 182.
1,3,6,8-Tetrahydroxyanthrachinon-2-carbonsäure 234.
$3\beta,7\alpha,16\alpha,26(27)$-Tetrahydroxy-5$\alpha$-cholestane → Myxinol 434.
9,10,12,13-Tetrahydroxy-docosansäure 223.
9,10,12,13-Tetrahydroxy-heneicosansäure 223.
1,5,6,8-Tetrahydroxy-3-methyl-anthrachinon 235.
Tetrahydroxytricosansäure 223.
Tetrahymanol 435.
Tetrahymnena pyriformis 435, 443.
α-Tetralone 194.
2,6,2',4'-Tetramethoxy-biphenyl 240.
Tetramethylammonium hydroxide 196.
3,5,3',5'-Tetramethyl-4,4'-diäthyl-pyrromethen 76.
Tetramethyldiguanylcystamine 455, 456.
Tetrapyrran → Bilan 62.
Tetrapyrren-Chromophor 74.
Tetrodotoxin 451, 452.
Thamnol 253.
Thea sinensis 4.
Thelephorsäure 269, 270, 284.
—, UV-Spektrum 212.
Thioäthanolamin 220, 284.
Thiobinupharidin 33.
Thioketalization 322.
Thionupharidin 32.
Thionyl chloride 161, 188, 381, 382, 385.
Thiophaninsäure 239.
Thiophansäure 237.
—, -6-methyläther 238.
—, Synthese 238.

Thornback ray *(Platyrhinoides triseriata)* 446.
Threit 275.
Threonin 123.
Thunbergene → Cembrene 403.
Thuringion 239.
—, UV-Spektrum 212.
Tocopherols 442, 470.
p-Toluenesulfonic acid 316, 339, 441.
p-Totuene-sulfonyl chloride 322.
Tolypothrix tenuis 117.
Tomatillidin 37.
Tora fugu 451.
Tosylation 380.
Totarol 399, 404.
Toxanthera natalensis 354.
Toxic bitter principles in *Cucurbitaceae* 308.
Toxic principles in amphibians 469.
Toxic species of marine origin 451.
Toxoplasma gondii RH 174.
Trachylobane 402, 408.
Trachylobium verrucosum 402.
Trebouxia 273, 274.
Trehalose 218.
Trentepohlia 274.
3,5,6-Tri-O-acetyl-1,2-o-methyl-orthoacetyl-α-D-galakto-furanose 218.
Triäthylamin s. Triethylamine.
Tricarbäthoxylecanorsäure 249.
Trichloressigsäure 77, 78.
2,5,7-Trichlornorlichexanthon 238.
4,5,7-Trichlorparietin 229.
Trichothecin 447.
Tricothecium roseum 404.
Triethylamine 195, 271.
Trifluoracetanhydrid 238, 247, 248, 255, 257.
Trifluoroacetic acid s. Trifluoressigsäure
Trifluoressigsäure 78, 382.
Trifolium repens 4.
Trigonelline 4, 456.
3α : 7α : 23-Trihydroxycholanic acid → β-Phocaecholic acid 433.
3β,7α,26(27)-Trihydroxy-5α-cholestane → 16-Deoxymyxinol 435.
2,3,7-Trihydroxy-6-acetyl-juglone 464.
2,6,7-Trihydroxy-3-ethyl-juglone 463.
6α,7α,22-Trihydroxyhopan 267.
3β,17α,20ξ-Trihydroxy-5α-lanosta-7 : 8,9 : 11-diene-18-carboxylic acid lactone 426.

2,3,4-Trihydroxy-6-n-propyl-benzoesäure 250, 251.
Triketone from elaterin 316.
—, IR-spectrum 317.
—, UV-spectrum 317.
2,3,6-Trimethoxy-7-hydroxy-naphthazarin 465.
Trimethylamin 217.
Trimethyl-(5-amino-5-carboxypentyl)-ammonium dioxalate 459.
2,3,6-Tri-O-methyl-D-glucopyranosid 220.
1,2,5-Trimethylnaphthalene 312.
1,2,8-Trimethylphenanthrene 309, 312.
4,4,14α-Trimethylpregn-8-ene-2,7,11,20-tetraene 336.
Triphenyltetrazolium chloride 314.
—, test for α-hydroxyketones 337.
Tripneustes gratilla 458, 463.
Tripyrren-Struktur, Elektronenstruktur 74.
Triterpene 211.
—, in Flechten 262.
—, mode of cyclization 397.
—, tetracyclic 307 f.
Tritium-markierte Verbindungen.
 Lysin 7, 8, 16.
 Orsellinsäure 278.
 Pipecolinsäure 13.
Tritium, retention of 408.
Trochomeria debilis 354.
Trochomeria sagittata 354.
Tropanbasen 17.
Tropantyp bei Alkaloiden 2.
Tropin 13, 14.
α-Truxillsäure 28, 29.
—, -chlorid 29.
Trypetheliopsis boninensis 260.
Tryptamin 34.
Tumor inhibitors 395.
Tumor necrotizing capacity 308.
Turf diseases 146.
Tyrosin 39.
Tyrosinase 273.

Ubiquinones 442.
—, distribution in marine species 443.
Ultrazentrifuge zur Molekulargewichts-Bestimmung 219.
Umbelliferae 9.
Umbelliprenin epoxide 373.
Umbilicaria angulata 219.
Umbilicaria caroliniana 219.

Umbilicaria hirsuta 219.
Umbilicaria polyphylla 219.
Umbilicaria pustulata 273.
Umbilicarsäure 273.
Umbilicin 218.
trans,cis,cis-Undeca-1,3,5,8-tetraene 440.
Urease 273.
Urobilin 62, 64, 94.
—, Alkali-Einwirkung 102, 103.
—, Chromsäureabbau 94.
—, $FeCl_3$-Reaktion 102.
—, katalytische Hydrierung 94.
—, Massenspektrum 94.
—, Totalsynthese 94, 95.
d-Urobilin 94.
—, -hydrochlorid, ORD-Spektrum 77, 78.
—, Konfiguration 78.
i-Urobilin 94, 96.
Urobilin-IX α-dimethylester, Massenspektrum 83.
Urobilinoide 94f.
Urobilinogen 95, 99.
—, Autoxydation in Eisessig 103.
Ursolsäure 268.
Urticaceae 10.
Urticaceen-Alkaloide 15.
Usnea aciculifera 254.
Usnea barbata 223.
Usnea diffracta 278, 279.
Usnea longissima 278.
Usninsäure 210, 211, 242f, 251, 284, 285.
—, Acetylderivate 243.
—, Analoga, Synthese 245.
—, antibiotische Wirkung 287.
—, Biosynthese 278, 279.
—, Hydrat 279.
—, Hydrierung 244.
—, IR-Spektrum 242.
—, Methylierung 243.
—, Natriumsalz 287.
—, NMR-Spektrum 242.
—, Racematspaltung 244.
—, Racemisierung 243.
—, Synthese 244, 245.
Usno 287.
UV-Spektren.
    Actiphenol 181.
    Anthrachinone, 1-hydroxylierte 212.
    Arthoniasäure 247.
    Chromone, 5-hydroxylierte 212.
    Ciguatoxin 453, 454.

UV-Spektren.
    Crustecdysone 425.
    Cucurbitacins 337, 346, 350, 356.
    Dehydrocycloheximide 149.
    22-Deoxocucurbitacin D 339.
    Elatericin A 316.
    Flechtenstoffe 211, 212.
    Gratiogenin derivatives 345.
    Holothurinogenins 426.
    trans-4-Hydroxy-4-methyl-pent-2-enoic acid 312.
    Intermediates in synthesis of cucurbitane skeleton 336.
    Protomycin 184.
    Reaction products from cucurbitacins 319, 321f, 325, 332, 333.
    Streptimidone 182.
    Triketone from elaterin 317.
    Urobiline 94.
    Xanthone 212.

Vandopsis longicaulis 3.
Variolaria sp. 228.
Venus mercenaria campechiensis 453.
Veralkamin 37.
Veratrum 37.
Vernamycin B_δ 40.
Verongia cauliformis 447.
Verongia fistularis 447.
Verongia thiona 449.
Vertebraten, Vorkommen von Biliverdin 104, 106.
Verticillol 403.
Vicanicin 258.
Vicenin type 467.
Victorina steneles 110.
Vinetorin 239.
Vinhaticoic acid 399.
Violaxanthin 268.
Virenssäure 257.
Virusinhibitoren 287.
Vitamin A 442.
Vitamin B_1 273.
Vitamin D_3 442.
Vitamine in Flechten 273.
D-Volemit 217.
Vulpinsäure 281, 285.
—, Biosynthese 280.
—, radioaktive 279.
—, Wirkung auf Blutzuckerregulation 287.

Wachstumsinhibitoren in *Lasallia*-Extrakten 287.
Wagner-Meerwein rearrangement 402.
Walker carcinoma (rats) 351.
Wasserstoffbrücken 78, 93, 242, 243.
—, in Biliproteiden 123.
Wasserstoffperoxid 93, 249.
Walleyed pike 443.
Whale oil 419.
White horehound *(Marrubium vulgare)* 398, 403.
White pine blister rust 146.
Wild cucumber *(Echinocystis macrocarpa)* 407, 408, 409, 410, 412.
Withania-Alkaloide 13, 16f.
Withania somnifera 12, 16, 17.
Wittig-Reaktion 242.
Wolff-Kishner-reduction 175, 265, 266, 267, 343, 380.
Woodward-Spaltung 39.

Xanthone 210, 211.
—, Alkalischmelze 215.
—, biogenese-artige Synthesen 240.
—, chlorsubstituierte 240.
—, Dünnschichtchromatographie 239.
—, Fluoreszenz 210, 211.
—, in Flechten 237f, 284.
—, IR-Spektren 239.
—, Massenspektren 215.
—, NMR-Spektren 213, 239.
—, UV-Spektren 212.
Xanthophyll 268, 444, 445.
Xanthophyta algae 444.
Xanthoria-Arten 231.
Xanthoria aureola 231, 274.
Xanthoria elegans 234.
Xanthoria parietina 217, 231, 268, 285.
Xanthorin 234.
Xiphigorgia anceps 434.
X-ray analysis s. Röntgenstrukturanalyse.
D-Xylose 218, 425.
Xylose-Konjugate 91.

Ylangene 438.

Zea mays 409.
Zeaxanthin 446.
—, -palmitate 446.
Zellpermeabilität, Erhöhung durch Flechtenstoffe 287.
Zeorin 211, 262, 264, 267.
Zeorinin 211.
Zeorininon 211.
Zeorinon 211, 262.
—, Dehydratisierung 263.
Zimtsäure 28, 29.
Zimtsäure-2-^{14}C 280.
Zimtsäurestoffwechsel 9.
Zinc 380.
Zinkacetat 66.
Zinkchlorid 237, 238.
Zink-Komplexe von Bilenen, Biladienen und Bilatrienen 66.
Zink-Komplexe von Gallenfarbstoffen 78.
Zoarces viviparus 106.
Zooanemonine 457.
Zuckeralkohole 284.
Zuckeralkohol-Glykoside in Flechten 218.
Zucker in Flechten 284.
Zygophyllaceae 34.
Zymase 273.
Zymosterol 421.

Fortschritte der Chemie organischer Naturstoffe
Progress in the Chemistry of Organic Natural Products

Bisher erschienen:

Erster Band: 41 Abbildungen. VI, 371 Seiten. 1938.
Gebunden DM 76,60, S 529,—

Zweiter Band: 24 Abbildungen. VII, 366 Seiten. 1939.
Gebunden DM 76,60, S 529,—

Dritter Band: 10 Abbildungen. VI, 252 Seiten. 1939.
Gebunden DM 58,50, S 404,—

Vierter Band: 47 Abbildungen. VIII, 499 Seiten. 1945.
Gebunden DM 105,—, S 725,—

Fünfter Band: 34 Abbildungen. VIII, 417 Seiten. 1948.
Gebunden DM 53,50, S 369,—

Sechster Band: 32 Abbildungen. VIII, 392 Seiten. 1950.
Gebunden DM 58,50, S 404,—

Siebenter Band: 12 Abbildungen. VII, 330 Seiten. 1950.
Gebunden DM 57,—, S 393,—

Achter Band: 47 Abbildungen. XI, 400 Seiten. 1951.
Gebunden DM 75,—, S 518,—

Neunter Band: 20 Abbildungen. XI, 535 Seiten. 1952.
Gebunden DM 87,50, S 604,—

Zehnter Band: 19 Abbildungen. IX, 529 Seiten. 1953.
Gebunden DM 88,—, S 607,—

Elfter Band: 67 Abbildungen. VIII, 457 Seiten. 1954.
Gebunden DM 79,50, S 549,—

Zwölfter Band: 15 Abbildungen. X, 550 Seiten. 1955.
Gebunden DM 88,—, S 607,—

Dreizehnter Band: 48 Abbildungen. XII, 624 Seiten. 1956.
Gebunden DM 114,—, S 787,—

Vierzehnter Band: 38 Abbildungen. VIII, 377 Seiten. 1957.
Gebunden DM 79,50, S 549,—

Fünfzehnter Band: 81 Abbildungen. VI, 244 Seiten. 1958.
Gebunden DM 43,50, S 300,—

Sechzehnter Band: 27 Abbildungen. VI, 226 Seiten. 1958.
Gebunden DM 42,50, S 293,—

Siebzehnter Band: 57 Abbildungen. X, 515 Seiten. 1959.
Gebunden DM 88,—, S 607,—

Achtzehnter Band: 65 Abbildungen. X, 600 Seiten. 1960.
Gebunden DM 109,—, S 752,—

Neunzehnter Band: 16 Abbildungen. VIII, 420 Seiten. 1961.
Gebunden DM 83,—, S 573,—

Weitere Bände siehe nächste Seite

Zwanzigster Band: 33 Abbildungen. XIII, 509 Seiten. 1962.
Gebunden DM 102,—, S 704,—

Generalregister / Cumulative Index / Index Général I—XX. 1938—1962.
XVI, 369 Seiten. 1964.   Gebunden DM 63,50, S 438,—

Einundzwanzigster Band: 14 Abbildungen. VII, 362 Seiten. 1963.
Gebunden DM 80,50, S 556,—

Zweiundzwanzigster Band: 8 Abbildungen. VII, 370 Seiten. 1964.
Gebunden DM 93,50, S 645,—

Dreiundzwanzigster Band: 58 Abbildungen. VIII, 397 Seiten. 1965.
Gebunden DM 99,50, S 687,—

Vierundzwanzigster Band: 25 Abbildungen. VIII, 475 Seiten. 1966.
Gebunden DM 118,—, S 814,—

Fünfundzwanzigster Band: 25 Abbildungen. VII, 348 Seiten. 1967.
Gebunden DM 84,—, S 580,—

Sechsundzwanzigster Band: 97 Abbildungen. IX, 456 Seiten. 1968.
Gebunden DM 132,—, S 911,—

Siebenundzwanzigster Band: 47 Abbildungen. VIII, 412 Seiten. 1969.
Gebunden DM 120,—, S 830,—

Über den Inhalt der Bände gibt der Verlag Auskunft

Achtundzwanzigster Band: 14 Abbildungen. XII, 503 Seiten. 1970.
Gebunden DM 155,—, S 1070,—

Inhalt: **E. Wong,** Structural and Biogenetic Relationships of Isoflavonoids. — **R. Eyjólfsson,** Recent Advances in the Chemistry of Cyanogenic Glycosides. — **D. Gross,** Naturstoffe mit Pyridinstruktur und ihre Biosynthese. — **E. W. Warnhoff,** Peptide Alkaloids. — **K. Eiter,** Insektensexuallockstoffe. — **H. Hikino** and **Y. Hikino,** Arthropod Molting Hormones. — **J. E. Pike,** Total Synthesis of Prostaglandins. — **R. B. Morin** and **B. G. Jackson,** Chemistry of Cephalosporin Antibiotics. — **H. Wiegandt** und **H. Egge,** Oligosaccharide der Frauenmilch. — **W. Bromer,** Glucagon: Chemistry and Action. — Namenverzeichnis. Author Index. — Sachverzeichnis. Subject Index.

Preisermäßigung für Subskribenten / Price reduction for subscribers: 10%

Vorzugspreis (20% Nachlaß) bei Bezug der Bände 1—20 inklusive Generalregister / Special price reduction (20% of the list price) for the set Vols. 1—20 plus Cumulative Index.

QD
241
F6
v.29
1971

SEP 18 1972